ANNUAL REVIEW OF ECOLOGY AND SYSTEMATICS

ANNUAL REVIEW OF ECOLOGY AND SYSTEMATICS

VOLUME 28, 1997

DAPHNE GAIL FAUTIN, *Editor*
University of Kansas

DOUGLAS J. FUTUYMA, *Associate Editor*
State University of New York at Stony Brook

FRANCES C. JAMES, *Associate Editor*
Florida State University

http://Annual Reviews.org   science@annurev.org   650-493-4400

ANNUAL REVIEWS INC.   4139 EL CAMINO WAY   P.O. BOX 10139   PALO ALTO, CALIFORNIA 94303-0139

ANNUAL REVIEWS INC.
Palo Alto, California, USA

International Standard Serial Number: 0066-4162
International Standard Book Number: 0-8243-1428-X
Library of Congress Catalog Card Number: 71-135616

⊗ The paper used in this publication meets the minimum requirements of American
National Standard for Information Sciences—Permanence of Paper for Printed Library
Materials, ANSI Z39.48-1992.

Annual Reviews Inc. and the Editors of its publications assume no responsibility for the
statements expressed by the contributors to this *Review*.

TYPESETTING BY TECHBOOKS, FAIRFAX, VA
PRINTED AND BOUND IN THE UNITED STATES OF AMERICA

Annual Review of Ecology and Systematics
Volume 28 (1997)

CONTENTS

(continued)   v

RELATED ARTICLES FROM OTHER *ANNUAL REVIEWS*

From the *Annual Review of Earth and Planetary Sciences*, Volume 25, 1997:

Hydrochemistry of Forested Catchments, M. R. Church

Stratigraphic Record of the Early Mesozoic Breakup of Pangea in the Laurasia-Gondwana Rift System, P. E. Olsen

Sediment Bacteria: Who's There, What Are They Doing, and What's New? K. H. Nealson

From the *Annual Review of Energy and the Environment*, Volume 21, 1996:

On the Concept of Industrial Ecology, T. E. Graedel

Restoration Ecology: The State of an Emerging Field, J. Cairns, Jr., John R. Heckman

Tropical Deforestation and the Global Carbon Budget, J. M. Melillo, R. A. Houghton, D. W. Kicklighter, A. D. McGuire

From the *Annual Review of Entomology*, Volume 42, 1997:

Ecology and Evolution of Galling Thrips and Their Allies, B. J. Crespi, D. A. Carmean, T. W. Chapman

Interactions Among Scolytid Bark Beetles, Their Associated Fungi, and Live Host Conifers, T. D. Paine, K. F. Raffa, T. C. Harrington

Physiology and Ecology of Dispersal Polymorphism in Insects, A. J. Zera, R. F. Denno

Systematics of Mosquito Disease Vectors (Diptera, Culicidae): Impact of Molecular Biology and Cladistic Analysis, L. E. Munstermann, J. E. Conn

Migratory Ecology of the Black Cutworm, W. B. Showers

Phylogeny of Trichoptera, J. C. Morse

Biology of Wolbachia, J. H. Werren

From the *Annual Review of Plant Physiology and Plant Molecular Biology*, Volume 48, 1997:

More Efficient Plants: A Consequence of Rising Atmospheric CO_2?, B. G. Drake, M. A. Gonzàlez-Meler, S. P. Long

ANNUAL REVIEWS INC. is a nonprofit scientific publisher established to promote the advancement of the sciences. Beginning in 1932 with the *Annual Review of Biochemistry*, the Company has pursued as its principal function the publication of high-quality, reasonably priced *Annual Review* volumes. The volumes are organized by Editors and Editorial Committees who invite qualified authors to contribute critical articles reviewing significant developments within each major discipline. The Editor-in-Chief invites those interested in serving as future Editorial Committee members to communicate directly with him. Annual Reviews Inc. is administered by a Board of Directors, whose members serve without compensation.

For the convenience of readers, a detachable order form/envelope is bound into the back of this volume.

Annu. Rev. Ecol. Syst. 1997. 28:1–25

MOLECULAR POPULATION GENETICS OF SOCIAL INSECTS

Pekka Pamilo, Pia Gertsch, Peter Thorén, Perttu Seppä

Department of Genetics, University of Uppsala, Box 7003, 750 07 Uppsala, Sweden; e-mail: Pekka.Pamilo@genetik.uu.se

KEY WORDS: dispersal, kin selection, polyandry, polygyny, relatedness

---

ABSTRACT

The life of social insects centers around sedentary colonies that can include individuals belonging to different patrilines or matrilines, with a turnover of reproductives. The colony is a scene for both cooperation and conflicts, and the conceptual framework for the evolution of social life and colony organization is provided by the kin selection theory. Variable molecular markers make it possible to dissect kinship within colonies, identifying patrilines and matrilines and estimating genetic relatednesses. Such markers have been used to test hypotheses on social conflicts between queens and workers (split sex ratio hypothesis), among workers (worker policing hypothesis), and among reproductive females (skew hypothesis). The data from several species of ants, bees, and wasps indicate that workers can obtain information on the genetic heterogeneity of their colonies and use that information to manipulate reproductive decisions. The social structure of colonies and the mode of colony founding affect the population-wide dispersal of sexuals. Populations with multi-queen colonial networks have limited dispersal; females stay in their natal colonies, and mating flights can be restricted. As a result, coexisting queens tend to be related to each other, maintaining the genetic integrity of colonies, and populations become spatially differentiated to an extent that can lead even to socially driven speciation.

INTRODUCTION

The origin of insect sociality, particularly the origin of the sterile caste in several phylogenetic lineages, has for decades been a central topic in evolutionary biology. Social evolution has continued in these lineages, creating a great diversity of societies and colonial life cycles (13, 21, 54). Colonies can be classified on

1

0066-4162/97/1120-0001$08.00

the basis of the mode of founding (with or without helping workers), the number of reproductive females (monogynous versus polygynous colonies), or the number of nest units forming a colony (monodomous colonies versus polydomous colonial networks). Many social bees and wasps have annual colonies, but ant queens can live years, even decades (54, 91). The life of social insects centers around the nest, and their populations are hierarchically organized. Dispersal and colony founding connect the two hierarchical levels, linking the social unit (the colony) and the breeding unit (the population) with each other; the two levels thus are intimately related. These features lead to two types of evolutionary and population genetic questions.

1. How do the societies function, and how are these functions controlled and coordinated? Individuals in a colony have different genotypes, and we can expect reproductive competition among the colony members as in any other organism. Such competition can arise among reproductive individuals, among nonreproductive individuals, and between these two castes. The individual conflicts concern both parentage and parental care of the offspring: Which members of the colony reproduce, and how are the resources allocated between reproductive and nonreproductive as well as between male and female functions? As the predicted solutions of the conflicts depend on specific kinship relationships of colony members, our understanding of the evolution of social life relies strongly on genetic analyses of colonial structures.

2. The sedentary life-style of most social insects means that dispersal and gene flow are restricted to a short period between the maturation of reproductives and colony founding. This gives the mode of colony founding a central position in determining the population structure. A simple colonial type has only one reproductive female, which means that the offspring must disperse and try to establish new colonies elsewhere. Many species have multiple egg-laying females, or supersedure of females in the nest, which immediately provides the option of not dispersing, as new females can stay in the natal colony and become reproductives there. As a result, social life affects dispersal within and connectedness between populations and so influences the pattern of geographic differentiation—even speciation.

The general conceptual framework for social evolution is provided by the kin selection theory, and kinship among interacting individuals is a key variable describing the relevant social structures (21, 46, 90, 112). The development of the genetic theory of social evolution has been recently followed by a revolutionary progress in the availability of molecular tools that allow identification of various patrilines and matrilines within colonies, estimation of the genetic structures of colonies and populations, and testing specific hypotheses derived from the theory (13, 21, 109, 121). As most of the theory has been developed for male-haploid insects, and as most empirical studies have been carried out with hymenopteran species (ants, bees, wasps), we focus on them unless otherwise mentioned.

KIN SELECTION AND INTRACOLONIAL CONFLICTS

The theory of kin selection was first formulated by Hamilton (46) who also pointed out the high relatedness among full sisters of a male-haploid insect (0.75). This led to the hypothesis that the social life of hymenopteran insects, which is basically sociality of females, results from kin selection and that the high relatedness among sisters contributed to the origin and maintenance of worker behavior (47). The first empirical studies on social insect population genetics therefore aimed to estimate genetic relatedness among nestmates in paper wasps (66, 71) and ants (19, 97). These studies showed that the relatedness of worker nestmates is often well below that of full sisters because of the existence of different patrilines and matrilines within colonies.

Trivers & Hare (155) pointed out that eusocial hymenopterans have strong queen-worker conflicts over the preferred population-wide sex investment ratios. The basis of this conflict can be formulated by using inclusive fitnesses. The inclusive fitness of an individual I can be written as (21, 90, 153)

$$W = N(g_f v_f f/F + g_m v_m m/M),$$
1.

where N is the number of offspring produced, g_f and g_m are the coefficients of genetic relatedness of female and male offspring to the individual I, v_f and v_m are the sex-specific reproductive values of females and males, and f and m are the investment proportions in the two sexes within a colony ($f + m = 1$) and F and M are the same proportions in the whole population. It should be noted that this simple linear equation (linear for f and m) does not hold when there is local mate or local resource competition. If worker behavior is to be explained by kin selection, we could expect workers to maximize their inclusive fitness. This can be done by adjusting the colonial sex ratio $m : f$. As the coefficients of genetic relatedness depend on the parentage of the individuals, it is evident that all individuals cannot maximize their inclusive fitnesses at the same time.

Recent developments of the kin selection theory have produced clear and testable predictions on various conflict situations that may arise in colonies of social insects. These are the hypothesis of split sex ratios (10), worker policing (111, 158), and reproductive skews (113). All these hypotheses rest on genetic relatednesses, and testing them requires that the relatednesses are estimated not only as means in the whole population but individually within single colonies or for specific individuals. Such estimates require suitable statistical methods and good genetic markers that provide enough information on the kinship. Both requirements are largely satisfied today. The coefficient of relatedness, defined as a genotypic correlation or regression (89), can be calculated for specific individuals when the rest of the population is kept as a reference (104). Genotypic data required for reliable estimates can be obtained by highly variable molecular markers. The method has also been expanded to band-sharing data produced by multilocus fingerprinting (114).

GENETIC TOOLS

Enzyme electrophoretic studies have generally shown low levels of heterozygosity in social insects (50a). Nevertheless, most genetic studies so far have been based on allozymes (21), but the need for more informative markers is obvious. Such markers are needed for examining the parentage of individuals within colonies, matrilines and patrilines, and for estimating dispersal of sexuals. Many questions can still be satisfactorily answered by applying allozymes (131).

Multilocus DNA fingerprinting is convenient for assessing parentage, and it has been used for detecting patrilines within colonies (6, 39), for other estimates of parentage (72, 76), for assessing the levels of relatedness among cofounding females (22, 128), and for estimating spatial population structures (103). Fingerprinting is based on individual variation in the number of tandemly repeated sequences, minisatellites (long repeats), or microsatellites (short repeats). Both types are often highly variable, but some populations may have very low levels of variation as in the bee *Halictus marginatum* (7). The complexity of banding patterns makes it commonly impossible to detect allelism, and the individual genotypes cannot be defined. This restricts the use of minisatellites.

Microsatellites (109) provide genotypic data when PCR-amplified using locus-specific primers, and they can be studied from minute amounts of material (for example, multiple inseminations can be estimated by amplifying DNA from the spermathecae) (31, 40a, 42, 102). Microsatellites have been estimated to occur every few thousand nucleotides in the genome of the social wasp *Vespula rufa* (154). As the current advances of the kin selection theory require good genotypic data for testing alternative hypotheses, it is no wonder that microsatellite markers have been vigorously developed for social insects including wasps (17, 55, 144, 147, 154), bees (27, 30, 99), and ants (13a, 16, 31, 35, 41, 45). The primers developed for one species work often in other species within the same genus.

Random amplified polymorphic DNA (RAPD) has been successfully used for studying parentage (37, 48, 115) and hybridization (136), and for constructing linkage maps in the honeybee (58). RAPD markers are based on amplification of DNA with PCR, and they are often dominantly inherited (failure of amplification), although some are codominant (57). The drawbacks in population studies include poor repeatability of results and the dominant inheritance, which prevents genotypic information.

Mitochondrial DNA is maternally inherited and can therefore be used for identifying matrilines and for estimating the dispersal of females. The whole mtDNA of the honeybee has been sequenced (20), and that sequence can be used for constructing primers for other hymenopteran species. Haplotypes can be identified (i) from sequences, (ii) from restriction fragments (RFLP) obtained either by hybridizing total DNA with known mitochondrial probes (140), or by amplifying mtDNA with PCR and separating the cleaved restriction fragments electrophoretically, or (iii) from nucleotide differences in PCR-amplified products detected as conformational polymorphisms (152). Haplotype variation in the fire ant *Solenopsis invicta* (135) and the meat ant *Iridomyrmex purpureus* (14) has been studied by amplifying a 4-kb sequence and cutting it with several restriction enzymes. Sequence information also preserves evolutionary history, for which reason sequence data from mitochondrial and nuclear genes are useful for estimating long-term evolutionary relationships of populations and species.

Genetic markers allow detecting matrilines and patrilines within colonies, and the detection probability depends on the level of genetic variation the markers have in the population (102). It is also clear that the absolute and genetically effective numbers of matrilines (or patrilines) can differ from each other when the reproductives contribute unequally (11).

SOCIOGENETICS

Mating and Dispersal

Mating flights of ants have been classified in two major categories, female calling syndrome and male aggregation syndrome (13, 21, 54). In the former, females on the ground attract males pheromonally, and in the latter males aggregate and attract females to these sites. Only a few studies have tried to examine the effective size of such aggregations or the neighborhood size of populations. Allozyme studies indicate that the continuous populations of the ants *Myrmica punctiventris*, *Leptothorax ambiguus*, *Formica sanguinea*, and *F. transkaucasica* are genetically structured, and the neighborhood sizes can be small (4, 52, 88). Inbreeding in the ant *Messor aciculatus* fits the assumption that the mating swarms contain sexuals from only a few colonies and that the dispersal distances of sexuals are less than 50 m (49). Restricted dispersal leads into local mate competition and female-biased sex ratios in this species. Contrasting patterns of spatial differentiation in mitochondrial and nuclear (allozyme) markers suggest large breeding units but restricted female dispersal in the ant *Leptothorax acervorum* (141, 142). Most genetic studies in social insects show no significant inbreeding, although diploid males have been observed with low frequencies in many species as a sign of inbreeding (21). The exceptions are found in socially parasitic species and in termites and some ants that have inbreeding cycles within colonies (21, 116).

Mating behavior and dispersal of new queens determine the genetic population structure, and they also influence the intracolonial genetic diversity. A hymenopteran female mates at the beginning of her life and stores the sperm in the spermatheca. The males are short-lived. Behavioral observations and dissections of spermathecae indicate that females of many species mate with several males (11, 21), but the success of various mates and the pattern of sperm use have remained unclear until estimated by molecular markers. Haploidy of the hymenopteran males simplifies this estimation. When p_i is the contribution by the ith mate, the effective number of matings m_e can be given as

$$m_e = 1/\Sigma p_i^2. \hspace{3cm} 2.$$

An interesting, although not completely unexpected, result is that the genetically observed effective number of inseminations is generally lower than the number of matings observed in behavioral studies (11, 21). It is also evident that the males contribute different amounts of sperm. For example, the most frequent patriline had an average frequency of 77% among progenies fathered by two males in the ant *Formica aquilonia* (92). Allozyme studies of paternity in 19 ant species show the values of m_e ranging from 1 to 1.48, even though some queens may mate up to seven times (11). Three major exceptions to the generally low mating numbers have been found in leaf-cutting ants, *Vespula* wasps, and honeybees. The effective number of matings exceeds two in the leaf-cutters (35, 115) and in *Vespula* (118; P Thorén, unpublished data). Honeybee queens mate normally with many drones. The maximal numbers of patrilines detected in colonies by using microsatellites are 20 in *Apis andreniformis*, 28 in *A. dorsata*, 14 in *A. florea*, 20 in *A. mellifera*, and 44 in *A. mellifera capensis* (28, 73, 74, 80, 83). Although microsatellites are highly polymorphic and detect different mates with a high probability, it is likely that all the patrilines were not detected because of limited sampling. The effective mean numbers calculated from the same data are 25.6 for *A. dorsata* and 5.7 for *A. florea*.

With few exceptions, polyandry has minor effects in increasing the genetic diversity within social insect colonies, at least when compared to the effects of multiple matrilines (11). But polyandry affects relatednesses and leads to variation among colonies. When the relatednesses vary, the workers have different optimal sex ratios depending on the type of colony they are in. This variation affects the queen-worker conflict and leads to split sex ratios under worker control (10). If polyandry is the main cause of the colony-level variation of relatedness asymmetries, the workers of a polyandrous queen should favor strongly male-biased sex allocation. Therefore, it seems that males should avoid mating with a female that has already mated, as her colony would (under worker control) produce mainly males that do not carry any paternal genes. The theory predicts a negative association between multiple mating by queens and sperm bias in monogynous colonies (9), but there are not yet data to test that prediction.

There are many hypotheses for explaining the evolution of polyandry (21), but most of them have not been rigorously tested and no single hypothesis is supported by existing data (11). It seems possible that the ant queens with large monogynous colonies simply need sperm from several males in order to produce enough workers. There is no significant increase in polyandry with decreasing polygyny of colonies, indicating that the hypotheses based on genetic diversity of colonies may not be general explanations for the evolution of polyandry (11)

Polygyny

FOUNDRESS ASSOCIATIONS   Mated females can join existing colonies or establish new nests without any help from old workers, either singly (haplometrosis) or in small groups (pleometrosis). Sociality in communally nesting females is restricted to female associations that have no workers and no cooperative brood care, but in eusocial species the pleometrotic founder associations will develop into mature colonies with separate queen and worker castes. The pleometrotic foundress associations in ants normally revert to single-queen colonies after the first workers emerge, as additional queens are killed, often by the workers, sometimes by other queens (54). In light of the kin selection theory, it is essential to ask whether founder associations are based on kinship or not.

Low—or an absence of—relatedness among cofoundresses has been observed in two groups of social insects, among cofounding ant queens and in communally nesting bees. The relatednesses in pleometrotic associations of ant queens have been estimated by using allozyme markers in *Acromyrmex versicolor* and *Veromessor pergandei* (43) and by analyzing band-sharing of DNA minisatellites in *Polyrhacis moesta* (128). All these species have small foundress associations (normally fewer than 5 females), and the estimates of relatedness do not differ significantly from zero. This is also supported by nongenetic studies, except when the nest densities are so low that the pleometrotic groups consist of sibs with a high probability (78). Even though kin groups seem to perform better in the founding stage in the ant *Lasius pallitarsis*, the queens associate on the basis of size rather than on kin (78). Incipient monogynous colonies of *Solenopsis invicta* have intense brood raiding among neighboring colonies and temporary multi-queen associations. Behavioral studies show that workers do not recognize their own mother but favor the most physogastric queen when eliminating extra queens (36). This decision would normally select a queen who is the mother of most workers in an incipient colony, but an analysis of matriline frequencies shows that this is not always the case (3). The queen that was eliminated was often the mother to the majority of workers, and the criteria used may be based on the size and condition rather than on the actual reproductive rate of the queen.

Communally nesting females also tend to cluster with unrelated individuals. Low relatedness has been estimated in a halictid bee *Lasioglossum hemichalceum* (64) and in the andrenid bees *Andrena jacobi* and *Perdita texana* (22, 99). An exception to this pattern of low relatedness in communal insects is given by the sphecid wasp *Cerceris antipodes*, whose coexisting females are closely related (estimates ranging from 0.25 to 0.64) (68). These estimates were associated with positive inbreeding coefficients that may boost relatednesses to some extent. Inbreeding has been observed also in *Andrena jacobi,* where 70% of the emerging females had already mated underground (98, 99).

The cofoundresses in eusocial bees and wasps appear to be closely related. The founding females arise commonly as nestmates with relatednesses of 0.6 in the allodapine bee *Exoneura bicolor* (129) and 0.3–0.8 in the wasp genera *Polistes* and *Mischocyttarus* (107, 146). Even though dominance hierarchies can develop among the foundresses, depending largely on the individual size and on the order of joining the founder association, the colonies often remain polygynous unlike the pleometrotic ant colonies.

The difference between pleometrotic ants and cofounding bees and wasps in queen relatedness does not reflect their abilities to recognize kin. Nestmate recognition has been documented in all major groups of social insects (21). Clear differences between the taxa exist in the dispersal of new queens, in the development of the behavioral relationships among the queens, and in the longevity of the colonies (143). Cofoundresses in annual wasp colonies develop strong dominance hierarchies, whereas pleometrotic ant associations do not show dominance until all the extra queens are killed. The ant queens participate in mating flights and disperse, and it can be impossible for them to locate relatives among other nest-founding females, unless they return to the natal nest site. The ant queens lose their wings after mating, and their establishment of new colonies in the vicinity of the parental nest would lead to local resource competition. Communal bees do not collaborate beyond nest-building and guarding, and kin associations may be less relevant. However, kin associations may arise because of philopatry without any particular kin recognition mechanism. Patterns of band-sharing in DNA minisatellites show that females in neighboring nest holes of the solitary cicada killer wasp (*Sphecius speciosus*) are, on average, more related than random females in the population and that the females are less aggressive toward their neighbors than toward other females (103). Whether the *Sphecius* females discriminate between related and unrelated neighbors has yet to be demonstrated.

RELATIONSHIPS AMONG QUEENS   Colonies established by several females can remain polygynous in bees and wasps, but tend to revert to single-queen, monogynous colonies in ants. Mature colonies can, however, accept new reproductive females, and monogynous colonies become secondarily polygynous. Polygynous colonies can also reproduce by fission or budding, when some of the queens and workers leave the old nest site and establish a new nest. The queens can be separated in the nest, and such a society can be seen either as a multi-queen colony or as a multi-nest aggregation, as in the ant *Lasius flavus* in which the workers of different matrilines mix in the nest galleries (12). Secondary polygyny implies that the life span of colonies can be much longer than that of individual queens, and the queens may benefit from posthumous reproduction when the colony is kept going by new, even unrelated, queens (34).

Even though ant cofoundresses are commonly unrelated, this does not apply to queens in secondarily polygynous colonies. Genetic relatedness among queens has been estimated mainly by allozyme markers in a number of ant species, and the estimates show that the queens are related to each other approximately as closely as the worker nestmates in the same populations (21, 61, 130). This finding agrees with the hypothesis that the females tend to join their natal nests as new reproductives. In a stable population with constant queen numbers per nest, that kind of queen recruitment would lead to a relatedness of $3/(3n+1)$ among diploid females, where n is the effective number of queens (97). When the number of queens increases, relatedness among nestmates decreases even though females stay in their natal colonies. Unrelated queens can also coexist in a colony, as demonstrated in the ants *Iridomyrmex purpureus*, *Myrmica tahoensis*, and *Leptothorax acervorum*, where some colonies have unrelated microsatellite genotypes or more than one mtDNA haplotype (14, 34, 140).

There are also cases in which the reproductive females are exceptionally closely related. Many ponerine ants have workers, called gamergates, that mate and lay diploid eggs. Some ponerine ants have completely lost the morphological queen caste and all reproduction is by workers, whereas other species have polymorphic colonial types, some run by queens (type A) and some by gamergates (type B). *Rhytidoponera chalybea* belongs to the latter type and its type A colonies are always monogynous. Type B colonies normally have several gamergates that are closer kin than the worker nestmates on average (157). It is evident that some of the type B colonies have shifted recently from type A to type B after losing the queen, and the gamergates in these colonies are full sisters.

Many genetic studies have shown an exchange of dominant queens in colonies within the lifetime of workers: matrilines are differently represented among the adult workers, and developing larvae and pupae, or workers include matrilines not found in present queens (13a, 51, 52, 102, 130). Unrelated matrilines may also result from intraspecific parasitism—egg-dumping by other females (77). Overall, there can be significant differences in how the various queens contribute in reproduction. Such a reproductive skew (113) can be different for worker production and sexual production, and the fitness of a queen is determined by her share of sexual production. Intraspecific parasitism with some queens trying to monopolize sexual production has been observed in the ants *Solenopsis invicta* and *Formica sanguinea* (95, 119).

Principles that apply to the origin of worker behavior apply also to the evolution of polygyny. The females can accept a subordinate status more readily if they are highly related to the dominant females. The theory of reproductive skew thus predicts that the amount of skew should be positively correlated with the relatedness of queens (113). From the point of view of the dominant queens the situation is different; they might allow subordinates to reproduce provided the subordinates are highly related to them. Critical tests of the skew are still wanting. Heinze (51) showed a positive association between relatedness and skew in *Leptothorax* ants, but remarked that high relatedness may not cause skew. Young queens are commonly readopted into their native colonies, and high relatedness and high skew may be mutually reinforcing. No relationship between the skew and relatedness was observed at the colonial level in the ant *Myrmica tahoensis* (33).

Genetic Diversity within Colonies

Social insect females can mate many times (polyandry of queens), and mature colonies can have many coexisting, physogastric females (polygyny of colonies). Genetic studies have confirmed that polyandry and polygyny are functional, and colonies do consist of mixtures of patrilines and matrilines. Estimates of genetic relatedness of workers or of sexual offspring within colonies have resulted in a whole spectrum of values from 0 to over 0.75 (21). The observed low values have led to a critical evaluation of the so-called haplodiploid hypothesis for the origin and maintenance of sociality. When the relatednesses are low, because of either polyandry or polygyny, the female workers do not get high fitness returns from sisters and should in fact produce their own daughters. Taking into account relatednesses and optimal sex ratios, Gadagkar (38) concluded that the role of haplodiploidy may have been overemphasized. However, the genetic structures of the present species may not reflect those at the dawn of sociality, and, all other things being equal, the ecological threshold for the evolution of worker behavior is lower in male-haploid than in diploid organisms (21).

Studies of intracolonial relatednesses have shown variation both among colonies within populations and among populations. The differences among populations can result from various levels either of polyandry as in the ant *Lasius niger* (50) or of polygyny as in the ants *Formica sanguinea* and *Leptothorax longispinosus* (53, 133). In extreme cases, there is clear social polymorphism and the societies (even populations) can be divided into two distinct categories, one with monogynous-monodomous colonies (M type) and the other with polygynous-polydomous colonial networks (P type) (94). Such social polymorphism has been detected in *Solenopsis invicta* (124) and several *Formica* species (67, 94, 149).

The coexistence of several to many matrilines and patrilines in a colony affects colony integrity by increasing the diversity of cues used for nestmate recognition. It is commonly observed that social insect workers can discriminate between nestmates and non-nestmates. This ability functions as a kin-recognition mechanism in the sense that workers can also identify related individuals they have not previously met (21). Cuticle hydrocarbons have been pointed out as one class of possible chemical labels for recognition, and the hydrocarbon profiles differ significantly among honeybee subfamilies that were identified by using microsatellite markers (2). The recognition cues seem not to be innate but are learned from nestmates. As a result, workers cannot normally show nepotism by discriminating between differently related individuals within a nest (24, 105).

Increased heterogeneity may have positive effects if colonial performance (survival and productivity) depends on genetic variation. Such selection might result from improved ability to cope with environmental variation (70), from disease resistance (137), or from a more efficient division of labor than in colonies with a single matriline and patriline. Workers do normally have an age-dependent task profile, and they move gradually to new tasks as they get older. Many studies have shown a genetic component in the division of labor among workers, and the differences seem to exist mainly in the speed with which the workers move to new tasks. These differences result in task profiles that can be interpreted as different task preferences, and genetically diverse honeybee colonies with many patrilines may have a selective advantage in responding to varying environmental conditions (86). Identifying the patrilines and matrilines of individual workers by various genetic markers has enabled the detection of such task specialization in honeybees (84), wasps (79), and ants (139, 148). Workers can specialize between major tasks such as guarding, cleaning the nest, and foraging, or between various items that they forage. Applying polymorphic DNA markers and crossing experiments have allowed mapping major quantitative trait loci (QTL) for task specialization in the honeybee, as the segregating marker alleles were nonrandomly distributed among worker bees that brought to the colony mainly pollen or nectar (59). DNA fingerprinting of workers in polygynous colonies of the termite *Schedorhinotermes lamanianus* show kin-biased behavior that may indicate task specialization, or kin recognition and discrimination within colonies, as foragers within the same gallery are on average more closely related to each other than to other nestmates (60).

Hybridization between species has generally negative effects, and the hybrids of *Solenopsis invicta* and *S. richteri* have greater fluctuating asymmetries than do the parental species (123). However, some positive effects at the level of colonies have been suggested in *Lasius* ants. Hymenopteran males are not affected by hybridization in the F1 generation, because they are haploid and are produced parthenogenetically. Genetic studies suggest unidirectional hybridization, females of the smaller species *L. alienus* mating with males of *L. niger* (100). As a result, the workers are larger than in normal *L. alienus* colonies, giving some competitive advantage that will be capitalized upon if the hybrid colonies produce mainly male sexuals.

Hymenopteran males develop from unfertilized eggs and are haploid, while fertilized eggs develop into females. More precisely, sex of the honeybee is determined by a single locus in such a way that heterozygotes develop into females (either fertile queens or practically sterile workers) and hemizygotes (i.e. haploid individuals) and homozygotes develop into males. Diploid males are nonviable or sterile and do not contribute to future generations. Most hymenopteran insects are believed to have a similar single-locus sex determination mechanism (18). The sex-determining locus requires variation within colonies as all the females, both queens and workers, must be heterozygous. This creates strong frequency-dependent selection for allelic diversity. The number of sex alleles has been estimated on the basis of the frequency of diploid males in several species, and the estimates go up to 86 in an Argentinian population of *Solenopsis invicta,* provided there is only one sex locus (126). High frequencies of diploid males in some colonial orchid bees (Euglossinae) have led to the hypothesis that the genetic sex-determination mechanism and inbreeding have prevented the evolution of eusociality, as the production of female workers becomes erratic in small and isolated populations (127). The sex locus has not yet been identified, but the locus has been mapped in the honeybee (5). When the sequence is known in the honeybee, the next tasks will include characterizing the allelic variation at the molecular level and confirming the function of the locus in other species.

Do Workers Rule?

One of the key goals in the biology and evolution of social insects is to understand how selection acts on workers in spite of their sterility or low fecundity, and how the workers can manipulate reproduction within the maternal colonies in order to increase the inclusive fitness. This leads to queen-worker conflicts over reproduction and resource allocation. The main conflicts concern the origin of haploid males, and the allocation of resources between sexual and worker production on one hand, and between female and male production on the other (90). It is important to note that the queen-worker conflict is not identical in all the colonies but may vary depending on the precise genetic structure of a colony (10). This colony-level variation gives powerful possibilities for testing conflicts, and it also emphasizes the importance of informative genetic markers in mapping genetic structures at the colonial level. Empirical studies increasingly show that workers do indeed strongly affect the colonial decisions in a way optimal for their own inclusive fitness, although they may not be able to control all the conflict situations (106).

WORKER REPRODUCTION AND WORKER POLICING   Many hymenopteran workers can lay haploid eggs that develop into males (21). Workers of some primitively eusocial bees can even mate and produce diploid offspring. The workers of *Halictus ligatus* first help the mother to reproduce, but while the queen produces most male-destined eggs, the workers lay many diploid eggs that develop into new queens (117). It occurs much more frequently that workers produce haploid males. When both queens and workers produce males, a worker can raise males with very different relatedness values. Workers should collectively prefer males that are most closely related to them, relatednesses that are affected by both polyandry and polygyny (21). In monogynous colonies, workers should prevent each other from laying eggs when the effective number of matings $m_e > 2$ (111, 158). This kind of worker policing has been observed in the highly polyandrous honeybee. Honeybee queens mark the eggs by specific pheromones, and workers eat nonmarked eggs. An allozyme analysis showed that workers laid 7% of all male eggs, but very few of them developed to adults (156). As perhaps might be expected, some workers seem to cheat the system and mark their eggs. Oldroyd et al (82) observed worker-produced males in a hive, and all these males belonged to a single patriline. In other words, all egg-laying workers were full sisters, indicating a clear genetic component in egg-marking. Similarly, reproductive dominance in the cape honeybee, *Apis mellifera capensis*, seems to have a genetic component, as the worker subfamilies are differently represented among the laying workers in dequeened colonies (73). The cape honeybee is notable for its thelytokous parthenogenesis, i.e. workers can produce diploid offspring parthenogenetically. Worker policing does not require that the mean number of matings exceeds two, but facultative policing can be expressed in those colonies where the effective number is high enough, if workers can detect the genetic heterogeneity of the colony.

POPULATION SEX RATIOS   Another major conflict between queens and workers concerns sex allocation. In monogynous colonies, the optimal population-wide sex investment ratio is 1:1 for the queens and is female biased for the workers, the precise worker optimum depending on the level of polyandry and on the proportion of worker-produced males (21, 155). The situation in polygynous colonies depends on the genetic relationships among the queens. If subordinate queens are daughters of the dominant queen, their preference approaches that of workers. The optimal sex ratio for the workers in polygynous societies is the same as in monogynous colonies if the queens are totally unrelated (8), but approaches 1:1 with the increasing number of related queens. Nevertheless, the general prediction from kin selection theory is that workers favor more female-biased population sex ratios than the queens do, the preferred allocation in males being (10, 90)

$$g_m v_m / (g_m v_m + g_f v_f), \qquad\qquad 3.$$

where the symbols are the same as in Equation 1. The worker-control hypothesis has been tested by collecting data on sex ratios from many species, and they are indeed more female biased in monogynous than in polygynous species (21, 155). Studies that have estimated both relatednesses and sex allocation in single ant and bee populations have shown rather good agreement between the observed sex ratio and that predicted under worker control (85, 93, 101).

SPLIT SEX RATIOS   A much more powerful way of testing worker control is to examine colony-level sex ratios (10). Workers should raise females in colonies where females are particularly highly related, i.e. the relatedness asymmetry is high. This hypothesis of split sex ratios requires that workers can measure the genetic heterogeneity of the colony and compare it to the population average. The hypothesis of split sex ratios furthermore predicts that the population mean sex allocation departs from that given by Equation 3 and depends on the frequencies of various colonial types in the population (10). The colonial sex ratios, particularly in ants, have a bimodal distribution (21, 93), and a few studies test the hypothesis that this bimodality results from optimal behavior of workers. The colonial sex ratios in monogynous forms of the ants *Formica exsecta* and *F. truncorum* have bimodal distributions, and the colonies with monandrous queens produce significantly more females than those with polyandrous queens (150, 151). Furthermore, sexing eggs by chromosome counts shows that haploid eggs are selectively eliminated in colonies headed by monandrous queens in *F. exsecta* (151). The results indicate that the workers behave in a way that increases their inclusive fitness, and that this behavior is conditional, depending on the genetic heterogeneity within colonies. In the ant *Myrmica tahoensis,* the relatedness asymmetry varies depending on polygyny. Interestingly, the colonial sex allocation was clearly associated with the colony-level relatedness of workers rather than with the number of queens, indicating that the workers are able to detect the genetic heterogeneity of the colony and use that information to bias sex allocation (32). In the ant *Leptothorax acervorum*, the sex ratios were significantly more female-biased in monogynous than in polygnous colonies (15). It will be interesting to see how widespread such worker control is. The colonial sex ratios in *Formica sanguinea* are bimodally distributed but show no association to intracolonial genetic heterogeneity (95).

Worker-manipulated split sex ratios have also been found in bees and wasps (76, 85, 110). The bee *Aucochlorella striata* has both eusocial colonies in which the reproductive female is the mother of both the workers and the sexual brood, and parasocial colonies in which the initial mother has been replaced by one of her daughters (76). Kin selection theory predicts that the workers should favor raising females (sisters) in eusocial and males (nephews) in parasocial colonies, and this is the case even when confounding factors such as worker-produced males, diploid males, and egg-dumping by other females have been taken into account (76, 77). The evidence from epiponine wasps is somewhat indirect but convincing: Males are produced in colonies that are more polygynous than average colonies (110). When females are produced in less polygynous colonies, the genetic integrity of new colonies remains high and the worker control of sexual investment helps to maintain the social cohesion of colonies. Similar cycles in polygyny, sex ratios, and relatednesses are postulated in *Leptothorax* ants (51).

FROM COLONIES TO POPULATIONS

Social insect populations have strongly hierarchical structures. A single family can make a colony, but often a nest includes several to many reproductive females. In extreme cases, hundreds of fertile females inhabit the same nest. In such situations we can rightly ask what is the relationship between the colony as a social unit and the population as a breeding unit. This relationship is affected by the mating flights, the mode of colony founding and philopatric behavior of females that may tend to stay in their natal colonies. The intracolonial social environment forms a factor that affects dispersal and gene flow within populations. At the same time, the population structure makes a framework for social life and social evolution.

Social Polymorphism in the Introduced Fire Ant

The fire ant *Solenopsis invicta* was introduced to northern America in 1930s, and it exists in two very different social forms, one being monogynous-monodomous (M type) and the other polygynous-polydomous (P type). This social polymorphism is found also in native populations in Argentina, although the level of polygyny there is not as high as in the introduced populations (125). The overall level of genetic variation is significantly lower in the introduced populations

than in the original area of the species (125). Particularly important consequences follow from the reduced number of sex alleles. The females mate only
with a single male, so each time when a male carries a sex allele present in
the female genome, half of the diploid offspring develop into males. If sex is
controlled by a single locus, the production of diploid males leads to estimates
of 10-13 sex alleles in the North American populations, whereas an Argentinian
population has more than 80 alleles (126). Diploid males form a heavy genetic
load, and the type M colonies that produce diploid males do not survive the
founding stage (121). However, diploid males are produced in type P colonies,
because the load is diluted by normal reproduction of other queens.

The two social forms live largely in separate populations. In general, allozyme studies indicate that the M and P types in the same geographical area
are related to each other, suggesting that the polygynous forms have originated
independently from the monogynous form locally (124). There is continuous, although weak, gene flow from the neighboring monogynous colonies to
the polygynous form (122). New mature queens produced in the P form lack
the reserves needed for independent founding of new monogynous colonies
(63). This inability to establish colonies is largely culturally transmitted. The
intrusion of females from monogynous to polygynous societies is also difficult. Genetic studies have pointed out a difference at the *Pgm-3* enzyme locus
between the two forms (120). Homozygous *Pgm-3a/-3a* females have particularly high fecundity, and the allele *Pgm-3a* is apparently selected for and has a
high frequency in the monogynous form. When homozygous females enter a
polygynous colony, they are actively attacked and killed by workers apparently
because of their high fecundity (62). The protein-coding locus *Gp-9* shows
similar differences between the two social forms (135). Restricted gene flow
by females has led to very different mitochondrial haplotype frequencies in the
M and P type colonies (135). Most males produced by type P colonies are
diploid, and the gene flow between the social types is mainly carried by males
from type M colonies mating with females in type P. For estimating gene flow,
it is important to examine genetic differentiation of the social forms at nuclear
loci that are not directly connected to the performance of queens. Populations
consisting of polygynous colonies are genetically more differentiated from each
other than are those of monogynous colonies within comparable geographical
areas (135). This observation is parallel to the general pattern of differentiation
among source and sink populations, where the sink populations show higher
genetic differentiation (25).

Population Structures in Formica Wood Ants

Variation of social forms similar to that in *S. invicta* is found in the ant genus
Formica, both as intraspecific social polymorphism and as a difference between

closely related species. Therefore, the genus offers excellent possibilities to test the association between intracolonial social structure and population-wide dispersal in the same way as in *Solenopsis*. Effective polyandry ranges between $m_e = 1$ and $m_e = 1.48$ in *Formica* species (87, 92, 149), the level of polygyny varies widely, and the relatedness among diploid nestmates ranges from zero to values close to 0.75 (21). As discussed above, coexisting queens are related about as closely as worker nestmates, suggesting that the queens commonly become reproductives in their natal polygynous colonies. This means that dispersal by females is restricted. Genetic differentiation in continuous populations of *F. transkaucasica* and *F. sanguinea* with high nest densities showed that the study populations are not panmictic, and that the neighborhood areas may consist of some tens of colonies. Populations of *F. exsecta* and *F. fusca* with commonly monogynous colonies show little geographic differentiation; F_{ST} among island populations within an area of a few square kilometres is below 0.05 (88). The populations of highly polygynous social forms in *F. truncorum* and *F. aquilonia* show strong genetic differentiation with F_{ST} about 0.2 (149; P Pamilo, unpublished data). Similarly high differentiation is found between two nearby populations of *F. cinerea* that differ also in level of polygyny (67; D Zhu, unpublished data). The results are concordant with those in *Solenopsis invicta* in that gene flow is restricted in the polygynous social forms, and apparently between the monogynous and polygynous forms (see also 132 for similar observations in *Myrmica* ants). The large values of F_{ST} indicate very weak gene flow, particularly when one takes into account that *Formica* queens have been estimated to live more than 20 years (91).

Isolation within sparse island populations, on islands too small to support but a few nests, leads to severe inbreeding and production of diploid males. About 15–40% of nests on the Baltic islands that produce diploid sexuals produce diploid males (96). In contrast to the situation in *Solenopsis*, monogynous nests of *F. rufa* that produce large numbers of diploid males (sometimes half of all diploid offspring) do survive the founding stage. The genetic load caused by diploid males is reduced in *Formica* because the queens start new nests by temporary parasitism, occupying nests of species of the subgenus *Serviformica*. Furthermore, diploid males are found only in the sexual brood, not when the colonies later in the season produce workers. The workers must identify and eliminate males when they occur at the wrong time of the year. Isolation must have contributed also to the strong allele frequency differences between the sexes in highly polygynous *F. aquilonia*. Three of six polymorphic allozyme loci are fixed for one allele in males, even though 20–50% of females are heterozygotes (92). The loci behave as if carrying recessive lethals, and no homozygotes are found for the alleles missing in males. The problem with the hypothesis of lethal alleles is that it makes it hard to understand the high frequencies of heterozygous queens, particularly when the loci segregate independently and are unlinked.

Colony Cycles in Polistinae

Polistinae are eusocial wasps with no consistent morphological differentiation between reproductive and nonreproductive female castes. The distinction between queens and workers is functional, and each female is capable of mating and becoming a queen in her own colony. Polistinae can be divided into two major groups based on of the mode of colony founding. Species with independent colony founding (e.g. the genera *Polistes* and *Mischochyttarus,* and some species of *Ropalidia*) have regular colony cycles. New nests are founded in the spring independently by one (haplometrosis) or more (pleometrosis) overwintered foundresses. A few cohorts of workers are produced during the season, and the last cohort in the autumn develops into sexuals, gynes and males. Some sexuals are also produced during the season. Nests have one to several egg layers, although one queen usually dominates. The other group of Polistinae wasps are the mostly neotropical swarm-founding wasps (Epiponinae), and they are typically highly polygynous. The kin structure of colonies has been studied in many species, first by allozymes and recently by using microsatellites. High relatednesses of some female nestmates (40, 146) and the analysis of spermathecal contents by microsatellites (42, 102) suggest that females are mainly monandrous in Polistinae. The colony relatednesses therefore reflect the existing matrilines.

Successful haplometrosis is rare in independently founding species, and colonies are started pleometrotically by nestmates from the previous autumn nests. Severe competition among cofoundresses leads to dominance of one of them, and assigning all the brood to the correct patrilines and matrilines by using microsatellites showed frequent turnovers of the dominant queens in the nests of *Polistes annularis* (102). As a result, female nestmates, and the cofoundresses of the following spring, are often full sisters. Wasps are powerful flyers, and genetic studies have not generally shown any population substructures, except between subpopulations less than 400 m apart in *Mischocyttarus immarginatus* (108) and in *Polistes exclamans* in a somewhat larger area (23). This differentiation has been attributed to inferior success of haplometrotic nests. Because pleometrotic nests are usually founded by nestmates from the previous autumn, a dispersing female may have a low probability to join any nest.

Epiponinae are often extremely polygynous, but relatednesses among workers are always higher than expected if the queens shared reproduction equally. This can be explained mainly by variable queen numbers across nests, relatedness among queens, and to a lesser degree by unequal reproduction by queens (56). The epiponine societies are characterized by a colony cycle called cyclical oligogyny (145). New queens are produced only immediately before swarming, when the queen number is at its lowest. This keeps queen relatedness high. Young nests are highly polygynous but produce only workers. New queens are produced again after several cohorts of worker brood, when the queen number falls low enough. Cyclical oligogyny seems to be associated with split sex ratios. In four Epiponinae wasps (*Polybia occidentalis, P. exigua, Protopolybia emaciata, Parachartergus colobopterus*), new queens are produced at the stage of the colony cycle when the queen number is lowest (and relatedness asymmetry highest). Males, however, are produced earlier in the cycle, when the colonies are still more polygynous (and the relatedness asymmetry is lower). As a consequence of this selfish worker behavior, queen and worker relatednesses stay high (110).

Cyclical oligogyny has some bearing on the population structure of Epiponinae, since it keeps the effective population sizes down and the swarms probably do not migrate very far. Such a demographic structure is expected to lead to spatial genetic differentiation, but differentiation has so far been studied only in *Protopolybia exigua* (40). Sampling sites 5 km apart showed only slight genetic differentiation ($F_{ST} = 0.06$).

Phylogeography in Apis

The honeybee *Apis mellifera* is one of the best studied insect species. Both morphometric and genetic studies suggest there are several genetically different evolutionary lineages around the Mediterranean area (1, 26). They form hybrid zones in these native populations and also in new areas where they have been introduced (81). The African and European lineages can be distinguished on the basis of nuclear and mitochondrial markers that have been used for studying the spreading of the African bee *A. mellifera scutellata* from southern to northern America (65, 69, 138). Both mitochondrial and nuclear markers of the African type are found in high frequencies in the neotropics, indicating that both females and males of the African type have a selective advantage that affects the local gene pool. The Africanized bees swarm readily, causing flow in genes carried by females. Paternal gene flow is also asymmetric, and introgression from the feral African-type population into European-type apiaries is greater than in the opposite direction (44). The diagnosis of hives and populations is made problematic by polymorphism within the source populations. A theoretical simulation study predicts that the hybridization of the African and European bees in America will lead to polymorphism of mtDNA rather than to complete replacement by the African type (75; see also 134). From the sociobiological point of view, the Africanized honeybees provide a model and a natural experiment for studying how the social characteristics (mating behavior, swarming) affect the genetic structure of populations and for separating the roles of males and females in causing gene flow. Clearly, gene flow in the honeybee, as well as in some bumblebees (29), differs radically from that seen in many other social insects in which local populations can differentiate over short distances to the extent that may promote speciation.

ACKNOWLEDGMENTS

Our work is supported by the Swedish Natural Science Research Council and the EU-TMR network on social evolution.

> Visit the *Annual Reviews home page* at
> http://www.annurev.org.

Literature Cited

1. Arias MC, Sheppard WS. 1996. Molecular phylogenetics of honey bee subspeies (*Apis mellifera* L.) inferred from mitochondrial DNA sequence. *Mol. Phyl. Evol.* 5:557–66
2. Arnold G, Quenet B, Cornuet JM, Masson C, De Schepper B, et al. 1996. Kin recognition in honeybees. *Nature* 379:498
3. Balas MT, Adams ES. 1996. The dissolution of cooperative groups: mechanisms of queen mortality in incipient fire ant colonies. *Behav. Ecol. Sociobiol.* 38:391–99
4. Banschbach VS, Herbers JM. 1996. Complex colony structure in social insects. I. Ecological determinants and genetic consequences. *Evolution* 50:285–97
5. Beye M, Moritz RFA, Crozier RH, Crozier YC. 1996. Mapping the sex locus of the honeybee (*Apis mellifera*). *Naturwissenschaften* 83:424–26
6. Blanchetot A. 1991. Genetic relatedness in honeybees as established by DNA fingerprinting. *J. Hered.* 82:391–96
7. Blanchetot A, Packer L. 1992. Genetic variability in the social bee *Lasioglossum marginatum* and a cryptic undescribed sibling species, as detected by DNA fingerprinting and allozyme electrophoresis. *Insect Mol. Biol.* 1:89–97
8. Boomsma JJ. 1993. Sex ratio variation in polygynous ants. In *Queen Number and Sociality in Insects,* ed. L Keller, pp. 86–109. Oxford: Oxford Univ. Press
9. Boomsma JJ. 1996. Split sex ratios and queen-male conflict over sperm allocation. *Proc. R. Soc. London Ser. B* 263:697–704
10. Boomsma JJ, Grafen A. 1991. Colony-level sex-ratio selection in the eusocial Hymenoptera. *J. Evol. Biol.* 4:383–407
11. Boomsma JJ, Ratnieks FLW. 1996. Paternity in eusocial Hymenoptera. *Philos. Trans. R. Soc. London B* 351:957–75

12. Boomsma JJ, Wright PJ, Brouwer AH. 1993. Social structure in the ant *Lasius flavus*: multi-queen nests or multi-nest mounds? *Ecol. Entomol.* 18:47–53
13. Bourke AFG, Franks NR. 1995. *Social Evolution in Ants.* Princeton, NJ: Princeton Univ. Press
13a. Bourke AFG, Green HAA, Bruford MW. 1997. Parentage, reproductive skew and queen turnover in a multiple queen ant analysed with microsatellites. *Proc. R. Soc. London Ser. B* 264:277–283
14. Carew ME, Tay WT, Crozier RH. 1997. Polygyny via unrelated queens indicated by mitochondrial DNA variation in the Australian meat ant *Iridomyrmex purpureus. Insectes Soc.* In press
15. Chan GL, Bourke AFG. 1994. Split sex ratios in a multiple-queen ant population. *Proc. R. Soc. London Ser. B* 258:261–66
16. Chapuisat M. 1996. Characterization of microsatellite loci in *Formica lugubris* B and their variability in other ant species. *Mol. Ecol.* 5:599–601
17. Choudhary M, Strassmann JE, Solis CR, Queller DC. 1993. Microsatellite variation in a social insect. *Biochem. Genet.* 31:87–96
18. Cook JM, Crozier RH. 1995. Sex determination and population biology in the Hymenoptera. *Trends Ecol. Evol.* 10:281–86
19. Craig R, Crozier RH. 1979. Relatedness in the polygynous ant *Myrmecia pilosula. Evolution* 33:335–41
20. Crozier RH, Crozier YC. 1993. The mitochondrial genome of the honeybee *Apis mellifera*: complete sequence and genome organization. *Genetics* 133:97–117
21. Crozier RH, Pamilo P. 1996. *Evolution of Social Insect Colonies.* Oxford: Oxford Univ. Press
22. Danforth BN, Neff JL, Barretto-Ko P.

1996. Nestmate relatedness in a communal bee, *Perdita texana* (Hymenoptera: Andrenidae), based on DNA fingerprinting. *Evolution* 50:276–84

23. Davis SK, Strassmann JE, Hughes C, Pletscher LS, Templeton AR. 1990. Population structure and kinship in *Polistes* (Hymenoptera, Vespidae): an analysis using ribosomal DNA and protein electrophoresis. *Evolution* 44:1242–53

24. DeHeer CJ, Ross KG. 1997. Lack of detectable nepotism in multiple-queen colonies of the fire ant *Solenopsis invicta* (Hymenoptera: Formicidae). *Behav. Ecol. Sociobiol.* 40:27–33

25. Dias PC. 1996. Sources and sinks in population biology. *Trends Ecol. Evol.* 11:326–30

26. Estoup A, Garnery L, Solignac M, Cornuet JM. 1995. Microsatellite variation in honey bee (*Apis mellifera* L.) populations: hierarchical genetic structure and test of the infinite allele and stepwise mutation models. *Genetics* 140:679–95

27. Estoup, A, Scholl A, Pouvreau A, Solignac M. 1995. Monoandry and polyandry in bumble bees (Hymenoptera: Bombinae) as evidenced by highly variable microsatellites. *Mol. Ecol.* 4:89–93

28. Estoup A, Solignac M, Cornuet JM. 1994. Precise assessment of the number of patrilines and of genetic relatedness in honeybee colonies. *Proc. R. Soc. London Ser. B* 258:1–7

29. Estoup A, Solignac M, Cornuet JM, Goudet J, Scholl A. 1996. Genetic differentiation and island populations of *Bombus terrestris* (Hymenoptera: Apidae) in Europe. *Mol. Ecol.* 5:19–31

30. Estoup A, Solignac M, Harry M, Cornuet JM. 1993. Characterization of (GT)n and (CT)n microsatellites in two insect species: *Apis mellifera* and *Bombus terrestris*. *Nucleic. Acids Res.* 21:1427–31

31. Evans JD. 1993. Parentage analyses in ant colonies using simple sequence repeat loci. *Mol. Ecol.* 2:393–97

32. Evans JD. 1995. Relatedness threshold for the production of female sexuals in colonies of a polygynous ant, *Myrmica tahoensis*, as revealed by microsatellite DNA analysis. *Proc. Natl. Acad. Sci. USA* 92:6514–17

33. Evans, JD. 1996. Competition and relatedness between queens of the facultatively polygynous ant *Myrmica tahoensis*. *Anim. Behav.* 51:831–40

34. Evans JD. 1996. Queen longevity, queen adoption, and posthumous indirect fitness in the facultatively polygynous ant *Myrmica tahoensis*. *Behav. Ecol. Sociobiol.* 39:275–84

35. Fjerdingstad EJ, Boomsma JJ, Thorén P. 1997. Polyandry in the leafcutter ant *Atta colombica*—a high resolution microsatellite DNA analysis. *Heredity.* In press

36. Fletcher DJC, Blum MS. 1983. Regulation of queen number by workers in colonies of social insects. *Science* 219:312–14

37. Fondrk MK, Page RE Jr, Hunt GJ. 1993. Paternity analysis of worker honeybees using random amplified polymorphic DNA. *Naturwissenschaften* 80:226–31

38. Gadagkar R. 1991. On testing the role of genetic asymmetries created by haplodiploidy in the evolution of eusociality in the Hymenoptera. *J. Genet.* 70:1–31

39. Gadau J, Heinze J, Hölldobler B, Schmid M. 1996. Population and colony structure of the carpenter ant *Camponotus floridanus*. *Mol. Ecol.* 5:785–92

40. Gastreich KR, Strassmann JE, Queller DC. 1993. Determinants of high genetic relatedness in the swarm-founding wasp, *Protopolybia exigua*. *Ethol. Ecol. Evol.* 5:529–39

40a. Gertsch PJ, Fjerdingstad EJ. 1997. Biased amplification and the utility of spermatheca-PCR for mating frequency studies in Hymenoptera. *Hereditas.* In press

41. Gertsch P, Pamilo P, Varvio SL. 1995. Microsatellites reveal high genetic diversity within colonies of *Camponotus* ants. *Mol. Ecol.* 4:257–60

42. Goodnight KF, Strassmann JE, Klingler CJ, Queller DC. 1996. Single mating and its implications for kinship structure in a multiple-queen wasp, *Parachartergus colobopterus*. *Ethol. Ecol. Evol.* 8:191–98

43. Hagen RH, Smith DR, Rissing SW. 1988. Genetic relatedness among cofoundresses of two desert ants, *Veromessor pergandei* and *Acromyrmex versicolor* (Hymenoptera: Formicidae). *Psyche* 95:191–201

44. Hall HG. 1990. Parental analysis of introgressive hybridization between African and European honeybees using nuclear DNA RFLPs. *Genetics* 125:611–22

45. Hamaguchi K, Ito Y, Takenaka O. 1993. GT dinucleotide repeat polymorphisms in a polygynous ant, *Leptothorax spinosior* and their use for measurement of relatedness. *Naturwissenschaften* 80:179–81

46. Hamilton WD. 1964. The genetical evolution of social behaviour. I. *J. Theor. Biol.* 7:1–16

47. Hamilton WD. 1964. The genetical evolution of social behaviour. II. *J. Theor. Biol.* 7:17–32

48. Hasegawa E. 1995. Parental analysis using RAPD markers in the ant *Colobopsis nipponicus:* a test of RAPD markers for estimating reproductive structure within social insect colonies. *Insectes Soc.* 42:337–46

49. Hasegawa E, Yamaguchi T. 1995. Population structure, local mate competition, and sex-allocation pattern in the ant *Messor aciculatus. Evolution* 49:260–65

50. van der Have T, Boomsma JJ, Menken SBJ. 1988. Sex investment ratios and relatedness in the monogynous ant *Lasius niger* (L.). *Evolution* 42:160–72

50a. Hedrick PW, Parker JD. 1997. Evolutionary genetics and genetic variation of haplodiploids and X-linked genes. *Annu. Rev. Ecol. Syst.* 28: SUPPLY

51. Heinze J. 1995. Reproductive skew and genetic relatedness in *Leptothorax* ants. *Proc. R. Soc. London Ser. B* 261:375–79

52. Herbers JM, Grieco S. 1994. Population structure of *Leptothorax ambiguus*, a facultatively polygynous and polydomous species. *J. Evol. Biol.* 7:581–98

53. Herbers JM, Stuart RJ. 1996. Patterns of reproduction in southern versus northern populations of *Leptothorax* ants (Hymenoptera: Formicidae). *Ann. Entomol. Soc. Am.* 89:354–60

54. Hölldobler B, Wilson EO. 1990. *The Ants.* Cambridge, MA: Harvard Univ. Press

55. Hughes CR, Queller DC. 1993. Detection of highly polymorphic microsatellite loci in a species with little allozyme polymorphism. *Mol. Ecol.* 2:131–37

56. Hughes CR, Queller DC, Strassmann JE, Solis CR, Negrón-Sotomayor JA, Gastreich KR. 1993. The maintenance of high genetic relatedness in multi-queen colonies of social wasps. In *Queen Number and Sociality in Insects,* ed. L Keller, pp. 153–170. Oxford: Oxford Univ. Press

57. Hunt GJ, Page RE Jr. 1992. Patterns of inheritance with RAPD molecular markers reveal novel types of polymorphism in the honey bee. *Theor. Appl. Genet.* 85:15–20

58. Hunt GJ, Page RE Jr. 1995. Linkage map of the honey bee *Apis mellifera*, based on RAPD markers. *Genetics* 139:1371–82

59. Hunt GJ, Page RE Jr, Fondrk MK, Dullum CJ. 1995. Major quantitative trait loci affecting honey bee foraging behavior. *Genetics* 141:1537–45

60. Kaib M, Hussender C, Epplen C, Epplen JT, Brandl R. 1996. Kin-biased foraging in a termite. *Proc. R. Soc. London Ser. B* 263:1527–32

61. Keller L. 1995. Social life: the paradox of multiple queen colonies. *Trends Ecol. Evol.* 9:355–60

62. Keller L, Ross KG. 1993. Phenotypic basis of reproductive success in a social insect: genetic and social determinants. *Science* 260:1107–10

63. Keller L, Ross KG. 1993. Phenotypic plasticity and 'cultural transmission' of alternative social organizations in the fire ant *Solenopsis invicta. Behav. Ecol. Sociobiol.* 33:121–29

64. Kukuk PF, Sage GK. 1994. Reproductivity and relatedness in a communal halictine bee *Lasioglossum (Chilalictus) hemichalceum. Insectes Soc.* 41:443–55

65. Lee ML, Hall HG. 1996. Identification of mitochondrial DNA of *Apis mellifera* (Hymenoptera: Apidae) subspecies groups by multiplex allele-specific amplification with competing fluorescent-labeled primers. *Ann. Entomol. Soc. Am.* 89:20–27

66. Lester LJ, Selander RK. 1981. Genetic relatedness and the social organization of *Polistes* colonies. *Am. Nat.* 117:147–66

67. Lindström K, Berglind SÅ, Pamilo P. 1996. Variation of colony types in the ant *Formica cinerea. Insectes Soc.* 43:329–32

68. McCorquodale DB. 1988. Relatedness among nestmates in a primitively social wasp, *Cerceris antipodes* (Hymenoptera: Sphecidae). *Behav. Ecol. Sociobiol.* 23:401–6

69. McMichael M, Hall HG. 1996. DNA RFLPs at a highly polymorphic locus distinguish European and African subspecies of the honey bee *Apis mellifera* L. and suggest geographical origins of New World honey bees. *Mol. Ecol.* 5:403–16

70. Messier S, Mitton JB. 1996. Heterozygosity at the malate dehydrogenase locus and developmental homeostasis in *Apis mellifera. Heredity* 76:616–22

71. Metcalf RA, Whitt GS. 1977. Intra-nest relatedness in the social wasp *Polistes metricus.* A genetic analysis. *Behav. Ecol. Sociobiol.* 2:339–51

72. Moritz RFA, Haberl M. 1994. Lack of meiotic recombination in thelytokous parthenogenesis of laying workers of *Apis mellifera capensis* (the Cape honeybee). *Heredity* 73:98–102

73. Moritz RFA, Kryger P, Allsopp MH. 1996. Competition for royalty in bees. *Nature* 384:31

74. Moritz RFA, Kryger P, Koeniger G, Koeniger N, Estoup A, Tingek S. 1995.

High degree of polyandry in *Apis dorsata* queens detected by DNA microsatellite variability. *Behav. Ecol. Sociobiol.* 37:357–63

75. Moritz RFA, Meusel MS. 1992. Mitochondrial gene frequencies in Africanized honeybees (*Apis mellifera* L.): theoretical model and empirical evidence. *J. Evol. Biol.* 5:71–81

76. Mueller UG. 1991. Haplodiploidy and the evolution of facultative sex ratios in a primitively eusocial bee. *Science* 254:442–44

77. Mueller UG, Eickwort GC, Aquadro CF. 1994. DNA fingerprinting analysis of parent-offspring conflict in a bee. *Proc. Natl. Acad. Sci. USA* 91:5143–47

78. Nonacs P. 1990. Size and kinship affect success of co-founding *Lasius pallitarsis* queens. *Psyche* 97:217–28

79. O'Donell S. 1996. RAPD markers suggest genotypic effects on forager specialization in a eusocial wasp. *Behav. Ecol. Sociobiol.* 38:83–88

80. Oldroyd BP, Clifton MJ, Wongsiri S, Rinderer TE, Sylvester HA, Crozier RH. 1997. Polyandry in the genus *Apis*, particularly *Apis andreniformis*. *Behav. Ecol. Sociobiol.* 40:17–26

81. Oldroyd BP, Cornuet JM, Rowe D, Rinderer TE, Crozier RH. 1995. Racial admixture of *Apis mellifera* in Tasmania, Australia: similarities and differences with natural hybrid zones in Europe. *Heredity* 74:315–25

82. Oldroyd BP, Smolenski AJ, Cornuet JM, Crozier RH. 1994. Anarchy in the beehive. *Nature* 371:749

83. Oldroyd BP, Smolenski AJ, Cornuet JM, Wongsiri S, Estoup A, Rinderer TE, Crozier RH. 1995. Levels of polyandry in intracolonial genetic relationships in *Apis florea*. *Behav. Ecol. Sociobiol.* 37:329–35

84. Oldroyd BP, Sylvester HA, Wongsiri S, Rinderer TE. 1994. Task specialization in a wild bee, *Apis florea* (Hymenoptera: Apidae), revealed by RFLP banding. *Behav. Ecol. Sociobiol.* 34:25–30

85. Packer L, Owen RE. 1994. Relatedness and sex ratio in a primitively eusocial halictine bee. *Behav. Ecol. Sociobiol.* 34:1–10

86. Page RE Jr, Robinson GE, Fondrk MK, Nasr ME. 1995. Effects of worker genotypic-diversity on honey bee colony development and behavior (*Apis mellifera* L.). *Behav. Ecol. Sociobiol.* 36:387–96

87. Pamilo P. 1982. Multiple mating in *Formica* ants. *Hereditas* 97:37–45

88. Pamilo P. 1983. Genetic differentiation within subdivided populations of *Formica* ants. *Evolution* 37:1010–22

89. Pamilo P. 1984. Genotypic correlation and regression in social groups: multiple alleles, multiple loci and subdivided populations. *Genetics* 107:307–20

90. Pamilo P. 1991. Evolution of colony characteristics in social insects. I. Sex allocation. *Am. Nat.* 137:83–107

91. Pamilo P. 1991. Life span of queens in the ant *Formica exsecta*. *Insectes Soc.* 38:111–19

92. Pamilo P. 1993. Polyandry and allele frequency differences between the sexes in the ant *Formica aquilonia*. *Heredity* 70:472–80

93. Pamilo P, Rosengren R. 1983. Sex ratio strategies in *Formica* ants. *Oikos* 40:24–35

94. Pamilo P, Rosengren R. 1984. Evolution of nesting strategies of ants: genetic evidence from different population types of *Formica* ants. *Biol. J. Linn. Soc.* 21:331–48

95. Pamilo P, Seppä P. 1994. Reproductive competition and conflicts in colonies of the ant *Formica sanguinea*. *Anim. Behav.* 48:1201–6

96. Pamilo P, Sundström L, Fortelius W, Rosengren R. 1994. Diploid males and colony level selection in *Formica* ants. *Ethol. Ecol. Evol.* 6:221–235

97. Pamilo P, Varvio-Aho SL. 1979. Genetic structure of nests in the ant *Formica sanguinea*. *Behav. Ecol. Sociobiol.* 6:91–98

98. Paxton RJ, Tengö J. 1996. Intranidal mating, emergence, and sex ratio in a communal bee *Andrena jacobi* Perkins 1921 (Hymenoptera: Andrenidae). *J. Insect Behav.* 9:421–40

99. Paxton RJ, Thorén PA, Tengö J, Estoup A, Pamilo P. 1996. Mating structure and nestmate relatedness in a communal bee *Andrena jacobi* (Hymenoptera, Andrenidae), using microsatellites. *Mol. Ecol.* 5:511–19

100. Pearson B. 1983. Hybridisation between the ant species *Lasius niger* and *Lasius alienus*: the genetic evidence. *Insectes Soc.* 30:402–11

101. Pearson B, Raybould AF, Clarke RT. 1995. Breeding behaviour, relatedness and sex-investment ratios in *Leptothorax tuberum* Fabricius. *Entomol. Exp. Appl.* 75:165–74

102. Peters JM, Queller DC, Strassmann JE, Solis CS. 1995. Maternity assignment and queen replacement in a social wasp. *Proc. R. Soc. London Ser. B* 260:7–12

103. Pfennig DW, Reeve HK. 1993. Nepotism in a solitary wasp as revealed by DNA fingerprinting. *Evolution* 47:700–4

104. Queller DC, Goodnight KF. 1989. Estimating relatedness using genetic markers. *Evolution* 43:258–75

105. Queller DC, Hughes CR, Strassmann JE. 1990. Wasps fail to make distinctions. *Nature* 344:388

106. Queller DC, Peters JM, Solis CR, Strassmann JE. 1997. Control of reproduction in social insect colonies: individual and collective relatedness preferences in the paper wasp, *Polistes annularis*. *Behav. Ecol. Sociobiol.* 40:3–16

107. Queller DC, Strassmann JE, Hughes CR. 1988. Genetic relatedness in colonies of tropical wasps with multiple queens. *Science* 242:1155–57

108. Queller DC, Strassmann JE, Hughes CR. 1992. Genetic relatedness and population structure in primitively eusocial wasps in the genus *Mischocyttarus* (Hymenoptera: Vespdae). *J. Hymenopterol. Res.* 1:115–45

109. Queller DC, Strassmann JE, Hughes CR. 1993. Microsatellites and kinship. *Trends Ecol. Evol.* 8:285–88

110. Queller DC, Strassmann JE, Solis CR, Hughes CR, DeLoach DM. 1993. A selfish strategy of social insect workers that promotes social cohesion. *Nature* 365:639–41

111. Ratnieks FLW. 1988. Reproductive harmony via mutual policing by workers in eusocial Hymenoptera. *Am. Nat.* 132:217–36

112. Ratnieks FLW, Reeve HK. 1992. Conflict in single-queen hymenopteran societies: the structure of conflict and processes that reduce conflict in advanced eusocial species. *J. Theor. Biol.* 158:33–65

113. Reeve HK, Ratnieks FLW. 1993. Queen-queen conflicts in polygynous societies: mutual tolerance and reproductive skew. In *Queen Number and Sociality in Insects*, ed. L Keller, pp. 45–85. Oxford: Oxford Univ. Press

114. Reeve HK, Westneat DF, Queller DC. 1992. Estimating average within-group relatedness from DNA fingerprints. *Mol. Ecol.* 1:223–32

115. Reichardt AK, Wheeler DE. 1996. Multiple mating in the ant *Acromyrmex versicolor*: a case of female control. *Behav. Ecol. Sociobiol.* 38:219–25

116. Reilly LM 1987. Measurements of inbreeding and average relatedness in a termite population. *Am. Nat.* 130:339–49

117. Richards MH, Packer L, Seger J. 1995. Unexpected patterns of parentage and relatedness in a primitively eusocial bee. *Nature* 373:239–41

118. Ross KG. 1986. Kin selection and the problem of sperm utilization in social insects. *Nature* 323:798–99

119. Ross KG. 1988. Differential reproduction in multiple-queen colonies of the fire ants, *Solenopsis invicta* (Hymenoptera: Formicidae). *Behav. Ecol. Sociobiol.* 23:341–55

120. Ross KG. 1992. Strong selection on a gene that influences reproductive competition in a social insects. *Nature* 355:347–49

121. Ross KG, Keller L. 1995. Ecology and evolution of social organization: insights from fire ants and other highly eusocial insects. *Annu. Rev. Ecol. Syst.* 26:631–56

122. Ross KG, Keller L. 1995. Joint influence of gene flow and selection on a reproductively important genetic polymorphism in the fire ant *Solenopsis invicta*. *Am. Nat.* 146:325–48

123. Ross KG, Robertson JL. 1990. Developmental stability, heterozygosity, and fitness in two introduced fire ants (*Solenopsis invicta* and *S. richteri*) and their hybrid. *Heredity* 64:93–103

124. Ross KG, Vargo EL, Fletcher DJC. 1987. Comparative biochemical genetics of three fire ant species in North America, with special reference to the two social forms of *Solenopsis invicta* (Hymenoptera: Formicidae). *Evolution* 41:979–90

125. Ross KG, Vargo EL, Keller L. 1996. Social evolution in a new environment: the case of introduced fire ants. *Proc. Natl. Acad. Sci. USA* 93:3021–25

126. Ross KG, Vargo EL, Keller L, Trager JC. 1993. Effect of a founder event on variation in the genetic sex-determining system of the fire ant *Solenopsis invicta*. *Genetics* 135:843–54

127. Roubik DW, Weigt LA, Bonilla MA. 1996. Population genetics, diploid males, and limits to social evolution in euglossine bees. *Evolution* 50:931–35

128. Sasaki K, Satoh T, Obara Y. 1996. Cooperative foundation of colonies by unrelated foundresses in the ant *Polyrhacis moesta*. *Insectes Soc.* 43:217–26

129. Schwarz MP. 1987. Intra-colony relatedness and sociality in the allodapine bee, *Exoneura bicolor*. *Behav. Ecol. Sociobiol.* 21:387–92

130. Seppä P. 1996. Genetic relatedness and colony structure in polygynous *Myrmica* ants. *Ethol. Ecol. Evol.* 8:279.90

131. Seppä P, Gertsch P. 1996. Genetic relatedness in the ant *Camponotus herculeanus*. A comparison of estimates from allozyme

and DNA microsatellite markers. *Insectes Soc.* 43:235–43

132. Seppä P, Pamilo P. 1995. Gene flow and population viscosity in *Myrmica* ants. *Heredity* 74:200–9

133. Seppä P, Sundström L, Punttila P. 1996. Facultative polygyny and habitat succession in boreal ants. *Biol. J. Linn. Soc.* 56:533–51

134. Sheppard WS, Rinderer TE, Mazzoli JA, Stelzer JA, Shimanuki H. 1994. Gene flow between African- and European-derived honey bee populations in Argentina. *Nature* 349:782–84

135. Shoemaker DD, Ross KG. 1996. Effects of social organization on gene flow in the fire ant *Solenopsis invicta*. *Nature* 383:613–16

136. Shoemaker DD, Ross KG, Arnold ML. 1994. Development of RAPD markers in two introduced fire ants, *Solenopsis invicta* and *S. richteri*, and their application to the study of a hybrid zone. *Mol. Ecol.* 3:531–39

137. Shykoff JA, Schmid-Hempel P. 1991. Genetic relatedness and eusociality: parasite-mediated selection on the genetic composition of groups. *Behav. Ecol. Sociobiol.* 28:371–76

138. Smith DR, Taylor OR, Brown WM. 1989. Neotropical African bees have African mitochondrial DNA. *Nature* 339:213–15

139. Snyder, LE. 1992. The genetics of social behavior in a polygynous ant. *Naturwissenschaften* 79:525–27

140. Stille M, Stille B. 1992. Intra-and internest variation in mitochondrial DNA in the polygynous ant *Leptothorax acervorum* (Hymenoptera: Formicidae). *Insectes Soc.* 39:335–40

141. Stille M, Stille B. 1993. Intrapopulation nestclusters of maternal mtDNA lineages in the polygynous ant *Leptothorax acervorum* (Hymenoptera: Formicidae). *Insect Mol. Biol.* 1:117–21

142. Stille M, Stille B, Douwes P. 1991. Polygyny, relatedness and nest founding in the polygynous myrmicine ant *Leptothorax acervorum* (Hymenoptera; Formicidae). *Behav. Ecol. Sociobiol.* 28:91–96

143. Strassmann JE. 1989. Altruism and relatedness at colony foundation in social insects. *Trends Ecol. Evol.* 4:371–74

144. Strassmann JE, Barefield K, Solis CR, Hughes CR, Queller DC. 1997. Trinucleotide microsatellite loci for a social wasp, *Polistes. Mol. Ecol.* 6:97–100

145. Strassmann JE, Gastreich KR, Queller DC, Hughes CR. 1992. Demographic and genetic evidence for cyclical changes in queen number in a Neotropical wasp, *Polybia emaciata. Am. Nat.* 140:363–72

146. Strassmann JE, Hughes CR, Queller DC, Turillazzi S, Cervo R, et al. 1989. Genetic relatedness in primitively eusocial wasps. *Nature* 342:268–69

147. Strassmann JE, Solis CR, Barefield K, Queller DC. 1996. Trinucleotide microsatellite loci in a swarm-founding neotropical wasp, *Parachartergus colobopterus* and their usefulness in other social wasps. *Mol. Ecol.* 5:459–61

148. Stuart RJ, Page RE Jr. 1991. Genetic component to division of labor among workers of a leptothoracine ant. *Naturwissenschaften* 78:375–77

149. Sundström L. 1993. Genetic population structure and sociogenetic organization in *Formica truncorum*. *Behav. Ecol. Sociobiol.* 33:345–54

150. Sundström L. 1994. Sex ratio bias, relatedness asymmetry and queen mating frequency in ants. *Nature* 367:266–68

151. Sundström L, Chapuisat M, Keller L. 1996. Conditional manipulation of sex ratios by ant workers: a test of kin selection theory. *Science* 274:993–95

152. Tay WT, Cook JM, Rowe DJ, Crozier RH. 1997. Migration between nests in the Australian arid-zone ant *Rhytidoponera* sp. 12 revealed by DGGE analyses of mitochondrial DNA. *Mol. Ecol.* 6:403–11

153. Taylor PD. 1988. Inclusive fitness models with two sexes. *Theor. Popul. Biol.* 34:145–68

154. Thorén PA, Paxton RJ, Estoup A. 1995. Unusually high frequency of (CT)n and (GT)n microsatellite loci in a yellowjacket wasp, *Vespula rufa* (L.) (Hymenoptera: Vespidae). *Insect Mol. Biol.* 5:141–48

155. Trivers RL, Hare H. 1976. Haplodiploidy and the evolution of the social insects. *Science* 191:249–63

156. Visscher PK. 1996. Reproductive conflict in honey bees: a stalemate of worker egg-laying and policing. *Behav. Ecol. Sociobiol.* 39:237–44

157. Ward PS. 1983. Genetic relatedness and colony organization in a species complex of ponerine ants. I. Phenotypic and genotypic composition of colonies. *Behav. Ecol. Sociobiol.* 12:285–99

158. Woyciechowski M, Lomnicki A. 1987. Multiple mating of queens and the sterility of workers among eusocial Hymenoptera. *J. Theor. Biol.* 128:317–27

Annu. Rev. Ecol. Syst. 1997. 28:27–54

EVOLUTION OF EUSOCIALITY IN TERMITES

Barbara L. Thorne
Department of Entomology, University of Maryland, College Park,
Maryland 20742-4454; e-mail: bt24@umail.umd.edu

KEY WORDS: termite, Isoptera, Dictyoptera, social insects, eusocial evolution

ABSTRACT

Eusociality in Isoptera (termites) converges along many lines with colony organization and highly social behavior in the phylogenetically distinct insect order
Hymenoptera (ants, bees, wasps). Unlike the haplodiploid Hymenoptera, however, both sexes of Isoptera are diploid. Termite families thus lack asymmetric
degrees of genetic relatedness generated by meiosis and fertilization, so explanations for eusocial evolution based on such asymmetries are not applicable to
Isoptera. The evolution of eusociality in termites likely occurred in small families in which most helpers retained developmental flexibility and reproductive
options. A suite of ecological and life-history traits of termites and their ancestors may have predisposed them toward eusocial evolution. These characteristics
include familial associations in cloistered, food-rich habitats; slow development;
overlap of generations; monogamy; iteroparity; high-risk dispersal for individuals; opportunities for nest inheritance by offspring remaining in their natal nest;
and advantages of group defense. Such life-history components are particularly
persuasive as fostering social evolution because many are present in a broad group
of eusocial taxa, including Hymenoptera, beetles, aphids, thrips, naked mole rats,
and shrimp. The evolution of eusociality in Isoptera likely evolved in response to
a variety of contributing elements and the selective pressures that they generated.

INTRODUCTION

The evolution of eusociality in termites has been a classic evolutionary conun-
drum since Darwin (25) recognized that the life histories of individuals in social
insect colonies posed "special difficulties" to his theory of natural selection because the majority of individuals in a colony never reproduce. The perplexing

case of eusociality in the termites is rendered even more conspicuous by the abundance of work on the social evolution of Hymenoptera that emphasizes the haplodiploid genetic system of that group (42, 43, 45). Both male and female termites are diploid: Hence, the asymmetric degree of relatedness inherent between hymenopteran brothers and sisters, and between their sisters and their offspring, is not generated by meiosis and fertilization in termites. The discovery and investigation of eusociality in other diploid animals such as aphids (4, 5), beetles (57), naked mole rats (2), and shrimp (30) have escalated interest in mechanisms of eusocial evolution in taxa that do not have skewed degrees of genetic relatedness within families.

There is active discussion over the precise definition of the term eusocial, and over which species have life-history patterns that fit within this "most highly social" category (e.g. 18, 20, 119). Despite differences and nuances of opinion with regard to some other taxa, all extant termites (Isoptera) are considered eusocial. By the conventional definition (75, 139), the key element of eusociality is markedly skewed reproduction among members of the society (a distinct reproductive division of labor). In eusocial groups, a limited number of individuals are fertile and fecund, but most have reduced reproductive capacities or, in the extreme, are completely and permanently sterile. The other defining components of eusociality are cooperation in the care of brood within the group, and overlap of adult generations. These features taken together result in workers helping to rear their siblings and/or the offspring of reproductives in their parents' generation.

Comparative studies of the social biology of a spectrum of solitary through eusocial species of bees and wasps have elucidated the evolution of complex societies, but parallel research on termites (or ants) is impossible because all living species are eusocial. Observations on the biology of extant taxa cannot be used to convincingly reconstruct ancestral states prior to the evolution of worker subfertility or sterility. Crozier (22, p. 8) states, "Many aspects of the biology of forms such as ants, honey bees, and termites seem scarcely relevant to evaluating theories on the origin of eusociality, because for them (especially those ants with sterile workers) there is no option open to workers for selfish behavior. . . ." Once prototermites evolved through the "bottleneck" of eusociality, life-history constraints, especially reproductive division of labor, are presumed to have been essentially irreversible. The new evolutionary dynamic involving the respective fitness interests of reproductives and workers likely shifted the adaptive pathway to a very different trajectory because the selective influences would be altered markedly once workers evolved lowered reproductive potential. Thus studies of living termite species can be used to generate ideas of possible evolutionary pressures and scenarios but cannot appropriately or convincingly be used to test hypotheses or predictions regarding the evolution of eusociality in Isoptera.

Thus a definitive, testable evolutionary scenario for the evolution of highly social behavior in termites is unrealistic. The best that can be developed is a comprehensive hypothesis, or set of hypotheses, each consistent both with the biology of termites and with principles of evolution. In this review I describe aspects of the biology of primitive termites, and insights that they provide into early social evolution in ancestral groups. I then present a synopsis of the major hypotheses previously presented to explain the evolution of eusociality in Isoptera. Ecological and life-history correlates of preconditions of eusociality are then discussed in a section that emphasizes general patterns recognized across phylogenetically disparate eusocial animals, including insects, naked mole rats, and shrimp. Termites have a striking number of these characteristics, and specific ecological factors are discussed as compelling forces fostering eusocial evolution in Isoptera. The ecological and life-history characteristics are then integrated into a hypothetical evolutionary scenario describing termite ancestors and possible selective influences in the transition to eusociality.

The Biology of Primitive Living Termites

Although extant termites cannot appropriately be used to test theories or predictions regarding the evolution of eusociality in Isoptera, knowledge of the biology of primitive living groups may be constructively used to generate insights and constructs regarding social evolution in their ancestors. In this section I present a synopsis of what is known about the habitat, castes, and reproductive biology of living termites considered to have retained the most primitive social organization and developmental traits. A robust phylogeny of the Isoptera has not yet been established (54, 59, 130). *Mastotermes darwiniensis* Froggatt (Mastotermitidae) is generally considered, based on morphological criteria, the most primitive living termite, but the Termopsidae, particularly the relictual himalayan termite *Archotermopsis wroughtoni* Desneux, are recognized as the most primitive socially and with respect to caste differentiation (37, 51, 95, 130). The following discussion is based on the biology of termopsids, detailing the relatively scant information available on *Archotermopsis,* supplemented by references to the more intensively studied genus *Zootermopsis.*

Termopsids live in decaying wood. *Archotermopsis wroughtoni* lives in small colonies (30–40 individuals) under the bark and within dead stumps and logs of fallen conifers (51, 108). Termopsids are "one piece" nesters (1, 93), living in and consuming their host log. They do not forage away from the nest, and colonies do not leave one log to colonize another. When the nest log, food, or space resources are depleted, many individuals within the colony differentiate into alates and disperse (102).

A soldier differentiates within the first brood of all termites in which development of incipient colonies has been studied (including *Zootermopsis*, but not yet *Archotermopsis*) (68). In *Z. angusticollis*, the first soldier in a young colony inhibits the differentiation of additional soldiers, but if it is removed, a replacement soldier develops (15, 68). Termopsid soldiers may be of either sex; females are distinguished from males by the enlarged 7th sternum as in imagoes (51, 126). In *Archotermopsis*, all soldiers have gonads that are as well developed as those of mature alates (51). Because there is no indication of inhibition or degeneration of soldier reproductive organs, Imms (51, p. 142) suggested that they may frequently be functional; however, this has not yet been demonstrated. Fertile soldier-like males and females with mature gonads ("reproductive soldiers" or "neotenic soldiers") are known in six species of the Termopsidae (reviewed in 78). Normal female pre-soldiers (callow soldiers that will molt into soldiers) of *Z. angusticollis* have oviducts, eggs, and a seminal receptacle. Mature female soldiers, however, are clearly infertile, with arrested development of various portions of the reproductive organs. The testes of mature male soldiers produce apparently normal sperm, but they have nonproductive seminal vesicles that probably render the sperm nonfunctional (126). Thompson (126, p. 524) concludes that normal male and female soldiers of *Z. angusticollis* are sterile, "although near the ancestral state of fertility."

Imms (51) reports that the "worker-like forms" (presumably 4th instar or older apterous individuals) of *Archotermopsis* have extensive gonad development, and that their fat body development is equivalent to that of alates. Imms (51) observed a captive "worker-like" *A. wroughtoni* lay seven eggs. The eggs did not develop, but whether due to sterility, lack of fertilization, or laboratory conditions is unknown.

The term neotenic reproductive refers to any termite reproductive that is not derived from an alate (129). Neotenic reproductives differentiate within their natal colony, never dispersing to outbreed. They can develop from a variety of instars from individuals with or without wingpads. True neotenic reproductives have not been reported in *A. wroughtoni* (51, 108). Neotenic reproductives of both sexes, up to several hundred per colony (68), are found in the field in *Zootermopsis* colonies that have lost the original king and/or queen. When the primary (alate-derived) founding pair is present, neotenic reproductives are normally absent (68). Healthy primary reproductives produce pheromones, distributed via the anus, that inhibit the differentiation of neotenic reproductives (69). If only the queen is present, only male neotenics develop, and vice versa (15, 68). When isolated from functional reproductives, neotenics differentiate in as little as 3–4 weeks in *Z. angusticollis* (39, 141). In some kalotermitids, neotenics develop in isolated groups in as little as 8–10 days (71, 83). Multiple neotenics function together in Mastotermitidae, Termopsidae, and Rhinotermitidae colonies (65), but excess neotenics are eliminated by fighting in Kalotermitidae, leaving only one of each sex (66, 81, 110). Alates remaining in their natal nest to replace dead or senescing primary reproductives ("adultoid" reproductives in the Termitidae) are not known to occur within the basal families of Isoptera (92, 97).

Thus colonies of living species of primitive termites differ from the general portrayal of termite societies. The king and queen do not live with a group of sterile helpers. Some individuals are near a state of fertility, and most colony members (all but soldiers) have options to differentiate into functional reproductives under certain circumstances. Developmental flexibility and retained reproductive options appear to be prominent components of the society.

Primitive Termite "Workers" Retain Developmental and Reproductive Options

In all species of termites there is a separation into two developmental pathways, the sexual (imaginal or nymphal) line identified by the presence of wing buds, and the apterous path, leading to individuals that function as workers. Distinction of these two lines appears at various instars, depending on the group (95, 96). Plasticity of developmental options also varies. Pathways are relatively more plastic in the Termopsidae, Kalotermitidae, and some Rhinotermitidae, and are relatively more rigid in *Mastotermes darwiniensis* (the single extant species of the Mastotermitidae), Hodotermitidae, some Rhinotermitidae, and Termitidae (95, 96, 99, 100). Termite castes are not genetically determined. All individuals carry developmental instructions for all castes such that hormonal and other stimuli induce particular pathways of differentiation (63, 96).

Several morphologically and/or developmentally distinct groups function as helpers or "workers" within termite colonies. In the Termopsidae, Kalotermitidae, and some Rhinotermitidae (e.g. *Prorhinotermes*), later instar "larvae" (individuals without wing pads) and nymphs (individuals with wing pads) perform "worker" tasks within their colonies. Apterous larvae may molt into nymphs, thus "switching" to the imaginal developmental path (95) (Figure 1). In these same termite groups (and in the rhinotermitid genus *Reticulitermes*), there may also be pseudergates, first defined by Grassé & Noirot (38). Pseudergates are nonreproductive, helper individuals that diverge from the imaginal line via a regressive or stationary molt at a relatively late instar (99). The principal morphological difference between a pseudergate and a nymph is the absence or regression of wing pads in the pseudergate; neither the brain nor the sex organs regress (95). Pseudergates retain the capacity to revert to the nymphal and then imaginal stages, or they can differentiate into soldiers or neotenic reproductives.

Thus larvae, nymphs, and pseudergates all function as helpers within colonies of termopsids, kalotermitids, and some rhinotermitids, and all of these individuals retain the capacity to differentiate into fertile alates or neotenic reproductives (or soldiers). Unfortunately, the circumstances that induce or contribute toward

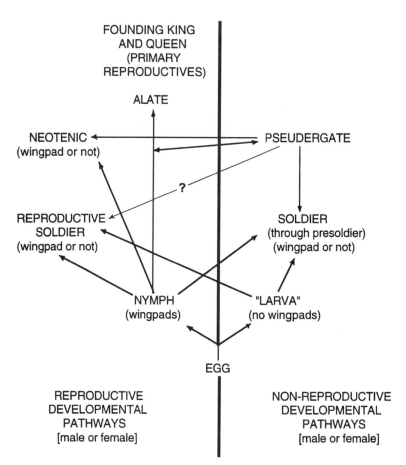

Figure 1   Developmental pathways in a primitive termite with flexible developmental options (e.g. *Zootermopsis*). Transitions can occur at various instars. Terminology as in Thorne (129).

differentiation along a particular pathway, and those that govern shifts from one path to another, are not fully understood (95).

A true *worker* termite is a nonreproductive, nonsoldier individual of the third or fourth (depending on the taxon) or a later instar that has diverged early and irreversibly from the imaginal line (94, 96, 99; irreversibly except in the rare cases that workers differentiate into replacement (ergatoid) reproductives—92, 97). True workers, a derived caste, are found in all living species of Mastotermitidae, Hodotermitidae, Serritermitidae (suggested, but not confirmed), and Termitidae, and in most species of Rhinotermitidae (*Prorhinotermes* is the only known rhinotermitid genus without true workers) (99, 100). Termites that forage away from their nest have a true worker caste (1, 50).

We lack a robust phylogeny of termite families but it appears that true workers evolved at least three times in the Isoptera (1, 50, 79, 82, 99, 100). Considering this along with the evolution of a monophyletic, nonreproductive soldier caste, Bourke & Franks (12) state that there is a minimum of four origins of reproductive altruism in termites (see also 32). A more parsimonious interpretation is that reproductive altruism in the form of a helper caste evolved only once in ancestral termites. The initial helpers directed some of their time and energy toward assistance within the colony (family), thereby at least delaying their own reproductive maturity and potential dispersal. The soldier caste evolved from the helper line with individuals first working and then in some cases becoming soldiers (98). Eventually, the highly constrained and essentially sterile true workers evolved, probably in several independent lines (as inferred from current phylogenetic hypotheses) (54) from species with developmentally and reproductively flexible helpers. Thus, eusociality in termites, characterized by reproductive altruism and division of reproductive labor, apparently evolved once. Evolution of sterile castes from already highly social ancestors followed.

A number of hypotheses have been proposed to explain the evolution of eusociality in termites. Five of the major hypotheses are summarized below along with an assessment of their merit as comprehensive evolutionary explanations.

PREVIOUS HYPOTHESES OF TERMITE SOCIAL EVOLUTION

Consensus on Subsociality

Two paths are recognized for the origin of social groups characterized by helpers and a reproductive division of labor. By the subsocial route societies originate from familial units initially composed of parents and offspring. Social groups may also form along the semisocial route as an association of individuals, related or not, from the same generation. Both pathways have been proposed for the Hymenoptera and vertebrates, although parental care may be a universal precursor of eusociality (2). In termite societies, however, both workers and soldiers are specialized juveniles, and there is no evidence for exchange of reproductives between different established colonies. Thus there is little doubt that termite sociality evolved via the subsocial route (2, 31, 56, 88, 106, 138). The wood diet of termites (and cryptocercid roaches), and reliance on cellulolytic protozoan symbionts rendered overlap of generations a necessity in these insects (17), thus providing opportunity for parental care and subsociality.

The Symbiont Transfer Hypothesis

Termites in the families Mastotermitidae, Kalotermitidae, Termopsidae, Hodotermitidae, and Rhinotermitidae (families collectively called the lower termites) (58) harbor distinctive groups of flagellate protists in their hindguts (13). These intestinal protozoa exist in mutualistic symbiosis with their hosts. The termites provide a habitat and mode of dispersal for the flagellates, and the protozoa digest cellulose consumed by the termites. Lower termites are thus obligately dependent on the intestinal symbionts for nutrition.

Each termite must acquire symbionts after hatching, and again following each molt. An initial innoculum of protozoa is passed to newly eclosed termites by feeding on the hindgut fluids of a nestmate. The gut and intestinal linings are shed during molts, and the symbionts are cast along with the exuvia (the protozoa do not encyst as occurs, for example, in the woodroach *Cryptocercus*) (17). Freshly molted termites may receive a new assemblage of symbionts by transfer via the hindgut fluids of a nestmate, or individuals may eat a freshly cast exoskeleton and reinstate the protozoa therein.

The impact of the intestinal symbionts on termite life history is fundamental because they are required for nutrition in all but the most derived family (Termitidae), and dependence on the protozoa precludes the option of solitary living in these hemimetabolous insects. Many authors, beginning with Cleveland et al (17) have noted that reliance on the intestinal flagellates predisposes termites to parental care and a social life, and that this symbiosis may have been one driving force in the evolution of sociality in this group (70, 84, 88). There is a broad consensus, however, that although the intestinal protozoa are a fundamental adaptation that requires group living and overlap of generations, this symbiotic relationship in termites does not by itself select for advanced components of eusociality such as reproductive division of labor (e.g. 3, 8, 60, 84, 121). Extended parent-offspring contact alone would facilitate transfer of symbionts. A similar symbiotic relationship with protozoa occurs in cryptocercid roaches within the context of a much simpler social system of extended parental care (84, 111). The symbiont transfer hypothesis, which has never been strongly advocated as a theory for the evolution of eusociality, is thus not persuasive except as an element predisposing termites to social living.

Hypotheses Based on Asymmetric Degrees of Relatedness

Many discussions regarding the evolution of eusociality in Hymenoptera are linked to the fact that their haplodiploid genetic system renders relatedness higher among sisters than between a queen and her female offspring. The high average degree of genetic relationship between sisters is seen as an explanation for the evolution of worker behaviors. Workers in ant, bee, and wasp colonies are female. By helping their mother raise their sisters (some of which will be fertile reproductives), workers likely increase their fitness (through inclusive fitness) over that which would be possible if they were solitary parents (45). Employing parallel reasoning, the cyclic-inbreeding and chromosome-linkage hypotheses have been proposed for the diplodiploid termites as mechanisms to generate similarly skewed degrees of relatedness among sibs in comparison to relatedness between parents and offspring. This would yield the consequences of haplodiploidy and, according to theory, facilitate kin selection as the driving force toward eusociality in termites.

CYCLIC-INBREEDING THEORY   Bartz (8, 9) constructs a hypothetical breeding pattern that would result in prototermite workers being more closely related to siblings than to offspring, thus favoring reproductive division of labor as reasoned for the Hymenoptera. This regime involves inbreeding, a condition that increases relatedness among family members and is therefore predicted to discourage selfishness within kin groups (45, 46, 77, 134). Bartz recognizes that although inbreeding increases an individual's relatedness to its siblings and to their offspring, it also boosts relatedness between individuals and their own offspring. Thus, if inbreeding continues, the extent to which inclusive fitness is increased by helping to raise siblings over offspring becomes limited, if present at all, and the selective force for such behavior is weakened (see also 60).

Bartz (8, 9) cleverly reasoned a pattern of cyclic inbreeding and outbreeding, staggered to generate and maintain asymmetries in relatedness that would favor helping behaviors. If new colonies are founded by a king and queen that are unrelated to one another but which each comes from inbred colonies, then their offspring will be relatively homogeneous and thus more closely related to one another than they would be to their own (outbred) offspring. If subsequent generations within the colony were inbred (products of parent-offspring or sib-sib matings within the nest), progeny would also be more closely related to siblings than to (outbred) offspring (see also 33, 104, 132). Hamilton (45, 46) noted that the ancestors of termites were likely restricted to the specialized habitat of decaying wood, leading to further opportunities for and predispositions toward inbreeding.

It is impossible to evaluate the reality of this type of breeding regime by studying extant termites because the model examines the case before establishment of a eusocial system, and evolutionary dynamics likely changed considerably once termites passed through the "bottleneck" of eusocial evolution. It is common, however, for termite colonies to replace a dead or senescing king or queen with one or more neotenic or adultoid reproductives that mature within the nest (95, 97, 129), thus providing opportunities for parent-offspring matings. If both original reproductives are gone, replacement reproductives often differentiate within the colony, leading to sibling-sibling, cousin-cousin, or other inbred matings. Fertile, dispersive progeny (alates) produced during this period would be inbred (and therefore relatively homozygous), and would fly to found a colony with a mate (potentially an outbred union). Thus, the fundamental life-history aspects of the model are plausible, but it is not possible to determine the extent to which ancestral termites fit the premises of the hypothesis. Kings and queens in extant species often live long enough to produce alates themselves (82). If at least some of the kings and queens are from inbred lineages, the asymmetries in relatedness will exist, but the degree of asymmetry will subside as the proportion of outbred reproductives increases (113).

CHROMOSOME-LINKAGE HYPOTHESIS   In some species of termites, a deviation from the expected relatedness structure of autosomal genes is caused by multiple reciprocal translocations among some of the chromosomes in males, apparently including the Y chromosome (73, 74, 122, 123, 133). These translocations result in the chromosomes forming rings instead of bivalents during meiosis, and segregating to the respective poles in linked clusters. Thus an entire set of translocated chromosomes moves together into Y-bearing gametes. In some species over half of the termite genome is involved in the translocation complex, and because it moves as a unit during meiosis and segregates like a single giant chromosome, the result may approach relatedness patterns typical of haplodiploidy. Lacy (60, 61) argued that these higher degrees of relationship could favor the evolution of altruistic behavior among siblings of the same sex.

In principle this "haplodiploid analogy" seems intriguing, but the translocations are absent in studied members of the primitive Mastotermitidae and Termopsidae, and the chromosome rearrangements appear to have arisen independently as a derived rather than ancestral condition (11, 23, 34). Further, no evidence has been found of sex-biased behaviors within termite colonies (41, 72). Leinaas (64) noted that if any move were made toward manipulation of the sex ratio of the brood, the male and female reproductives would generate a conflict of interest due to the linked genes within the translocation complex. In short, there is widespread consensus that multiple sex-linked reciprocal translocations were not a major factor, if any, in the evolution of eusociality in termites (23, 24).

The other two major hypotheses to explain eusocial evolution in termites involve intrafamilial dynamics and kin selection but do not depend upon generation of asymmetric degrees of genetic relatedness within a colony.

Shift-in-Dependent-Care Hypothesis

Nalepa (85, 88) discusses a possible scenario for describing how a prototermite familial group might make a transition to cooperative brood care and reproductive division of labor. The need for symbiont transfer to neonates, the relatively poor nutritional quality of wood, and life within a log habitat together selected for subsociality, monogamy, and slow development. Within this context, Nalepa postulates a behavioral shift within a young family: The older brood makes a transition from food recipients to food donors, thus shifting the responsibility for care of dependents from parents to older offspring. Because of the limiting nutrients that are directed toward neonates (and parents), individuals functioning as workers have depleted nutritional reserves and spend a prolonged period in the juvenile stages, although they may eventually become fertile, winged imagoes.

This scenario may reflect what occurred during the transition from a subsocial to a eusocial life history, but it does not address the selective dynamic favoring such a switch in responsibilities from parents to elder offspring. Why is it in the best interests of the "workers" to make this behavioral change, delaying or possibly foregoing their own reproductive potential to care for younger siblings and parents? Perhaps this is due to kin selection if a helper's inclusive fitness is boosted by this behavioral transition, or the change might be postulated as a result of parental manipulation (still kin selection) (12). The shift of care from parents to older offspring is important because it allows the parents to invest more rapidly in their next clutch, but their production of first fertile offspring is apparently delayed in this scenario because the older offspring spend longer periods in the juvenile stages. Thus, the fundamental evolutionary question remains: What were the selective forces that favored this shift in dependent care and other features of the life histories of termite reproductives and helping offspring? What evolutionary forces resulted in termite eusociality, whereas *Cryptocercus* family groups, living with similar life histories in similar habitats, remain subsocial?

Intragroup Conflict and Selection for a Helper Phenotype Hypothesis

Roisin (106) proposes that intragroup competition among late instar larvae or nymphs to reach the alate stage would result in conflicts that could lead to a helper phenotype in prototermites. He suggests that some individuals are deflected from alate development because of wing pad injuries inflicted as bites due to competition among sibs vying to become alates. Individuals with damaged wing pads became "lower status" helpers with reduced chances of future dispersal. Roisin cites as evidence of such conflicts the wing bud damage (generally interpreted as mutilations by other colony members) on nymphs of kalotermitids, termopsids, and some rhinotermitids. Nymphs with damaged wing buds cannot molt directly into alates but instead undergo regressive or stationary molts (to become pseudergates). Subsequent wing bud regeneration and formation of a normal alate is possible, but it requires a delay and additional molts (105, 107, 115, 120, 142). Roisin suggests that the mutilated "losers" in intracolony conflicts formed the original helpers in termites. At first only the losers would express the helper phenotype, but as helpers became increasingly efficient and effective in contributing to the colony, the phenotype was expressed by undamaged individuals because of indirect fitness benefits. This scenario would result in a caste of helpers that became fixed from a "condition sensitive" origin. This line of reasoning has been applied to the evolution of helpers in some bees (76) and wasps (137), and Roisin extended this theory to termites.

This is an intriguing hypothesis because it addresses some of the various interests among individuals within termite colonies, and how these may have contributed to the evolution of helpers. Three issues need to be considered in evaluating the likelihood of Roisin's scenario. The first is that although the "losers" that become helpers have a reduced or delayed chance of future dispersal as alates, they are positioned to be reproductive winners if the opportunity arises to become a replacement reproductive within the colony. Roisin (106) recognizes that reproduction within the established natal nest is less risky than a dispersal flight and colony initiation, but in the model the "losers" are viewed as helpers, rather than as hopeful reproductives. A second consideration regarding Roisin's theory pertains to developmental pathways. According to Roisin's constructs, the original helper phenotypes derived from nymphs deflected from alate development that were forced to regress to pseudergates. This occurs relatively late in the ontogeny of an individual. In modern termites individuals from at least the 4th instar on function as workers, generally without first differentiating into a wingbud form. Thus Roisin's earliest helpers develop indirectly and as older individuals. How the behavior would be expressed directly in young individuals is not explained. The third aspect of Roisin's theory that is difficult to interpret is the fact, acknowledged by Roisin (106 p. 757), that wing bud scars in the Termopsidae appear to be due to self-induced abcission, not mutilation by colony members as apparently occurs in kalotermitids (29, 49, 51, 106, 142; Thorne et al, personal observation). Work in our lab on complete colonies suggests that wing bud abcission in *Zootermopsis* occurs in the context of opportunities to become a replacement reproductive (BL Thorne et al, unpublished data), as had been noticed by Lenz & Runko (67) in the rhinotermitid *Coptotermes lacteus*. Behaviors and dynamics in modern termites are difficult to interpret, but the case of the termopsids is of interest because they are socially, morphologically, and developmentally more primitive than the kalotermitids and rhinotermitids. The reproductive biology of termites, and the contexts under which primitive termites lose wing buds, need to be better understood before this hypothesis can be fully evaluated.

None of these five hypotheses can be summarily dismissed, but based on current knowledge, none is completely persuasive. It should also be noted that the hypotheses are not mutually exclusive. The symbiont hypothesis is not overly compelling, although the symbiosis with protists enabled termites to specialize on a wood diet, a nutritionally weak food source that may have had its own implications, and the symbiosis required an overlap of generations for transfer of an innoculum of protozoa to juveniles. The chromosome linkage hypothesis was interesting when first proposed, but as data have accumulated on the phylogenetic distribution of the translocation complexes, and on lack of sex-biased behavior in termites, this hypothesis has been broadly discarded (23, 24).

It seems probable that eusociality in Isoptera evolved in response to a variety of contributing elements and the selective pressures that they generated. In addition to the hypotheses described above, a number of frequently overlooked factors warrant consideration as affecting the evolutionary dynamic of eusociality in termites. These pertain to the ecology and life history of termites and their ancestors, and these are particularly persuasive characteristics because many are present in a broad group of eusocial taxa, including Hymenoptera, beetles, aphids, thrips, naked mole rats, and shrimp (2, 19, 30, 42, 43, 57).

ECOLOGICAL AND LIFE HISTORY CORRELATES OF EUSOCIALITY

As eusociality is discovered in more and more animals and details of their life histories are unveiled, it is clear that a comparative approach may be productive in gaining insights into the evolution of highly social groups. Each evolutionary context was obviously unique, but there are enough characteristics shared among phylogenetically diverse eusocial animals to make a compelling case for the existence of suites of ecological and life-history characters that are correlates of eusociality, and in various clusters may serve as factors that foster the evolution of eusociality. A synopsis of some of these traits is presented in Table 1. The list of animal groups bearing each characteristic is likely incomplete; as more is learned about each eusocial species the table can be developed. Termites have each of the life-history components listed in Table 1 except for haplodiploidy. The following discussion expands upon each of those ecological characteristics as represented in primitive termites, along with a brief explanation about how these life-history features relate to eusocial evolution.

Primitive termites nest under the bark of large dead trees or logs. These environments provide a protected, food-rich habitat. The initial nest is a small, cloistered cavity, but is readily expanded into adjacent parts of the wood as a colony grows. In their confined nesting area, families live together for more than one generation, providing opportunities for kin-selected reproductive altruism and rendering it feasible for parents to capitalize on the food-gathering behaviors of their offspring (2, 3, 8, 16, 42, 43, 46, 102, 124). Parental care (subsociality) exists through the first several instars in young colonies, with brood

Table 1 Ecological and life history preconditions/correlates of eusociality

Precondition/correlate	Eusocial group	References
NESTING HABITAT		
Claustral Familial Associations		
• Safe, initially small, long-lasting (multigenerational), expandable, food-rich habitats; nesting in protected cavities keeps relatives together, thus providing opportunities for kin-selected reproductive altruism; Nesting aggregations make it easy for parents to parasitize the food-gathering behavior of their offspring	Eusocial Hymenoptera, termites, naked mole rats, shrimp, ambrosia beetles, gall-dwelling aphids, thrips	(2, 3, 8, 16, 19, 42, 43, 46, 52, 57, 102, 114, 124)
PARENTAL CARE (SUBSOCIALITY)		
• Family (kin groups)	Eusocial Hymenoptera, termites, aphids, ambrosia beetle, thrips, naked mole rat shrimp	(2, 4, 5, 17, 19, 30, 52, 57, 138)
DEVELOPMENT		
Slow Development, Long Generation Time, Overlap of Generations		
• Long life span, especially of reproductives (parents evolve to live longer than their helper offspring)	Eusocial Hymenoptera, termites, naked mole rat	(2, 46, 62, 112)
• Gradual metamorphosis (individuals begin helping as immatures, and can improve in helping ability as they age)	Termites, naked mole rat, shrimp	(2, 55, 98)
GENETICS		
Haplodiploidy		
• Genetic asymmetries increase the reproductive advantage of a female tending full siblings rather than producing offspring	Hymenoptera, thrips	(3, 19, 42, 43)
High Chromosome Numbers		
• Relatively high chromosome numbers would reduce the ability of sibs to differentiate among each other based on relatedness, and would reduce inclusive fitness variance among sibs, thus facilitating social evolution	Most social Hymenoptera, termites	(112, 118, 125; but see 3, 12, 26)

(Continued)

Table 1 *(Continued)*

MATING SYSTEM: SINGLE FATHER (MONOGAMY)		
• In haploid-diploid groups monogamy assures that sisters share all the genes from their father and thus, on average, 3/4 of their genes are identical by immediate descent	Primitive condition in social Hymenoptera	(42, 43)
• In diploid groups monogamy renders siblings genetically indifferent as to whether they rear fertile siblings or their own offspring	Termites, naked mole rat?, possibly shrimp	(16, 45)
REPRODUCTIVE CYCLE: ITEROPAROUS		
• Older offspring care for younger siblings	All eusocial animals	(2, 24, 112)
HIGH-RISK AND/OR TIME INVESTMENT IN DISPERSAL AND FOUNDING OF NEW NESTS		
• Remaining as a helper in the parental nest may be substantially safer and more efficient than attempting dispersal and successful development of a new nest	Hymenoptera, termites, naked mole rat	(27, 51, 89, 103)
OPPORTUNITIES FOR SUBSTANTIAL INHERITANCE		
Replacement Reproductive Opportunities		
• Possibility of maturing and reproducing within the group, either as a replacement or supplementary reproductive, and thus inheriting group resources	Termites, wasps (ants & bees produce male eggs, naked mole rat	(2, 3, 79, 139)
DEFENSE		
• Need for group defense against predators	Social insects	(12, 44, 70)
• Need for group defense against intra- and interspecific competitors	Termites, shrimp	(30, 131)
• Specialized defense: sting (facilitated the evolution of eusociality among Hymenoptera in exposed locations); major or soldier caste (ants, termites, aphids, thrips); major claw (shrimp)	Hymenoptera, aphids, thrips, shrimp	(2, 4, 6, 19, 30, 139)

care responsibilities largely transferred to older siblings as the colony grows (17, 31, 88, 109, 116, 138).

Termites have slow and gradual development, with relatively great longevity, especially of reproductives. The extended development time is thought to be due, at least in part, to the nutritionally impoverished diet of decayed wood (46, 62, 88, 117, 135). Life span of reproductives in the field is not known, and modern termites offer only inferences about ancestral characteristics, but a king of the primitive termite *Mastotermes darwiniensis* is known to have lived for 17 years in the laboratory (136). Termites have hemimetabolous development; thus, in contrast to Hymenoptera, they engage in social activities and assistance as soon as they pass the early instars (55, 98). A further consequence of gradual development in termites is that under certain circumstances larval and nymphal individuals can differentiate into neotenic reproductives, a widespread phenomenon in primitive termites (80, 82, 91, 98).

Termites generally have higher chromosome numbers than do related taxa, a characteristic that reduces the variance in the proportion of genes shared by siblings, thus making them less able to favor more closely related sibs and perhaps facilitating care and altruism among all siblings (112, 118, 125). The importance of high chromosome numbers is uncertain, however, because it is difficult to model how the trait would spread (3, 26). Also, it is dissuasive that a species in the primitive ant genus *Myrmecia* has a haploid chromosome number of one (21).

A monogamous reproductive pair normally cloisters to found a termite colony. Monogamy in a diploid organism results, on average, in both male and female siblings sharing one half of their genes. Because this degree of relatedness by descent is identical to the relatedness that termites share with their offspring, it is genetically (and fitness) neutral whether an individual termite produces fertile offspring or, by helping in its parents' colony, it boosts the production of fertile siblings by an equivalent number (16, 45). Alate dispersal and successful founding of new colonies are risky in termites (27, 51, 89, 102), thus there are clear safety advantages to remaining as a helper in the natal nest (79, 106). Temporal efficiency may also select for nondispersive helpers in termites because it takes time for queens to develop fully productive ovaries and a brood of helpers to support rearing her eggs. Individuals remaining in their parents' nest likely help their mother produce more offspring (including fertile progeny) than she would without their assistance, and workers partake of inclusive fitness benefits earlier, and with less risk, than if they sought direct fitness through independent colony development.

All eusocial insects are iteroparous. This leads to staggered age classes within colonies, enabling older offspring to rear younger siblings. It is notable that the woodroach *Cryptocercus* is semelparous (86), and not eusocial.

Termite societies offer several opportunities for substantial inheritance of re-sources to heirs within the colony. In the case of primitive termites, replacement reproductives differentiate among offspring following senescence or death of the original king and/or queen. In a decaying log, the presence of more than one colony will eventually result in intercolony interactions, which can lead to death of reproductives and opportunities for replacement (BL Thorne et al, unpublished data). Heirs also receive an established nest, food, a work- and defensive force of laborers and soldiers, and, possibly, close relatives that will differentiate into fertile alates. These possible inheritances constitute a poten-tially high payoff to some of the individuals that remain in the parent colony (79, 80). The fact that mechanisms exist for colony reproductive succession suggests that all individuals, both helpers and reproductives, are better assured persistence and thus an eventual pay-off of their investments.

A well-documented advantage to each individual living in a group is the benefit of cooperative defense against predators, parasites, and competitors (2, 32, 70, 139). In addition to the advantage of numbers, most eusocial groups have a specialized mode of defense. This may be weaponry present on all in-dividuals, such as the sting in eusocial Hymenoptera (females) or the major chela (claw) in shrimp (2, 30). Alternatively, some eusocial taxa have a mor-phologically specialized defensive subgroup, or caste, such as major ants or soldier termites, aphids, and thrips (4, 6, 19, 139). The soldier caste in termites is considered to be monophyletic, and to have evolved early in the evolution of the order, but probably after or concurrently with the evolution of eusociality (90, 100). Although the Hymenoptera sting and the shrimp claw may have been preadaptations favoring social options, group defensive behaviors and special-ized castes evolved along with social living. The risks of solitary living and the advantages of group defense may, however, have served as selective pressures for advanced social life in some or all of these organisms.

Thus a wide array of ecological characteristics of other eusocial animals, that are therefore viewed as potentially predisposing taxa toward eusociality, are also found in termites (Table 1). The fact that Isoptera have so many of these life-history characteristics lends credibility to the suggestion that some or all of these factors, probably acting in concert, served to channel prototermites toward evolution of eusociality.

TERMITE ANCESTORS AND THE TRANSITION TO EUSOCIALITY

There is a broadly accepted notion that termites are derived from a *Cryptocercus*-like ancestor (17, 85, 88, 121, 139), although the phylogenetic relationship be-tween Cryptocercidae and Isoptera remains the subject of investigation (7, 35,

54, 130). Similarities in life histories between *Cryptocercus* and termites may have been due to convergence rather than common ancestry (46). Because discussions of social evolution so frequently link *Cryptocercus* and termites, insight can be gained by comparing and contrasting the biology of the two groups, at least based on modern representatives. Both *Cryptocercus* and termites nest in decaying wood, do not forage out of their nest galleries, live in families with overlapping generations, and rely on similar intestinal symbionts for digestion. However, it is not clear whether the symbionts were acquired in each taxon through inheritance or by transfer (36, 87, 127, 128).

Given the similarities in claustral habitat, diet, and family groupings, one can ask why one of these dictyopteran groups evolved eusociality whereas the other has not. In contrast to termites, *Cryptocercus* are semelparous, with high parental investment in small broods that average about 20 individuals in *C. punctulatus* (84–86, 111). Termites are iteroparous and have higher fecundity, although colony sizes of the primitive termite *Archotermopsis wroughtoni* are thought to be relatively small, typically consisting of 30–40 individuals (108). In termites offspring are cared for by other offspring whereas wood roaches rely on parental care. Another major difference is that *Cryptocercus* are apterous whereas termites have a winged, dispersive reproductive form. Termites also have non-winged reproductive forms (neotenics) that differentiate when the original founders die, thus enabling an established colony to persist for many generations.

Therefore, despite several similarities, the social systems and life histories of *Cryptocercus* and termites are significantly different, and the former does not necessarily represent an intermediate evolutionary transition to the latter. It is thus important to consider possible step-by-step scenarios for the evolution of eusociality in termites, beginning with a solitary ancestor and following the transition through to eusocial groups. Such a sequence is obviously impossible to reconstruct or to test, but plausible alternatives, with evolutionary justification for each step, may be constructed and evaluated. Below I describe one such scenario for the evolution of eusociality in termites, based in large part on the composite of life-history characteristics identified in Table 1.

There is a consensus that eusociality in termites evolved in dead trees rather than evolving elsewhere with a subsequent transition into decaying wood (46). Thus solitary ancestors of termites likely fed upon decayed vegetation. They began to consume decayed wood, first facultatively and then, with the symbiosis with cellulolytic protists firmly established, as specialists and permanent residents with occasional bouts of dispersal. They probably first inhabited the cambial layers (cambium plus phloem) of dead trees, a relatively rich and well-balanced food source in modern trees (40, 46, 117), and probably in Mesozoic gymnosperms as well (10). After reproductive pairs colonized the layers directly under the bark, extended development times (that resulted from a wood diet) and the confined nest cavities of excavated wood favored sequestering of the monogamous family groups. Individuals did not leave the nest area to forage. All surviving offspring eventually matured into winged alates and dispersed. The claustral nests, relatively slow development, and need for reinfaunation of symbionts selected for parental care and long-term associations of small family groups (subsocial "colonies"). Iteroparous parents produced a brood of staggered age classes all living together.

As prototermites became increasingly specialized in habitat and with symbiotic protists, and as the cambial tissues of a log became crowded with competitors, the insects may have fed upon adjacent decaying sapwood. Colonizing the sapwood and heartwood layers would have expanded resources and improved protection from predators, but the nutritional quality of those tissues is poorer than that of the cambial region (40, 46, 117, 135) and may have further slowed individual growth rates. Slow growth resulted in older sibs remaining in the nest for a relatively long time before maturing into alates. They would thus have been poised to assist as adolescent helpers feeding and grooming dependent instars, defending the group if necessary, and perhaps feeding and grooming parents. The costs of intermittent sharing of food in terms of delayed growth rates of the helpers might have been relatively low because no transit energy would have been required to acquire food. Feeding capabilities of juveniles are conceivable because both king and queen termites feed the earliest instars of their first brood; thus parenting behaviors could reasonably be expressed in male and female immatures (2, p. 20). Potential costs of nest defense by immatures would have been high (i.e. injury or death), but defense risks are high independent of group or solitary nesting, and the cost per individual might well be lower in a group (44). Thus circumstances that might have facilitated offspring remaining in the nest and assisting in sibling brood care were present as a result of the primitive termite habitat and diet.

A significant transition in this dynamic may have begun as a shift toward a larger and more consistent commitment toward brood care by older instar sibs (88). An extended maturation time for older offspring in the colony (due to energetic allocations to brood care) delayed the age of first reproduction of F1 (helper) individuals, the inclusive fitness that F1 individuals gained from reproduction by sibs, and production of fertile progeny by the parental generation. These costs were presumably balanced by higher fecundity of reproductives, made possible because the parents were relieved of some dependent care responsibilities, freeing time and energy for additional egg production. Parents also had assistants to support care of a larger brood. In competition with other colonies, increased colony size may have been a strong advantage. Thus in its earliest stages, with relatively small colonies, delay in reproductive output as a result of sibling brood care was ultimately compensated by a higher probability of survival and production of a larger number of alates by the group as a whole.

The transition toward helping and delayed maturity would have required the evolution of a longer life span, but selection for longevity was already in progress as a result of slowed development due to the wood diet. The apparent altruism of older sibs delaying their own maturity to care for younger sibs can be explained by simple kin selection, plausible because in this monogamous, diploid system individuals share, on average, exactly as many genes by descent with their siblings as with their own offspring. Given the low probability that any individual termite alate would survive dispersal and successfully rear a brood to maturity, the fitness payoff may well have been higher and more assured to termites that remained to help boost sibling production in their already established parental nest.

In this evolutionary scenario thus far, the prototermite colonies have no morphological castes, and all surviving progeny eventually become dispersing alates. Two changes occurred in the next step of the social transition: the development of lifetime helpers and of neotenic reproductives within a colony. Because many offspring had delayed maturation and continued to assist within the natal nest, not all F1 individuals matured into alates, and a portion of the colony thus became lifetime helpers. It was at this point that the society could be considered fully eusocial (139), with overlapping generations, cooperative brood care, and a reproductive division of labor. The helpers were non-reproductive, but not sterile. All individuals could become neotenic reproductives even in relatively early instars, but inferences based on modern species suggest that neotenic differentiation probably occurred only after senescence or death of the primary reproductive of that sex. Sometime in this early evolution of termites, hormones became important in suppressing and releasing gonad development among individuals within the brood. The hormones were produced by functional reproductives and spread by trophallaxis (proctodeal trophallaxis had already evolved for transfer of hindgut symbiotic protists; oral trophallaxis was an established behavior for feeding dependent brood and, possibly, reproductives).

It need not have been altruism or parental manipulation that led to lifetime helpers in termite colonies. The evolutionary trade-off for F1 individuals was high-risk dispersal and improbable production of mature offspring versus the low risk of remaining in an established colony to rear siblings, coupled with a chance at the high pay-off of inheriting reproductive status as a neotenic should a parent senesce or die. Eusocial evolution occurred in small colonies. Primary reproductives may not have lived as long as they do in modern species, so opportunities for reproductive (and nest) inheritance may have been relatively high for individual helpers even in young colonies [a similar dynamic to "helpers at the nest" in birds (14)]. Differentiation of neotenics as replacement reproductives would be an advantage to the whole family in carrying on the colony. Each helper would "prefer" to become a reproductive, but at worst reproductive siblings would inherit the colony. Suppose, for example, that the original queen survived but that the king was replaced by one of his offspring. A helper in the colony would then be an offspring of the queen and a sibling of the neotenic replacement male. That helper would still share, on average, one half of its genes with progeny of that inbred union. Even if both primary reproductives died and were replaced by offspring, a helper sib of the new reproductives would rear inbred nieces and nephews, with potential rewards of substantial inclusive fitness.

Individuals that became extended helpers retained the option of eventual dispersal as alates. Primitive termites have remarkably flexible development, maintaining the ability to molt from apterous to brachypterous lines, and vice versa (99). In certain circumstances, such as depletion of host log resources or presence of numerous competitor colonies, a low probability of colony survivorship might induce a large percentage of individuals within the colony to become alates rather than to remain as helpers.

There is one exception to the options for flexible development, and another feature of termite evolution that must be addressed: the soldier caste. Soldiers apparently evolved early because they appear to be monophyletic among extant termites (48, 95, 99). The soldier caste is terminal; soldiers do not molt and therefore lack further developmental flexibility (92, 95, 99, 100). Their mandibles are clearly derived, extended versions of nymphal mandibles, with homologous dentition (48). In primitive living termites, soldiers differentiate from a variety of instars [normally beginning with the 4th (95)], and from both apterous and brachypterous individuals (126). All soldiers of *Archotermopsis wroughtoni* have fully developed gonads (51), and fertile reproductive soldiers (or "soldier neotenics") differentiate in some Termopsidae (78). Reproductive soldiers are normally the first replacement reproductives to differentiate in young, orphaned colonies of the primitive dampwood termite *Zootermopsis nevadensis* (BL Thorne et al, unpublished data). The first termite soldiers may have retained reproductive capability, but it is unclear whether the soldier morph appeared and was selected for as a defensive caste or as replacement or supplementary reproductives. If the latter occurred, then subsequent evolution of gonad degeneration of most individuals differentiating along that soldier pathway would result in what is now a typically sterile defensive caste.

As termite species radiated, derived traits such as construction of nests, foraging away from the nest, and constrained developmental pathways (e.g. true workers) evolved. An overwhelming majority of the 2000+ extant species of termites have this highly derived colony structure, with discrete and canalized castes; a distinct division of labor; vestigial gonads and effective sterility of all soldiers and, except under very unusual circumstances, all workers; a long-lived king and physogastric queen; large colony population size; and highly organized foraging behaviors. Many of these derived species build complex nest structures that provide a relatively homeostatic environment for reproductives and brood, and serve as a headquarters for foragers. Termites are ecologically important and conspicuous members of many temperate and tropical communities. Their success may be in large part due to the fact that they are eusocial, with the advantage of efficient allocation of tasks and resources among cooperating individuals within flexible and resilient societies (140).

CONCLUSIONS

It is unlikely that eusociality in termites arose as a result of evolutionary forces acting on any one dynamic or on any single life-history component. Circumstances such as cyclic inbreeding (8), confined, subsocial groups with a poor diet (2, 46, 88), or intragroup competition (106) may have all provided impetus toward eusociality in termites, but at this point no single condition can be identified as the dominant driving force. The additional ecological and life-history attributes that termites share with other eusocial animals as apparent correlates of eusociality (Table 1) are a particularly compelling ensemble because termites have all of these except haplodiploidy. Against this framework of favorable preconditions one must still define evolutionary dynamics that would have promoted the most extreme eusocial characteristic, highly skewed reproductive division of labor within a colony.

In termites, the "colonies" in which eusociality evolved were small families. Individual prototermites in a young family would have faced three options. First, they could spend no time or energy helping, and instead develop directly into a winged alate. Dispersal and colony initiation would have been risky, with no direct fitness pay-off until a successful colony produced fertile offspring. A second choice would be for offspring within a family to kill their parents and take over reproduction in the excavated nest galleries. Such behavior would not be favored by natural selection because it is in the interest of offspring to have their parents (the "king and queen") keep producing their siblings, especially given the neutrality of genetic relatedness between offspring and siblings (one half in a diploid system). Further, parents would likely evolve mechanisms to supplant mutinies among progeny.

A final choice for offspring developing slowly within a monogamous, iteroparous family living in a confined cavity within an expandable resource would be to remain in the nest, for at least a while, to help rear siblings. In animals like termites this might have been an especially productive strategy because fertile siblings provide an identical fitness pay-off (genetic relatedness = one half) as offspring. Helping a colony to expand gives it a higher probability of survival, facilitating its persistence and continued production of fertile relatives. Further, the ability of termites to develop into neotenic reproductives offers the possibility that helper individuals may become replacement reproductives, which confers a fitness advantage augmented by inheritance of the nest, labor, and food resources. In primitive, developmentally flexible termites the helping alternative might have been relatively low risk because, except for soldiers and reproductives, all individuals retained the option of differentiating into an alate.

Thus the trade-off faced by individuals within small prototermite families was no helping, no boost in inclusive fitness, and early, high-risk dispersal as an alate versus temporary helping, potential replacement reproductive opportunities, and the cost of delayed high-risk dispersal as an alate. Ultimately, or perhaps immediately, some temporary helpers served their entire lifetime within the colony, thus becoming permanent helpers in a eusocial system.

Although we may never definitively identify and prove the driving forces behind the evolution of eusociality in termites, the probable life-history characteristics of their immediate ancestors suggest some compelling possible scenarios for eusocial evolution based on individual selection. Eusociality in Isoptera was probably fostered by a suite of contributing factors and the concurrent and cumulative selective pressures that they generated.

ACKNOWLEDGMENTS

I express sincere appreciation to NL Breisch, RF Denno, and ME Suàrez for insightful and constructive comments on earlier versions of this manuscript. Discussions with many distinguished isopterists and social insect biologists during the course of my career have helped me to formulate and synthesize ideas presented in this paper. In the regard, I particularly acknowledge the mentoring and insights of Bert Hölldobler, Michael Lenz, Charles Noirot, Bill Nutting, Yves Roisin, and Ed Wilson.

> Visit the *Annual Reviews home page* at
> http://www.annurev.org.

Literature Cited

1. Abe T. 1990. Evolution of worker caste in termites. In *Social Insects and the Environment,* ed. GK Veeresh, B Mallik, CA Viraktamath, pp. 29–30. New Delhi: Oxford & IBH
2. Alexander RD, Noonan KM, Crespi BJ. 1991. The evolution of eusociality. In *The Biology of the Naked Mole Rat,* ed. PW Sherman, JUM Jarvis, RD Alexan-

der, pp. 1–44. Princeton, NJ: Princeton Univ. Press
3. Andersson M. 1984. The evolution of eusociality. *Annu. Rev. Ecol. Syst.* 15:165–89
4. Aoki S. 1977. *Colophina clematis* (Homoptera, Pemphigidae), an aphid species with soldiers. *Kontyu* 45:276–82
5. Aoki S. 1982. Soldiers and altruistic

dispersal in aphids. In *Biology of Social Insects,* ed. MD Breed, CD Michener, HE Evans, pp. 154–58. Boulder: Westview

6. Aoki S. 1987. Evolution of sterile soldiers in aphids. In *Animal Societies: Theories and Facts,* ed. Y Ito, JL Brown, J Kikkawa, pp. 53–66. Tokyo: Jpn. Sci. Soc. Press

7. Bandi C, Sironi M, Damiani G, Magrassi L, Nalepa CA, et al. 1995. The establishment of intracellular symbiosis in an ancestor of cockroaches and termites. *Proc. R. Soc. London Ser. B* 259:293–99

8. Bartz SH. 1979. Evolution of eusociality in termites. *Proc. Natl. Acad. Sci. USA* 76:5764–68

9. Bartz SH. 1980. Correction. *Proc. Natl. Acad. Sci. USA* 77:3070

10. Beck CB, Wight DC. 1988. Progymnosperms. In *Origin and Evolution of Gymnosperms,* ed. CB Beck, pp. 1–84. New York: Columbia Univ. Press

11. Bedo D. 1987. Undifferentiated sex chromosomes in *Mastotermes darwiniensis* Froggatt (Isoptera: Mastotermitidae) and the evolution of eusociality in termites. *Genome* 29:76–79

12. Bourke AFG, Franks NR. 1995. *Social Evolution in Ants.* Princeton, NJ: Princeton Univ. Press. 529 pp.

13. Breznak JA. 1982. Intestinal microbiota of termites and other xylophagous insects. *Annu. Rev. Microbiol.* 36:323–43

14. Brown JL. 1987. *Helping and Communal Breeding in Birds: Ecology and Evolution.* Princeton, NJ: Princeton Univ. Press

15. Castle GB. 1934. The damp-wood termites of western United States, genus *Zootermopsis* (formerly, *Termopsis*) In *Termites and Termite Control,* ed. CA Kofoid, pp. 273–310. Berkeley: Univ. Calif. Press

16. Charnov EL. 1978. Evolution of eusocial behavior: offspring choice or parental parasitism? *J. Theoret. Biol.* 75:451–65

17. Cleveland LR, Hall SR, Sanders EP, Collier J. 1934. The wood-feeding roach *Cryptocercus,* its protozoa, and the symbiosis between protozoa and roach. *Mem. Am. Acad. Arts Sci.* 17:185–342

18. Costa J, Fitzgerald TD. 1996. Developments in social terminology: semantic battles in a conceptual war. *Trends Ecol. Evol.* 11:285–89

19. Crespi BJ. 1992. Eusociality in Australian gall thrips. *Nature* 359:724–26

20. Crespi BJ, Yanega D. 1995. The definition of eusociality. *Behav. Ecol.* 6:109–15

21. Crossland MWJ, Crozier RH. 1986. *Myrmecia pilosula,* an ant with only one pair of chromosomes. *Science* 231:1278

22. Crozier RH. 1982. Of insects and insects: twists and turns in our understanding of the evolution of eusociality. In *Biology of Social Insects,* ed. MD Breed, CD Michener, HE Evans, pp. 4–9. Boulder, CO: Westview

23. Crozier RH, Luykx P. 1985. The evolution of termite eusociality is unlikely to have been based on a male-haploid analogy. *Am. Nat.* 126:867–69

24. Crozier RH, Pamilo P. 1996. *Evolution of Social Insect Colonies.* Oxford: Oxford Univ. Press

25. Darwin C. 1859. *On the origin of species by means of natural selection, or the preservation of favoured races in the struggle for life.* London: John Murray

26. Dawkins R. 1982. *The Extended Phenotype.* Oxford: Freeman

27. Deligne J, Quennedy A, Blum MS. 1981. The enemies and defense mechanisms of termites. In *Social Insects,* ed. HR Hermann, pp. 2–76. New York: Academic

28. DeSalle R, Gatesy J, Wheeler W, Grimaldi D. 1992. DNA sequences from a fossil termite in Oligo-Miocene amber and their phylogenetic implications. *Science* 257:1933–36

29. Desneux J. 1906. The Kashmir termite, *Termopsis wroughtoni. J. Bombay Nat. Hist. Soc.* 17:293–98

30. Duffy JE. 1996. Eusociality in a coral-reef shrimp. *Nature* 381:512–14

31. Emerson AE. 1958. The evolution of behavior among social insects. In *Behavior and Evolution,* ed. A Roe, GG Simpson, pp. 311–35. New Haven, CT: Yale Univ. Press

32. Evans HE. 1977. Extrinsic versus intrinsic factors in the evolution of insect sociality. *BioScience* 27:613–17

33. Flesness NR. 1978. Kinship asymmetry in diploids. *Nature* 276:495–96

34. Fontana F. 1991. Multiple reciprocal translocations and their role in the evolution of eusociality in termites. *Ethol. Ecol. Evol.* 1:15–19

35. Grandcolas P. 1996. The phylogeny of cockroach families: a cladistic appraisal of morpho-anatomical data. *Can. J. Zool.* 74:508–27

36. Grandcolas P, Deleporte P. 1996. The origin of protistan symbionts in termites and cockroaches: a phylogenetic perspective. *Cladistics* 12:93–98

37. Grassé PP. 1986. *Termitologia.* Vol. III. Paris: Masson

38. Grassé PP, Noirot C. 1947. Le poly-morphism social du termite à cou jaune (*Calotermes flavicollis* F.). Les faux-ouvriers ou pseudergates et les mues regressives. *C. R. Acad. Sci. Paris* 224:219–21

39. Greenberg SLW, Stuart AM. 1982. Precocious reproductive development (neoteny) by larvae of a primitive termite *Zootermopsis angusticollis* (Hagen). *Insectes Soc.* 29:535–47

40. Haack RA, Slansky F Jr. 1987. Nutritional ecology of wood-feeding Coleoptera, Lepidoptera, and Hymenoptera. In *Nutritional Ecology of Insects, Mites, and Spiders*, ed. F Slansky, JG Rodriguez, pp. 449–86. New York: Wiley & Sons

41. Hahn PD, Stuart AM. 1987. Sibling interactions in two species of termites: a test of the haplodiploid analogy (Isoptera: Kalotermitidae; Rhinotermitidae). *Sociobiology* 13:83–92

42. Hamilton WD. 1964. The genetical evolution of social behavior I. *J. Theoret. Biol.* 7:1–16

43. Hamilton WD. 1964. The genetical evolution of social behavior II. *J. Theoret. Biol.* 7:17–52

44. Hamilton WD. 1971. Geometry for the selfish herd. *J. Theoret. Biol.* 31:295–311

45. Hamilton WD. 1972. Altruism and related phenomena, mainly in social insects. *Annu. Rev. Ecol. Syst.* 3:193–232

46. Hamilton WD. 1978. Evolution and diversity under bark. In *Diversity of Insect Faunas*, ed. LA Mound, N Waloff. *Symp. R. Entomol. Soc. London* 9:154–75. New York: Halsted

47. Hamilton WD, May RM. 1977. Dispersal in stable habitats. *Nature* 269:578–81

48. Hare L. 1937. Termite phylogeny as evidenced by soldier mandible development. *Ann. Entomol. Soc. Am.* 30:459–86

49. Heath H. 1927. Caste formation in the termite genus *Termopsis. J. Morphol. Physiol.* 43:387–425

50. Higashi M, Yamamura N, Abe T, Burns TP. 1991. Why don't all termite species have a sterile worker caste? *Proc. R. Soc. London Ser. B* 246:25–9

51. Imms AD. 1919. On the structure and biology of *Archotermopsis*, together with descriptions of new species of intestinal protozoa and general observations on the Isoptera. *Phil. Trans. R. Soc. London.* 209:75–180

52. Ito Y. 1989. The evolutionary biology of sterile soldiers in aphids. *Trends Ecol. Evol.* 4:69–73

53. Kambhampati S. 1996. Phylogenetic relationship among cockroach families inferred from mitochondrial 12S rRNA gene sequence. *Syst. Entomol.* 21:89–98

54. Kambhampati S, Kjer KM, Thorne BL. 1996. Phylogenetic relationship among termite families based on DNA sequence of mitochondrial 16S ribosomal RNA gene. *Insect Mol. Biol.* 5:229–38

55. Kennedy JS. 1947. Child labor of the termite society versus adult labor of the ant society. *Sci. Monthly* 65:309–24

56. Kennedy JS. 1966. Some outstanding questions in insect behavior. *Symp. R. Entomol. Soc. London* 3:97–112

57. Kent DS, Simpson JA. 1992. Eusociality in the beetle *Austroplatypus incompertus* (Coleoptera: Curculionidae). *Naturwissenschaften* 79:86–87

58. Krishna K. 1969. Introduction. In *The Biology of Termites*, ed. K Krishna, FM Weesner, 1:1–18. New York: Academic

59. Kristensen NP. 1995. Forty years' insect phylogenetic systematics. *Zool. Beitr. N.F.* 36:83–124

60. Lacy RC. 1980. The evolution of eusociality in termites: a haplodiploid analogy? *Am. Nat.* 116:449–51

61. Lacy RC. 1984. The evolution of termite eusociality: reply to Leinaas. *Am. Nat.* 123:876–78

62. LaFage JP, Nutting WL. 1978. Nutrient dynamics of termites. In *Production Ecology of Ants and Termites,* ed. MV Brian, pp. 165–244. Cambridge, Engl. Cambridge Univ. Press

63. Lefeuve P, Thorne BL. 1984. Nymph-soldier intercastes in *Nasutitermes lujae* and *N. columbicus* (Isoptera; Termitidae). *Can. J. Zool.* 62:959–64

64. Leinaas HP. 1983. A haplodiploid analogy in the evolution of termite eusociality? Reply to Lacy. *Am. Nat.* 121:302–04

65. Lenz M. 1985. Is inter- and intraspecific variability of lower termite neotenic numbers due to adaptive thresholds for neotenic elimination? Considerations from studies on *Porotermes adamsoni* (Froggatt) (Isoptera: Termopsidae). See Ref. 136a, pp. 125–46

66. Lenz M, Barrett RA, Williams ER. 1985. Reproductive strategies in Cryptotermes: Neotenic production in indigenous and "tramp" species in Australia (Isoptera: Kalotermitidae). See Ref. 136a, pp. 147–62

67. Lenz M, Runko S. 1993. Long-term impact of orphaning on field colonies of *Coptotermes lacteus* (Froggatt)

(Isoptera: Rhinotermitidae). *Insectes Sociaux* 40:439–56

68. Light SF. 1943. The determination of caste of social insects. *Q. Rev. Biol.* 18:46–63

69. Light SF. 1944. Experimental studies on ectohormonal control of the development of supplementary reproductives in the termite genus *Zootermopsis* [formerly *Termopsis*]. *Univ. Calif. Publ. Zool.* 43:413–54

70. Lin N, Michener CD. 1972. Evolution of sociality in insects. *Q. Rev. Biol.* 47:131–59

71. Lüscher M. 1952. Die Produktion und Elimination von Ersatz-Geschlechtstieren bei der Termite *Kalotermes flavicollis* (Fabr.) *Z. Vergl. Physiol.* 34:123–41

72. Luykx P, Michel J, Luykx J. 1986. The spatial distribution of the sexes in colonies of the termite *Incisitermes schwarzi* Banks (Isoptera: Kalotermitidae). *Insectes Sociaux* 33:406–21

73. Luykx P, Syren RM. 1979. The cytogenetics of *Incisitermes schwarzi* and other Florida termites. *Sociobiology* 4:191–209

74. Luykx P, Syren RM. 1981. Multiple sex-linked reciprocal translocations in a termite from Jamaica. *Experientia* 37:819–20

75. Michener CD. 1969. Comparative social behavior of bees. *Annu. Rev. Entomol.* 14:299–342

76. Michener CD. 1985. From solitary to eusocial: Need there be a series of intervening species? In *Experimental Behavioral Ecology and Sociobiology,* ed. B Hölldobler, M Lindauer. *Fortschritte der Zoologie.* 31:293–305. Stuttgart: Fischer

77. Michod RE. 1982. The theory of kin selection. *Annu. Rev. Ecol. Syst.* 13:23–55

78. Myles TG. 1986. Reproductive soldiers in the Termopsidae (Isoptera). *Pan-Pac. Entomol.* 62:293–99

79. Myles TG. 1988. Resource inheritance in social evolution from termites to man. In *The Ecology of Social Behavior,* ed. CN Slobodchikoff, pp. 379–423. New York: Academic

80. Myles TG. 1994. Causal factors in the origin of termite eusociality. *Proc. Congr. Int. Union Stud. Soc. Insect,* 12th, p. 50. Paris: Univ. Paris Nord

81. Myles TG, Chang F. 1984. The caste system and caste mechanisms of *Neotermes connexus* (Isoptera: Kalotermitidae). *Sociobiology* 9:163–321

82. Myles TG, Nutting WL. 1988. Termite eusocial evolution: a re-examination of Bartz's hypothesis and assumptions. *Q. Rev. Biol.* 63:1–23

83. Nagin R. 1972. Caste determination in *Neotermes jouteli* (Banks). *Ins. Soc.* 19:39–61

84. Nalepa CA. 1984. Colony composition, protozoan transfer and some life history characteristics of the woodroach *Cryptocercus punctulatus* Scudder (Dictyoptera: Cryptocercidae). *Behav. Ecol. Sociobiol.* 14:273–79

85. Nalepa CA. 1988a. Cost of parental care in the woodroach *Cryptocercus punctulatus* Scudder (Dictyopotera: Cryptocercidae). *Behav. Ecol. Sociobiol.* 23:135–40

86. Nalepa, C.A. 1988b. Reproduction in the woodroach *Cryptocercus punctulatus* Scudder (Dictyoptera: Cryptocercidae): mating, oviposition, and hatch. *Ann. Entomol. Soc. Am.* 81:637–41

87. Nalepa CA. 1991. Ancestral transfer of symbionts between cockroaches and termites: an unlikely scenario. *Proc. R. Soc. Lond., Ser. B* 246:185–89

88. Nalepa CA. 1994. Nourishment and the origin of termite eusociality. In *Nourishment and Evolution in Insect Societies,* ed. JH Hunt, CA Nalepa, pp. 57–104. Boulder, CO: Westview

89. Nalepa CA, Jones SC. 1991. Evolution of monogamy in termites. *Biol. Rev.* 66:83–97

90. Noirot C. 1955. Recherches sur le polymorphisme des termites supérieurs (Termitidae). *Ann. Sci. Nat. Zool.* 17:399–595

91. Noirot, C. 1956. Les sexués de remplacement chez les termites supérieurs (Termitidae). *Insectes Sociaux* 3:145–58

92. Noirot C. 1969. Formation of castes in the higher termites. In *The Biology of Termites,* ed. K Krishna, FM Weesner, 1:311–50. New York: Academic

93. Noirot, C. 1970. The nests of termites. In *The Biology of Termites,* ed. K Krishna, FM Weesner, 2:73–125. New York: Academic

94. Noirot C. 1982. La caste des ouvriers, élément majeur du succès évolutif des termites. *Rivista di Biologia* 75:157–95

95. Noirot C. 1985. Pathways of caste development in the lower termites. See Ref. 136a, pp. 41–57

96. Noirot C. 1985. The caste system in higher termites. See Ref. 136a, pp. 75–86

97. Noirot C. 1985. Differentiation of reproductives in higher termites. See Ref. 136a, pp. 177–86
98. Noirot, C. 1989. Social structure in termite societies. *Ethol. Ecol. Evol.* 1:1–17
99. Noirot C, Pasteels JM. 1987. Ontogenetic development and evolution of the worker caste in termites. *Experientia* (Basel) 43:851–60
100. Noirot C, Pasteels JM. 1988. The worker caste is polyphyletic in termites. *Sociobiology* 14:15–20
101. Noirot C, Thorne BL. 1988. Ergatoid reproductives in *Nasutitermes columbicus* (Isoptera, Termitidae). *J. Morphol.* 195:83–93
102. Nutting WL 1969. Flight and colony foundation. In *The Biology of Termites,* ed. K Krishna, FM Weesner, 1:233–82. New York: Academic
103. O'Riain MJ, Jarvis JUM, Faulkes CG. 1996. A dispersive morph in the naked mole rat. *Nature* 380:619–21
104. Pamilo P. 1984. Genetic relatedness and evolution of insect sociality. *Behav. Ecol. Sociobiol.* 15:241–48
105. Roisin Y. 1988. Morphology, development and evolutionary significance of the working stages in the caste system of *Prorhinotermes* (Insecta, Isoptera). *Zoomorphology* 107:339–47
106. Roisin Y. 1994. Intragroup conflicts and the evolution of sterile castes in termites. *Am. Nat.* 143:751–65
107. Roisin Y, Pasteels JM. 1991. Polymorphism in the giant cocoa termite, *Neotermes papua* (Desneux). *Insectes Sociaux* 38:263–72
108. Roonwal ML, Bose G, Verma SC. 1984. The Himalayan termite, *Archotermopsis wroughtoni* (synonyms *radcliffei* and *deodarae*). Identity, distribution and biology. *Rec. Zool. Surv. India* 81:315–38
109. Rosengaus RB, Traniello JFA. 1991. Biparental care in incipient colonies of the dampwood termite *Zootermopsis angusticollis* Hagen (Isoptera: Termopsidae). *J. Insect Behav.* 4:633–48
110. Ruppli E. 1969. Die elimination überzähliger ersatzgeschlichtstiere bei der termite *Kalotermes flavicollis* (Fabr.). *Insectes Sociaux* 16:235–48
111. Seelinger G, Seelinger U. 1983. On the social organization, alarm and fighting in the primitive cockroach *Cryptocercus punctulatus* Scudder. *Zeitschrift für Tierpsychol.* 61:315–33
112. Seger J. 1983. Conditional relatedness, recombination, and the chromosome number of insects. In *Advances in Herpetology and Evolutionary Biology,* ed.

AGJ Rhodin, K Miyata, pp. 596–612. Cambridge, MA: Harvard Univ. Press
113. Seger J. 1991. Cooperation and conflict in social insects. In *Behavioral Ecology: An Evolutionary Approach,* ed. JR Krebs, NB Davies, 2:338–73. Oxford: Blackwell
114. Seger J, Moran NA. 1996. Snappling social swimmers. *Nature* 381:473–74
115. Sewell JJ, Watson JAL. 1981. Developmental pathways in Australian species of *Kalotermes* Hagen (Isoptera). *Sociobiology* 6:243–324
116. Shellman-Reeve JS. 1990. Dynamics of biparental care in the dampwood termite, *Zootermopsis nevadensis* (Hagen): response to nitrogen availability. *Behav. Ecol. Sociobiol.* 26:389–97
117. Shellman-Reeve JS. 1994. Limited nutrients in a dampwood termite: nest preference, competition and cooperative nest defence. *J. Anim. Ecol.* 63:921–32
118. Sherman PW. 1979. Insect chromosome number and eusociality. *Am. Nat.* 113:924–35
119. Sherman PW, Lacey EA, Reeve HK, Keller L. 1995. The eusociality continuum. *Behav. Ecol.* 6:102–8
120. Springhetti A. 1969. Il controllo sociale della differenziazione degli alati In *Kalotermes flavicollis* (Isoptera). *Ann. dell' Univ. di Ferrara,* (Sezione 3), 3:73–96
121. Starr CK. 1979. Origin and evolution of insect sociality: a review of modern theory. In *Social Insects,* ed. HR Hermann, 1:35–79. New York: Academic
122. Syren RM, Luykx P. 1977. Permanent segmental interchange complex in the termite *Incisitermes schwarzi. Nature* 266:167–68
123. Syren RM, Luykx P. 1981. Geographic variation of sex-linked translocation heterozygosity in the termite *Kalotermes approximatus* Snyder (Insecta: Isoptera). *Chromosoma* 82:65–88
124. Taylor VA. 1978. A winged élite in a subcortical beetle as a model for a prototermite. *Nature* 276:73–75
125. Templeton A. 1979. Chromosome number, quantitative genetics and eusociality. *Am. Nat.* 113:937–54
126. Thompson CB. 1922. The castes of *Termopsis. J. Morphol.* 36:495–531
127. Thorne BL. 1990. A case for ancestral transfer of symbionts between cockroaches and termites. *Proc. R. Soc. Lond., Ser. B* 241:37–41
128. Thorne BL. 1991. Ancestral transfer of symbionts between cockroaches and

termites: an alternative hypothesis. *Proc. R. Soc. Lond., Ser. B* 246:191–95

129. Thorne BL. 1996. Termite terminology. *Sociobiology* 28:253–63

130. Thorne BL, Carpenter JM. 1992. Phylogeny of the Dictyoptera. *Syst. Entomol.* 17:253–68

131. Thorne BL, Haverty MI. 1991. A review of intracolony, intraspecific, and interspecific agonism in termites. *Sociobiology* 19:115–45

132. Tyson JJ. 1984. Evolution of eusociality in diploid species. *Theoret. Pop. Biol.* 26:283–95

133. Vincke PP, Tilquin JP. 1978. A sex-linked ring quadrivalent in Termitidae (Isoptera). *Chromosoma* 67:151–56

134. Wade MJ, Breden F. 1981. Effect of inbreeding on the evolution of altruistic behavior by kin selection. *Evolution* 35:844–58

135. Waller DA, LaFage JP. 1987. Nutritional ecology of termites. In *Nutritional Ecology of Insects, Mites, and Spiders*, ed. F Slansky, JG Rodriguez, pp. 487–532. New York: Wiley

136. Watson JAL, Abbey HM. 1989. A 17-year old primary reproductive of *Mastotermes darwiniensis* (Isoptera). *Sociobiology* 15:279–84

136a. Watson JAL, Okot-Kotber BM, Noirot C, eds. *Caste Differentiation in Social Insects.* Oxford: Oxford Pergamon

137. West-Eberhard MJ. 1987. Flexible strategy and social evolution. In *Animal Societies: Theories and Facts*, pp. 35–51. Jpn. Scientific Soc., Tokyo

138. Wheeler WM. 1930. Societal evolution. In *Human Biology and Racial Welfare*, ed. EV Cowdry, pp. 139–55. New York: Hoeber

139. Wilson EO. 1971. *The Insect Societies.* Cambridge, MA: Belknap

140. Wilson EO. 1992. *The Diversity of Life.* Cambridge: Belknap

141. Yin C-M, Gillot C. 1975. Endocrine control of caste differentiation In *Zootermopsis angusticollis* Hagen (Isoptera). *Can. J. Zool.* 53:1701–708

142. Zimmerman RB. 1983. Sibling manipulation and indirect fitness in termites. *Behav. Ecol. Sociobiol.* 12:143–45

Annu. Rev. Ecol. Syst. 1997. 28:55–83

EVOLUTIONARY GENETICS AND GENETIC VARIATION OF HAPLODIPLOIDS AND X-LINKED GENES

Philip W. Hedrick and Joel D. Parker

Department of Zoology, Arizona State University, Tempe, Arizona 85287-1501

KEY WORDS: effective population size, heterozygosity, inbreeding, selection, X-linked genes

ABSTRACT

The evolutionary genetics of haplodiploids and X-linked genes share many features and are different from diploid (autosomal) genes in many respects. For example, the conditions for a stable polymorphism, the amount of genetic load, and the effective population size are all expected to be quite different between haplodiploids or X-linked genes and diploids. From experimental data, the genetic load for X-linked genes is much less than autosomal genes and appears less for haplodiploids than for diploids. The observed amount of molecular variation for haplodiploids is much less than that for diploids, even more so than predicted from the differences in effective population size. Extensive recently published data suggest that the differences in variation for X-linked and autosomal genes for *Drosophila*, mice, and humans are consistent with the differences predicted theoretically based on the relative effective population sizes.

INTRODUCTION

The evolutionary genetics of all genes in haplodiploid species and X-linked genes in diploid organisms have many similarities and differ from autosomal genes in diploid organisms in a number of ways. In this review, we examine the commonalities of the evolutionary genetics of these two systems, particularly in light of recent molecular, theoretical, and experimental developments. Evolutionary factors such as gene flow and genetic drift should be identical in autosomal and X-linked genes in the same organism, except for some intrinsic differences between the two groups of loci, so that autosomal genes

55

act as a "control" for the amount of genetic variation and the impact of other evolutionary factors, such as selection and mutation, in X-linked genes. For haplodiploids, no such direct comparison is possible, and for evaluation of levels of genetic variation, diploid insects, which may have many other differences from haplodiploids, are the best standard.

More than 15% of all animal species are haplodiploid (129), most of them in the Hymenoptera (ants, bees, and wasps), but this number includes some other species, such as insects in the Thysanoptera (thrips), some Coleoptera (beetles), and some Homoptera (whiteflies and scales), as well as some arachnids (mites) and rotifers (34, 67). In most organisms with X chromosomes, less than 10% of the genes are X-linked and the remainder are autosomal, but in some species the proportion is much larger, e.g. about 38% of the euchromatin (and presumably the expressed genes) of *Drosophila robusta* is X-linked (18), while in *D. melanogaster* approximately 20% of the genes are on the X chromosome (36).

Several major similarities exist for genes in haplodiploids and X-linked genes.

1. For autosomal genes in diploids, both sexes inherit alleles equally from both parents. This is also true for females in haplodiploids and for females for X-linked genes. (We assume the homogametic sex is the female and the heterogametic sex is the male but recognize that in birds, snakes, and butterflies, the chromosomal constitution of the sexes are reversed; in those cases, the same conclusions for X-linked genes would hold, but the sexes would be reversed.) However, unlike diploids (the term diploids is used synonymously with autosomal genes in diploids hereafter), males for haplodiploids or X-linked genes inherit all their genes from their mother. For haplodiploids, this occurs because haploid males are the consequence of unfertilized eggs, and for X-linked genes, it is the consequence of not receiving an X-chromosome from the male parent. Thus, males in haplodiploids have no father and, although males do not inherit X-linked genes from their father, they do inherit autosomal and Y-linked genes from him.

2. Female haplodiploids or females with X-linked genes have twice as many copies of the genes as do males, a factor important in determining overall allelic frequencies, mutation rates, and rates of recombination.

3. Because females are diploid in haplodiploids and for X-linked genes and, as a result, can be heterozygous, selection is modeled in the same manner as for diploid genes with varying levels of dominance. On the other hand, because males have only one copy of a gene, selection in males is modeled in the same manner as haploid or gametic selection.

We do not have space to discuss some interesting aspects of the evolutionary genetics of haplodiploids or X-linked genes. Some important topics in haplodiploids, most related to social behavior, omitted here, include the evolution of sociality and sex ratios (12, 39a, 65, 123, 137), measures of relatedness (129), evolution of social organization (142), colony-level selection (113, 125, 126), sex-determining loci and production of diploid males (31, 32, 120, 130), and selection on worker-produced males (119). Similarly, we do not discuss some important topics related to X-linked genes, such as reproductive isolation and speciation (77, 159, 168, 172), meiotic drive (41, 89, 97), male-driven evolution (93, 149), distribution of transposable elements over chromosomes (24, 48, 103), and X-linked human genetic diseases (99). In addition, dosage compensation (6, 94) and quantitative genetics (79, 92, 116), topics related to understanding the evolutionary genetics of both haplodiploids and X-linked genes, are not discussed.

We first introduce some basic evolutionary genetic theory relating to a particular evolutionary factor (or factors) and give references for more detailed theoretical coverage. We also give pertinent experimental examples and molecular data, when available, that are relevant to the evolutionary genetic theory. Finally, we discuss the amount of molecular variation observed in various groups of haplodiploids and that observed for X-linked and autosomal genes. We then discuss possible evolutionary genetic explanations for these observations.

EVOLUTIONARY GENETICS

Allelic and Genotypic Frequencies

The estimation of allelic frequencies and their variances in females and males for a haplodiploid or a X-linked gene is straightforward and follows estimation for genes in diploid and haploid organisms, respectively (92). Because two thirds of the genes are in females and one third in males, the estimate of the mean allele frequency must be weighted accordingly. However, if the initial frequencies in the two sexes differ (127) and there are nonoverlapping generations, the allele frequencies in the two sexes will oscillate above and below the mean frequency and dampen to the mean frequency in a few generations (the difference in allelic frequency between the sexes is halved each generation) (81, 92). However, if there are overlapping generations, then allelic differences in frequency between the sexes generally dampen more quickly (33, 106). Clegg & Cavener (29) examined oscillations between the sexes in allele frequency for two X-linked allozyme loci 64 map units apart in *Drosophila melanogaster* and found their observations generally consistent with these predictions.

If there are different allelic frequencies in the two sexes, this leads to an expected excess of heterozygotes and a deficiency of homozygotes compared to Hardy-Weinberg proportions. With an observed heterozygosity of H, two alleles A_1 and A_2 with mean allelic frequencies of $\overline{p}$ and $\overline{q}$, and with frequencies of A_2 in females and males of q_f and q_m, respectively, the expected excess of heterozygotes is $H - 2\overline{pq} = 2(4q_f - q_m)(q_f - q_m)/9$ (72). For example, if the difference in allelic frequencies between the two sexes is 0.2, the excess of heterozygotes is 0.031. When the allelic frequencies are near 0.0 or 1.0, the expected heterozygosity is low, and there can be a considerable proportional excess of heterozygotes. If there is substructure to the population, a deficiency of heterozygotes may result (72, 92), which could cover up any excess of heterozygotes from different allelic frequencies in the sexes.

Selection

With directional selection and relatively small selective differences, allelic frequency differences in the two sexes converge rather quickly (58, 107). Assuming that the allelic frequencies in the two sexes are similar and that dosage compensation (defined here as when the difference in fitness between the two male genotypes is same as the difference in fitness between the two homozygotes in females), then the rate and pattern of response is approximately one third faster in haplodiploids or X-linked genes than in diploids (68) and is faster for all levels of dominance (2). This increased response results because all alleles are exposed to selection in males for haplodiploids and X-linked genes and not protected from selection in heterozygotes as in diploids. Further, a recessive detrimental allele is never covered up in males by a dominant allele, suggesting that the frequency of detrimentals for haplodiploids or X-linked genes should be lower than for diploid genes (see below). Also, a favorable allele should increase in frequency faster in a haplodiploid or an X-linked gene than in a diploid, in both an infinite and a finite population (2, 23; see also 70). Fisher's fundamental theorem, i.e. the rate of change in fitness is equal to the variance in fitness, holds well in haplodiploids or X-linked genes when there are relatively small selective effects and dosage compensation (68).

However, when there is an initial difference in the allelic frequency between the sexes and nonoverlapping generations, these differences may be perpetuated during selection. An experimental example is given in Figure 1, where the observed and expected change in allelic frequency for the X-linked mutant *w* (white) in *D. melanogaster* with equal initial allelic frequency in the two sexes (Figure 1*a*) is contrasted with the initial frequencies being different (Figure 1*b*) (69). In both examples, the *w* allele is rapidly eliminated, but the elimination is slightly slower when the initial frequencies are different. Both the oscillations in allelic frequency and the difference in rate of elimination are predicted theoretically using independently estimated fitness components (69).

Many researchers have determined the conditions for a stable polymorphism for genes in haplodiploids or X-linked genes (40, 60, 67, 124, 128, see other references in 67). When there is no selection at a particular gene in the males, i.e. the trait is sex-limited to females (83), the condition for a polymorphism in haplodiploids or X-linked genes is heterozygous advantage (in females), the

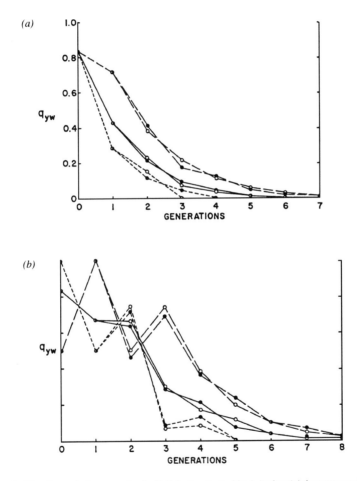

Figure 1   The change in frequency for the X-linked mutant white (*w*) when (*a*) there are equal initial frequencies in the two sexes and (*b*) when the initial frequencies are different. The *open* and *closed circles* indicate experimental results and theoretical predictions, respectively. The *solid, short-broken*, and *long-broken lines* refer to the mean, female, and male allelic frequencies, respectively (from 69).

same as for a diploid gene. Although these conditions are equivalent to those in diploids, the extent of selection (only in one sex) is lower so that, in combination with genetic drift or gene flow, the impact may be less.

With selection also acting in males, heterozygous advantage in females is neither a necessary nor a sufficient condition for a stable polymorphism. Figure 2a (after 124) gives the example for haplodiploids or X-linked genes where there is heterozygous advantage in females and differential selection in males (the relative fitness values are $1 - s_1$, 1, and $1 - s_2$ for genotypes A_1A_1, A_1A_2, and A_2A_2, respectively, in females (both haplodiploid or X-linked and diploid) and diploid males, and $1 - s_1$ and $1 - s_2$ for haplodiploid or X-linked males with genotypes A_1 and A_2, respectively). For diploids under this fitness regime, the whole region allows a stable polymorphism (s_1, $s_2 < 1.0$). However, for haplodiploids and X-linked genes, the region (given between the broken curved lines) is much smaller, only 58% that for diploids. Intuitively, this can be understood by realizing that when selection is very different between the two male genotypes (the upper left and lower right corners of Figure 2a), this strong directional selection overcomes the heterozygote advantage in females.

In addition, a stable polymorphism can be maintained if there is selection in opposite directions in the two sexes even though there is no heterozygote advantage in either sex (Figure 2b; the relative fitness values here are 1, $1 - s_1/2$, and $1 - s_1$ for genotypes A_1A_1, A_1A_2, and A_2A_2, respectively, in both haplodiploid or X-linked and diploid females, $1 - s_2$, $1 - s_2/2$, and 1 for diploid males with genotypes A_1A_1, A_1A_2, and A_2A_2, respectively, and $1 - s_2$ and 1 for haplodiploid or X-linked males with genotypes A_1 and A_2, respectively). When the two sexes have selection in opposite directions, then a stable polymorphism in a diploid can be maintained if the harmonic means of the homozygotes are less than those of the heterozygote (solid curved lines in Figure 2b, 85). For haplodiploids or X-linked genes, the region of stability is smaller (broken curved lines)—63% that in diploids (124). When there is differential selection between the sexes, the allelic frequencies at equilibrium may be greatly different (see 72 for an example) and, as a result, before selection there may be an excess of heterozygotes over Hardy-Weinberg expectations.

In both these theoretical cases, dosage compensation is assumed, i.e. the same difference occurs between the fitnesses in the two male genotypes as between the two female homozygotes. If there is no dosage compensation, and the difference between the fitness of the male genotypes is one half that of the female homozygotes, the region of a stable polymorphism is intermediate between those given for haplodiploids and diploids in Figure 2a and is smaller than that given for haplodiploids in Figure 2b.

Perhaps the most convincing example of a balanced polymorphism for a haplodiploid or an X-linked gene is that for the *Pgm*-3 locus in the fire ant

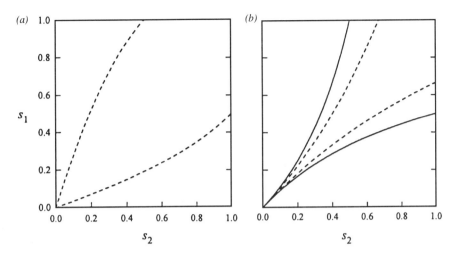

Figure 2 Regions of stability for a haplodiploid or X-linked gene (between the *curved broken lines*) or an autosomal gene [the *whole square* or between the *solid curved lines* for (*a*) and (*b*) (adapted from 124)] when there is heterozygote advantage in the females (*a*) and when there is selection in different directions in the two sexes (*b*). See the text for the definition of the selection coefficients (s_1 and s_2 values) for the two different situations.

(*Solenopsis invicta*), in which there is strong selection against homozygous *Pgm*-3[a] queens in polygynous populations and selection for this genotype in monogynous populations (141, 143). However, to describe this situation completely, a model with both selection and gene flow simultaneously is necessary (for a general theoretical introduction to selection and gene flow in haplodiploids or X-linked genes, see 114). Significant differences between the sexes in allelic frequency have been described for three of six allozyme loci examined in red wood ants; these differences may be the result of differential selection although it is not known if there is a stable polymorphism (127). Selection appears to significantly influence the maintenance of polymorphism for the X-linked allozyme loci G6PD and 6PDG in *D. melanogaster* (10). G6PD deficiency in humans, also X-linked, has long been thought to be associated with resistance to malaria (95, 105). Recently, an approximately 50% reduction in risk for severe malaria has been shown for both female heterozygotes and males with G6PD deficiency (144). There may have been a counterbalancing selective disadvantage of G6PD deficiency in the past, but at present in areas with endemic malaria, there appears to be overall directional selection favoring G6PD variants.

Effective Population Size

A haplodiploid or an X-linked gene has fewer copies of a given gene (only one copy in males) in a population than does a diploid gene. If random sampling of gametes is assumed, the effective population size for diploids is $N_{e.d} = 4N_f N_m/(N_f + N_m)$ and that for haplodiploids or X-linked genes is $N_{e.hd-X} = 9N_f N_m/(2N_f + 4N_m)$, where N_f is the number of breeding females and N_m is the number of breeding males (170). With equal sex ratios, the effective population size for haplodiploids or X-linked genes is $3/4$ that of diploids.

Crozier (37) observed that the ratio of the haplodiploid or X-linked to diploid effective population sizes varies for different sex ratios and that the ratio approaches $9/16$ with an extreme male-biased ratio and $9/8$ with an extreme female-biased sex ratio (the two effective populations sizes are equal at $7N_m/2$ when $N_f = 7N_m$). Several other authors (111, 168) have pointed out that differences in effective population size (and the resulting heterozygosity under the neutral model) between haplodiploids or X-linked genes and diploids depend on the mating system. The effective population sizes may be small at these sex-ratio limits: $N_{e.d}$ approaches four times the number of the less common sex at both extremes, and $N_{e.hd-X}$ approaches $9N_f/4$ for the extreme male-biased sex ratio and approaches $9N_m/2$ for extreme female-biased sex ratio.

An intuitive explanation for these ratios at the sex-ratio limits is as follows. When there are many more males than females, the diploids have a larger effective size because the contribution from the haploid males approaches one half that of diploid males; hence the low ratio value of $9/16$. When there are many more females than males so that females have a larger effective size, the female contribution is two thirds for the haplodiploid and only one half for the diploid; hence the ratio approaches the value greater than unity of $9/8$. Sex ratios in haplodiploids vary from high female-to-male in some parasitoids to high male-to-female in honey bees (65, 137). However, the effective number of matings for a given female may be generally small (12) so that the effective sex ratio may not deviate greatly from unity, thereby making the expected ratio of effective population sizes for haplodiploids and X-linked genes to that for diploids, all else being similar, close to $3/4$.

Several recent detailed theoretical treatments on the effective population size in haplodiploids and X-linked genes have included consideration of unequal family sizes and the sex of both the parent and offspring (16, 109, 134, 161). When there are overlapping generations, the equations in these references still hold if standardized by time to first reproduction and by letting N_f and N_m be equal to the number of females and males entering the breeding age in that time interval (75, 136). We discuss two of the more interesting predictions from these theoretical treatments.

Effective population size is a function of the variance in progeny number per parent and it is generally assumed to have a Poisson distribution, the expectation from random sampling of gametes. For a diploid gene, if the variances in progeny numbers are zero (all families have the same number of progeny), then $N_e = 2N$, where N is the number of parents (170). For a haplodiploid or a X-linked gene, the effect is slighter greater: $N_e = 9N/4$, when the variances in progeny number are zero (16). The higher value for haplodiploids or X-linked genes appears to occur because there is no segregation in these males. However, the two sexes have greatly different influences on this effect for haplodiploids or X-linked genes. When the variance in male progeny is Poisson and that for female progeny is zero, $N_e = 3N/2$; with the variance in male progeny zero and the variance in female progeny Poisson, $N_e = 9N/10$ (165). In other words, N_e is increased much more by a smaller variation in the number of female progeny per family than that in male progeny.

When there is inbreeding in a diploid population, the effective population size is $N_{e.d} = 4N_f N_m/[(1 + \alpha)(N_f + N_m)]$ (17), where α is the equilibrium inbreeding coefficient (73). When α is at its maximum of unity, $N_{e.d}$ is reduced to half its value when there is no inbreeding ($\alpha = 0$). For a haplodiploid or an X-linked gene with inbreeding, $N_{e.hd-X} = 9N_f N_m/[2N_f + 4(1 + \alpha)N_m]$ (163). When α is unity, $N_{e.hd-X}$ is reduced to 0.60 its value when there is no inbreeding. For a haplodiploid or an X-linked gene, the effect of inbreeding on effective population size is also a function of the sex ratio. When there is a high female-to-male sex ratio, there is no effect of different inbreeding levels on the effective population size, which is $9N_m/2$ for all α values. For a high male-to-female sex ratio, the effective population size is reduced from $9N_f/4$ with random mating to $9N_f/8$ when $\alpha = 1$, similar to the effect for diploids. At these extreme sex ratios, the effective population sizes could be small if the numbers of the rarer sex are low.

Inbreeding

Our discussion of inbreeding (and the related concept, "relatedness") will not be lengthy because relatedness is the subject of a review in this volume (129). Close inbreeding, particularly high levels of brother-sister mating, appears to be common in some wasps (66, 161) and ants (134). Several aspects of inbreeding distinguish haplodiploids or X-linked genes from those of diploids.

Because males are haploid, they cannot be inbred (identity by descent of two alleles can occur only in diploids). When one assumes the inbreeding coefficient in males is zero and averages inbreeding over both sexes, then haplodiploids and X-linked genes have lower expected inbreeding than do autosomal genes. The inbreeding coefficient calculated from a pedigree for haplodiploid or X-linked females may be different from that calculated for an autosomal gene. A procedure to calculate the inbreeding coefficient (f) from a pedigree for a haplodiploid or an X-linked gene is to use the expression $f = \Sigma[(1/2)^{N_f}(1 + f_{CA})]$, where the summation is over all common ancestors (CA), N_f represents the number of females in the chain linking the parents of the inbred individual to the common ancestor, and f_{CA} is the inbreeding coefficient in the common ancestors (35, 92, 170). Two important aspects differentiate this from the inbreeding coefficient for diploid genes, namely, only female ancestors are counted and if there are two or more consecutive male ancestors in the chain (for X-linked genes; haplodiploid males do not have a male parent), there can be no inbreeding. For example, if two half sibs mate and their common parent is a male, then there is no possibility of inbreeding in their offspring for an X-linked gene, even for female offspring.

In some human populations, the differences between sexes of the ancestors of two related individuals may actually make the inbreeding coefficient for an X-linked gene higher than for an autosomal gene. For example, first-cousin matings, which give an inbreeding coefficient of 1/16 for autosomal genes, give different inbreeding coefficients for females for the four different types of first-cousin matings, depending upon the sex of the parents of the first cousins. A first-cousin mating can result in an inbreeding coefficient of 0 for the two types of matings in which the parent of the male first cousin is a male, 1/8 when the parent of the male first cousin is a female and the parent of the female first cousin is a male, and 3/16 when both parents of the first cousins are female (the last is called a matrilineal or matrilateral cross-cousin mating, 145). Estimates of the inbreeding coefficients in a large number of couples of different ages, religions, castes, etc from South India were quite high for human populations (138, 145). Further, the inbreeding coefficient was significantly higher for X-linked than for autosomal genes: 0.051 versus 0.023 in rural Andhra Pradesh (145), 0.0411 versus 0.0371 in rural Tamil Nadu (138), and 0.0228 versus 0.0205 for urban Tamil Nadu (138). These differences were attributed to a cultural preference for matrilineal cross-cousin marriages over the other types of first-cousin marriages.

For diploids, the coefficient of relatedness in the parents of an inbred individual is twice the inbreeding coefficient for an inbred offspring from that mating (92, 169). This same relationship holds for two sisters in haplodiploids, as used in the development of kin selection theory (63, 64). However, for individuals of different ploidy levels, the direction of the comparison influences the level of relatedness. In haplodiploids and X-linked genes, there is an asymmetry in the relatedness of individuals of different sexes. For example, sister-to-brother relatedness is 1/2 while brother-to-sister relatedness is only 1/4 (for further discussion, see 128 and references therein).

Chapman & Stewart (20) surveyed seven polymorphic allozyme loci in a solitary wasp and used deviations from Hardy-Weinberg proportions to estimate the level of inbreeding. For all the loci, there was an extreme deficiency of heterozygotes; the estimated proportion of full-sib mating at inbreeding equilibrium (73) necessary to cause these deviations was 0.913. Because the inbreeding equilibrium is reached only after approximately five generations from full-sib mating, these high estimates indicate that similar levels of full-sib mating appear to have been present, without interruption, for a number of generations.

Two or More Loci

The estimation of gametic frequencies for two or more loci in diploids is complicated because it is not possible to observe directly the phase of the gametes in genotypes heterozygous at two or more loci, making it necessary to use a maximum likelihood procedure (74, 154). However, for haplodiploids or X-linked genes in males, the phase of all gametes can be observed directly, and the estimation procedure is identical to that for haploids or observed gametes (72; see 122 for an application in a bee population). The estimation of gametic frequencies over both sexes for a haplodiploid or an X-linked gene, if it assumed that there is no difference in gametic frequencies between the sexes, can use the estimates in males in the maximum likelihood estimation for females (154).

Because there is no recombination in males for both haplodiploids and X-linked genes ($c_m = 0$), and two thirds of the gametes are produced by females, the mean recombination rate is $\bar{c} = 2c_f/3 + c_m/3 = 2c_f/3$, or one third less than the rate for autosomal genes. (An interesting contrast is in *Drosophila* species that have no recombination in males for any genes so that the rate of recombination between autosomal genes, $c_f/2$, is lower than that for X-linked genes.) Therefore, the rate of decay of gametic or linkage disequilibrium (nonrandom association of alleles at different loci) for two loci, assuming equal gametic frequencies in the two sexes, is $D_t = D_0(1 - 2c_f/3)^t$, where D_t is the disequilibrium in the tth generation. If the gametic frequencies in the two sexes differ, the treatment is substantially more complicated (7, 29, 171). However, a difference in gametic frequencies in the two sexes increases the proportion of double heterozygotes somewhat and consequently increases the number of new recombinant gametes and the rate of decay of gametic disequilibrium. Further, with overlapping generations, the predicated rate of decay of disequilibrium is even further complicated (29, 135), as illustrated by the observations in experiments for two X-linked allozyme loci in *D. melanogaster* (29).

Hedrick (69) examined the change in the frequency of two tightly linked genes ($c_f = 0.015$), yellow (y) and white (w), on the X chromosome in *D. melanogaster* (Figure 3). Starting in maximum gametic disequilibrium and using estimates of the relative fecundity and mating ability of the different genotypes, the change in gamete frequencies and amount of disequilibrium over the 21 generations of the experiment was quite closely predicted, using independently estimated fitness components. In particular, the observed increase in the frequency of the wild-type gamete ($++$), which was produced by recombination during the experiment, was close to that predicted.

The maintenance of polymorphism, with or without disequilibrium, by two-locus selection depends on the maintenance of polymorphisms at both loci, a

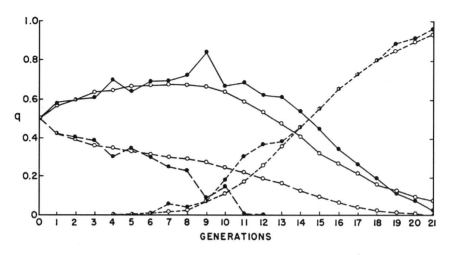

Figure 3 Frequency of gametes in males over generations when the population was initiated with equal numbers of +w and y+ gametes where w is the mutant white allele, y is the mutant yellow allele, and + is the wild type for the two genes. The *solid, long broken,* and *short broken lines* indicate the frequencies in males of +w, y+, and ++ gametes, respectively, and the *closed* and *open circles* indicate observed and simulated values (from 69).

condition that is more restrictive in haplodiploids or X-linked genes than in diploids, as discussed above. With selection limited to females, the conditions reduce to that in diploids except that the recombination is two thirds that for diploid genes the same distance apart, and the overall amount of selection acting is less than that for a diploid gene. The lowered recombination makes it easier to maintain a two-locus polymorphism with disequilibrium but the lowered amount of selection has the opposite effect. Overall it appears more difficult to maintain a two-locus polymorphism for haplodiploids or X-linked genes than for diploids (115). In addition, a certain level of epistasis is critical to maintain disequilibrium, an amount and type that probably is less likely to occur in haplodiploids or X-linked genes than in diploids.

Genetic hitchhiking may be more likely for haplodiploids or X-linked genes than for diploids, primarily because the overall rate of recombination is lower than in diploids. Theoretically this appears to be true, but Owen (115) has shown that for diploids without recombination in males, like *Drosophila*, the likelihood of hitchhiking is stronger than for haplodiploids or X-linked genes. In addition, haplodiploids or X-linked genes may have a smaller effective population size, which could generate disequilibrium, and a faster rate of change for favorable variants, which creates a better situation for genetic hitchhiking to occur (39, 71, 157).

Molecular data in *Drosophila* suggest that the amount of nucleotide variation varies greatly among genes and that there is generally a direct relationship between the level of recombination and the amount of variation observed (1, 5). Different regions of the X chromosome vary considerably in the amount of genetic variation and, for example, the telomeric region of the X chromosome in *D. melanogaster* has both low genetic variation and very low recombination (4). Currently two hypotheses, both based on genetic hitchhiking, seek to explain this association; they are termed selective sweeps (9) and background selection (22, 25). Briefly, the selective sweep hypothesis suggests that when an advantageous mutant arises and goes to fixation, it fixes neutral alleles in disequilbrium with it during the process. The background selection hypothesis suggests that detrimental alleles are spread over all chromosomes and are destined to be lost. When they decline in frequency, they reduce variation at neutral genes in disequilibrium with them. For both processes, the tighter the linkage, the larger the region influenced.

Mutation-Selection Balance and Inbreeding Depression

The input of mutations, which reduce fitness, and purifying selection, which eliminates these mutants, are thought to be two major factors influencing the frequency of detrimental or lethal alleles in a population. Because of the continuous exposure of detrimental or lethal alleles to selection in haploid males for haplodiploids and X-linked genes, it is generally assumed in a random-mating population that the frequency of detrimental alleles and the consequent genetic load would be lower than in diploids. If a population segregating for such fitness variation is inbred, the variants hidden as heterozygotes in a random-mating population are expressed, resulting in inbreeding depression. If there is less hidden variation, then the reduction in fitness upon inbreeding should be less, suggesting that inbreeding depression should be less in haplodiploids or X-linked genes than in diploids.

For example in random-mating diploids, the equilibrium frequency of a detrimental allele is $q_{e.d} = (u/s)^{1/2}$, and it is a function of the mutation rate (u) to this allele and of the amount of selection (s) against the allele, given that fitnesses of genotypes AA, Aa, and aa are 1, 1, and $1 - s$, respectively ($s = 1$ for lethals). For haplodiploids or X-linked genes, given that the fitnesses of male genotypes A and a are 1 and $1 - s$, respectively, then $q_{e.hd-X} = 3u/s$, a much lower level (59). To illustrate the difference, the ratio of these values is $q_{e.hd-X}/q_{e.d} = (9u/s)^{1/2}$, quite small for any reasonable values of s and u.

The equilibrium frequency of individuals with the lowered fitness is u/s for diploids. For haplodiploids or X-linked genes, this equilibrium is approximately $3u/2s$ in males and approaches zero in females, so the overall value, assuming an equal sex ratio, is $3u/4s$, or 25% lower than for diploids. In other words, the expected genetic load is less for a detrimental allele in haplodiploids or X-linked genes than in diploids. These calculations make a number of assumptions, including that the allelic frequencies are the same in the two sexes, the detrimental allele is recessive, and selection is equivalent in both sexes.

Werren (166) relaxed these assumptions to determine, in general, the expected differences in genetic load between haplodiploids or X-linked genes and diploids. His theoretical conclusions were that in random-mating populations at equilibrium, the genetic load in haplodiploid or X-linked females is significantly lower than that in haploid males of a haplodiploid species or diploids, which generally have similar loads. For many parameter combinations, the genetic load for haplodiploids or X-linked genes is approximately 25% less than that of diploids (see 166, Figure 3.1), although the apportionment between sexes varies for different parameter combinations. In addition, for female-limited load, both the equilibrium allele frequency and genetic load are still lower for haplodiploids or X-linked genes than for female-limited diploids.

Inbreeding can take two general forms (166): chronic, as in a population that has a continuous level of inbreeding, or acute, as when a random-mating population undergoes a sudden high level of inbreeding (this, of course, occurs when the effect of inbreeding on fitness is determined experimentally). For chronic inbreeding, a new level of equilibrium genetic load is reached that is reduced more for diploids than for haplodiploids or X-linked genes. For acute inbreeding, the fitness is reduced in the first generations much more for diploids because the genetic load in a random-mating diploid population is higher.

Several types of data are relevant to these predictions about genetic load and inbreeding depression. First, the genetic load appears to be much less for the X chromosome than for the autosomes in *Drosophila*. For example, the median proportion of autosomal chromosomes in *D. melanogaster, D. pseudoobscura*, and *D. persimilis* that have a semilethal or lethal gene in natural populations is approximately 25% (45). In two samples of X chromosomes, the overall frequency of semilethal or lethal chromosomes was only 2.2% (47, 54), much less than for autosomes even when the smaller size of the X chromosome is taken into account. Overall, the total genetic load for viability (about half the load for the autosomal chromosomes is from detrimentals and half from lethals and semilethals) estimated for autosomes in *D. melanogaster, D. pseudoobscura*, and *D. willistoni* is 0.609 (151), and only 0.046 or less for the X chromosome in *D. melanogaster* (47)—a very large difference.

Second, the amount of genetic load in haplodiploids also appears to be less than that in the autosomes in the *Drosophila* studies, although, the number of studies are smaller. Lowered fitness or more bilateral asymmetry upon inbreeding in honey bees has been reported (14, 15, 83), but more recent studies have not found such effects (27, 28, 82). In studies on both *Drosophila* (46, 54, 84) and honey bees (83), there is the suggestion that some of the observed viability load is limited to females. However, some of this effect may be artifactual (38) or due to lack of proper controls (47); the most extensive study of X-linked genetic load in *D. melanogaster* (47) showed no sex-limited genetic load for viability. On the other hand, the fitness reduction for X chromosomes observed in laboratory equilibrium experiments (167) may be the result of female-limited fertility effects.

Third, wild populations of *D. melanogaster* were surveyed in North Carolina and Great Britain for allozyme-null alleles, i.e., alleles that show no catalytic activity in their gel staining assay, at both autosomes and the X chromosome (90, 160). At 20 autosomal loci, 58 null alleles were found of 24,678 alleles screened, for an average frequency of 0.0024 (the North Carolina and British samples were quite similar). However, for these samples and a sample from Japan for five X-linked genes, no null alleles were found in 8,209 alleles screened, a highly significant difference. All but one of the null alleles were both viable and fertile (44). If these null alleles are assumed to be in mutation-selection equilibrium, the lower frequency of null alleles on the X chromosome than on the autosomes is consistent with theoretical predictions given above.

Neutrality

Under the neutrality theory, the amount of genetic variation is a function of the effective population size, variation being reduced through genetic drift, and mutation producing new variation. The expected heterozygosity for a diploid is $H_{e.d} = 4N_{e.d}u/(4N_{e.d}u + 1) = \theta_d/(\theta_d + 1)$, where u is the mutation rate to a new allele (87), the infinite-allele model. For a haplodiploid or X-linked gene, $H_{e.hd-X} = 4N_{e.hd-X}u/(4N_{e.hd-X}u + 1) = \theta_{hd-X}/(\theta_{hd-X} + 1)$ (108). Assuming a sex ratio of unity (and all other factors equivalent), $N_{e.hd-X} = 3N_{e.d}/4$, and $H_{e.hd-X} = 3N_{e.d}u/(3N_{e.d}u+1) = 0.75\theta_d/(0.75\theta_d+1)$ (98). For microsatellite loci, the stepwise-mutation model may be more appropriate (151, although see 50), and we will also give the ratio of θ values from the formula for the expected heterozygosity in this model, $H_e = 1 - 1/(1 + 8N_eu)^{1/2}$.

Similarly, the expectation at equilibrium of the amount of nucleotide diversity and the proportion of polymorphic sites for diploids is $\theta_d = \pi_d = 4N_{e.d}u$ (86) and for haplodiploids or X-linked genes is $\theta_{hd-X} = \pi_{hd-X} = 4N_{e.hd-X}u$. In this case, with a sex ratio of unity, $\theta_{hd-X} = \pi_{hd-X} = 3N_{e.d}u = 0.75\theta_d$ (104).

Avery (2) suggested that the observed variance of allozyme heterozygosity over species could be compared to its expectation under neutrality to determine if the pattern of allozyme variation is consistent with neutrality. This approach for both haplodiploid species within social categories (2) or haplodiploid species within genera (118) has shown good general agreement between the observed and theoretical variances, suggesting that the pattern of allozyme variation is consistent with neutrality.

There is some evidence that mutation rate in males is higher than that in females in primates (149) and rodents (19). If so, this results in a higher average mutation rate for autosomes than for X-linked genes because half of the genes are in males for an autosomal gene and only one third are for a X-linked gene. The average mutation rate for an autosomal (diploid) gene is $u_d = u_f/2 + u_m/2$ and that for a X-linked gene is $u_X = 2u_f/3 + u_m/3$. Assuming that the mutation rate in males is α times that in females, then $u_X/u_d = [2(2 + \alpha)]/[3(1 + \alpha)]$. As expected, this ratio is unity if there is no difference in mutation rate between the sexes and approaches two thirds as α becomes large (the estimates of α in primates and rodents are appproximately six and two, respectively). Using the expression above, when there is an equal sex ratio but assuming that $u_X = 2u_d/3$ then $H_{e.X} = 2N_{e.d}u_d/(2N_{e.d}u_d + 1)$ and $\theta_X = \pi_X = 2N_{e.d}u$, lowering the expected heterozygosity even further. Recently, there is the suggestion that mutation rates may be lower for X chromosomes than autosomes in rodents (99a), potentially resulting in a lower heterozygosity for X-linked than autosomal genes. Below we use estimates of heterozygosity and diversity to determine the difference in θ values for haplodiploids and diploids, and for X-linked and autosomal genes.

GENETIC VARIATION

Comparison of Haplodiploid and Diploid Insects

The earliest searches for allozyme variation in haplodiploids were largely unsuccessful (101, 133, 155). As a result, a variety of hypotheses, based primarily on evolutionary genetics, were invoked to explain this apparent lack of genetic variation. The most prominent explanation was that the expected effective population size for haplodiploids is three fourths that in diploids when there is an equal sex ratio, resulting in a proportionately lower heterozygosity in haplodiploids. There were also three hypotheses related to differences in selection between haplodiploids and diploids: The conditions for a stable polymorphism from selection were significantly more restrictive in haplodiploids so less variation would be maintained by selection in haplodiploids; allozyme alleles with a detrimental effect would have a lower equilibrium heterozygosity from mutation-selection balance for haplodiploids; and favorable variants, as they go to fixation, would be polymorphic for a shorter time in haplodiploids. Genetic hitchhiking has also been proposed as more important for haplodiploids than diploids, as mentioned above.

In addition, a number of studies have attempted to compare haplodiploids across levels of sociality (8, 34, 57, 87, 112, 131, 139, 140, 148). In general, eusocial haplodiploids were found to have reduced genetic variation relative to haplodiploids with a solitary life history. Several hypotheses were suggested

Table 1 Mean expected allozyme heterozygosity (H) and the standard error of the mean over species (in parentheses) for all studies and for studies with 15 or more loci divided by social category and species group for the given number of species

Group	All studies		≥15 loci		References
	Species	H	Species	H	
Haplodiploid insects	119	0.046 (0.004)	66	0.047 (0.005)	
I. Hymenoptera	115	0.046 (0.004)	6	0.046 (0.005)	
A. Eusocial	73	0.036 (0.004)	36	0.024 (0.004)	
1. Advanced	31	0.045 (0.006)	12	0.030 (0.006)	
a. Ants	26	0.041 (0.005)	10	0.036 (0.005)	8, 61, 132, 150, 164, 165
b. Bees	4	0.069 (0.041)	2	0.000	30, 162
c. Wasps	1	0.000	0		132
2. Primitive	42	0.031 (0.005)	24	0.021 (0.005)	
a. Bees	38	0.027 (0.005)	22	0.020 (0.006)	13, 88, 118, 121, 132, 140, 155, 160
b. Wasps	4	0.062 (0.010)	3	0.053 (0.005)	91
B. Social parasites					
a. Bees	5	0.010 (0.002)	2	0.008 (0.004)	118, 131, 132
C. Solitary	37	0.069 (0.009)	26	0.076 (0.008)	
a. Bees	10	0.035 (0.008)	3	0.057 (0.010)	91, 101, 131, 155
b. Wasps	13	0.065 (0.019)	9	0.051 (0.005)	20, 91, 100, 101, 131
c. Sawflies	14	0.096 (0.012)	14	0.096 (0.012)	11, 131, 148, 150
II. Thysanoptera					
A. Solitary	4	0.058 (0.011)	4	0.058 (0.011)	34
Diploid insects	151	0.117 (0.006)	151	0.117 (0.006)	
a. Drosophila	31	0.135 (0.060)	31	0.135 (0.060)	34, 57

to explain this association, including that selection in eusocial species from an environmentally constant nest was less likely to maintain genetic variation (131, 155), and that the effective population size in eusocial haplodiploids is smaller than that in solitary haplodiploids due to inbreeding (8).

In Table 1, we summarize data on allozyme variation from studies known to us. We divide the data into two categories: all studies, defined as those in which at least six loci were examined, and those in which ≥15 loci were examined (we also examined data for ≥20 loci, as suggested by 34, but the results were nearly identical to those for ≥15 loci). Some of these data are from earlier summaries, but we used the newer data if newer data on the same species were available. The expected heterozygosities (110) for each species were calculated and the unweighted mean and standard error were calculated over all species in a given category. The Hymenoptera were subdivided into categories based on social status as suggested by 139, i.e. advanced eusocial (large perennial nests and sterile worker caste), primitive eusocial (small or annual nests, potentially fertile workers), and solitary. In addition, we separated the social parasites from the eusocial and solitary categories because they live in a eusocial environment but lack a worker caste.

Eusocial Hymenoptera demonstrate significantly lower heterozygosity compared to the solitary Hymenoptera, as found by others, both for all studies and for studies with ≥15 loci. However, the advanced eusocial species have somewhat higher heterozygosity than the primitive eusocial species, so the trend of higher heterozygosity in solitary species is not complemented with the lowest heterozygosity in advanced eusocial species. [A great part of the reason that the solitary species have a high mean heterozygosity is the high heterozygosity observed in sawflies.] Interestingly, the other haplodiploid insects for which there are data, four species of solitary Thysanoptera (thrips), have a heterozygosity similar to that of the hymenoptera.

Diploid insects have a higher heterozygosity than do haplodiploid insects, even when *Drosophila*, a group with high allozyme heterozygosity, is removed. The loci in diploid insects include those on all chromosomes, including the X, so that the expectation would be that if the X-linked loci were removed (the chromosomal location of the allozymes surveyed in most of these species is not known), the mean value would increase somewhat. However, heterozygosity levels at X-linked genes in *Drosophila* are similar to those for autosomal genes (2), and the diploid insect species with the largest sample of known X and autosomal loci, *D. melanogaster*, has a heterozygosity of 0.154 for both the 9 X-linked loci and the 54 autosomal loci examined (152).

If we use the mean heterozygosities of haplodiploids (0.047) and diploids (0.117) to estimate θ_{hd} and θ_d from the equations above, then $\theta_{hd}/\theta_d = 0.372$. This ratio is half that expected based only on the differences in effective population size, suggesting other differences between the two groups that result in lower heterozygosity in haplodiploids. Such factors in the haplodiploids could be a smaller overall effective population size, higher inbreeding, lower mutation rate, or some effect of sociality. Note that the inclusion of monomorphic loci reduces the expected difference between haplodiploids and diploids because these loci cannot decrease in heterozygosity further due to lower effective population size.

For the comparisons between groups to be more meaningful, it would be good to use the same number of homologous, polymorphic loci in two groups that differ only by a given characteristic. We know of no such related pairs of haplodiploids and diploids, but the eight species of naked mole rats show different levels of sociality (80). Eusocial, colonial, and solitary species have heterozygosities of 0.314 (one species), 0.215 (four species), and 0.281 (three

Table 2   Mean expected microsatellite heterozygosity (H) and the standard error over species (in parentheses) in haplodiploid insects and for microsatellites in *D. melanogaster*, with the mean number of loci per species

Group	Species	Loci	H	References
Haplodiploid insects				
A. Eusocial	12	5.2	0.459 (0.069)	
1. Advanced	4	4.0	0.700 (0.092)	
a. Ants	3	3.0	0.729 (0.092)	21, 52, 55
b. Bees	1	7	0.614	50
2. Primitive	8	5.4	0.338 (0.055)	
a. Bees	5	2.0	0.283 (0.080)	51
b. Wasps	3	12.3	0.340 (0.018)	26, 78, 156
Diploid insects				
a. Drosophila	1	81 (19*)	0.504 (0.715*)	49, 56, 102, 146, 147

*Dinucleotide microsatellite loci with 10 repeats or more.

species) respectively (80); the eusocial species had the highest heterozygosity, but the intermediate level of sociality, colonial, had the lowest.

Microsatellite loci, which are generally repeats of two, three, or four nucleotides, with different alleles at a locus having different numbers of repeats, have recently become the preferred type of polymorphic locus because they are codominant, often have high variability, and generally are consistent with neutrality. A summary of the studies to date of microsatellites in haplodiploid insects and *D. melanogaster* is given in Table 2. The lack of variability in allozyme loci that has plagued studies of social insects does not appear in microsatellite loci, as studies cited in Table 2 and others (62, 158) have found many variable loci. Because the mutation rate for microsatellite loci is generally higher than for allozyme loci, this finding is not unexpected.

In this preliminary summary of microsatellite variation in haplodiploids, the advanced eusocial species have a higher heterozygosity than do the primitive eusocial species, but the number of species and number of loci are quite small. The mean heterozygosity for all the haplodiploids (they are all eusocial here) is 0.459, somewhat lower than the mean of 0.504 for the 81 loci located to chromosome in *D. melanogaster*. Using these heterozygosity values, θ_{hd}/θ_d is 0.787 and 0.835 for the stepwise and infinite allele models. The amount of variation for many microsatellite loci in *D. melanogaster* is fairly low, possibly due to a low mutation rate (146), the small size number of repeats at many of these loci (147), and the location of many loci in this sample near transcribed genes. If the heterozygosity is calculated for the 19 known dinucleotide loci with more than ten repeats in *D. melanogaster* (as were nearly all the microsatellite loci in the human and mouse studies discussed below), the average heterozygosity is much higher—0.715.

Comparison of X-Linked and Autosomal Genes

As mentioned above, the levels of heterozygosity for X-linked and autosomal allozyme loci in *D. melanogaster* appear similar (2, 152). A survey of RFLPs in humans showed about one third as much variation on X-linked as on autosomal markers (76), and a survey of nucleotide diversity for 49 human genes showed quite low variation overall (94), with some variation at two of the six X-linked loci and at 17 of the 46 autosomal loci. These studies have been superseded by extensive reports or summaries of X-linked and autosomal variation for nucleotide diversity (104) and microsatellites (147) in *D. melanogaster*, for nucleotide diversity in *D. simulans* (104), for marker polymorphisms [(89% were microsatellite loci and the remainder RFLPs) in the house mouse (42)], and for microsatellite loci in humans (43) (Table 3).

Nucleotide variation in *Drosophila* is evaluated in (104). It appears that excluding the loci in regions of low recombination differs little between X chromosomes and autosomes (104). However, the extensive data in mice and humans in Table 3 are part of the effort to map the genomes in these two species, so we briefly discuss some aspects of those data. The heterozygosities for the 216 microsatellite loci on the human X chromosome are given by location in Figure 4 (43, 53) along with the means for all 216 X-linked loci (0.65) and

Table 3 Variation at X-linked and autosomal loci and the ratio of θ_X/θ_A. "Number" for nucleotide diversity is the number of base pairs, and that for the other data is the number of loci. The amount of variation for nucleotide diversity and for the polymorphic loci in mice is θ, and that for the microsatellite loci is heterozygosity. The two values of θ_X/θ_A for microsatellite loci are for the stepwise and infinite allele models, respectively

Species	X-linked		Autosomal		θ_X/θ_A	References
	Number	Variation	Number	Variation		
Nucleotide diversity						
D. melanogaster	8,774	0.00274	14,757	0.00441	0.621	104
D. simulans	7,741	0.00470	7325	0.00971	0.484	104
Polymorphic loci						
M. musculus	230	0.330	6106	0.486	0.679	42
Microsatellite loci						
Humans	216	0.65	5048	0.70	0.708, 0.796	43
D. melanogaster	19	0.602	62	0.474	2.032, 1.679	49, 56, 102, 146, 147
	7*	0.670	12*	0.741	0.588, 0.710	

*Dinucleotide microsatellite loci with 10 repeats or more.

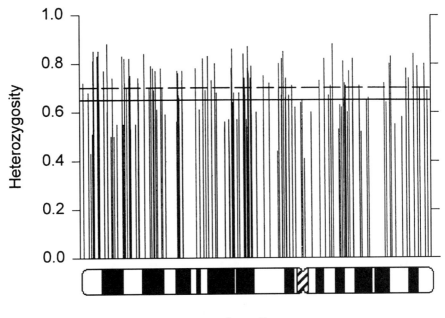

Location

Figure 4 Heterozygosity for 216 microsatellite loci on the human X chromosome show by map location (adapted from 43, 53). The horizontal solid and broken lines indicate the mean heterozygosity for the X-linked genes and the 5048 autosomal genes, respectively.

all 5,048 autosomal loci (0.70). Notice that although there appears to be some clustering of variable sites along the chromosome, the loci near the ends of the chromosome do not appear to have significantly lower heterozygosity, as they do on the *D. melanogaster* X chromosome. This is not unexpected because recombination at the ends of the human X chromosome is not reduced. The range of heterozygosity over the 22 autosomes is fairly narrow, from 0.69 to 0.73, so that the heterozygosity for the X chromosome is significantly lower than that for autosomes. Similarly, the proportion of polymorphism for the 230 genes on the mouse X chromosome (0.33) was lower than that observed for any of the 19 autosomes (from 0.35 to 0.57). Neither of these studies are perfect population samples, with some of the variation in the mouse survey potentially coming from ancestry from *Mus domesticus* (42) and the human sample being composed of families, so that not all individuals are independent observations (43) (M Nachman, personal communication). However, these biases may not be too important, both because so many loci were studied and because we are interested in the amounts of variation for X-linked relative to autosomal genes.

The ratio of the estimated values in Table 3 of θ_X/θ_A are similar to, or slightly below, the 0.75 neutrality prediction for the difference in effective population size for all the comparisons, except in *D. simulans*, for which the X-linked nucleotide variation is only 0.484 that of the autosomal variation (see 104 for a discussion of this observation, and also 3). Particularly impressive is the fact that the variation at 6336 marker loci in mice and 5264 microsatellite loci in humans show ratios of 0.679 and 0.708 (stepwise) or 0.796 (infinite), values quite similar to the 0.75 value predicted under neutrality. These data do not appear consistent with the hypothesis that the primate mutation rate is higher in males than in females (or that the mutation rate is lower for X-linked than autosomal genes). If the male mutation rate was sixfold larger, then the predicted ratio for humans would be approximately 0.57, using the expression above, quite different from 0.708 or 0.796.

CONCLUSIONS

Evolutionary genetic theory predicts a number of differences between hap-lodiploids or X-linked genes and diploids. For example, the conditions for a polymorphism, the amount of genetic load, and the rate of change of new variants are all expected to be different. It is difficult to evaluate experimentally or empirically the differences between the two systems for the amount of selectively maintained polymorphism (there are few such documented polymorphisms for either system) or the rate of change of new variants. However, there does appear to be substantial experimental support for lower genetic load and inbreeding depression for X-linked than for autosomal genes and some support for less inbreeding depression in haplodiploids than diploids.

For neutral molecular variation, the expected effective population size is smaller for haplodiploids or X-linked genes than for diploids and should result in less genetic variation. For comparisons of allozymes between haplodiploids and diploids, and for comparisons between X-linked variants and autosomes for nucleotide diversity in *Drosophila*, and for microsatellites in mice, humans, and *D. melanogaster* (for dinucleotide repeats of 10 or more), the observations are quite consistent with these predictions. However, we caution that such consistency does not necessarily mean that we have the correct explanation for these observations. There are several discrepancies, i.e., the level of variation for haplodiploids relative to diploids and the variation for X-linked to autosomal genes in *D. simulans* are even less than predicted by theory. As we have mentioned, other differences, such as a smaller overall effective population size for haplodiploids than for diploids, or an indirect effect of sociality on effective population size or level of inbreeding, could contribute to the difference between haplodiploids and diploids.

Several observations suggest that loci may have quite varied properties. First, and as often noted, the allozyme loci sampled vary greatly between species and may include loci with different levels of variability or different proportions of polymorphic loci, factors that could dilute any predicted differences. Second, in *Drosophila*, the amount of variation at a given locus appears to be directly correlated with the level of recombination in that region, suggesting that more detailed linkage information would be preferable for the markers used in such comparisons. Third, the amount of variation for microsatellite loci appears to be a function of the size of the repeat, the number of repeats, and probably, for some unknown reason at this point, the species examined (147).

Finally, researchers interested in the population genetics of haplodiploids and those working on X-linked genes are generally two quite separate groups. Obviously, knowledge of the population genetics for these two very similar systems is significant and sophisticated. We hope that we have succeeded here in demonstrating the usefulness of population genetics theory and data from haplodiploids for application to X-linked genes and for X-linked genes to haplodiploids.

ACKNOWLEDGMENTS

We thank numerous colleagues for sending us their published and unpublished research, and Ross Crozier, Michael Nachman, Robin Owen, and Pekka Pamilo for comments on the manuscript.

Visit the *Annual Reviews home page* at
http://www.annurev.org.

Literature Cited

1. Aquadro CF, Begun BJ, Kindahl EC. 1994. Selection, recombination, and DNA polymorphism in *Drosophila*. In *Non-Neutral Evolution, Theories and Molecular Data*, ed. B Golding, pp. 46–56. New York: Chapman & Hall

2. Avery PJ. 1984. The population genetics of haplo-diploids and X-linked genes. *Genet. Res.* 44:321–41

3. Begun DJ. 1996. Population genetics of silent and replacement variation in *Drosophila simulans* and *D. melanogaster*: X/autosome differences? *Mol. Biol. Evol.* 13:1405–7

4. Begun DJ, Aquadro CF. 1991. Molecular population genetics of the distal portion of the X chromosome in *Drosophila*: evidence for genetic hitchhiking of the yellow-achaete region. *Genetics* 129:1147–58

5. Begun DJ, Aquadro CF. 1992. Levels of natural occurring DNA polymorphism correlate with recombination rate in *Drosophila melanogaster*. *Nature* 356:519–20

6. Belmont JW. 1996. Genetic control of X inactivation and processes leading to X-inactivation skewing. *Am. J. Hum. Genet.* 58:1101–8

7. Bennett JH, Oertel CR. 1965. The approach to random association of genotypes with random mating. *J. Theor. Biol.* 9:67–76

8. Berkelhamer RC. 1983. Intraspecific genetic variation and haplodiploidy, eusociality, and polygyny in the Hymenoptera. *Evolution* 37:540–45

9. Berry AJ, Ajiola JW, Kreitman M. 1991. Lack of polymorphism in the *Drosophila* fourth chromosome resulting from selec-

tion. *Genetics* 129:1111–17

10. Bijlsma R. 1982. Biochemical and physiological basis for fitness differences at allozyme loci in *Drosophila melanogaster.* In *Advances in Genetics, Development, and Evolution of Drosophila,* ed. S Lakovaara, pp. 297–308. New York: Plenum

11. Boato A, Battisti A. 1996. High genetic variability despite haplodiploidy in primitive sawflies of the genus *Cephalcia* (Hymenoptera, Pamphiliidae). *Experientia* 52:516–21

12. Boomsma JJ, Ratnieks FLW. 1996. Paternity in eusocial Hymenoptera. *Philos. Trans. R. Soc. Lond. B.* 351:947–75

13. Bregazzi V, Laverty T. 1992. Enzyme gene variation in five species of bumble bees (Hymenoptera: Apidae). *Can. J. Zool.* 70:1263–66

14. Bruckner C. 1976. The influence of genetic variation on wing symmetry on honeybees (*Apis mellifera*). *Evolution* 30:100–8

15. Bruckner C. 1978. Why are their inbreeding effects in haplodiploid systems? *Evolution* 32:230–33

16. Caballero A. 1995. On the effective size of populations with separate sexes, with particular reference to sex-linked genes. *Genetics* 139:1007–11

17. Caballero A, Hill WG. 1992. Effective size of non-random mating populations. *Genetics* 130:909–16

18. Carson HL. 1955. Variation in genetic recombination in natural populations. *J. Cell. Compar. Physiol.* 45:221–35

19. Chang H-J, Shimmin LC, Shyue S-K, Hewett-Emmett D, Li W-H. 1994. Weak male-driven molecular evolution in rodents. *Proc. Natl. Acad. Sci. USA* 91:827–31

20. Chapman TW, Stewart SC. 1996. Extremely high levels of inbreeding in a natural population of the free-living wasp *Ancistrocerus antislope* (Hymenoptera: Vespidea: Eumeninae). *Heredity* 76:65–69

21. Chapuisat M. 1996. Characterization of microsatellite loci in *Formica lugubris* B and their variability in other ant species. *Mol. Ecol.* 5:599–601

22. Charlesworth B. 1996. Background selection and patterns of genetic diversity in *Drosophila melanogaster. Genet. Res.* 68:131–49

23. Charlesworth B, Coyne JA, Barton NH. 1987. The relative rates of evolution of sex chromosomes and autosomes. *Am. Nat.* 130:113–46

24. Charlesworth B, Langley CH. 1989. The population genetics of *Drosophila* transposable elements. *Annu. Rev. Genet.* 23:251–87

25. Charlesworth B, Morgan MT, Charlesworth D. 1993. The effect of deleterious mutations on neutral molecular variation. *Genetics* 134:1289–303

26. Choudary M, Strassmann JE, Solis CR, Queller DC. 1993. Microsatellite variation in a social insect. *Biochem. Genet.* 31:87–96

27. Clarke GM, Brand GW, Whitten MJ. 1986. Fluctuating asymmetry: a technique for measuring developmental stress caused by inbreeding. *Aust. J. Biol. Sci.* 39:143–53

28. Clarke GM, Oldroyd BP, Hunt P. 1992. The genetic basis of developmental stability in *Apis mellifera*: heterozygosity versus genic balance. *Evolution* 46:753–62

29. Clegg MT, Covener DR. 1982. Dynamics of correlated genetic systems. VII. Demographic aspects of sex-linked transmission. *Am. Nat.* 120:108–18

30. Comparini A, Biasiolo A. 1991. Genetic discrimination of Italian bee, *Apis mellifera ligustica* versus Carniolan bee, *Apis mellifera carnica* by allozyme variability analysis. *Biochem. Syst. Ecol.* 19:189–94

31. Cook JM. 1993. Sex determination in the Hymenoptera: a review of models and evidence. *Heredity* 71:421–35

32. Cook JM, Crozier RH. 1995. Sex determination and population biology in the Hymenoptera. *Trends Ecol. Evol.* 10:281–86

33. Cornette JL. 1978. Sex-linked genes in age-structured populations. *Heredity* 40:291–97

34. Crespi BJ. 1991. Heterozygosity in the haplodiploid Thysanoptera. *Evolution* 45:458–64

35. Crow JF, Roberts WC–1950. Inbreeding and homozygosity in bees. *Genetics* 35:613–21

36. Crozier RH. 1970. On the potential for genetic variability in haplo-diploidy. *Genetica* 41:551–56

37. Crozier RH. 1976. Counter-intuitive property of effective population size. *Nature* 262:384

38. Crozier RH. 1976. Why male-haploid and sex-linked genetic systems seem to have unusually sex-limited mutational genetic loads. *Evolution* 30:613–24

39. Crozier RH. 1985. Adaptive consequences of male-haploidy. In *Spider Mites: Their Biology, Natural Enemies and Control,* ed. W Helle, MW Sabelis, pp. 201–22. Amsterdam: Elsevier

39a. Crozier RH, Pamilo P. 1996. *Evolution of Social Insect Colonies. Sex Allocation and Kin Selection.* Oxford, UK: Oxford Univ. Press. 320 pp.

40. Curtsinger JW. 1980. On the opportunity for polymorphism with sex-linkage or haplodiploidy. *Genetics* 96:995–1006

41. Curtsinger JW. 1984. Components of selection in X chromosome lines of *Drosophila melanogaster*: sex ratio modification by meiotic drive and viability selection. *Genetics* 108:941–52

42. Deitrich WF, Miller J, Steen R, Merchant MA, Damron-Boles D, et al. 1996. A comprehensive genetic map of the mouse genome. *Nature* 380:149–52

43. Dib C, Faure S, Fizames C, Samson D, Drouot N, et al. 1996. A comprehensive map of the human genome based on 5,264 microsatellites. *Nature* 380:152–54

44. Dickson Burkhart B, Montgomery E, Langley CH, Voelker RA. 1984. Characterization of allozyme null and low activity alleles from two natural populations of *Drosophila melanogaster*. *Genetics* 107:295–306

45. Dobzhansky T. 1970. *Genetics of the Evolutionary Process.* New York: Columbia Univ. Press. 505 pp.

46. Drescher W. 1964. The sex limited genetic load in natural populations of *Drosophila melanogaster*. *Am. Nat.* 98:167–81

47. Eanes WF, Hey J, Houle D. 1985. Homozygous and hemizygous viability variation on the X chromosome of *Drosophila melanogaster*. *Genetics* 111:831–44

48. Eanes WF, Wesley C, Hey J, Houle D, Ajioka JW. 1988. The fitness consequences of *P* element insertion in *Drosophila melanogaster*. *Genet. Res.* 52:17–26

49. England PR, Briscoe DA, Frankham R. 1996. Microsatellite polymorphisms in a wild population of *Drosophila melanogaster*. *Genet. Res.* 67:285–90

50. Estoup A, Garney L, Solignac M, Cornuet JM. 1995. Microsatellite variation in Honey Bee (*Apis mellifera* L.) populations: hierarchical genetic structure and test of the infinite allele and stepwise mutation models. *Genetics* 140:679–95

51. Estoup A, Tailliez C, Cornuet JM, Solignac M. 1995. Size homoplasy and mutational processes of interrupted microsatellites in two bee species, *Apis mellifera* and *Bombus terrestris* (Apidae). *Mol. Biol. Evol.* 12:1074–84

52. Evans JD. 1993. Parentage analyses in ant colonies using simple sequence repeat loci. *Mol. Ecol.* 2:393–97

53. Fain PR, Kort EN, Chance PF, Nguyen K, Redd DF, et al. 1995. A 2D crossover–based map of the human X chromosome for map integration. *Nat. Genet.* 9:261–66

54. Gallo AJ. 1978. Genetic load of the X-chromosome in natural populations of *Drosophila melanogaster*. *Genetica* 49:153–57

55. Gertsch P, Pamilo P, Varvio SL. 1995. Microsatellites reveal high genetic diversity within colonies of *Camponotus* ants. *Mol. Ecol.* 4:257–60

56. Goldstein DB, Clark AG. 1995. Microsatellite variation in North American populations of *Drosophila melanogaster*. *Nucleic Acids Res.* 23:3882–86

57. Graur D. 1985. Gene diversity in Hymenoptera. *Evolution* 39:190–99

58. Haldane JBS. 1926. A mathematical theory of natural and artificial selection. Part III. *Proc. Cambridge Philos. Soc.* 23:363–72

59. Haldane JBS. 1935. The rate of spontaneous mutation of a human gene. *J. Genet.* 31:317–26

60. Haldane JBS, Jayakar SD. 1964. Equilibria under natural selection at a sex-linked locus. *J. Genet.* 59:29–36

61. Halliday RB. 1981. Heterozygosity and genetic distance in sibling species of meat ants (*Iridomyrmex purpureus* group). *Evolution* 35:234–42

62. Harnaguchi K, Ito Y. 1993. GT dinucleotide repeat polymorphisms in a polygynous ant, *Leptothorax spinosior* and their use for measurement of relatedness. *Naturwissenschaften* 80:179–81

63. Hamilton WD. 1964. The genetical evolution of social behavior. I. *J. Theor. Biol.* 7:1–16

64. Hamilton WD. 1964. The genetical evolution of social behavior. II. *J. Theor. Biol.* 7:17–52

65. Hamilton WD. 1967. Extraordinary sex ratios. *Science* 156:477–88

66. Hamilton WD. 1979. Wingless and fighting males in fig wasps and other insects. In *Sexual Selection and Reproductive Competition in Insects*, ed. MS Blum, NA Blum, pp. 167–220. New York: Academic

67. Hartl DL. 1971. Some aspects of natural selection in arrhenotokous populations. *Am. Zool.* 11:309–25

68. Hartl DL. 1972. A fundamental theorem of natural selection for sex linkage or arrhenotoky. *Am. Nat.* 106:516–24

69. Hedrick PW. 1976. Simulation of X-linked selection in Drosophila. *Genetics* 83:551–71

70. Hedrick PW. 1980. Selection in finite populations. III. An experimental examination. *Genetics* 96:297–313

71. Hedrick PW. 1982. Genetic hitchhiking: a new factor in evolution. *Bioscience* 32:845–53
72. Hedrick PW. 1985. *Genetic of Populations.* Boston: Jones & Bartlett. 629 pp.
73. Hedrick PW, Cockerham CC. 1986. Partial inbreeding: equilibrium heterozygosity and the heterozygosity paradox. *Evolution* 40:856–61
74. Hill WG. 1974. Estimation of linkage disequilibrium in randomly mating populations. *Heredity* 33:229–39
75. Hill WG. 1979. A note on effective population size with overlapping generations. *Genetics* 92:317–22
76. Hofker MH, Skaastad MI, Bergen AAB, Wapenaar MC, Bakker E, et al. 1986. The X chromosome shows less genetic variation at restriction sites than the autosomes. *Am. J. Hum. Genet.* 39:438–51
77. Hollocher H, Wu C-1. 1996. The genetics of reproductive isolation in the *Drosophila simulans* clade: X vs. autosomal effects and male vs. female effects. *Genetics* 143:1243–55
78. Hughes CR, Queller DC. 1993. Detection of highly polymorphic microsatellite loci in a species with little allozyme polymorphism. *Mol. Ecol.* 2:131–37
79. Hunt GJ, Page RE, Fondrk MK, Dullum CJ. 1995. Major quantitative trait loci affecting honey bee foraging behavior. *Genetics* 141:1537–45
80. Janecek LL, Honeycutt IL, Rautenbach BH, Erasmus SR, Schitter DA. 1992. Allozyme variation and systematics of African mole-rats (Rodentia: Bathyergidae). *Biochem. Syst. Ecol.* 20:401–16
81. Jennings HS. 1916. The numerical results of diverse systems of breeding. *Genetics* 1:53–89
82. Keller L, Passera L. 1993. Incest avoidance, fluctuation asymmetry, and the consequences of inbreeding in *Iridomyrmex humilis*, an ant with multiple queen colonies. *Behav. Ecol. Sociobiol.* 233:191–99
83. Kerr WE. 1976. Population genetic studies in bees. 2. sex-limited genes. *Evolution* 30:94–99
84. Kerr WE, Kerr LS. 1952. Concealed variability in the X-chromosome of *Drosophila melanogaster. Am. Nat.* 86:405–07
85. Kidwell JF, Clegg MT, Stewart FM, Prout T. 1977. Regions of stable equilibria for models of differential selection in the two sexes under random mating. *Genetics* 85:171–83
86. Kimura M. 1969. The number of het-

erozygous nucleotide sites maintained in a finite population due to steady flux of mutation. *Genetics* 61:893–903
87. Kimura M, Crow JF. 1964. The number of alleles that can be maintained in a finite population. *Genetics* 49:725–38
88. Kukuk PF, May B. 1985. A reexamination of genetic variability in *Dialictus zephyrus* (Hymenoptera: Halictidae). *Evolution* 39:226–28
89. LaMunyon CW, Ward S. 1997. Increased competitiveness of nematode sperm bearing the male X chromosome. *Proc. Natl Acad. Sci. USA* 94:185–89
90. Langley CH, Voelker RA, Leigh Brown AJ, Ohnishi S, Dickson B, Montgomery E. 1981. Null allele frequencies at allozyme loci in natural populations of *Drosophila melanogaster. Genetics* 99:151–56
91. Lester LJ, Selander RK. 1979. Population genetics of haplodiploid insects. *Genetics* 92:1329–43
92. Li CC. 1976. *First Course in Population Genetics.* Pacific Grove, CA: Boxwood. 631 pp.
93. Li W-H, Ellsworth DL, Krushkal J, Chang BH-J, Hewett-Emmett D. 1996. Rates of nucleotide substitution in primates and rodents and the generation-time effect hypothesis. *Mol. Phylogenet. Evol.* 5:182–87
94. Li W-H, Sadler LA. 1991. Low nucleotide diversity in man. *Genetics* 129:513–23
95. Lisker R, Motulsky AG. 1967. Computer simulation of evolutionary trends in an X linked trait, application to glucose-6-phosphate dehydrogenase deficiency in man. *Acta Genet.* 17:465–74
96. Lyon MF. 1992. Some milestones in the history of X-chromosome inactivation. *Annu. Rev. Genet.* 26:17–28
97. Lyttle TW. 1991. Segregation distorters. *Annu. Rev. Genet.* 25:511–37
98. Mayo 0. 1976. Neutral alleles at X-linked loci: a cautionary note. *Hum. Hered* 26:49–63.
99. McKusick V. 1994. *Mendelian Inheritance in Man.* Baltimore, MD: Johns Hopkins Univ. Press, 3009 pp. 11 ed.
99a. McVean GT, Hurst LD. 1997. Evidence for a selectively favorable reduction in the mutation rate of the X chromosome. *Nature* 386:388–92
100. Menken SJB. 1982. Enzymatic characterization of nine endoparasite species of small Ermine moths (Yponomeutidae). *Experientia* 38:1461–62
101. Metcalf RA, Marlin JC, Whitt GS. 1975. Low levels of genetic heterozygosity in Hymenoptera. *Nature* 257:792–93

102. Michalakis Y, Veuille M. 1996. Length variation of CAG/CAA trinucleotide repeats in natural populations of *Drosophila melanogaster* and its relation to the recombination rate. *Genetics* 143:1713–25

103. Montgomery E, Charlesworth B, Langley CH. 1987. A test for the role of natural selection in the stabilization of transposable element copy number in a population of *Drosophila melanogaster. Genet. Res.* 49:31–41

104. Moriyama EN, Powell JR. 1996. Intraspecific nuclear DNA variation in Drosophila. *Mol. Biol. Evol.* 13:261–77

105. Motulsky AG. 1960. Metabolic polymorphism and the role of infectious disease in human evolution. *Hum. Biol.* 32:28–62

106. Nagylaki T. 1975. A continuous selective model for a X-linked locus. *Heredity* 34:273–78

107. Nagylaki T. 1979. Selection in dioecious populations. *Ann. Hum. Genet.* 43:143–50

108. Nagylaki T. 1981. The inbreeding effective population number and the expected homozygosity for an X-linked locus. *Genetics* 47:731–37

109. Nagylaki T. 1995. The inbreeding effective population number in dioecious populations. *Genetics* 139:473–85

110. Nei M. 1978. Estimation of average heterozygosity and genetic distance from a small number of individuals. *Genetics* 89:583–90

111. Nunney L. 1993. The influence of mating system and overlapping generations on effective population size. *Evolution* 47:389–92

112. Owen RE. 1985. Difficulties with the interpretation of patterns of genetic variation in the eusocial Hymenoptera. *Evolution* 39:205–10

113. Owen RE. 1986. Colony-level selection in the social insects: single locus additive and nonadditive models. *Theor. Pop. Biol.* 29:198–234

114. Owen RE. 1986. Gene frequency clines at X-linked or haplodiploid loci. *Heredity* 57:209–19

115. Owen RE. 1988. Selection at two sex-linked loci. *Heredity* 60:415–25

116. Owen RE. 1989. Differential size variation of male and female bumblebees. *J. Hered.* 80:39–43

117. Owen RE. 1993. Genetics of parasitic hymenoptera. In *Applications of Genetics to Arthropods of Biological Control Significance,* ed. SK Narang, AC Bartlett, RM Faust, pp. 69–89. Boca Raton, FL: CRC

118. Owen RE, Mydynnski LJ, Packer L, McCorquodale DB. 1992. Allozyme variation in Bumble Bees (Hymenoptera: Apidae). *Biochem. Syst.* 30:443–53

119. Owen RE, Owen ARG. 1989. Effective population size in social Hymenoptera with worker-produced males. *Heredity* 63:59–65

120. Owen RE, Packer L. 1994. Estimation of the proportion of diploid males in populations of Hymenoptera. *Heredity* 72:219–27

121. Packer L, Owen RE. 1989. Allozyme variation in *Halictus rubicundus* (Christ): a primitively social Halictine Bee (Hymenoptera: Halictidae). *Can. Entomol.* 121:1049–57

122. Packer L, Owen RE. 1990. Allozyme variation, linkage disequilibrium and diploid male production in a primitively social bee *Augochorella striate* (Hymenoptera; Halictidae). *Heredity* 65:241–48

123. Page RE. 1986. Sperm utilization in social insects. *Annu. Rev. Ecol. Syst.* 31:297–320

124. Pamilo P. 1979. Genic variation at sex-linked loci: quantification of regular selection models. *Hereditas* 91:129–33

125. Pamilo P. 1991. Evolution of colony characteristics in social insects. 1. Sex allocation. *Am. Nat.* 137:83–107

126. Pamilo P. 1991. Evolution of colony characteristics in social insects. II. Number of reproductive individuals. *Am. Nat.* 138:412–33

127. Pamilo P. 1993. Polyandry and allele frequency differences between the sexes in the ant *Formica aquilonia. Heredity* 70:472–80

128. Pamilo P, Crozier RH. 1981. Genic variation in male haploids under deterministic selection. *Genetics* 98:199–214

129. Pamilo P, Gertsch P, Thorén P, Seppä J. 1997. Molecular population genetics of social insects. *Annu. Rev. Ecol. Syst.* 28:1–26

130. Pamilo P, Sundstrom L, Fortelius W, Rosengren R. 1994. Diploid males and colony-level selection in *Formica* ants. *Ethol. Ecol. Evol.* 6:221–35

131. Pamilo P, Varvio-Aho S, Pekkarinen A. 1978. Low enzyme gene variability in Hymenoptera as a consequence of haplodiploidy. *Hereditas* 88:93–99

132. Pamilo P, Varvio-Aho S, Pekkarinen A. 1984. Genetic variation in bumblebee s (*Bombus, Psithyrus*) and putative sibling species of *Bombus lucorum. Hereditas* 101:245–51

133. Pamilo P, Vepsalainen K, Rosengren R. 1975. Low allozymic variability in Formica ants. *Hereditas* 80:293–96

134. Passera L, Keller L, Suzzoni J. 1988. Queen replacement in dequeened colonies of the Argentine ant *Iridomyrmex humilis* (Mayr). *Psyche* 95:59–63

135. Pollak E. 1984. Gamete frequencies at two sex-linked loci in random mating age-structured populations. *Math. Biosci.* 70:217–35

136. Pollak E. 1990. The effective population size of an age-structured population with a sex-linked locus. *Math. Biosci.* 101:121–30

137. Queller DC. 1993. Worker control of sex ratio and selection for extreme multiple mating by queens. *Am. Nat.* 142:346–51

138. Rao SSR. 1983. Religion and intensity of inbreeding in Tamil Nadu, South India. *Soc. Biol.* 30:413–22

139. Reeve HK, Sherman-Reeve J, Pfennig DW. 1985. Eusociality and genetic variability: a re-evaluation. *Evolution* 39:200–1

140. Rosemeier L, Packer L. 1993. A comparison of genetic variation in two sibling species pair of Haplodiploid insects. *Biochem. Genet.* 31:185–200

141. Ross KG. 1992. Strong selection on a gene that influences reproductive competition in a social insect. *Nature* 355:347–49

142. Ross KG, Keller L. 1995. Ecology and evolution of social organization: insights from fire ants and other highly eusocial insects. *Annu. Rev. Ecol. Syst.* 26:631–56

143. Ross KG, Keller L. 1995. Joint influence of gene flow and selection on a reproductively important genetic polymorphism in the fire ant *Solenopsis invicta. Am. Nat.* 146:325–48

144. Ruwende C, Khoo SC, Snow RW, Yates SNR, Kwiatowski D, et al. 1995. Natural selection of hemizygotes for G6PD deficiency in Africa by resistance to severe malaria. *Nature* 376:246–49

145. Sanghvi LD. 1966. Inbreeding in India. *Eugen. Q.* 13:291–301

146. Schug MD, Mackay TFC, Aquadro CF. 1997. Low mutation rates of microsatellite loci in *Drosophila melanogaster. Nat. Genet.* 15:99–102

147. Schug MD, Wettstrand KA, Gaudette MS, Lim RH, Hutter CM, Aquadro CF. 1997. The distribution and frequency of microsatellite loci in *Drosophila melanogaster. Nucleic Acids Res.* In press

148. Sheppard WS, Heydon SL. 1986. High levels of genetic variability in three male-haploid species (Hymenoptera: Argidae, Tenthredinidae). *Evolution* 40:1350–53

149. Shimmin LC, Chang BH-W, Li W-H. 1993. Male driven evolution of DNA sequences. *Nature* 362:745–47

150. Shoemaker DD, Ross KG. 1992. Estimates of heterozygosity in two social insects using a large number of electrophoretic markers. *Heredity* 69:573–82

151. Simmons MJ, Crow JF. 1977. Mutations affecting fitness in *Drosophila* populations. *Annu. Rev. Genet.* 11:49–78

152. Singh RS, Rhomberg LR. 1987. A comprehensive study of genic variation in natural populations of *Drosophila melanogaster.* II. Estimates of heterozygosity and patterns of geographic differentiation. *Genetics* 117:255–71

153. Slatkin N. 1995. A measure of population subdivision based on microsatellite allele frequencies. *Genetics* 139:457–62

154. Slatkin M, Excoffier L. 1996. Testing for linkage disequilibrium in genotypic data using the Expectation-Maximization algorithm. *Heredity* 76:377–83

155. Snyder TP. 1974. Lack of allozymic variability in three bee species. *Evolution* 28:687–89

156. Strassmann JE, Barefield K, Solis CR, Hughes CR, Queller DC. 1997. Trinucleotide microsatellite loci for a social wasp, *Polistes. Mol. Ecol.* 6:97–100

157. Thomson G. 1977. The effects of a selected locus on linked neutral loci. *Genetics* 85:753–88

158. Thorén PA, Paxton RJ, Estoup A. 1995. Unusually high frequency of (CT)n and (GT)n microsatellite loci in a yellowjacket wasp, *Vespula rufa* (L.) (Hymenoptera: Vespidae). *Biochem. Genet.* 4:141–48

159. Turelli M, Orr HA. 1995. The dominance theory of Haldane's rule. *Genetics* 140:389–402

160. Voelker RA, Langley CH, Leigh Brown AJ, Ohnishi S, Dickson B, et al. 1980. Enzyme null alleles in natural population of *Drosophila melanogaster:* frequencies in a North Carolina population. *Proc. Natl. Acad. Sci. USA* 77:1091–95

161. Waage JK. 1986. Family planning in parasitoids: adaptive patterns of progeny and sex allocation. In *Insect Parasitoids,* ed. JK Waage, D Greathead, pp. 63–95. London: Academic

162. Wagner AE, Briscoe DA. 1983. An absence of enzyme variation within two species of *Trigona* (Hymenoptera). *Heredity* 50:97–103

163. Wang J. 1996. Inbreeding coefficient and effective size for an X-linked locus in nonrandom mating populations. *Heredity* 76:569–77

164. Ward PS. 1980. Genetic variation and population differentiation in the *Rhytidoponera impressa* group, a species complex of ponerine ants (Hymenoptera: Formicidae). *Evolution* 34:1060–76

165. Ward PS, Taylor RW. 1981. Allozyme variation, colony structure and genetic relatedness in the primitive ant *Nothomyrmecia macrops* Clark (Hymenoptera: Formicidae). *J. Aust. Entomol. Soc.* 20:177–83

166. Werren JH. 1993. The evolution of inbreeding in haplodiploid organisms. In *The Natural History of Inbreeding and Outbreeding: Theoretical and Empirical Perspectives,* ed. NW Thornhill, pp. 42–59, Chicago: Univ. Chicago Press

167. Wilton AD, Sved JA. 1979. X chromosomal heterosis in *Drosophila melanogaster. Genet. Res.* 34:303–15

168. Whitlock MC, Wade MJ. 1995. Speciation: founder events and their effects on X–linked and autosomal genes. *Am. Nat.* 145:676–85

169. Wright S. 1922. Coefficients of inbreeding and relationship. *Am. Nat.* 56:330–38

170. Wright S. 1933. Inbreeding and homozygosis. *Proc. Natl. Acad. Sci. USA* 19:411–20

171. Wright S. 1969. *Evolution and Genetics of Populations.* Vol. II. *The Theory of Gene Frequencies.* Chicago: Univ. Chicago Press. 511 pp.

172. Wu C-1, Davis AW. 1993. Evolution of postmating reproductive isolation: the composite nature of Haldane's rule and its genetic bases. *Am. Nat.* 142:187–212

Annu. Rev. Ecol. Syst. 1997. 28:85–104

DISSECTING GLOBAL DIVERSITY PATTERNS: Examples from the Ordovician Radiation

Arnold I. Miller

Department of Geology, P. O. Box 210013, University of Cincinnati, Cincinnati, Ohio 45221-0013; e-mail: arnold.miller@uc.edu

KEY WORDS: Ordovician Radiation, global biodiversity, diversity trends, evolutionary paleoecology, Ordovician biogeography

---

ABSTRACT

Although the history of life has been characterized by intermittent episodes of radiation that can be recognized in global compilations of biodiversity, it does not necessarily follow that these episodes are caused by processes that occurred uniformly around the world. Major diversity increases could be generated by the cumulative effects of different mechanisms operating simultaneously at several geographic or environmental scales. The purpose of this review is to describe ongoing research on the manifestations, at several scales, of the Ordovician Radiation, which was among the most extensive intervals of diversification in the history of life. Through much of the period, diversity was concentrated most heavily near regions of active mountain building and volcanism; differences in diversity patterns from continent to continent, and among regions within continents, reflect this overprint. While this suggests a linkage of the Radiation and tectonic activity, this is by no means the only mediating agent. Outcrop-based research in North America has demonstrated that tectonic activity was detrimental to some biotic elements, in contrast to its effects on other organisms. Moreover, in the Great Basin of North America where the local stratigraphic record is of particularly high quality, biotic transitions characteristic of the period occurred far more rapidly than observed in global compilations of diversity, suggesting that the global rate of transition may represent the aggregate sum of transitions that occurred abruptly, but at different times, around the world. Finally, it has been demonstrated that, in concert with an increase in average age, the environmental and geographic ranges of Ordovician genera both increased significantly through

0066-4162/97/1120-0085$08.00

the period, indicating a role for intrinsic factors in producing Ordovician biotic patterns.

———————————

INTRODUCTION

Researchers from a spectrum of subdisciplines, with shared interests in ecology and evolution, have come to appreciate the importance of comparing and reconciling biotic diversity patterns across spatio-temporal scales (14, 41, 70). For example, if we want to know why a particular gastropod is found living near the top of a seagrass blade in a Caribbean lagoon, we must recognize that the "explanation" derives from an interaction of processes operating at scales ranging from the zonation of resources on a seagrass blade to the long-term, Phanerozoic evolution of gastropods.

On a broad scale, the Phanerozoic history of global biodiversity is itself the consequence of interaction among physical and biological processes at several hierarchical levels, played out over millions of years. While a thorough inventory of these processes might seem a daunting task, an ever-improving understanding of the physical history of the earth and the growing sophistication of analyses conducted on fossil data are permitting an unprecedented level of insight into major ecological/evolutionary trends exhibited throughout the history of life. In this context, on the heels of more than two decades of intensive research into the calibration of global biodiversity trends through the Phanerozoic (90), an emerging focus of paleobiological research has been the dissection of these synoptic trends into more local, geographic, and environmental components. Global diversity trends do not result simply from processes that occur uniformly around the world (see below), and this kind of treatment is therefore a prerequisite for understanding what causes them.

Here, I review the ongoing investigation of perhaps the most extensive global diversity increase in the history of marine animal life: the Ordovician Radiation. Because the Ordovician Period (505 to 440 MYA) was characterized by a three- to fourfold increase in the number of marine animal families and genera worldwide (79, 81; but see 53) and was also an unusually dynamic time in the earth's physical history, it continues to capture the attention of diverse researchers and provides perhaps an unparalleled opportunity to understand the kinetics of a major, global-scale diversification (22).

THE ORDOVICIAN WORLD

A central theme of this review is the relationship between Ordovician diversity trends and contemporaneous physical patterns and transitions. Thus, as a backdrop to consideration of biological patterns during the period, it is useful to review physical attributes of the Ordovician world that were likely to have affected marine biotas.

Positions of Paleocontinents

Scotese & McKerrow illustrated the positions of paleocontinents throughout the Ordovician in a series of four reconstructions (see Figures 6–9 in 73); a quick perusal of these maps, and others provided for the rest of the Paleozoic, reveals that the Paleozoic world bore little resemblance to that of the present day. In fact, many present-day continents are agglomerations of smaller paleocontinental fragments that collided with one another after the Ordovician, whereas others are the remnants of much larger paleocontinents that broke apart.

Although some Ordovician fossils have been found in ancient deep sea and oceanic island environments, most were associated with shallow seas and fore-land basins tied to individual paleocontinents. As they traveled in concert with Ordovician plate motion, paleocontinents carried along marine biotas that were, in some instances, limited to individual paleocontinents or groups of paleocon-tinents near one another. For this reason biogeographic evidence can play a significant role in determining the relative positions of paleocontinents. This is particularly true for Paleozoic reconstructions, which are more problematic than those for the post-Paleozoic; post-Paleozoic reconstructions are facilitated by tracing the oceanographic "tracks" of continental movement subsequent to the breakup of the Late Paleozoic-Early Mesozoic supercontinent of Pangea. Such tracking is not possible for Paleozoic reconstructions because much of the Paleozoic oceanographic record has been subducted.

Along with paleomagnetic and lithologic evidence, biogeographic data are the foci of an ongoing debate concerning the positions of paleocontinents dur-ing the Early Paleozoic. At stake are the positions of several paleocontinents surrounding the ancient Iapetus Ocean (Figure 1), the existence and locations of several oceanic islands within the Iapetus Ocean (35, 56, 57, 99), and, perhaps most prominently, the position of what is now South America with respect to Laurentia, a paleocontinent that encompasses much of North America. It has been suggested (16–18) that South America, which was part of the Paleozoic supercontinent of Gondwana, collided with Laurentia during the Ordovician, suggested in part by the alleged biogeographic affinity of the fossil biota from an Argentine region known as the Precordillera with contemporaneous biotas of North America. However, others (95) support an alternative scenario: that the Precordillera was a microcontinent that rifted from Laurentia during the Cambrian and collided with Gondwana during the Ordovician (95).

Regardless of the specifics of this debate, there is little disagreement on two points of particular importance: (*a*) the positions of the six paleocontinents and

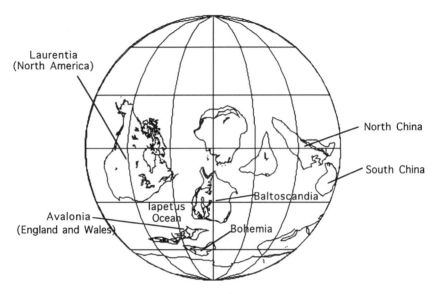

Figure 1   Global reconstruction showing the approximate positions of six paleocontinents during the Ordovician Period. For further discussion, see text.

microcontinents labeled in Figure 1 appear to be reasonably well-constrained and were arrayed over a fair portion of the Ordovician world; and (*b*) despite uncertainty about the precise positions of these paleocontinents, their individual geological histories can be understood and contrasted.

Sea Level

Another important hallmark of the Ordovician world was a global sea level that significantly exceeded the level observed today. Although recent reevaluations (1) have reduced previous estimates of differences between the sea levels of Ordovician and the present-day (34, 96), Ordovician seas were sufficiently extensive to cover large portions of many Ordovician paleocontinents and microcontinents, including those highlighted here. However, the degree of marine invasion varied substantially from one paleocontinent to the next because of regional differences in the extent of tectonic activity and other geological attributes (see Figure 9 of Ref. 1).

Tectonic Activity

Although proposed scenarios for the closing of the Iapetus Ocean differ, there is little doubt that plate collisions took place with increasing intensity through the period, with a concomitant increase in global tectonic activity, including orogeny (mountain building) and volcanism (42, 43). Ordovician volcanic eruptions, for example, may have been of an extent that dwarfed anything recognized during human history. Huff & colleagues (37–39, 44) have now recognized the altered volcanic ash layer (known as a K-bentonite) from one such Middle Ordovician eruption over several paleocontinents, and they have estimated that it was perhaps the largest eruption to have occurred during the Phanerozoic. Erosion of regions newly uplifted by orogeny and volcanism contributed large quantities of sediment that came to dominate paleoenvironmental gradients in several parts of the world as the Ordovician progressed.

Geochemistry and Nutrient Levels

Perhaps as a direct consequence of heightened tectonic uplift and erosion of source areas, ocean geochemistry changed markedly through the Ordovician, as described by Martin (Figure 3 in Ref. 45, and references cited therein). Although Martin argued that oceanographic conditions during the Early Paleozoic as a whole were oligotrophic relative to most of the remaining Phanerozoic, the trend in the Ordovician was clearly away from oligotrophy. Relative to the Cambrian, the Ordovician was characterized by a growing $^{87}Sr/^{86}Sr$ ratio, increased levels of phosporite and $\delta^{13}C$, and decreased $\delta^{34}S$ levels. Taken together, these changes indicate increased nutrient input from continental weathering, increased photosynthesis/primary productivity, and increased overturn/oxygenation of marine waters.

ORDOVICIAN GLOBAL BIODIVERSITY TRENDS

Raw, temporal trajectories of Ordovician global marine biodiversity, compiled at the genus and family levels, reveal a sharp increase that apparently continued, nearly unabated, through most of the period (Figure 2; 76, 79, 81). The degree of increase at the class level and higher was not as great as that exhibited during the preceding Cambrian Explosion (25), suggesting that the Ordovician Radiation was not characterized so much by the origination of major body plans associated with new phyla as it was by the elaboration and diversification of those that had already become established. However, this is not to downplay the biological significance of the Ordovician Radiation: there was considerable origination at the class level (25), and at lower taxonomic levels, diversification rates far outstripped those of the Cambrian.

In addition, the Ordovician Radiation involved a global-scale transition from a marine biota dominated by trilobites and other faunal elements of Sepkoski's (76) Cambrian Evolutionary Fauna to one with an increasing diversity of taxa belonging both to Sepkoski's Paleozoic (e.g. articulate brachiopods, stalked crinoids, stenolaemate bryozoans, rugose and tabulate corals) and Modern (e.g. gastropod and bivalve molluscs) Evolutionary Faunas (see also 80). Thus,

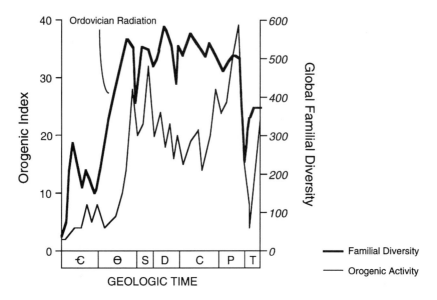

Figure 2   A comparison of global familial diversity and levels of global orogenic activity during the Paleozoic Era and the Triassic Period of the Mesozoic Era. The curve depicting orogenic index is adapted from a histogram in Khain & Seslavinskii (42); familial diversity is after Sepkoski (79). Figure after Miller (50).

while the Cambrian Explosion and Ordovician Radiations both encompassed profound biotic transitions, they differed profoundly on a global scale, and each was likely associated with unique causal mechanisms. Despite their temporal juxtaposition, it might therefore not be appropriate to lump them together as one Early Paleozoic "event" that can be explained by a single, unifying mechanism (e.g. 97).

A potential problem with the interpretation of raw temporal diversity trajectories, like those cited here, is their being compromised by variations in sample size from one stratigraphic interval to the next. Raup (67, 68) illustrated a close relationship between the broad, Phanerozoic trend of global species-level diversity and that of sedimentary rock volume (see Figure 10.6 and 10.7 in 70). Sepkoski et al (82) sought to overcome this problem by demonstrating that several datasets all revealed the same general Phanerozoic trajectory, even though they were collected with different methodologies at various taxonomic levels, and some were independent of rock volume. However, even the latter data were not demonstrably free of other potential sampling biases, leaving lingering doubts about the biological significance of raw diversity trajectories. In the case of the Ordovician Radiation, Miller & Foote (53), applying rarefaction techniques (40, 66) to a database of individual occurrences of Ordovician genera worldwide, found that, after accounting for sample size differences among the six major stratigraphic subdivisions of the system, the increase in global diversity may have been over by the end of the first half of the period (see Figure 2 of 53). This is not to suggest that the Ordovician Radiation was concluded by the middle of the period: the transitions among evolutionary faunas, described earlier, were apparently protracted through the entire period (but see GEOGRAPHIC DISSECTION: THE INTRACONTINENTAL LEVEL, below). There is little doubt that standing diversity at the genus and family levels was substantially higher in the Ordovician than it was in the Cambrian, but the Radiation need not have been associated with a period-long increase in total diversity, as implied by the raw curve.

GEOLOGICAL DISSECTION

In a series of papers focusing on the kinetics of Phanerozoic global diversification (74, 75, 77), Sepkoski used coupled versions of an equation for logistic (equilibrial) diversity growth to demonstrate that the three evolutionary faunas were characterized by successively higher levels of equilibrium diversity. This was elaborated on by Bambach (6, 7), who suggested an ecological explanation for these differences: the advent and diversification of taxa with novel life habits in successive evolutionary faunas that could exploit previously underutilized habitats at, above, and below the sediment-water interface (see Figures of 7). Thus, on a global scale, the taxonomic transitions that accompanied the Ordovician Radiation were associated with a proliferation of new life habits and the occupation of new habitats. However, while these are important diagnostic symptoms of the Radiation, they do not, in themselves, explain why the Radiation took place.

Vermeij (97), recognizing two major intervals of Phanerozoic diversification, suggested that they were linked to broad cycles of increased tectonic activity described previously by Fischer (26) and others. The earlier interval encompassed both the Cambrian Explosion and the Ordovician Radiation; Vermeij suggested that increased contemporaneous submarine volcanism may have led to elevated global temperatures in concert with growing levels of atmospheric CO_2. Because of these and other volcanic effects, global nutrient and energy levels may have been augmented, which would have permitted, for the first time, the subsistence and diversification of organisms with relatively high metabolic rates. Similarly, Bambach (8) noted that members of the Modern Evolutionary Fauna, on average, had greater individual biomasses, in terms of "fleshiness" and size, than their counterparts in the Paleozoic Evolutionary Fauna, with a concomitant need for more food. He suggested that the transition from the

Paleozoic to the Modern Fauna was tied to a secular, global increase in the available food supply.

As an explanation for the Ordovician Radiation, Vermeij's hypothesis is appealing, especially given the independent evidence, described earlier, that the Ordovician was a time of increasing volcanism and growing nutrient supply. However, it is not sufficient by itself, because it does not account for an important attribute of the Radiation: the predominant diversification of the Paleozoic Fauna, relative to the Modern Fauna. Articulate brachiopods and other members of the Paleozoic fauna were probably energy minimalists, in comparison to bivalve molluscs and other members of the Modern Fauna (8). Bivalves, in particular, were well-represented by several "modern-looking" life habit-types by the Middle Ordovician (48, 62–65, 92–94), and their initial, Ordovician diversification may, indeed, have been food-related (see below). However, if food and other energy-related transitions "controlled" the Ordovician Radiation by themselves, why, then, did the diversity levels of bivalves and other elements of the Modern Fauna not quickly exceed those of the Paleozoic Fauna, instead of the other way around? This is not to say that changes in nutrient supply did not play a role in the Radiation, but perhaps they acted in concert with other causal agents that fueled a diversification dominated more appreciably by organisms with lesser food requirements.

As noted earlier, the Ordovician was a time not only of increased volcanism, but also of growing orogenic activity. A quantification of levels of orogenic activity through the entire Paleozoic Era shows a trajectory that is tantalizingly similar to the global familial diversity trend through that interval (Figure 2). However, this similarity could be entirely artifactual: increased orogeny could result directly in increased sedimentary rock volume because of erosion of newly uplifted source areas. This, in turn, would favor the preservation of fossils during times when orogenic activity was at high levels (see earlier discussion of 67, 68).

Thus, a more definitive test of any hypothesized relationship between increased orogenic activity and the Ordovician Radiation is required. One way to accomplish this is with a geologic dissection of the Radiation, to determine directly whether diversifying biotas were concentrated in regions of orogenic activity. To this end, Miller & Mao (54) mapped Ordovocian biodiversity with respect to probable centers of orogeny, using a literature-derived database of some 6600 individual occurrences of Ordovician genera mainly from the paleocontinents shown in Figure 1; these genera belonged to a representative cross-section of major higher taxa from each of Sepkoski's three evolutionary faunas. Even after accounting for sample size differences of the kind that could be generated by differences in sedimentary rock volume, genus diversity was shown to be significantly greater through most of the period in regions of likely orogenic (and other tectonic) activity than it was in regions remote from those effects.

Thus, there may have also been a relationship between orogenic activity and the Ordovician Radiation. Miller & Mao (54) suggested three possible explanations for this relationship, all of which might have acted in concert. First, they noted the possible importance of nutrient input from newly uplifted areas in the enhancement of food supplies in Ordovician oceans and epicontinental seas (see 45); this dovetails with Vermeij's hypothesis (97), described earlier. Second, there is considerable evidence from North America and elsewhere that the erosion of uplifted areas inundated large regions with mud and other siliciclastic sediments (as a consequence of both orogeny and volcanism); while this may have induced the extinction of some organisms (see below), it would have favored the diversification of higher taxa that thrived in muddy conditions, regardless of their metabolic rate or their membership in a particular evolutionary fauna. For example, many phacopid trilobites, strophomenid brachiopods, and bivalve molluscs, all of which diversified through the Ordovician, appeared to exhibit a shared affinity for mud. A third effect of orogenic and other tectonic activity may have been an increasing likelihood of allopatric speciation, owing to an increased disruption of the Ordovician sea floor. In general, Cracraft (15) suggested a direct relationship between rates of speciation and habitat partitioning associated with tectonism, which would have served to increase diversity in a variety of taxa.

GEOGRAPHIC DISSECTION: COMPARISONS AMONG PALEOCONTINENTS

The significance of tectonism and other potential factors in the Ordovician Radiation can be evaluated further by contrasting diversification patterns among paleocontinents arrayed throughout the Ordovician world. A comparison of this kind was conducted (51, 55) using the database of genus occurrences that had been dissected previously in a geological context (54). Diversification patterns, in aggregate and for individual higher taxa, varied substantially from one paleocontinent to the next (see Figures 1–4 of 51), corroborating the impression, based on the earlier geological dissection, that the Ordovician Radiation did not transpire uniformly around the world. Moreover, some of these intercontinental differences can be interpreted in light of tectonic overprints and related paleoenvironmental factors. For example, as the Ordovician progressed, Laurentia became dominated increasingly by a molluscan biota; in fact, by the Late Ordovician, the aggregate genus diversity of bivalves, gastropods, and monoplacophorans appears to have exceeded that of articulate brachiopods; a preponderance of molluscs can be found, for example, in Middle and Upper

Ordovician fossil assemblages of the central Appalachian Mountains and the Tug Hill Plateau region of New York State (11–13), and in shale-rich facies of the Cincinnati area (19, 32). By contrast, South China appeared to harbor a more limited benthic molluscan biota, the diversity of which was far exceeded by articulate brachiopods throughout the period (51).

It is likely that the difference between South China and Laurentia was a consequence of the differences in levels of tectonic activity and associated siliciclastic sediments derived from uplifted source regions. Erosion of areas uplifted in eastern Laurentia during tectonic events related to the Taconic Orogeny (69) produced a large quantity of siliciclastic sediment that reached central regions of the paleocontinent late in the period. For a combination of reasons, including nutrient needs and substrate preferences described earlier, the presence of siliciclastic sediments appears to have fostered this diverse and abundant molluscan biota (11–13, 46, 47). South China, on the other hand, was relatively quiet throughout the period (36) and was dominated by carbonate sedimentation.

With respect to differences among paleocontinents and the timing of diversification, it is instructive to consider the Ordovician history of bivalve diversification in the geographic context described by Babin (3, 4). The earliest Ordovician radiation of bivalves apparently took place in siliciclastic sediments marginal to Gondwana; they were virtually absent from Laurentia at that time. It was not until later, with the onset of widespread siliciclastic sedimentation, that bivalves began to flourish over a large portion of eastern Laurentia.

Similarly, Lower to Middle Ordovician fossil biotas in siliciclastic sediments of western Argentina were dominated by molluscs and trilobites (72), whereas contemporaneous carbonate-rich sediments in the Precordillera were rich in articulate brachiopods and other elements of the Paleozoic Fauna. In describing these differences, Sánchez & Waisfeld (72) noted that the Precordillera was a microcontinent in the vicinity of Laurentia at that time (see Positions of Paleocontinents), whereas the western Argentinean siliciclastics were deposited along the margin of Gondwana.

GEOGRAPHIC DISSECTION: THE INTRACONTINENTAL LEVEL

Because of the diversity differences within geological regimes and among paleocontinents, it can be concluded with some confidence that the Ordovician Radiation did not take place uniformly around the world. Rather, the transitions apparent in global-scale compilations are best viewed as synoptic summaries that must be dissected to be understood. Further, while it is useful to describe the summed attributes of one paleocontinent relative to another, variations at the intracontinental scale can also be substantial. Intracontinental dissections of diversity patterns, with a high degree of spatial and stratigraphic resolution, yield signals relevant to global biodiversity at the same time that they provide insight into organismal response in the face of significant regional perturbations and transitions.

Patzkowsy & Holland (59–61) have evaluated geographic-temporal diversity patterns of Middle and Upper Ordovician brachiopods in eastern and central North America (southeastern Laurentia). By mapping distributional patterns in a high-resolution lithostratigraphic framework, they found that brachiopod biotas in eastern North America (Tennessee, Virginia, Indiana, Ohio, and Kentucky) declined abruptly across a stratigraphic boundary in the Middle Ordovician as a probable consequence of the combined effects of a major rise in sea level and the Taconic Orogeny. Environmental effects associated with this confluence of events, including upwelling and sedimentary influx, are thought to have produced cold, turbid, nutrient-rich waters, thereby decimating carbonate-producing organisms that had thrived in that region but were intolerant of the new conditions. Patzkowsky & Holland demonstrated that this biota persisted in warmer-water refugia to the north and west, and subsequently re-emerged over part of southeastern Laurentia when favorable conditions returned midway through the Late Ordovician.

Patzkowsy & Holland's recognition that the Taconic orogeny may have been detrimental to brachiopods and, in fact, may have contributed to a regional decimation of that biota, would seem to challenge the earlier assertion that tectonic activity helped to foster the Ordovician Radiation. However, this apparent contradiction can be resolved by recalling that the fauna in southeastern Laurentia became dominated increasingly by molluscs, phacopid trilobites, and brachiopods different from those studied by Patzkowsky & Holland. In fact, environmental conditions detrimental to the pre-Taconic biota may have been quite hospitable to the faunal elements that dominated subsequently in southeastern Laurentia, especially after the hypothesized upwelling abated. Late Ordovician biotas of southeastern Laurentia were broadly divided into those that apparently favored carbonate-rich environments and others that flourished in the widespread siliciclastic conditions that came to dominate much of the region (2, 11–13, 91).

Ordovician strata are also well-exposed in the Great Basin region of California, Nevada, and Utah (western Laurentia). Over the past several years, these carbonate-dominated rocks have been investigated intensively by Droser et al (20–24), who have now documented several rather abrupt transitions:

1. There was a sudden shift at the base of the Middle Ordovician in the compositions of shell beds from a biota dominated by trilobites to one in which brachiopods and a greater diversity of other faunal elements were most

abundant. It also appears that shell beds in the Middle Ordovician were thicker and more frequent stratigraphically than in the Lower Ordovician.

2. The trilobite fauna in the Middle Ordovician was rather different than that in Lower Ordovician rocks, reflecting both an indigenous radiation of new taxa and an immigration of other taxa previously found only in association with other paleocontinents.

3. Trace fossils, which record the burrowing habits of infaunal, soft-bodied organisms, indicate a substantial increase in the depth of bioturbation within Upper Ordovician strata, relative to the Middle Ordovician.

The extent to which these abrupt shifts are manifested elsewhere remains to be seen, but they have the potential to alter dramatically our view of the global Ordovician transition from the Cambrian Fauna to the Paleozoic fauna. As noted earlier, despite evidence (53) that aggregate faunal diversity peaked much earlier in the Ordovician than previously recognized, the transition from the Cambrian fauna to the Paleozoic fauna is thought to have continued through most of the period. However, if it can be shown that abrupt stratigraphic transition from the Cambrian Fauna to Paleozoic faunal elements took place at localities worldwide, but that the timing of this transition varied from locality to locality because of geological/environmental differences among paleocontinents, then appearance of a gradual global transition would actually represent the sum of many abrupt transitions (87).

Faunal data from Ordovician strata of Utah and Nevada were brought to bear by Guensberg & Sprinkle (33) on the issue of the larger-scale Ordovician transition among evolutionary faunas. They found that a major, Early Ordovician radiation of stalked crinoids, which were important components of the Paleozoic Fauna, had an extrinsic trigger: a major, global rise in sea level flooded siliciclastic source areas, producing a shift to dominantly carbonate sedimentation. Combined with certain geochemical transitions, this substantially increased the availability of hard substrates on the Early Ordovician sea floor, thereby providing a much expanded habitat area for crinoids and other attached echinoderms.

Taken together, the analyses described in this section demonstrate the sensitivity of benthic marine biotas to various kinds of environmental changes, and they serve to bolster suggestions that similar kinds of transitions, if occurring simultaneously in a sufficient number of regions around the world, could substantially affect global diversity. They also suggest that the record of biotic transition in any area is likely to be governed by numerous extrinsic factors, ranging from global to local, the effects of which may be superimposed complexly, but perhaps not intractably, upon one another (52).

ALPHA AND BETA DIVERSITY TRENDS

Ecologists and paleoecologists alike have often found it useful to summarize the attributes of biotic assemblages by measuring α ("within-community") and β ("between-community") diversity. For evolutionary paleoecologists, the calibration of trends in these attributes through geologic time permits direct, numerical assessment of the relationship between diversity at hierarchical scales ranging from individual communities to the entire world. In a period-by-period analysis of median species richness (α diversity) among open marine paleocommunities from Laurentia and Avalonia, Bambach (5) documented a Phanerozoic trend that was subsequently found to closely approximate that exhibited in synoptic compilations of global diversity (82).

More recently, Sepkoski (78) focused on Paleozoic assemblages of Laurentia and reported α and β diversity values at the genus level for each period. α diversity was measured as the mean genus richness of assemblages within each stratigraphic interval, and β diversity was measured by calculating the mean value of the Jaccard similarity coefficient for pairwise comparisons of aggregate faunal lists compiled for each of six environmental zones arrayed from onshore to offshore; a decrease in the Jaccard coefficient, reflecting reduced similarity among assemblages, would indicate that β diversity was increasing. Sepkoski found that mean α diversity and β diversity both increased from the Cambrian to the Ordovician, and then stabilized for the remainder of the Paleozoic, suggesting that these two metrics paralleled one another in accordance with the theoretical ecological expectations that he described.

However, because Sepkoski presented data only at the period level of temporal resolution, it is possible that most of the increase from the Cambrian to the Ordovician resulted from a transition that took place near the start of the Ordovician, and that there was no actual trend of increasing α or β diversity through most of the period. Indeed, an evaluation of α and β diversity trends within South China and Laurentia throughout the Ordovician suggests that α diversity may have increased, but that β diversity did not parallel this trend (49, 55, 58); in fact, β diversity may have actually decreased during the Ordovician Radiation.

At first glance, the possibility of an inverse relationship between α and β diversity trends appears surprising, given expectations based on ecological principles (78). However, further investigation of environmental and paleogeographic attributes of the Radiation reveals that, as the period progressed, the mean geographic and environmental ranges of Ordovician genera both increased significantly, apparently in concert with a significant aging of the biota (49; AI Miller, unpublished observations). If genera became more widespread, dissimilarities among individual assemblages should have decreased, thereby reducing β diversity.

Thus, it appears that, during the Ordovician Radiation, diversity trends at different hierarchical levels did not simply parallel one another; they were perhaps governed individually by mechanisms unique to a particular level (55). In addition, the recognition of a relationship between the aging of the biota and the environmental/paleogeographic expansion of genera suggests that intrinsic factors, perhaps independent of the extrinsic, physical mechanisms highlighted in most of this review, also played an important role in the Radiation (see 83, 84). In fact, Ordovician diversity patterns almost certainly resulted from interaction between intrinsic and extrinsic factors (33). For example, there was a major mass extinction at the end of the Ordovician Period; among Phanerozoic marine mass extinctions, it was second only to the Late Permian event in severity. A full discussion of the causes of this extinction, which may have resulted from a complex interplay of sea-level fluctuations and a major glaciation (9, 10), is beyond the scope of this review. What is relevant here is the likelihood that, while the extinction was caused by extrinsic perturbations, its effect on the marine biota was mediated by the intrinsic factors that were likely responsible for the mean geographic range increases of genera through the Ordovician. This is demonstrably true because the mass extinction selectively eliminated taxa with limited geographic ranges (85, 86, 88, 89). Had the range expansion that accompanied the Ordovician Radiation not taken place, the nature of the subsequent extinction might have been substantially different.

MORPHOLOGICAL DIVERSITY

Even if they are not strictly monophyletic, the grouping of taxa on an ecological basis may be the most appropriate and effective means of evaluating biotic transitions in response to extrinsic mechanisms (71, 72). The calibration and interpretation of taxonomic diversity trends described in this review are dependent on the assumption that the taxonomic groupings considered are ecologically meaningful units at some scale with members that respond in similar fashion to particular physical or biological stimuli.

Rather than assuming that higher taxa are adequate proxies for assessing the role of hypothesized forcing mechanisms in mediating the history of life, particular morphologies or morphological diversity can be evaluated directly in light of these mechanisms. While it may be reasonable, in many cases, to infer that comparative diversity trends within higher taxa adequately mirror the long-term fates of morphological attributes, there is growing evidence that, following the initial diversification of a higher taxon, taxonomic diversity trends may exhibit patterns that are clearly at variance with patterns of morphological diversity. For example, Foote (27–30) statistically analyzed the morphologies of trilobite cranidia to describe morphological diversity trends among trilobites during the Paleozoic Era. He found that, among trilobites as a whole, as well as within certain groups of trilobites, morphological diversity was either stable or increased significantly during several intervals when taxonomic diversity was declining (see Figures 4–9 of 30); one such interval encompassed the Ordovician. While this does not negate the importance of the decline in taxonomic diversity, it does suggest, among other things, that a failure to analyze morphology directly may impede the recognition of important episodes of morphological diversification or decline that perhaps transcended taxonomy and were ecologically mediated.

Compared to the calibration of taxonomic diversity, the quantitative analysis of morphological diversity is a relatively new enterprise, and the initial analyses of Paleozoic trends have focused rather effectively on global scale patterns in comparison with taxonomic diversity. However, in the same way that it has been fruitful to dissect taxonomic diversity in the context of geologic, geographic, and environmental parameters, it would be valuable, as a next step, to do likewise with morphologic diversity (cf. 71). Given that morphological attributes have now been quantified for many species or genera within several higher taxa (e.g. 27–31, 98), it should be possible, in the near future, to merge these data, at several scales, with the other kinds of data presented in this review.

CONCLUDING REMARKS

In the study of mass extinctions, it has long been appreciated that any attempt to understand the causes of extinction should include rigorous assessment of biological selectivity. Evaluations of environmental, geographic, taxonomic, and ecological patterns of extinction may well reflect the mechanisms that produced them, and the literature is now replete with studies that either consider the selectivity of individual mass extinctions or compare patterns among two or more events to help evaluate suggestions of similarity in causal processes. A central goal of this review has been to demonstrate that a similar approach is also vital to the investigation of longer-term Phanerozoic global diversity trends, including radiations.

In perusing global curves of Phanerozoic marine diversity (e.g. Figures 1–2 in 77) one cannot help but be struck by the apparent simplicity of the summed pattern, which may, indeed, reflect global-scale "rules" of diversification associated with the characteristic turnover rates of taxa and the aggregate space limitations of marine habitats (74, 75, 77). However, this is not to say that Phanerozoic diversity trends are "caused" by global-scale processes. It is possible for the overall shape of the global curve, and perhaps the rates at which biotic transitions take place among faunal elements, to be mediated globally, at the same time that the actual inducements for change are local or regional in scope and vary from place to place around the world. Whether this is, indeed, the case throughout the Phanerozoic can only be determined through the kind of analysis of selectivity that is already ongoing in studies of the Ordovician Radiation. Results thus far for the Ordovician, summarized herein, suggest strongly that the global signal is, in large measure, an aggregation of patterns and processes unique to particular regions or scales.

More generally, the data required to tease apart patterns of faunal diversity throughout the Phanerozoic are largely available in the worldwide literature in the form of descriptions and lists of taxa contained in local strata. However, they must be assembled into databases, including stratigraphic, environmental, and geographic data, that can be sorted to investigate and dissect diversity patterns based on any of a number of possible parameters, either individually or in combination. If these can be combined with local and regional outcrop-based studies targeted to address issues for which the literature data are insufficient or ambiguous (e.g. possible worldwide variation in the timing of major transitions like those recognized by Droser et al (22, 23, 24) in the Ordovician of the Great Basin), the results would undoubtedly include a much-improved understanding of Phanerozoic diversification and, just as importantly, a better sense of how present-day biotas would be expected to respond in the face of ongoing physical transitions and stresses.

ACKNOWLEDGMENTS

My research on the Ordovician Radiation has been facilitated by a grant from the National Aeronautics and Space Administration, Program in Exobiology (Grant NAGW-3307).

> Visit the *Annual Reviews home page* at
> http://www.annurev.org.

Literature Cited

1. Algeo TJ, Seslavinsky KB. 1995. The Paleozoic world: continental flooding, hypsometry, and sea level. *Am. J. Sci.* 295:787–822
2. Anstey RL. 1986. Bryozoan provinces and patterns of generic evolution and extinction in the Late Ordovician of North America. *Lethaia* 19:33–51
3. Babin C. 1993. Rôle des plates-formes gondwaniennes dans les diversifications des mollusques bivalves durant l'Ordivicien. *Bull. Soc. Géol. Fr.* 164: 141–53
4. Babin C. 1995. The initial Ordovician bivalve mollusc radiations on the western Gondwanan shelves. See Ref. 14a, pp. 491–98
5. Bambach RK. 1977. Species richness

in marine benthic habitats through the Phanerozoic. *Paleobiology* 3:152–67

6. Bambach RK. 1983. Ecospace utilization and guilds in marine communities through the Phanerozoic. In *Biotic Interactions in Recent and Fossil Benthic Communities*, ed. MJS Tevesz, PL McCall, pp. 719–46. New York: Plenum

7. Bambach RK. 1985. Classes and adaptive variety: the ecology of diversification in marine faunas through the Phanerozoic. In *Phanerozoic Diversity Patterns: Profiles in Macroevolution*, ed. JW Valentine, pp. 191–253. Princeton, NJ: Princeton Univ Press

8. Bambach RK. 1993. Seafood through time: changes in biomass, energetics, and productivity in the marine ecosystem. *Paleobiology* 19:372–97

9. Brenchley PJ. 1988. The Late Ordovician extinction. In *Mass Extinctions: Processes and Evidence*, ed. SK Donovan, pp. 104–32. New York: Columbia Univ. Press

10. Brenchley PJ, Marshall JD, Carden GAF, Robertson DBR, Long DGF, et al. 1994. Bathymetric and isotopic evidence for a short-lived Late Ordovician glaciation in a greenhouse period. *Geology* 22:295–98

11. Bretsky PW Jr. 1969. Central Appalachian Late Ordovician communities. *Geol. Soc. Am. Bull.* 80:193–202

12. Bretsky PW Jr. 1970. Upper Ordovician ecology of the central Appalachians. *Peabody Mus. Nat. Hist. Bull. 34.* 150 pp.

13. Bretsky PW Jr. 1970. Late Ordovician benthic communities in north-central New York. *New York St. Mus. Sci. Serv. Bull. 414* 34 pp.

14. Brown JH. 1995. *Macroecology.* Chicago: Univ. Chicago Press 269 pp.

14a. Cooper JD, Droser ML, Finney SC, eds. 1995. *Ordovician Odyssey: Short Papers for the Seventh International Symposium on the Ordovician System.* Fullerton, CA: Pacific Section SEPM

15. Cracraft J. 1985. Biological diversification and its causes. *Missouri Bot. Gard. Ann.* 72:794–822

16. Dalla Salda LH, Cingolani CA, Varela R. 1992. Early Paleozoic orogenic belt of the Andes in southwestern South America: result of Laurentia-Gondwana collision? *Geology* 20:617–20

17. Dalla Salda LH, Dalziel IWD, Cingolani CA, Varela R. 1992. Did the Taconic Appalachians continue into soutern South America? *Geology* 20:1059–62

18. Dalziel IWD. 1995. Earth before Pangea. *Sci. Am.* 272:58–63

19. Dattilo BF. 1996. A quantitative paleoecological approach to high-resolution cyclic and event stratigraphy: the Upper Ordovician Miamitown Shale in the type Cincinnatian. *Lethaia* 28:21–37

20. Droser ML, Bottjer DJ. 1989. Ordovician increase in extent and depth of bioturbation: implications for understanding early Paleozoic ecospace utilization. *Geology* 17:850–52

21. Droser ML, Bottjer DJ. 1993. Trends and patterns of Phanerozoic ichnofabrics. *Annu. Rev. Earth Planet. Sci.* 21:205–25

22. Droser ML, Fortey RA, Li X. 1996. The Ordovician Radiation. *Am. Scientist* 84:122–32

23. Droser ML, Sheehan PM. 1995. Paleoecology of the Ordovician Radiation and the Late Ordovician extinction event: evidence from the Great Basin. In *Ordovician of the Great Basin: Fieldtrip Guidebook and Volume for the Seventh International Symposium on the Ordovician System*, ed. J. Cooper, pp. 64–106. Fullerton, CA: Pacific Section SEPM

24. Droser ML, Sheehan PM, Fortey RA, Li X. 1995. The nature of diversification and paleoecology of the Ordovician Radiation with evidence from the Great Basin. See Ref. 14a, pp. 405–8

25. Erwin DH, Valentine JW, Sepkoski JJ Jr. 1987. A comparative study of diversification events: the early Paleozoic versus the Mesozoic. *Evolution* 41:1177–86

26. Fischer AG. 1985. The two Phanerozoic supercycles. In *Catastrophes in Earth History: The New Uniformitarianism*, ed. WA Berggren, JA Van Couvering, pp. 129–50. Princeton, NJ: Princeton Univ. Press

27. Foote M. 1989. Perimeter-based Fourier analysis: a new morphometric method applied to the trilobite cranidium. *J. Paleontol.* 63:880–85

28. Foote M. 1990. Nearest-neighbor analysis of trilobite morphospace. *Syst. Zool.* 39:371–82

29. Foote M. 1991. Morphologic patterns of diversification: examples from trilobites. *Palaeontology* 34:461–85

30. Foote M. 1993. Discordance and concordance between morphological and taxonomic diversity. *Paleobiology* 19:185–204

31. Foote M. 1995. Morphological diversification of Paleozoic crinoids. *Paleobiology* 21:273–99

32. Frey RC. 1987. The paleoecology of a Late Ordovician shale unit from southwest Ohio and southeastern Indiana. *J. Paleontol.* 61:242–67

33. Guensberg TE, Sprinkle JS. 1992. Rise of echinoderms in the Paleozoic evolutionary fauna: significance of paleoenvironmental controls. *Geology* 20:407–10
34. Hallam A. 1992. *Phanerozoic Sea-Level Changes.* New York: Columbia Univ. Press. 266 pp.
35. Harper DAT, Mac Niocaill C, Williams SH. 1996. The palaeogeography of early Ordovician Iapetus terraces: an integration of faunal and paleomagnetic constraints. *Palaeogeogr. Palaeoclimatol. Palaeoecol.* 121:297–312
36. Hsü K, Jiliang L, Haihong C, Qinchen W, Shu S, Sengör AMC. 1990. Tectonics of South China: key to understanding West Pacific geology. *Tectonophysics* 183:9–39
37. Huff WD, Bergström SM, Kolata DR. 1992. Gigantic Ordovician volcanic ash fall in North America and Europe: biological, tectonomagmatic, and eventstratigraphic significance. *Geology* 20:875–78
38. Huff WD, Kolata DR. 1990. Correlation of the Ordovocian Deicke and Millbrig K-bentonites between the Mississippi Valley and the southern Appalachians. *Am. Assoc. Petrol. Geol. Bull.* 74:1736–47
39. Huff WD, Kolata DR, Bergström SM, Zhang Y-S. 1996. Large-magnitude Middle Ordovician volcanic ash falls in North America and Europe: dimensions, emplacement and post-emplacement characteristics. *J. Volc. Geoth. Res.* 73:285–301
40. Hurlbert SH. 1971. The nonconcept of species diversity: a critique and alternative parameters. *Ecology* 52:577–86
41. Jablonski D, Sepkoski JJ Jr. 1996. Paleobiology, community ecology, and scales of ecological pattern. *Ecology* 77:1367–78
42. Khain VE, Seslavinskii KB. 1991. *Historical Geotectonics.* Moscow: Nedra. 400 pp.
43. Khain VE, Seslavinskii KB. 1994. Global rhythms of the Phanerozoic endogenic activity of the earth. *Stratigr. Geol. Correl.* 2:520–41
44. Kolata DR, Huff WD, Bergström SM. 1996. Ordovician K-Bentonites of Eastern North America. *Geol. Soc. Am. Spec. Pap.* 313. 84 pp.
45. Martin RE. 1995. Cyclic and secular variation in microfossil biomineralization: clues to the biogeochemical evolution of Phanerozoic oceans. *Global Planetary Change* 11:1–23
46. Miller AI. 1988. Spatio-temporal transitions in Paleozoic Bivalvia: an analysis of North American fossil assemblages. *Hist. Biol.* 1:251–73
47. Miller AI. 1989. Spatio temporal transitions in Paleozoic Bivalvia: a field comparison of Upper Ordovician and upper Paleozoic bivalve-dominated fossil assemblages. *Hist. Biol.* 2:227–60
48. Miller AI. 1990. Bivalves. In *Evolutionary Trends,* ed. KJ McNamara, pp. 143–61. London: Belhaven
49. Miller AI. 1996. Beta diversity trends during the Ordovician Radiation: "unexpected" patterns and new perspectives. *Geol. Soc. Am. Abstr. Prog.* 28(7):A176
50. Miller AI. 1996. Subsection on evolution. In *1997 McGraw-Hill Yearbook of Science and Technology,* pp. 181–83. New York: McGraw-Hill
51. Miller AI. 1997. Comparative diversification dynamics among palaeocontinents during the Ordovician Radiation. *Geobios Mem.* In press
52. Miller AI. 1997. Coordinated stasis or coincident relative stability? *Paleobiology* 23: In press
53. Miller AI, Foote M. 1996. Calibrating the Ordovician Radiation of marine life: implications for Phanerozoic diversity trends. *Paleobiology* 22:304–9
54. Miller AI, Mao S. 1995. Association of orogenic activity with the Ordovician Radiation of marine life. *Geology* 23:305–8
55. Miller AI, Mao S. 1997. Scales of diversification and the Ordovician Radiation. In *Biodiversity Dynamics: Turnover of Populations, Taxa, and Communities,* ed. ML McKinney. New York: Columbia Univ. Press. In press
56. Neuman RB. 1984. Geology and paleobiology of islands in the Ordovician Iapetus Ocean: review and implications. *Geol. Soc. Am. Bull.* 95:1188–1201
57. Neuman RB, Harper DAT. 1992. Paleogeographic significance of Arenig-Llanvirn Toquima-Table Head and Celtic brachiopod assemblages. In *Global Perspectives on Ordovician Geology,* ed. BD Webby, J Laurie, pp. 241–54. Rotterdam: Balkema
58. Patzkowsky ME. 1995. Ecologic aspects of the Ordovician radiation of articulate brachiopods. See Ref. 14a, pp. 413–14
59. Patzkowsky ME. 1995. Gradient analysis of Middle Ordovician brachiopod biofacies: biostratigraphic, biogeographic, and macroevolutionary implications. *Palaios* 10:154–79
60. Patzkowsky ME, Holland SM. 1993. Biotic response to a Middle Ordovician paleoceanographic event in eastern North America. *Geology* 21:619–22

61. Patzkowsky ME, Holland SM. 1996. Extinction, invasion, and sequence stratigraphy: patterns of faunal change in the Middle and Upper Ordovician of the eastern United States. *Geol. Soc. Am. Spec. Pap.* 306:131–42

62. Pojeta J. 1971. Review of Ordovician Pelecypods. *US Geol. Surv. Prof. Pap.* 695, 46 pp.

63. Pojeta J. 1978. The origin and early taxonomic diversification of pelecypods. *Philos. Trans. R. Soc., Ser. B* 284:225–46

64. Pojeta J. 1987. Class Pelecypoda. In *Fossil Invertebrates,* ed. RS Boardman, AH Cheetham, AJ Rowell, pp. 386–435. Oxford: Blackwell Sci.

65. Pojeta J. 1987. The first radiation of the Class Pelecypoda. *Geol. Soc. Am. Abstr. Prog.* 19:238–9

66. Raup DM. 1975. Taxonomic diversity estimation using rarefaction. *Paleobiology* 1:333–42

67. Raup DM. 1976. Species diversity in the Phanerozoic: a tabulation. *Paleobiology* 2:279–88

68. Raup DM. 1976. Species diversity in the Phanerozoic: an interpretation. *Paleobiology* 2:289–97

69. Rodgers J. 1971. The Taconic Orogeny. *Geol. Soc. Am. Bull.* 82:1141–78

70. Rosenzweig ML. 1995. *Species Diversity in Space and Time.* Cambridge/New York: Cambridge Univ. Press. 436 pp.

71. Roy K. 1994. Effects of the Mesozoic Marine Revolution on the taxonomic, morphologic, and biogeographic evolution of a group: aporrhaid gastropods during the Mesozoic. *Paleobiology* 20:274–96

72. Roy K. 1996. The roles of mass extinction and biotic interaction in large-scale replacements: a reexamination using the fossil record of stromboidean gastropods. *Paleobiology* 22:436–52

72a. Sánchez TM, Waisfeld BG. 1995. Benthic assemblages in the northwest of Gondwana: a test of the Ordovician evolutionary radiation model. See Ref. 14a, pp. 409–12

73. Scotese CR, McKerrow WS. 1990. Revised world maps and introduction. In *Palaeozoic Palaeogeography and Biogeography,* ed. WS McKerrow, CR Scotese, pp. 1–21. London: Gel. Soc. Mem. 12

74. Sepkoski JJ Jr. 1978. A kinetic model of Phanerozoic taxonomic diversity: I. Analysis of marine orders. *Paleobiology* 4:223–51

75. Sepkoski JJ Jr. 1979. A kinetic model of Phanerozoic taxonomic diversity: II. Early Phanerozoic families and multiple equilibria. *Paleobiology* 5:222–51

76. Sepkoski JJ Jr. 1981. A factor analytic description of the marine fossil record. *Paleobiology* 7:36–53

77. Sepkoski JJ Jr. 1984. A kinetic model of Phanerozoic taxonomic diversity: III. Post-Paleozoic marine families and mass extinctions. *Paleobiology* 10:246–57

78. Sepkoski JJ Jr. 1988. Alpha, beta, or gamma: Where does all the diversity go? *Paleobiology* 14:221–34

79. Sepkoski JJ Jr. 1993. Ten years in the library: new data confirm paleontological patterns. *Paleobiology* 19:246–57

80. Sepkoski JJ Jr. 1995. The Ordovician Radiations: diversification and extinction shown by global genus-level data. See Ref. 14a, pp. 393–96

81. Sepkoski JJ Jr. 1997. Biodiversity: past, present, future. *J. Paleontol.* 71: In press

82. Sepkoski JJ Jr, Bambach RK, Raup DM, Valentine JW. 1981. Phanerozoic marine diversity and the fossil record. *Nature* 293:435–37

83. Sepkoski JJ Jr, Miller AI. 1985. Evolutionary faunas and the distribution of Paleozoic benthic communities in space and time. In *Phanerozoic Diversity Patterns: Profiles in Macroevolution,* ed. JW Valentine, pp. 153–90. Princeton, NJ: Princeton Univ Press

84. Sepkoski JJ Jr, Sheehan PM. 1983. Diversification, faunal change, and community replacement during the Ordovician radiations. In *Biotic Interactions in Recent and Fossil Benthic Communities,* ed. MJS Tevesz, PL McCall, pp. 673–717. New York: Plenum

85. Sheehan PM. 1973. The relation of Late Ordovician glaciation to the Ordovician-Silurian changeover in North American brachiopod faunas. *Lethaia* 6:147–54

86. Sheehan PM. 1975. Brachiopod synecology in a time of crisis (Late Ordovician-Early Silurian). *Paleobiology* 1:205–12

87. Sheehan PM. 1996. A new look at Ecological Evolutionary Units (EEUs). *Palaeogeogr. Palaeoclimatol. Palaeoecol.* 127: 21–32

88. Sheehan PM, Coorough PJ. 1990. Brachiopod zoogeography across the Ordovician-Silurian extinction event. In *Palaeozoic Palaeogeography and Biogeography,* ed. WS McKerrow, CR Scotese, pp. 181–87. London: Gel. Soc. Mem. 12

89. Sheehan PM, Coorough PJ, Fastovsky DE. 1996. Biotic selectivity during the K/T and Late Ordovician events. *Geol. Soc. Am. Spec. Pap.* 307:477–89

90. Signor PW. 1990. The geologic history of diversity. *Annu. Rev. Ecol. Syst.* 21:509-39

91. Springer DA, Bambach RK. 1985. Gradient versus cluster analysis of fossil assemblages: a comparison from the Ordovician of southwest Virginia. *Lethaia* 18:181–98

92. Stanley SM. 1972. Functional morphology and evolution of byssally attached bivalve mollusks. *J. Paleontol.* 46:165–212

93. Stanley SM. 1977. Trends, rates, and patterns of evolution in the Bivalvia. In *Patterns of Evolution,* ed. A Hallam, pp. 209–50. Amsterdam: Elsevier

94. Stanley SM. 1979. *Macroevolution: Pattern and Process.* San Francisco: Freeman

95. Thomas WA, Astini, RA. 1996. The Argentine Precordillera: a traveler from the Ouachita Embayment of North America Laurentia. *Science* 273:752–7

96. Vail PR, Mitchum RM Jr, Thompson S III. 1977. Global cycles of relative changes of sea level. *Am. Assoc. Petrol. Geol. Mem.* 26:83–97

97. Vermeij G. 1995. Economics, volcanoes, and Phanerozoic revolutions. *Paleobiology* 21:125–52

98. Wagner PJ. 1995. Testing evolutionary constraint hypotheses with early Paleozoic gastropods. *Paleobiology* 21:248–72

99. Williams SH, Harper DAT, Neuman RB, Boyce WD, Mac Niocaill C. 1995. Lower Paleozoic fossils from Newfoundland and their importance in understanding the history of the Iapetus Ocean. *Geol. Assoc. Can. Spec. Pap.* 41:115–26

Annu. Rev. Ecol. Syst. 1997. 28:105–28

A COMPARISON OF ALTERNATIVE STRATEGIES FOR ESTIMATING GENE FLOW FROM GENETIC MARKERS[1]

Joseph E. Neigel

Department of Biology, University of Southwestern Louisiana, Lafayette, Louisiana 70504; e-mail: jneigel@usl.edu

KEY WORDS: gene flow, migration, dispersal, F statistics, population structure

---

ABSTRACT

The estimation of gene flow from the distribution of genetic markers in populations requires an indirect approach. Gene flow parameters are defined by demographic models, and population genetic models provide the link between these parameters and the distributions of genetic markers. Following the introduction of allozyme methods in the 1960s, a standard approach to the estimation of gene flow was developed. Wright's island model of population structure was used to relate the distribution of allozyme alleles in populations to $N_e\,m$, the product of the effective population size and the rate of migration.

Alternative strategies for the estimation of gene flow have been developed using different genetic markers, different models of demography and population genetics, and different methods of parameter estimation. No alternative strategy now available is clearly superior to the standard approach based on Wright's model and allozyme markers. However, this may soon change as methods are developed that fully utilize the genealogical relationships of DNA sequences. At present, alternative strategies do fill important needs. They can provide independent estimates of gene flow, measure different components of gene flow, and detect historical changes in population structure.

---

[1] The US government has the right to retain a nonexclusive, royalty-free license in and to any copyright covering this paper.

105

INTRODUCTION

In the late 1960s, surveys of allozyme variation in natural populations set a new direction for studies of gene flow (61–65). Wright's analysis of nonrandom mating provided a quantitative relationship between his measure of population subdivision, F_{ST}, and the two parameters: population size, N, and rate of migration, m (127, 130–133). Because F_{ST} can be interpreted as the standardized variance in the frequency of an allele among populations (132), it could be readily estimated from allozyme data. A standard approach to quantifying gene flow among populations was put into practice, using allozyme markers to estimate F_{ST} and Wright's island model (127, 129) to transform estimates of F_{ST} into estimates of Nm (95, 97). This approach is considered an indirect method (95) for the estimation of gene flow, because population genetic models are required to infer the magnitude of gene flow from its effects on the distributions of genetic markers. Throughout this review, I refer to F_{ST} analysis of allozyme data as the F_{ST}-allozyme approach.

The F_{ST}-allozyme approach has been widely applied to natural populations of plants, animals, and microorganisms. Its many advantages explain its continued use. Allozyme surveys require relatively little expense, time, or specialized laboratory equipment, and they can be successfully carried out for most species (63). The theoretical basis of this approach is well established, which has led to progressive refinements in how F_{ST} is estimated (21, 102, 124). However, there are compelling reasons for investigating alternative strategies. First, it is difficult to directly test the reliability of Nm estimates. Our confidence in them rests largely on their robustness under diverse conditions in theoretical models (70, 95, 97, 102). Empirical tests such as comparisons with mark-and-recapture estimates of dispersal are at best ambiguous. Indirect methods estimate the cumulative effects of gene flow, acting over all temporal and spatial scales. In contrast, direct estimates of gene flow apply only to the interval of time and space over which observations are made (95). Alternative strategies are also needed to better utilize new types of data now available to population geneticists. A shift in emphasis from classical Mendelian alleles to DNA sequences has changed the way population genetic variation is described. DNA sequences represent a more diverse set of genetic markers than allozymes, and we now have the tools to survey them in natural populations (68). It is also possible to infer genealogical relationships among DNA sequences (4, 62). Interpretation of these relationships has led to the development of coalescent models (43, 53, 54, 115) and to more powerful methods for the estimation of population genetic parameters (34, 37). We can now develop new strategies to estimate gene flow from alternative DNA-based genetic marker systems, population genetic models, and statistical methods. It is not yet clear which, if any, single new approach will replace the F_{ST}-allozyme approach as the standard method for gene flow estimation.

Although the efficiency of methods for DNA analysis has greatly improved, allozyme methods are still more accessible, and methods for the analysis of DNA sequence data are still under development. However, alternative strategies have progressed sufficiently to yield some interesting results. They have allowed us to examine aspects of gene flow distinct from those that can be examined with allozyme data alone. And in a few cases, results with DNA-based markers appear to contradict those obtained with allozyme data and thus have led us to reevaluate F_{ST}-allozyme based estimates of gene flow.

In this review, I first identify the basic methodological, theoretical, and statistical components necessary for the indirect estimation of gene flow from genetic markers. I examine alternatives for each of these components and how they may be used in gene flow estimates. Finally, I consider the assumptions made about the scale of relevant parameters in the estimation of gene flow, and how scale should influence our interpretations of these estimates.

COMPONENTS OF AN INDIRECT APPROACH

Indirect methods of gene flow estimation characterize the spatial distribution of genotypes by some parameter and then apply a population genetic model to ask what level of gene flow would produce a distribution with the same parameter value (95, 97). The logic of indirect methods is thus more complex than estimation from direct measurements, a point first carefully articulated by Weir & Cockerham (124). Organisms are sampled from populations, and a statistic is calculated from the distribution of a genetic marker in the samples. This statistic is then used to estimate a parameter of the genetic marker's distribution in the populations. Estimates of this distribution parameter from one or more genetic markers are combined and used with a population genetic model to estimate a gene flow parameter. Four components are combined by this logic: a demographic model, a genetic marker system, a population genetic model, and a parameter estimator. Alternative methods for the estimation of gene flow have introduced changes in each of these components, and the significance of these modifications should be considered individually.

TERMINOLOGY

Before gene flow can be measured, it must be defined, and any quantitative definition of gene flow must be based upon a model of population structure. The terminology applied to gene flow in population models is potentially confusing because the more general meanings of the terms *migration* and *dispersal* do not match their meanings when applied to gene flow. In a general sense, migration can be defined to include any movements, including cyclical movements that may regularly return an organism to its original location (7). In contrast, the term dispersal is more precisely restricted to movements that increase the distances between organisms, gametes, or propagules (7). Gene flow is defined as the movement of genes in populations, and thus it includes all movements of gametes, propagules, and individuals that are effective in changing the spatial distributions of genes (95). The distinction between movements of organisms and movements of genes is clearly important and should not be forgotten when interpreting estimates of gene flow. However, the terms migration and dispersal have been given secondary meanings in the population genetic literature. When we consider population genetic models, we usually refer to gene flow between discrete subpopulations as migration, and to gene flow within a continuous population as dispersal. I employ this useful division of the concept of gene flow for this review.

The terms deme, subpopulation, population, total population, and metapopulation refer to different scales of population structure, but these scales are not defined precisely or consistently. For this review, I use only two terms: subpopulation and total population. I consider a subpopulation to be a unit of population structure that exchanges migrants with other subpopulations. I consider a total population to be the set of all subpopulations that exchange migrants.

F_{ST}-ALLOZYME APPROACH

Demographic Models

Most indirect estimates of gene flow have been based on models in which migration occurs between discrete subpopulations. Each subpopulation is conceived of as an "island," with no internal spatial structure, and with gene flow via migration of individuals. These models represent a restricted subset of diverse demographic models that have been developed without a population genetic component (see for example 79, 85). The population genetic consequences of demographic models and their use in estimates of gene flow have been the subject of several reviews (32, 95, 97, 102).

The simplest migration model is the infinite island model (95), which is equivalent to the continent-island model introduced by Wright (127, 129). It features an infinitely large source population, and one or more finite subpopulations ("islands") that receive migrants from the source at a constant rate. The source population can either be a single infinitely large population (a "continent"), or an infinite number of finite subpopulations (the total population) that both produce and receive migrants ("infinite islands"). In either case the source population is infinite and therefore not subject to genetic drift. Subpopulations are filled by N zygotes each generation, of which the fraction m are migrants from the source population. The number and relative locations of the island subpopulations are unspecified because each is dependent only on the source population.

The infinite island model leaves out much of what we expect to be true of migration in nature. However, the simplicity of this model has led to its widespread use for estimates of gene flow. Its only parameters are those of concern here, N and m. A more general model would specify a matrix of migration rates between all subpopulations, but such a model would be too complex to be useful for the indirect estimation of gene flow (32). However, some special cases with simple migration matrices have been considered.

In the finite island model or n-Island model (95), migration occurs between n finite subpopulations of size N (60). The migration rate, m, represents the fraction of each subpopulation derived from migration, with all subpopulations equally likely to serve as the source of a migrant. In *stepping stone* models (52), migration occurs only between neighboring subpopulations. The subpopulations may be arrayed in one or more dimensions. Predictions of the model are made with reference to the number of steps between pairs of subpopulations. Although direct gene flow occurs only between neighboring subpopulations, the "stepping stone" effect allows gene flow to occur between nonneighboring subpopulations via the subpopulations that connect them.

Allozyme Markers

The literature on allozymes as genetic markers is extensive. Lewontin's book (61), written less than a decade after the first applications of allozyme methods to population genetics, clearly lays out the rationale for studying allozyme variation and summarizes the early findings. Lewontin has also provided thoughtful retrospectives on how allozyme studies have influenced the development of population genetics (62, 63). A general overview of allozymes and other molecular markers in studies of natural history and evolution is provided by Avise (2).

Allozymes have several desirable properties as genetic markers for indirect estimates of gene flow. They can be assayed without formal genetic analysis and therefore can be surveyed in nearly any species. The nature of allozyme variation and the electrophoretic methods used for its detection provide markers that fit a simple Mendelian model. In higher plants and animals, most allozymes are encoded by autosomal loci that exhibit standard biparental Mendelian inheritance. Multiple polymorphic loci can be surveyed, and typically two or three alleles are detected at each. The rates of mutations that produce detectable differences in allozyme alleles are on the order of 10^{-6} (119).

Genetic Models

Wright's work on nonrandom mating provided the theoretical foundation for estimates of gene flow from allozyme data (127, 130–133). He defined a set of correlation coefficients, which he called F-statistics, to partition departures from random mating into components due to nonrandom mating within populations and to population subdivision (132). For the continent island model of population structure with discrete, nonoverlapping generations of diploid organisms, Wright (127) found the following often-cited relationship between F_{ST}, and $N_e m$, the product of effective population size and migration rate: $F_{ST} \simeq 1/(4N_e m - 1)$. Following Slatkin (99), the quantity $N_e m$ can be considered "the amount of gene flow," and be represented by the single parameter M.

Parameter Estimation

There has been much confusion over what F-statistics actually represent, and how they should be estimated (21, 95, 124, 133). It is useful to consider three meanings. First, F-statistics can be defined as demographic parameters, without reference to any specific form of genetic variation. This corresponds to Wright's definition of an inbreeding coefficient (126). In this sense, F-statistics are parameters that define how gametes are sampled to form zygotes. The quantity relevant to gene flow is F_{ST} (132).

F-statistics may also be defined as parameters of an allele's frequency distribution among individuals and subpopulations. In the absence of selection, the expectations for these distributions correspond to those of the parameters that define how gametes are sampled. In this sense, F_{ST} is the standardized variance of p, the frequency of an allele, among subpopulations: $F_{ST} \simeq \sigma_p^2 / p(1 - p)$ (132). However, because allele frequencies are subject to stochastic variation, the value of F_{ST} is expected to vary among alleles at different loci, among alleles at a multiallelic locus, and over time for any single allele. The stochastic variance imposes a lower limit on the variance of any estimate of F_{ST} as a demographic parameter if that estimate is based on a single locus. In addition to this stochastic variance, realized values of F_{ST} may reflect the effects of selection and mutation.

F-statistics have also been defined as true statistics, quantities calculated from data. These statistics are used to estimate parametric values of F_{ST}. Weir & Cockerham (124) have recommended that parameters be given precise definitions and that they be clearly distinguished from statistics. This is a critical distinction, because the expectation of F_{ST} as a statistic is not equal to F_{ST} as a parameter. F_{ST} as a statistic adds components of variance due to sampling individuals and subpopulation to the actual variance in allele frequencies among subpopulations (74, 124).

F_{ST} can also be related to the process that generates genealogical relationships of genes, known as the coalescent (53). The coalescence time of two genes is the number of generations in the past that separates them from a common ancestral gene. Slatkin (98) derived the following relationship between Wright's F_{ST} and coalescence times, as a limit for mutation rates that are too small to affect F_{ST}: $F_{ST} = (\bar{t} - \bar{t}_0)/\bar{t}$ where $\bar{t}_0$ is the average coalescence time of two genes drawn from the same subpopulation, and $\bar{t}$ is the average coalescence time of two genes drawn from the same total population. This relationship is important in comparisons of alternative methods of estimating gene flow, because it separates the demographic processes of migration and genetic drift from the process of mutation.

In addition to F_{ST}, some related quantities have been introduced that correspond to the different meanings that have been attached to F_{ST}. G_{ST} was originally introduced by Nei (72) to provide an explicit procedure for combining information from multiple loci and alleles in a single measure of population subdivision. Because G_{ST} was defined for samples of populations and individuals, it has been considered a statistic that can be used as an estimator of F_{ST} (124). However, the meaning of G_{ST} has broadened with subsequent usage. Crow & Aoki (23) defined it as a parameter rather than a statistic, and it has been used as a measure of population subdivision for animal mtDNA (112). N_{ST} was first introduced as a statistic based on average differences between DNA sequences in populations (64), and can be considered an estimator of F_{ST} (98) In general, the relationships among these quantities have been determined, and it has been useful to consider them as alternative estimators of the same underlying parameter (21, 98, 102, 124).

Several approaches have been taken to the problem of estimating F_{ST} and related quantities from allozyme data. As discussed above, the problem is not trivial because the statistics calculated from allozyme data are removed by several logical steps from the parameters defined in population genetic models. Weir & Cockerham (124) emphasized that F_{ST} as a parameter cannot be estimated unless it is adequately defined. They developed such a definition for a specific model in which a finite number of subpopulations act as independent realizations of the same process. With this parametric definition of F_{ST}, it is possible to define estimators that, under the conditions of the model, are unbiased. However, if the underlying model is not valid, it is unclear what is being estimated or whether the estimator's properties are preserved. For this reason, Cockerham & Weir (21) did not consider estimation of F_{ST} to be equivalent to an estimation of gene flow. Others (72, 74, 102) have focused on the theoretical link between F_{ST} and M, and the development of workable methods for the estimation of M from population genetic data. Estimators are sought that are robust under a range of conditions, rather than well defined under restricted conditions. The standard approach for estimation of M is based on Wright's island model (127), which suggests the following estimator:

$$\hat{M} = (1 - F_{ST})/4F_{ST}.$$

A few methods have been explored that bypass the estimation of F_{ST} by using other parameters of allele frequency distributions that can be linked to M in population genetic models. Maximum likelihood estimates of M based on Wright's island model (128) assume a beta distribution for allele frequencies among subpopulations (8, 123). The parameters of the beta distribution can be expressed in terms of M and average allele frequencies among subpopulations. Slatkin introduced the "private alleles" method for the estimation of M from the frequencies of rare alleles that are found in only one or a few subpopulations (96). Although this method does not involve explicit estimation of F_{ST}, it is based on the same population genetic models and the properties of allele frequency distributions that correspond to F_{ST} (9). Spatial autocorrelation methods can be used to describe patterns of allele frequency distributions (106), detect population structure (107), and estimate gene flow parameters (105). However, migration parameters are expected to be only weakly correlated with the quantities estimated by spatial autocorrelation analysis (29, 95, 101), which are also subject to mutation, stochastic variance, and nonequilibrium conditions (101).

The process of gene flow in natural populations may be far more complex than its representation in population models. However, if a model is to be used for indirect estimation of gene flow, it may be of little value to add additional parameters if their values cannot be determined. There are differing opinions on the robustness of the methods used to estimate population genetic parameters. Lewontin (62) expressed the opinion that parameter estimates would be too sensitive to such unaccounted influences as natural selection and population history to be reliable. Slatkin (95) acknowledged Lewontin's concerns but argued that, in the case of migration parameter estimates, a more optimistic view was warranted.

Models too complex to be used for estimation of gene flow may still be useful for testing the robustness of conclusions drawn from relatively simple models. For example, Slatkin & Barton (102) used a combination of analytical models and numerical simulations to show that gene flow estimates may be robust with respect to unknown details of population structure and low levels of selection or mutation. They concluded that the F_{ST}-allozyme approach can be generalized to more complex models of population structure.

ALTERNATIVE APPROACHES

Demographic Models

METAPOPULATION MODELS   There has been much interest in the ecological and evolutionary consequences of metapopulation dynamics (see, for example, 40 and others in that volume). Recolonization can maintain an assemblage of subpopulations in which individual subpopulations are subject to frequent extinction. The term metapopulation has been applied to such assemblages, although the term is sometimes used more broadly to include almost any process that involves multiple populations (including conventional migration). The parameters of a metapopulation model define rates of population replacement and the number and source of propagules or individuals that recolonize populations.

To my knowledge, no general method has been proposed for indirect estimation of gene flow parameters in metapopulations. There may be little basis for such a method, because the effect of gene flow on the distribution of genetic variation in metapopulations is expected to depend on the details of the recolonization process, including the founding number, probability of common origin, and kin structure (94, 120), and the outcome may also be critically affected by nonequilibrium dynamics (125). If any general conclusion can be drawn from these theoretical studies, it may be that naive application of a standard method for the estimation of gene flow to populations that are subject to frequent extinctions could yield misleading results. However, it is conceivable that population genetic data could be used in conjugation with other information to estimate parameters within the framework of a metapopulation model.

SPATIAL MODELS   For organisms of relatively limited vagility, gene flow may be restricted by distance alone within a single population. Wright called this effect isolation by distance (130, 131). Although consideration of gene flow in spatial models is not new, these models have not been developed as fully as subpopulation models and are less often used for indirect estimates of gene flow. Two basic types of spatial models have been considered for estimates of gene flow. In continuum models, the locations of individuals are not specified by an array but are generated by the distributions of dispersal movements, births, and deaths within a continuous space. In the special case of every individual leaving exactly one offspring, a random distribution of individuals will develop, such as expected for particles subject to Brownian movement (92). However, if individuals leave multiple offspring, siblings will form clusters that reflect the distribution of dispersal distances, and their descendants in turn will form larger clusters that represent patterns of multigeneration dispersal. This process leads to spatial distributions with infinitely large clumps (87), which are biologically unrealistic, and severely complicates the analysis of the population genetic consequences of these models (31, 32).

In natural populations, it can be assumed that there are mechanisms that distribute individuals more evenly than predicted by simple continuum models (1). Lattice models (102) represent the most rigid form of local density regulation. Individuals are distributed in regular arrays of one or two dimensions, and in the simplest case, exactly one individual occupies each position in the array. This restriction provides a constant population density and size, and greatly simplifies mathematical analysis. Lattice models are formally similar to island and stepping stone models, because each point on the lattice may be treated as a subpopulation. Other forms of density regulation cannot be easily treated analytically and are likely to be explored primarily with numerical simulations.

DNA Markers

ANIMAL MITOCHONDRIAL DNA   The first DNA-based genetic marker system that could be routinely applied to surveys of genetic variation in natural populations was animal mitochondrial DNA (mtDNA). This molecule was amenable to analysis by methods that were widely available in the early 1980s for manipulating plasmid DNA. Animal mtDNA could be easily purified by ultracentrifugation, and sequence differences could be detected as restriction fragment length polymorphisms (RFLPs) (15, 59). More recently, the introduction of the polymerase chain reaction (PCR) has nearly eliminated the need to directly purify specific DNA sequences (86). Improvements in DNA sequencing methods have made it feasible to determine the exact nucleotide sequence of amplified regions of mtDNA for large numbers of individuals (69).

In many important respects, mtDNA polymorphisms are quite different from allozyme alleles. Animal mtDNA is generally inherited maternally as a single linkage unit of about 15 kb (5, 12, 13, 59). Because the mitochondrial genome is transmitted without recombination as a single linkage unit, mtDNA sequence variants are usually referred to as haplotypes, rather than alleles. The inheritance of mtDNA is thus formally similar to a single haploid locus. Mutation rates are often higher for animal mtDNA than for nuclear genes (14, 118), and there are generally many more detectable polymorphisms than there are for single allozyme loci (3, 5). Perhaps most significantly, detailed characterization of sequence differences can be used to infer genealogical relationships and estimate divergence time (reviewed in 4, 13).

CHLOROPLAST DNA   The chloroplast genome is a circular molecule, typically about 160 kb in size (80). In most higher plants, chloroplast DNA (cpDNA) is uniparentally inherited—maternally in angiosperms and paternally in conifers (22, 41, 90). Variation in cpDNA sequences can be detected by either restriction site analysis or sequencing. In these respects, cpDNA is similar to animal mtDNA. However, variation in cpDNA is more often used to infer relationships among species than as a population genetic marker. This is primarily because in land plants, cpDNA has a relatively slow rate of evolution, and consequently low levels of intraspecific polymorphism (20, 80). However, intraspecific polymorphism has been found in some species (19, 27, 36, 42, 51, 65, 71, 81, 108, 109), and so it has been possible to use cpDNA variation to investigate gene flow (27, 66).

NUCLEAR DNA   Nuclear DNA (nDNA) variation has been used to estimate gene flow in a few studies. Early studies of nDNA polymorphisms were either limited to surveys of RFLPs detected by Southern hybridization (58, 88), or they required laborious molecular cloning methods to isolate genes for nucleotide sequencing (57). The advent of the PCR (polymerase chain reactions), a growing database of sequences that can be used to design PCR primers, and automated sequencing (39, 46) have greatly increased the feasibility of nDNA sequence surveys (93). Two forms of nuclear sequence variation are widely used as genetic markers: base substitutions and variable numbers of tandem repeats (VNTRs). Base substitutions are more difficult to survey in populations but offer a greater possibility of inferring genealogical relationships among sequences. VNTRs are relatively easy to survey in populations. They are often highly polymorphic, and length variation can be detected by simple electrophoretic methods. VNTR sequences are classified by size as microsatellites or minisatellites. Microsatellites consist of up to 50 copies of tandemly repeated sequences that are each 1–10 basepairs (bp) in length. Because their total length is usually less than a few hundred bp, it is possible to analyze microsatellite length variation by direct size measurements of PCR-amplified sequences on electrophoretic gels (113a). Minisatellite sequences contain up to several hundred copies of repeat units that range from 10–200 bp in length, with total lengths up to 50 kb. Because of their large size, minisatellite sequences often cannot be amplified by PCR, and their length polymorphisms are usually detected as RFLPs with Southern hybridization (48a).

Two mechanisms are expected to generate variation in the number of tandem repeats. Recombination is expected to generate a broad distribution of length changes, at rates that may be as high as 5×10^{-2} (16, 47). Thus, in sequence comparisons, little correlation is expected between the magnitude of accumulated length differences and the number of unequal crossing over events that produced them. Replication slippage that occurs during DNA replication appears to favor small stepwise changes in the number of tandem repeats (25, 91, 117). The rates at which variants are generated by this process appears to be between 10^{-4} and 10^{-3} (24, 28, 122). In contrast to variation generated by recombination, a correlation is expected between the accumulated length differences and the number of generating events, provided the number of these events is not too large (117). Although the mechanisms that generate length variation in VNTRs are not well understood, the distribution of length variation in microsatellite sequences is most consistent with replication slippage (89, 114, 117). Recombination and related gene conversion processes appear to be responsible for variation in minisatellite sequences (47, 48).

Genetic Models

There is an obvious temptation to apply estimators based on familiar population genetic models developed for the F_{ST}-allozyme approach to other forms of gene flow and other genetic marker systems. In some cases, the fit would not be bad, and the models would be expected to yield reasonable gene flow estimates. In other cases, the assumptions of these models would be severely violated and could result in very inaccurate estimates of gene flow. However, even where the standard models are not inappropriate, new models may provide more effective methods for the estimation of gene flow.

COALESCENT MODELS   The introduction of nucleotide sequence data to population genetics has been accompanied by the development of genealogical, or coalescent, models (43, 53, 54, 115), which have led to new approaches for both testing evolutionary hypotheses (49) and estimating population genetic parameters (35, 37). Two kinds of information in a genealogy of DNA sequences can be used for estimates of gene flow. First, the lengths of the branches that connect sequences in the genealogical tree, which correspond to coalescence times, are sometimes referred to as the distances between the sequences. Second, the relative order of branches in the tree, which corresponds to the order of coalescence events, constitute the cladistic relationships of the sequences. Both types of information can be used to estimate gene flow, and an ideal method would make the fullest use of both.

The results of standard population genetic models of gene flow can be recast in genealogical terms because of the close relationship between coalescence time and allelic identity (98). However, because these models consider only average pairwise relationships between sequences, they do not fulfill the potential of genealogical models. Progress in the development of genealogical models that incorporate gene flow (49, 111, 113) is expected to lead to the further development of methods for estimating gene flow and other population genetic parameters from DNA sequence data (35, 45).

RAPIDLY MUTATING LOCI   If the mutation rate at a locus is extremely high, individual alleles generated by mutation will appear and be dispersed only briefly in a population before they are altered by subsequent mutations. Under these conditions, the spatial distributions of alleles in a population can be interpreted as short-term traces of gene flow. O'Connell & Slatkin (78) used both analytic and simulation models to determine expected spatial distributions of alleles at rapidly mutating loci. The results of their analyses cannot be simply summarized, but they suggest that if the mutation rate is known, it should be possible to estimate neighborhood size from the spatial distributions of alleles at rapidly mutating loci. These models could be applied to microsatellite or minisatellite loci, although the predictions of these models are sensitive to details of the mutation process (78).

ORGANELLAR GENOMES   Specific models have been developed for the population genetics of organellar genomes (mtDNA and cpDNA) (29a, 81a, 112). These models indicate that rates of both gene flow and genetic drift should depend strongly on the mode of transmission of organellar genomes (maternal, paternal or biparental), the effective population sizes of haploid and uniparentally transmitted genomes, and the consequences of multiple modes of dispersal (e.g. pollen vs seed). A common prediction of these models is that organellar genomes should often be subject to greater genetic drift than nuclear genomes, because both haploidy and uniparental transmission may reduce effective population size. A relative reduction in gene flow is also expected for maternally inherited organellar genomes in plants, because both seed and pollen movements result in gene flow for nuclear genes, while only seed dispersal results in gene flow for maternally inherited genes.

Parameter Estimation

CLADISTIC MEASURE OF GENE FLOW   In comparison with allozyme studies, most surveys of DNA sequence variation sample fewer individuals and loci but may provide information that can be used to infer genealogical relationships among sequences. In these respects, DNA sequence data are not well suited for analysis by F_{ST}, which is best applied to large samples and does not make use of cladistic relationships. The cladistic measure of gene flow was developed as an alternative (103). DNA sequences are sampled from subpopulations, and sequence variation is used to infer a phylogenetic tree by any of several standard methods (for example, 110). Each sequence on the tree is assigned a multistate unordered character that indexes the subpopulation from which it was sampled. A parsimony criterion is then used to determine the minimum number of character state transitions required for the phylogenetic distribution of this character to be consistent with the tree. A robust relationship between s and M was found in simulations of island subpopulations over a range of conditions. A table can be used to convert values of s to estimates of M. From additional simulations, a simple relationship was also found between M and the geographic distance between samples in stepping stone and lattice models (104).

The performance of the cladistic measure of gene flow was compared to F_{ST}-based estimates in simulations of a finite island demographic model, with sequences subject to recombination and an infinite-sites model of mutation (45). Performance criteria were bias, variance, and the accuracy of values for the median and other percentiles of the estimator distributions. For low migration rates, F_{ST} provided the best estimates. At moderate to high rates of migration, the two methods were differentially affected by recombination rate. Low rates of recombination favored the cladistic method, while F_{ST} based estimates improved with higher rates of recombination. Contrary to initial expectations (103), the cladistic method performed nearly as well as F_{ST} under conditions of moderate to high migration and high recombination. At present, the cladistic measure of gene flow is probably best applied to animal mtDNA sequences. Unlike nDNA, mtDNA is not subject to recombination, and unlike cpDNA, there are usually sufficient polymorphic sites in mtDNA to fully resolve genealogies.

NEIGHBORHOOD SIZE   Two parameters can be related to gene flow in spatial models. One corresponds to single generation dispersal distances, the other is the effective neighborhood size, N_b which is analogous to M in subpopulation models. The relationship between these parameters can be understood by considering a symmetrical distribution of single-generation dispersal movements. With respect to a single spatial coordinate, this distribution includes both positive and negative movements, and it has a mean of zero. A typical dispersal distance can be characterized by the standard deviation of this distribution, σ_D. This quantity may be referred to as the standard dispersal distance (77). Together with the population density, D, σ_D determines N_b. If selfing can occur, N_b is the expected number of individuals within a range of $2\sigma_D$ (130).

Underlying similarities between spatial models and stepping-stone island models suggest that F_{ST} analysis can be used to estimate N_b (102). It is encouraging that similar population genetic results have been reached by analyses of different spatial models (32, 78, 102). However, the validity of the approximations used for continuum models has been questioned (32), and these models may not represent the full range of effects that local density regulation could have on gene flow. Simulation models of gene flow, while not on a par with analytic models (21), do allow the investigation of different mechanisms of density regulation on gene flow in continuous populations (76, 77).

DISPERSAL DISTANCE   Striking patterns are often observed when the genealogical relationships of animal mtDNA sequences are overlaid on their geographic locations. These patterns can be interpreted by a combination of population genetic, systematic, and biogeographic principles, and the synthesis of these principles is referred to as phylogeography (4). We investigated phylogeographic patterns that could develop from isolation by distance alone (76, 77) with extensions of models that had been developed for the phylogenetic relationships of mtDNA lineages within species (6) and between species (75). We first considered a continuum model in which mtDNA lineages were dispersed by a simple random-walk process. If two members of the same lineage coalesce n generations in the past and the distribution of single generation dispersal distances of females has variance σ_F^2, the probability distribution for the distance between them will have variance $\sigma_F^2 = 2n\sigma_F^2$. This suggests that it should be possible to estimate the standard dispersal distance, σ_F^2, from an appropriate sample of pairs of mtDNA sequences with estimates of n based on sequence divergence. An attractive feature of this approach is that it is based on a process that is not dependent on genetic drift and provides an estimate of σ_F^2, the standard dispersal distance, which could be directly compared with mark-and-recapture based estimates of dispersal.

There are two problems in applying the random walk model to pairwise data. The first is obtaining pairs of individuals that represent independent realizations of the random walk process. In branching genealogies, many individuals will share portions of their paths of descent and thus also the paths of dispersal that led to their present locations. Data from such pairs of individuals are not independent. There are several possible solutions to this problem, including simply paring down a sample to eliminate nonindependent pairs.

The second and more difficult problem is obtaining distributions of pairwise distances that are conditional on coalescence times (10, 77). The variances of these spatial distributions would provide a direct estimate of σ_G^2. However, in general, we cannot randomly sample individuals with respect to their spatial locations. Rather, we select the locations from which the samples are taken, which imposes the distribution of these locations on the sample.

We developed a workable solution to these problems by using the variance of the spatial distribution of all members of a lineage, σ_H^2, as an approximation of the variance for independent pairs of individuals, σ_G^2 (77). Because members of the same lineage are not independent, σ_H^2 tends to be smaller than σ_G^2, and thus we should anticipate some bias in estimates based on σ_H^2. The second problem now becomes sampling the spatial distributions of entire lineages, rather than pairs of individuals. Ideally, a uniform spatial distribution of sampling locations would be used, but few mtDNA surveys provide such data. However, samples from nonuniform distributions can be used by weighting them appropriately, as we have described. In simulations, this sampling method provided a distribution of estimates for σ_F with the expected bias of σ_H/σ_G and a standard deviation of $0.14\sigma_F$ (77). An advantage of sampling entire lineages is that separate estimates of σ_F^2 can be based on individual lineages or lineages grouped by location or age. This raises the interesting possibility of analyzing spatial or temporal heterogeneity in the dispersal process.

A simple random walk model of lineage dispersion assumes that the process is uniform and unconstrained. This is unlikely to be true for natural populations. For most species, local population density regulation and barriers to dispersal, including range limits, would violate these assumptions. Furthermore, over long time scales, range limits and population sizes would be expected to change. Therefore, it is important to consider the robustness of the random walk model and to identify conditions under which it may fail.

Range limits will impose an upper limit on σ_H^2. This range saturation effect can be predicted by comparing the expected spatial distribution of a lineage with the size of the available range. As range saturation occurs, it is expected that the rate at which σ_H^2 increases with time will decline, and the correlation between σ_H^2 and σ_F^2 will weaken (76). In additional simulations, with local population density regulation and cyclical range contractions, the relationship $\sigma_G^2 = 2n\sigma_F^2$ was robust for lineages in which range saturation had not occurred (76).

The predictions of the random walk model were tested with published mtDNA surveys (76, 77). As expected, the best fits were found for species with broad geographic ranges and low vagility. In these cases, strong correlations were observed between lineage age and σ_H^2. For the deer mouse (*Peromyscus maniculatus*), our bias-corrected estimate of σ_F^2 was 225 m (77), which corresponds closely to mark-and-recapture estimates of 230 m for females (11) and 264 m for both sexes (26). Also as expected, there was evidence of range saturation for highly vagile species of birds and marine fishes. For these cases, there was no significant correlation between lineage age and σ_H^2, and the observed values of σ_H^2 were not significantly different from those obtained with geographically randomized data (76).

CONSIDERATIONS OF SCALE AND INTERPRETATION

Population Size

Assumptions about population size are often implicit in gene flow models. Estimates of M are really measures of the relative strengths of genetic drift and migration. In small populations, the effects of genetic drift on allele frequencies will be strong relative to other forces. Equilibrium between genetic drift and migration will be quickly established (23) and will be mostly insensitive to relatively weaker forces such as selection or mutation (102). However, in large populations, genetic drift is a relatively weak force. If the migration rate is also small, equilibria that are determined primarily by genetic drift will be reached slowly (23), and the distributions of genetic markers will be more susceptible to mutation and selection. At very low migration rates, the expected time in generations to equilibrium is on the order of the effective population size. If this approaches the ages of populations, historical relationships among populations, rather than gene flow, may be the primary determinant of the distribution of genetic markers (33).

In large populations, selection acting on genetic markers may bias estimates of gene flow. A persistent question about allozymes is to what extent their polymorphisms are subject to natural selection. Some allozyme data is at least consistent with the effects of selection (18, 56, 57a, 83, 121). Balancing selection may be of particular relevance to gene flow estimation. It has long been suggested that either heterosis or frequency dependent selection maintains some allozyme polymorphisms (44, 55, 56). Estimates of F_{ST} based on allozymes subject to balancing selection would lead to overestimates of gene flow (102).

A thought-provoking comparison of allozyme, mtDNA, and nDNA markers has been provided by studies of the American oyster (*Crassostrea virginica*). Its life history is typical of many benthic marine invertebrates, with planktonic larvae that are capable of long-distance dispersal. Buroker (17) surveyed allozymes in American oyster populations from the Atlantic coast of North America and the Gulf of Mexico. Allele frequencies at five polymorphic loci were generally similar among all locales, indicating a high rate of gene flow. In contrast, Reeb & Avise (84) found Atlantic and Gulf populations were distinguished by two very divergent groups of mtDNA. The distributions of these haplotypes implies that Atlantic and Gulf of Mexico populations have long been completely isolated with respect to gene flow. Several alternative explanations can generally be proposed to account for differences in gene flow estimates from allozymes and mtDNA. By conventional wisdom, an unlikely explanation would be that selection has acted on all five allozyme loci to maintain similar gene frequencies in isolated populations. However, based on subsequent evidence, this appears to be the most probable explanation. A survey of four randomly cloned nDNA markers revealed a consistent geographic pattern similar to the distribution of mtDNA haplotypes (50). These nuclear sequences should be subject to the same forces of gene flow and genetic drift as allozyme loci, but, because they are not protein-coding loci, they are not likely subject to the same selective forces. It appears that in the two large regional populations of oysters, balancing selection at allozyme loci may have overcome the relatively weak force of genetic drift and thereby created the appearance of high gene flow.

Mutation Rate

The relationship between F_{ST} and M predicted by standard population genetic models is based on the assumption that mutation rates are much lower than migration rates. If mutation rates are higher than assumed, estimates of F_{ST}

will be downwardly biased. The exact form and magnitude of this bias depends on the mutation process. For an infinite alleles model of mutation and an island model of gene flow: $F_{ST} \simeq 1/(1 + 4N_e m + 4N\mu)$ (23). Thus, mutation has the same effect on F_{ST} as gene flow. For a stepwise mutation model, which may be appropriate for microsatellite loci (89, 114, 117), the bias is greater (100). This dependence of estimates of F_{ST} on mutation rates should not be important for allozyme markers, which have very low mutation rates. However, for DNA sequences with much higher mutation rates, this problem could be quite severe.

If DNA sequence data are used, the problem of mutation saturation can be avoided by using any of several measures that effectively weight pairwise comparisons of sequences by the number of mutations that differentiate them (64, 73, 112). More generalized methods for the use of weighted pairwise data in analysis of population structure have also been developed (30, 82). However, for mtDNA data, a cladistic measure of gene flow (103, 104) may be preferable to an F_{ST}-based estimate (45).

It is less obvious how to correct estimates of F_{ST} for the high mutation rates of microsatellite and minisatellite loci. A weighted pairwise distance measure (38) and a measure of population structure that is analogous to F_{ST} (100) have been proposed for microsatellites that undergo a stepwise mutation process. These measures perform well in simulations, but further testing of the stepwise mutation model is needed before their reliability can be assumed. For minisatellites, an infinite alleles model may provide a reasonable approximation of random unequal crossover events and a large number of potential length states. However, estimators that use corrections based solely on multiple event probabilities would be subject to the high variance of these corrections, and the loss of information that occurs when only a small proportion of alleles remain identical in state.

Time

Barton & Wilson (10) raised a general criticism of any attempt to estimate gene flow from genealogical data, such as mtDNA genealogies. With coalescent models of isolation by distance and neighborhood sizes typical of most species, average coalescence times were sufficiently long for the dispersal of lineages to reach range saturation and thus randomize most information about gene flow. Because in many species mtDNA genealogies are not geographically randomized, they concluded that this long-term expectation of isolation by distance is never realized and that mtDNA distributions are generated primarily by historical processes.

This is a serious criticism not only of gene flow estimates from genealogical data but of any method of gene flow estimation based on a process that may be strongly influenced by population history (33). The problem is not so much the models as the scale on which they are applied. For example, the time in generations, t, required for F_{ST} to approach equilibrium is approximately $1/[2m + 1/(2N_e)]$, where m is the rate of migration and N_e the effective population size (23). While it is often stated that F_{ST} approaches equilibrium quickly, clearly this is not the case for large N_e and small m. Does this invalidate estimates of gene flow for broad relatively isolated regional populations, which are likely to have been influenced by historical processes as well as gene flow?

The answer to this may be in part a matter of perspective. Large-scale patterns of population structure, whether described by allele frequencies or DNA sequence genealogies have almost certainly been influenced by history. However, throughout the history of a population, the mechanisms behind broad transformations of its structure must necessarily include finer-scale processes of dispersal, migration, and genetic drift, albeit in a more dynamic context than is generally represented in models. The challenge is to develop tools that can resolve these processes and not to assume that simple measurements necessarily require simplistic interpretations.

Gene flow estimates are usually made without reference to a specific timeframe; however, it is possible to adapt them to detect historical changes in patterns of gene flow. For stepping-stone models of population structure, the relationship between estimates of M and the distance between samples can provide evidence for a history that has prevented the attainment of equilibrium (99). Similar analyses can be based on cladistic measures of gene flow (104). Comparisons of population structure measures for nuclear vs. organellar genomes (67, 81a), or for rapidly and slowly mutating loci (100), also have the potential to reveal whether gene flow and genetic drift have reached equilibrium. Estimates of dispersal distance based on the geographical distribution of mtDNA lineages have been viewed as particularly sensitive to historical effects (10). However, this sensitivity could be used to examine the history of dispersal by comparing gene flow estimates among lineages of different ages (76, 77). Finally, it may be possible to identify the processes that generate complex phylogeographic patterns (4) by testing the fit of identifiable elements of these patterns to alternative models that represent both historical and dispersal processes (2, 116). Progress in these areas should benefit from the availability of genetic markers that represent different components of gene flow, from models that are built from different assumptions, and from fresh perspectives on gene flow as a complex process operating on many spatial and temporal scales.

ACKNOWLEDGMENTS

I am especially indebted to M Slatkin for his contributions in this area, and for his patience with my own efforts and for helpful comments on the manuscript. I also thank J Avise for sharing his ideas and encouraging the development of my own. This review was supported by NSF grant DBI-9630311. This work is a result of research sponsored by NOAA National Sea Grant College Program Office, Department of Commerce, under Grant R/CFB-21.

Visit the *Annual Reviews home page* at
http://www.annurev.org.

Literature Cited

1. Andrewartha HG, Birch LC. 1954. *The Distribution and Abundance of Animals.* Chicago: Univ. Chicago Press

2. Avise JC. 1994. *Molecular Markers, Natural History and Evolution.* New York: Chapman & Hall

3. Avise JC. 1986. Mitochondrial DNA and the evolutionary genetics of higher animals. *Philos. Trans. R. Soc. London* B312:325–42

4. Avise JC, Arnold J, Ball RM, Bermingham E, Lamb T, et al. 1987. Intraspecific phylogeography: the mitochondrial DNA bridge between population genetics and systematics. *Annu. Rev. Ecol. Syst.* 18:489–522

5. Avise JC, Lansman RA. 1983. Polymorphism of mitchondrial DNA in populations of higher animals. In *Evolution of Genes and Proteins,* ed. M Nei, RK Koehn, pp. 147–64. Sunderland, MA: Sinauer

6. Avise JC, Neigel JE, Arnold J. 1984. Demographic influences on mitochondrial DNA lineage survivorship in animal populations. *J. Mol. Evol.* 20:99–105

7. Baker RR. 1978. *The Evolutionary Ecology of Animal Migration.* New York: Holmes & Meier

8. Barton NH, Halliday RB, Hewitt GM. 1983. Rare electrophoretic variants in a hybrid zone. *Heredity* 50:139–46

9. Barton NH, Slatkin M. 1986. A quasi-equilibrium theory of the distribution of rare alleles in a subdivided population. *Heredity* 56:409–15

10. Barton NH, Wilson I. 1995. Genealogies and geography. *Philos. Trans. R. Soc. Lond. B* 349:49–59

11. Blair WF. 1940. A study of prairie deer-mouse populations in southern Michigan. *Am. Midland Nat.* 24:273–305

12. Brown WM. 1983. Evolution of animal mitochondrial DNA. In *Evolution of Genes and Proteins,* ed. M Nei, RK Koehn, pp. 62–88. Sunderland, MA: Sinauer

13. Brown WM. 1985. The mitochondrial genome of animals. In *Molecular Evolutionary Genetics,* ed. RJ MacIntyre. New York: Plenum

14. Brown WM, George JM, Wilson AC. 1979. Rapid evolution of animal mitochondrial DNA. *Proc. Natl. Acad. Sci. USA* 76:1967–71

15. Brown WM, Goodman HM. 1979. Quantification of intrapopulation variation by restriction analysis of human mitochondrial DNA. In *Extrachromosomal DNA,* ed. DJ Cummings, P Borst, IB Dawid, SM Weismann, CF Fox. New York: Academic

16. Bruford MW, Hanotte O, Brookfield JFY, Burke T. 1992. Multi- and single-locus DNA fingerprinting. In *Molecular Analysis of Populations: A Practical Approach,* ed. AR Hoelzel, pp. 225–69. Oxford: IRL Press

17. Buroker NE. 1983. Population genetics of the American oyster *Crassostrea virginica* along the Atlantic coast and the Gulf of Mexico. *Mar. Biol.* 75:99–112

18. Burton RS, Feldman MW. 1983. Physiological effects of an allozyme polymorphism: glutamate-pyruvate transaminase and response to hyperosmotic stress in the copepod *Tigriopus californicus. Biochem. Genet.* 21:239–51

19. Byrne M, Moran GF. 1994. Population divergence in the chloroplast genome of *Eucalyptus nitens. Heredity* 73:18–28

20. Clegg MT, Gaut BS, Learn GH, Morton BR. 1994. Rates and patterns of chloroplast DNA evolution. *Proc. Natl. Acad. Sci. USA* 91:6795–801

21. Cockerham CC, Weir BS. 1993. Estimation of gene flow from F-Statistics. *Evolution* 47:855

22. Corriveau JL, Coleman AW. 1988. Rapid screening method to detect potential biparental inheritance of plastid DNA and results for over 200 angiosperm species. *Am. J. Bot.* 75:1443–58

23. Crow JF, Aoki K. 1984. Group selection for a polygenic behavioral trait:

estimating the degree of population subdivision. *Proc. Natl. Acad. Sci. USA* 81:6073–77

24. Dallas JF. 1992. Estimation of microsatellite mutation rates in recombinant inbred strains of mouse. *Mammal. Genome* 5:32–38

25. Di Rienzo A, Peterson AC, Garza JC, Valdes AM, Slatkin M, et al. 1994. Mutational processes of simple-sequence repeat loci in human populations. *Proc. Natl. Acad. Sci. USA* 91:3166–70

26. Dice LR, Howard WE. 1951. Distances of dispersal by prairie deer mice from birthplaces to breeding sites. *Cont. Lab. Vert. Biol. Univ. Mich.* 50:1–15

27. Dong J, Wagner DB. 1994. Paternally inherited chloroplast polymorphism in *Pinus*: estimation of diversity and population subdivision, and tests of disequilibrium with a maternally inherited mitochondrial polymorphism. *Genetics* 136:1187–94

28. Edwards A, Hammond HA, Jin L, Caskey CT, Chakraborty R. 1992. Genetic variation at five trimeric and tetrameric tandem repeat loci in four human population groups. *Genomics* 12:241–53

29. Endler JA. 1977. *Geographic Variation, Speciation and Clines.* Princeton, NJ: Princeton Univ. Press

29a. Ennos RA. 1994. Estimating the relative rates of pollen and seed migration among plant populations. *Heredity* 72:250–59

30. Excoffier L, Smouse PE, Quattro JM. 1992. Analysis of molecular variance inferred from metric distances among DNA haplotypes: application to human mitochondrial DNA restriction data. *Genetics* 131:479–91

31. Felsenstein J. 1975. A pain in the torus: some difficulties with models of isolation by distance. *Am. Nat.* 109:359–68

32. Felsenstein J. 1976. The theoretical population genetics of variable selection and migration. *Annu. Rev. Genet.* 10:253–80

33. Felsenstein J. 1982. How can we infer geography and history from gene frequencies? *J. Theor. Biol.* 96:9–20

34. Felsenstein J. 1992. Estimating effective population size from samples of sequences: a bootstrap monte carlo integration method. *Gen. Res.* 60:209–20

35. Felsenstein J. 1992. Estimating effective population size from samples of sequences: inefficiency of pairwise and segregating sites as compared to phylogenetic estimates. *Genet. Res. Camb.* 59:139–47

36. Fenster CB, Ritland K. 1992. Chloroplast DNA an isozyme diversity in two *Mimulus* species (Scrophulariaceae) with contrasting mating systems. *Am. J. Bot.* 79:1440–47

37. Fu YX. 1994. Estimating effective population size or mutation rate using the frequencies of mutations of various classes in a sample of DNA sequences. *Genetics* 138:1375–86

38. Goldstein DB, Ruiz Linares A, Cavalli-Sforza LL, Feldman MW. 1995. An evaluation of genetic distances for use with microsatellite loci. *Genetics* 139:463–71

39. Gyllensten UB, Erlich HA. 1988. Generation of single-stranded DNA by the polymerase chain reaction and its application to direct sequencing of the HLA-DQA locus. *Proc. Natl. Acad. Sci. USA* 85:7652–56

40. Hanski I, Gilpin M. 1991. Metapopulation dynamics—brief history and conceptual domain. *Biol. J. Linn. Soc.* 42:3–16

41. Harris H, Hopkinson DA. 1976. *Handbook of Enzyme Electrophoresis in Human Genetics.* Oxford, UK: North Holland

42. Hong Y, Hipkins VV, Strauss SH. 1993. Chloroplast DNA diversity among trees, populations and species in the California closed-cone pines (*Pinus radiata, Pinus muricata* and *Pinus attenuata*). *Genetics* 135:1187–96

43. Hudson RR. 1990. Gene genealogies and the coalescent process. In *Oxford Surveys in Evolutionary Biology*, ed. DJ Futuyma, J Antonovics

44. Hudson RR, Kreitman M, Aguade M. 1987. A test of neutral molecular evolution based on nucleotide data. *Genetics* 116:153–59

45. Hudson RR, Slatkin M, Madison WP. 1992. Estimation of levels of gene flow from DNA sequence data. *Genetics* 132:583–89

46. Hunkapiller T, Kaiser RJ, Koop BF, Hood L. 1991. Large-scale and automated DNA sequence determination. *Science* 254:59–68

47. Jeffreys AJ, Royle NJ, Wilson V, Wong Z. 1988. Spontaneous mutation rates to new length alleles at tandem-repetitive hypervariable loci in human DNA. *Nature* 332:278–81

48. Jeffreys AJ, Tamaki K, Macleod A, Monckton DG, Neil DL, et al. 1994. Complex gene conversion events in germline mutations at human minisatellites. *Nature Genet.* 6:136–45

48a. Jeffreys AJ, Wilson V, Thein SL. 1985. Hypervariable "minisatellite" regions in human DNA. *Nature* 314:67–73

49. Kaplan N, Hudson RR, Izuka M. 1991. The coalescent process in models with selection, recombination and geographic subdivision. *Genet. Res. Camb.* 57: 83–91

50. Karl SA, Avise JC. 1992. Balancing selection at allozyme loci in oysters: implications from nuclear RFLPs. *Science* 256:100–2

51. Kim K-J, Jansen RK, Turner BL. 1992. Evolutionary implications of intraspecific chloroplast DNA variation in dwarf dandelions (Krigia; Asteraceae). *Am. J. Bot.* 79:708–15

52. Kimura M. 1953. "Stepping-stone" model of population. *Annu. Rep. Natl. Inst. Genet. Japan* 3:62–63

53. Kingman JFC. 1982. The coalescent. *Stochast. Proc. Theor. Appl.* 13:235–48

54. Kingman JFC. 1982. On the genealogy of large populations. *J. Appl. Prob.* 19A:27–43

55. Koehn RK, Gaffney PM. 1984. Genetic heterozygosity and growth rate in *Mytilus edulis. Mar. Biol.* 82:1–7

56. Kohen RR, Hilbish TJ. 1987. The adaptive importance of genetic variation. *Am. Sci.* 75:134–41

57. Kreitman M. 1983. Nucleotide polymorphism at the alcohol dehydrogenase locus of *Drosophila melanogaster. Nature* 304:412–17

57a. Kreitman M, Akashi H. 1995. Molecular evidence for natural selection. *Annu. Rev. Ecol. Syst.* 26:403–22

58. Kreitman ME, Aguade M. 1986. Excess polymorphism at the *Adh* locus in *Drosophila melanogaster. Genetics* 114:93–110

59. Lansman RA, Shade RO, Shapira JF, Avise JC. 1981. The use of restriction endonucleases to measure mitochondrial DNA sequence relatedness in natural populations. III. Techniques and potential applications. *J. Mol. Evol.* 17:214–26

60. Latter BDH. 1973. The island model of population differentiation: a general solution. *Genetics* 73:147–57

61. Lewontin RC. 1974. *The Genetic Basis of Evolutionary Change.* New York: Columbia Univ. Press

62. Lewontin RC. 1985. Population genetics. In *Evolution: Essays in Honour of John Maynard Smith,* ed. PJ Greenwood, PH Harvey, M Slatkin, pp. 3–18. Cambridge: Cambridge Univ. Press

63. Lewontin RC. 1991. Twenty-five years ago in Genetics. Electrophoresis in the development of evolutionary genetics: milestone or millstone? *Genetics* 128:657–62

64. Lynch M, Crease TJ. 1990. The analysis of population survey data on DNA sequence variation. *Mol. Biol. Evol.* 7:377–94

65. Mason-Gamer RJ, Holsinger KE, Jansen RK. 1995. Chloroplast DNA haplotype variation within and among populations of *Coreopsis grandiflora* (Asteraceae). *Mol. Biol. Evol.* 12:371–81

66. McCauley DE. 1994. Contrasting the distribution of chloroplast DNA an allozyme polymorphism among local populations of *Silene alba*: implications for studies of gene flow in plants. *Proc. Natl. Acad. Sci. USA* 91:8127–31

67. McCauley DE. 1995. The use of chloroplast DNA polymorphism in studies of gene flow in plants. *Trends Ecol. Evol.* 10:198–202

68. Mitton JB. 1994. Molecular approaches to population biology. *Annu. Rev. Ecol. Syst.* 25:45–69

69. Murray V. 1989. Improved double-stranded DNA sequencing using the linear polymerase chain reaction. *Nucleic Acids Res.* 17:8889

70. Nagylaki T. 1978. The geographical structure of populations. In *Studies in Mathematics,* ed. SA Levin, pp. 588–624

71. Neale DB, Saghai-Maroof MA, Allard RW, Zhang Q, Jorgensen RA. 1986. Chloroplast DNA diversity in populations of wild and cultivated barley. *Genetics* 120:1105–10

72. Nei M. 1973. Analysis of gene diversity in subdivided populations. *Proc. Natl. Acad. Sci. USA* 70:3321–23

73. Nei M. 1982. Evolution of human races at the gene level. In *Human Genetics, Part A: The Unfolding Genome,* ed. B Bohhe-Tamir, P Cohen, RN Goodman, pp. 167–181. New York: Alan R. Liss

74. Nei M, Chesser RK. 1983. Estimation of fixation indices and gene diversities. *Ann. Hum. Genet.* 47:253–59

75. Neigel JE, Avise JC. 1986. Phylogenetic relationships of mitochondrial DNA under various demographic models of speciation. In *Evolutionary Processes and Theory,* ed. S Karlin, E Nevo. New York: Academic

76. Neigel JE, Avise JC. 1993. Application of a random walk model to geographic distributions of animal mitochondrial DNA variation. *Genetics* 135:1209–20

77. Neigel JE, Ball RM, Avise JC. 1991. Estimation of single generation migration distances from geographic variation in animal mitochondrial DNA. *Evolution* 45:423–32

78. O'Connell N, Slatkin M. 1993. High mutation rate loci in a subdivided population. *Theor. Pop. Biol.* 44:110–27

79. Okubo A. 1980. *Diffusion and Ecological Problems: Mathematical Models.* Berlin: Springer-Verlag

80. Palmer JD. 1991. Plastid chromosomes: structure and evolution. In *The Molecular Biology of Plastids,* ed. IK Vasil. 7A:5–53. New York: Academic

81. Petit RJ, Kremer A, Wagner DB. 1993. Geographic structure of chloroplast DNA polymorphisms in European oaks. *Theor. Appl. Genet.* 87:122–28

81a. Petit RJ, Kremer A, Wagner DB. 1993. Finite island model for organelle and nuclear genes in plants. *Heredity* 71:630–41

82. Pons O, Petit RJ. 1996. Measuring and testing genetic differentiation with ordered versus unordered alleles. *Genetics* 144:1237

83. Powers DA, DiMichele L, Place AR. 1983. The use of enzyme kinetics to predict differences in cellular metabolism, developmental rate and swimming performance between LDH-B genotypes of the fish *Fundulus heteroclitus. Curr. Top. Biol. Med. Res.* 10:147–70

84. Reeb CA, Avise JC. 1990. A genetic discontinuity in a continuously distributed species: mitochondrial DNA in the American Oyster, *Crassostrea virginica. Genetics* 124:397–406

85. Renshaw E. 1991. *Modelling Biological Populations in Space and Time.* Cambridge: Cambridge Univ. Press

86. Saiki RK, Gelfand DH, Stoffel S, Scharf SJ, Higuchi R, et al. 1988. Primer-directed enzymatic amplification of DNA with a thermostable DNA polymerase. *Science* 239:487–91

87. Sawyer S. 1967. Branching diffusion processes in population genetics. *Adv. Appl. Prob.* 8:659–89

88. Schaeffer SW, Aquadro CF, Anderson WW. 1987. Restriction-map variation in the alcohol dehydrogenase region of *Drosophila pseudoobscura. Mol. Biol. Evol.* 4:254–65

89. Schlotter C, Tautz D. 1993. Slippage synthesis of microsatellites. *Nucleic Acids Res.* 20:211–15

90. Sears BB. 1980. The eliminatino of plastids during spermatogenesis and fertilization in the plant kingdom. *Plasmid* 4:233–55

91. Shriver MD, Jin L, Chakraborty R, Boerwinkle E. 1993. VNTR allele frequency distributions under the stepwise mutation model: a computer simulation approach. *Genetics* 134:983–93

92. Skellam JG. 1951. Random dispersal in theoretical populations. *Biometrika* 38:196–218

93. Slade RW, Moritz C, Heideman A, Hale PT. 1993. Rapid assessment of single copy nuclear DNA variation in diverse species. *Mol. Ecol.* 2:359–73

94. Slatkin M. 1977. Gene flow and genetic drift in a species subject to frequent local extinction. *Theor. Popul. Biol.* 12:253–62

95. Slatkin M. 1985. Gene flow in natural populations. *Annu. Rev. Ecol. Syst.* 16:393–430

96. Slatkin M. 1985. Rare alleles as indicators of gene flow. *Evolution* 39:53–65

97. Slatkin M. 1987. Gene flow and the geographic structure of natural populations. *Science* 236:787–92

98. Slatkin M. 1991. Inbreeding coefficients and coalescence times. *Genet. Res. Camb.* 58:167–75

99. Slatkin M. 1993. Isolation by distance in equilibrium and non-equilibrium populations. *Evolution* 47:264–79

100. Slatkin M. 1995. A measure of population subdivision based on microsatellite allele frequencies. *Genetics* 139:457–62

101. Slatkin M, Arter H. 1991. Spatial autocorrelation methods in population genetics. *Am. Nat.* 138:499–517

102. Slatkin M, Barton NH. 1989. A comparison of three indirect methods for estimating average levels of gene flow. *Evolution* 43:1349–68

103. Slatkin M, Maddison WP. 1989. A cladistic measure of gene flow inferred from the phylogenies of alleles. *Genetics* 123:603–13

104. Slatkin M, Maddison WP. 1990. Detecting isolation by distance using phylogenies of genes. *Genetics* 126:249–60

105. Sokal RR, Jacquez GM, Wooten MC. 1989. Spatial autocorrelation analysis of migration and selection. *Genetics* 121:845–55

106. Sokal RR, Oden NL. 1978. Spatial autocorrelation in biology. II. Some biological implications and four applications of evolutionary and ecological interest. *Biol. J. Linn. Soc.* 10:229–49

107. Sokal RR, Wartenberg DE. 1983. A test of spatial autocorrelation analysis using

128 NEIGEL

an isolation-by-distance model. *Genetics* 105:219–37

108. Soltis DE, Mayer MS, Soltis PS, Edgerton M. 1991. Chloroplast-DNA variation in *Tellima grandifliora. Am. J. Bot.* 78:1379–90

109. Soltis DE, Soltis PS, Ranker TA, Ness BD. 1989. Chloroplast DNA variation in a wild plant, *Tolmiea menziesii. Genetics* 121:819–26

110. Swofford DL, Olsen GJ, Waddell PJ, Hillis DM. 1996. Phylogenetic inference. In *Molecular Systematics,* ed. DM Hillis, C Moritz, BK Mable, pp. 407–514. Sunderland, MA: Sinauer

111. Takahata N. 1988. The coalescent in two partially isolated diffusion populations. *Genet. Res.* 52:213–22

112. Takahata N, Palumbi SR. 1985. Extranuclear differentiation and gene flow in the finite island model. *Genetics* 109:441–57

113. Takahata N, Slatkin M. 1990. Genealogy of neutral genes in two partially isolated populations. *Theor. Popul. Biol.* 38:331–50

113a. Tautz D. 1989. Hypervariability of simple sequences as a general source for polymorphic markers. *Nucleic Acids Res.* 17:6463–571

114. Tautz D, Schlotterer C. 1994. Simple sequences. *Curr. Opin. Genet. Dev.* 4:832–37

115. Tavarae S. 1984. Line-of-descent and genealogical processes, and their applications in population genetic models. *Theor. Popul. Biol.* 26:119–64

116. Templeton AR, Routman E, Phillips CA. 1995. Separating population structure from population history: a cladistic analysis of the geographical distribution of mitochondrial DNA haplotypes in the Tiger Salamander, *Ambystoma tigrinum. Genetics* 140:767–82

117. Valdes AM, Slatkin M, Friemer NB. 1993. Allele frequencies at microsatellite loci: the stepwise mutation model revisited. *Genetics* 133:737–49

118. Vawter L, Brown WM. 1986. Nuclear and mitochondrial DNA comparisons reveal extreme rate variation in the molecular clock. *Science* 234:194–96

119. Voelker RA, Schaffer HE, Mukai T.

1980. Spontaneous allozyme mutations in *Drosphila melanogaster*: rate of occurrence and nature of the mutants. *Genetics* 94:961–68

120. Wade MJ, McCauley DE. 1988. Extinction and recolonization: their effects on the genetic differentiation of local populations. *Evolution* 42:995–1005

121. Watt WB. 1983. Adaptation at specific loci. II. Demographic and biochemical elements in the maintenance of the Colias PGI polymorphism. *Genetics* 103:691–724

122. Weber JL, Wong C. 1993. Mutation of human short tandem repeats. *Hum. Mol. Genet.* 2:1123–28

123. Wehrhahn CF, Powell R. 1987. Electrophoretic variation, regional differences, and gene flow in the coho salmon (*Onchorhynchus kisutch*) of southern British Columbia. *Can. J. Fish. Aqaut. Sci.* 44:822–31

124. Weir BS, Cockerham CC. 1984. Estimating F-Statistics for the analysis of population structure. *Evolution* 38:1358–70

125. Whitlock MC, McCauley DE. 1990. Some population genetic consequences of colony formation and extinction: genetic correlations within founding groups. *Evolution* 44:1717–24

126. Wright S. 1921. Systems of mating. I. The biometric relations between parent and offspring. *Genetics* 6:111–23

127. Wright S. 1931. Evolution in Mendelian populations. *Genetics* 16:97–159

128. Wright S. 1938. The distribution of gene frequencies under irreversible mutation. *Proc. Natl. Acad. Sci. USA* 245:253–59

129. Wright S. 1940. Breeding structure of populations in relation to speciation. *Am. Nat.* 74:232–48

130. Wright S. 1943. Isolation by distance. *Genetics* 28:114–38

131. Wright S. 1946. Isolation by distance under diverse systems of mating. *Genetics* 28:139–56

132. Wright S. 1951. The genetical structure of populations. *Ann. Eugen.* 15:323–53

133. Wright S. 1965. The interpretation of population structure by F-statistics with special regard to systems of mating. *Evolution* 19:395–420

Annu. Rev. Ecol. Syst. 1997. 28:129–52

THE EVOLUTION OF MORPHOLOGICAL DIVERSITY

Mike Foote

Geophysical Sciences Department, University of Chicago, Chicago, Illinois 60637

KEY WORDS: ecomorphology, evolutionary radiation, extinction, macroevolution, theoretical morphology

ABSTRACT

The diversity of organismic form has evolved nonuniformly during the history of life. Quantitative morphological studies reveal profound changes in evolutionary rates corresponding with the generation of morphological disparity at low taxonomic diversity during the early radiation of many clades. These studies have also given insight into the relative importance of genomic and ecological factors in macroevolution, the selectivity of extinction, and other issues. Important progress has been made in the development of morphological spaces that can accommodate highly disparate forms, although this area still needs more attention. Other future directions include the relationship between morphological and ecological diversification, geographic patterns in morphological diversity, and the role of morphological disparity as a causal factor in macroevolution.

MORPHOLOGICAL DIVERSITY IN SYSTEMATIC BIOLOGY

How has the diversity of organic form, living and extinct, come to be? This question, perennially at the heart of systematic biology, has been addressed in paleobiology with a focus on quantitative approaches and large-scale evolutionary questions. The goals of this review are to discuss some recent developments, mainly paleobiological, in the study of morphological diversity [disparity (55, 56, 57, 112, 172)] as opposed to taxonomic richness, and to outline some promising future directions in disparity studies. Many of the pressing questions, such as the pattern of divergence early in clade history and the filling of morphological space, have a long pedigree (54, 108, 116, 122). Although

129

some macroevolutionary studies mainly provide documentation of evolutionary patterns and mechanisms that have been proposed without quantitative
approaches, nevertheless a unique perspective exists that reflects an interest in
how distributions of form evolve over the fullness of geologic time (58) and
at different hierarchical levels. This discussion also entails some consideration
of methodological problems, such as the measurement of disparity, the comparison of results from different approaches, the use of evolutionary models
in understanding disparity, and the effect of paleontological incompleteness on
perceived evolutionary patterns.

APPROACHES TO STUDYING DISPARITY

Although taxonomic richness is the most common measure of biological diversity (81), the distinction between variety and numbers of species is essential
(11, 29, 30, 31, 35–38, 40, 41, 43). For descriptive purposes, it may not be
necessary to assess form quantitatively (64, 65), but many substantive issues in
disparity studies involve comparing levels of morphological diversity among
taxa (12, 16, 18, 32, 39, 109, 110, 164, 165) or among different periods in the
history of a single clade. Therefore, some quantification of disparity is necessary. In this brief survey of disparity measures, I do not consider advantages and
disadvantages exhaustively (partly because this has been done elsewhere, e.g.
155, 162–165, 172, and partly because the relative strengths and weaknesses
for various questions are not yet fully understood). Instead, I comment on some
of the more salient features of each approach.

Indirect Measures

The extent of morphological divergence among taxa generally increases with
taxonomic rank (e.g. 16–18, 36, 42, 44, 59, 173, 175), so secular changes in the
number of higher taxa (generally, phyla, classes, and orders) provide an obvious
index of disparity (3, 28, 142–144). Using higher taxonomic richness to assess
disparity has a number of advantages. The pertinent data are relatively easy to
compile, the common currency of taxa can be summed across disparate biological groups, and a reasonable concordance with more direct measures may be
obtained (45). Although taxonomic proxies may be criticized on the grounds
that taxa are artificial, subjective, nonmonophyletic, or erected on the basis of
criteria other than morphological distinctiveness (124–127), these criticisms
miss the point somewhat. The relevant issue is not the biological meaning, but
rather the information content, of higher taxa. Simply, is there a reasonable empirical concordance between taxonomic richness and more direct measures of
disparity? If so, the heterogeneity and comparability of higher taxa, while still
interesting questions (153), may be of secondary importance. Moreover, even

direct measures of disparity are not free of subjectivity, because they rely upon the choice of a finite number of organismic traits. In a limited analysis of three major clades of Paleozoic marine invertebrates (trilobites, crinoids, and blastozoans), the number of higher taxa (orders and suborders) captured many, but not all, features of the history of morphological disparity revealed by more direct measures (45). Some discrepancies are substantial enough to recommend direct morphological analysis. Nevertheless, higher taxonomic data have suggested a number of important evolutionary patterns that have later been corroborated (see below); thus, ignoring this rich source of information would be shortsighted.

Morphotypes, which are recognized sometimes by eye and sometimes biometrically (23, 26, 53, 66, 72, 113, 114), present many of the same advantages and disadvantages as taxonomic proxies. However, morphotypes are generally quite deliberately conceived in a way that cuts across phylogenetic lines. Thus, compared to simple taxonomic proxies, they facilitate the study of iterative and convergent evolution. For example, Fortey & Owens's study (53) of trilobite morphotype and family diversity suggests the rise and fall of blind trilobites as one component of the rise and fall of morphological diversity through the early Paleozoic.

Morphological Measures

The study of ecomorphology has focused on the expansion of the concept of diversity to include morphological similarity (31), the relationship between morphological and ecological differences within species (152a) and among species (104, 166), and the extent to which greater morphological (ecological) packing and expansion of morphological (ecological) variance among species result from an increase in the number of species in a community (104–106, 157, 158, 160). Although paleontological studies of secular patterns in disparity have had somewhat different goals, many of the methods of measuring morphological diversity are similar to those used in ecomorphology. Starting with a sample of species represented in a morphological space of discrete or continuous variables, disparity can be measured as the variance, range, average pairwise distance between species, number of discrete character states in the sample, number of character-state combinations, as well as related measures and multivariate extensions (2, 35–38, 40, 111, 117, 132, 136, 140, 154, 155, 164, 165). Several studies have been concerned with the effect of sampling on measures such as the range and volume, which increase monotonically with sample size (38, 172). This problem is especially relevant in paleontological studies; the incompleteness of the fossil record implies that substantial changes in range can result simply from changes in the quality of preservation or sampling (38, 172). When volume is measured as the product of variances or standard deviations (e.g. 106), the sampling issue is not as relevant.

A point of contention has been whether morphological distances between species should be measured along the branches of an estimated genealogy (patristic dissimilarity) or not (phenetic dissimilarity) (43, 45, 126, 162, 165). There is no single, correct approach. Nonphylogenetic measures have the advantage of not relying on an estimate of genealogy that may be inaccurate. On the other hand, phylogenetic measures may allow a more direct assessment of transition magnitudes, something that is often of interest in inferring mechanisms in the evolution of disparity. It should always be kept in mind that morphological differences among species include autapomorphies and symplesiomorphies. Even though such traits are not informative for cladistic branching sequence, they are informative for other aspects of genealogy, such as ancestor-descendant relationships (163), and are essential for assessing disparity (56, 57). The crucial question is whether one is interested in net evolutionary change (how dissimilar two species end up, regardless of the evolutionary pathways) or total evolutionary change (how long the evolutionary pathways are, regardless of where they end up) (43, 45). Of course, in the absence of homoplasy, the two measures are identical. It is worth considering both approaches, since in some cases they corroborate each other, and in some cases the disagreement between them is evolutionarily informative (162, 164, 165). This is discussed in more detail below.

"Predicted" Character Diversity

A curious hybrid between direct and indirect approaches stems from the incorporation of phylogenetic analysis into studies of diversity, especially in the context of conservation biology. Since we can consider only a small, potentially biased sample of the indefinitely large number of organismic traits, the measurement of patristic dissimilarity based on observed traits may provide an inadequate proxy for total character diversity (the same, of course, is true of any measure of disparity). A number of authors have advocated predicting (estimating) character diversity based on an estimate of genealogy (or, in the absence of a genealogy, taxonomic structure) and a presumed model of character evolution (29, 30, 171). This raises the obvious questions of how sensitive estimated character diversity is to the presumed model and to the accuracy of the genealogy, and whether a genealogy estimated from a biased subset of characters could yield an unbiased estimate of character diversity. Although some models of character evolution have been advocated over others (30), a more thorough analysis of the robustness of this approach, based on extensive simulation or on jackknifing of observed characters, for example, is essential.

Choice of Traits

It should go without saying that a limited set of traits allows measurement not of "overall" disparity (30, 78), but rather of the diversity of form in the chosen traits. Most studies have attempted to include a broad range of anatomical features, or, when necessary, those features that are sufficiently well preserved to be measured. A complementary approach is to select those traits of particular importance for the question at hand, for example functional morphology (12, 65) or ecology (166). In this vein, some authors have advocated a greater consideration of the developmental and architectural significance of characters as a prerequisite for understanding the Cambrian explosion of animal disparity at a deeper level than permitted by a tabulation of the number and size of evolutionary transitions (56, 57, 88, 107). Although his interpretation of characters may be controversial, Wagner (163) has made a step in this direction by categorizing gastropod shell traits as related to trophic demands versus fundamental architecture.

Factors of Uncertain Relation to Disparity

Morphological diversity is commonly discussed in connection with the complexity and "bizarreness" of organisms. If we think of complexity as a property of individual organisms, there need be no correspondence with disparity, which is a property of distributions of organisms. We may observe a wide spectrum of simple forms or a limited array of complex forms (90). Schemes such as the Skeleton Space (139–141), the quantification of tagmosis in arthropods (20), the number of cell types (150), or the differentiation among serial elements such as vertebrae (87, 89) assess complexity in terms of the disparity among parts within the same organism, thereby allowing the evolution of average complexity and variance in complexity to be studied together. Unlike complexity (20, 87, 89, 90, 114, 150), the notion of bizarreness has not been properly operationalized; currently, the concept is highly subjective. Most workers seem to regard bizarre forms as those that are morphologically extreme (e.g. 73, 93), but at least one author (63) has suggested that flatness is a key element of bizarreness! It is also common to regard forms of uncertain genealogical relationship as "weird" (161), but taxa whose phylogenetic relationships are understood can nevertheless be quite disparate morphologically; disparity and branching sequence are logically distinct (8–10, 56, 96, 147, 161).

PROGRESS IN DISPARITY STUDIES

A primary contribution of disparity studies has been the simple description of evolutionary history, the "kinetics " of biological diversity (119). Documenting the morphological exuberance of clades through their history, and the ways that different taxa contribute to overall morphological diversity (12, 39), helps hone specific evolutionary questions and hypotheses. For example, I initiated a largely exploratory study of crinoid disparity through the Paleozoic with no intention of testing for constraints on form. Yet the striking pattern that maximal disparity (measured as mean pairwise phenetic distance) was attained early, and that the morphological extremes reached early in the group's history were scarcely exceeded over the next 200 My, despite the proliferation of hundreds of new genera, together suggested some severe limits on the evolution of crinoid form (41–44). (See 113 and 137 for a similar assessment in Carboniferous ammonoids, and 8, 9, 49, and 77 for a discussion of this issue in arthropods.) Here I outline some of the principal substantive issues that have been addressed in the paleobiological analysis of disparity.

Testing for Adaptive Radiations

The common conception of adaptive radiations concerns both a proliferation in numbers of taxa and a diversification of form (122). For example, in an influential treatment of macroevolution in the fossil record, Stanley (131) cited the case of Cambrian trilobites (among other groups), partly using the increase in number of families as evidence for adaptive radiation. However, morphometric data show that the diversification of trilobite form (increase in morphological range and variance) was actually rather limited in the Cambrian, and that the greater proliferation of morphological diversity followed in the Ordovician, during a decline in family-level taxonomic diversity (34, 36, 40). The point here is not to criticize Stanley's example, but to illustrate the nature of the test. Likewise, Cambrian biomeres (repeated stratigraphic sequences apparently marked by iterative evolutionary patterns in trilobites) have been discussed as examples of adaptive radiation, but most analyses have focused on taxonomic data (60, 133, 134). Sundberg (136) analyzed morphometric data on trilobites through one of these biomeres, verifying that the evolutionary sequence involves a substantial diversification of form. Implicit in these tests is the ecomorphological assumption that a diversification of form is likely to reflect an ecological diversification (see discussion below).

Patterning in Morphospace

Although many disparity studies have focused on the extent of morphospace occupation and variance among species, other aspects of pattern in morphospace have also provided insight. For example, Cambrian trilobites seem to exhibit pronounced homeomorphy, and they have been notoriously difficult to group into higher taxa (families and superfamilies) (135, 169), a pattern that may be related to developmental flexibility in the Cambrian (67, 68, 86; but see 128). In contrast, Ordovician trilobites are, for the most part, easier to sort into morphologically distinct groups (135, 169). Morphometric analysis suggests that higher taxa in the Ordovician occupy a greater range of morphological space, but with less overlap (36). Ignoring higher taxa altogether, the analysis of morphological nearest-neighbor distances has supported the pattern of greater clustering in the Ordovician (34). Tabachnick & Bookstein (138) suggested that, at a smaller scale, specimens of the Miocene planktonic foraminiferan *Globorotalia* are spread continuously through morphospace, so that named taxa and morphotypes do not correspond to clusters or modes in the distribution. Moreover, this pattern itself has evolved; some periods in foraminiferan history are apparently marked by greater clustering (RE Tabachnick, personal communication). Although some preliminary simulations (34; see 103) suggest that an increase in morphological clustering could result from simple diffusive evolution and extinction of intermediates, the evolutionary processes responsible for changes in the clustering of morphological distributions need to be explored in greater depth.

Ecomorphology

Determining the extent of community convergence is a principal question in ecomorphological studies. Do ecologically similar communities in different places show similar patterns of morphological similarity and morphological diversity among species (104)? Van Valkenburgh has explored this issue in a temporal context and found that a number of mammalian paleocommunities show similar guild structure (based on extent of morphospace occupation and nearest-neighbor distances), despite the passing of tens of millions of years and substantial taxonomic turnover (157–160). This suggests that ecological interactions may be strong enough to outweigh historical influences on community structure (157, 158). Another study noted similarities in Pleistocene and Recent community structure in vultures (61), based on the relationship between body size and feeding strategy. The difficulty of ecomorphological analysis of fossil taxa is discussed briefly below.

Responses to Extinction

Because variance in form is generally unbiased by sample size, a morphologically random culling of taxa should, on average, leave variance unchanged. Thus, the question whether morphological variance among species changes as taxonomic diversity declines has been used as a test for extinction selectivity (19, 35, 40, 43, 46, 84, 109–111, 158; cf. 152). Because many of these studies compare disparity before an extinction event to disparity at some period of time afterwards, they confound the change in the morphological distribution attributable to extinction with that attributable to subsequent diversification. However, Churchill (19) overcame this problem by comparing all taxa before the event to the subset of taxa surviving the event. McGhee (84) documented an interesting pattern in articulate brachiopods (specifically, the subset of them having two convex shells). He found that reduction in diversity generally resulted in a reduction of the distribution of forms to a morphological mode which he interpreted as advantageous in allowing a high ratio of body volume to surface area (82).

An unsolved problem in extinction studies concerns the sensitivity of tests for selectivity. What combinations of intensity of selectivity, pattern of selectivity, and sample size allow departures from random survivorship to be detected? In principle, this is easily addressed with simulation studies, but, to my knowledge, the necessary work has not yet been carried out.

Replacements and Successive Diversifications

Extinction of incumbent taxa commonly opens new opportunities for the diversification of other groups (4, 5, 121, 122). Although studies of replacement have generally focused on taxonomic diversity, morphological analysis has shown that the replacing taxon sometimes occupies vacated morphological space [Ward (167) on Tertiary nautilids versus Mesozoic ammonoids, and Roy (110) on Tertiary strombids versus Cretaceous aporrhaids]. To the extent that species in the two groups in question coexisted spatially and thus were capable of interaction, and to the extent that morphological traits are ecologically significant, the colonization of morphological space provides stronger evidence for the role of ecological interaction (competition) in macroevolution than do data on taxonomic richness alone.

The pattern of morphological diversification during successive intervals of taxonomic diversification provides some evidence bearing on the role of ecological versus genomic and developmental changes in macroevolution. The origin of higher taxa is concentrated early in the history of many groups (e.g. 22, 108, 116, 122, 142, 143, 146–148). Although some might argue that this says more about taxonomic practice than morphological divergence (21, 24, 125, 127), available morphological data support a rapid, early proliferation of morphological diversity (see discussion below). Two leading explanations for this pattern are that ecological opportunites were greater in the early history of many clades, diminishing as the world became ecologically saturated, and that genetic and developmental systems were less canalized early on (27, 28, 88, 142, 143, 146–149, 151). Erwin (27) suggested a test of these alternatives involving the analysis of disparity. If ecological opportunity were responsible, then one would expect later radiations of a clade following extinction events to involve a rapid proliferation of morphological diversity (perhaps to the same high level attained earlier), whereas an increase in genomic and developmental canalization might severely limit morphological diversification later on. In support of the ecospace model, Wagner (164) noted that some subclades of Paleozoic gastropods exhibited accelerated morphological diversification following the Late Ordovician extinction event. Similarly, Foote (46) documented a rapid increase in disparity in early Mesozoic crinoids (the same pattern as seen in the

Paleozoic), following the end–Paleozoic extinction event that had drastically reduced the taxonomic diversity of this group.

Rapid Filling of Morphological Space During Evolutionary Radiations

Perhaps the most common theme in disparity studies so far has been the asymmetric deployment of morphological diversity early in the radiation of major clades. This pattern has long been advocated (e.g. 22, 54, 97, 98, 108, 116, 129, 130, 143, 176), but it has been disputed because it is sometimes discussed in terms of taxonomic proxies (8, 21, 24, 124–127). However, most authors who describe the pattern in terms of higher taxa use this description as a shorthand to express what are perceived as profound morphological differences (e.g. 55, 129, 130).

Although we can always benefit from more examples, it seems safe to say at this point that studies of disparity within class- and higher-level taxa have documented more cases of accelerated morphological diversification early in a clade's history, at relatively low taxonomic diversity, than of a more gradual unfolding of morphological and taxonomic diversity together. Evidence for acceleration of morphological evolution early in history has taken a number of forms: peak disparity early; more rapid proliferation of disparity versus diversity; secular decline in the rate of increase of morphological disparity; secular decline in the dissimilarities between sister taxa (i.e. decline in estimated magnitude of evolutionary transitions); failure of taxa to converge morphologically toward their time of phylogenetic splitting (11); and failure of later-evolving subclades, throughout their entire history, to generate as much morphological diversity as the more inclusive clade did just in its initial phase of diversification (77, 88, 164). Because the pattern in question concerns the magnitude of morphological differences, arguments based purely on cladistic branching order (8–10, 96, 161) are not immediately relevant to the question of early disparity (56, 147).

Examples of pronounced early increase in disparity include Cambrian marine arthropods (9, 49, 77, 172), Paleozoic gastropods (163, 176), Paleozoic rostroconch molluscs (165), Paleozoic stenolaemate bryozoans (1), Paleozoic seeds (123), Cretaceous angiosperms (based on pollen; R Lupia, personal communication), Cenozoic ungulates (72), Carboniferous ammonoids (113, 114, 137), Paleozoic articulate brachiopods (13, 84), Ordovician trilobites (but not Paleozoic trilobites as a whole) (91), early-mid Paleozoic tracheophytes (75), Paleozoic crinoids (41, 42, 43, 44), Mesozoic crinoids (46), Paleozoic blastozoans (37, 47, 162), and Cambrian Metazoa (141). Counterexamples include Early Jurassic ammonites (23), Paleozoic trilobites (40), Paleozoic blastoids (35, 40), Paleozoic cladid and flexible crinoids (43), and, apparently, insects from the mid-Paleozoic to the Recent (based on number of mouthpart morphotypes; 76). Several of these studies suggest that disparity within large clades may increase rapidly early in history, while disparity within concurrently diversifying constituent subclades increases more gradually. If this pattern proves to be more general, it will provide important support for hierarchical views of evolution that regard patterns at different scales as distinct qualitatively rather than just quantitatively (71, 142). Interestingly, at least one author has considered the pattern of early morphological diversification to be such a robust evolutionary generality that he has assumed, for the sake of phylogenetic analysis, that the groups showing greater disparity in the basal part of the cladogram are more primitive (79)!

At first glance, one might suspect that, since many of the foregoing studies are based on discrete morphological characters, the pattern of early maximal disparity is an artifact of this type of data. There are a number of reasons to think the pattern is not a simple artifact, however. First, not all clades that have been analyzed with discrete character data show this pattern (e.g. 37). In fact, the very same set of discrete characters showed a rapid increase to maximal disparity within crinoids as a whole but not within a major subclade of crinoids (43, 44). Second, in at least one case, the same pattern was found when characters used to differentiate the higher taxa were omitted (41, 42). Third, in the cases I have studied (37, 41–44, 46), no pair of species exhibits a morphological dissimilarity approaching the theoretically maximal value (i.e. the clade is not up against the theoretical limits of the morphospace imposed by the choice of characters). Fourth, different sets of discrete characters sometimes yield different evolutionary patterns in a single taxon (e.g. 41, 42), which we would not expect if the discrete nature of characters were itself responsible for perceived patterns. Finally, at least one study of the same group comparing landmark-based, continuous measures to discrete characters found that both kinds of data show a long-term increase in morphological diversity over time, although the patterns differ in detail (38). Thus, it is more reasonable that the common pattern of maximal early disparity reflects true early divergence to extremely dissimilar forms, such that it is necessary to use discrete characters to quantify them (limited homologies make biometric or landmark approaches problematic).

The combination of disparity data and simple evolutionary models has had limited success in explaining the proximate mechanisms underlying the early diversification of form (40, 47). Although some potential explanations for a rapid, early increase in disparity seem unlikely (e.g. logistic taxonomic diversification—47), many other factors, such as a secular decline in taxonomic turnover rates, a secular decline in the size of morphological transitions, and boundaries in morphological space, can yield this particular evolutionary pattern (47). Nevertheless, the fact that this pattern is soundly grounded in morphological analysis rules out the possibility that it is an artifact of taxonomic practice, such as the erection of higher taxa that diverge early phylogenetically, but only later morphologically (21, 24). Comparison of results from different approaches to disparity has helped narrow down evolutionary mechanisms in particular cases (see below).

Sensitivity of Patterns to Approaches Adopted

In the foregoing survey of approaches to measuring disparity, I deliberately avoided advocating one method over others. With so many ways to quantify form and measure the differences between forms, it would be pointless to argue that one approach is best in principle, since this depends on the organisms studied and the kinds of evolutionary patterns one hopes to detect, among other factors. The more fruitful approach has been to explore various methods and to test for consistency of evolutionary patterns. When a number of methods of quantifying form and measuring disparity converge on a similar result, we can be more confident in that result. [I should point out that by consistency I do not mean, strictly, the same temporal pattern in two or more disparity metrics, but rather a pattern in the metrics that has the same evolutionary implication. For example, if disparity is measured as patristic dissimilarity between sister-species on the one hand and by mean phenetic distance among all species on the other hand, a pattern of constant patristic dissimilarity would be consistent with a steady increase in phenetic distance, since constant step size in a diversifying clade yields an increase in variance among forms (47, 150).]

Wills et al (172) showed that disparity metrics including range, variance, and distance from basal node on the cladogram all point to comparable disparity in Cambrian and Recent arthropods. Jernvall et al (72) found an early increase in morphological diversity of Cenozoic ungulates, whether based on number of morphotypes or pairwise phenetic distances. Wagner (162) considered the case of blastozoan echinoderms, in which an early increase in disparity at low taxonomic diversity had been used to infer that morphological transitions were larger early on (37). This result had been disputed because the disparity metric was not phylogenetic (126), but Wagner showed that, using patristic dissimilarity per branch on a cladogram to estimate transition magnitudes, the original, indirect, inference was supported. Wagner also found a similar concordance in an analysis of rostroconch molluscs (165). Many other examples could be cited in which different disparity metrics, different morphological traits, different sampling protocols, different estimates of phylogeny, or different methods of character weighting were used to test the robustness of evolutionary patterns (34, 41–44, 46, 74, 165).

Of course, different approaches need not yield the same patterns. Such discordances can be interesting in their own right, rather than suggesting that one approach is right or wrong. It can be very informative to break away from arguments about phenetic versus phylogenetic metrics (162, 163, 165). For example, in rostroconch molluscs, Wagner (165) found that patristic dissimilarity increased much more than phenetic distance during the Ordovician. This suggests substantial evolutionary transitions with a high degree of homoplasy (165). Likewise, a discordance between abundant character change and limited expansion of the morphological range may suggest either extreme convergence or boundaries in morphospace (43, 49, 77).

Considering what we can learn from the agreements and disagreements among various methods, it would be wise, especially during this expansive phase of the history of disparity studies, to take a lesson from evolutionary radiations by practicing early experimentation. Whether and how we settle into patterns of later standardization is an open question.

Effects of Incompleteness on Temporal Patterns of Disparity

The incompleteness of the fossil record affects various measures of disparity in different ways. Simple average phenetic distance and variance have an advantage relative to measures of extremes such as the range in that the former measures are less sensitive to completeness, provided that sampling is representative (35, 38, 48). Thus, contrary to some suggestions (73), incomplete preservation will not bias average distances unless the species preserved are systematically more or less extreme morphologically than those not preserved (see discussion below). Of course, if a clade is not preserved at all for a substantial part of its early history, then the temporal patterns of morphological and taxonomic diversity will be biased; it remains to be seen how common a problem this is (52).

Sister-species differences generally increase as the record becomes less complete, since there are, on average, more missing intermediates. Wagner (163) has addressed this problem by estimating sampling intensity. He found that a temporal decrease in the morphological distance between gastropod sister–species (or ancestors and descendants) was not matched by an increase in sampling intensity, and that the evolutionary pattern was therefore probably not an artifact of incompleteness. While this is an important first step, it is also worth estimating completeness (proportion of taxa preserved in an interval) in addition to sampling intensity. This is because, if extinction rate is higher (taxonomic durations are shorter), the same intrinsic preservability and the same intensity of sampling will still yield lower completeness (50, 128a), and thus artificially greater dissimilarities between sister species. Therefore, a decline in taxonomic turnover rates, which is sometimes found during the diversification of clades (47, 156), could bias perceived patterns in the magnitude of evolutionary transitions. Wagner (165), estimating the proportion of rostroconch taxa preserved, found that temporal changes in sister-species differences could not be explained by changes in completeness (165). Since quantitative methods for estimating completeness are not yet fully developed, this problem deserves further consideration (6, 48a, 50, 80, 99, 128a).

OUTSTANDING PROBLEMS

In contrast with the almost routine documentation of taxonomic diversity in the history of biologic groups, there is still a relatively small (but rapidly growing) number of case studies quantifying secular patterns of disparity. I do not dwell on the obvious need for more data to establish, for example, how common it is for subclades to show qualitatively different evolutionary patterns than their more inclusive clades. Rather, I consider several unanswered but important questions (some of which have already been discussed at length by others).

Broadening the Taxonomic Scope of Disparity Studies

Without suggesting that taxonomic ranks have a consistent meaning, we can note that it is common to be able to accommodate species within the same class with a consistent biometric scheme (e.g. 13, 31, 35, 36, 84, 100, 105). On the other hand, studies that span several classes within a phylum tend to be stymied by difficulties in establishing measurable homologies, and therefore they often must rely upon discrete character data. Generative and architectural models have transcended taxonomic boundaries to some extent. Of these, models of shell coiling (83, 100, 101, 115) and branching growth (14, 15, 51, 85, 94) have been most common in paleobiological studies. Other models, such as those that view organisms as fluid-filled sacks taking on a shape that balances internal and external forces, have been successful as heuristic tools in constructional morphology (102, 118), but there have been only limited attempts to establish model parameters and estimate these parameters on observed organisms (25, 102). A general advantage of theoretical models of growth and form over simple empirical descriptions is that the former allow a comparison between the observed spectrum of form and the theoretically conceivable spectrum (65, 83). This is important in assessing the fullness of morphological space.

Perhaps the most significant development in broadening the taxonomic scope of morphological diversity studies is the application of a combinatorial system for describing some of the principal features of skeletal structures—the Skeleton Space of Thomas & Reif (139–141). This is a bold attempt to distill the skeletal elements of animals to their most salient constructional features: for example, location (internal or external), number of parts, mechanical properties, mode of growth, and nature of contact or articulation. The Skeleton Space is of greatest utility in assessing the morphological diversity of very high-level taxa such as phyla or kingdoms, since taxa below this level tend to be relatively invariant in their skeletal structures (e.g. the common skeletal formula for gastropods is a single, external, tube-like, accreted, rigid, self-produced shell; but then, of course, there are opercula, slugs, and the occasional bivalved snail). Thomas & Reif stated that there are over 1500 possible combinations of features in their space, but, to keep their analysis tractable, they considered pairwise combinations of traits (e.g. skeletal elements that are internal and accreted, external and accreted, internal and remodeled, external and remodeled, and so on). With this approach, nearly all pairwise combinations have been exploited in living and extinct animals, and thus the space seems quite richly occupied. It would be worth extending this analysis to include the full spectrum of combinations, not just the features taken in pairs.

One complication that must be kept in mind when comparing the Skeleton Space to other approaches is that a single organism can occupy many loci in the Skeleton Space, whereas in nearly all other morphospaces, each organism is considered to occupy a single point. This is not just because features are considered pairwise (e.g. vertebrate long bones occupy the rigid-remodeled locus, the rigid-articulated locus, the internal-rod-shaped locus, and so on), but, more importantly, it is because different parts of an organism may have fundamentally different structures. For example, the long bones and the cranium of vertebrates have different skeletal formulae.

To some extent, the apparent fullness of the space depends on the way features are decomposed (e.g. the number of elements has only three states: one, two, or greater than two). Although one may criticize the space on such grounds, it is the first scheme to allow a quantitative assessment of the morphological diversification of all (skeletonized) animals. Applying the Skeleton Space to the Middle Cambrian Burgess Shale fauna, Thomas & Stewart (141) found that about 90% of designs ultimately used by animals (considered as pairwise combinations of skeletal features) had already been exploited rather early in animal history [or at least early in their preserved history (174)]. Thus, we have direct morphological documentation for a broad diversification of skeletal designs during the Cambrian explosion.

Comparing the Fullness of Different Morphospaces

Do snails, based on parameters of shell coiling, occupy more of the morphospace available to them than do arborescent bryozoans, based on the parameters of branching growth? In morphospaces that lack theoretical maxima and/or minima for at least some parameters (such as the coiling space), this question may be intractable. If we somewhat less ambitiously restrict ourselves to the observed extremes (observed maximal and minimal values of quantitative traits), then average differences between species can be expressed as a proportion of the maximal possible difference (as is commonly done with discrete character data). Clearly, how fully a morphospace is occupied is potentially very sensitive to the choice of traits (83). A common approach in interpreting secular patterns of morphological diversity is to consider disparity at any time relative to the maximal disparity reached by a group in its history (e.g. 40, 47). Even if this does not allow us to say that diatoms are more diverse morphologically than dinosaurs, it does allow us to address whether morphospace was filled more gradually or abruptly in one group versus another.

Sampling and Preservation of Morphological Diversity

One measure of the robustness of evolutionary patterns concerns their sensitivity to sampling (48, 120, 168, 170). Blackburn & Gaston (7) found that smaller-bodied species in several groups of living animals have been discovered at an increasing rate toward the present day. Thus, apparent geographic and ecological patterns based on body size may not be robust to sampling. On the other hand, a study of several large groups of fossil marine invertebrates showed that, based on multivariate measures of morphology, there is no appreciable preference for morphologically extreme or modal species to be described earlier or later in the history of systematic paleontology (48). This suggests effectively random sampling of preserved forms at the large scale, although certain details of evolutionary patterns of disparity within the studied groups have changed as more material has been discovered and described (48).

Whether a sample of fossil species is biased with respect to morphological disparity (relative to the entire statistical population of preserved species) is a different question than whether those preserved species are a representative sample of all the species that lived in some group. Some organisms are more likely to be preserved than others. For example, all else being equal, thick skeletons are more likely to enter the fossil record than are thin skeletons, and single-element skeletons are more likely to preserve than those consisting of unfused sclerites. But this bias does not imply that disparity itself will be biased; the crucial question is whether the average morphological dissimilarity among preserved species is the same as that among all species that could have been preserved. To my knowledge, this question has not been addressed in detail, but a simple test is possible (similar to that which Valentine performed to assess the completeness of the fossil record of marine molluscs in the Californian province—145). Take a large group of Recent species that can be divided into a number of groups, say taxonomically or geographically. Quantify the form of each species and measure the disparity of the entire group and each subgroup. Now consider only those living species that are known from the fossil record, and measure the disparity of the entire fossil sample and the fossil samples of subgroups. If disparity is unbiased by preservation, then the values of disparity for the entire fossil sample and fossil subsets should be statistically indistinguishable from the values for the entire living sample and its corre-sponding subsets. This test requires that disparity be measured in a way that is sensitive not to sample size alone but to the representativeness of the sample; variance of morphology, for example, would be preferable to the range.

Ecomorphology

The success of ecomorphology rests on the ecological or functional significance of measured morphological features (104, 159). In many cases the correlation between form and ecology or function seems sufficiently strong (12, 104), but this may not always be the case. For example, is it reasonable to suppose that the diversification of form as represented by the outline shapes of trilobite heads reflects ecological diversification (36)? Considering the ubiquity of character correlations (95), the diversity of a large array of haphazardly selected mor-phological features may provide a good proxy for ecological diversity, but this should be tested extensively with living species. For example, we could quantify trophic and functional differences among a large number of species and compare these to the morphological differences among these same species based on traits that are not deliberately selected for their presumed ecological or functional sig-nificance. If many comparisons of this kind revealed a general correspondence between morphological and ecological dissimilarities, then the inference of eco-logical diversification from morphological diversification would be reliable.

Geographic Context of Morphological Diversification

Latitudinal, provincial, and bathymetric patterns of taxonomic diversity have revealed interesting patterns in life's history. For example, Jablonski & Bottjer (71) showed that higher taxa (orders) tend to originate preferentially in near-shore environments, in contrast to lower taxa (genera), and Jablonski (70) doc-umented the preferential origin of higher taxa in the tropics. Miller & Mao (92) found that global patterns of taxonomic diversity in the Ordovician were not matched by patterns within provinces, implying that explanations for global diversification could not simply be extrapolated up from smaller-scale explana-tions, but might, for example, involve changes in faunal differentiation among provinces. In contrast, geographic patterns of morphological diversity in the fos-sil record have scarcely been explored (111). Do those areas that generate evo-lutionary novelties (nearshore environments, tropics) also accumulate greater morphological diversity, or are they simply a source of novelties, with the net disparity accumulating elsewhere? Do areas with higher taxonomic diversity tend to have a greater diversity of form, or do they reflect numerous trivial variations on the same themes? How is global morphological diversity bro-ken down into provincial patterns? In some preliminary analyses, AI Miller & M Foote (unpublished) found that the global increase in morphological diversity of Ordovician trilobites is matched by an increase in disparity of endemic genera, but that the disparity of cosmopolitan genera scarcely changes through the Ordovician. This suggests that global diversification of form is mainly attributable to endemic radiations.

Evolutionary Models and Data Analysis

I stated earlier that simple branching models of evolution have been of limited success in isolating the mechanisms of morphological diversification, since a range of different parameters can yield very similar patterns of taxonomic and morphological diversity. The range of patterns that could result from stochastic variation about a constant set of parameters has mainly been addressed by simulation (35, 39, 47, 103). The variance in taxonomic diversity trajectories is understood analytically, and it would be worth developing a similar analytic distribution for disparity trajectories. This would give clues as to how different two observed disparity histories need to be before we have some confidence that the difference does not simply reflect sampling error or stochastic variation.

In addition to incorporating phylogenetic information to assess the size of transitions and to understand in more detail how morphological space is filled (33, 62, 65, 126, 163–165), it is also crucial to consider alternative ways of analyzing diversity and disparity data. For example, Wagner has suggested that the comparison of cumulative diversity and cumulative disparity (total number of taxa that have lived up to some point in time and the disparity among them) may help address whether a rapid proliferation of morphological diversity is attributable to many smaller evolutionary transitions or to a few larger transitions (164). (There are certainly difficulties here. For example, a clade may wander around in morphospace with the result that cumulative disparity increases substantially even though standing disparity may change little; such a pattern would be difficult to distinguish from one in which a clade continually expands its diversity of form. But such complications would thwart interpretations only if we relied on just one mode of analysis.) I suggested earlier that using a variety of methods to study disparity is generally more enlightening than attempting to select the single most appropriate approach. Considering the diversity of methods that have been developed, a serious effort to understand the theoretical and empirical relationships among these approaches, and how these relationships depend on particular models of evolution, could result in significant advances.

CONCLUSION: THE ROLE OF DISPARITY
IN MACROEVOLUTION

This review and most of the work discussed here have focused on morphological disparity essentially as a passive response variable. How is disparity affected by extinction events? How do changes in taxonomic and morphological rates of evolution affect disparity? How do new ecological opportunities allow a clade to diversify morphologically? How do limits of form check the increase in disparity? However, just as higher-level properties of taxa (such as species richess of genera and geographic range of species) may affect the risk of extinction (69), the morphological diversity of a clade may also affect its evolution. For example, Wagner compared the evolutionary histories of two concurrently evolving clades of rostroconch molluscs to determine whether the group that survived the end-Ordovician extinction also showed a more substantial early diversification of form (165). This might be expected, since a wider range of form could represent a greater range of ecological and functional modes, which would enhance the probability that at least some lineages would survive. In this case, a concordance between early diversification of form and later resistance to extinction was not found. In contrast, in a preliminary study (unpublished) of variation in body size among species within families of Late Ordovician trilobites, I found that the families with a greater variance in size preferentially survived the end-Ordovician extinction event, even when the effect of species richness was factored out statistically.

The study of disparity in the fossil record is in some ways still young, yet it has already enhanced our understanding of large-scale heterogeneities in the history of life (such as the early generation of substantial morphological diversity at low taxonomic diversity and the characteristically different pattern of morphological diversification within major clades versus their constituent subclades), the nature of evolutionary radiations and biotic replacements, selectivity of extinction, and the role of ecological interactions in shaping macroevolution and community structure. The future success of disparity studies will continue to rest upon a pluralistic attitude and a willingness to consider morphological diversity, not just as a characteristic of evolving systems, but also as a causative agent in macroevolution.

ACKNOWLEDGMENTS

The ideas discussed here were influenced by many people. I especially thank B Chernoff, DC Fisher, SJ Gould, D Jablonski, DW McShea, DJ Miller, AI Miller, AR Solow, DM Raup, K Roy, JJ Sepkoski Jr, HJ Sims, RE Tabachnick, JW Valentine, and PJ Wagner. I thank R Lupia, AI Miller, AR Solow, and PJ Wagner for sharing unpublished work. HJ Sims and PJ Wagner kindly read the manuscript. Research was supported by the National Science Foundation (DEB-9207577, DEB-9496348, and EAR-9506568) and by the Donors of the Petroleum Research Fund, administered by the American Chemical Society.

Visit the *Annual Reviews home page* at
http://www.annurev.org.

Literature Cited

1. Anstey RL, Pachut JF. 1995. Phylogeny, diversity history, and speciation in Paleozoic bryozoans. In *New Approaches to Speciation in the Fossil Record,* ed. DH Erwin, RL Anstey, pp. 239–84. New York: Columbia Univ. Press. 342 pp.
2. Ashton JH, Rowell AJ. 1975. Environmental stability and species proliferation in Late Cambrian Trilobite faunas: a test of the niche-variation hypothesis. *Paleobiology* 1:161–74
3. Bambach RK, Sepkoski JJ Jr. 1992. Historical evolutionary information in the traditional Linnean hierarchy. *Paleontol. Soc. Spec. Publ.* 6:16
4. Benton MJ. 1987. Progress and competition in macroevolution. *Biol. Rev.* 62:305–38
5. Benton MJ. 1996. On the nonprevalence of competitive replacement in the evolution of tetrapods. In *Evolutionary Paleobiology,* ed. D Jablonski, DH Erwin, JH Lipps, pp. 185–210. Chicago: Univ. Chicago Press. 484 pp.
6. Benton MJ, Hitchin R. 1996. Testing the quality of the fossil record by groups and by major habitats. *Hist. Biol.* 12:111–57
7. Blackburn TM, Gaston KJ. 1994. Animal body size distributions change as more species are described. *Proc. R. Soc. Lond.* B 257:293–97
8. Briggs DEG, Fortey RA. 1989. The early radiation and relationships of the major arthropod groups. *Science* 246:241–43
9. Briggs DEG, Fortey RA, Wills MA. 1992. Morphological disparity in the Cambrian. *Science* 256:1670–73
10. Briggs DEG, Fortey RA, Wills MA. 1993. Cambrian and Recent morphological disparity. *Science* 258:1817–18
11. Campbell KSW, Marshall CR. 1987. Rates and modes of evolution among Palaeozoic echinoderms. In *Rates of Evolution,* ed. KSW Campbell, MF Day, pp. 61–100. London: Allen & Unwin. 314 pp.
12. Carlson SJ. 1989. The articulate brachiopod hinge mechanism: morphological and functional variation. *Paleobiology* 15:364–86
13. Carlson SJ. 1992. Evolutionary trends in the articulate brachiopod hinge mechanism. *Paleobiology* 18:344–66
14. Cheetham AH, Hayek LC. 1983. Geometric consequences of branching growth in adeoniform Bryozoa. *Paleobiology* 9:240–60
15. Cheetham AH, Hayek LC, Thomsen E. 1981. Growth models in fossil arborescent cheilostome bryozoans. *Paleobiology* 7:68–86
16. Cherry LM, Case SM, Kunkel JG, Wilson AC. 1979. Comparison of frogs, humans, and chimpanzees. *Science* 204:435
17. Cherry LM, Case SM, Kunkel JG, Wyles JS, Wilson AC. 1982. Body shape metrics and organismal evolution. *Evolution* 36:914–33
18. Cherry LM, Case SM, Wilson AC. 1978. Frog perspective on the morphological difference between humans and chimpanzees. *Science* 200:209–11
19. Churchill LL. 1996. Testing for differences in selectivity during mass and background extinctions using the relationship between size and extinction intensity in Trilobita. *Paleontol. Soc. Spec. Pub.* 8:70
20. Cisne JL 1974. Evolution of the world fauna of aquatic free-living arthropods. *Evolution* 22:337–66
21. Derstler KL. 1981. Morphological diversity of early Cambrian echinoderms. In *Short Papers for the Second Int. Symp. on the Cambrian System,* ed. ME Taylor, pp. 71–75. *US Geol. Survey Open File Report* 81-743
22. DiMichele WA, Bateman RM. 1996. Plant paleoecology and evolutionary inference: two examples from the Paleozoic. *Rev. Palaeobot. Palynol.* 20:223–47
23. Dommergues J-L, Laurin B, Meister C. 1996. Evolution of ammonoid morphospace during the Early Jurassic radiation. *Paleobiology* 22:219–40
24. Dzik, J. 1993. Early metazoan evolution and the meaning of its fossil record. *Evol. Biol.* 27:339–89
25. Ellers O. 1993. A mechanical model of growth in regular sea urchins: predictions of shape and a developmental morphospace. *Proc. R. Soc. Lond.* B 254:123–29
26. Erwin DH. 1990. Carboniferous-Triassic gastropod diversity patterns and the Permo-Triassic mass extinction. *Paleobiology* 16:187–203
27. Erwin DH. 1994. Early introduction of major morphological innovations. *Acta Palaeontol. Polonica* 38:281–94
28. Erwin DH, Valentine JW, Sepkoski JJ Jr. 1987. A comparative study of diversification events: the early Paleozoic versus the Mesozoic. *Evolution* 41:1177–86

29. Faith DP. 1992. Systematics and conservation: on predicting the feature diversity of subsets of taxa. *Cladistics* 8:361–73

30. Faith DP. 1994. Phylogenetic pattern and the quantification of organismal biodiversity. *Philos. Trans. R. Soc. Lond.* **B** 345:45–58

31. Findley JS. 1973. Phenetic packing as a measure of faunal diversity. *Am. Nat.* 107:580–84

32. Findley JS. 1979. Comparison of frogs, humans, and chimpanzees. *Science* 204:434–35

33. Fisher DC. 1991. Phylogenetic analysis and its application in evolutionary paleobiology. In *Analytical Paleobiology,* ed. NL Gilinsky, PW Signor, pp. 103–22. Knoxville, TN: Paleontol. Soc. 216 pp.

34. Foote M. 1990. Nearest-neighbor analysis of trilobite morphospace. *Syst. Zool.* 39:371–82

35. Foote M. 1991. Morphologic and taxonomic diversity in a clade's history: the blastoid record and stochastic simulations. *Univ. Mich. Mus. Paleontol. Contrib.* 28:101–40

36. Foote M. 1991. Morphologic patterns of diversification: examples from trilobites. *Palaeontology* 34:461–85

37. Foote M. 1992. Paleozoic record of morphological diversity in blastozoan echinoderms. *Proc. Natl. Acad. Sci. USA* 89:7325–29

38. Foote M. 1992. Rarefaction analysis of morphological and taxonomic diversity. *Paleobiology* 18:1–16

39. Foote M. 1993. Contributions of individual taxa to overall morphological disparity. *Paleobiology* 19:403–19

40. Foote M. 1993. Discordance and concordance between morphological and taxonomic diversity. *Paleobiology* 19:185–204

41. Foote M. 1994. Morphological disparity in Ordovician-Devonian crinoids and the early saturation of morphological space. *Paleobiology* 20:320–44

42. Foote M. 1994. Morphology of Ordovician-Devonian crinoids. *Univ. Mich. Mus. Paleontol. Contrib.* 29:1–39

43. Foote M. 1995. Morphological diversification of Paleozoic crinoids. *Paleobiology* 21:273–99

44. Foote M. 1995. Morphology of Carboniferous and Permian crinoids. *Univ. Mich. Mus. Paleontol. Contrib.* 29:135–84

45. Foote M. 1996. Perspective: evolutionary patterns in the fossil record. *Evolution* 50:1–11

46. Foote M. 1996. Ecological controls on the evolutionary recovery of post–Paleozoic crinoids. *Science* 274:1492–95

47. Foote M. 1996. Models of morphological diversification. In *Evolutionary Paleobiology,* ed. D Jablonski, DH Erwin, JH Lipps, pp. 62–86. Chicago: Univ. Chicago Press. 484 pp.

48. Foote M. 1997. Sampling, taxonomic description, and our evolving knowledge of morphological diversity. *Paleobiology.* 23:181–206

48a. Foote M. 1997. Estimating taxonomic durations and preservation probability. *Paleobiology.* In press

49. Foote M, Gould SJ. 1992. Cambrian and Recent morphological disparity. *Science* 258:1816

50. Foote M, Raup DM. 1996. Fossil preservation and the stratigraphic ranges of taxa. *Paleobiology* 22:121–40

51. Fortey RA, Bell A. 1987. Branching geometry and function of multiramous graptoloids. *Paleobiology* 13:1–20

52. Fortey RA, Briggs DEG, Wills MA. 1996. The Cambrian evolutionary 'explosion': decoupling cladogenesis from morphological disparity. *Biol. J. Linn. Soc.* 57:13–33

53. Fortey RA, Owens RM. 1990. Evolutionary radiations in the Trilobita. In *Major Evolutionary Radiations,* ed. PD Taylor, GP Larwood, pp. 139–64. Oxford: Clarendon. 437 pp.

54. Gould SJ. 1970. Evolutionary paleontology and the science of form. *Earth Sci. Rev.* 6:77–119

55. Gould SJ. 1989. *Wonderful Life.* New York: Norton. 347 pp.

56. Gould SJ. 1991. The disparity of the Burgess Shale arthropod fauna and the limits of cladistic analysis: Why we must strive to quantify morphospace. *Paleobiology* 17:411–23

57. Gould SJ. 1993. How to analyze Burgess Shale disparity—a reply to Ridley. *Paleobiology* 19:522–23

58. Gould SJ. 1996. *Full House.* New York: Harmony Books. 244 pp.

59. Hafner MS, Remsen JV, Lanyon SM. 1984. Bird versus mammal morphological diversity. *Evolution* 38:1154–56

60. Hardy MC. 1985. Testing for adaptive radiation: the ptychaspid (Trilobita) biomere of the Late Cambrian. In *Phanerozoic Diversity Patterns,* ed. JW Valentine, pp. 379–97. Princeton, NJ: Princeton Univ. Press. 441 pp.

61. Hertel F. 1994. Diversity in body size and feeding morphology within past and

present vulture assemblages. *Ecology* 75:1074–84

62. Hickman CS. 1980. Gastropod radulae and the assessment of form in evolutionary paleontology. *Paleobiology* 6:276–94

63. Hickman CS. 1988. Analysis of form and function in fossils. *Am. Zool.* 28:775–93

64. Hickman CS. 1993. Biological diversity: elements of a paleontological agenda. *Palaios* 8:309–10

65. Hickman CS. 1993. Theoretical design space: a new program for the analysis of structural diversity. *Neues Jb. Geol. Paläont. Abh.* 190:169–82

66. Hirsch P, Rades-Rohkohl E, Kölbel-Boelke J, Nehrkorn A. 1992. Morphological and taxonomic diversity of groundwater microorganisms. In *Progress in Hydrogeochemistry,* ed. G Matthes, F Frimmel, P Hirsch, HD Schultz, H-E Usdowski, pp. 311–25. Berlin: Springer-Verlag. 544 pp.

67. Hughes NC. 1991. Morphological plasticity and genetic flexibility in a Cambrian trilobite. *Geology* 19:913–16

68. Hughes NC, Chapman RE. 1995. Growth and variation in the Silurian proetide trilobite *Aulacopleura konincki* and its implications for trilobite palaeobiology. *Lethaia* 28:333–53

69. Jablonski D. 1986. Background and mass extinctions: the alternation of macroevolutionary regimes. *Science* 231:129–33

70. Jablonski D. 1993. The tropics as a source of novelty through geological time. *Nature* 364:142–44

71. Jablonski D, Bottjer DJ. 1991. Environmental patterns in the origin of higher taxa: the post-Paleozoic fossil record. *Science* 252:1831–33

72. Jernvall J, Hunter JP, Fortelius M. 1996. Molar tooth diversity, disparity, and ecology in Cenozoic ungulate radiations. *Science* 274:1489–92

73. Kaiser HE, Boucot AJ. 1996. Specialisation and extinction: Cope's Law revisited. *Hist. Biol.* 11:247–65

74. Kendrick DC. 1996. Morphospace filling in flexible crinoids. *Paleontol. Soc. Spec. Pub.* 8:208

75. Knoll AH, Niklas KJ, Gensel PG, Tiffney BH. 1984. Character diversification and patterns of evolution in early vascular plants. *Paleobiology* 10:34–47

76. Labandeira CC, Sepkoski JJ Jr. 1993. Insect diversity in the fossil record. *Science* 261:310–15

77. Lee MSY. 1993. Cambrian and Recent morphological disparity. *Science* 258:1816–17

78. MacLeod N. 1996. Empirical shape space representations and shape modeling of fossils from landmark-registered 2D outlines, 3D outlines, and 3D surfaces, with a comment on the indeterminacy of empirical "mono-morphospace" analysis. *Paleontol. Soc. Spec. Pub.* 8:254

79. Mamkaev YV. 1986. Initial morphological diversity as a criterion in deciphering turbellarian phylogeny. *Hydrobiologia* 132:31–33

80. Marshall CR. 1991. Estimation of taxonomic ranges from the fossil record. In *Analytical Paleobiology,* ed. NL Gilinsky, PW Signor, pp. 19–38. Knoxville, TN: Paleontol. Soc. 216 pp.

81. May RM. 1994. Conceptual aspects of the quantification of the extent of biological diversity. *Philos. Trans. R. Soc. Lond. B* 345:13–20

82. McGhee GR Jr. 1980. Shell form in the biconvex articulate Brachiopoda: a geometric analysis. *Paleobiology* 6:57–76

83. McGhee GR Jr. 1991. Theoretical morphology: the concept and its applications. In *Analytical Paleobiology,* ed. NL Gilinsky, PW Signor, pp. 87–102. Knoxville, TN: Paleontol. Soc. 216 pp.

84. McGhee GR Jr. 1995. Geometry of evolution in the biconvex Brachiopoda: morphological effects of mass extinction. *Neues Jb. Geol. Paläont. Abh.* 197:357–82

85. McKinney FK, Raup DM. 1982. A turn in the right direction: simulation of erect spiral growth in the bryozoans *Archimedes* and *Bugula. Paleobiology* 8:101–12

86. McNamara KJ. 1986. The role of heterochrony in the evolution of Cambrian trilobites. *Biol. Rev.* 61:121–56

87. McShea DW. 1992. A metric for the study of evolutionary trends in the complexity of serial structures. *Biol. J. Linn. Soc.* 45:39–55

88. McShea DW. 1993. Arguments, tests, and the Burgess Shale—a commentary on the debate. *Paleobiology* 19:399–402

89. McShea DW. 1993. Evolutionary change in the morphological complexity of the mammalian vertebral column. *Evolution* 47:730–40

90. McShea DW. 1996. Perspective: Metazoan complexity and evolution: Is there a trend? *Evolution* 50:477–92

91. Miller AI, Foote M. 1996. Calibrating the Ordovician radiation of marine life: implications for Phanerozoic diversity trends. *Paleobiology* 22:304–9

92. Miller AI, Mao S. 1997. Scales of diversification and the Ordovician radiation. In *Biodiversity Dynamics: Turnover of Populations, Taxa, and Communities,* ed. ML McKinney. New York: Columbia Univ. Press. In press

93. Mooi R, David B. 1996. Phylogenetic analysis of extreme morphologies: deepsea holasteroid echinoids. *J. Nat. Hist.* 30:913–53

94. Niklas, KJ 1982. Computer simulations of early land plant branching morphologies: canalization of patterns during evolution? *Paleobiology* 8:196–210

95. Olson EC, Miller RL. 1958. *Morphological Integration.* Chicago: Univ. Chicago Press. 317 pp.

96. Padian K, Lindberg DR, Polly PD. 1994. Cladistics and the fossil record: the uses of history. *Annu. Rev. Earth Planet. Sci.* 22:63–91

97. Paul CRC. 1977. Evolution of primitive echinoderms. In *Patterns of Evolution,* ed. A Hallam, pp. 123–57. Amsterdam: Elsevier. 591 pp.

98. Paul CRC. 1979. Early echinoderm radiation. In *The Origin of Major Invertebrate Groups,* ed. MR House, pp. 415–34. London: Academic. 515 pp.

99. Paul CRC. 1982. The adequacy of the fossil record. In *Problems of Phylogenetic Reconstruction,* ed. KA Joysey, AE Friday, pp. 75–117. London: Academic. 442 pp.

100. Raup DM. 1966. Geometric analysis of shell coiling: general problems. *J. Paleontol.* 40:1178–90

101. Raup DM. 1967. Geometric analysis of shell coiling: coiling in ammonoids. *J. Paleontol.* 41:43–65

102. Raup DM. 1968. Theoretical morphology of echinoid growth. *J. Paleontol.* 42(Suppl.):50–63

103. Raup DM, Gould SJ. 1974. Stochastic simulation and evolution of morphology—towards a nomothetic paleontology. *Syst. Zool.* 23:305–22

104. Ricklefs RE, Miles DB. 1994. Ecological and evolutionary inferences from morphology: an ecological perspective. See Ref. 166, pp. 13–41

105. Ricklefs RE, O'Rourke K. 1975. Aspect diversity in moths: a temperate-tropical comparison. *Evolution* 29:313–24

106. Ricklefs RE, Travis J. 1980. A morphological approach to the study of avian community organization. *The Auk* 97: 321–38

107. Ridley M. 1993. Analysis of the Burgess Shale. *Paleobiology* 19:519–21

108. Romer AS. 1949. Time series and trends in animal evolution. In *Genetics, Paleontology, and Evolution,* ed. GL Jepsen, E Mayr, GG Simpson, pp. 103–20. Princeton, NJ: Princeton Univ. Press. 474 pp.

109. Roy K. 1994. Effects of the Mesozoic Marine Revolution on the taxonomic, morphologic, and biogeographic evolution of a group: aporrhaid gastropods during the Mesozoic. *Paleobiology* 20:274–96

110. Roy K. 1996. The roles of mass extinction and biotic interaction in large-scale replacements: a reexamination using the fossil record of stromboidean gastropods. *Paleobiology* 22:436–52

111. Roy K, Foote M. 1997 Morphological diversity as a biodiversity metric. *Trends Ecol. Evol.* 12: In press

112. Runnegar B. 1987. Rates and modes of evolution in the Mollusca. In *Rates of Evolution,* ed. KSW Campbell, MF Day, pp. 39–60. London: Allen & Unwin. 314 pp.

113. Saunders WB, Swan ARH. 1984. Morphology and morphologic diversity of mid-Carboniferous (Namurian) ammonoids in time and space. *Paleobiology* 10:195–228

114. Saunders WB, Work DM. 1996. Shell morphology and suture complexity in Upper Carboniferous ammonoids. *Paleobiology* 22:189–218

115. Savazzi E. 1990. Biological aspects of theoretical shell morphology. *Lethaia* 23:195–212

116. Schindewolf OH. 1950 [1993]. *Basic Questions in Paleontology.* [transl. J Schaeffer]. Chicago: Univ. Chicago Press. 467 pp.

117. Schram FR. 1981. On the classification of Eumalacostraca. *J. Crustac. Biol.* 1:1–10

118. Seilacher A. 1991. Self-organizing mechanisms in morphogenesis and evolution. In *Constructional Morphology and Evolution,* ed. N Schmidt-Kittler, K Vogel, pp. 251–71. Berlin: Springer-Verlag. 409 pp.

119. Sepkoski JJ Jr. 1978. A kinetic model of Phanerozoic taxonomic diversity. I. Analysis of marine orders. *Paleobiology* 4:223–51

120. Sepkoski JJ Jr. 1993. Ten years in the library: new data confirm paleontological patterns. *Paleobiology* 19:43–51

121. Sepkoski JJ Jr. 1996. Competition in macroevolution: the double wedge revisited. In *Evolutionary Paleobiology*, ed. D Jablonski, DH Erwin, JH Lipps, pp. 211–55. Chicago: Univ. Chicago Press. 484 pp.

122. Simpson GG. 1953. *The Major Features of Evolution*. New York: Columbia Univ. Press. 434 pp.

123. Sims HJ. 1996. Morphological diversification of Late Devonian and Carboniferous seeds. *Paleontol. Soc. Spec. Pub.* 8:361

124. Smith AB. 1988. Patterns of diversification and extinction in early Palaeozoic echinoderms. *Palaeontology* 31:799–828

125. Smith AB. 1990. Evolutionary diversification of echinoderms during the early Palaeozoic. In *Major Evolutionary Radiations*, ed. PD Taylor, GP Larwood, pp. 265–86. Oxford: Clarendon. 437 pp.

126. Smith AB. 1994. *Systematics and the Fossil Record*. Oxford: Blackwell Sci. 223 pp.

127. Smith AB, Patterson C. 1988. The influence of taxonomic method on the perception of patterns of evolution. *Evol. Biol.* 23:127–216

128. Smith LH. 1996. Developmental integration in trilobites. *Geol. Soc. Am. Abstr. Progr.* 28:A53

128a. Solow AR, Smith W. 1997. On fossil preservation and the stratigraphic ranges of taxa. *Paleobiology.* In press

129. Sprinkle J. 1980. Patterns and problems in echinoderm evolution. *Echinoderm Stud.* 1:1–18

130. Sprinkle J. 1992. Radiation of Echinodermata. In *Origin and Early Evolution of the Metazoa*, ed. JH Lipps, PW Signor, pp. 375–98. New York: Plenum. 570 pp.

131. Stanley SM. 1979. *Macroevolution*. San Francisco: Freeman. 332 pp.

132. Stebbins GL Jr. 1951. Natural selection and the differentiation of angiosperm families. *Evolution* 5:299–324

133. Stitt JH. 1971. Repeating evolutionary pattern in Late Cambrian trilobite biomeres. *J. Paleontol.* 45:178–81

134. Stitt JH. 1975. Adaptive radiation, trilobite paleoecology, and extinction, Ptychaspid biomere, Late Cambrian of Oklahoma. *Fossils Strata* 4:381–90

135. Stubblefield CJ. 1960. Evolution in trilobites. *Q. J. Geol. Soc. London* 115:145–62

136. Sundberg FA. 1996. Morphological diversification of Ptychopariida (Trilobita) from the Marjumiid biomere (Middle and Upper Cambrian). *Paleobiology* 22:49–65

137. Swan ARH, Saunders WB. 1987. Function and shape in late Paleozoic (mid-Carboniferous) ammonoids. *Paleobiology* 13:297–311

138. Tabachnick RE, Bookstein FL. 1990. The structure of individual variation in Miocene *Globorotalia. Evolution* 44:416–34

139. Thomas RDK, Reif W-E. 1991. Design elements employed in the construction of animal skeletons. In *Constructional Morphology and Evolution*, ed. N Schmidt-Kittler, K Vogel, pp. 283–94. Berlin: Springer-Verlag. 409 pp.

140. Thomas RDK, Reif W-E. 1993. The skeleton space: a finite set of organic designs. *Evolution* 47:341–60

141. Thomas RDK, Stewart GW. 1995. Extent and pattern of exploitation of skeletal design options by Middle Cambrian Burgess Shale organisms. *Geol. Soc. Am. Abstr. Progr.* 27:A269–70

142. Valentine JW. 1969. Patterns of taxonomic and ecological structure of the shelf benthos during Phanerozoic time. *Palaeontology* 12:684–709

143. Valentine JW. 1980. Determinants of diversity in higher taxonomic categories. *Paleobiology* 6:444–50

144. Valentine JW. 1986. Fossil record of the origin of Baupläne and its implications. In *Patterns and Processes in the History of Life*, ed. DM Raup, D Jablonski, pp. 209–22. Berlin: Springer-Verlag. 447 pp.

145. Valentine JW. 1989. How good was the fossil record? Clues from the Californian Pleistocene. *Paleobiology* 15:83–94

146. Valentine JW. 1991. Major factors in the rapidity and extent of the metazoan radiation during the Proterozoic-Phanerozoic transition. In *The Early Evolution of Metazoa and the Significance of Problematic Taxa*, ed. A Simonetta, S Conway Morris, pp. 11–13. Cambridge, UK: Cambridge Univ. Press. 296 pp.

147. Valentine JW. 1992. The macroevolution of phyla. In *Origin and Early Evolution of the Metazoa*, ed. JH Lipps, PW Signor, pp. 525–53. New York: Plenum. 570 pp.

148. Valentine JW. 1995. Why no new phyla after the Cambrian? Genome and ecospace hypotheses revisited. *Palaios* 10:190–94

149. Valentine JW, Campbell CA. 1975. Genetic regulation and the fossil record. *Am. Sci.* 63:673–80

150. Valentine JW, Collins AG, Meyer CP. 1994. Morphological complexity increase in metazoans. *Paleobiology* 20: 131–42

151. Valentine JW, Erwin DH. 1987. Interpreting great developmental experiments: the fossil record. In *Development as an Evolutionary Process,* ed. RA Raff, EC Raff, pp. 71–107. New York: Alan R. Liss. 329 pp.

152. Valentine JW, Walker TD. 1987. Extinctions in a model taxonomic hierarchy. *Paleobiology* 13:193–207

152a. Van Valen L. 1965. Morphological variation and width of ecological niche. *Am. Nat.* 99:377–90

153. Van Valen L. 1973. Are categories in different phyla comparable? *Taxon* 22:333–73

154. Van Valen L. 1974. Multivariate structural statistics in natural history. *J. Theoret. Biol.* 45:235–47

155. Van Valen L. 1978. The statistics of variation. *Evol. Theory* 4:33–43

156. Van Valen L. 1985. How constant is extinction? *Evol. Theory* 7:93–106

157. Van Valkenburgh B. 1985. Locomotor diversity within past and present guilds of large predatory mammals. *Paleobiology* 11:406–28

158. Van Valkenburgh B. 1988. Trophic diversity in past and present guilds of large predatory mammals. *Paleobiology* 14:156–73

159. Van Valkenburgh B. 1994. Ecomorphological analysis of fossil vertebrates and their paleocommunities. See Ref. 166, pp. 140–66

160. Van Valkenburgh B. 1994. Extinction and replacement among predatory mammals in the North American Late Eocene and Oligocene: tracking a paleoguild over twelve million years. *Hist. Biol.* 8:129–50

161. Waggoner BM. 1996. Phylogenetic hypotheses of the relationships of arthropods to Precambrian and Cambrian problematic fossil taxa. *Syst. Biol.* 45:190–222

162. Wagner PJ. 1995. Systematics and the fossil record. *Palaios* 10:383–88

163. Wagner PJ. 1995. Testing evolutionary constraint hypotheses with early Paleozoic gastropods. *Paleobiology* 21:248–72

164. Wagner PJ. 1996. Patterns of morphological diversification during the initial radiation of the "Archaeogastropoda." In *Origin and Evolutionary Radiation of the Mollusca,* ed. JD Taylor, pp. 161–69. Oxford: Oxford Univ. Press. 392 pp.

165. Wagner PJ. 1997. Patterns of morphologic diversification among the Rostroconchia. *Paleobiology.* 23:115–50

166. Wainwright PC, Reilly SM, eds. 1994. *Ecological Morphology.* Chicago: Univ. Chicago Press. 367 pp.

167. Ward PD. 1980. Comparative shell shape distributions in Jurassic-Cretaceous ammonites and Jurassic-Tertiary nautilids. *Paleobiology* 6:32–43

168. Weishampel DB. 1996. Fossils, phylogeny, and discovery: a cladistic study of the history of tree topologies and ghost lineage durations. *J. Vert. Paleontol.* 16:191–97

169. Whittington HB. 1966. Phylogeny and distribution of Ordovician trilobites. *J. Paleontol.* 40:696–737

170. Williams A. 1957. Evolutionary rates of brachiopods. *Geol. Mag.* 94:201–11

171. Williams PH, Humphries CJ. 1996. Comparing character diversity among biotas. In *Biodiversity,* ed. KJ Gaston, pp. 54–76. Oxford: Blackwell Sci. 396 pp.

172. Wills MA, Briggs DEG, Fortey RA. 1994. Disparity as an evolutionary index: a comparison of Cambrian and Recent arthropods. *Paleobiology* 20:93–130

173. Wilson AC, Kunkel JG, Wyles JS. 1984. Morphological distance: an encounter between two perspectives in evolutionary biology. *Evolution* 38:1156–59

174. Wray GA, Levinton JS, Shapiro, LH. 1996. Molecular evidence for deep Precambrian divergences among metazoan phyla. *Science* 274:568–73

175. Wyles JS, Kunkel JG, Wilson AC. 1983. Birds, behavior, and anatomical evolution. *Proc. Natl. Acad. Sci. USA* 80:4394–97

176. Yochelson EL. 1979. Early radiation of Mollusca and mollusc-like groups. In *The Origin of Major Invertebrate Groups,* ed. MR House, pp. 323–58. London: Academic. 515 pp.

Annu. Rev. Ecol. Syst. 1997. 28:153–93

INSECT MOUTHPARTS: Ascertaining the Paleobiology of Insect Feeding Strategies

Conrad C. Labandeira

Department of Paleobiology, Smithsonian Institution, National Museum of Natural History, Washington, DC 20560; e-mail: labandeira.conrad@simnh.si.edu

KEY WORDS: mouthpart class, functional feeding group, dietary guild, morphological data, ecological attributes, mouthpart classes, fossil record, plant-insect interactions

---

ABSTRACT

One of the most intensively examined and abundantly documented structures in the animal world is insect mouthparts. Major structural types of extant insect mouthparts are extensive, consisting of diverse variations in element structure within each of the five mouthpart regions—labrum, hypopharynx, mandibles, maxillae, and labium. Numerous instances of multielement fusion both within and among mouthpart regions result in feeding organs capable of ingesting in diverse ways foods that are solid, particulate, and liquid in form. Mouthpart types have a retrievable and interpretable fossil history in well-preserved insect deposits. In addition, the trace-fossil record of insect-mediated plant damage, gut contents, coprolites, and insect-relevant floral features provides complementary data documenting the evolution of feeding strategies during the past 400 million years.

From a cluster analysis of insect mouthparts, I recognize 34 fundamental mouthpart classes among extant insects and their geochronological evolution by a five-phase pattern. This pattern is characterized, early in the Devonian, by coarse partitioning of food by mandibulate and piercing-and-sucking mouthpart classes, followed by a rapid rise in herbivore mouthpart types for fluid- and solid-feeding during the Late Carboniferous and Early Permian. Mouthpart innovation during the Late Triassic to Early Jurassic added mouthpart classes for fluid and aquatic particle-feeding. This ecomorphological expansion of mouthpart design was associated with the radiation of holometabolous insects, especially Diptera. The final phase of mouthpart class expansion occurred during the Late Jurassic and Early Cretaceous, with addition of surface-fluid-feeding mouthpart classes that

153

subsequently became important during the ecological expansion of angiosperms. Conclusions about the evolution of mouthpart design are based on the mapping of phenetic mouthpart classes onto (ideally) cladistic phylogenies of lineages bearing those same mouthpart classes. The plotting of phenetic and associated ecological attributes onto baseline phylogenies is one of the most important uses of cladistic data.

INTRODUCTION

One of the most intensively studied structures among animals is insect mouthparts. Although reasons for the extensive literature addressing insect mouthpart structure are as varied as insects themselves, three major aspects are central. First, because insects are ubiquitous residents of virtually all terrestrial and freshwater habitats and have elevated taxonomic diversity and ecomorphologic disparity, their mouthparts represent a broad spectrum of feeding modes that are ideal for comparative studies. Second, insect mouthparts represent one of the most externally complex, yet structurally integrated and homologous morphologies known (67, 206, 317), such that detailed studies can be made of element and multielement innovations in the conversion of one mouthpart type to another (65, 44) or in the convergence toward a mouthpart type among unrelated lineages (3, 185, 239). Third, considerable effort in understanding economically related consequences of insect feeding, particularly in agricultural fields such as pest control, crop pollination, and the transmission of insect-vectored diseases, historically has required a fundamental understanding of mouthpart structure and function. However, only recently have this complex structural system and its ecological correlates been placed in a phylogenetic context. In this review I provide a synopsis of the geochronologic deployment of insect mouthpart types and, in particular, detail the role that paleobiology has to offer in recapturing the pattern of the ecological partitioning of food resources by insects during the past 400 million years.

INSECT MOUTHPARTS IN PERSPECTIVE

Basic Patterns of Insect Mouthparts

The head capsule of insects appears to be subdivided into six regions that correspond to embryonic segments (215, but see 271), of which five bear mouthpart appendages that are relevant for documentation of insect mouthpart structure in the fossil record and the evolution of insect feeding strategies (43, 172). In a generalized mandibulate insect such as a grasshopper, the head capsule is characterized by two dorsolaterally placed compound eyes, three median ocelli

located frontally at the vertices of a small inverted triangle, and the anatomically ventral mouthparts borne on five regions. The most anteriorly positioned
mouthpart is a median flap known as the labrum, which is the "upper lip"
that contains an inner membranous surface rich in sensilla, the epipharynx.
Posterior of the labrum is the centrally positioned, medial, and tongue-like hypopharynx, which is laterally encompassed by articulating mandibles. The two
similar posteriormost regions consist of a pair of proximal and distal sclerites
that are separate in the laterally positioned maxillae but fused in the posteriormost labium into a mesal "lower lip." Both the maxillae and the labium have
laterally attached, multisegmented palps, fleshy inner lobes, and often sclerotized, sometimes elongated outer lobes. For the maxillae, these lobes are known
respectively as the galeae and laciniae, whereas in the labium they are termed
the glossae and paraglossae. Placement of mouthparts on the head ranges from
the hypognathous condition described above to those that are angulated pos-
teriorly, as in the opisthognathous mouthparts of cicadas, and those that are
directed forward, such as prognathous ground beetles.

Interesting Deviations

The generalized mandibulate conditon, found in orthopteroid insects such as
crickets and termites, or in holometabolous insects such as beetles and sawflies,
is frequently modified so that mouthpart elements are co-opted in the formation of multielement complexes necessary for fluid feeding (43, 172, 319).
Co-optation may include transformation and association of the labrum, hypopharynx, mandibles, and maxillary laciniae into interlocking stylets in extinct and nonhaustellate Paleozoic paleodictyopteroids (166), or into stylets
housed within an enclosing labial sheath, or haustellum, formed by medial fusion of the labial palps as in nematocerous flies (43, 130). Other examples of
the haustellate-stylate condition include thrips, homopterans, bugs, and fleas.
In other taxa, such as proturans and certain small beetles, stylate mouthparts
have originated from the prolongation of mouthpart elements, concomitant with
a rotation of musculature to produce protracting/retracting stylets rather than
mesially adducting/abducting mandibles or laciniae (65, 241). Both nonhaustellate and haustellate types of stylate mouthparts function by piercing and sucking
and are involved overwhelmingly in invasive fluid feeding, principally of algal
and fungal protoplasts, plant sap, and animal blood.

Noninvasive types of fluid feeding involve mouthpart modifications for surface feeding of exposed fluids such as honeydew, nectar, and exudates from
animal wounds. In advanced flies, fused labial palpi form a sponging organ
consisting of a fleshy labellum with a feeding surface of orally-directed pseu-
dotracheae and miniscule sclerites for directing the capillary flow of fluid food
(85, 106, 291). In advanced moths, fluid feeding is accomplished by medial

fusion of elongate maxillary galea to form a coiled siphon that is extended by hydraulic action (83, 146). By contrast, in advanced Hymenoptera, principally bees, the maxillae and labium form a joint labiomaxillary complex, character-ized by prolongation of the labial glossa into a hirsute and often long organ for lapping fluids during extension. Upon glossal retraction, cupped lateral galeae and adpressed labial palps form an enclosed structure containing the glossa, on which fluid food ascends by capillary action and then is suctioned into the mouth by a pharyngeal pump (219, 319).

Mouthparts of subadult insects are equally varied, particularly those of holo-metabolous larvae (172). The unique labial mask of dragonfly naiads is folded into a Z-shaped configuration under the head when at rest (242) but can be protracted instantaneously by abdominal contraction (233) to extend a signifi-cant distance anteriorly beyond the head. The labial mask terminates in labial palps modified into clawed pincers (285, 331) that are responsible for securing and impaling prey. A similar structure with identical function occurs in some staphylinid beetle larvae (13), but instead of terminal pincers, adhesives on labial appendages stick to prey for its rapid retrieval into the pharynx. The lar-val mouthparts of flies, particularly Culicidae, Simuliidae, and Chironomidae (188, 232, 299), and of other orders such as mayflies (84) and beetles (282), are modified into elaborate setiferous fans, pilose brushes, grooved rakes, and other devices for passively filtering and actively sieving suspended organic mat-ter (347). Terrestrial larvae of planipennians, by contrast, bear falcate, acumi-nate mandibles that are longitudinally channeled on the ventral surface and can be covered by the adjacent lacinial plate which, when articulated over the mandible, results in a closed canal for releasing proteolytic enzymes into a prey item (150, 284). After the prey is secure, a reversal of the direction of fluid flow results in imbibition of the liquefied contents of the prey item. An analogous feeding mechanism exists for several lineages of larval beetles, except that the tubular channel is contained within individual mandibles (44, 345).

Modifications of mandibulate mouthparts other than for the familiar deep-tissue and surface fluid feeding have occurred in some lineages. For example, adult beetle mouthparts are typically characterized as mandibulate, with power-ful mandibles occurring especially in large species. However, several lineages of scarab beetles have evolved mouthparts incapable of chewing solid food (227) and have been modified into flexible flaps for procurement of honey (90), for pollen consumption (31), and for the likely absence of feeding altogether (210). Typical pseudotracheate and fleshy labella of flies have been modified into a rasping organ by conversion of typically nonabrasive prestomal scle-rites into batteries of circumoral teeth. The armature of these teeth is used for predation on other insects, abrasion of fruit, and scarification of vertebrate in-tegument for feeding on exuded fluids, including blood (364). This labellum type has originated independently at least several times in muscomorph flies

(85, 131), including the Muscidae (161), Scathophagidae (228), Glossinidae (250), and Tephritidae (80). A last example is the transformation of the noctuid moth siphon into a stylet, primarily by sclerotization of the siphon tip and tubular stiffening (9), a process that occurred independently in Southeast Asia and Africa (9, 322, 361). These stylate moths consume fruit or they pierce humans and the thinner integumentary regions of large ungulates.

PALEOBIOLOGY AND EVOLUTIONARY BIOLOGY OF INSECT MOUTHPARTS

Approaches Centered on the Fossil Record

The insect fossil record historically has been considered poor—overwhelmingly dependent on descriptions of wings (40, 41). However, this is a caricature. At the family level, 63% of modern families are represented as fossils (174), surpassing many invertebrate and vertebrate groups. Additionally, recent studies have revealed rich morphological detail from soft-part material in both sedimentary compresssion (10, 28, 97, 108, 196, 263) and amber (160, 245, 369) deposits. Moreover, investigations of the fossil history of plant-insect associations, including the record of insect-mediated plant damage (176, 179, 301, 302, 323), increasingly have been placed in an evolutionary context (177, 236), as have examination of plant tissues in coprolites (180, 283) and insect guts (34, 166, 265, 266), and understanding the origin and timing of floral features for the evolution of insect pollination syndromes (52, 53, 230, 355). These developments provide considerable and mutually independent data addressing the history of insect mouthpart design and, more broadly, the macroevolutionary history of insect feeding strategies.

MOUTHPART STRUCTURE PRESERVED IN THE FOSSIL RECORD   Even in deposits with well-preserved, soft-part anatomy, detailed preservation of insect mouthparts is uncommon. However, during the past 15 years, particularly Mesozoic deposits (108, 127, 135, 203, 261, 263, 269, 350) have yielded sufficient mouthpart detail that direct evidence now exists for the occurrence of major mouthpart types in the pre-Cenozoic fossil record. In this section a brief summary is provided of some notable occurrences of fossil insect mouthparts, from which certain deductions can be made about feeding strategies and diets (23, 355). Most of the record of mouthpart structures is driven by exceptionally well-preserved deposits that are spatiotemporally scarce but reveal details of feeding mechanisms for many species.

Paleozoic   From the Early Devonian, two deposits, at approximately 390 Ma, indicate that collembolans (125) and archaeognathans (175) bore distinctive mouthpart types, including primitive entognathous and ectognathous mandibulate and possibly piercing-and-sucking types (175). The earliest known stratum for which there is abundant, well-preserved material occurs in the Late Pennsylvanian (Late Moscovian) age (307 Ma) ironstone nodules of Mazon Creek in north-central Illinois (273). This deltaic deposit contains mouthpart and head detail of primitively wingless forms, including large entognathous diplurans (165), and a monuran and thysanuran with broadly articulating, dicondylic mandibles and leg-like palps (165). Of the extinct paleodictyopteroids, paleodictyopterans have been described with long, robust beaks containing five unsheathed and interlocking stylets (165, 166), variously interpreted as imbibing small particulate matter such as lycopod spores (281, 310), fluidized endosperm from cordaite or pteridosperm seeds (307, 310), or sap from the vascular tissue of tree ferns (179). Megasecopterans with smaller, truncate beaks (179) are also known. Also documented are the ancestral hemipteroids *Eucaenus* with long and slender maxillary palps and a prominent clypeal dome (42; CC Labandeira, personal observation), and *Gerarus* (32, 165, 168), also with a combination of orthopteroid and primitive hemipteroid features. An undescribed endopterygote larva has been figured, apparently bearing mouthparts resembling an externally-feeding caterpillar (166). In slightly younger Late Pennsylvanian (Kasimovian) ironstone deposits at Commentry, France, the raptorial mouthparts of the protodonatan dragonfly *Meganeura monyi*, one of the largest known insects, had mandibles and maxillary lobes with sharp, terminal incisors for cutting prey (310). Equally impressive is the large paleodictyopteran *Eugereon boeckingii* from the lowermost Permian of Germany (224), with a beak 3.1 cm long (310), that probably fed on plant sap (179).

From younger, Lower Permian (264 Ma) marginal marine deposits at Elmo, Kansas, head and associated mouthpart structures are documented for several insects. These include a monuran (81), a diaphanopterodean adult with a bulging clypeal pump and transverse linear depressions indicating dilator muscles (39), the rostrate psocopteran *Dichentomum* with terminally located mouthparts, including small, biting mandibles (37, 38, 166), and the prominent clypeus and beak with an enclosed stylet of an archescytinid homopteran (36). The Chekarda insect fauna, of Early Permian age (Kungurian; 258 Ma), near Perm, Russia, represents the most extensive documentation of Paleozoic insect mouthparts. From compressions in these fine-grained deposits, the mouthpart and head structure has been documented for a monuran (305) resembling that of modern archaeognathans. For paleodictyopteroids, documentation and reconstructions are available for the heads and mouthparts of paleodictyopterans (281), megasecopterans with more gracile beaks compared to their Carboniferous predecessors (169, 281), diaphanopterodeans, especially *Permuralia* (281, 163, 167), and the small, delicate, and acuminate beaks of a near dipterous permothemistid (166). Generalized orthopterans with mandibulate mouthparts similar to those of modern grasshoppers are known (Sharov 1968). Ancestral hemipteroids such as *Synomaloptilia* bear a rostrate head with exserted, apically bifid lacinial blades and a clypeal dome (259, 281, 310)—features that have been interpreted as a strategy for feeding on gymnosperm megaspores (259). One of the earliest known homopterans, *Permocicada*, possessed a three-segmented opisthognathous beak, and a modest clypeal bulge on which occurred a battery of transverse ridges representing pronounced cibarial dilator muscles (16), indicating phloem feeding (309). An early endopterygote larva is known with head capsule and mandibular detail (21, 304). Two major Paleozoic mouthpart types—in paleodictyopteroids the unsheathed beak with interlocking stylets, and in ancestral hemipteroids the combination of mandibulate mouthparts, exposed lacinial blades, and a prominent clypeal dome (166)—have no modern analogs. In addition, the distinctive rostrate head bearing terminal mouthparts in the psocopteran *Dichentomum* may be unique to the Paleozoic.

Triassic and Jurassic   Little is known of insect mouthparts or even head structures of well-preserved Triassic insects. The first documented appearances of modern taxa correspond with the near absence of Paleozoic-aspect taxa in better preserved Triassic deposits (97, 235). An assumption that the mouthpart structures of these Triassic lineages are similar to those of their modern descendants is supported by more extensive, better-preserved, and intensively collected Jurassic material containing the earliest known occurrences of many modern mouthpart types. From the Lower Jurassic (Hettangian, 203 Ma) shales of Issyk Kul in Kirghizstan (278) and better preserved material from various Early and Middle Jurassic localities in Siberia (142), several mouthpart types first occur in association with the radiation of nematoceran flies (278, 360). These include plumose and setose mouthparts modified for filtration and sieving of particulate matter by aquatic larvae (56, 142, 360), the cephalopharyngeal complex and associated, dorsoventrally articulating mouthhooks used for piercing by terrestrial larvae (142), various multistylate proboscides for piercing integument in adult blood feeding (140, 141, 151), a sponging labellum for adult surface fluid feeding (152, 153), and adults with rudimentary mouthparts indicating aphagy (155). Additionally, this interval marks the appearance of weevil mouthparts (370), generally used for chewing through indurated substrates (58), and the distinctive labiomaxillary complex associated with the dominantly parasitic apocritan Hymenoptera (262).

Perhaps the most celebrated and best-preserved Jurassic insect deposit is the lacustrine shale at Karatau, Kazakhstan, dated as Oxfordian to Kimmeridgian in age (≈152 to 157 Ma). Well-preserved insects from this deposit were initially described by Martynov in the 1920s, and subsequently specialists have described a wealth of insect taxa (7, 279), including structural details of mouthparts. Prominent are cockroaches, which include ovipositor-bearing mesoblattinids with typical orthopteroid mouthparts (342), and an apparently oöthecate lineage of snakefly-mimics, possessing elongate, forwardly oriented heads and prolonged mandibulate mouthparts associated with long, gracile maxillary palps (343), evidently used in probing and feeding in concealed places. Also present are protopsyllidiid and other homopteran taxa (17, 18) with modest clypeal expansions that are similar to those of Paleozoic forms; some of the earliest fossils of extant phytophagous beetle lineages, such as alleculids (212), mordellids (58), and chrysomelids with robust, toothed mandibles (204, 211); the earliest diverse suite of weevils, representing several major lineages with rostra of various lengths and dorsoventral orientations (6, 170, 204, 369); orthophlebiid scorpionflies with rostrate heads and hypognathous mouthparts; and apparent nemestrinid flies with long proboscides (108, 280). Many of these mouthpart types indicate interaction with plants, which, judging from the Karatau floral record, consisted principally of ferns, caytonialeans, bennettitaleans, cycads, ginkgoaleans, and coniferaleans (73). From the Late Jurassic/Early Cretaceous boundary (≈145 Ma), Kozlov (157) described the earliest known siphonate proboscis, possibly belonging to a ditrysian moth (177). If true, this find precedes the earliest known angiosperms by approximately 15 my. Nepticuloid leaf mines are documented in penecontemporaneous deposits from Australia (287), indicating that primitive siphonate mouthparts may have been present in another lepidopteran lineage (177).

Cretaceous   The Cretaceous record, once considered depauperate (40, 41), has improved dramatically because of the discovery and description of insect taxa from several compression deposits of exceptional quality outside Europe and North America, and a significant rise in the number of insect taxa described from amber (174). During the Early Cretaceous, the coarse morphological spectrum of insect mouthparts converged on the Recent, and by approximately 80 million Ma, the modern spectrum of mouthparts was already deployed. Although the detail of Early Cretaceous compression fossils from fine-grained lacustrine shales and carbonates is not as impressive as that in Late Cretaceous amber from forested settings, the Early Cretaceous insect material provides insights into insect-bearing deposits that did not reappear until the Cenozoic.

The lithographic limestone of Lérida, Spain is one of the earliest, insect-bearing Cretaceous deposits, of Berriasian age (143 Ma). In this deposit are found the head and rostra of nemonychid weevils (350), heptageniid mayfly naiads with prominent maxillae and mandibular tusks (350), and one of the earliest known adult termites, bearing asymmetric but robust mandibles with conspicuously hardened incisors (183). Also known are dipteran larvae bearing a cephalopharyngeal apparatus with parallel ventral impressions of mouthhooks (350).

Several uppermost Jurassic to mostly Lower Cretaceous localities ($\approx$150 to 130 Ma) in Transbaikalian Russia have revealed siphlonurid mayfly naiads with prominent labra and mandibles, an anthocorid hemipteran with trisegmented beak (252), aquatic hydrophilid beetles with prognathous mouthparts (248), an orthophlebiid scorpionfly with conspicuous rostrate and hypognathous mouthparts (327), and well-preserved nematocerous flies with both labellate and stylate proboscides (141, 154). A rare, externally feeding sawfly larva, *Kuengilarva,* bears a typical head capsule with mandibulate, albeit indistinct, mouthparts (264). The most enigmatic insect is *Saurophthirus*, evidently a panorpoid holometabolan (246) but ordinally unplaced and interpreted by Ponomarenko (246, 247) as ectoparasitic on pterosaurs and related to fleas. It possesses an opisthognathous head bearing a pronounced cheek region and an elongated, palp-bearing proboscis that abruptly tapers distally into a terminus with exserted stylets.

Perhaps less enigmatic but equally controversial are the four taxa of presumed fleas from the Koonwarra Insect Bed, in Victoria, Australia, of Late Aptian age, dated at approximately 118 Ma (135). Two of these taxa have been assigned to the extant family Pulicidae (135, 275), and a third, *Tarwinia*, has a body facies and prolonged head shape with stylate mouthparts that converges on *Saurophthirus* and extant insects ectoparasitic on warm-blooded vertebrates (e.g. 189). The Koonwarra specimens possess heads that are modestly compressed laterally, and display mouthpart specializations that include elongate palps and long styliform structures interpreted as epipharynges and laciniae (135). Unlike other recently discovered Lower Cretaceous compression faunas, Koonwarra insects are of interest because of their unique taxonomic character and abundance of immature stages. Additional examples include a siphlonurid mayfly naiad bearing spinose maxillary appendages and detail of mandibular dentition; mesophlebiid and coenagrionid dragonfly naiads with distinctive labial masks, one with stiff, parallel hairs on the labial palps; and exceptionally well-preserved mouthparts in a hydrophilid beetle larva, including a convex clypeal border, edentate, falcate mandibles and maxillae with a long palpus (135).

The Santana Formation, from northeastern Brazil, also of Late Aptian age, has yielded a diverse insect fauna (108, 109). It includes a dragonfly nymph with robust, inwardly curving antennae and a robust labial mask (35), an earwig with typical mandibulate mouthparts (249), and a diverse spectrum of homopterous Hemiptera (109, 115, 116) mostly preserved in lateral aspect, with evident but variously developed clypeal bulges and beak lengths indicating a variety of feeding strategies. Large insect-feeding robber flies with dagger-like, monostylate beaks have also been described (108, 109).

An insect fauna of Aptian to Albian age (124 to 100 Ma) from northwestern Mongolia taxonomically parallels that of Transbaikalia mentioned above (261). It includes a hexagenitid mayfly naiad (316) with distal mandibular and maxillary surfaces that indicate a scraping feeding strategy, a predatory odonatan naiad (255) with an unextended labial mask 0.8 cm long, a cicadelloid (308) exhibiting a pronounced clypeal expansion and muscle insertion chevrons indicating a powerful pump for imbibing deep-seated vascular fluids, and the aquatic corixid bug *Velocorixa* (251) that bears prothoracic legs with elongate, inwardly-directed, filiform hairs, and a short, truncate, three-segmented beak—structures indicating algivory. Also documented are nematocerous flies, some with prominent, elongate labial palps for piercing (Chaoboridae) and others with short, broader labellae for surface fluid feeding (Anisopodidae) (139), and the enigmatic *Saurophthiroides*, less well-preserved than its presumed Transbaikalian relative, *Saurophthiris* (247). Additionally the seeds of an extinct gingkophyte were used by caddisfly naiads to construct cases (159). At Orapa, Botswana, dated as Cenomanian (93 Ma) in age, a glimpse into an early Late Cretaceous insect fauna reveals weevils with well-preserved, decurved beaks (171), staphylinid beetles with typical mandibulate mouthparts (269), and a tipulid fly (268) housing an elongate rostrum 0.3 cm long, the tip bearing elongate, five-segmented palps and a small labellum, interpreted as a modification for nectarivory (267, 268).

The record of mouthpart structures from predominately Late Cretaceous amber is equally informative, and complements the dominantly Early Cretaceous occurrences of compression deposits. The earliest documented amber with insect inclusions is Hauterivian in age ($\approx$133 Ma; 297), and contains thrips with a mouthcone, labial palps, and exserted mandibular stylets (370), phytophagous whitefly labial sheaths with enclosed stylets (296), a monotrisian lepidopteran larva bearing a head capsule (DA Grimaldi personal communication), and nematoceran dipterans (297) including an phlebotomine fly with extruded stylate mouthparts (120). From New Jersey amber of Campanian age ($\approx$80 Ma), an early ant with a wasplike head and short, bidentate mandibles is known (359), as is a labellate phorid dipteran (107). Also present is the earliest known bee glossa, belonging to a meliponine bee, which reveals the distal margin of an apically dentate mandible, although other mouthpart structures are obscured (220). From known correlates between stylet dentition and palpal sensillae in modern ceratopogonid midges and their host preferences (208, 209, 286), Borkent (23) deduced that two and perhaps three ceratopogonid species from New Jersey and slightly younger Canadian ambers fed on exposed regions of the bodies of large vertebrates, specifically hadrosaurs, with accessible, vascularized integument (see also 76, 112, 141).

Siberian amber from the Taimyr Peninsula of Russia, of Santonian age (85 Ma) (369) has revealed the most diverse suite of Cretaceas insect mouthparts.

Included in this assemblage are a variety of aphidoids, particularly several extinct family-level lineages similar to modern phylloxerans (149), some with beaks exceeding the body in length and undoubtedly used for deep-seated vascular tissue feeding and probably capable of penetrating wood. Also present are adult caddisflies possessing well-preserved fluid-feeding haustella and associated labial and maxillary palps (25), ceratopogonids with robust, elongate proboscides and adjacent palps (270), bethylid and trigonalid wasps with unelongated mouthparts bearing prominent, apically dentate mandibles (92, 260) probably used for insect predation, and bombyliids with short labella for nectar feeding on bowl-shaped flowers (365), unlike subsequent Tertiary descendants with significantly more elongated labella for nectaring in flowers with tubular corollas.

Cenozoic   The post-Cretaceous insect record improves dramatically, especially with reference to highly collected, specimen-rich, and taxonomically diverse faunas such as Baltic Amber (160) and the Green River shales of the western United States. However, this record is uninformative with regard to major mouthpart innovation, and it collectively mirrors the Recent. Essentially all major mouthpart types (172; 182) and feeding strategies (172) had occurred by the end of Cretaceous, and overwhelmingly during the earlier Mesozoic. For example Szadziewski (330) discussed the life habits of ceratopogonid flies, found in Baltic amber ($\approx$37 Ma), some of which pursued large mammalian hosts in ways similar to vertebrate-seeking Late Cretaceous ancestors (23). The Cenozoic record of insect mouthparts does reveal minor variations within previously established mouthpart types, such as modifications involved in faithful pollination of certain angiosperms (51, 54, 355) or fine-tuned ectoparasitic specializations for mammalian hosts (145, 198, 213, 340).

THE RECORD OF INSECT-MEDIATED PLANT DAMAGE   Insects consume virtually all live and dead organic matter, including substances that are minimally rewarding nutritionally such as xylem sap (30) or those that are nonnutritive, such as wood, but are used indirectly as substrates for fungi or other saprophytic organisms to be consumed (12, 353). Virtually the entire fossil record of insect-organismic associations focuses on the interactions between insects and plants, preserved as trace fossils. Direct evidence of insect-fungal interactions (128, 352) are very rare but may be inferred from the presence of fossil taxa (134, 294) whose modern representatives are intimately coevolved with wood and fungi. Evidence for insect predation on animals has been derived principally from mouthpart morphology (23, 25, 141, 255).

The trace-fossil record of insect-plant interactions is largely decoupled from the record of insect body fossils (14), attributable to differing taphonomic circumstances in which insects and plants occur in the fossil record. Thus, inferences regarding plant associations have been made from insect body fossils in one of three ways: by tight correspondences between the taxonomy and ecological attributes of relevant modern insect taxa (60), functional interpretation of unique or otherwise well-preserved mouthpart structure (268, 309), or, ideally, assignment of botanically identifiable gut or coprolite contents to known fossil insect taxa (197, 265). Similarly, from the often well-preserved trace-fossil record of insect-mediated plant damage, it is often difficult to circumscribe an insect culprit with any taxonomic reliability. In instances of excellent preservation of insect-mediated damage on known plant hosts, the taxonomic identities of insect herbivores can be made at the level of the family or genus, either by reference to highly stereotyped damage patterns inflicted by ancestors of modern lineages whose interactions with plants are well characterized in the agricultural and entomological literature (113, 191, 192, 243, 328), or for extinct lineages by functional inference from mouthpart-inflicted damage seen on anatomically preserved plant host tissue and cellular structure (179, 180).

In this context, the Paleozoic poses special challenges because it records plant damage attributable almost exclusively to extinct lineages (15), requiring dependence on the functional morphology of plant damage and the presence of requisite insects with appropriate mouthparts (8, 144). By contrast, in Cenozoic floras, functional-feeding-groups such as leaf-miners (236, 323, 324), gallers (69, 184, 222, 302), and highly stereotyped external feeders (102, 293, 323) have been assigned to modern taxa. Increasingly this appears possible, perhaps with a relaxing of the taxonomic level, for Cretaceous angiosperm-dominated floras (177, 178, 323). [But see Jarzembowski (134) and Rozefelds (287) for highly stereotyped feeding patterns on Late Jurassic to Early Cretaceous nonangiospermous hosts.]

In many instances the historical record of insect-plant associations parallels mouthpart evolution as reflected in the body-fossil records of insect herbivores. Ironically, in some instances the plant damage record can provide a better assessment of a clade's ecological presence and plant-host specificity than body fossils (156, 177, 323), particularly for poorly sclerotized taxa of low preservational potential that lack a significant mouthpart history. Major examples include endophytic larvae that leave highly distinctive feeding patterns in woody plants, especially lepidopteran leaf miners (122, 257, 320), cecidomyiid, cynipid, and other gallers (1, 313, 357), and some phytophagous coleopterans whose larvae and adults feed in highly patterned and recognizable ways (138, 240).

GUT CONTENTS AND COPROLITES   The most direct way of establishing the diets of fossil insects is examination of well-preserved specimens with gut contents. This approach has not been used extensively, but recent examination of insects from a variety of Late Carboniferous to Miocene settings, including those in ironstone nodules, lacustrine shales, and amber have established that ingested spores and pollen frequently are well preserved and identifiable to source plant taxa. By contrast, insect coprolites containing inclusions that are anatomically preserved with histological and cellular detail are commonly encountered only in permineralized coal-balls from the Euramerican Pennsylvanian (283, 303).

Documented gut contents of Paleozoic insects occur in the Late Carboniferous Mazon Creek locality in Illinois (274) and at Chekarda in Russia (265). Mazon Creek insects that possess gut content palynomorphs include a thysanuran (301), a diaphanopterodean nymph (165), the adult ancestral hemipteroid *Eucaenus* (303), and an unidentified "protorthopteran" (303), all of which consumed spores and probably pollen, but only arborescent lycopod spores have been positively identified (303). In penecontemporaneous coal-swamp deposits 225 km from the clastic-swamp delta at Mazon Creek, abundant spore-bearing coprolites are found amid permineralized plant tissues (173). From the younger deposits at Chekarda, a more diverse assemblage of palynomorphs has been identified in the guts of an ancestral hemipteroid, a grylloblattid, and an unassigned species, collectively containing gastric residues of conifer, peltasperm, and glossopterid pteridosperm, and probably gnetalean pollen (217, 265, 266). Of these, only the unassigned species contains a monospecific pollen assemblage; the others were eclectic consumers of several gymnospermous pollen types.

It is notable that the two extinct mouthpart types associated with the Paleozoic insect fauna are also associated with pollinivory—the robust beak of palaeodictyopteroids and the laciniate mouthparts of ancestral hemipteroids. The mouthpart structure and inferred feeding mechanism of these insects has been discussed by several authors (168, 265, 281) who note the presence of a prolonged hypognathous head, rod-shaped maxillary laciniae, and a pronounced clypeal dome that indicate pollinivorous feeding habits. Modern psocopteran mouthparts (43, 205, 206) exhibit the closest similarities to Paleozoic hypoperlids. Both mechanisms of pollinivory independently have originated in modern taxa, including mandibulate beetles (31, 295) and adult syrphid flies (101, 126, 300) with tubular suction. A third major mechanism of pollen ingestion is the punch-and-suck mechanism of certain thrips (148), in which individual pollen grains are punctured by the single stout mandible, followed by adpression of the mouth-cone over the pollen grain and suction of the internal contents; this method may have Paleozoic antecedents (344).

Post-Paleozoic examples of insect gut contents center on three major mandibulate taxa: beetles (268, 293, 355), sawfly hymenopterans (34, 158), and a katydid orthopteran (266). The earliest occurrence, a Late Jurassic katydid from Karatau, Russia, contains pollen from a cheirolepidaceous conifer (266). Early Cretaceous sawflies from Santana, Brazil, consumed winteraceous angiospermous pollen (34, 79), whereas at a slightly earlier locality in Transbaikalia,

Russia, three xyelid sawflies consumed pollen of pinaceous and other conifers (158). From Eocene deposits, buprestid beetles have been recovered at Messel, Germany, with well-preserved pollen (293), and a scraptiid beetle from Baltic amber was coated by a diverse assemblage of pollen types (354). Related examples of insects with body surfaces bearing various palynomorphs include an early Late Cretaceous beetle at Orapa, Botswana, with undetermined pollen on its abdomen (268), an apid bee from the Middle Eocene of Germany bearing *Tricolporopollenites* grains on its sternites and metatarsal brush (197), and, from Early Miocene Dominican amber, ambrosia beetles bearing ascomycetous spores in mycangial cavities as well as a meliponine bee covered with *Hymenaea* and with unidentified pollen on its abdomen (110).

After an intestinal bolus exits the insect body as a fecal pellet, it can be fossilized immediately into a coprolite, albeit with loss of the original taxonomic context of its fabricator. However, coprolite assemblages can provide unparalleled ecological information if they are sufficiently abundant, compositionally diverse, and well preserved, such as those present in coal-ball floras from Late Carboniferous Euramerican wetlands. Such assemblages are identified by coprolite types distinctive in size, shape, composition, and plant associations (14, 303), and they often are linked to insect taxa at higher levels (180, 181, 283, 301). Thus, evidence from coprolites containing anatomically pristine tissues—equivalent in histological detail to modern embedded and sectional fecal pellets—can be used to document and understand insect dietary strategies (173). For an insect body fossil record for Late Carboniferous coal-swamp deposits (14), detailed identification of coprolite contents offers much promise for documenting early dietary diversity of mandibulate insects.

FLORAL STRUCTURE AND POLLINATION SYNDROMES   Much suggestive evidence indicates insect pollination of diverse Paleozoic gymnospermous plants, including pteridosperms, conifers, and cycadophytes (70). Among some Late Carboniferous medullosan pteridosperms, pollen is sufficiently large (70) that prepollen transfer was achieved probably by an insect vector (333). However, if sufficiently buoyant, *Medullosa* prepollen could move distances as much as 200 m with modest air currents (229). Certainly insects such as some large ancestral hemipteroids, orthopteroids bearing robust mandibles with shearing incisors, or paleodictyopteroids with stout, short beaks were capable of fragmenting and consuming medullosan prepollen of large size. Supplemental evidence of plant damage occurs in the form of a 1.2 cm-long beak probe, terminal feeding chamber, and associated reaction tissue in a bell-shaped medullosan pollen organ (298), wherein a paleodictyopteroid consumed prepollen and intercalated tissue while the organ was still alive (179, 272). The Early Permian conifer *Fergliocladus* of Argentina bore delicate, deliquescing tissue at the micropylar apex of each platyspermic ovule, indicating formation of liquid that rose up the micropylar tube and into a circumapertural depression at the external surface (2). This condition may indicate insect pollination. Mamay (201) proposed a function for characteristic glandular bodies occurring on cycadophyte leaves and between the seeds of associated reproductive structures, in coeval strata from Texas. He proposed that such glands were attractants for fluid-feeding insects that would acquire pollen in the feeding process and pollinate adjacent or distant ovules (but see 15). If this interpretation is correct, likely pollinators would be mandibulate insects possessing hypognathous heads and pectinate or hirsute mouthparts for imbibing fluid and possibly pollen.

As illustrated above, it is difficult to evaluate from fossil material evidence for the existence of pollination syndromes. The evidence almost always is indirect, requiring evaluation of a suite of floral or other reproductive characters and the likely presence, with or without co-ordinate paleoentomological evidence, of suitable insects vector to effect pollination. Such is the case for well-reasoned assessments inferring insect pollination of a Late Triassic gnetophyte and the angiosperm-like *Sanmiguelia* (46, 47). Although several modern gnetophytes are associated with pollinating insects (143, 214, 253, 341), the indirectness of this approach centers on specification of an insect to achieve pollination in return for a reward by the pollinated plant that benefits the insect (93, 314, 346). Alternatively, the presence of certain mouthpart structures known to effect pollination occur in a bewildering variety of modern insects (132), and similar mouthpart types present during the Middle and Late Jurassic can be interpreted as consuming plant substances that result in pollination. One example in the Late Jurassic Karatau insect fauna is beetles with mouthparts resembling extant boganiids, chrysomelids, and nemonychids (5, 6, 58, 59, 211, 212). Additionally, it is suggested that nemonychid weevils from these deposits (170) were inhabiting cycads in ways similar to some modern primitive weevils that occur in cycads (5, 6, 58), a conclusion supported by paleobotanical evidence documenting the presence of five cycad and allied genera (73).

Recent discoveries documenting extant beetle associations, in particular pollination of several cycad species (59, 95, 231, 332, 346), led to the conclusion that consumption by primitive beetles of cone axes, microsporophylls, pollen, and occasionally ovuliferous tissue is probably an ancient association (58, 60, 89, 223, 225). Evidence from the Early Cretaceous of a more or less related clade, the Cycadeoidea, reveals patterns of damage in the reproductive tissues of several specimens (50, 52, 58, 105) indicating that robust insects, undoubtedly beetles (58), were capable of entering the concealed interstices of a cycadeoid strobilus (cf 332) to consume interseminal scales and adjacent tissues (50). This resulted in a more extreme version of the "mess-and-soil" or cantharophilous pollination syndrome common to some extant beetles on magnolialean angiosperms (105, 336). A conclusion from these direct and indirect examples suggests that primitive, beetle pollination syndromes existed during the Late Jurassic and Early Cretaceous on nonangiospermous seed plants (71, 105).

Some of the earliest, best-documented evidence for angiosperm-centered pollination syndromes comes from the Early Cenomanian (97 Ma) Dakota Formation, in which several flower morphotypes and dispersed pollen have been described and documented (11, 72). One of these, the "Rose Creek flower" is a bisexual, actinomorphic rosid' with pentamerous and showy petals, an abbreviated style, and a robust and hypogynous receptacular disk possibly with nectaries. Judging by the pollinator spectra of equivalent modern flowers, this and other functionally similar magnoliid flowers with significant visual impact were probably pollinated predominantly by small to large mandibulate beetles (4, 63, 88, 99, 100) and medium-sized flies with sponging labella (78, 351). However, some early angiospermous flowers were small, simple, either sessile or closely attached to axes (53, 104), and characteristic of the myophilous pollination syndrome. These flowers resembled modern Winteraceae (86, 338), Chloranthaceae (87), and related families that are pollinated by small insects, especially flies (339, 351) and to a lesser extent thrips (86, 337, 338). Early Cretaceous lineages of nematocerous and plesiomorphic brachycerous Diptera occur in many insect faunas, although most modern representatives are generally considered hematophagous or saprophagous. These same clades sporadically have pollinating nectarivorous members today, including the Tipulidae (207, 277), Chironomidae (77, 195), Ceratopogonidae (33, 75, 292), Culicidae (114, 186, 335), Cecidomyiidae (94, 234, 363), Sciaridae (348) and Tabanidae (200), indicating that stylate mouthpart structure can be associated with nectarivory.

The radiation of the parasitoid Hymenoptera during the Jurassic (262) provided another pollinating insect group, since the labiomaxillary apparatus of small parasitoid wasps often is involved in nectaring (123, 136, 190, 256). Members of these clades entered a pollination syndrome, melittophily, involving further modification of the labiomaxillary complex into the glossate mouthparts of bees and its functional equivalent in some advanced wasps (172, 289, 290), Melittophily probably originated during the later Early Cretaceous (220). The phalaenophily of large moths and butterflies—but including long-tongued labellate flies that often hover, such as tabanids, nemestrinids, vermeleonids, and bombyliids (132, 200, 202, 326, 334)—was a Late Cretaceous or later development (82, 256) and is indicated by the fine-tuning of floral features to accommodate pollinator behavior and structure, such as zygomorphy, sympetaly, tubular corollas, and the presence of elaiophores (53, 355). Faithful modes of pollination, such as flowers with deep funnelform corollas and insects with significantly long mouthparts (200, 202), are probably post-Cretaceous (51, 52).

Mapping Mouthpart Classes onto Fossil-Calibrated Cladograms

A conservative estimate of the number of primary literature papers published since 1880 on insect mouthparts is 2100. These articles can be divided into standard descriptions of head and mouthparts of selected insect species (68, 83, 137), examinations of specific homologous structures or mouthpart regions across a wide range of insect taxa (49, 91, 318), monographic comparisons of head and mouthpart structure within an insect order (55, 244, 321, 325), and biomechanical analyses of mouthparts, such as those involving labial mask extension in dragonfly naiads (233, 254, 285) or aquatic filter feeding (26, 27, 48, 299). A sample of 1200 mouthpart and head sources from the literature up to 1989 was used in a phenetic analysis of the structural diversity and fundamental types of modern insect mouthparts (Figure 1). This literature contains a wealth of data on the structure of insect heads and mouthparts in an area of research that has been largely abandoned (but see 19, 80, 193) except as a source of data for supplying characters and character-states for cladistic analyses (20, 117, 238, 315).

This phenetic summary of the insect mouthpart literature was accomplished by a cluster analysis of 1365 modern insect species, including noninsectan apterygotes (Figure 1; 172). Criteria for inclusion of literature-based sources included rejection of all groundplan or otherwise idealized or generalized abstractions, citation of a documented taxon to genus or species level, repeatability of data among investigators, and multiplicity of data sources. Variables consisted of qualitative descriptions and nonmeristic size and shape assessments of mouthpart elements and mouthpart complexes and head-associated structures that meaningfully could be compared across all insects. Clusters were determined objectively by dendrogram branch lengths, similarity levels, overall dendrogram topology, and absence of chaining, although cluster 1-3, expressed as a single cluster in Figure 1, consisted of three discrete clusters in subsequent analyses (172). After criteria were used for cluster delimitation, a mouthpart class was designated for each cluster. Post hoc justifications for each cluster included demonstration that they were internally consistent with common mouthpart features, structurally unique when compared with the 33 other clusters, and, in some instances, that they had been previously recognized in the mouthpart literature. The resulting 34 fundamental mouthpart clusters are thus defined by membership of a clade or, more often, multiple unrelated clades that converge on a particular suite of head and mouthpart features forming a discrete structural unit in nature. Most of these mouthpart classes were previously recognized informally in the mouthpart literature (67, 133, 216, 317), and a subsequent summary employing a classical morphological approach (318, 319) independently repeated many of the results (43).

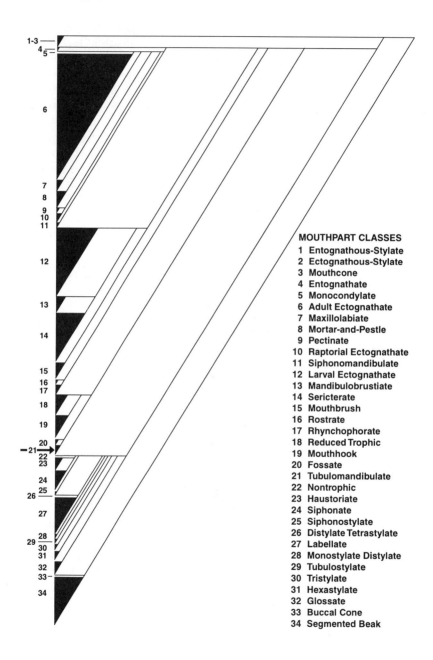

MOUTHPART CLASSES

1 Entognathous-Stylate
2 Ectognathous-Stylate
3 Mouthcone
4 Entognathate
5 Monocondylate
6 Adult Ectognathate
7 Maxillolabiate
8 Mortar-and-Pestle
9 Pectinate
10 Raptorial Ectognathate
11 Siphonomandibulate
12 Larval Ectognathate
13 Mandibulobrustiate
14 Sericterate
15 Mouthbrush
16 Rostrate
17 Rhynchophorate
18 Reduced Trophic
19 Mouthhook
20 Fossate
21 Tubulomandibulate
22 Nontrophic
23 Haustoriate
24 Siphonate
25 Siphonostylate
26 Distylate Tetrastylate
27 Labellate
28 Monostylate Distylate
29 Tubulostylate
30 Tristylate
31 Hexastylate
32 Glossate
33 Buccal Cone
34 Segmented Beak

The purpose of summarizing the extensive mouthpart literature by a cluster analysis was not only to determine the fundamental number and taxal composition of major extant mouthpart designs, but also to map each mouthpart class onto its constituent evolutionary lineages, calibrated by fossil occurrences (172). Thus patterns of taxal origination and diversification were assessed for the appearance and timing of mouthpart innovations. In addition, given an appropriate level of taxonomic analysis that ranges from the order to the genus, depending on the taxonomic scope of the mouthpart class, an assessment can be made of the degree of major mouthpart innovation within a monophyletic clade (e.g. 194), or alternatively of convergence toward a given mouthpart class in unrelated lineages (e.g. 3, 239). Thus, events such as the radiation of the Diptera during the Triassic and Early Jurassic (121, 278) is associated with the origin of six mouthpart classes; by contrast, the same mouthpart themes were repeated in two diverse clades, illustrated in the tubulomandibulate example of Figure 2. For the tubulomandibulates, which had arisen by the Late Triassic and radiated during the Jurassic, characteristic tubular mandibles originated from two to seven times within two major beetle clades (44, 66, 345, 366) as evaluated at the family level. It is possible that both clades may reveal additional convergences on the tubulomandibulate condition if the level of taxonomic resolution were increased to the genus or species levels. The defining feeding structure in tubulomandibulates is sickle-shaped, piercing mandibles that house tubular canals for ejection of proteolytic enzymes to liquefy insectan and invertebrate prey, as opposed to typical cutting mechanisms in mandibulate sister-group taxa.

A GEOCHRONOLOGIC HISTORY OF INSECT FEEDING STRATEGIES

When the method of analysis described above is applied to all 34 mouthpart classes, the result is a distinctive geochronological pattern of mouthpart class ap-

←

Figure 1   Resulting dendogram from a cluster analysis of recent hexapod mouthparts. Literature-derived data for this analysis consisted of 49 qualitative variables with characters ranging from 3 to 19 states, and representing all mouthpart regions, major mouthpart elements, and associated cephalic structures. 1365 cases were used, representing a diversity of modern hexapod mouthpart structure, including all 33 conventional orders and 70% of all recognized families. A matrix inversion of the BMDP1M cluster analysis program (118) was used with a Jaccard similarity index and an average linkage clustering algorithm. The root of the dendrogram is at the upper right, and the horizontal positions of nodes that link clusters indicate phenetic similarity. The vertical length of each blackened cluster is proportional to the number of included species; for example, cluster 6 contains 291 species, whereas cluster 32 contains 29 species. *Arrow* designates cluster 21, the Tubulomandibulate Mouthpart Class, detailed in Figure 2. Additional details and elucidation of clusters are provided in Labandeira (172).

TUBULOMANDIBULATE MOUTHPARTS

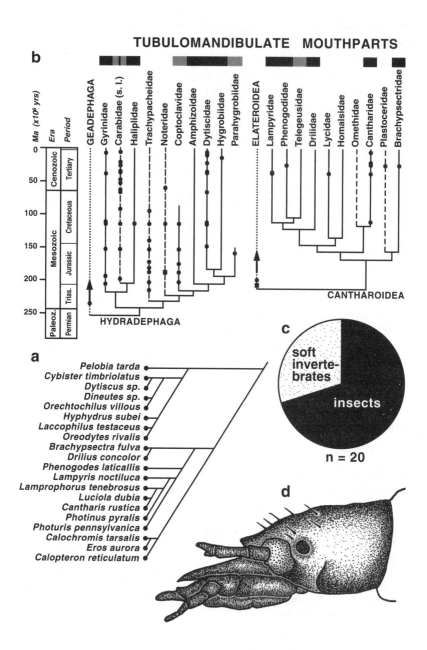

pearances (Figure 3). These data are combined with occurrences from the physical mouthpart and plant-insect association records to yield a five-phase sequence of insect mouthpart innovation and associated feeding strategies (Figure 4, top). This history is based upon minimal dates of mouthpart class appearance, and occasionally on indirect inference regarding mouthpart presence attributable to sister-group relationships of mouthpart-bearing taxa and types of fossil evidence, indicated by the cross-hatched portions of the vertical bars. Two intervals of the insect fossil record provide no data or minimal data for resolution: a 60 my interval from the Middle Devonian to Early Carboniferous, and a 13 my interval during the Early and Middle Triassic.

In Figure 4, morphological innovation of mouthpart types, expressed by the five-phased temporal pattern of mouthpart class diversity, is compared with family-level taxonomic diversity for the same geochronologic interval (174, 176, 182). The different forms of these two curves suggest that the evolution of distinctive mechanisms for processing food predates a rise in overall taxonomic diversity. Preliminary estimates of insect diversity at the genus level merely repeat the pattern observed at the family level.

Phase 1: Early Devonian

Primitive modes of mandibulate feeding and undoubted piercing-and-sucking extend to the Early Devonian, indicating the coarsest possible partitioning of food resources. Mandibulate feeding was accomplished by three mouthpart types borne by small hexapods (175): the entognathate (103) and monocondylate (22) types of detritivory with weakly abducting, milling mandibles, and probably an unelaborated version of adult ectognathate mouthparts (98) characterized by dicondylic and muscularly more powerful mandibles (175). Entognathous-stylate mouthparts are found generally in proturans (96) and

---

Figure 2   The Tubulomandibulate Mouthpart Class. (*a*) Membership and inter-relationships of taxa comprising cluster 21 in Figure 1. Tubulomandibulates constitute those larval beetles with tubular mandibles used for fluid-feeding on live prey. Details are provided by Labandeira (172). (*b*) Phylogeny, calibrated by fossil occurrences, of family-level lineages comprising the taxa in (*a*). The fossil occurrences are represented by *circles* for reliable assignments and *squares* for less secure identifications; they are not necessarily a complete inventory of all known, relevant fossils. The *bar* at top indicates membership in the Tubulomandibulate Mouthpart Class (*black*) or probable membership for fossils lacking in mandibular detail or for extant taxa with undocumented mouthpart structure (*grey*). For the Carabidae, the tubulomandibulate condition is documented in one species (366). The phylogenies are after Beutel & Roughley (21) for Hydradephaga, and Lawrence (187) and Crowson (58) for the Cantharoidea. Abbreviations: Paleoz. = Paleozoic, Trias. = Triassic. (*c*) Dietary spectrum of taxa in (*a*). Soft invertebrates are primarily gastropods. (*d*) A representative member, the cantharid *Chauliognathus* sp. Note that this mouthpart class originated from two to seven times, when evaluated at the family level.

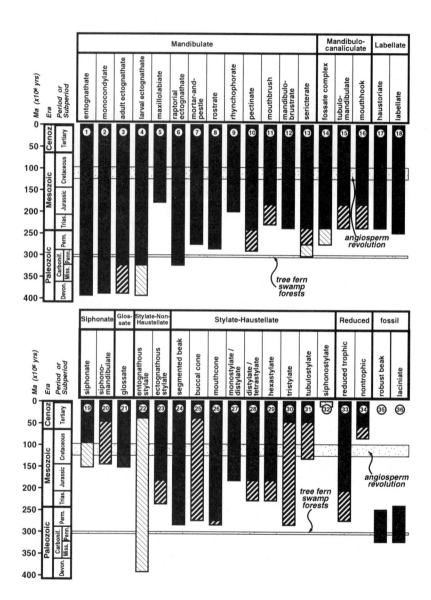

some collembolans, and evidence indicates that piercing-and-sucking occurred on primitive land plant stems (8, 179).

Phase 2: Pennsylvanian

There is no insect record for the Late Devonian and Mississippian; apparently, several major additional mouthpart classes originated not long before the earliest Pennsylvanian. Raptorial ectognathate mouthparts (312) of adult protodonatan predators (29), possibly including adult ephemeropterans (164), represented active pursuit predation of other insects. In another paleopterous lineage, the distinctive robust beak mouthparts of paleodictyopteroids (28, 166) were an effective piercing-and-sucking mechanism for tissue penetration and stylet maneuverability. Evidence exists for plant damage caused by this mouthpart type (179, 272, 307). Laciniate mouthparts occurred in some ancestral hemipteroid lineages (168, 259, 281), providing a combination of lacinial puncturing of indurated food, mandibulate chewing capability, and a clypeal suction pump for ingestion. Perhaps dating to the Early Pennsylvanian (166), sericterate mouthparts of larval holometabolans (311) provided multiple feeding strategies, including external feeding of foliar material by caterpillar-like larvae (166) and consumption of stem and petiole parenchyma by endophytic apodous gallers (180). Meanwhile, there was a continual expansion of lineages bearing adult ectognathate mouthparts, particularly orthopteroids.

Phase 3: Early Permian

During the Early Permian, mouthpart classes increased from six to nine, attributable to the earliest documented appearances of true hemipteroids and holometabolans. Most of these mouthpart classes were innovations for fluid-feeding, whereas others represented new ways of consuming aquatic or terrestrial detritus. Among hemipteroids, the mortar-and-pestle mouthpart class of psocopterans (205) was a major modification of the laciniate condition.

←——

Figure 3 The geochronologic distribution of modern mouthpart classes elucidated in Figure 1, including two extinct mouthpart classes not analyzed in Figure 1 but sufficiently distinctive for inclusion herein. Data is updated from Labandeira (172). *Solid black segments* of vertical bars indicate presence of a mouthpart class as body fossils in well-preserved deposits; *heavy slashed segments* indicate presence based on sister-group relationships when one lineage of a pair occurs as fossils and the sister lineage, whose modern representatives bear the mouthpart class in question, is inferred to have been present. The *lightly slashed segments* indicate more indirect evidence for presence, including trace fossil evidence and the documented occurrence of a mouthpart class in one life-stage of a species (e.g. larva) when the mouthpart class of interest is inferred to have been present in another life-stage (e.g. adult) that lacks a fossil record. Abbreviations: Devon. = Devonian, Carbonif. = Carboniferous, Miss. = Mississippian, Penn. = Pennsylvanian, Perm. = Permian, Trias. = Triassic, Cenoz. = Cenozoic.

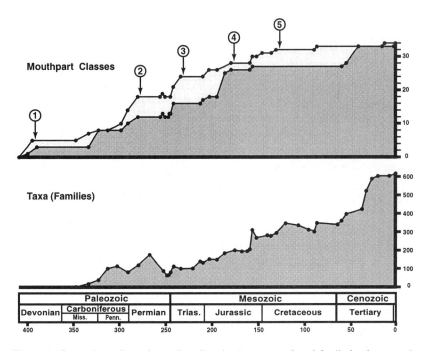

Figure 4 Comparison of mouthpart class diversity (*upper panel*) and family-level taxonomic diversity (*lower panel*) since the Early Devonian. Data for both graphs are resolved at the stage level, shown at bottom sequentially as data points but included in the *top panel* only to reflect diversity changes. For mouthpart diversity, *darker pattern* indicates strong evidence for presence (i.e *black portion* of bars in Figure 3) and *lighter pattern* indicates less reliable evidence (*slashed patterns* of bars in Figure 3). Numbers attached to arrows refer to the five phases of mouthpart class diversification described in the text. Mouthpart data are an updated version of Labandeira (172), and taxonomic data are from Labandeira & Sepkoski (182), documented in Labandeira (174). Abbreviations: Miss. = Mississippian, Penn. = Pennsylvanian, Trias. = Triassic.

Mouthcone mouthparts (221) are found in lophioneurid thysanopterans, which bore an asymmetrical, hypognathously drawn head that apparently bore stylets (344) used for feeding on shallow plant tissues. Prosbolid and archescytinid homopterans with opisthognathous, segmented beaks (16) mark the appearance of segmented beak mouthparts; they probably were fluid feeders on more nutritionally rewarding and deeper-seated plant sap. Larvae of mandibulate planipennians, overwhelmingly fluid feeders on other insects, probably possessed the fossate complex mouthparts of recent forms (362) and were present during the Lower Permian (304). Adult mecopterans or their stem-group (358), which occur in Lower Permian deposits, bore a characteristic hypognathous rostrum with mouthparts attached posterioventrally to the head capsule (119); modern forms are scavengers on detritus and dead insects. Indirect, sister-group evidence is present for buccal cone (124) and tristylate (349) mouthparts, both involved in the ectoparasitic extraction of blood from vertebrates, although it is equally likely that both lineages are more recently derived (121). The evidence for reduced trophic mouthparts (147) is indirect but centers on the presence of Permian coccoid homopterans, of which most modern adult males are nonfeeding. Lastly, presence of aquatically filtering pectinate mouthparts (842) is assumed from Early Permian occurrences (162), although little is known of preadult ephemeropterans during the Permian, and the single well-documented Pennsylvanian form probably was predaceous (164).

Phase 4: Late Triassic to Early Jurassic

During the Late Triassic to Early Jurassic, there was an explosion in mouthpart innovation and feeding strategies, with approximately one third of all modern mouthpart classes coming into existence. Half of these 12 new mouthpart classes are attributable to the origin and ecological diversification of nematocerous Diptera (142, 278). This proliferation of new mouthpart designs included distinctive larval feeding apparatuses and in adults the combination of an elongate protractile labium surmounted by a fleshy labellum that enclosed piercing or cutting stylets of varying number and length (130, 244). Although most of these mouthpart types were designed for fluid feeding, ingestion was accomplished by sponging with a pseudotracheate labellum (291), by piercing and sucking with haustellate stylets (208, 329), or by laterally-to-dorsoventrally occluding falcate mouthhooks in larvae of some advanced lineages (276). Adult diets included vertebrate blood or insect haemolymph for those with hexastylate, distylate/monostylate, and monostylate/distylate mouthparts (112, 141). Labellates imbibed exposed plant, animal, and fungal fluids, including honeydew—polysaccharide containing secretions from extrafloral nectaries (78)—and protein- and lipid-rich exudates from carrion or exposed wounds. Mouthhook-bearing larvae were scavengers or predators principally on other insects. However, dipteran feeding strategies were successful not only on land—in freshwater habitats, modification occurred of labra, mandibles, and maxillary appendages into the penicillate or setate filtering structures of mouthbrush mouthparts (48, 64, 258)—and allowed for exploitation of particulate organic material suspended in the water or as epibenthic detritus (61). Early trichopteran lineages paralleled dipterans in that adults developed a fleshy, labellum-like sponging organ (55) in the form of haustoriate mouthparts, and naiads similarly possessed seta- and hair-bearing mandibulobrustiate mouthparts designed for sieving or filtering particulate organic matter (24). In freshwater environments, a revolution occurred in the trophic partitioning of the water and upper benthic zones, with several diverse lineages of filterers, scrapers, gatherers, and shredders (61, 62) appearing during this time interval (360).

The presence of new food resources was probably attributable to the emergence of several seed plant clades during the Triassic, leading to increased interactions with insects, examples of which include the earliest occurrence of leaf mining (as plant damage—288), surface fluid feeding and undoubtedly pollination. Nondipteran clades that participated in the exploitation of new food resources included the maxilollabiates, mostly diminutive adults of parasitoid Hymenoptera that possessed mandibles and a labiomaxillary apparatus (237) for chewing solid food and lapping fluids. Rhynchophorates appeared during the Late Triassic (368), becoming abundant during the Late Jurassic, and bore a characteristic decurved snout with terminal mandibulate mouthparts (74) designed for masticating hard substrates such as seeds, stem sclerenchyma, and wood. Both maxillolabiates and rhynchophorates could have fed on and pollinated Jurassic seed plants. The tubulomandibulate mouthpart class (345), consisting of specialized fluid-feeding predators on other invertebrates, originated in water initially during the Late Triassic and re-evolved during the Late Jurassic within the terrestrial Cantharoidea (57, 58). During the Triassic and Early Jurassic, several separate origins of small beetles characterized the ectognathous-stylate mouthpart class in which typical adducting/abducting mandibles become transformed into protracting/retracting stylets, with accompanying modifications in labral, maxillary, and labial structures (65, 241).

Phase 5: Late Jurassic to Early Cretaceous

Almost all of the innovation in major mouthpart design occurred before the Cretaceous. Phase 4 of the earlier Mesozoic had broadened the repertoire of feeding to include noninvasive feeding on surface fluids by several mouthpart types. This was extended during the later Mesozoic to include the siphonate (157, 177) and glossate (220) mouthpart classes, which involved novel mechanisms in the imbibition of surface fluids (83, 218), becoming crucial during the Late Cretaceous by contributing members to major pollination syndromes with flowering plants (53, 256). By the Eocene, faithful modes of pollination were principally centered on these two speciose mouthpart types. The documentation of nontrophic mouthparts in the fossil record is spotty because it is difficult to ascertain even in modern insects whether vestigial mouthparts are capable of securing food, whether adults are truly nonfeeding or simply have not been observed to feed. Vestigial mouthparts occur sporadically in ephemeropterans (226), lepidopterans, and dipterans, and circumstantial evidence indicates that this mouthpart class originated during the Late Cretaceous, although it could have occurred significantly earlier for ephemeropterans.

Three other mouthpart classes have inadequately understood histories because their fossil records are missing or poorly documented. Siphonomandibulate mouthparts, in which maxillary galeae or palps are fused, originated independently among several genera, including meloid (111) and rhipiphorid (356) beetles, and probably certain bees with tubular maxillary proboscides (129, 185). Among three mid-Cenozoic occurrences of rhipiphorids (172), it is not known if they are siphonomandibulates; based on other evidence, this mouthpart class probably extends to the mid-Cretaceous. The fossil record of the tubulostylate mouthpart class is better, since glossinids occurred during the latest Eocene (45), and an Early Miocene hippoboscid is known (199). Siphonostylates consisting of multiple innovations of skin-piercing noctuid moths that puncture fruit (322, 361) and the integument of large, even pachydermous, mammals (9) are the only mouthpart class that probably originated during the Cenozoic.

CONCLUSIONS

Insect mouthparts and associated head structures consist of distinctive, highly integrated structural units containing single and multiple elements that are coordinated in processing food. For foods different than those of generalized, solid-food feeding mandibulate insects, many subsequently originating mouthpart types broadened the spectrum of food available for consumption, such as organic particles and both exposed and tissue-bound fluids. These innovations involved significant multielement co-optation and modification between two or more mouthpart regions, resulting in major mouthpart types engaged in aquatic filter-feeding, piercing-and-sucking, sponging, siphoning, and lapping, among others.

The fossil insect mouthpart record occurs sporadically in approximately 40 exceptionally well-preserved deposits. This record provides detail for minimal dates of origin and the timing of diversification for the vast majority of modern mouthpart types; it also includes documentation of unique Paleozoic mouthpart types lacking modern analogs. Supplementing this body-fossil record is the trace-fossil record of mouthpart-mediated plant damage, insect gut and coprolitic contents, and floral and other plant features indicative of particular insect feeding styles.

In addition to the above approach centered on the fossil record, an objective classification of modern insect mouthparts into formal mouthpart classes was made, in which each mouthpart class was mapped onto its constituent lineage or lineages, calibrated by fossil occurrences. From a summary cluster analysis of modern insect mouthparts, 34 structurally defined mouthpart classes of insects and their apterygote relatives were recognized; most of these mouthpart classes have multiple origins among clades not sharing common ancestors.

The geochronologic record of mouthpart innovation, when supplemented by data on the effects of mouthparts on plants, indicates a fivefold phase in the evolution of feeding strategies, with all documentable mouthpart classes in the fossil record antedating the mid-Cretaceous. These five phases of mouthpart class origination—Early Devonian, Mississippian, Early Permian, Late Triassic to Early Jurassic, and Late Jurassic to Early Cretaceous—are characterized by successive partitioning of freshwater and especially terrestrial food resources, and a trend from solid-food feeding toward exposed- and tissue-bound fluid feeding and particle feeding. Major mouthpart innovations include a proliferation of mostly herbivorous mouthpart classes during the Pennsylvanian, expansion of fluid-feeding strategies during the Early Permian, a dramatic increase in several fluid-feeding and particle-capturing mouthpart classes associated with the radiation of the Holometabola, particularly Diptera, during the Late Triassic and Early Jurassic, and further expansion of surface-fluid feeding strategies during the Late Jurassic and Early Cretaceous. Many earlier Mesozoic mouthpart classes that were engaged in fluid-feeding subsequently became important as associates of angiosperms during the mid-Cretaceous.

An important consequence of a phenetic summary of the profuse literature of modern insect mouthparts is condensation of a wealth of morphological data into 34 basic entities. The macroevolutionary distribution of these 34 mouthpart classes is made meaningful only when they can be plotted on cladistic phylogenies of lineages bearing those mouthpart classes. This review suggests that one of the most important uses of cladograms is to provide an objective phylogenetic framework by which ecological attributes can be compared, particularly when both phylogenies and ecological data are anchored in a known fossil record.

ACKNOWLEDGMENTS

Special thanks go to Finnegan Marsh for generating the figures accompanying this article. This article benefitted from review by several colleagues including Roy Crowson, Jarmila Kukalová-Peck and James Pakaluk. During the past decade the interlibrary loan staffs of the University of Chicago, University of Illinois at Urbana-Champaign, and the National Museum of Natural History were indefatigable in securing articles for my dissertation and this review. The Scholarly Studies Program of the Smithsonian Institution is acknowledged for contributing to this research. This is contribution No. 54 from the Evolution of Terrestrial Ecosystems at the National Museum of Natural History.

Literature Cited

1. Ananthakrishnan TN, ed. 1984. *Biology of Gall Insects.* New Delhi: Oxford & IBH. 362 pp.
2. Archangelsky S, Cuneo R. 1987. Feugliocladaceae, a new conifer family from the Permian of Gondwana. *Rev. Palaeobot. Palynol.* 51:3–30
3. Arens W. 1994. Striking convergence in the mouthpart evolution of stream-living algae grazers. *J. Zool. Syst. Evol. Res.* 32:319–43
4. Armstrong JE, Irvine AK. 1990. Functions of staminodia in the beetle-pollinated flowers of *Eupomatia laurina. Biotropica* 22:429–31
5. Arnol'di LV 1977a. Introduction. See Ref. 7, pp. 1–18
6. Arnol'di LV 1977b. Eobelidae. See Ref. 7, pp. 195–241
7. Arnol'di LV, Zherikhin VV, Nikritin LM, Ponomarenko AG. 1977. *Mesozoic Coleoptera.* Transl. NJ Vandenberg, 1992 (from Russian). Washington, DC: Smithson. Inst. Libraries, Natl. Sci. Found. 285 pp.
8. Banks HP, Colthart BJ. 1993. Plant-animal-fungal interactions in Early Devonian trimerophytes from Gaspé, Canada. *Am. J. Bot.* 80:992–1001
9. Bänziger H. 1980. Skin-piercing blood-sucking moths. III: Feeding act and piercing mechanism of *Calyptra eustrigata* (Hmps.) (Lep., Noctuidae). *Mitt. Schweiz. Entomol. Ges.* 53:127–42
10. Barthel KW, Swinburne NHM, Conway Morris S. 1994. *Solnhofen: A Study in Mesozoic Palaeontology.* Cambridge, UK: Cambridge Univ. Press. 236 pp.
11. Basinger JF, Dilcher DL. 1984. Ancient bisexual flowers. *Science* 224:511–13
12. Batra LR, ed. 1978. *Insect-Fungus Symbiosis, Nutrition, Mutualism, and Commensalism.* New York: Wiley. 276 pp.
13. Bauer T, Pfeiffer M. 1991. "Shooting" springtails with a sticky rod: the flexible hunting behaviour of *Stenus comma* (Coleoptera; Staphylinidae) and the counter-strategies of its prey. *Anim. Behav.* 41:819–28
14. Baxendale RW. 1979. Plant-bearing coprolites from North American Pennsylvanian coal balls. *Palaeontology* 22:537–48
15. Beck AL, Labandeira CC, Mamay SH. 1996. Host spectrum and intensity of insect herbivory on a Lower Permian riparian flora: implications for the early sequestering of vascular plant tissues. *Geol. Soc. Am. Abstr. Prog.* 28:105 (Abstr.)

16. Becker-Migdisova EE. 1940. Fossil Permian cicadas of the family Prosbolidae from the Soyana River. *Trans. Paleontol. Inst.* 11:1–79 (In Russian)
17. Becker-Migdisova EE. 1968. Protopsyllidiids and their morphology (Homoptera, Protopsyllidiidae). See Ref. 279, pp. 87–98
18. Becker-Migdisova EE. 1985. Fossil insect Psyllomorpha. *Trans. Paleontol. Inst.* 206:1–93 (In Russian)
19. Bennett A, Toms RB. 1995. Sexual dimorphism in the mouthparts of the king cricket *Libanasidus vittatus* (Kirby) (Orthoptera: Mimnermidae). *Ann. Transvaal Mus.* 36:205–14
20. Beutel RG. 1993. Phylogenetic analysis of Adephaga (Coleoptera) based on characters of the larval head. *Syst. Entomol.* 18:127–47
21. Beutel RG, Roughley RE. 1988. On the systematic position of the family Gyrinidae (Coleoptera: Adephaga). *Z. Zool. Syst. Evol.-forsch.* 26:380–400
22. Bitsch J. 1963. Morphologie cephalique des machilides (Insecta, Thysanura). *Ann. Sci. Nat. (Zool.)* 5:585–706
23. Borkent A. 1995. *Biting Midges in the Cretaceous Amber of North America (Diptera, Ceratopogonidae).* Leiden: Backhuys. 237 pp.
24. Botosaneanu L. 1956. Contributiona à la connaissance des stades aquatiques des Trichoptères crénobiontes: *Rhyacophila laevis* Pict., *Wormaldia triangulifera* MacLachl., *Drusus romanicus* Murg. & Bots., *Silo varipilosa* Bots. (Trichoptera). *Beitr. Entomol.* 6:590–624
25. Botosaneanu L, Wichard W. 1983. Upper Cretaceous Siberian and Canadian amber caddisflies (Insecta: Trichoptera). *Bijdr. Dierk.* 53:187–217
26. Braimah SA. 1987. Mechanisms of filter feeding in immature *Simulium vittatum* Malloch (Diptera: Simuliidae) and *Isonychia campestris* McDunnough (Ephemeroptera: Oligoneuriidae). *Can. J. Zool.* 65: 504–13
27. Braimah SA. 1987. Pattern of flow around filter feeding structures of immature *Simulium bivittatum* Malloch (Diptera: Simuliidae) and *Isonychia campestris* McDunnough (Ephemeroptera: Oligoneuriidae). *Can. J. Zool.* 65:514–21
28. Brauckmann C. 1988. Hagen-Vorhalle, a new important Namurian Insecta-bearing locality (Upper Carboniferous; FR Germany). *Entomol. Gen.* 14:73–79

29. Brauckmann C, Zessin W. 1989. Neue Meganeuridae aus dem Namurium von Hagen-Vorhalle (BRD) und die Phylogenie der Meganisoptera (Insecta, Odonata). *Deut. Entomol. Z.* 36:177–215

30. Brues CT. 1972. *Insect, Food, and Ecology*. New York: Dover. 466 pp.

31. Bürgis H. 1981. Beitrag zur Morphologie des Kopfes der Imago von *Cetonia aurata* L. (Coleoptera, Insecta). *Zool. Jb. (Anat.)* 106:186–220

32. Burnham L. 1986. *Revisionary Studies on Upper Carboniferous Insects in the Order Protorthoptera*. PhD diss. Cornell Univ., Ithaca, NY. 178 pp.

33. Bystrak PG, Wirth WW. 1978. The North American species of *Forcipomyia*, subgenus *Euprojoannisia* (Diptera: Ceratopogonidae). *US Dept. Agric. Tech. Bull.* 1591:1–51

34. Caldas MB, Martins-Neto RG, Lima-Filho FP. 1989. *Afropollis* sp. (polém) no trato intestinal de vespa (Hymenoptera: Apocrita: Xyelidae) no Cretáceo da Bacia do Araripe. *Actas Simp. Geol. Nordeste* 13:195–96 (Abstr.)

35. Carle FL, Wighton DC. 1990. Odonata. See Ref. 108, pp. 51–68

36. Carpenter FM. 1931. The Lower Permian insects of Kansas. Part 4. The order Hemiptera, and additions to the Paleodictyoptera and Protohymenoptera. *Am. J. Sci.* (5)22:113–30

37. Carpenter FM. 1933. The Lower Permian insects of Kansas. Part 6. Delopteridae, Protelytroptera, Plectoptera and a new collection of Protodonata, Odonata, Megasecoptera, Homoptera, and Psocoptera. *Proc. Am. Acad. Arts Sci.* 68: 411–503

38. Carpenter FM. 1938. Structure of Permian Homoptera and Psocoptera. *Nature* 141:164–65

39. Carpenter FM. 1971. *Proc. North Am. Paleontol. Conv., 1st, 1968,* 2:1236–51. Lawrence, KS: Allen Press

40. Carpenter FM. 1992. Volume 3: *Superclass Insecta*. In *Treatise on Invertebrate Paleontology, Part R, Arthropoda 4*, ed. RC Moore, RL Kaesler, E Brosius, J Keim, J Priesner. Boulder, CO: Geol. Soc. Am.; Lawrence, KS: Univ. Kansas Press. 665 pp.

41. Carpenter FM, Burnham L. 1985. The geological record of insects. *Annu. Rev. Earth Planet. Sci.* 13:297–314

42. Carpenter FM, Richardson ES Jr. 1976. Structure and relationships of the Upper Carboniferous insect, *Eucaenus ovalis* (Protorthoptera: Eucaenidae). *Psyche* 83: 223–42

43. Chaudonneret J. 1990. *Le Pièces Buccales des Insectes: Thème et Variations.* Dijon: Berthier. 256 pp.

44. Cicero JM. 1994. Composite, haustellate mouthparts in netwinged beetle and firefly larvae (Coleoptera, Cantharoidea: Lycidae, Lampyridae). *J. Morphol.* 219:183–92

45. Cockerell TDE. 1907. A fossil tsetse fly in Colorado. *Nature* 76:414

46. Cornet B. 1989. The reproductive morphology and biology of *Sanmiguelia lewisii*, and its bearing on angiosperm evolution in the Late Triassic. *Evol. Trends Plants* 3:25–51

47. Cornet B. 1996. A new gnetophyte from the Late Carnian (Late Triassic) of Texas and its bearing on the origin of the angiosperm carpel and stamen. In *Flowering Plant Origin, Evolution and Phylogeny*, ed. DW Taylor, LJ Hickey, pp. 32–67. New York: Chapman & Hall

48. Craig DA, Chance MM. 1982. Filter feeding in larvae of Simuliidae (Diptera: Culicomorpha): aspects of functional morphology and hydrodynamics. *Can. J. Zool.* 60:712–24

49. Crampton GC. 1925. A phylogenetic study of the labium of holometabolous insects, with particular reference to the Diptera. *Proc. Entomol. Soc. Washington* 27:68–91

50. Crepet WL. 1972. Investigations of North American cycadeoids: pollination mechanisms in *Cycadeoidea*. *Am. J. Bot.* 59: 1048–56

51. Crepet WL. 1979. Some aspects of the pollination biology of Middle Eocene angiosperms. *Rev. Palaeobot. Palynol.* 27: 213–38

52. Crepet WL, Friis EM. 1987. The evolution of insect pollination in angiosperms. In *The Angiosperms and their Biological Consequences*, ed. EM Friis, WG Chaloner, PR Crane, pp. 181–201. Cambridge, UK: Cambridge Univ. Press

53. Crepet WL, Nixon KC. 1996. The fossil history of stamens. In *The Anther: Form, Function and Phylogeny*, ed. WG D'Arcy, RC Keating, pp. 25–57. Cambridge, UK: Cambridge Univ. Press

54. Crepet WL, Taylor DW. 1985. Diversification of the Leguminosae: first fossil evidence of the Mimosoideae and Papilionoideae. *Science* 228:1087–89

55. Crichton MI. 1957. The structure and function of the mouthparts of adult caddis flies (Trichoptera). *Philos. Trans. R. Soc. London (B)* 241:45–91

56. Crosskey RW. 1994. The fossil pupa *Simulimima* and the evidence it provides

for the Jurassic origin of the Simuliidae (Diptera). *Syst. Entomol.* 16:401–06

57. Crowson RA. 1972. A review of the classification of Cantharoidea (Coleoptera), with the definition of two new families, Cneoglossidae and Omethidae. *Rev. Univ. Madrid* 21:35–77

58. Crowson RA. 1981. *The Biology of the Coleoptera.* New York: Academic. 802 pp.

59. Crowson RA. 1990. A new genus of Boganiidae (Coleoptera) from Australia, with observations on glandular openings, cycad associations and geographical distribution in the family. *J. Austral. Entomol. Soc.* 29:91–99

60. Crowson RA. 1991. The relations of Coleoptera to Cycadales. In *Advances in Coleopterology,* ed. M Zunino, X Bellés, M. Blas, pp. 13–28. Barcelona: Eur. Assn. Coleopterol.

61. Cummins KW, Klug MJ. 1979. Feeding ecology of stream invertebrates. *Annu. Rev. Ecol. Syst.* 10:147–72

62. Cummins KW, Merritt RW. 1984. Ecology and distribution of aquatic insects. In *An Introduction to the Aquatic Insects,* ed. RW Merritt, KW Cummins, pp. 59–65. Dubuque, IA: Kendall-Hunt. 2nd ed.

63. Dafni A, Bernhardt P, Shmida A, Ivri Y, Greenbaum S, et al. 1990. Red bowl-shaped flowers: convergence for beetle pollination in the Mediterranean Region. *Israel J. Bot.* 39:81–92

64. Dahl C, Widahl LE, Nilsson C. 1988. Functional analysis of the suspension feeding system in mosquitoes (Diptera: Culicidae). *Ann. Entomol. Soc. Am.* 81:105–27

65. Dajoz R. 1976. Les Coléoptères Cerylonidae. Etude des espèces de la fauna Paléartique. *Bull. Mus. Natl. Hist. Nat. (3)* 360:249–81

66. De Marzo L. 1978. Studio per fini sistematici del comportamento dei caratteri delle mandibole nelle larve di alcune specie della subf. Colymbetinae (Coleoptera, Dytiscidae). *Atti Congr. Naz. Ital. Entomol.* 11:73–78

67. Denis JR, Bitsch J. 1973. Morphologie de tête des insectes. In *Traité de Zoologie: Anatomie, Systematique, Biologie,* ed. P-P Grassé. 8(1):1–593. Paris: Masson

68. Dennell R. 1942. The structure and function of the mouth-parts, rostrum and foregut of the weevil *Calandra granaria* L. *Philos. Trans. R. Soc. London (B)* 231:247–91

69. Diéguez C, Nieves-Aldrey JL, Barrón E. 1996. Fossil galls (zoocecids) from the Upper Miocene of La Cerdaña (Lérida, Spain). *Rev. Palaeobot. Palynol.* 94:329–43

70. Dilcher DL. 1979. Early angiosperm reproduction: an introductory report. *Rev. Palaeobot. Palynol.* 27:291–328

71. Dilcher DL. 1996. Plant reproductive strategies: using the fossil record to unravel current issues in plant reproduction. *Univ. Fla. Contr. Paleobiol.* 425. In press

72. Dilcher DL, Crane PR. 1984. *Archaeanthus*: an early angiosperm from the Cenomanian of the western interior of North America. *Ann. Missouri Bot. Gard.* 71:351–83

73. Doludenko MP, Orlovskaya ER. 1976. Jurassic floras of the Karatau Range, southern Kazakhstan. *Palaeontology* 19: 627–40

74. Dönges J. 1954. Der Kopf von *Cionus scrophulariae* L. (Curculionidae). *Zool. Jb. (Anat.)* 74:1–76

75. Downes JA. 1955. The food habits and description of *Atrichopogon pollinivorus* sp. n. (Diptera: Ceratopogonidae). *Trans. R. Entomol. Soc. London* 106:439–53

76. Downes JA. 1971. The ecology of blood-sucking Diptera: an evolutionary perspective. In *Ecology and Physiology of Parasites,* ed. AM Fallis, pp. 232–58. London: Univ. Toronto Press

77. Downes JA. 1974. The feeding habits of adult Chironomidae. *Entomol. Tidskr.* 95 Suppl:84–90

78. Downes WL Jr, Dahlem GA. 1987. Keys to the evolution of Diptera: role of Homoptera. *Environ. Entomol.* 16:847–54

79. Doyle JA, Hotton CL, Ward JV. 1990. Early Cretaceous tetrads, zonasulculate pollen, and Winteraceae. II. Cladistic analysis and implications. *Am. J. Bot.* 77:1558–68

80. Driscoll CA, Condon MA. 1994. Labellar modifications of *Blepharoneura* (Diptera: Tephritidae): neotropical fruit flies that damage and feed on plant surfaces. *Ann. Entomol. Soc. Am.* 87:448–53

81. Durden CJ. 1978. A dasyleptid from the Permian of Kansas, *Lepidodasypus sharovi* n. gen., n. sp. (Insecta: Thysanura: Monura). *Pearce-Sellards Ser., Texas Mem. Mus.* 30:1–9

82. Durden CJ, Rose H. 1978. Butterflies from the middle Eocene: the earliest occurrence of fossil Papilionoidea (Lepidoptera). *Pearce-Sellards Ser., Texas Mem. Mus.* 19:1–25

83. Eastham LES, Eassa YEE. 1955. The feeding mechanism of the butterfly *Pieris brassicae* L. *Philos. Trans. R. Soc. London (B)* 239:1–43

84. Elpers C, Tomka I. 1992. Struktur der Mundwerkzeuge und Nahrungsaufnahme bei den Larven von *Oligoneuriella rhenana* Imhoff (Ephemeroptera: Oligoneuriidae). *Mitt. Schweiz. Entomol. Ges.* 65:119–39

85. Elzinga RJ, Broce AB. 1986. Labellar modifications of Muscomorpha flies (Diptera). *Ann. Entomol. Soc. Am.* 79:150–209

86. Endress PK. 1986. Reproductive structures and phylogenetic significance of extant primitive angiosperms. *Plant Syst. Evol.* 152:1–28

87. Endress PK. 1987. The Chloranthaceae: reproductive structures and phylogenetic position. *Bot. Jahrb. Syst.* 109:153–226

88. Endress PK. 1990. Evolution of reproductive structures and functions in primitive angiosperms (Magnoliidae). *Mem. New York Bot. Gard.* 55:5–34

89. Endrödy-Younga S, Crowson RA. 1986. Boganiidae, a new beetle family for the African fauna (Coleoptera: Cucujoidea). *Ann. Transvaal Mus.* 34:253–73

90. Evans AV, Nel A. 1989. Notes on *Macrocyphonistes kolbeanus* Ohaus and *Rhizoplatys auriculatus* (Burmeister), with comments on their melittophilous habits (Coleoptera: Melolonthidae: Dynastinae: Phileurini). *J. Entomol. Soc. So. Afr.* 52:45–50

91. Evans AW. 1921. On the structure and occurrence of maxillulae in the orders of insects. *J. Linn. Soc., Zool.* 34:429–56

92. Evans HE. 1973. Cretaceous aculeate wasps from Taimyr, Siberia (Hymenoptera). *Psyche* 80:166–78

93. Faegri K, Van der Pijl L. 1980. *The Principles of Pollination Ecology.* Oxford: Pergamon. 244 pp. 3rd ed.

94. Feil JP. 1992. Reproductive ecology of dioecious *Siparuna* (Monimiaceae) in Ecuador—a case of gall midge pollination. *Bot. J. Linn. Soc.* 110:171–203

95. Forster PI, Machin PJ, Mound L, Wilson GW. 1994. Insects associated with reproductive structures of cycads in Queensland and northeast New South Wales, Australia. *Biotropica* 26:217–22

96. François J. 1969. Anatomie et morphologie cephalique des protoures (Insecta Apterygota). *Mem. Mus. Natl. Hist. Nat.* 59:1–144

97. Fraser NC, Grimaldi DA, Olsen PC, Axsmith B. 1996. A Triassic Lagerstätte from eastern North America. *Nature* 380:615–20

98. Gapud VP. 1968. The external morphology of the head and mouthparts of some Philippine Orthoptera. *Philip. Entomol.* 1:11–32

99. Gazit S, Galon I, Podoler H. 1982. The role of nitidulid beetles in natural pollination of *Annona* in Israel. *J. Am. Soc. Hort. Sci.* 107:849–52

100. Gibbs PE, Semir J, Da Cruz ND. 1977. The development of the flower in *Talauma ovata* and the behaviour of pollinating beetles. *Ciên. Cult.* 29:1437–41

101. Gilbert FS. 1981. Foraging ecology of hoverflies: morphology of the mouthparts in relation to feeding on nectar and pollen in some common urban species. *Ecol. Entomol.* 6:245–62

102. Givulescu R. 1984. Pathological elements on fossil leaves from Chuizbaia (galls, mines and other insect traces). *Dari Seama Sedint.* 68:123–33

103. Goto HE. 1972. On the structure and function of the mouthparts of the soil-inhabiting collembolan *Folsomia candida. Biol. J. Linn. Soc.* 4:147–68

104. Gottsberger G. 1974. The structure and function of the primitive angiosperm flower—a discussion. *Acta Bot. Neerl.* 23:461–71

105. Gottsberger G. 1988. The reproductive biology of primitive angiosperms. *Taxon* 37:630–43

106. Graham-Smith GS. 1930. Further observations on the anatomy and function of the proboscis of the blow-fly, *Calliphora erythrocephala* L. *Parasitology* 22:47–115

107. Grimaldi DA. 1989. The genus *Metopina* (Diptera: Phoridae) from Cretaceous and Tertiary ambers. *J. N.Y. Entomol. Soc.* 97:65–72

108. Grimaldi DA, ed. 1990. Insects from the Santana Formation, Lower Cretaceous, of Brazil. *Bull. Am. Mus. Nat. Hist.* 195:1–191

109. Grimaldi DA. 1991. The Santana Formation insects. In *Santana Fossils: An Illustrated Atlas,* ed. JG Maisey, pp. 378–406. Neptune City, NJ: TFH Publ.

110. Grimaldi DA, Bonwich E, Dellanoy M, Doberstein W. 1994. Electron microscopic studies of mummified tissues in amber fossils. *Am. Mus. Novit.* 3097:1–31

111. Grinfel'd EK. 1975. Anthophily in beetles (Coleoptera) and a critical evaluation of the cantharophilous hypothesis. *Entomol. Rev.* 54:18–22

112. Grogan WL Jr, Szadziewski R. 1988. A new biting midge from Upper Cretaceous (Cenomanian) amber of New Jersey (Diptera, Ceratopogonidae). *J. Paleontol.* 62:808–12

113. Guo S-x. 1991. A Miocene trace fossil of insect from Shanwang Formation in Linqu, Shandong. *Acta Palaeontol. Sinica* 30:739–42

114. Haeger JS. 1955. The non-blood feeding habits of *Aedes taeniorhynchus* (Diptera, Culicidae) on Sanibel Island, Florida. *Mosquito News* 15:21–26

115. Hamilton KGA. 1990. Homoptera. See Ref. 108, pp. 82–122

116. Hamilton KGA. 1996. Cretaceous Homoptera from Brazil: implications for classification. In *Studies on Hemipteran Phylogeny,* ed. CW Schaefer, pp. 89–110. Lanham, MD: Entomol. Soc. Am.

117. Hannappel U, Paulus HF. 1991. Some undetermined Helodidae larvae from Australia and New Zealand: fine structure of mouthparts and phylogenetic position. In *Advances in Coleopterology,* ed. M Zunino, X Bellés, M Blas, pp. 89–127. Barcelona: Eur. Assoc. Coleopterol.

118. Hartigan J. 1988. Program 1M. Cluster analysis of variables. In *BMDP Statistical Software Manual,* ed. WJ Dixon, MB Brown, L Engelman, MA Hill, RI Jennrich, 2:745–54. Los Angeles: Univ. Calif. Press. 2nd ed.

119. Heddergott H. 1938. Kopf und Vorderdarm von *Panorpa communis* L. *Zool. Jb. (Anat.)* 64:229–94

120. Hennig W. 1972. Insektenfossilien aus der unteren Kreide. IV. Psychodidae (Phlebotominae), mit einer kritischen Ubersicht über das phylogenetische System der Familie und die bisher beschriebenen Fossilien (Diptera). *Stutt. Beit. Naturk.* 241:1–69

121. Hennig W. 1981. *Insect Phylogeny.* New York: Wiley. 514 pp.

122. Hering EM. 1951. *Biology of the Leaf Miners.* 's-Gravenhage: W. Junk. 420 pp.

123. Hespenheide HA. 1985. Insect visitors to extrafloral nectaries of *Byttneria aculeata* (Sterculiaceae): relative importance and roles. *Ecol. Entomol.* 10:191–204

124. Hirsch F. 1986. Die Mundwerkzeuge von *Phthirus pubis* L. (Anoplura). *Zool. Jb. (Anat.)* 114:167–204

125. Hirst S, Maulik S. 1926. On some arthropod remains from the Rhynie Chert (Old Red Sandstone). *Geol. Mag.* 63:69–71

126. Holloway BA. 1976. Pollen-feeding in hover-flies (Diptera: Syrphidae). *N. Zeal. J. Zool.* 3:339–50

127. Hong Y-c, Wang W-l. 1990. Fossil insects from the Laiyang Basin, Shandong Province. In *The Stratigraphy and Palaeontology of Laiyang Basin, Shandong Province,* ed. Regional Geol. Surveying Team, pp. 44–189. Shandong: Bur. Geol. Mineral Resourc.

128. Hotton CL, Hueber FM, Labandeira CC. 1996. Plant-arthropod interactions from early terrestrial ecosystems: two Devonian examples. *Paleontol. Soc. Spec. Publ.* 8:181 (Abstr.)

129. Houston TF. 1983. An extraordinary new bee and adaptation of palpi for nectarfeeding in some Australian Colletidae and Pergidae (Hymenoptera). *J. Aust. Entomol. Soc.* 22:263–70

130. Hoyt CP. 1952. The evolution of the mouth parts of adult Diptera. *Microentomology* 17:61–125

131. Iwasa M. 1983. A comparative study on the mouth parts of medically and veterinarily important flies, with special reference to the development and origin of the prestomal teeth in cyclorrhaphous Diptera. *Jpn. J. Sanit. Zool.* 34:177–206

132. Jaeger P. 1976. Les rapports mutuels entre fleurs et insectes. In *Traité de Zoologie,* ed. P-P Grassé, 8(4):673–933. Paris: Masson

133. James MT, Harwood RF. 1969. The feeding apparatus of insects and Acarina. In *Herm's Medical Entomology,* pp. 24–38. London: Macmillan. 6th ed.

134. Jarzembowski EA. 1990. A boring beetle from the Wealden of the Weald. In *Evolutionary Paleobiology of Behavior and Coevolution,* ed. AJ Boucot, pp. 373–76. Amsterdam: Elsevier

135. Jell PA, Duncan PM. 1986. Invertebrates, mainly insects, from the freshwater, Lower Cretaceous, Koonwarra fossil bed (Korumburra Group), South Gippsland, Victoria. *Mem. Assoc. Australas. Paleontol.* 3:111–205

136. Jervis MA, Kidd NAC, Fitton MG, Huddleston T, Dawah HA. 1993. Flower-visiting by hymenopteran parasitoids. *J. Nat. Hist.* 27:67–105

137. Jobling B. 1933. A revision of the structure of the head, mouth-part and salivary glands of *Glossina palpalis* Rob.-Desv. *Parasitology* 24:449–90

138. Johnson WT, Lyon HH. 1991. *Insects That Feed on Trees and Shrubs.* Ithaca, NY: Cornell Univ. Press. 560 pp.

139. Kalugina NS. 1986. Muscida (= Diptera). Tipulomorpha et Culicomorpha. See Ref. 261, pp. 112–25

140. Kalugina NS. 1989. New Mesozoic psychodomorph dipteran insects from Siberia (Diptera Eoptychopteridae, Ptychopteridae). *Palaeontol. Zh.* 1989(1):65–77 (In Russian)

141. Kalugina NS. 1992. New Mesozoic Simuliidae and Leptoconopidae and the origin

of bloodsucking in the lower dipteran insects. *Paleontol. J.* 1991(1):66–77

142. Kalugina NS, Kovalev VG. 1985. *Dipterous Insects of Jurassic Siberia.* Moscow: Acad. Sci. 198 pp.

143. Kato M, Inoue T. 1994. Origin of insect pollination. *Nature* 368:195

144. Kevan PG, Chaloner WG, Savile DBO. 1975. Interrelationships of early terrrestrial arthropods and plants. *Palaeontology* 18:391–417

145. Kim KC. 1988. Evolutionary parallelism in Anoplura and eutherian mammals. *Syst. Assn. Spec. Vol.* 37:91–114

146. Kingsolver JG, Daniel TL. 1979. On the mechanics and energetics of nectar feeding in butterflies. *J. Theor. Biol.* 76:167–79

147. Kinzelbach RK. 1966. Zur Kopfmorphologie der Facherflügler (Strepsiptera, Insecta). *Zool. Jb. (Anat.)* 84:559–684

148. Kirk WDJ. 1984. Pollen-feeding in thrips (Insecta: Thysanoptera). *J. Zool.* 204:107–17

149. Kononova EL. 1977. New aphid species (Homoptera, Aphidinea) from Upper Cretaceous deposits of the Taymyr. *Entomol. Rev.* 56:72–80

150. Korn W. 1943. Die Muskulatur des Kopfes und des Thorax von *Myrmeleon europaeus* und ihre Metamorphose. *Zool. Jb. (Anat.)* 68:273–330

151. Kovalev VG. 1982. The oldest representatives of the Diptera with short antennae from the Jurassic in Siberia. *Paleontol. J.* 1981(3):84–100

152. Kovalev VG. 1982. Some Jurassic Diptera-rhagionids (Muscida, Rhagionidae). *Paleontol. J.* 1982(3):87–99

153. Kovalev VG. 1988. New Mesozoic mycetophiloid Diptera of the family Pleciofungivoridae. *Paleontol. J.* 1987(2):67–79

154. Kovalev VG. 1990. Dipterans. Muscida. See Ref. 263, pp. 123–77

155. Kovalev VG. 1990. Eremochaetidae, the Mesozoic family of brachycerous dipterans. *Paleontol. J.* 1990(2):100–05

156. Kozlov MV. 1988. Paleontology of lepidopterans and problems in the phylogeny of the order Papilionida. In *The Cretaceous Biocoenotic Crisis in the Evolution of Insects,* ed. AG Ponomarenko, pp. 16–69. Moscow: Acad. Sci. (In Russian)

157. Kozlov MV. 1989. New Upper Jurassic and Lower Cretaceous Lepidoptera (Papilionida). *Paleontol. J.* 23:34–39

158. Krassilov VA, Rasnitsyn AP. 1983. A unique find: pollen in the intestine of Early Cretaceous sawflies. *Paleontol. J.* 1982(4):80–95

159. Krassilov VA, Sukacheva ID. 1979. Cad-

disfly cases from *Karkenia* seeds (Gingkophyta) in Lower Cretaceous deposits of Mongolia. *Trans. Soil Biol. Inst.* 53:119–21 (In Russsian)

160. Krzeminska E, Krzeminski W. 1992. *Les Fantomes de l'Ambre: Insectes Fossiles dans l'Ambre de la Baltique.* Neuchâtel: Musée d'Histoire Naturelle de Neuchâtel. 142 pp.

161. Kühne KS. 1992. Morphologische Beschreibung und Funktionsmechanismus des Proboscis von *Coenosia tigrina* Fab. (Muscidae). *Zool. Beit.* 34:113–25

162. Kukalová-Peck J. 1968. Permian mayfly nymphs. *Psyche* 75:310–27

163. Kukalová-Peck J. 1983. New Homoiopteridae (Insecta: Paleodictyoptera) with wing articulation from Upper Carboniferous strata of Mazon Creek, Illinois. *Can. J. Zool.* 61:1670–87

164. Kukalová-Peck J. 1985. Ephemeroid wing venation based upon new gigantic Carboniferous mayflies and basic morphology, phylogeny, and metamorphosis of pterygote insects (Insecta, Ephemeroptera). *Can. J. Zool.* 63:933–55

165. Kukalová-Peck J. 1987. New Carboniferous Diplura, Monura, Thysanura, the hexapod ground plan, and the role of thoracic side lobes in the origin of wings (Insecta). *Can. J. Zool.* 65:2327–45

166. Kukalová-Peck J. 1991. Fossil history and the evolution of hexapod structures. In *The Insects of Australia,* ed. ID Naumann, PB Carne, JF Lawrence, ES Nielsen, JP Spradbery, et al, 1:141–79. Ithaca, NY: Cornell Univ. Press. 2nd ed.

167. Kukalová-Peck J. 1992. The "Uniramia" do not exist: the ground plan of the Pterygota as revealed by Permian Diaphanopterodea from Russia (Insecta: Paleodictyopteroidea). *Can. J. Zool.* 70:236–55

168. Kukalová-Peck J, Brauckmann C. 1992. Most Paleozoic Protorthoptera are ancestral hemipteroids: major wing braces as clues to a new phylogeny of Neoptera (Insecta). *Can. J. Zool.* 70:2452–73

169. Kukalová-Peck J, Sinichenkova ND. 1992. The wing venation and systematics of the Lower Permian Diaphanopterodea from the Ural Mountains, Russia (Insecta: Paleoptera). *Can. J. Zool.* 70:229–35

170. Kuschel G. 1983. Past and present of the relict family Nemonychidae (Coleoptera, Curculionoidea). *Geojournal* 7:499–504

171. Kuschel G, Oberprieler RG, Rayner RJ. 1994. Cretaceous weevils from southern Africa, with description of a new genus and species and phylogenetic and zoogeographical comments (Coleoptera:

Curculionoidea). *Entomol. Scand.* 25: 137–49

172. Labandeira CC. 1990. *Use of a phenetic analysis of Recent hexapod mouthparts for the distribution of hexapod food resource guilds in the fossil record.* PhD diss. Univ. Chicago. 1186 pp.

173. Labandeira CC. 1991. The trophic basis for the terrestrial radiation of insects. *Geol. Soc. Am., Abstr. Prog.* 23:23 (Abstr.)

174. Labandeira CC. 1994. A compendium of fossil insect families. *Milwaukee Publ. Mus. Contr. Biol. Geol.* 88:1–71

175. Labandeira CC, Beall BS, Hueber FM. 1988. Early insect diversification: evidence from a Lower Devonian bristletail from Québec. *Science* 242:913–16

176. Labandeira CC, Beall BS. 1990. Arthropod terrestriality. In *Arthropods: Notes for a Short Course,* ed. D Mikulic, pp. 214–56. Knoxville: Univ. Tenn. Press

177. Labandeira CC, Dilcher DL, Davis DR, Wagner DL. 1994. 97 million years of angiosperm-insect association: paleobiological insights into the meaning of coevolution. *Proc. Natl. Acad. Sci. USA* 91:12278–82

178. Labandeira CC, Nufio CR, Wing SL, Davis DR. 1995. Insect feeding strategies from the Late Cretaceous Big Cedar Ridge flora: comparing the diversity and intensity of Mesozoic herbivory with the present. *Geol. Soc. Am., Abstr. Prog.* 27:447 (Abstr.)

179. Labandeira CC, Phillips TL. 1996. Insect fluid-feeding on Upper Pennsylvanian tree ferns (Palaeodictyoptera, Marattiales) and the early history of the piercing-and-sucking functional feeding group. *Ann. Entomol. Soc. Am.* 89:157–83

180. Labandeira CC, Phillips TL. 1996. A Late Carboniferous petiole gall and the origin of holometabolous insects. *Proc. Natl. Acad. Sci. USA* 93:8470–74

181. Labandeira CC, Phillips TL, Norton RA. 1997. Oribatid mites and decomposition of plant tissues in Paleozoic coal-swamp forests. *Palaios* 12:317–51

182. Labandeira CC, Sepkoski JJ Jr. 1993. Insect diversity in the fossil record. *Science* 261:310–15

183. Lacasa Ruiz A, Martínez Delclòs X. 1986. *Meiatermes*: Nuevo género fósil de insecto isóptero (Hodotermitidae) de las calizas Neocomienses del Montsec (Provincia de Lérida, España). Lleida: Institut d'Estudis Llerdencs. 65 pp.

184. Larew HG. 1992. Fossil galls. See Ref. 313, pp. 50–59

185. Laroca S, Michener CD, Hofmeister RM. 1989. Long mouthparts among "short-tongued" bees and the fine structure of the labium in *Niltonia* (Hymenoptera, Colletidae). *J. Kans. Entomol. Soc.* 62:400–10

186. Larsen EB. 1948. Observations on the activity of some culicids. *Entomol. Meddel.* 25:263–77

187. Lawrence JF. 1987. The family Pterogeniidae, with notes on the phylogeny of the Heteromera. *Coleopt. Bull.* 31:25–56

188. Lee RMKW, Craig DA. 1983. Maxillary, mandibulary, and hypopharyngeal stylets of female mosquitoes (Diptera: Culicidae); a scanning electron microscope study. *Can. Entomol.* 115:1503–12

189. Lehane MJ. 1991. *Biology of Blood-Sucking Insects.* London: Harper Collins. 288 pp.

190. Leius K. 1960. Attractiveness of different foods and flowers to the adults of some hymenopterous parasites. *Can. Entomol.* 92:369–76

191. Lewis SE. 1994. Evidence of leaf-cutting bee damage from the Republic sites (Middle Eocene) of Washington. *J. Paleontol.* 68:172–73

192. Lewis SE, Carroll MA. 1991. Coleopterous egg deposition on alder leaves from the Klondike Mountain Formation (Middle Eocene), northeastern Washington. *J. Paleontol.* 65:334–35

193. Liu N, Nagatomi A. 1995. The mouthpart structure of Scenopinidae (Diptera). *Jap. J. Entomol.* 63:181–202

194. Livingstone D, Kumarswami NS. 1990. Adaptive radiation of mandibles of *Poecilocerus pictus* (Fabricius) in relation to the feeding strategies. *J. Soil Biol. Ecol.* 10:41–47

195. Lloyd DG, Wells MS. 1992. Reproductive biology of a primitive angiosperm, *Pseudowintera colorata* (Winteraceae), and the evolution of pollination systems in the Anthophyta. *Plant Syst. Evol.* 181:77–95

196. Lutz H. 1987. Die Insekten-Thanatocoenose aus dem Mittel-Eozän der "Grube Messel" bei Darmstadt: Erste Ergebnisse. *Courier Forsch. Inst. Senck.* 91:189–201

197. Lutz H. 1993. *Eckfeldapis electrapoides* nov. gen. n. sp., eine "Honigbiene" aus dem Mittel-Eozän des "Eckfelder Maares" bei Manderscheid/Eifel, Deutschland (Hymenoptera: Apidae, Apinae). *Mainz Naturwiss. Archiv* 31:177–99

198. Lyal CHC. 1987. Co-evolution of trichodectid lice (Insecta: Phthiraptera) and their mammalian hosts. *J. Nat. Hist.* 21:1–28

199. Maa TC. 1966. Studies in Hippoboscidae (Diptera). Part 2. *Pac. Ins. Mon.* 10:1–148
200. Mackerras, I. 1954. The classification and distribution of Tabanidae (Diptera). 1. General review. *Aust. J. Zool.* 2:431–54
201. Mamay SH. 1976. Paleozoic origin of the cycads. *US Geol. Surv. Prof. Pap.* 934:1–48
202. Manning JC, Goldblatt P. 1996. The *Prosoeca peringueyi* (Diptera: Nemestrinidae) pollination guild in southern Africa: long-tongued flies and their tubular flowers. *Ann. Missouri Bot. Gard.* 83:67–86
203. Martínez-Delclòs, X. 1991. Insectes de les calcàries litogràfiques de la Serra del Montsec. Cretaci Inferiour de Catalunya Espanya. In *Les Calcàries Litografiques del Cretaci Inferiour del Montsec. Deu Anys de Campanyes Paleontològiques*, ed. X Martínez-Delclòs, pp. 91–110. Barcelona: Inst. d'Estudis Llerdencs
204. Martynov AV. 1926. To the knowledge of fossil insects from Jurassic beds in Turkestan. 5. On some interesting Coleoptera. *Ezheg. Russ. Paleontol. Obshch.* 5:1–38
205. Masumoto M, Nagashima T. 1993. Head structure of the psocid *Psococerastis nubila* (Enderlein) (Psocoptera, Psocidae). *Jpn. J. Entomol.* 61:671–78
206. Matsuda R. 1965. Morphology and evolution of the insect head. *Mem. Am. Entomol. Inst.* 4:1–334
207. May EM. 1979. Adaptations of crane fly mouthparts for nectar feeding. *J. Kans. Entomol. Soc.* 564 (Abstr.)
208. McKeever S, Hagan DV, Grogan WL. 1991. Comparative study of mouthparts of ten species of predaceous midges of the tribe Ceratopogonini (Diptera: Ceratopogonidae). *Ann. Entomol. Soc. Am.* 84:93–106
209. McKeever S, Wright MD, Hagan DV. 1988. Mouthparts of females of four *Culicoides* species (Diptera: Ceratopogonidae). *Ann. Entomol. Soc. Am.* 81:332–41
210. McQuillan PB, Semmens TD. 1990. Morphology of antenna and mouthparts of adult *Adoryphorous couloni* (Burmeister) (Coleoptera: Scarabaeidae: Dynastinae). *J. Aust. Entomol. Soc.* 29:75–79
211. Medvedev LN. 1968. Leaf-beetles from the Jurassic of Karatau. See Ref. 279, pp. 155–65
212. Medvedev LN. 1969. New Mesozoic Coleoptera (Cucujoidea) of Asia. *Paleontol. J.* 1969(1):108–113
213. Medvedev SG. 1990. Peculiarities of the evolution of fleas, parasites of Chiroptera.

Parazitologiya 24:457–65 (In Russian with English abstract)
214. Meeuse ADJ, De Meijer AH, Mohr OWP, Wellinga SM. 1990. Entomophily in the dioecious gymnosperm *Ephedra aphylla* Forsk. (= *E. alte* C.A. Mey.), with some notes on *Ephedra campylopoda* C.A. Mey. III. Further anthecological studies and relative importance of entomophily. *Israel J. Bot.* 39:113–23
215. Melnikov OA, Rasnitsyn AP. 1984. Zur Metamerie des Arthropoden-Kopfes: Das Acron. *Beitr. Entomol.* 34:3–90
216. Metcalf CL. 1929. The mouthparts of insects. *Trans. Ill. Acad. Sci.* 21:109–35
217. Meyen SV. 1987. *Fundamentals of Paleobotany.* London: Chapman & Hall. 432 pp.
218. Michener CD. 1944. Comparative external morphology, phylogeny, and a classification of the bees (Hymenoptera). *Bull. Am. Mus. Nat. Hist.* 82:155–301
219. Michener CD, Greenberg L. 1985. The fate of the lacinia in the Halictidae and Oxaeidae (Hymenoptera-Apoidea). *J. Kansas Entomol. Soc.* 58:137–41
220. Michener CD, Grimaldi DA. 1988. The oldest fossil bee: apoid history, evolutionary stasis, and antiquity of social behavior. *Proc. Natl. Acad. Sci USA* 85:6424–26
221. Mickoleit E. 1963. Untersuchungen zur Kopfmorphologie der Thysanopteren. *Zool. Jb. (Anat.)* 81:101–50
222. Möhn E. 1960. Eine neue Gallmücke aus der niederrheinischen Braunkohle, *Sequoiomyia kraeuseli* n. g., n. sp. (Diptera, Itonidae). *Senck. Leth.* 41:513–22
223. Monrós F. 1954. Revision of the chrysomelid subfamily Aulacoscelinae. *Bull. Mus. Comp. Zool.* 112:321–60
224. Müller AH. 1978. Zur Entomofauna des Permokarbon. über die Morphologie, Taxonomie und Ökologie von *Eugereon boeckingi* (Palaeodictyoptera). *Freib. Forsch. (C)* 334:7–20
225. Muñiz R, Barrera A. 1969. *Rhopalotria dimidiata* Chevrolat, 1878: estudio morfológico del adulto y descripción de la larva (Ins. Col. Curcul.: Oxycoryninae). *Rev. Soc. Méx. Hist. Nat.* 30:205–22
226. Needham JG, Traver JR, Hsu Y-C. 1935. *The Biology of Mayflies, with a Systematic Account of North American Species.* Ithaca, NY: Comstock. 351 pp.
227. Nel A, De Villiers WM. 1988. Mouthpart structure in adult scarab beetles (Coleoptera: Scarabaeoidea). *Entomol. Gener.* 13:94–114
228. Nelson JM. 1991. Further observations on the prestomal teeth of dung-flies (Dipt., Scathophagidae) with special reference

to Nearctic species. *Entomol. Mon. Mag.* 127:39–42

229. Niklas KJ. 1992. *Plant Biomechanics: An Engineering Approach to Plant Form and Function.* Chicago: Univ. Chicago Press. 607 pp.

230. Nixon KC, Crepet WL. 1993. Late Cretaceous fossil flowers of ericalean affinity. *Am. J. Bot.* 80:616–23

231. Norstog KJ, Stevenson DM, Niklas KJ. 1986. The role of beetles in the pollination of *Zamia furfuracea* L. fil. (Zamiaceae). *Biotropica* 18:300–6

232. Olafsson JS. 1992. A comparative study on mouthpart morphology of certain larvae of Chironomini (Diptera: Chironomidae) with reference to the larval feeding habits. *J. Zool.* 228:183–204

233. Olesen J. 1979. Prey capture in dragonfly nymphs (Odonata, Insecta): labial protraction by means of a multi-purpose abdominal pump. *Vidensk. Medd. Dansk Naturh. Foren.* 141:81–96

234. Ollerton J. 1996. Interactions between gall midges (Diptera: Cecidomyiidae) and inflorescences of *Piper novae-hollandiae* (Piperaceae) in Australia. *Entomologist* 115:181–84

235. Olsen PE, Remington CL, Cornet B, Thomson KS. 1978. Cyclic change in Late Triassic lacustrine communities. *Science* 201:729–33

236. Opler PA. 1973. Fossil lepidopterous leaf mines demonstrate the age of some insect-plant relationships. *Science* 179:1321–23

237. Osten T. 1982. Vergleichend-funktionsmorphologische Untersuchungen der Kopfkapsel und der Mundwerkzeuge ausgewählter "Scolioidea" (Hymenoptera, Aculeata). *Stutt. Beit. Naturk. (A)* 354:1–60

238. Osten T. 1988. Die Mundwerkzeuge von *Proscolia spectator* Day (Hymenoptera: Aculeata). Ein Beitrag zur Phylogenie der "Scolioidea". *Stutt. Beit. Naturk. (A)* 414:1–30

239. Osten T. 1991. Konvergente Entwicklung der Mundwerkzeuge von *Epomidiopteron* (Tiphiidae) und den Scoliidae (Hymenoptera). *Stutt. Beit. Naturk. (A)* 466:1–7

240. Packard AS. 1890. *Insects Injurious to Forest and Shade Trees.* Washington, DC: US Govt. Print. Off. 957 pp.

241. Pakaluk J. 1987. Revision and phylogeny of the Neotropical genus *Hoplicnema* Matthews (Coleoptera: Corylophidae). *Trans. Am. Entomol. Soc.* 113:73–116

242. Parry DA. 1983. Labial extension in the dragonfly larva *Anax imperator.* *J. Exp. Biol.* 107:495–99

243. Pemberton RW. 1992. Fossil extrafloral nectaries, evidence for the ant-guard antiherbivore defense in an Oligocene *Populus. Am. J. Bot.* 79:1242–46

244. Peterson A. 1916. The head-capsule and mouth-parts of Diptera. *Ill. Biol. Mon.* 3(2):1–112

245. Poinar GO Jr. 1992. *Life in Amber.* Stanford, CA: Stanford Univ. Press. 350 pp.

246. Ponomarenko AG. 1976. A new insect from the Cretaceous of Transbaikalia, a possible parasite of pterosaurians. *Paleontol. J.* 1976(3):339–43

247. Ponomarenko AG. 1986. Scarabaeiformes incertae sedis. See Ref. 261, pp. 110–12

248. Ponomarenko AG, Ryvkin AB. 1990. Beetles. Scarabaeida. See Ref. 263, pp. 39–87

249. Popham EJ. 1990. Dermaptera. See Ref. 108, pp. 69–75

250. Popham EJ, Abdillahi M. 1979. Labellar microstructure in tsetse flies (Glossinidae). *Syst. Entomol.* 4:65–70

251. Popov YA. 1986. Cimicida (= Homoptera + Heteroptera). Peloridiina (= Coleorrhyncha) et Cimicina (= Heteroptera). See Ref. 261, pp. 50–83

252. Popov YA. 1990. Bugs. Cimicina. See Ref. 263, pp. 20–39

253. Porsch O. 1910. *Ephedra campylopoda* C. A. Mey., eine entomophile Gymnosperme. *Ber. Deut. Bot. Ges.* 28:404–12

254. Pritchard G. 1976. Further observations on the functional morphology of the head and mouthparts of dragonfly larvae (Odonata). *Quaest. Entomol.* 12:89–114

255. Pritykina LN. 1986. Libellulida (= Odonata). See Ref. 261, pp. 165–66

256. Proctor M, Yeo P, Lack A. 1996. *The Natural History of Pollination.* Portland, OR: Timber. 479 pp.

257. Puplesis R. 1994. *The Nepticulidae of Eastern Europe and Asia.* Leiden: Backhuys. 748 pp.

258. Rashid SS, Mulla MS. 1990. Comparative functional morphology of the mouth brushes of mosquito larvae (Diptera: Culicidae). *J. Med. Entomol.* 27:429–39

259. Rasnitsyn AP. 1977. New Paleozoic and Mesozoic insects. *Paleontol. J.* 11:60–72

260. Rasnitsyn AP. 1977. New Hymenoptera from the Jurassic and Cretaceous of Asia. *Paleontol. J.* 1977(3):349–357

261. Rasnitsyn AP, ed. 1986. Insects in the Early Cretaceous ecosystems of [sic] west Mongolia. *Trans. Joint Sov.-Mong. Palaeont. Exped.* 28:1–216 (In Russian)

262. Rasnitsyn AP. 1988. An outline of evolution of the hymenopterous insects (Order Vespida). *Oriental Insects* 22:115–45
263. Rasnitsyn AP, ed. 1990. Late Mesozoic insects of eastern Transbaikalia. *Trans. Paleontol. Inst.* 239:1–224 (In Russian)
264. Rasnitsyn AG. 1990. Hymenopterans. Vespida. See Ref. 263, pp. 177–205
265. Rasnitsyn AP, Krassilov VA. 1996. First find of pollen grains in the gut of Permian insects. *Paleontol. J.* 30:484–90
266. Rasnitsyn AP, Krassilov VA. 1996. Pollen in the gut contents of fossil insects as evidence of coevolution. *Paleontol. J.* 30:716–22
267. Rayner RG, Waters SB. 1990. A Cretaceous crane-fly (Diptera: Tipulidae): 93 million years of stasis. *Zool. J. Linn. Soc.* 99:309–18
268. Rayner RG, Waters SB. 1991. Floral sex and the fossil insect. *Naturwissenschaften* 78:280–82
269. Rayner RJ, Waters SB, McKay IJ, Dobbs PN, Shaw AL. 1991. The mid-Cretaceous palaeoenvironment of central Southern Africa (Orapa, Botswana). *Palaeogeog. Palaeoclim. Palaeoecol.* 88:147–56
270. Remm KY. 1976. Late Cretaceous biting midges (Diptera, Ceratopogonidae) from fossil resins of the Khatanga Depression. *Paleontol. J.* 1976(3):344–51
271. Rempel JG. 1975. The evolution of the insect head: the endless dispute. *Quaest. Entomol.* 11:9–25
272. Retallack GJ, Dilcher DL. 1988. Reconstructions of selected seed ferns. *Ann. Missouri Bot. Gard.* 75:1010–57
273. Richardson ES Jr. 1956. Pennsylvanian invertebrates of the Mazon Creek area, Illinois. *Fieldiana: Geology* 12:1–76
274. Richardson ES Jr. 1980. Life at Mazon Creek. In *Middle and Late Pennsylvanian Strata of [the] Margin of [the] Illinois Basin*, ed. RL Langenheim Jr, CJ Mann, pp. 217–24. Urbana, IL: Univ. Illinois Press
275. Riek EF. 1970. Lower Cretaceous fleas. *Nature* 227:746–47
276. Roberts MJ. 1971. The structure of the mouthparts of some calypterate dipteran larvae in relation to their feeding habits. *Acta Zool.* 2:171–88
277. Rogers JS. 1926. Some notes on the feeding habits of adult crane-flies. *Fla. Entomol.* 10:5–8
278. Rohdendorf BB. 1964. *Historical Development of the Diptera.* Transl. JE Moore, 1974 (From Russian). Edmonton: Univ. Alberta Press. 360 pp.
279. Rohdendorf BB, ed. 1968. *Jurassic Insects of Karatau.* Moscow: Acad. Sci. 252 pp.
280. Rohdendorf BB. 1968. New Mesozoic nemestrinids (Diptera, Nemestrinidae). See Ref. 279, pp. 180–89
281. Rohdendorf BB, Rasnitsyn AP. 1980. Historical development of the class Insecta. *Trans. Paleontol. Inst.* 175:1–258 (In Russian)
282. Rothmeier G, Jach MA. 1986. *Proc. Eur. Congr. Entomol., 3rd.* Amsterdam, 1986, pp. 133–37. Amsterdam: Nederlandse Entomol. Vereniging
283. Rothwell GW, Scott AC. 1983. Coprolites within marattiaceous fern stems (*Psaronius magnificus*) from the Upper Pennsylvanian of the Appalachian Basin U.S.A. *Palaeogeog., Palaeoclim., Palaeoecol.* 41:227–32
284. Rousset A. 1966. Morphologie cephalique des larves de Planipennes (Insectes Neuropteroides). *Mem. Mus. Natl. Hist. Nat.* 42:1–199
285. Rowe RJ. 1987. Predatory versatility in a larval dragonfly, *Hemianax papuensis* (Odonata: Aeshnidae). *J. Zool. London* 211:193–207
286. Rowley WA, Cornford M. 1972. Scanning electron microscopy of the pit of the maxillary palp of selected species of *Culicoides. Can. J. Zool.* 50:1207–10
287. Rozefelds AC. 1988. Lepidoptera mines in *Pachypteris* leaves (Corystospermaceae: Pteridospermophyta) from the Upper Jurassic/Lower Cretaceous Battle Camp Formation, North Queensland. *Proc. R. Soc. Queensland* 99:77–81
288. Rozefelds AC, Sobbe I. 1987. Problematic insect leaf mines from the Upper Triassic Ipswich Coal Measures of southeastern Queensland, Australia. *Alcheringa* 11:51–57
289. Saini MS, Dhillon SS. 1979. Comparative morphology of galea and lacinia in Hymenoptera. *Entomon* 4:149–55
290. Saini MS, Dhillon SS. 1979. Glossal and paraglossal transformation in order Hymenoptera. *Entomon* 4:355–60
291. Satô M. 1993. Comparative morphology of the mouthparts of the family Dolichopodidae (Diptera). *Ins. Matsum.* 45:49–75
292. Saunders LG. 1959. Methods for studying *Forcipomyia* midges, with special reference to cacao-pollinating species (Diptera, Ceratopogonidae). *Can. J. Zool.* 37:33–51
293. Schaarschmidt F. 1992. The vegetation: fossil plants as witnesses of a warm climate. In *Messel: An Insight into the*

History of Life and of the Earth, ed. S Schaal, W Ziegler, pp. 29–52. Oxford, UK: Clarendon

294. Schedl KE. 1947. Die Borkenkäfer des baltischen Bernsteins. Zentral. Gesamt. Entomol. 2:12–45

295. Schicha E. 1967. Morphologie und Funktion der Malachiiden-mundwerkzeuge unter besonderer Berücksichtigung von Malachius bipustulatus L. (Coleopt., Malacodermata). Z. Morphol. Ökol. Tiere 60:376–433

296. Schlee D. 1970. Insektenfossilien aus der unteren Kreide. 1. Verwandtschaftsforschung an fossilen und rezenten Aleyrodina (Insecta, Hemiptera). Stutt. Beitr. Naturk. 213:1–72

297. Schlee D, Dietrich H-G. 1970. Insektenführender Bernstein aus der Unterkreide des Libanon. Neues Jb. Geol. Paläont. Monat. 1:40–50

298. Schopf JM. 1948. Pteridosperm male fructifications: American species of Dolerotheca, with notes regarding certain allied forms. J. Paleontol. 22:681–724

299. Schröder P. 1987. Labral filter fans of blackfly larvae: differences in fan area and fan ray number and the consequences for utilization and particle selection. Zool. Beitr. 31:365–94

300. Schuhmacher H, Hoffmann H. 1982. Zur Funktion der Mundwerkzeuge von Schwebfliegen bei der Nahrungsaufnahme (Diptera: Syrphidae). Entomol. Gen. 7:327–42

301. Scott AC, Stephenson J, Chaloner WG. 1992. Interaction and coevolution of plants and arthropods during the Palaeozoic and Mesozoic. Philos. Trans. R. Soc. London (B) 335:129–65

302. Scott AC, Stephenson J, Chaloner WG. 1994. The fossil record of leaves with galls. See Ref. 357, pp. 447–470

303. Scott AC, Taylor TN. 1983. Plant/animal interactions during the Upper Carboniferous. Bot. Rev. 49:259–307

304. Sharov AG. 1953. The first discovery of a Permian larva of Megaloptera from Kargala. Dokl. Acad. Nauk USSR 1953(4):731–32 (In Russian)

305. Sharov AG. 1966. Basic Arthropodan Stock with Special Reference to Insects. Oxford: Pergamon. 271 pp.

306. Sharov AG. 1968. Phylogeny of the Orthopteroidea. Transl. J. Salkind, 1971 (From Russian). Jerusalem: Keter. 251 pp.

307. Sharov AG. 1973. Morphological features and mode of life of the Palaeodictyoptera. In Readings in the Memory of Nikolaj Aleksandrovich Kholodkovskij, ed. GY

Bei-Benko, pp. 49–63. Leningrad: Sci. Publ. (In Russian)

308. Shcherbakov DE. 1986. Cimicida (= Homoptera + Heteroptera). Cicadina (= Auchenorrhyncha). See Ref. 261, pp. 47–50

309. Shcherbakov DE. 1996. Origin and evolution of the Auchenorrhyncha as shown by the fossil record. In Studies on Hemipteran Phylogeny, ed. CW Schaefer, pp. 31–45. Lanham, MD: Entomol. Soc. Am.

310. Shear WA, Kukalová-Peck J. 1990. The ecology of Paleozoic terrestrial arthropods: the fossil evidence. Can. J. Zool. 68:1807–34

311. Short JRT. 1952. The morphology of the head of larval Hymenoptera with special reference to the head of the Ichneumonoidea, including a classification of the final instar larvae of the Braconidae. Trans. R. Entomol. Soc. London 103:27–66

312. Short JRT. 1955. The morphology of the head of Aeshna cyanea (Müller) (Odonata: Anisoptera). Trans. R. Entomol. Soc. London 106:197–211

313. Shorthouse JD, Rohfritsch O, eds. 1992. Biology of Insect-Induced Galls. New York: Oxford Univ. Press. 285 pp.

314. Simpson BB, Neff JL. 1983. Evolution and diversity of floral rewards. In Handbook of Experimental Pollination Biology, ed. CE Jones, RJ Little, pp. 142–159. New York: Van Nostrand Reinhold

315. Sinclair BJ. 1992. A phylogenetic interpretation of the Brachycera (Diptera) based on the larval mandible and associated mouthpart structures. Syst. Entomol. 17:233–52

316. Sinichenkova, N. 1986. Order Ephemeroptera (= Ephemeroptera). See Ref. 261, pp. 45–47

317. Smith JJB. 1985. Feeding mechanisms. In Comprehensive Insect Physiology, Biochemistry, and Pharmacology, ed. GA Kerkut, LI Gilbert, 4:33–85. Oxford: Pergamon

318. Snodgrass RE. 1928. Morphology and evolution of the insect head and its appendages. Smithson. Misc. Collns. 81:1–158

319. Snodgrass RE. 1935. Principles of Insect Morphology. New York: McGraw-Hill. 667 pp.

320. Spencer KA. 1990. Host Specialization in the World Agromyzidae (Diptera). Dordrecht: Kluwer. 444 pp.

321. Spooner CS. 1938. The phylogeny of the Hemiptera based on a study of the head capsule. Ill. Biol. Mon. 16(3):1–102

322. Srivastava RP, Bogawat JK. 1969. Feeding mechanism of a fruit-sucking moth *Othreis materna* (Lepidoptera: Noctuidae). *J. Nat. Hist.* 3:165–81

323. Stephenson J. 1991. *Evidence of plant/insect interactions in the Late Cretaceous and Early Tertiary*. PhD diss. Univ. London. 378 pp.

324. Straus A. 1977. Gallen, Minen und andere Frass-spuren im Pliokän von Willershausen am Harz. *Verhand. Bot. Ver. Prov. Brandenburg* 113:41–80

325. Strenger A. 1942. Funktionelle Analyse des Orthopterenkopfes, eine systematisch-funktionsanatomische Studie. *Zool. Jb. (Syst.)* 75:1–72

326. Stuckenberg BR. 1996. A revised generic classification of the wormlion flies of Southern Africa previously placed in *Lampromyia* Macquart, with reinstatement of *Leptynoma* Westwood 1876, and descriptions of a new subgenus and two new species (Diptera, Vermileonidae). *Ann. Natal Mus.* 37:239–66

327. Sukacheva ID. 1990. Scorpion flies. Panorpida. See Ref. 263, pp. 88–94

328. Süss H, Müller-Stoll WR. 1980. Das fossile Holz *Pruninium gummosum* Platen emend. Süss aus dem Yellowstone Nationalpark und sein Parasit *Palaeophytobia prunorum* sp. nov. nebst Bemerkungen über Markflecke. In *100 Jahre Arboretum Berlin (1879–1979), Jubiläumsschrift*, ed. W Vent, pp. 343–64. Berlin: Humboldt-Universität zu Berlin

329. Sutcliffe JF, Deepan PD. 1988. Anatomy and function of the mouthparts of the biting midge, *Culicoides sanguisuga* (Diptera: Ceratopogonidae). *J. Morphol.* 198:353–65

330. Szadziewski R. 1988. Biting midges (Diptera, Ceratopogonidae) from Baltic amber. *Pol. Pismo Entomol.* 57:3–283

331. Tanaka Y, Hisada M. 1980. The hydraulic mechanism of the predatory strike in dragonfly larvae. *J. Exp. Biol.* 88:1–19

332. Tang W. 1987. Insect pollination in the cycad *Zamia pumila* (Zamiaceae). *Am. J. Bot.* 74:90–99

333. Taylor TN, Millay MA. 1979. Pollination biology and reproduction in early seed plants. *Rev. Palaeobot. Palynol.* 27:329–55

334. Tetley H. 1918. The structure of the mouth-parts of *Pangonia longirostris* in relation to the probable feeding-habits of the species. *Bull. Entomol. Res.* 8:253–67

335. Thien LB. 1969. Mosquito pollination of *Habenaria obtusata* (Orchidaceae). *Am. J. Bot.* 56:232–37

336. Thien LB. 1974. Floral biology of *Magnolia. Am. J. Bot.* 61:1037–45

337. Thien LB. 1980. Patterns of pollination in the primitive angiosperms. *Biotropica* 12:1–13

338. Thien LB. 1982. Fly pollination in *Drimys* (Winteraceae), a primitive angiosperm. *Monogr. Biol.* 42:529–33

339. Thien LB, White DA, Yatsu LY. 1983. The reproductive biology of a relict—*Illicium floridianum* Ellis. *Am. J. Bot.* 70:719–27

340. Traub R. 1980. The zoogeography and evolution of some fleas, lice and mammals. In *Fleas*, ed. R Traub, H Starcke, pp. 93–172. Rotterdam: Balkema

341. Van der Pijl L. 1953. On the flower biology of some plants with general remarks on fly-traps (species of *Annona, Artocarpus, Typhonium, Gnetum, Arisaema* and *Abroma*). *Ann. Bogor.* 1:77–99

342. Vishniakova VN. 1968. Mesozoic cockroaches with an external ovipositor and the specific relations of their reproduction (Blattodea). See Ref. 279, pp. 55–86

343. Vishniakova VN. 1973. New cockroaches (Insecta: Blattodea) from the Upper Jurassic deposits of the Karatau Mountains. In *Readings in the Memory of Nikolaj Aleksandrovich Kholodkovskij*, ed. GY Bei-Benko, pp. 64–77. Leningrad: Sci. Publ. (In Russian)

344. Vishniakova VN. 1981. New Paleozoic and Mesozoic lophioneurids (Thripida, Lophioneuridae). *Trans. Paleontol. Inst.* 183:43–63 (In Russian)

345. Vogel R. 1915. Beitrag zur Kenntnis des Baues und der Lebensweise der larve von *Lampyris noctiluca*. *Z. Wiss. Zool.* 112:291–432

346. Vovides AP. 1991. Cone idioblasts of eleven cycad genera: morphology, distribution, and significance. *Bot. Gaz.* 152:91–99

347. Wallace JB, Merritt RW. 1980. Filter-feeding ecology of aquatic insects. *Annu. Rev. Entomol.* 25:103–32

348. Webber AC. 1981. Alguns aspectos da biologia floral de *Annona sericea* Dun. (Annonaceae). *Acta Amazon.* 11:61–65

349. Wenk P. 1953. Der Kopf von *Ctenocephalus canis* (Curt.) (Aphaniptera). *Zool. Jb. (Anat.)* (73):103–64

350. Whalley PES, Jarzembowski EA. 1985. Fossil insects from the lithographic limestone of Montsech (late Jurassic—early Cretaceous), Lérida Province, Spain. *Bull. Br. Mus. Nat. Hist. (Geol.)* 38:381–412

351. White DA, Thien LB. 1985. The pollination of *Illicium parviflorum* (Illiciaceae). *J. Elisha Mitchell Sci. Soc.* 101:15–18

352. White JF Jr, Taylor TN. 1989. A trichomycete-like fossil from the Triassic of Antarctica. *Mycologia* 81:643–46

353. Wilding N, Collins NM, Hammond PM, Webber JF, eds. 1989. *Insect-Fungus Interactions*. London: Academic. 344 pp.

354. Willemstein SC. 1980. Pollen in Tertiary insects. *Acta Bot. Neerl.* 29:57–58

355. Willemstein SC. 1987. An evolutionary basis for pollination biology. *Leiden Bot. Ser.* 10:1–425

356. Williams IW. 1938. The comparative morphology of the mouth-parts of the order Coleoptera treated from the standpoint of phylogeny. *J. N. Y. Entomol. Soc.* 41:1–34

357. Williams MAJ, ed. 1994. *Plant Galls: Organisms, Interactions, Populations*. Oxford: Clarendon

358. Willmann R. 1989. Evolution und Phylogenetisches System der Mecoptera (Insecta: Holometabola). *Abh. Senckenberg. Naturforsch. Ges.* 544:1–153

359. Wilson EO, Carpenter FM, Brown WL Jr. 1967. The first Mesozoic ants. *Science* 157:1038–40

360. Wootton RJ. 1988. The historical ecology of aquatic insects: an overview. *Palaeogeog., Palaeoclimat., Palaeoecol.* 62: 477–92

361. Wu J-t, Chou T-j. 1985. Studies on the cephalic endoskeleton, musculature and proboscis of citrus fruit-piercing noctuid moths in relation to their feeding habits. *Acta Entomol. Sinica* 28:165–72

362. Wundt H. 1961. Der Kopf der Larve von *Osmylus chrysops* L. (Neuroptera, Planipennia). *Zool. Jb. (Anat.)* 79:557–662

363. Young AM. 1985. Studies of cecidomyiid midges (Diptera: Cecidomyiidae) as cocoa pollinators (*Theobroma cacao* L.) in Central America. *Proc. Entomol. Soc. Wash.* 87:49–79

364. Zaitzev VF. 1984. Microstructure of the labella of the fly proboscis. II. Pseudotracheal framework; structure and evolution. *Entomol. Rev.* 63:33–41

365. Zaitzev VF. 1987. New species of Cretaceous fossil bee flies and a review of paleontological data on the Bombyliidae (Diptera). *Entomol. Rev.* 66:150–60

366. Zetto Brandmayr T, Marano I, Paarmann W. 1994. *Graphipterus serrator*: a myrmecophagous carabid beetle with mandibular suctorial tube in the larva (Coleoptera, Carabidae, Graphipterini). In *Carabid Beetles: Ecology and Evolution*, ed. K Desender, et al., pp. 87–91. Dordrecht: Kluwer

367. Zherikhin VV. 1977. Curculionidae. See Ref. 7, pp. 244–50

368. Zherikhin VV, Gratshev VG. 1993. Obrieniidae, fam. nov., the oldest Mesozoic weevils (Coleoptera, Curculionoidea). *Paleontol. J.* 27(1A):50–69

369. Zherikhin VV, Sukacheva ID. 1973. On Cretaceous insects from "amber" (retinites) of northern Siberia. In *Readings in the Memory of Nikolaj Aleksandrovich Kholodkovskij*, ed. GY Bei-Benko, pp. 3–48. Leningrad: Sci. Publ. (In Russian)

370. Zur Strassen R. 1973. Insektenfossilien aus der unteren Kreide. 5. Fossile Fransenflüger aus Mesozoïschem Bernstein des Libanon (Insecta: Thysanoptera). *Stutt. Beitr. Naturk. (A)* 256:1–51

Annu. Rev. Ecol. Syst. 1997. 28:195–218

HALDANE'S RULE

H. Allen Orr

Department of Biology, University of Rochester, Rochester, New York 14627;
e-mail: haorr@darwin.biology.rochester.edu

KEY WORDS: hybrid inviability, hybrid sterility, postzygotic isolation, reproductive isolation, speciation

ABSTRACT

Haldane's rule—the preferential sterility or inviability of hybrids of the heterogametic (XY) sex—characterizes speciation in all known animals. Over the past decade, an enormous amount of experimental and theoretical work has been devoted to explaining this pattern. This work has falsified several once-popular theories and, more important, has produced a strong consensus on the likely causes of Haldane's rule. Experiments show that the dominance theory, which posits that "speciation genes" act as partial recessives in hybrids, can explain Haldane's rule for hybrid inviability. Dominance likely also contributes to Haldane's rule for sterility. Recent experiments further show that faster evolution of hybrid male steriles plays an important role. Faster evolution of X-linked loci may also contribute, though the evidence here is weaker. Evolutionary geneticists now largely agree that the simultaneous action of these forces explains Haldane's rule.

INTRODUCTION

Progress in evolutionary biology often reflects the discovery of new or unexpected phenomena, e.g. the protein molecular clock. But now and then, progress results from the just-as-unexpected finding that our confidently held explanations of old phenomena are wrong. The recent burst of work in the genetics of speciation falls into this second category.

Since 1985, the causes of hybrid sterility and inviability have been intensively studied, and our understanding of speciation has grown dramatically. Ironically, the problem that triggered—and that has largely sustained—this modern renaissance of speciation work is very old, predating Dobzhansky's *Genetics and the Origin of Species* (23) and even Fisher's *The Genetical Theory of Natural Selection* (27). The problem is, of course, Haldane's rule.

195

0066-4162/97/1120-0195$08.00

In 1922, JBS Haldane (31) observed that:

> When in the offspring of two different animal races one sex is absent, rare, or sterile, that sex is the heterozygous [heterogametic or XY] sex.

Haldane showed that this rule is obeyed in several animal groups (see below), and more recent surveys suggest that it holds in all animals known to possess sex chromosomes.

The significance of Haldane's rule is obvious. It suggests that speciation—or at least speciation by postzygotic isolation—proceeds in a similar way among very different kinds of animals. This implied universality explains the attention paid to Haldane's rule. It is not that the pattern itself is deemed so inherently interesting, but that it implies some fundamental similarity in the genetic events causing speciation in all animals. The question is: Just what is the same in bird, bug, mammal, and moth speciation? What shared genetic process gives rise to this pattern in group after group?

Although evolution texts invariably offered up some pat explanation of Haldane's rule, work in the mid-1980s showed surprisingly that all was not simple. It quickly became clear that we did not, in fact, understand why animals conform to Haldane's rule, nor therefore what genetic processes underlie animal speciation. A wave of experiment and theory ensued.

Given the lavish attention paid Haldane's rule, it may not be obvious that yet another review is needed. Three good reasons justify such a review. First, Haldane's rule has had an extraordinarily tortuous history. Theories that once seemed viable—gracing textbooks as well as many issues of *Nature*—have been slain; theories that once seemed dead (or at least dying) have been resurrected. The result of all this is considerable confusion. Second, many of these unexpected developments have occurred in the past several years. Consequently, reviews only a few years old are already badly dated.

Most important, a remarkable consensus has recently emerged on the causes of Haldane's rule. It would be premature to read this consensus as a flat resolution—the Haldane's rule story is not over, and in several key places, the data remain sufficiently flimsy that we could all yet prove wrong. But it would also be a mistake to downplay the significance of this consensus. For the first time since the "new work" on Haldane's rule began, most workers (e.g., Coyne, Davis, Laurie, Orr, Turelli, and Wu; see references below) appear to agree that the likely causes have been found. It does not seem too naive to hope that we have finally arrived at the correct answer.

My approach here will be simple. I first summarize the phenomenology of Haldane's rule: How strong is the pattern, where is it seen, and how many times has it evolved independently? I then review background information on the genetics of postzygotic isolation; Haldane's rule cannot be understood without this material. I devote the remainder of the review to the genetic causes of Haldane's rule. I emphasize those explanations that now appear correct and, because of space limitations, only briefly allude to those theories that have been rejected. I also emphasize places where more critical experiments are needed.

THE PHENOMENON

Strength of the Pattern

No theorist predicted Haldane's rule. The pattern was simply noticed by Haldane (31), who presented data from Lepidoptera, birds, flies, mammals, Anoplura, and Cladocera. Since then, several reviewers have tabulated the frequency with which Haldane's rule is obeyed in various taxa (12, 16, 80). Table 1 is adapted from Coyne (12), but similar results were obtained in all of the above reviews.

Haldane's rule—despite claims to the contrary (76)—is well obeyed in all the taxa surveyed. In *Drosophila*, for instance, a remarkable 112 out of 114 species crosses that produce sterile hybrids of one sex only obey Haldane's rule. But the most important fact emerging from Table 1 is this: The rule is obeyed in taxa in which males are heterogametic (e.g., *Drosophila*, mammals) and in taxa in which females are heterogametic (e.g., Lepidoptera, birds). Thus, the correct explanation of Haldane's rule must be tied not to sex per se, but to the sex chromosomes (16).

Another important fact is not apparent from Table 1: that Haldane's rule represents an early stage in speciation. Although this had been widely assumed, it is not necessarily true. As Coyne & Orr (15) emphasized, there could be two paths to speciation. In the first, the heterogametic sex would become sterile or

Table 1   The strength of Haldane's rule. Modified from Coyne (12)

Group	Phenotype	Asymmetric hybridizations	Number obeying Haldane's rule
Heterogametic males			
Drosophila	Sterility	114	112
	Inviability	17	13
Mammals	Sterility	25	25
	Inviability	1	1
Heterogametic females			
Lepidoptera	Sterility	11	11
	Inviability	34	29
Birds	Sterility	23	21
	Inviability	30	30

inviable early on, with homogametic hybrids being affected later. In the second, both hybrid sexes would become sterile or inviable simultaneously. Even if this second path were common, Haldane's rule would be obeyed: When only one sex is afflicted, it is the heterogametic one.

To see if there are multiple paths to speciation, Coyne & Orr (15) collected data on the age and strength of reproductive isolation between 119 pairs of *Drosophila* species. The age of a species pair was measured by Nei's genetic distance, D, which is roughly linear with divergence time, at least early on. The strength of isolation was measured as the proportion of the "four" hybrid sexes (males and females in both reciprocal crosses) that were sterile or inviable. The results showed that the severity of hybrid sterility and inviability increases with time, as expected (Figure 1). The correlation is strong enough that, of 47 young species pairs (Nei's $D < 0.5$), 43 have an isolation index of 0–0.50, i.e., no more than half of the four hybrid sexes are sterile or inviable. More important, these

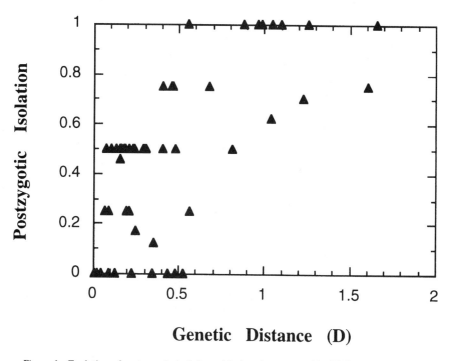

Figure 1 Evolution of postzygotic isolation with time (as measured by Nei's genetic distance, D). Data have been phylogenetically corrected for independence (18). Almost all young species pairs (those in the lower left quadrant) represent cases of Haldane's rule. (Modified from Coyne & Orr—18).

weakly isolated taxa almost always conform to Haldane's rule: Of 43 species pairs that have an isolation index of 0.25 or 0.50, 37 are cases of Haldane's rule. The lower left quadrant of Figure 1 is, then, made up almost entirely of cases of Haldane's rule. These results prove that Haldane's rule is a nearly obligatory first step in the evolution of postzygotic isolation (15). These findings have been confirmed by a more recent survey (18) that includes data from 171 pairs of *Drosophila* species.

Is Haldane's Rule Significant?

Because so many species pairs obey Haldane's rule—242 out of 255 crosses in Table 1—it may seem absurd to ask if the pattern is significant. But Read & Nee (66) argued that the association between postzygotic isolation and heterogamety may be illusory. Their argument was straightforward: The relevant sample size may not be the number of hybridizations tested. Instead, to prove an association between heterogamety and the sex most afflicted by hybridization, one must look only at those cases in which the sex that is heterogametic has evolved independently. The reason is simple: Birds and *Drosophila* differ in many ways, not just in heterogamety. Thus, the fact that all *Drosophila* (male heterogamety) and all birds (female heterogamety) obey Haldane's rule shows only that something all flies share makes them conform to the rule, and similarly, something all birds share makes them conform. This "something" may or may not be heterogamety. Read & Nee argued that there were only four independent associations between heterogamety and the sex afflicted in species crosses: Lepidoptera (female heterogamety), all other members of class Insecta (male heterogamety), Aves (female heterogamety), and Mammalia (male heterogamety).

As Coyne et al pointed out (14), Read & Nee's analysis overlooked cases of Haldane's rule in at least one other group known to have independently evolved male heterogamety, salamanders of the genus *Triturus* (68). More cases of Haldane's rule have since been described in previously uncharacterized taxa. In species crosses among lizards of the genus *Lacerta*, hybrid sterility occurs nearly exclusively in females, which are heterogametic (67). Among nematodes, crosses between two species of the genus *Caenorhabditis* produce lethal hybrid males, which are heterogametic (2). Among stick insects, crosses between *Bacillus rossius* and *B. grandii benazzii* produce sterile hybrid males, which are XO (42). The first two cases appear to represent phylogenetically independent instances of Haldane's rule, although the latter probably does not. In any case, the range of taxa known to obey Haldane's rule has grown considerably, and it seems likely that Haldane's rule is obeyed in all animal taxa possessing heteromorphic sex chromosomes—and we now know of at least seven independent associations between heterogamety and the sex most afflicted by hybridization.

But we need not rely on such comparative data to demonstrate an association between postzygotic isolation and heterogamety. The notion that Haldane's rule is due to heterogamety can be tested by direct genetic analysis (14). If, for instance, genetic analysis showed that the factors causing Haldane's rule were always X-linked or always recessive (and thus only expressed in the heterogametic sex), we would understand causally why heterogametic hybrids suffer disproportionately low fitness. As we will see, there is now an overwhelming body of direct genetic data supporting a causal connection between heterogamety and Haldane's rule.

I assume, then, that Haldane's rule is "significant." The real question is why the heterogametic sex disproportionately suffers the effects of hybridization. To understand the reasons, I first review several principles underlying the genetics of speciation. I then attempt to derive the cause of Haldane's rule from these principles.

GENETICS OF POSTZYGOTIC ISOLATION

Any attempt to understand the evolution of hybrid sterility and inviability confronts a simple problem: Both phenotypes seem patently maladaptive. How then could natural selection allow the routine evolution of progeny that are infertile or inviable? This question once represented the most fundamental problem in speciation and, though solved, it still represents one of the most commonly misunderstood. Thus, before considering the evolution of sex-specific hybrid problems, we must understand how hybrid problems afflicting any sex can evolve.

Species that produce sterile or inviable hybrids can be pictured as residing on two adaptive peaks, with the hybrid genotype residing in a valley between the parental peaks. In Wrightian language, the problem of speciation is to explain how two species could come to reside on separate peaks without either lineage passing through the adaptive valley (which would not be allowed by selection). Imagine the simplest possible genetic scenario: One species has genotype AA, the other aa, and the sterile or inviable hybrid Aa. Now consider the evolution of these two genotypes from a common ancestor, say AA: Beginning with two allopatric populations, one must remain AA, while the other must evolve from AA to aa. But how can it? The a mutation arises in heterozygous state (Aa) which, unfortunately, corresponds to the sterile or inviable genotype. Speciation, it seems, requires passing through an adaptive valley.

This fundamental problem was first solved by William Bateson in 1909 (3), although his precedent was only recently discovered (56). Credit is usually given instead to Dobzhansky (23) and Muller (44, 45), who independently offered the same solution early in the modern synthesis. The "Dobzhansky-Muller model" of speciation forms the basis of all modern work on the genetics of hybrid sterility and inviability, including that on Haldane's rule.

The Dobzhansky-Muller Model

Consider two allopatric populations, each experiencing many substitutions through evolutionary time. If an allele from one population is introduced into the genome of the other, would it function? It is easy to imagine that this allele would function reasonably well or not at all. But it is difficult to imagine that it would function better than on its "own" genetic background. This simple asymmetry forms the foundation of Dobzhansky's and Muller's view of speciation: Alleles from different loci but from the same species have been "tested" together, while those from different species have not. On average, then, we expect a mixture of genes from two species to be less well adjusted than those from a single species. Hybrid sterility or inviability could, therefore, be a simple by-product of the divergence of two genomes in geographical isolation.

Dobzhansky and Muller formalized this idea with a simple model. Suppose a species is split into two allopatric populations, each of which begins with an *aa* genotype at one locus and *bb* at another. An *A* allele appears and is fixed in one of the populations; the *Aabb* and *AAbb* genotypes are viable and fertile. In the other population, a *B* mutation appears and is fixed; the *aaBb* and *aaBB* genotypes are also viable and fertile. The critical point is that, although the *B* allele is compatible with *a*, it has not been tested with the *A* allele. It is entirely possible that *B* has a deleterious effect that appears only when *A* is present. If the two populations meet and hybridize, the resulting *AaBb* hybrid may be inviable or sterile. Genes showing this kind of deleterious epistatic interaction are known as "complementary genes." The important point of the Dobzhansky-Muller model is obvious: Two taxa separated by an adaptive valley can evolve even though no genotype ever passed through the valley.

In the Dobzhansky-Muller model, the *A* and *B* genes need not have a devastating effect on hybrid fitness; any particular incompatibility might cause only slight hybrid inviability or sterility. Complete postzygotic isolation might then result from the cumulative effect of many small incompatibilties. Similarly, there is no reason to think that any particular incompatibility must involve only a pair of genes. As Muller (45) emphasized, it is possible that a hybrid must carry the "correct" alleles at three or more loci for any hybrid sterility or inviability to occur. Indeed, there is good evidence for such complex interactions (5, 45). Last, one can imagine many variations on the scenario sketched above. Both substitutions might occur in one population (*aabb* $\longrightarrow$ *AAbb* $\longrightarrow$ *AABB*), for instance, and none in the other. The point remains: There is no guarantee that alleles from the two species can be mixed.

The Dobzhansky-Muller model has been formalized mathematically (55, 58). The most important result of this analysis is that postzygotic isolation is expected to "snowball" with the time of divergence between taxa, i.e., the severity of hybrid sterility and inviability, as well as the number of loci causing postzygotic isolation, will increase much faster than linearly with time.

The Evidence

Recent work shows that the Dobzhansky-Muller model is almost certainly correct. Indeed, this finding may represent the single most important result of the past decade of work in the genetics of speciation: Despite frequent disagreement over details, there is a strong consensus that hybrid sterility and inviability in animals are caused by between-locus incompatibilties. Within species, alleles that are fit on one genetic background but unfit on another are well documented. Complementary genes have been found segregating in natural populations (24, 25, 41, 72). Moreover, it is well known that some laboratory mutations, while viable and fertile by themselves, act as lethals or steriles when combined (44, 72), e.g., *pn* and *Killer-of-pn* (71).

The evidence showing that sterility or inviability of species hybrids is caused by complementary interactions is now overwhelming. Most of the data derive from *Drosophila*. Pantazidis & Zouros and Pantazidis et al (61, 62), for instance, mapped an autosomal factor in *Drosophila mojavensis* that causes hybrid male sterility when brought together with the Y chromosome of *D. arizonae*. Similarly, X-linked genes from *D. persimilis* interact with Y-linked genes from *D. pseudoobscura* to cause hybrid male sterility (49), whereas hybrid sterility in the *D. simulans* subgroup often involves interactions between X-linked and autosomal genes (39, 83). Additional examples from the *Drosophila* literature could be multiplied almost indefinitely (19, 22, 57, 64, 65).

Such between-locus interactions are not limited to *Drosophila*. They have, for example, been studied in platyfish-swordtail hybrids (77). Nor are they limited to animals. Indeed, the first complementary system was found in *Crepis* (32). Between-locus interactions are also responsible for the lethality of certain cotton hybrids (30), as well as that of hybrids between populations of the yellow monkey flower *Mimulus* (10).

Evidence for the Dobzhansky-Muller model is stronger than the mere observation of between-locus incompatibilities. The Dobzhansky-Muller model makes the more specific prediction that incompatibilities should be asymmetric: If allele *A* from one species is incompatible with allele *B* from the other, then their allelomorphs *a* and *b* must be compatible; this reciprocal combination must represent an ancestral genotype (i.e., either the initial common ancestor or, in more complicated scenarios, some intermediate step in the divergence of taxa). This ancestor, by definition, must be viable and fertile. The data reveal that hybrid incompatibilities are, in fact, asymmetric. Perhaps the best data come from a study of *Drosophila pseudoobscura–D. persimilis* hybrid male sterility (79), although similar results have been obtained in other *Drosophila* hybridizations (51, 75).

In sum, overwhelming evidence confirms that hybrid sterility and inviability in animals are caused by Dobzhansky-Muller incompatibilities. While cases of single gene speciation (underdominance) and of chromosomal speciation (e.g., involving large rearrangements) may occur now and then, they clearly do not represent the stuff of most animal speciation.

THE CAUSES OF HALDANE'S RULE

Preliminary Comments

Knowing that speciation is caused by between-locus incompatibilities does not explain Haldane's rule. But the Dobzhansky-Muller model remains immensely important for two reasons. First, any theory of Haldane's rule must work within its confines. We need not entertain theories of Haldane's rule that are inconsistent with the Dobzhansky-Muller mechanism. (And many such theories are possible, e.g., that speciation reflects heterozygote disadvantage.) Second, one of the leading explanations of Haldane's rule—the dominance theory—is a simple extension of the Dobzhansky-Muller model.

A dizzying number of hypotheses have been offered to explain Haldane's rule. Most have, however, been falsified. It is now clear that Haldane's rule is not invariably caused by incompatibilities between X- and Y-linked genes (38, 39, 51, 57, 83; indeed the rule is obeyed in XO taxa, 42), or by meiotic drive (9, 17, 29, 35, 40; despite 34), or by a disruption of dosage compensation (50), or by species-specific translocations between the X chromosome and autosomes (16). Each of these mechanisms may well act now and then, but the evidence strongly suggests that none accounts for the ubiquity of Haldane's rule.

Only three hypotheses—the dominance theory, the "faster-male" theory, and the "faster-X" theory—remain viable. There is little doubt that both the dominance and faster-male theories play an important role in Haldane's rule. Although the evidence for the faster-X theory is weaker, it may also contribute. I devote the remainder of this review to these leading theories. I discuss each in a fair amount of detail, emphasizing evidence pro and con, as well as places where better data are needed.

The Dominance Theory

INTRODUCTION   HJ Muller (44, 45) was the first to consider how the Dobzhansky-Muller model might give rise to Haldane's rule. Although his insight was simple enough, the history of his theory and its modern descendant—the dominance theory—is complex. The theory was first sketched verbally by Muller, then rejected 50 years later, then experimentally resurrected, and finally formalized mathematically. (To make matters worse, the later mathematics revealed that the original verbal incarnation was slightly incorrect.) It is important to see why the dominance theory has gone from being pronounced dead to being proclaimed a (if not the) leading theory of Haldane's rule.

MULLER'S THEORY   Consider two complementary genes that interact to cause hybrid inviability. For clarity, we assume that males are the heterogametic sex (as in *Drosophila*) and change our nomenclature, writing one species's genotype as $A_1A_1B_1B_1$, and the other as $A_2A_2B_2B_2$. Allele A_1 from the first species is incompatible with B_2 from the second species.

First, consider the case in which both genes are autosomal. If females of species 1 hybridize with males of species 2, the resulting hybrids have genotype $A_1A_2B_1B_2$. Because A_1 and B_2 occur as heterozygotes, hybrids will be inviable only if both factors are fairly dominant. [Dominance here refers to the dominance of an allele for fitness when it occurs on a foreign genetic background, i.e., when it occurs with its complementary partner (45). Nothing is assumed about the dominance of alleles on their normal within-species genetic background.] Note that we have no reason to expect a difference in the fitness of hybrid males vs. females as both sexes have the same genotype.

Now consider the case where one locus (A) is X-linked and the other (B) is autosomal. If females of species 1 hybridize with males of species 2, male hybrids have genotype $A_1B_1B_2$ (where I ignore the Y chromosome), and female hybrids have genotype $A_1A_2B_1B_2$. As before, females will be inviable only if the complementary partners A_1 and B_2 are both fairly dominant. If, for instance, B_2 is dominant, but A_1 is recessive, hybrid females remain perfectly fit. But under the same conditions, hybrid males will be lethal: Because the X is hemizygous in males, hybrid males are lethal regardless of the dominance of A_1.

Muller's point is intuitively obvious: Hybrid males are afflicted by all X-linked "speciation genes," dominant and recessive, whereas hybrid females are afflicted only by that subset of genes that are fairly dominant. As long as some fraction of the genes causing hybrid problems is recessive, hybrid males should fare worse than females, giving rise to Haldane's rule. Despite the confusing terminology, note that Muller's "dominance" theory hinges on the existence of alleles that act recessively in hybrids.

REJECTION OF MULLER'S THEORY   Although Muller consistently claimed that Haldane's rule reflects recessivity (44, 45), his attempts to couch his idea in biochemical terms were confusing. In 1942, he pointed out that the genes

incapacitating hybrids are "apt to be recessive" (45) and hinted that this recessivity may be due to alleles that, when introduced into hybrids, act as hypo-
or amorphs. But earlier (44), he speculated that male (heterogametic) hybrids
suffer an "X-autosomal imbalance": While females carry an X and a haploid set
of autosomes from each species, males lack an X from one species (see also 31).
Because levels of expression of X-linked and autosomal genes from the same
species are "balanced," X-linked genes in hybrid males may find the expression
of their autosomal partners inappropriate. This imbalance is especially severe,
Muller argued, because many X-linked genes are recessive; their inappropriate
doses are thus "as fully expressed as in the parent species itself."

Because of these differences in biochemical emphasis, Muller's model is
sometimes referred to as the "recessivity theory" or the "X-autosomal imbalance theory." For our purposes, these distinctions are of little significance. All
versions of Muller's model hinge on the full expression of X-linked "speciation
genes," and all make more or less the same predictions.

Perhaps the simplest of these predictions was identified by Coyne (13), who
pointed out that, if hybrid males are sterile or inviable because they express
recessive X-linked genes causing hybrid problems, then hybrid females that
are homozygous for the same X should also be sterile or inviable (Figure 2).
These females, after all, experience the same degree of X-autosomal imbalance
as F1 males. Coyne tested this prediction in two *Drosophila* hybridizations
obeying Haldane's rule for hybrid sterility: *D. simulans–D. mauritiana* and

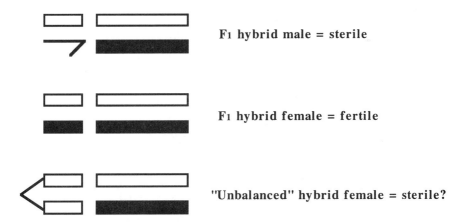

F1 hybrid male = sterile

F1 hybrid female = fertile

"Unbalanced" hybrid female = sterile?

Figure 2   The "unbalanced" female test. The short horizontal bars represent sex chromosomes
(the Y is shown with a hook). The long horizontal bars represent haploid sets of autosomes. One
species's chromosomes are shown in black and the other in white. Unbalanced hybrid females carry
an attached-X chromosome.

D. simulans–D. sechellia. Using an attached-X chromosome from *D. simulans*, he produced "unbalanced" hybrid females that carry a *D. simulans* attached-X on an otherwise hybrid genetic background (Figure 2). Despite being homozygous for their X chromosomes, these unbalanced females remained fertile in both species crosses.

Although this result was unexpected—indeed so unexpected that it single-handedly reignited interest in the genetics of speciation—it was quickly confirmed by analogous tests with other *Drosophila* species. Tests were performed on at least five evolutionarily independent hybridizations, and in all cases unbalanced females remained perfectly fit (16, 49, 51, 57, 74). In the *D. melanogaster* subgroup, Coyne (13) further presented evidence that hybrid male sterility is caused by incompatibilities between X- and Y-linked loci, not by X-autosomal imbalance. These findings appeared to falsify Muller's theory.

RESURRECTION OF MULLER'S THEORY   As Wu and colleagues (78, 80) emphasized, the above tests suffered one limitation: All involve hybrid sterility. It thus remained possible that Muller's theory explains Haldane's rule for inviability. Unbalanced female tests for hybrid inviability had not been performed for purely technical reasons: Instances of Haldane's rule for inviability are rarer than for sterility in *Drosophila* (Table 1), and the required attached-X stocks were not available in the relevant species.

In 1993, Orr (53) found two *Drosophila* species crosses in which the required tests could be performed: *D. simulans–D. teissieri* and *D. melanogaster–D. simulans.* [Descriptions of the latter cross (69, 70) typically describe females as lethal in one direction of the hybridization, suggesting that female viability involves a maternal effect that might compromise unbalanced female tests. Fortunately, strains were found in which all hybrid females are viable (53); with these strains, the species cross neatly obeys Haldane's rule.] Remarkably, Orr found that females who are homozygous for an X on an otherwise hybrid background are lethal in both species crosses. Moreover, these females die at the same developmental stage as F1 males (53), suggesting the same genes cause hybrid male and unbalanced female lethality. In *D. melanogaster–D. simulans,* this interpretation is essentially proved by the finding that unbalanced female viability is restored by mutations known to rescue hybrid males, e.g. *Hmr* and *In(1)AB* (36, 37). Taken together, these results provide strong support for Muller's theory of Haldane's rule: Heterozgygous females are viable, but homozygotes are lethal.

In retrospect, it is clear why the hybrid inviability and sterility results differ (53, 80). Mutation studies in *D. melanogaster* show that lethals almost always affect both sexes (the exceptions being lesions at genes belonging to the sex determination cascade, e.g., *Sxl*), while steriles typically affect one sex only (reviewed in 1). If the same pattern holds in species hybrids, an X that carries a recessive allele killing hybrid males will, when made homozygous, kill hybrid females. But an X that carries a recessive allele sterilizing hybrid males may, when made homozygous, not sterilize females: A female is none the worse for being homozygous for a hybrid male sterile (16). Indeed, there is now good evidence for sex-limited expression of hybrid steriles: Several backcross (57) and introgression (33, 73) experiments show that chromosome regions causing male sterility between two species differ from those causing female sterility.

Although unbalanced female tests provide strong support for the dominance theory for inviability, it is unclear if the repeated recovery of fertile unbalanced females tells us anything about hybrid sterility. The question is difficult because the answer depends on the genetic architecture of speciation. Consider two cases. In the first, hybrid sterility involves few genes. If these genes are recessive, Haldane's rule is obeyed: Female steriles, unlike male, are invariably heterozygous. But unbalanced females will often remain fertile as an X that bears a hybrid male sterile will often not bear a female sterile (74). Thus, unbalanced female tests lead us to incorrectly reject the dominance theory. In the second case, hybrid sterility involves many genes. If these genes are recessive, Haldane's rule is again obeyed. But now unbalanced females should often be sterile: Because the X is saturated with hybrid male steriles, it likely also carries female steriles (74). If, therefore, unbalanced females usually remain fertile, we correctly reject the dominance theory—something else must cause Haldane's rule.

In reality, we do not usually know which of these genetic architectures holds (if we knew such details, we would have little need for unbalanced female tests). Indeed, we still do not know if the genetic basis of postzygotic isolation is usually simple or complex (5, 6, 10, 20, 21, 43, 47, 48, 52, 60, 63, 74, 77, 82, 83). Given this uncertainty—as well as that over whether hybrid male and female steriles evolve at the same rate (see below)—unbalanced female tests cannot be used, as they once were, to reject the dominance theory for sterility. Evidence for or against the dominance theory for hybrid sterility must be found elsewhere.

FORMALIZING THE DOMINANCE THEORY   Muller's theory of Haldane's rule has been formalized in several mathematical models (54, 59, 74). Although confirming the gist of his argument, this work has revealed an error in Muller's intuition. While Muller was correct that hybrid females, unlike males, partly mask the effects of X-linked speciation genes, he overlooked the fact that females, by carrying twice as many X chromosomes, suffer twice as many hybrid incompatibilities involving the X (54). If the alleles causing hybrid problems act additively in hybrids (no dominance), these two forces balance, and hybrid males and females suffer equally. But if the alleles causing hybrid problems act partially recessively, hybrid males suffer more than females. Thus,

Muller's theory of Haldane's rule requires more than the mere existence of some recessive alleles—it requires that speciation genes are, on average, partially recessive.

Although Orr's (54) original model assumed that speciation is due to many genes of small effect each and that these factors interact multiplicatively in hybrids, Turelli & Orr (74) relaxed both assumptions. Even when hybrid fitness declines as an arbitrary function of the number of Dobzhansky-Muller incompatibilities, and even when individual incompatibilities can have large effects, the conditions for Haldane's rule remain essentially unchanged: Haldane's rule holds whenever the alleles causing hybrid problems are partially recessive, $d < 1/2$. [The parameter d incorporates both dominance (h) and any covariance between deleterious effects in hybrids and dominance. As before, dominance refers solely to dominance for hybrid fitness.]

The above calculations naively assume that genes afflicting hybrid males vs. females accumulate at the same evolutionary rate. Although this is presumably true for hybrid inviability, it is not necessarily true for sterility: Because different sets of loci cause male vs. female sterility, we have no guarantee that both types of steriles appear at the same rate. Indeed, there is good evidence that male and female steriles accumulate at different rates (see below). To take this possibility into acount, Turelli & Orr (74) presented a formal "composite model" of Haldane's rule that simultaneously incorporates the effects of dominance and differential accumulation of male vs. female genes. They showed that Haldane's rule occurs when

$$d < \frac{\tau p_x}{2[1 - \tau(1 - p_x)]},$$
1.

where p is the proportion of all Dobzhansky-Muller incompatibilities that involve X-linked loci, and τ reflects the ratio of male to female hybrid steriles at any point in time (if hybrid male steriles accumulate faster than female, $\tau > 1$). Note that, if male and female steriles accumulate at the same rate ($\tau = 1$), Equation 1 reduces to $d < 1/2$, as expected.

The more interesting cases occur when $\tau \neq 1$, i.e., when both forces, dominance and differential accumulation of sex-specific factors, act. The qualitative effect is straightforward: As the rate of male (heterogametic) evolution begins to exceed female, it becomes easier to obtain Haldane's rule, i.e., the mean dominance, d, consistent with Haldane's rule increases above $d = 1/2$. This is intuitively reasonable: In the extreme case in which only male steriles accumulate, Haldane's rule clearly holds for any dominance. Similarly, rearrangement of Equation 1 shows that, as the alleles causing speciation become more and more recessive, Haldane's rule holds even if male steriles accumulate more slowly than do female steriles. In sum, either force alone may be sufficient to yield Haldane's rule (59, 74). Their joint action merely makes it easier to obtain Haldane's rule.

FURTHER EVIDENCE   Additional support for the dominance theory has recently emerged. Most of this work involves tests of one of the theory's simplest predictions: If speciation genes are recessive, it should be possible to find autosomal alleles that, while masked in F1 hybrids, cause postzygotic isolation when made homozygous on a hybrid background. Such hidden factors have been found in backcross experiments involving nearly intact chromosomes (20, 46, 52) and in introgression experiments involving smaller chromosome regions (33, 73). This work—which involves both hybrid inviability and sterility—provides strong support for the dominance theory. Hollocher & Wu (33), for example, in their study of 18 autosomal introgressions from *D. mauritiana* and *D. sechellia* into *D. simulans*, found that "[a]lthough individuals heterozygous for second chromosome introgressions generally do not suffer sterility or inviability, individuals homozygous for these same regions show dramatic increases in both" (see also 73). Similarly, recent work in the haplodiploid wasp *Nasonia* shows that F2 males (haploid) are more severely afflicted in species crosses than F2 females (diploid) and that this difference reflects the partial recessivity of the factors causing hybrid inviability (4). It is important to note, however, that such data often remain qualitative. We typically lack quantitative measures of dominance for the factors causing postzygotic isolation (but see 52). More rigorous estimates of dominance are needed.

M Turelli and D Begun (unpublished data) have recently performed a novel test of the dominance theory: All else being equal, taxa with large Xs should, under the dominance theory, evolve Haldane's rule faster than taxa with small Xs. This prediction was confirmed in a comparison of the time (in units of genetic distance) to Haldane's rule for both inviability and sterility in pairs of *Drosophila* species having 40% vs. 20% of their genome X-linked.

A final line of evidence for the dominance theory is provided by the "large X-effect." Backcross analyses typically reveal that substitution of a hemizygous X has a much larger effect on hybrid sterility or inviability than substitution of a heterozygous autosome (16). Interpretation of this effect has, however, proved difficult. Early on, it was thought to reflect rapid evolution of X-linked speciation genes (8, 16). This now seems unlikely (see below). Wu and colleagues have instead argued that the large X-effect reflects an observational bias as backcrosses only allow one to compare hemizygous-X with heterozygous-autosome substitutions (33, 80, 81). But it is important to draw a distinction. If speciation genes act additively (no dominance), one might expect hemizygous X substitutions to have twice the effect of same-sized heterozygous autosome substitutions; such a difference might be legitimately branded an artifact of hemizygosity. But if speciation genes are very recessive, hemizygous X substitutions might have much larger effects than autosomal. In this case, the large X-effect would not be an artifact of, but evidence for dominance.

A simple observational bias also cannot explain one of the best known facts about the X-effect: that it is seen only with hybrid sterility and inviability, not with morphological species differences (8, 16). If the X-effect is an artifact of contrasting hemizygous and heterozygous substitutions, why isn't it seen with all characters? The likely answer is that morphological species differences often involve alleles that act more or less additively, while postzygotic isolation involves alleles that act fairly recessively. Although the X-effect provides qualitative support for the dominance theory, quantitative estimates of dominance from hybrid backcross data are not yet available. The problem is that any chromosome substitution simultaneously changes many kinds of hybrid interactions, e.g., a second chromosome substitution changes X-2, 2-3, 2-4, etc interactions. Some interactions are improved and others worsened. Consequently, simple predictions about how a substitution affects hybrid fitness as a function of dominance are hard to come by. Very recent theory suggests, however, that it may be possible to estimate the dominance of speciation genes from hybrid backcross data (M Turelli, HA Orr, unpublished data).

In sum, recent work has resulted in a strong consensus that dominance alone can explain Haldane's rule for inviability (59, 73, 74, 80, 81). The data further suggest—but do not prove—that hybrid steriles are, on average, partially recessive.

The Faster-Male Theory

INTRODUCTION   Haldane's rule would obviously arise if genes afflicting heterogametic hybrids evolved at a faster rate than those afflicting homogametic hybrids. Wu (80, 81) has championed a version of this idea: Hybrid male steriles, he suggests, accumulate faster than hybrid female steriles. Two factors might cause such "faster male" evolution: 1. Spermatogenesis might be an inherently sensitive process that is easily perturbed in hybrids (80). If so, male-expressed genes may evolve at the same rate as female-expressed genes, but a larger fraction of the former ultimately cause problems in hybrids. 2. Sexual selection might cause faster evolution of genes expressed in males than in females (80). Indeed there is good evidence that male reproductive characters evolve faster than female, at least in insects: Male genitalia are the most rapidly evolving of all morphological characters in insects (26). Moreover, two-dimensional electrophoresis reveals that proteins from the male reproductive tract diverge between *Drosophila* species faster than proteins from most other tissues (11). Under the Dobzhansky-Muller model, it stands to reason that rapidly diverging male proteins would be incompatible with each other more often than slowly evolving female proteins.

EVIDENCE   The faster-male theory makes a simple prediction: Chromosome regions introgressed from one species into another should contain hybrid male steriles more often than female. This prediction has been confirmed in what surely ranks as the most important recent work on Haldane's rule: True et al (73) marked the *D. mauritiana* genome with P element inserts at 87 locations distributed over all four chromosomes (the P elements bore a w+ eye marker). They then separately introgressed (with replication) each of their *D. mauritiana* markers into a *D. simulans* background by backcrossing P[w+]-bearing hybrid females to *D. simulans* for 15 generations. This scheme should produce "*D. simulans*" lines that carry, on average, a 9 cM introgressed region from *D. mauritiana*. When autosomal introgressions were made homozygous, True et al found that 36% of introgression sublines were male-sterile, but only 7% were female-sterile. As these data reflect an enormous sample size—185 homozygous viable sublines—there can be little doubt of the reality of the result.

Interpretation of the result is, though, somewhat complicated by one feature of True et al's design: Because they repeatedly backcrossed through females, the excess of hybrid male over female steriles might reflect selection against female steriles during the course of introgression. Fortunately, True et al's results were qualitatively confirmed by an independent experiment in which selection against hybrid female steriles was relatively mild: Hollocher & Wu (33) introgressed material from *D. mauritiana* and *D. sechellia* into *D. simulans*. Although they produced far fewer introgression lines and restricted their study to the second chromosome, Hollocher & Wu's introgression scheme involved more generations of selection in hybrid males than females. Nonetheless, they observed a large excess of hybrid male over female steriles. Although data from additional species are badly needed (the above results may reflect a single evolutionary event as *D. simulans–D mauritiana* and *D. simulans–D. sechellia* are not necessarily independent), these experiments provide strong support for the faster male theory. There can be little doubt that faster accumulation of hybrid male steriles contributes to Haldane's rule in *Drosophila*.

The faster male theory does, however, suffer two limitations (59). First, it cannot explain Haldane's rule for hybrid inviability, as there is essentially no such thing as a male vs. a female lethal. Indeed, both of the recent introgression experiments tested the viability effects of introgressed regions and, as expected, found little or no sex-specific hybrid lethality (33, 73).

Second, the faster male theory cannot explain Haldane's rule for sterility in taxa with female heterogamety (59). In birds and Lepidoptera, for instance, faster evolution of male steriles due either to sexual selection or to the inherent sensitivity of spermatogenesis would work against Haldane's rule: It is hybrid females, after all, that are preferentially sterile (Table 1). In these taxa, other forces must give rise to Haldane's rule, and these forces must be strong enough to overcome any faster male effect. In this context, it is worth noting that

Hollocher & Wu (33) found that heterozygous introgressions have no detectable fitness effect in hybrids, while homozygous introgressions often do. Based on such findings, Wu and colleagues (80, 81) acknowledged that, in birds and butterflies, dominance must be invoked to explain Haldane's rule for sterility.

The Faster-X Theory

INTRODUCTION   Charlesworth et al (8) have offered yet another explanation of Haldane's rule. They argue that the pattern is an epiphenomenon of the large X-effect: If X-linked genes have a disproportionate effect on hybrid fitness, it is hardly surprising that heterogametic hybrids suffer more than homogametic. The problem, then, is to understand why the X plays such a large role in postzygotic isolation. Charlesworth et al speculated that X-linked genes simply evolve faster than autosomal: They showed that, when evolution is driven by natural selection, the ratio of substitution rates at autosomal vs. X-linked genes is $A/X = 4h/(2h + 1)$. Faster X evolution thus occurs when favorable mutations are, on average, partially recessive ($h < 1/2$), where dominance now refers to the dominance of alleles on their normal species genetic background; nothing is assumed about dominance in hybrids.

The faster X theory differs in an important way from the dominance and faster male theories: It cannot, by itself, explain Haldane's rule. The problem is that, if genes affecting males and females evolve at the same rate ($\tau = 1$) and act additively in hybrids ($d = 1/2$), male and female hybrids are equally fit regardless of the rate of evolution on the X (see Equation 1 above). The reason, as noted earlier, is that, while the heterogametic sex suffers the full brunt of X-linked genes, homogametic hybrids suffer from twice as many incompatibilities involving the X. With additivity, these forces balance (54, 74). The faster-X theory can, however, be modified in two ways to yield Haldane's rule.

First, we can assume that the genes causing hybrid problems also act as partial recessives in hybrids (8). While the faster-X theory now depends on the dominance theory, faster-X evolution exaggerates the effects of dominance: When speciation genes are partially recessive in hybrids and are concentrated on the X, heterogametic hybrids will suffer disproportionately (74). Alternatively, we can assume that many of the genes that ultimately afflict hybrids are expressed in one sex only (16). If so, X-linked genes expressed in the heterogametic sex will evolve faster than those expressed in the homogametic sex when favorable mutations are partially recessive (for their effects within species): The former, after all, are fully expressed in hemizygotes. This faster divergence of heterogametic-sex genes on the X could, then, account for the disproportionate problems suffered by heterogametic hybrids.

EVIDENCE FROM INTROGRESSION EXPERIMENTS   Introgression experiments provide a potentially powerful test of faster-X evolution: By comparing hemizygous-X with homozygous-autosomal introgressions, one can assess the density of X vs autosomal speciation genes without bias. Unfortunately, the two recent introgression studies appear to give mixed results. Hollocher & Wu (33) argued that hybrid steriles are no more common on the X than second chromosome in the *D. simulans* clade. Their results were compromised, however, by a small sample of introgression lines (6 from *D. mauritiana* and 12 from *D. sechellia)* and by the fact that their X vs. autosomal data were collected in different experiments. Moreover, their introgressions were often large, further reducing statistical power (i.e., two large introgressions are both likely to be sterile, although one may include more steriles than the other). The work of True et al (73) provides more precise data. They found that X-linked regions from *D. mauritiana* were 50% more likely than (homozygous) autosomal regions to cause male sterility when introgressed into *D. simulans.* The effect was statistically significant. Nonetheless, True et al urged caution in interpreting this result given some uncertainty about the sizes of X-linked vs. autosomal introgressions.

The faster-X theory as extended by Coyne & Orr (16) predicts not only more X-linked than autosomal steriles, but more heterogametic than homogametic steriles (as only heterogametic-expressed genes evolve at an accelerated rate on the X). Although True et al (73) observed both patterns, they rejected the faster-X theory, noting that it cannot explain the excess of male steriles seen on the autosomes. Unfortunately this is incorrect. Although the faster-X theory predicts that favorable mutations affecting males vs. females will be fixed at the same rate on the autosomes (8, 16), it does not follow that autosomal hybrid male vs. female steriles will be equally common: Because autosomal genes, when introgressed, interact with many more diverged X-linked genes affecting males than females, a greater proportion of autosomal "male genes" will act as hybrid steriles even when autosomal genes affecting the two sexes evolve at the same rate. (The expected ratio of hybrid male to female steriles on the autosomes under the faster-X theory is $m/f = 1 + x(\tau - 1)$, where x is the fraction of the genome that is X-linked and τ is the ratio of male-to-female substitution rates on the X (HA Orr, unpublished). Although evidence for the faster-X theory is less than overwhelming (see below), faster-X evolution remains a potential contributor to the excess of male steriles in *Drosophila.* Fortunately, the faster-X theory makes a prediction that, in principle, distinguishes it from the faster-male alternatives (sexual selection and "spermatogenesis is special"): Only the faster-X theory predicts an excess of hybrid female over male steriles in birds and butterflies.

OTHER EVIDENCE   The faster-X theory makes two predictions that can be tested without studying speciation. First, if favorable mutations are partially recessive, we should observe more rapid molecular evolution at X-linked than autosomal genes. Although some evidence points to such an effect in the *Drosophila athabasca* complex (28), we currently possess far too little molecular population genetic data to allow any confident conclusions. Worse, this population genetic test is one-sided: If we consistently find faster sequence divergence at X-linked genes, the faster-X theory is supported; if we do not, the substitutions observed may have been effectively neutral and thus unsuitable for testing the faster-X hypothesis. We can largely circumvent this problem by studying morphological species differences (where it seems far more likely that natural selection has acted). If favorable mutations are typically partially recessive, we should observe large X-effects in genetic analyses of both postzygotic isolation and morphological species differences. As noted, the data here are clear: Morphological differences do not show large X-effects (8, 16). Although post hoc explanations of this difference are possible—e.g., postzygotic isolation involves a few major factors that, for metabolic reasons, are likely to act non-additively, while morphology involves many nearly additive genes (9)—the lack of X-effects for morphology poses an awkward problem for the faster-X theory.

The second prediction is that adaptive evolution in inbreeding species—in which the favorable effects of recessives are quickly expressed—should involve recessives more often than in outbreeding species (7). Charlesworth (7) tested this prediction by comparing the frequency of recessive vs. dominant derived states in partially selfing plants. He found that the derived state among inbreeders is often recessive; this result appears to differ profoundly from that in outbreeding animals. Although this finding is suggestive, it is not conclusive for several reasons. First, the direction of evolution was, in several cases, uncertain (7). Second, inbreeding species will fix recessive alleles more often than outbreeders even if favorable mutations are not, on average, partially recessive. Third, most of the characters studied were morphological and likely polygenic (7). But the argument made above asserts that polygenic morphological characters are not suitable for testing the faster-X theory as they likely involve additive polygenes. We cannot have it both ways: Polygenic morphological characters are either suitable material or they are not. And if they are, genetic analyses of morphology should reveal large X-effects; they do not.

Last, the faster-X theory suffers from two theoretical problems, though neither is fatal. First, the claim that X-linked loci evolve faster than autosomal may depend on the assumption that adaptation uses newly arising unique mutations. If adaptation uses variation previously maintained at mutation-selection balance, it is unclear if X-linked loci will evolve faster than autosomal. Second, the notion that most favorable mutations are partially recessive is biochemically counterintuitive: Metabolic theory and genetic data show that recessive alleles are typically loss-of-function lesions. The faster-X theory thus requires us to believe that most adaptive evolution involves substitution of loss-of-function alleles, which is far from obvious.

CONCLUSIONS

After a decade of intensive study, a consensus has emerged that two forces, dominance and faster-male evolution, cause Haldane's rule. A third force, faster-X evolution, may also play a role, though the evidence here is considerably weaker.

One of these forces—dominance—differs from the others in several important ways. First, only the dominance theory can explain Haldane's rule for both hybrid phenotypes (inviability and sterility) and in all taxa (those with heterogametic males or females). Second, the other theories rely, in one way or another, on dominance: The faster-male theory requires dominance to explain Haldane's rule in half the taxa surveyed (those with heterogametic females), while the faster-X theory requires dominance to explain more than the large X-effect. An argument can be made, therefore, that dominance plays a more universal and thus fundamental role in Haldane's rule (59).

There can be no doubt, however, that faster-male evolution also contributes to Haldane's rule for sterility. Thus, at least in some taxa, Haldane's rule probably reflects the simultaneous action of dominance and faster-male evolution. It is reassuring to note that this conclusion—reached by sometimes-arcane genetic studies—makes sense of one of the simplest patterns associated with Haldane's rule. For if both forces play a role, one might expect more instances of Haldane's rule for sterility than inviability (81): With sterility, both dominance and faster-male evolution act, whereas with inviability, dominance alone acts. Moreover, this excess should be limited to those taxa having heterogametic males as both forces act in the same direction in these taxa only. This is, of course, the pattern observed (Table 1): There are many more cases of hybrid sterility than inviability in mammals and *Drosophila*, but not in birds and Lepidoptera (81).

This pattern, together with the results of detailed genetic experiments, strongly suggests that the causes of Haldane's rule have at last been identified. Future experiments—especially those dissecting the genetics of speciation in taxa with heterogametic females—will reveal if the explanation of Haldane's rule championed here fares any better than its many predecessors.

ACKNOWLEDGMENTS

I thank J Coyne, C Laurie, and M Turelli for very helpful comments. This work was supported by NIH grant GM51932.

Literature Cited

1. Ashburner M. 1989. *Drosophila: A Laboratory Handbook.* New York: Cold Spring Harbor Lab. Press. 1331 pp.
2. Baird SE, Sutherlin ME, Emmons SW. 1992. Reproductive isolation in Rhabditidae (Nematoda: Secernentea); mechanisms that isolate six species of three genera. *Evolution* 46:585–94
3. Bateson W. 1909. Heredity and variation in modern lights. In *Darwin and Modern Science,* ed. AC Seward, pp. 85–101. Cambridge, UK: Cambridge Univ. Press.
4. Breeuwer JAJ, Werren JH. 1996. Hybrid breakdown between two haplodiploid species: the role of nuclear and cytoplasmic genes. *Evolution* 49:705–17
5. Cabot EL, Davis AW, Johnson NA, Wu C-I. 1994. Genetics of reproductive isolation in the *Drosophila simulans* clade: complex epistasis underlying hybrid male sterility. *Genetics* 137:175–89
6. Carvajal AR, Gandarela MR, Naveira HF. 1996. A three-locus system of interspecific incompatibility underlies male inviability between *Drosophila buzzatii* and *D. koepferae. Genetica* 98:1–19
7. Charlesworth B. 1992. Evolutionary rates in partially self-fertilizing species. *Am. Nat.* 140:126–48
8. Charlesworth B, Coyne JA, Barton N. 1987. The relative rates of evolution of sex chromosomes and autosomes. *Am. Nat.* 130:113–46
9. Charlesworth B, Coyne JA, Orr HA. 1993. Meiotic drive and unisexual hybrid sterility: a comment. *Genetics* 133(2):421–424
10. Christie P, Macnair MR. 1984. Complementary lethal factors in two North American populations of the yellow monkey flower. *J. Hered.* 75:510–11
11. Coulthart MB, Singh RS. 1988. High level of divergence of male reproductive tract proteins between *Drosophila melanogaster* and its sibling species, *D. simulans. Mol. Biol. Evol.* 5:182–91
12. Coyne J. 1992. Genetics and speciation. *Nature* 355:511–15
13. Coyne JA. 1985. The genetic basis of Haldane's rule. *Nature* 314:736–38
14. Coyne JA, Charlesworth B, Orr HA. 1991. Haldane's rule revisited. *Evolution* 45:1710–14
15. Coyne JA, Orr HA. 1989. Patterns of speciation in *Drosophila. Evolution* 43:362–81
16. Coyne JA, Orr HA. 1989. Two rules of speciation. In *Speciation and Its Consequences,* ed. D Otte, J Endler, pp. 180–207. Sunderland, MA: Sinauer. 679 pp.
17. Coyne JA, Orr HA. 1993. Further evidence against meiotic-drive models of hybrid sterility. *Evolution* 47:685–87
18. Coyne JA, Orr HA. 1997. Patterns of speciation revisited. *Evolution* 51:295–303
19. Crow JF. 1942. Cross fertility and isolating mechanisms in the *Drosophila mulleri* group. *Univ. Texas Publ.* 4228:53–67
20. Davis AW, Noonburg EG, Wu C-I. 1994. Evidence for complex genic interactions between conspecific chromosomes underlying hybrid female sterility in the *Drosophila simulans* clade. *Genetics* 137:191–99
21. Davis AW, Wu C-I. 1996. The broom of the sorcerer's apprentice: the fine structure of a chromosomal region causing reproductive isolation between two sibling species of *Drosophila. Genetics* 143:1287–98
22. Dobzhansky T. 1936. Studies on hybrid sterility. II. Localization of sterility factors in *Drosophila pseudoobscura* hybrids. *Genetics* 21:113–35
23. Dobzhansky T. 1937. *Genetics and the Origin of Species.* New York: Columbia Univ. Press. 364 pp.
24. Dobzhansky T. 1946. Genetics of natural populations. XIII. Recombination and variability in populations of *Drosophila pseudoobscura. Genetics* 31:269–90
25. Dobzhansky T, Levene H, Spassky B, Spassky N. 1959. Release of genetic variability through recombination. III. *Drosophila prosaltans. Genetics* 44:75–92
26. Eberhard WG. 1985. *Sexual Selection and Animal Genitalia.* Cambridge, MA: Harvard Univ. Press. 244 pp.
27. Fisher RA. 1930. *The Genetical Theory of Natural Selection.* Oxford, UK: Oxford Univ. Press. 291 pp.
28. Ford MJ, Aquadro CF. 1996. Selection on X-linked genes during speciation in the *Drosophila athabasca* complex. *Genetics* 144:689–703
29. Frank SH. 1991. Divergence of meiotic drive-suppressors as an explanation for sex-biased hybrid sterility and inviability. *Evolution* 45:262–67
30. Gerstel DU. 1954. A new lethal combination in interspecific cotton hybrids. *Genetics* 39:628–39
31. Haldane JBS. 1922. Sex ratio and unisexual sterility in animal hybrids. *J. Genet.* 12:101–9
32. Hollingshead L. 1930. A lethal factor in *Crepis* effective only in interspecific hybrids. *Genetics* 15:114–40
33. Hollocher H, Wu C-I. 1996. The genetics

of reproductive isolation in the *Drosophila simulans* clade: X vs. autosomal effects and male vs. female effects. *Genetics* 143:1243–55

34. Hurst LD, Hurst GG. 1996. Genomic revolutionaries rise up. *Nature* 384:317–18

35. Hurst LD, Pomiankowski A. 1991. Causes of sex ratio bias may account for unisexual sterility in hybrids: a new explanation of Haldane's rule and related phenomena. *Genetics* 128:841–58

36. Hutter P, Ashburner M. 1987. Genetic rescue of inviable hybrids between *Drosophila melanogaster* and its sibling species. *Nature* 327:331–33

37. Hutter P, Roote J, Ashburner M. 1990. A genetic basis for the inviability of hybrids between sibling species of *Drosophila*. *Genetics* 124:909–20

38. Johnson NA, Hollocher H, Noonburg E, Wu C-I. 1993. The effects of interspecific Y chromosome replacements on hybrid sterility within the *Drosophila simulans* clade. *Genetics* 135:443–53

39. Johnson NA, Perez DE, Cabot EL, Hollocher H, Wu C-I. 1992. A test of reciprocal X-Y interactions as a cause of hybrid sterility in *Drosophila*. *Nature* 358:751–53

40. Johnson NA, Wu C-I. 1992. An empirical test of the meiotic drive models of hybrid sterility: sex ratio data from hybrids between *Drosophila simulans* and *Drosophila sechellia*. *Genetics* 130:507–11

41. Krimbas CB. 1960. Synthetic sterility in *Drosophila willistoni*. *Proc. Natl. Acad. Sci. USA* 46:832–33

42. Mantovani B, Scali V. 1992. Hybridogenesis and androgenesis in the stick-insect *Bacillus rossius-grandii benazzii* (Insecta, Phasmatodea). *Evolution* 46:783–96

43. Marin I. 1996. Genetic architecture of autosome-mediated hybrid male sterility in *Drosophila*. *Genetics* 142:1169–80

44. Muller HJ. 1940. Bearing of the *Drosophila* work on systematics. In *The New Systematics,* ed. JS Huxley, pp. 185–268. Oxford: Clarendon. 583 pp.

45. Muller HJ. 1942. Isolating mechanisms, evolution, and temperature. *Biol. Symp.* 6:71–125

46. Muller HJ, Pontecorvo G. 1942. Recessive genes causing interspecific sterility and other disharmonies between *Drosophila melanogaster* and *simulans*. *Genetics* 27:157

47. Naveira H, Fontdevila A. 1986. The evolutionary history of *Drosophila buzzatii*. XII. The genetic basis of sterility in hybrids between *D. buzzatii* and its sibling *D. serido*

from Argentina. *Genetics* 114:841–57

48. Naveira H, Fontdevila A. 1991. The evolutionary history of *Drosophila buzzatii*. XXI. Cumulative action of multiple sterility factors on spermatogenesis in hybrids of *D. buzzatii* and *D. koepferae*. *Heredity* 67:57–72

49. Orr HA. 1987. Genetics of male and female sterility in hybrids of *Drosophila pseudoobscura* and *D. persimilis*. *Genetics* 116:555–63

50. Orr HA. 1989. Does postzygotic isolation result from improper dosage compensation? *Genetics* 122:891–94

51. Orr HA. 1989. Genetics of sterility in hybrids between two subspecies of *Drosophila*. *Evolution* 43:180–89

52. Orr HA. 1992. Mapping and characterization of a "speciation gene" in *Drosophila*. *Genet. Res.* 59:73–80

53. Orr HA. 1993. Haldane's rule has multiple genetic causes. *Nature* 361:532–33

54. Orr HA. 1993. A mathematical model of Haldane's rule. *Evolution* 47:1606–11

55. Orr HA. 1995. The population genetics of speciation: the evolution of hybrid incompatibilities. *Genetics* 139:1805–13

56. Orr HA. 1996. Dobzhansky, Bateson, and the genetics of speciation. *Genetics* 144:1331–35

57. Orr HA, Coyne JA. 1989. The genetics of postzygotic isolation in the *Drosophila virilis* group. *Genetics* 121:527–37

58. Orr HA, Orr LH. 1996. Waiting for speciation: the effect of population subdivision on the time to speciation. *Evolution* 50:1742–49

59. Orr HA, Turelli M. 1996. Dominance and Haldane's rule. *Genetics* 143:613–16

60. Palopoli MF, Wu C-I. 1994. Genetics of hybrid male sterility between *Drosophila* sibling species: a complex web of epistasis is revealed in interspecific studies. *Genetics* 138:329–41

61. Pantazidis AC, Galanopoulos VK, Zouros E. 1993. An autosomal factor from *Drosophila arizonae* restores normal spermatogenesis in *Drosophila mojavensis* males carrying the *D. arizonae* Y chromosome. *Genetics* 134:309–18

62. Pantazidis AC, Zouros E. 1988. Location of an autosomal factor causing sterility in *Drosophila mojavensis* males carrying the *Drosophila arizonensis* Y chromosome. *Heredity* 60:299–304

63. Perez DE, Wu C-I, Johnson NA, Wu M-L. 1993. Genetics of reproductive isolation in the *Drosophila simulans* clade: DNA-marker assisted mapping and characterization of a hybrid-male sterility gene, *Odysseus* (*Ods*). *Genetics* 134:261–75

64. Pontecorvo G. 1943. Hybrid sterility in artificially produced recombinants between *Drosophila melanogaster* and *D. simulans*. *Proc. R. Soc. Edinburgh B* 61:385–97

65. Pontecorvo G. 1943. Viability interactions between chromosomes of *Drosophila melanogaster* and *Drosophila simulans*. *J. Genet.* 45:51–66

66. Read A, Nee S. 1991. Is Haldane's rule significant? *Evolution* 45:1707–9

67. Rykena S. 1991. Hybridization experiments as tests for species boundaries in the genus *Lacerta* sensu stricto. *Mitt. Zool. Mus. Berl.* 67:55–68

68. Spurway H. 1953. Genetics of specific and subspecific differences in European newts. *Symp. Soc. Exp. Biol.* 7:200–37

69. Sturtevant AH. 1920. Genetic studies on *Drosophila simulans*. I. Introduction. Hybrids with *Drosophila melanogaster*. *Genetics* 5:488–500

70. Sturtevant AH. 1929. The genetics of *Drosophila simulans*. *Carneg. Inst. Wash. Publ.* 399:1–62

71. Sturtevant AH. 1956. A highly specific complementary lethal system in *Drosophila melanogaster*. *Genetics* 41:118–23

72. Thompson V. 1986. Synthetic lethals: a critical review. *Evol. Theory* 8:1–13

73. True JR, Weir BS, Laurie CC. 1996. A genome-wide survey of hybrid incompatibility factors by the introgression of marked segments of *Drosophila mauritiana* chromosomes into *Drosophila simulans*. *Genetics* 142:819–37

74. Turelli M, Orr HA. 1995. The dominance theory of Haldane's rule. *Genetics* 140:389–402

75. Vigneault G, Zouros E. 1986. The genetics of asymmetrical male sterility in *Drosophila mohavensis* and *Drosophila arizonensis* hybrids: interactions between the Y chromosome and autosomes. *Evolution* 40:1160–70

76. White MJD. 1978. *Modes of Speciation*, ed. CI Davern. San Francisco: WH Freeman. 455 pp.

77. Wittbrodt J, Adam D, Malitschek B, et al. 1989. Novel putative receptor tyrosine kinase encoded by the melanoma-inducing *Tu* locus in *Xiphophorus*. *Nature* 341:415–21

78. Wu C-I. 1992. A note on Haldane's rule: hybrid inviability versus hybrid sterility. *Evolution* 46:1584–87

79. Wu C-I, Beckenbach AT. 1983. Evidence for extensive genetic differentiation between the sex-ratio and the standard arrangement of *Drosophila pseudoobscura* and *D. persimilis* and identification of hybrid sterility factors. *Genetics* 105:71–86

80. Wu C-I, Davis AW. 1993. Evolution of postmating reproductive isolation: the composite nature of Haldane's rule and its genetic bases. *Am. Nat.* 142:187–212

81. Wu C-I, Johnson NA, Palopoli MF. 1996. Haldane's rule and its legacy: Why are there so many sterile males? *Trends Ecol. Evol.* 11:411–13

82. Zeng L-W, Singh RS. 1993. A combined classical genetic and high resolution two-dimensional electrophoretic approach to the assessment of the number of genes affecting hybrid male sterility in *Drosophila simulans* and *Drosophila sechellia*. *Genetics* 135:135–47

83. Zeng L-W, Singh RS. 1993. The genetic basis of Haldane's rule and the nature of asymmetric hybrid male sterility among *Drosophila simulans*, *Drosophila mauritiana* and *Drosophila sechellia*. *Genetics* 134:251–60

Annu. Rev. Ecol. Syst. 1997. 28:219–41

ECHINODERM LARVAE AND PHYLOGENY

Andrew B. Smith
Department of Palaeontology, The Natural History Museum, Cromwell Road,
London SW7 5BD, United Kingdom; e-mail: A.Smith@nhm.ac.uk

KEY WORDS: cladistics, molecular phylogeny, congruence, metamorphosis, life history

ABSTRACT

New robust phylogenies for echinoderms, based on congruent patterns derived from multiple data sets, provide a sound foundation for plotting the evolution of life-history strategies and comparing rates and patterns of larval and adult morphological change. This approach demonstrates that larval morphology has been evolving independently of adult morphology, that larval morphology displays more homoplasy than adult morphology, and that early developmental patterns are remarkably flexible. Larval morphology on its own can mislead phylogenetic analysis, not because of lateral gene transfer among distantly related taxa, but because of massive convergence in the form of nonfeeding larvae brought about by the loss of complex structures and the strong functional constraints on feeding larvae. The degree to which larval tissue is resorbed at metamorphosis is believed to be important in determining adult body plan. Although the correspondence is not precise, it does provide a model for understanding skeletal homologies among the classes.

INTRODUCTION

Echinoderms, like many other marine invertebrates, have a complex life-history in which an adult benthic phase alternates with a planktonic larval stage. The metamorphosis that occurs between larva and adult is extreme, usually involving extensive tissue resorption and the transformation from bilateral to pentaradial symmetry as the adult form develops from a rudiment on the larval body (8, 26).

Ever since the work of Müller (62), it has been recognized that the feeding (planktotrophic) larvae of each extant class of echinoderm has a distinctive

219

body plan. The echinopluteus and ophiopluteus of echinoids and ophiuroids, respectively, have long arms supported by skeletal rods, which greatly extend the ciliary bands, while the shorter-armed auricularia and bipinnaria character- ize holothurians and asteroids, respectively. Clades within each class also have distinct larval morphologies (22, 98). Consequently larval form has generally been thought to be a useful indicator of phylogenetic relationship. However, many lineages have either modified or lost their larval stage altogether, devel- oping directly from the embryo through a prolonged metamorphosis during which feeding does not occur (Figure 1). All five extant classes of echino- derms contain nonplanktotrophic members that come in a bewildering variety of forms (19, 58, 102). So larvae of closely related species whose adult mor- phology is almost identical can appear different, as in the case of the echinoids *Heliocidaris erythrogramma* and *H. tuberculata* (99), or larvae from adults of very different morphology, such as crinoids and holothurians, can appear very similar. Indeed, changes in larval morphology often appear to correlate more closely with shifts in life-history strategy than with adult morphology and phylogenetic history.

Recently, a much clearer appreciation of the link between larval morphology, life-history strategy, and phylogeny has started to emerge, thanks largely to the development of robust phylogenies.

RECENT ADVANCES IN CONSTRUCTING ECHINODERM PHYLOGENIES

New Molecular Data Bases

Phylogenetic hypotheses have traditionally been based on classical morpholog- ical criteria, but during the 1980s biochemical data started to be applied to the question of echinoderm relationships. Early biochemical phylogenies covered small groups and applied distance methods to immunological data (49–54), DNA-DNA hybridization data (55, 57, 79, 87), or amino acid sequences (94) to estimate relationships. More recently, phylogenies based on sequence data of nuclear and mitochondrial ribosomal genes have been constructed for echinoids (23, 47, 82), asteroids (43, 96), and ophiuroids (86). Although a few sequences for crinoids and holothurians currently exist (48), coverage is still too sparse for constructing a broad phylogenetic scheme for either class.

The relationship of the five extant echinoderm classes has also been tackled using molecular sequence data (46, 48, 56, 71, 73, 80, 95). For three of the five classes (echinoids, asteroids, and ophiuroids), sampling is now sufficiently broad that small sample size, a major source of error in phylogenetic analysis (14, 45, 68), is probably no longer a significant problem. In addition to gene

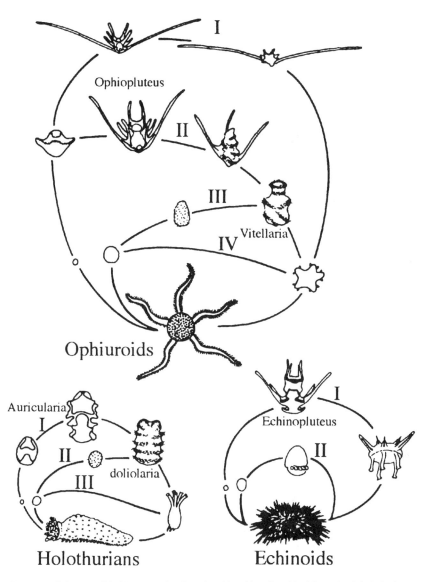

Figure 1 Schematic life-history cycles for (*a*) ophiuroids, (*b*) echinoids, and (*c*) holothurians, indicating some of the different pathways that can be taken (not to scale). Ophiuroid pathways: I, indirect development through planktotrophic ophiopluteus with suspended rudiment in Ophiothricidae; II, indirect development through planktotrophic ophiopluteus and vitellaria in Ophiocomidae; III, indirect development through nonfeeding vitellaria in Ophionereidae; IV, direct development in Amphiuridae. Echinoid pathways: I, indirect development through planktotrophic echinopluteus; II, direct development in *Heliocidaris*. Holothurian pathways: I, indirect development through planktotrophic auricularia (Stichopodidae); II, indirect development through nonfeeding doliolaria (Cucumariidae); III, direct development.

sequence data, gene order around the mitochondrion has been used to examine relationships both at class level (88) and within echinoids (27).

Congruence Between Independent Estimates

The reliability of single data sets for estimating phylogenetic relationships can be tested only by perturbing the original data matrix (e.g. bootstrapping, jack-knifing, or PTP tests) or relaxing the parsimony criterion (e.g. branch decay support). None of these methods actually tests for the correctness of the homology statements hypothesized or the topology found; they simply test whether there are sufficient data (of unknown quality) in the matrix to be confident that the clades identified represent more than just chance associations. Real confidence in hypothesized phylogenetic relationships can come only when similar topologies are derived from effectively independent data sets (7, 13, 59). Congruence studies are most advanced for echinoid and interclass relationships.

ECHINOIDS   Four independent data sets covering a broad taxonomic range of echinoids have been compiled for adult morphology, larval morphology, and gene sequences of large (partial) and small (complete) subunit ribosomal RNA (47, 85). Each provides an estimate of the true phylogenetic relationships, and each, unfortunately, generates a different topology. However, much similarity appears in the strongly supported parts of the four cladograms, and much of the difference among estimates may simply reflect error introduced through small sample size. One way to determine whether the four topologies can be considered estimates (with sampling error) of the same underlying topology is to apply Templeton's test of congruence (44). When this is done three of the four topologies turn out to be statistically indistinguishable from the single most parsimonious topology derived from all data combined (83). The fourth data set, derived from partial sequences of the LSU rRNA gene, does show a strong and significant conflict in the topology supported. This is due solely to the placement of one taxon, *Arbacia*, which groups with high bootstrap support among irregular echinoids (a highly unexpected placement based on morphologiccal and SSU rRNA evidence). The authenticity of this sequence urgently needs confirmation. When *Arbacia* is removed from the LSU rRNA data, the resultant topology is congruent with the other three, and Templeton's test is passed.

Since all four lines of evidence can be considered congruent, they are probably providing estimates of the same underlying phylogeny. Under these circumstances, combining all the data should produce the best working hypothesis of echinoid relationships (13, 59). The result is a highly robust topology that is identical whether *Arbacia* is included or excluded from the LSU rRNA data set (Figure 2).

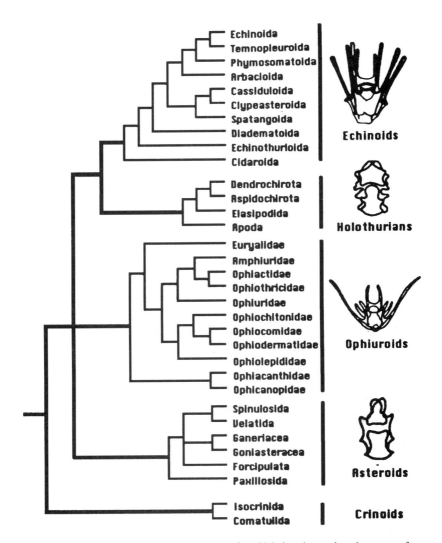

Figure 2 Cladogram of major echinoderm groups for which there is corroborative support from multiple independent data sets. Echinoid relationships are derived from the combined analysis of two molecular and two morphological data sets (47, 85). Holothurian relationships represent congruent SSU and LSU rRNA-based topologies (data from 48). Ophiuroid relationships are derived from LSU rRNA data (data from 48) which, except for the positions of Ophiacanthidae and Ophiocomidae, are identical to those derived from morphology (86). Asteroid relationships are based on molecular and morphological data of Lafay et al (43) with additional taxa from Wada et al (96). Class relationships are as derived from the morphological and molecular analysis of Littlewood et al (48). A representative planktotrophic larva of each class is shown.

ASTEROIDS   Incongruence exists among estimates of asteroid relationships based on mitochondrial rDNA (96), nuclear LSU rRNA (43), and morphological (6, 25) evidence. The two morphological analyses share much topological similarity but are rooted at different positions. The two molecular data sets also differ from one another, most significantly in the position of Solasteridae, but identify a similar rooting position. In the Wada et al analysis (96), the solasterid *Crossaster* groups within Asterina under parsimony and maximum likelihood analyses, making the genus *Asterina* paraphyletic, whereas in the Lafay et al analysis (43) *Crossaster* is sister group to the Echinasteridae. The latter is independently supported by adult morphology (6, p. 523) and embryological (40, p. 72) evidence; indeed a close relationship of *Crossaster* to *Asterina* has never previously been proposed, and no one has ever considered it to belong within *Asterina*. I suspect therefore that the Wada et al sequence may be a laboratory contaminant.

Collapsing branches with less than 50% bootstrap support in the Wada et al topology (i.e. branches that can be considered near random aassociations on the basis of their data) and ignoring the placement of *Crossaster* produce a topology congruent with that of Lafay et al. Both place paxillosids as sister group to other asteroids, contradicting Blake's (6) proposal that paxillosids are derived. Unfortunately, although a number of higher taxa are well supported within the Asteroidea, their interrelationships cannot be deduced with any confidence on current molecular information (Figure 2), and they presumably diverged over a short time interval.

HOLOTHURIANS   Though only a small number of taxa have been sampled, both LSU and SSU ribosomal RNA gene sequences identify the same topology, namely (outgroup (Elasipoda (Aspidochirotida, Dendrochirottida))). Only LSU rRNA sequence data are available for Synaptida, but these identify it as sister taxon to the other holothurians with high bootstrap support.

OPHIUROIDS   LSU rRNA gene sequence data (Figure 2) and morphology (86) generate rather weakly supported but almost identical topologies, differing primarily in the placement of Ophiacanthids. Morphology places *Ophiocanops* alone as sister group to all others, whereas the molecular data pair it with ophiacanthids. Additional data are needed to resolve its placement.

ECHINODERM CLASS RELATIONSHIPS   There has always been controversy as to the precise relationships of the five extant echinoderm classes (46, 56), with different kinds of data and different methods of analysis having generated many rival hypotheses. However, as groups have become more widely sampled and larger data bases have been constructed, just two rival topologies have emerged as serious contenders (48). A high degree of corroboration now exists among independent data sets of adult morphology and the gene sequences of LSU and SSU rRNA (Figure 2). This combined data set provides overwhelming support for crinoids as sister group to the other four classes, and for pairing echinoids and holothurians as sister taxa (=Echinozoa). However, morphological and molecular data are inconclusive about whether ophiuroids alone are sister group to the Echinozoa, or whether asteroids and ophiuroids themselves form a clade that is sister taxon to the Echinozoa. Clearly the three clades are separated by very short internodes and presumably arose over a short period of time.

Additional evidence from mitochondrial gene order (88) strongly supports an unrooted topology of (asteroids, ophiuroids) (echinoids, holothurians), but until the crinoid mitochondrial gene order is successfully recovered, which pattern represents the primitive arrangement remains speculative (56).

IS LARVAL MORPHOLOGY PHYLOGENETICALLY IRRELEVANT?

Larval morphology has generally been assumed to provide a phylogenetic signal at both low (98, 100) and high (37, 63, 64) taxonomic levels. Yet striking similarities in larval morphology exist between echinoderm groups that apparently have no close relationship (as, for example, between the doliolaria of crinoids and holothurians). Fell (22) believed that homoplasy among larval forms was so rampant that larval similarity provided little clue to the higher relationships of echinoderms. Strathmann (91) pointed out that no matter what phylogeny of echinoderms eventually proves correct, there must be striking convergence in larval features, while Wray (102) has shown larval morphology of echinoids can be strongly misleading in phylogenetic reconstruction if direct and indirect developing forms are treated together.

Similarities between larval form have generally been ascribed to convergent evolution by natural selection or parallel evolution under relaxed constraints (39, 69, 92). However, Williamson (97) proposed a radical explanation, invoking horizontal gene transfer of the entire larval phenotype during hybridization between members of different higher taxonomic groups. If this were true, larval form would provide no guide to phylogeny because life-history stages could be transferred and inserted at any time in the evolution of a group.

Williamson's evidence came from an experiment in which eggs of a tunicate (*Ascidia*) were fertilized with sperm from a sea urchin (*Echinus*). Larvae found in the aquaria subsequently proved to be plutei, like those of echinoids. This was taken as evidence that tunicate eggs had been fertilized by sea urchin sperm and had developed as viable sea-urchin-like larvae.

Williamson's hypothesis was subsequently tested by Hart (30) using molecular techniques. Since mitochondrial genes are inherited through the maternal line only, those from the hybrid larva should be tunicate in origin, whereas nuclear genes should show a mixed tunicate and echinoid origin. Hart found that mitochondrial genes of the putative hybrids were unambiguously echinoid in origin, with no trace of tunicate mitochondrial genes. Transgeneric transfer of larval form thus remains unproven, and Williamson's experiment appears to have been a laboratory artefact. Larval morphology can therefore continue to be treated as a valid source of information for reconstructing echinoderm phylogeny.

APPROACHES TO LIFE-HISTORY STUDIES

The many similarities in feeding larvae across echinoderm classes strongly imply common ancestry (89). Indeed, all deuterostomes share as a primitive character a pelago-benthic life cycle with a filter-feeding planktonic larva (63, 64). Yet clearly life-history strategies have become highly modified many times independently. In documenting when and why these changes have occurred, two approaches can be employed.

Fossil Evidence for Changes in Life History

The most direct line of evidence is to use the fossil record of echinoderms to identify the first occurrences of recognizable nonplanktotrophic development. Evidence for nonplanktotrophic development comes from three sources—morphological modifications associated with brooding (78), sexual dimorphism in the size of gonopores (17), and for echinoids, apical disk crystallographic pattern (15, 17). This last technique reflects the fact that during metamorphosis two of the skeletal arm rods in an echinopluteus are not fully resorbed but act as templates for crystal growth of genital plates 3 and 5 in the adult. This imparts a near tangential c-axis orientation to these two plates, in contrast to the perpendicular c-axis orientation of genital plates formed de novo. In nonplanktotrophic lecithotrophs, where no larval skeleton is formed, genital plates of the apical disc all have identical and perpendicular c-axes. Unfortunately, this correlation applies only to echinoids. Although ophioplutei have larval skeletal rods, these are completely resorbed during metamorphosis and leave no c-axis signature on the adult skeleton (17, 32).

The geological record unexpectedly lacks substantiated reports of sexual dimorphism or brood structures in any Palaeozoic echinoderm group, whether extant or extinct. Even among echinoids in which larval form is easiest to recognize in fossils, nonplanktotrophic development represents a relatively new phenomenon. Prior to the Cretaceous (ca 100 My ago), no echinoids show evidence of nonplanktotrophic development. Furthermore, there is good evidence that the switch from planktotrophic to nonplanktotrophic development evolved over a very short period of time in eight lineages of echinoid between 70 and 60 My ago (81). This rapid and parallel switch, which immediately predates the much vaunted K-T extinction event strongly suggests that external forcing agents of some kind were involved. Nonplanktotrophic development was adopted by other echinoid clades subsequently during the Tertiary, but the precise timing and environmental correlates of these events have yet to to be studied. Today some 14 clades include taxa with nonplanktotrophic larval development (18).

Whether this pattern holds true for other echinoderm classes is unknown but seems unlikely. All extant crinoids, for example, undergo nonplanktotrophic development and presumably have done so since at least the early Mesozoic when the extant groups diverged. Brooding is widespread today in ophiuroids and asteroids, but because of their poor fossil record and the difficulty of recognizing life-history strategy from the adult skeleton, no direct evidence exists for when this developed.

Optimization of Life-History Traits onto Cladograms

Given a robust phylogeny for a group, it is possible to optimize life-history patterns onto the cladogram of modern taxa to discover when and how often planktotrophy has been lost. This has been done for echinoids by Wray (98, 102). Crown group echinoids began to diverge ca 250 My ago and undoubtedly had planktotrophic development originally. Groups in which nonplanktotrophy has evolved are widely scattered over the cladogram and predominantly located toward the termini. Since at least some retain vestiges of structures unique to planktotrophic forms, such as remnant pluteal skeletons (20), nonplanktotrophy is clearly the derived condition.

Asteroids show a pattern similar to that of echinoids. Luidiids, astropectinids, and benthopectinids, which represent the deepest branches, are predominantly planktotrophic, whereas scattered throughout the crown group are clades with both planktotrophic and lecithotrophic life history. Although McEdward & Janies (58) argued on the basis of Blake's (6) cladogram that the ancestral state for asteroid larvae was to have had a bipinnarian stage followed by a brachiolarian stage, more recent molecular data clearly point to the brachiolaria being a derived larval stage within Asteroidea (43, 96). The bipinnaria larva is thus primitive for asteroids, as suggested by Ogura (65). Optimizing larval life-history strategy onto a working cladogram implies that nonplanktotrophy has evolved at least eight times within the post-Palaeozoic asteroids. However, it is probably considerably more frequent since a recent detailed analysis of just one seastar family (Asterinidae) suggests at least four independent evolutions of nonplanktotrophy (M Hart, M Byrne, MJ Smith, submitted). Furthermore, compared to echinoids, a much higher proportion of asteroid species whose larval development is known are nonplanktotrophs (21), possibly implying a more ancient origin for nonplanktotrophic life history in asteroids.

For ophiuroids, the situation is somewhat different. All of the deep branches in the Smith et al (86) cladogram have only nonplanktotrophic development (data from 34). Only in one major derived clade, the suborder Ophiurina, are taxa with planktotrophic larvae found. Nonplanktotrophic larval stages are also developed among Ophiurina, but at least some of these retain vestigial structures that are found only in planktotrophs (32), showing again that nonplanktotrophy is a secondary modification. However, remnant structures have yet to be reported in the larvae of more primitive groups.

Are ophiuroids really nonplanktotrophic primitively? This seems inherently unlikely, because the morphological similarity between the various stages in indirect developers (e.g. 60) and in other echinoderm classes would imply remarkable convergence (89). Hotchkiss (35), therefore, made the argument that the ancestral ophiuroid was a ciliated feeding planktotroph. Indeed, if ophiuroids prove to be sister group to echinoids and holothurians then it is equally parsimonious to treat plutei with larval skeletons as plesiomorphic (48).

It seems clear, therefore, that change in life-history strategy has been all in one direction and that echinoderms could not secondarily evolve planktotrophic development once it has been lost (90). Where robust phylogenies exist, as in echinoids, there are many clades in which planktotrophic development has been lost, but not a single example of the reverse, i.e. a planktotrophic clade nested within a larger nonplanktotrophic clade. If loss is irreversible, there is a strong argument for much of the transformation being accomplished by deleterious mutation under relaxed selection pressure (69, 98) rather than by positive selection toward nonfeeding.

There is, however, one apparent exception to this rule. Most species of the seastar *Pteraster* brood their young through to metamorphosis within a special internal aboral chamber. *Pteraster tesselatus*, by contrast, releases its developing embryos from the brood chamber at an early stage, and the larvae are dispersed to metamorphose in the water (40). Although *P. tesselatus* has not reevolved a feeding larval stage, it is unique in secondarily having reverted to planktonic dispersal.

RATES AND PATTERNS OF LARVAL EVOLUTION

Given robust phylogenies, we can start to explore how larval evolution has progressed in comparison to the more familiar patterns based on adult morphology.

How Conservative Are Larval Developmental Patterns?

It is often thought that early developmental features are highly conserved and fundamental to the development of later adult body plan. Yet there is growing evidence that early developmental patterns are remarkably flexible. Wray (101) studied evolution of cell lineage patterns among echinoderms, concentrating on the pattern and orientation of early cleavage, and the time, order, number, and location of founder cell appearances. Although he found some parameters remained highly conserved, Wray also found that switches in life-history strategy could produce profound changes in cell lineage patterns. Not only did these changes take place over a geologically short time (106), they also had no effect on adult morphology and are thus independent of post-larval body patterning. The position and timing of protein expression also vary greatly among echinoid taxa (104), showing that nonmorphological aspects of early development are equally flexible.

Have Larval and Adult Forms Evolved in Parallel or Independently?

Given that larvae and adults are adapted for very different life-styles, it seems likely that they have been subjected to independent selection pressures. Alternatively, there are various mechanisms by which concerted evolution of larva and adult might occur. For example, the effects of a genetic bottleneck will affect all life-history stages simultaneously. Both Fell (22) and Wray (98) argued for a decoupled view of evolution in which larval and adult phenotypes evolve largely independently. This was tested for echinoids by Smith et al (84, 85). If larval characters have been evolving independently of adult characters, there should be little correlation between the relative amounts of change in larval and adult characters along the branches of the cladogram. If, on the other hand, changes in larval and adult morphology were linked, individual branches with proportionally more change in larval characters should also show proportionally larger change in adult characters. No significant correlation was found between larval and adult character changes over the cladogram (84, 85), indicating that larval and adult morphologies have been evolving independently at least over the past 250 Ma.

Does Larval Morphology Show Unusually High Levels of Homoplasy?

There is little doubt that much of the convergence seen in larval morphology reflects convergence of life-history strategy. Similarity among direct-developing larvae is largely due to their having undergone overall simplification whereby derived features of the planktotrophic larva have become degraded or lost in the transformation. This is most clearly seen in echinoids, in which the morphological similarity of some distantly related nonplanktotrophic larvae results in their being grouped together in cladistic analysis based on larval characters alone (102). Nonplanktotrophy has evolved in echinoids at least 14 times and probably many more than that. In each case parallel changes have occurred in egg size, cleavage geometry, larval morphogenesis, formation of the adult rudiment, time to metamorphosis, and gene expression (103). Many of these changes simply represent repeated suppression or loss of complex features that are essential for feeding and maintaining posture in the water. Relaxation of stabilizing selection and the accumulation of mutations have been suggested as sufficient to explain the convergence (105), with the failure to initiate complex later structures, or with the acceleration of development to speed up the acquisition of adult traits. Thus, evolutionary changes in developmental mechanism appear to be driven primarily by changes in life-history strategy (101). The same seems to be true for asteroids in which a progressive loss of derived characters in the transformation of indirect planktotrophic to direct lecithotrophic development has also occurred (58).

However, not all convergence can be explained simply as the loss of derived characters; some must be directly related to functional considerations. Furthermore, even if only the morphologically complex planktotrophic forms are considered, there is still strong evidence that larval form is more prone to homoplasy than is the adult form. When pluteal and adult characters are scored for an identical suite of echinoid species and optimized onto the best estimate of the phylogenetic tree, the consistency and retention indices of larval characters are considerably lower than those of adult characters (84). This could arise because of the structural simplicity of larval characters or because of a stronger functional constraint on larval structure.

STRUCTURES ARE MUCH SIMPLER IN LARVAE THAN ADULTS AND ARE MORE READILY MISTAKEN AS HOMOLOGOUS THAN ARE COMPLEX FEATURES   Clearly ample numbers of characters can be scored for planktotrophic larval forms with complex skeletons. Wray (98, 102) coded more than 60 discrete characters for echinoplutei. Unfortunately, because of the microscopic size of larvae, most of these characters are relatively simple, relating to features such as the presence or absence of specific unbranched larval rods. This severely hampers the chances of distinguishing homoplasy from homology because the standard criterion of similarity is more stringent for characters with structural complexity.

Smith et al (85) tested whether the size and simplicity of larval characters being scored was a significant factor by comparing homoplasy levels with those of similarly sized and low-complexity structures found in adults, namely pedicellariae. Pedicellariae are tiny appendages on the tests of echinoids and can be scored for a suite of skeletal and soft-tissue characters that refer to small structural features very comparable to those of echinoplutei. For an identical suite of taxa, larval characters displayed higher homoplasy levels than did the pedicellariae of adults. This suggests that simplicity of characters alone is not sufficient to explain why larval morphology displays higher levels of homoplasy.

STRONG FUNCTIONAL CONSTRAINTS PROMOTE PARALLEL EVOLUTION OF LAR-
VAL FORM   The alternative explanation for heightened homoplasy in feeding
larvae is that the limited number of ways in which the echinopluteus can function have been repeatedly re-evolved. The requirements of feeding place severe
constraints on the configuration of the ciliary band and therefore on body form
of an echinoderm larva (92). The ciliary band in an echinoderm feeding larva
must be looped in a very precise way to enable it to fulfil both locomotory
and particle-capture roles. This contrasts with the situation in the tornaria of
enteropneusts, in which feeding and locomotory ciliated bands are completely
separate.

Yet, despite a widespread belief that echinoderm larvae are strongly constrained functionally, little in the way of direct experimental evidence has been
forthcoming. By far the best work in this field is that of Emlet (19). Using
a comparative approach, he found that body form, ciliation pattern, and buoyancy are all strongly linked with functional requirements for maintenance of
swimming performance. The switch from planktotrophy to nonfeeding larvae
is generally accompanied by an increase in larval body size (21). Planktotrophs
produce small eggs with insufficient nutritional reserves to support larval development through to metamorphosis, whereas lecithotrophs have much larger
eggs that contain reserves sufficient that feeding during larval development is
unnecessary. However, this increase in size has consequences for swimming
and buoyancy, and Emlet showed that the pattern of ciliation is size-related and
independent of phylogeny. Therefore strong functional reasons exist for differences in ciliation pattern, and the fact that crinoids, holothurians, and ophiuroids
all develop similar nonfeeding barrel-shaped larvae with four or five transverse
ciliary bands (doliolariae or vitellariae) may be simply a functional convergence
associated with the need for efficient locomotion.

If echinoplutei and ophioplutei independently evolved larval skeletal rods, as
Hotchkiss (35) believed, can a functional significance be found? They cannot
have evolved simply for support because some larvae have long arms without
skeletal support (luidiid asteroids) and others have skeletal rods without arms
(holothurians). Pennington & Strathmann (67) investigated the functional significance of larval skeletons by experimentally removing skeletons from both
living and dead larvae. The skeleton assisted in, but was not essential to, maintaining passive vertical orientation, with the larva's anterior end upwards. The
cost of this in terms of added density and faster sinking rates was very slight
(>1% total aerobic energy supply). However, the role of skeletal rods in passive
vertical orientation was probably not sufficient to explain their evolution in the
first place.

Feeding larvae of echinoids are highly flexible in their timing of development and allocation of resources. Indeed, an increased ratio of endogenous

to exogenous food may trigger the switch to nonfeeding development (93), although probably in conjunction with other factors (105). Even the change to nonfeeding development may involve strong selective pressure toward minimizing the time required to complete metamorphosis. Thus in developing lecithotrophic larvae, there may be a strongly correlated shift toward deferred skeletogenesis and early adult ectoderm specification (100, 104), though supressed rather than deferred skeletogenesis in the larva is possibly more common than previously supposed (20).

LARVAL METAMORPHOSIS AND THE EVOLUTION OF ADULT BODY PLANS

The Fate of Larval Tissues at Metamorphosis

Metamorphosis comprises the complex set of changes that transforms a bilateral larva into a pentaradiate adult (8). Typically during this transformation much of the larval tissue is resorbed and adult structures form from a rudiment on the side of the larval body. The extent to which larval tissue is resorbed or becomes incorporated into the adult varies considerably among echinoderm classes (9) and has recently been seen as having fundamental importance in determining adult body plan (11).

David & Mooi (11, 61) recognized two systems of skeletal plating in adult echinoderms, an axial portion of the skeleton generated by tissue of the rudiment and an extraxial portion derived from original larval tissue. Axial skeleton comprises the ambulacral skeleton and associated elements generated at the growing tip of the radial water vessel, while extraxial skeleton comprises the rest of the body wall and has plates added at random over the surface. Axial skeleton is thus defined as any skeletal elements that are derived terminally at the radii, according to what Mooi et al (61) termed the ocular plate rule.

This hypothesis, based partially on the pattern of skeletal formation in adults and partially on mapping tissue fates during metamorphosis from larva to juvenile, has implications for echinoderm relationships. Differences in body shape are seen as correlated to the degree to which the rudiment evaginates and larval tissue is resorbed during metamorphosis (11). In camarodont echinoids, the rudiment develops within a deep ectodermal invagination termed the vestibule. At metamorphosis the vestibule is everted so that the rudiment, which represents a disc comprising the oral half of the juvenile, comes to project. At the same time larval tissue withdraws to a small mound at the aboral surface so that the test ends up almost entirely composed of tissue derived from the rudiment. By contrast, the rudiment in asteroids and ophiuroids only partially everts, and axial and extraxial elements are equally developed in the adult body.

Holothurians retain much of their larval tissue during metamorphosis and undergo no evagination (76) so that most of the adult body is derived primarily from the pluteal portion of the metamorphosing larva (11, 12).

Some Problems

This new view of echinoderm plating certainly has merit, particularly in inferring the fundamentally different embryological origins of the two systems. However, some outstanding problems remain to be addressed.

THE EXTENT OF EVAGINATION AT METAMORPHOSIS IS INDEPENDENT OF ADULT BODY FORM   David & Mooi (11) explain echinoderm morphology in terms of the relative degree of evagination that takes place at metamorphosis, echinoids showing the most extensive evagination, which manifests itself in an almost complete inversion of the vestibular floor and the retraction of larval tissues to a small apical region. However, this is the case for only one of the two subclasses of echinoids, the euechinoids. Emlet (16) has documented that the other subclass, cidaroids, undergo a simpler metamorphism that involves no evagination and no vestibular cavity. Furthermore, much of the larval epidermis in the body of the cidaroid larva remains intact into the juvenile. The absence of an amniotic invagination in cidaroids may therefore be primitive (16), implying that evagination of the rudiment is a derived characteristic of only a subset of modern echinoids. As cidaroids and euechinoids have identical adult body plans, the degree of evagination at metamorphosis and the adult form cannot be directly correlated.

NONAXIAL SKELETAL ELEMENTS FIT THE OCULAR PLATE RULE   The ocular plate rule works well for echinoids, in which the entire corona comprises axial skeleton, and both ambulacral and adjacent interambulacral columns of plates are generated subterminally along the margin of an ocular plate. Furthermore, developmental abnormalities clearly demonstrate that the ambulacrum and the immediately adjacent column of interambulacral plates on each side are intimately linked (e.g. 38, p. 491 et seq). Although Palaeozoic echinoids such as *Aulechinus* and *Ectinechinus* were supposed to differ from crown-group (i.e. post-Palaeozoic) echinoids in having extraxial rather than axial interambulacra (61), interambulacral plates in these extinct forms almost certainly formed along the ocular margins and later became displaced laterally, as plates were added, to form multiple columns in exactly the same way as multiple ambulacral columns were generated in Palaeozoic echinoids (42).

The presence of an interambulacral series conforming to the ocular plate rule has been proposed as unique to echinoids (61). Yet both ophiuroids and asteroids have a series of plates, the lateral arm plates (adambulacrals) that form subterminally at the border of the oculars (28) and abut the ambulacral plates. Marginal ossicles in asteroids also form at the termini (24) and follow the ocular plate rule. Consequently, interambulacra of echinoids do not constitute a new plating system in echinoderms but have possible homologues among asteroids and ophiuroids (i.e. they may correspond to marginals of Palaeozoic asterozoans). Furthermore, even some plates of the aboral surface originate subterminally adjacent to the terminal plate. This includes the dorsal arm plates of ophiuroids and the dorsal carinal plates of asteroids (28). Thus, elements of oral, marginal, and aboral plating series in asterozoans are serially added at the growing tip of rudiment's hydrocoelic lobes.

SOME AXIAL ELEMENTS ARE DERIVED FROM LARVAL STRUCTURES   Mooi et al (61) interpreted ocular plates as axial elements and genital plates as extraxial elements. Yet the origin of the two sets of plates at metamorphosis is very similar. Four of the five genital plates originate directly from larval skeletal rods, but so do two of the five ocular plates (15). The remaining ocular and genital plates form de novo. Consequently, either the correspondence between larval or rudiment tissue and axial/extraxial skeleton is imprecise, or ocular plates are part of the extraxial system.

The Origins of Adult Body Plans

Although there remain some outstanding problems with David & Mooi's theory, it is certain that insight into the origin of new body plans will come from a better appreciation of larval metamorphosis.

ECHINOIDS AND HOLOTHURIANS   Echinoids are easily derived from an aster-ozoan body plan by relative reduction of the aboral (extraxial) plating and expansion of the oral (axial) skeleton (78). The required change in the rate of generation of extraxial relative to axial plating has occurred several times (e.g. in the evolution of the asteroids *Podosphaeraster* (74) and *Sphaeraster* (5). This change must be initiated at metamorphosis and, despite the anomalous situation with cidaroids mentioned above, is presumably connected to the retraction of larval skeletal tissue (from which ocular and genital plates derive) to a small aboral region.

Whereas the echinoid body plan represents the accelerated development of the oral (axial) skeleton component compared to aboral (extraxial) skeleton component at metamorphosis, the holothurian body plan is fundamentally dif-ferent, being totally dominated by larval tissue (extraxial), with only a small axial component (11). The structures making up the rudiment do not develop beyond the region within the oral tentacular circle and most larval tissues are directly incorporated into the adult (76). Furthermore, holothurians retain the larval axis of symmetry through metamorphosis into the adult (76). This has led to the interpretation of holothurians as strongly pedomorphic, retaining a suite of essentially larval traits into adulthood (11, 12). Such extreme pedomorphosis explains why holothurian relationships have been problematic to resolve from morphological data. Luckily, molecular studies identify holothurians with very high support as sister group to echinoids. So a change in the relative development of rudiment versus larval tissue instigated at metamorphosis offers the most likely mechanism for this transformation.

Reinterpreted this way, the tentacles of holothurians become the direct homologues of the radial water vessels in other echinoderms, and the later-forming radial extensions along the main body axis are a novel feature with no direct homologue (11). The proposed homology between the brachiolar podia of asteroids and the buccal tube-feet of holothurians (77) becomes untenable because the brachiolar podia of asteroids are derived entirely from axocoel extensions and the buccal tube-feet are derivatives of the hydrocoel ring alone (4).

XYLOPLAX AND ITS DERIVATION FROM ASTEROIDS   *Xyloplax* is a small disc-shaped echinoderm with such distinct anatomy that it was considered a new class when first described (3, 75). It has an aboral plated surface and three concentric rings of skeletal ossicles: an inner ring composed of modified mouth-frame skeleton, a middle ring of perforate ambulacral ossicles associated with tube-feet, and an outer ring of marginal ossicles bearing spines. Unique features include the unpaired and circumferential nature of the ambulacra and the double water-vascular ring.

Rowe et al (75) derived the *Xyloplax* body by transformation of an adult asteroid body plan. However, Janies & McEdward (41) proposed that the origin of the *Xyloplax* body plan could be better understood if it were derived progenetically from a late metamorphic stage seen in certain velatid asteroids. In velatid metamorphosis, larval and adult axes of symmetry are parallel, and the water vascular ring forms by lateral fusion of five enterocoel pouches (40). Janies & McEdward (41) postulated that duplication of the lateral evagination and fusion of the enterocoels could produce the unique water vascular arrangement of *Xyloplax*.

Although this model has some appeal, it fails to explain the position of the terminal plates in *Xyloplax*. Terminal plates, by definition, form the radial tips of the radial water vessels and thus cannot be interradial in position, as required by the Janies & McEdward hypothesis (41). Thus, although progenesis is almost certainly involved in the origin of *Xyloplax*, the detailed mechanism proposed seems implausible.

CRINOIDS   Views on the origin and relationships of crinoids have radically changed recently. The oldest supposed crinoid, the Middle Cambrian *Echmatocrinus*, has now been shown almost certainly to be an octocoral (2). The most likely sister group to crinoids are fistuliporite rhombiferans (1), based on stem, cup, and arm morphological similarities. If crinoids evolved from paedomorphic fistuliporite cystoids, left somatocoel elements were presumably housed in the deep oral groove of the arm plates. In crinoids, these tissues subsequently became enclosed in the internal canal running through the ossicles. Extension of the genital rachis into the pinnules of commatulid and isocrinid crinoids is probably a derived feature of this large clade of modern crinoids, since extant cyrtocrinids, like cystoids, have internal gonads (31).

Crinoid arms are therefore homologous with asteroid arms, in as much as they both represent ambulacral structures associated with radial growth of the water vascular system. However, arms did not simply evolve once (i.e. armless edrioasteroids and cystoids gave rise to armed forms that were common ancestors to crinoids, asteroids, and ophiuroids—61). Although some cystoids (e.g. blastoids, diploporite cystoids) certainly had no extension of somatic components into their free appendages, others (e.g. paracrinoids, fistuliporite and rhombiferan cystoids) had large arms fully comparable to those of crinoids (66).

Pentamery

Pentameral symmetry is one of the most characteristic features of adult echinoderms. Pentamery is initiated when the hydrocoel puts out five radial lobes during metamorphosis. However, the fossil record shows that adult echinoderms were primitively unrayed. Solutes are among the most primitive known echinoderms, having just a single ambulacrum and no trace of pentamery (10). In solutes, the anterior corresponds to the tip of the ambulacrum and the posterior of the body to the posterior attached stalk. Since juveniles attach to this stalk (10), it is probably homologous to the pterobranch stolon, i.e. it is an axocoel-derived attachment structure, in marked contrast to the crinoid stalk, which has no axocoel component (29).

The evolution of pentamery involved reduction or loss of the post-oral body and serial repetition of the ambulacrum at metamorphosis. Of key importance to understanding the origin of pentamery is Lovén's law of heterotopy, which recognizes that there is an asymmetry to the pattern in which ambulacral plates are laid down in the metamorphosing larva of many echinoderms (35). In the adult skeleton this is manifest in whether the left-(A) or right-hand (B) plate in each ambulacral biseries is larger at the peristome. Echinoids, ophiuroids, holothurians, edrioasteroids, and ophiocistioids all share the pattern BAABA, whereas cystoids with biserial ambulacra show the pattern AAAAA or BBBBB (35, 36). Lovén's law is too unusual to have arisen multiply by chance, and Hotchkiss explains its presence by invoking Bateson's rule of reduplication. This not only explains why Lovén's law applies; it also provides an explanation for why pentamery became strongly fixed in echinoderms. The observation that eleutherozoans and cystoids have different expressions of pentamery strongly supports the placement of edrioasteroids as Eleutherozoa, rather than as sister group to cystoids and all extant classes as suggested by Mooi et al (61).

How pentamery arose is currently under intense investigation through the study of *hox*-gene expression. Since *hox* genes in bilaterally symmetrical animals are expressed along the anterior-posterior axis, discovering how they are expressed in echinoderms may shed light on how a pentameral body plan has arisen. Initial findings are consistent with the idea that echinoderms originally had a unirayed condition with anterior at the arm tip (70, p. 421-3; 72), but research is at an early stage, and major advances can be expected in this field over the next few years.

ACKNOWLEDGMENTS

I should like to thank my recent collaborators, Tim Littlewood, Benedicte Lafay, and Richard Christen for willingly sharing their molecular data with me.

> Visit the *Annual Reviews home page* at
> http://www.annurev.org.

Literature Cited

1. Ausich, WI. 1997. Origin of the Crinoidea. In *Echinoderms, San Francisco: Proceedings of the 9th Int. Echinoderms Conf.,* ed. R Mooi. Rotterdam: AA Balkema. In press
2. Ausich WI, Babcock, LE. 1997. Phylogenetic position of *Echmatocrinus brachiatus* Sprinkle, 1973. *Palaeontology* 40: In press
3. Baker AN, Rowe FWE, Clark HES. 1986. A new class of Echinodermata from New Zealand. *Nature* 321:862–4
4. Balser EJ, Ruppert EE, Jaeckle WB. 1993. Ultrastructure of the coeloms of auricularia larvae (Holothuroidea: Echinodermata): evidence for the presence of an axocoel. *Biol. Bull.* 185:86–96
5. Blake DB. 1984. Constructional morphology and life habits of the Jurassic sea star *Sphaeraster* Quenstedt. *N. Jahrb. Geol. Pal. Abhandl.* 169:74–101
6. Blake DB. 1987. A classification and phylogeny of post-Palaeozoic sea stars (Asteroidea: Echinodermata). *J. Nat. Hist.* 21:481–528
7. Bull JJ, Huelsenbeck JP, Cunningham CW, Swofford DL, Waddell PJ. 1995. Partitioning and combining data in phylogenetic analysis. *Syst. Biol.* 42:384–97
8. Burke RD. 1989. Echinoderm metamorphosis: comparative aspects of the change in form. *Echinoderm Studies* 3:81–108
9. Chia FS, Burke RD. 1978. Echinoderm metamorphosis: fate of larval structures. In *Settlement and Metamorphosis of Marine Invertebrate Larvae,* ed. FS Chia, ME Rice, pp. 219–34. New York: Elsevier
10. Daley PEJ. 1996. The first solute which is attached as an adult: a mid-Cambrian fossil from Utah with echinoderm and chordate affinities. *Zool. J. Linn. Soc.* 117:405–40
11. David B, Mooi R. 1996. Embryology supports a new theory of skeletal homologies for the phylum Echinodermata. *C. R. Acad. Sci. Paris. Sci. Vie* 319:577–84
12. David B, Mooi R. 1997. Major events in the evolution of echinoderms viewed in the light of embryology. In *Echinoderms, San Francisco: Proc. of the 9th Int. Echinoderms Conference,* ed. R. Mooi. Rotterdam: AA Balkema. In press
13. De Queiroz A, Donoghue MJ, Kim J. 1995. Separate versus combined analysis of phylogenetic evidence. *Annu. Rev. Ecol. Syst.* 26:657–81

14. De Rijk P, Van de Peer Y, Van den Broek I, De Wachter R. 1995. Evolution according to large ribosomal subunit RNA. *J. Mol. Evol.* 41:366–75

15. Emlet RB. 1985. Crystal axes in Recent and fossil adult echinoids indicate trophic mode in larval development. *Science* 230:937–40

16. Emlet RB. 1988. Larval form and metamorphosis of a "primitive" sea urchin, *Eucidaris thouarsi* (Echinodermata: Echinoidea: Cidaroida), with implications for developmental and phylogenetic studies. *Biol. Bull.* 174:4–19

17. Emlet RB. 1989. Apical skeletons of sea urchins (Echinodermata: Echinoidea): two methods for inferring mode of larval development. *Paleobiology* 15:223–54

18. Emlet RB. 1990. World patterns of developmental mode in echinoid echinoderms. In *Advances in Invertebrate Reproduction,* ed. M Hoshi, O Yamishita, pp. 329–35. Amsterdam: Elsevier

19. Emlet RB. 1994. Body form and patterns of ciliation in non-feeding larvae of echinoderms: functional solutions to swimming in the plankton. *Am. Zool.* 34:570–85

20. Emlet RB. 1995. Larval spicules, cilia, and symmetry as remnants of indirect development in the direct developing sea urchin *Heliocidaris erythrogramma. Dev. Biol.* 167:405–15

21. Emlet RB, McEdward LR, Strathmann RR. 1987. Echinoderm larval ecology viewed from the egg. *Echinoderm Stud.* 2:55–136

22. Fell HB. 1948. Echinoderm embryology and the origin of chordates. *Biol. Rev.* 23:81–107

23. Féral JP, Derelle E. 1991. Partial sequence of the 28S ribosomal RNA and the echinoid taxonomy and phylogeny. Application to the Antarctic brooding schizasterids. In *Echinoderm Biology,* ed. T Yanagisawa, I Yasumasu, C Oguro, N Suzuki, T Motokawa, pp. 331–37. Rotterdam: AA Balkema

24. Fewkes JW. 1888. On the development of the calcareous plates of Asterias. *Bull. Mus. Comp. Zool., Harvard Univ.* 17:1–56

25. Gale AS. 1987. Phylogeny and classification of the Asteroidea (Echinodermata). *Zool. J. Linn. Soc.* 89:107–32

26. Geise AC, Pearse VB, Pearse JS. 1991. *Reproduction of Marine Invertebrates.* Vol. 6. *Lophophorates and Echinoderms.* Pacific Grove, CA: Boxwood. 767 pp

27. de Giorgi C, Martiradonna A, Lanave C, Saccone C. 1996. Complete sequence of the mitochondrial DNA in the sea urchin *Arbacia lixula*: conserved features of the echinoid mitochondrial genome. *Mol. Phylog. Evol.* 5:323–32

28. Gordon I. 1929. Skeletal development in *Arbacia, Echinarachnius* and *Leptasterias. Philos. Trans. R. Soc. Lond. B* 217:289–334

29. Grimer JC, Holland ND, Kubota H. 1984. The fine structure of the stalk of the pentacrinoid larva of a feather star, *Comanthus japonica* (Echinodermata: Crinoidea). *Acta Zool.* 65:41–58

30. Hart MW. 1996. Testing cold fusion of phyla: maternity in a tunicate x sea urchin hybrid determined from DNA comparisons. *Evolution* 50:1713–8

31. Heinzeller T, Fechter H. 1995. Microscopic anatomy of the cyrtocrinid *Cyathidium meteorensis* (sive *foresti*) (Echinodermata, Crinoidea). *Acta Zool.* 76:25–34

32. Hendler G. 1978. Development of *Amphioplus abditus* (Verrill) (Echinodermata: Ophiuroidea). II. Description and discussion of ophiuroid skeletal ontogeny and homologies. *Biol. Bull.* 154:79–95

33. Hendler G. 1982. An echinoderm vitellaria with a bilateral larval skeleton: evidence for the evolution of ophiuroid vitellariae from ophioplutei. *Biol. Bull.* 163:431–7

34. Hendler G. 1991. Echinodermata: Ophiuroidea. In *Reproduction of Marine Invertebrates,* Vol. 6. *Echinoderms and Lophophorates,* ed. AC Geise, JS Pearse, VB Pearse, pp. 355–511. Pacific Grove, CA: Boxwood

35. Hotchkiss FHC. 1995. Lovén's law and adult ray homologies in echinoids, ophiuroids, edrioasteroids and ophiocistioid (Echinodermata: Eleutherozoa). *Proc. Biol. Soc. Washington* 108:401–35

36. Hotchkiss FHC. 1997. Discussion on pentamerism: the five-part pattern of Stromatocystites, Asterozoa and Echinozoa. In *Echinoderms San Francisco: Proceedings of the 9th Int. Echinoderms Conf.,* ed. R Mooi. Rotterdam: AA Balkema. In Press

37. Hyman LH. 1955. *The Invertebrates: Echinodermata.* New York: McGraw-Hill. 763 pp

38. Jackson RT. 1927. Studies of *Arbacia punctulata* and allies, and of non-pentamerous echini. *Mem. Boston Soc. Nat. Hist.* 8:435–565

39. Jägersten G. 1972. *Evolution of the Metazoan Life Cycle.* London: Academic Press

40. Janies DA, McEdward LR. 1993. Highly derived coelomic and water-vascular

morphogenesis in a starfish with pelagic direct development. *Biol. Bull.* 185:56–76

41. Janies DA, McEdward LR. 1994. A hypothesis for the evolution of the concentricycloid water-vascular system. In *Reproduction and Development of Marine Invertebrates,* ed. WH Wilson, SA Stricker, GL Shinn, pp. 246–57. Baltimore, MD: Johns Hopkins Univ. Press

42. Kesling RV, Strimple HL. 1966. Suggested growth pattern in the Mississippian (Chester) echinoid *Lepidesthes formosa* Miller. *J. Paleontol.* 40:1167–77

43. Lafay B, Smith AB, Christen R. 1995. A combined morphological and molecular approach to the phylogeny of asteroids (Asteroidea: Echinodermata). *Syst. Biol.* 44:190–208

44. Larson A. 1994. The comparison of morphological and molecular data in phylogenetic systematics. In *Molecular Ecology and Evolution: Approaches and Applications* ed. B Schierwater, B Streit, GP Wagner, R DeSalle, pp. 372–90. Basel: Birkhauser

45. Lecointre G, Philippe H, Le HVL, Guyader H le. 1993. Species sampling has a major impact on phylogenetic inference. *Mol. Phylog. Evol.* 3:205–24

46. Littlewood DTJ. 1995. Echinoderm class relationships revisited. In *Echinoderm Research 1995,* ed. RH Emson, AB Smith, AC Campbell, pp. 19–28. Rotterdam: AA Balkema

47. Littlewood DTJ, Smith AB. 1995. A combined morphological and molecular phylogeny for echinoids. *Philos. Trans. R. Soc. Lond. B* 347:213–34

48. Littlewood DTJ, Smith AB, Clough KA, Emson RH. 1997. The interrelationships of echinoderm classes: morphological and molecular evidence. *Biol. J. Linn. Soc.* 61: In press

49. Matsuoka N. 1985. Biochemical phylogeny of the sea-urchins of the family Toxopneustidae. *Comp. Biochem. Physiol.* 80B:767–71

50. Matsuoka N. 1986. Further immunological study on the phylogenetic relationships among sea-urchins of the order Echinoida. *Comp. Biochem. Physiol.* 84B:465–8

51. Matsuoka N. 1987. Biochemical studies on the taxonomic situation of the sea-urchin *Pseudocentrotus depressus.* *Zool. Sci.* 4:339–47

52. Matsuoka N. 1989. Biochemical systematics of four sea-urchin species of the family Diadematidae from Japanese waters. *Biochem. Syst. Ecol.* 17:423–9

53. Matsuoka N, Suzuki H. 1987. Elec-

trophoretic study on the taxonomic relationship of the two morphologically very similar sea-urchins *Echinostrephus aciculatus* and *E. molaris. Comp. Biochem. Physiol.* 88B:637–41

54. Matsuoka N, Suzuki H. 1989. Electrophoretic study on the phylogenetic relationships among six species of sea-urchins of the family Echinometridae found in the Japanese waters. *Zool. Sci.* 6:589–98

55. Marshall CR. 1992. Character analysis and the integration of molecular and morphological data in an understanding of sand dollar phylogeny. *Mol. Biol. Evol.* 9:309–22

56. Marshall CR. 1994. Molecular approaches to echinoderm phylogeny. In *Echinoderms Through Time,* ed. B David, A Guille, JP Feral, M Roux, pp. 63–71. Rotterdam: AA Balkema

57. Marshall CR, Swift H. 1992. DNA-DNA hybridization phylogeny of sand dollars and highly reproducible extent of hybridization values. *J. Mol. Evol.* 34:31–44.

58. McEdward LR, Janies DA. 1993. Life cycle evolution in asteroids: What is a larva? *Biol. Bull.* 184:255–68

59. Miyamoto MM, Fitch WM. 1995. Testing species phylogenies and phylogenetic methods with congruence. *Syst. Biol.* 44:64–76

60. Mladenov PV. 1985. Development and metamorphosis of the brittle star *Ophiocoma pumila:* evolutionary and ecological implications. *Biol. Bull.* 168:285–95

61. Mooi R, David B, Marchand D. 1994. Echinoderm skeletal homologies: classical morphology meets modern phylogenetics. In *Echinoderms Through Time,* ed. B David, A Guille, JP Feral, M Roux, pp. 87–95. Rotterdam: AA Balkema

62. Müller J. 1850. Ueber die Larven und die Metamorphose der Echiniodermen. Dritte Abhandlungen. *Abhandl. Konigl. Preuss. Akad. Wiss. Berlin* 1849:35–72

63. Nielsen C. 1994. Larval and adult characters in animal phylogeny. *Am. Zool.* 34:492–501

64. Nielsen C. 1994. *Animal Evolution: Interrelationships of the Living Phyla.* Oxford: Oxford Univ. Press. 467 pp

65. Ogura C. 1989. Evolution of the development and larval types in asteroids. *Zool. Sci.* 6:199–210

66. Paul CRC, Smith AB. 1984. The early radiation and phylogeny of echinoderms. *Biol. Rev.* 59:443–81

67. Pennington JT, Strathmann RR. 1990. Consequences of the calcite skeletons of planktonic echinoderm larvae for

orientation, swimming, and shape. *Biol. Bull.* 179:121–33

68. Philippe H, Douzery E. 1994. The pitfalls of molecular phylogeny based on four species, as illustrated by the Cetacea/Artiodactyla relationships. *J. Mammal. Evol.* 2:133–52

69. Raff RA. 1987. Constraint, flexibility, and phylogenetic history in the evolution of direct development in sea urchins. *Dev. Biol.* 119:6–19

70. Raff RA. 1996. *The Shape of Life: Genes, Development, and the Evolution of Animal Form.* Chicago: Univ. Chicago Press

71. Raff RA, Field KG, Ghiselin MT, Lane DJ, Olsen GJ, et al. 1988. Molecular analysis of distant phylogenetic relationships in echinoderms. In *Echinoderm Phylogeny and Evolutionary Biology,* ed. CRC Paul, AB Smith, *Current Geological Concepts 1,* pp. 29–41. Oxford: Oxford Sci. Publ. and Liverpool Geol. Soc.

72. Raff RA, Popodi EM. 1996. Evolutionary approaches to analyzing development. In *Molecular Zoology: Advances, Strategies and Protocols,* ed. JD Ferrais, SR Palumbi, pp. 245–65. New York: Wiley-Liss

73. Ratto A, Christen R. 1990. Molecular phylogeny of echinoderms as deduced from partial sequences of 28S ribosomal RNA. *C. R. Acad. Sci. Paris* 310:169–73

74. Rowe FWE. 1985. On the genus *Podosphaeraster* A.M. Clark & Wright (Echinodermata, Asteroidea), with description of a new species from the North Atlantic. *Bull. Mus. Hist. Nat. Paris, 4,* 7(A):309–25

75. Rowe FWE, Baker AN, Clark HES. 1988. The morphology, development and taxonomic status of *Xyloplax* Baker, Rowe & Clark, 1986 (Echinodermata: Concentricycloidea), with the description of a new species. *Proc. R. Soc. London B* 223:431–9

76. Smiley S. 1986. Metamorphosis of *Stichopus californicus* (Echinodermata: Holothuroidea) and its phylogenetic implications. *Biol. Bull.* 171:611–31

77. Smiley S. 1988. The phylogenetic relationships of holothurians: a cladistic analysis of the extant echinoderm classes. In *Echinoderm Phylogeny and Evolutionary Biology,* ed. CRC Paul, AB Smith, *Current Geological Concepts 1,* pp. 69–84. Oxford: Oxford Sci. Publ. and Liverpool Geol. Soc.

78. Smith AB. 1984. *Echinoid Palaeobiology.* London: Allen & Unwin

79. Smith AB. 1988. Phylogenetic relationship, divergence times, and rates of molecular evolution for camarodont sea urchins. *Mol. Biol. Evol.* 5:345–65

80. Smith AB. 1989. RNA sequence data in phylogenetic reconstruction: testing the limits of its resolution. *Cladistics* 5:321–44

81. Smith AB, Jeffery CH. 1997. Changes in the diversity, taxic composition and lifestyle of echinoids over the past 145 million years. In *Biotic Response to Global Change: The Last 145 Million Years,* ed. SJ Culver, P Rawson. London: Chapman & Hall. In press

82. Smith AB, Lafay B, Christen R. 1992. Comparative variation of morphological and molecular evolution through geological time: 28S ribosomal RNA versus morphology in echinoids. *Philos. Trans. R. Soc. London B* 338:365–82

83. Smith AB, Littlewood DTJ. 1997. Molecular and morphological evolution during the post-Paleozoic diversification of echinoids. In *Molecular Evolution and Adaptive Radiation,* ed. TJ Givnish, KJ Sytsma. Cambridge: Cambridge Univ. Press. In press

84. Smith AB, Littlewood DTJ, Wray GA. 1995. Comparing patterns of evolution: larval and adult life history stages and small subunit ribosomal RNA of post-Palaeozoic echinoids. *Philos. Trans. R. Soc. London B* 349:11–18

85. Smith AB, Littlewood DTJ, Wray GA. 1996. Comparative evolution of larval and adult life history stages and small subunit ribosomal RNA amongst post-Palaeozoic echinoids. In *New Uses for New Phylogenies,* ed. PH Harvey, AJ Leigh Brown, J Maynard Smith, S Nee, pp. 234–54. Oxford: Oxford Univ. Press

86. Smith AB, Paterson GL, Lafay B. 1995. Ophiuroid phylogeny and higher taxonomy: morphological, molecular and palaeontological perspectives. *Zool. J. Linn. Soc.* 114:213–43

87. Smith MJ, Nicholson R, Stuerzul M, Lui A. 1982. Single copy DNA homology in sea stars. *J. Mol. Evol.* 18:92–101

88. Smith MJ, Arndt A, Gorski S, Fajber E. 1993. The phylogeny of echinoderm classes based on mitochondrial gene arrangements. *J. Mol. Evol.* 36:365–82

89. Strathmann RR. 1974. Introduction to function and adaptation in echinoderm larvae. *Thalassia Jugoslav.* 10:321–39

90. Strathmann RR. 1978. The evolution and loss of feeding larval stages of marine invertebrates. *Evolution* 32:894–906

91. Strathmann RR. 1988. Larvae, phylogeny and von Baer's Law. In *Echinoderm Phylogeny and Evolutionary Biology,* ed.

CRC Paul, AB Smith, *Current Geological Concepts 1*, pp. 53–68. Oxford: Oxford Sci. Publ. and Liverpool Geol. Soc.

92. Strathmann RR. 1988. Functional requirements and the evolution of developmental patterns. In *Echinoderm Biology*, ed. RD Burke, PV Mladenov, P Lambert, RL Parsley, pp. 53–68. Rotterdam: AA Balkema

93. Strathmann RR, Fenaux L, Strathmann MF. 1992. Heterochronic developmental plasticity in larval sea urchins and its implications for evolution of non-feeding larvae. *Evolution* 46:972–86

94. Suzuki N, Yoshino KI. 1992. The relationship between amino acid sequences of sperm-activating peptides and the taxonomy of echinoids. *Comp. Biochem. Physiol.* 102B:679–90

95. Wada H, Satoh N. 1994. Phylogenetic relationships among extant classes of echinoderms, as inferred from sequences of 18S rDNA, coincide with relationships deduced from the fossil record. *J. Mol. Evol.* 38:41–9

96. Wada H, Komatsu M, Satoh N. 1996. Mitochondrial rDNA phylogeny of the Asteroidea suggests the primitiveness of the Paxillosida. *Mol. Phyl. Evol.* 6:97–106

97. Williamson DI. 1992. *Larvae and Evolution: Toward a New Zoology*. London: Chapman & Hall

98. Wray GA. 1992. The evolution of larval morphology during the post-Paleozoic radiation of echinoids. *Paleobiology* 18:258–87

99. Wray GA. 1994. Punctuated evolution of embryos. *Science* 267:1115–6

100. Wray GA. 1994. Larval morphology and echinoid phylogeny. In *Echinoderms Through Time*, ed. B David, A Guille, JP Feral, M Roux, p. 921. Rotterdam: AA Balkema

101. Wray GA. 1994. The evolution of cell lineage in echinoderms. *Am. Zool.* 34:353–63

102. Wray GA. 1996. Parallel evolution of nonfeeding larvae in echinoids. *Syst. Biol.* 45:308–22

103. Wray GA, Bely AE. 1994. The evolution of echinoderm development is driven by several distinct factors. *Dev. 1994* (Suppl.):97–106

104. Wray GA, McClay DR. 1989. Molecular heterochronies and heterotropies in early echinoid development. *Evolution* 43:803–13

105. Wray GA, Raff RA. 1991. The evolution of developmental strategy in marine invertebrates. *TREE* 6:45–50

106. Wray GA, Raff RA 1991. Rapid evolution of gastrulation mechanisms in a sea urchin with lecithotrophic larvae. *Evolution* 45:1741–50

Annu. Rev. Ecol. Syst. 1997. 28:243–68

PRESERVING THE INFORMATION CONTENT OF SPECIES: Genetic Diversity, Phylogeny, and Conservation Worth

R. H. Crozier

School of Genetics and Human Variation, La Trobe University, Bundoora, Victoria 3083, Australia; e-mail: genrhc@gen.latrobe.edu.au

KEY WORDS: genetic distance, genomic information content, conservation worth, conservation philosophy, ecological economics, molecular phylogeny, gene number, species richness, genetic diversity

---

ABSTRACT

A variety of phylogenetic measures have been proposed to quantify distinctiveness, often held to mark species of high conservation worth. However, distinctiveness of species and their numbers have different implications for conservation policy, depending on whether moral, esthetic, or utilitarian reasons are accepted as justifying conservation. The utilitarian position values species according to increasing numbers, and as they are more, as opposed to less, distinctive. The view is taken that conservation should seek to maximize the preserved information of the planet's biota, best expressed in terms of genetic information held in genes and not in portions of the genome of uncertain or no function. Gene number is thus an important component of assessing conservation value. Phylogenetic measures are better indicators of conservation worth than species richness, and measures using branch-lengths are better than procedures relying solely on topology. Distance measures estimating the differences between genomes are preferable to substitution distances. Higher-taxon richness is a promising surrogate for branch-length measures. Complete enumeration of biotas in terms of phylogeny is desirable to avoid uncertainties in the use of indicator groups, and this is achievable now for bacteria. Phylogenetic measures are already important for management of sets of populations within species and are applicable for sets of species. Measures incorporating extinction probabilities and decision costs are being developed, and these, in conjunction with the use of confidence limits on the conservation worth of alternative reserves, are vital for conservation decision-making.

0066-4162/97/1120-0243$08.00

When sequencing becomes cheap enough, and reading genetic codes is as routine as counting feathers and molar teeth, we will be technically prepared in full to address the question of how much biodiversity exists on earth. Wilson (1992, p. 161).

BIODIVERSITY AS INFORMATION CONTENT

Interaction with Reasons Justifying the Preservation of Biodiversity

Biological conservation has tended to be a matter of the heart. For example, conservation effort by the US government correlates strongly with the anthropocentric appeal of species (122), and this may also be so for some conservationists (197). But for conservation biology to be most effective in setting priorities for conserving species and ecosystems, an objective framework is desirable. Much progress has been made in practical aspects of biological conservation, especially with regard to the preservation of populations endangered due to their small population size (30, 69) (i.e., considering genetic diversity at the within- rather than between-species level), as well as in general principles for ecosystem management (115), but attention must still be paid to an overall theoretical framework. A signal advance in this regard came from the suggestion (205) that the aim should be the preservation of the information content contained in the DNA of all species on the Earth. Developments relevant to this aim are explored in this review. A further consideration is the estimate of the amount of biodiversity to be conserved. There is no estimate yet of this in terms of information content, but estimates of the number of species have risen sharply over the last two decades, e.g. from 1.7 million in 1978 (206) to between 3 and 30 million in 1995 (118).

The information content view of biodiversity and increased estimates of the numbers of species have different implications depending on the reasons accepted for the conservation of biodiversity, organized under three, or possibly two, headings: moral (other species have a right to exist), esthetic (species are like works of art, and it would be foolish to destroy them), and utilitarian (humans derive material benefit from the existence of other species).[1] The cause of conservation is strengthened by adherents, whatever their reasons, but each of these reasons has difficulties that should be appreciated.

The moral reason raises the difficulty that rights are human constructs and hence have no validity outside human belief systems. Strongly contradictory conclusions in philosophical analyses of human rights (137, 161) reflect the non-universality of rights. Reasoning from a moral perspective may also seem

---

[1]There are many terminologies and classifications of reasons for biological conservation. The same three justifications advanced here can also be termed intrinsic, transformative, and human demand values (135). Authors may stress (22, 31) or omit (15) the moral reason.

incongruous because, under the biological species concept, species may be separated by very few gene substitutions (25, 143) or none at all (as in the case of symbiont-related speciation—154, 179), whereas under other species concepts, species may not be distinct evolutionary entities.

EO Wilson has described (personal communication) eliminating species for perceived economic gain as being akin to burning paintings by Rembrandt to keep the house warm. However, an esthetic reason for conservation faces the obvious difficulty that not everyone would ascribe beauty to natural phenomena. The concept that people possess an innate liking for other species and may benefit psychologically from access to them (204) can be supported (100), but it is uncertain that sufficient numbers of people can be convinced of this to yield political support for conservation.

Utilitarian arguments can be divided into the view that human survival is enhanced by the existence of large numbers of other species and that direct material benefits flow from other species. The former "ecosystem services" view may be correct and is supported at least for small numbers of species (scores) (131, 132, 181–183, but see 77), but the continuance of this effect for large numbers of species per ecosystem so far lacks the support of controlled experimental evidence and may not occur (130). Finnish forests, for example, are notably species-poor and yet appear quite stable (123), although the small number of tree generations since records began impels caution.

The bioprospecting industry rests on biodiversity as the source of novel products (39). Major drug companies have active programs for assessing natural plant and microbial products from biodiverse areas in return for payment to the source countries (23, 185); given international conventions favoring repayment, biodiverse countries are responding by formulating procedures to protect their natural wealth of biodiversity (8, 94, 95, 112, 162, 172). Based on the known rate of discoveries of pharmacologically active compounds from bioprospecting, the expected annual loss to the United States alone in terms of sales value foregone when plants become extinct before they could be assayed has been estimated at US $3.5 billion. Thus expenditure on conservation would be justified if it prevented the loss of plant species (160). While species might lose their value when assayed, this seems unlikely to affect biodiversity swiftly; assay methods change with advances in knowledge, so species not previously known to provide valuable products might later be found to do so. Moreover, the proportion of species examined is as yet very small.

In the moral view, the information concept and estimates of species numbers are irrelevant: Every species should be conserved out of moral necessity, and each is equally worthy of preservation. Under the other two views, however, the worth of a species is determined by its phylogenetic distinctness from other species (but see below). These positions are contrasted in Figure 1.

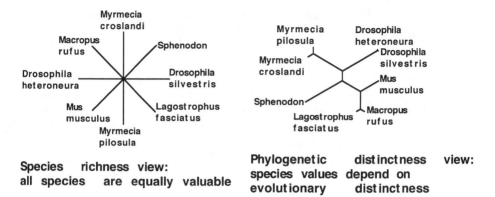

Species richness view: all species are equally valuable

Phylogenetic distinctness view: species values depend on evolutionary distinctness

Figure 1 Alternative views of a set of species chosen purely for illustrative purposes. The *Drosophila* species chosen are very similar genetically but extremely interesting scientifically because of their great morphological divergence, apparently based on few substitutions (178). Two macropod marsupials (*Macropus rufus* and *Lagostrophus fasciatus*) are included and also the common mouse, *Mus musculus*. The two *Myrmecia* bulldog ant species include one with the lowest possible chromosome number (41, 43). *Sphenodon* is the tuatara, a single genus comprising an extremely divergent group. Distances in the right-hand tree roughly correspond to times of separation.

The increasing size of the estimated number of species has a potentially deleterious impact on the justification for preserving biodiversity under the esthetic and utilitarian views. To understand the problem: If you have only a solitary $1 note, it will be very precious to you. If you gain a million more such notes, even though each is unique, your original note becomes much less valuable to you. The same point can be made about species: The more there are, the less any one of them will be worth. However, under both the esthetic and utilitarian views, the more distinctive species will have a higher conservation worth, less distinctive ones a lower. Wilson (205) made this point in relation to the possible rapid replacement of extinct species by speciation from a few survivors: The new species would be very similar to existing ones, and by comparison with those lost would be "cheap species."

Opposing Wilson's (205) position is the argument that rapidly speciating groups should be given priority over phylogenetically isolated taxa because biodiversity would then be more rapidly recreated after extinction (54, 108). However, this "picking winners" strategy ignores Wilson's point and the long-held conclusion of evolutionary biology that in fact picking winners is very difficult (103, 120). In any case the distinction is a false one: If faced with choosing either the highly divergent genus *Sphenodon* (tuataras, with two species) or all of the lizards, one would keep all the lizards. But the choice will more likely be the equivalent of whether to keep *Sphenodon* or a lizard species, and in that case it would be hard to justify ruling against *Sphenodon* under any scheme.

Although increasingly large estimates of numbers of species may seem to weaken the value of biodiversity, this challenge does not extend to the sum total of species. In the search for pharmaceutical products, for example, the choice is between focusing on chemical space, the universe of imaginable compounds (50), or bioprospecting. As the number of species increases, so does the likely payoff of bioprospecting compared with searching the chemical space. To put it another way, the best evolutionary algorithm to discover pharmaceutical and other products (26) is evolution itself. Given that more divergent species will present more divergent properties in all aspects, phylogenetic distinctness—a criterion stemming from the information content approach—should also be a predictor of utilitarian value. Reviewing the numbers of species in plant families and the occurrence of plants with useful products (35, 55, 81) shows that "useful" plant species are widely scattered across families. It remains to be determined whether a negative relationship exists between family size and number of "useful" species (i.e., divergent species are more likely to yield novel products), and whether families differ in the likelihood of a species being "useful" (i.e., families differ on a per-species basis in the probability of yielding a novel product).

A precautionary principle clearly operates under the utilitarian view: Given that nature has actual and potential value to human welfare, it is foolish to allow destruction of nature without knowing what it is worth. There is a strong and perhaps controversial form of this principle: Not only will it be a long time before the natural world is even sketchily surveyed, but also, given changing human needs and technology, the task is impossible to finish.

It is unclear whether, under the esthetic view, the worth of nature as a whole increases with increasing numbers of species, given that these will often be at least outwardly similar. Such an increase of value with increasing numbers of species may however occur if species are given an existence value.

GENOMIC INFORMATION CONTENT

The information content of a genome is, for the purposes of biodiversity, best restricted to coding portions of the DNA, roughly speaking, the genes coding for proteins, tRNAs and ribosomal RNAs, and controlling elements in the flanking regions of genes (107). Variation in the numbers of genes between organisms, while considerable, is much less than variation in the total amount of DNA per genome. This difference in range stems from the fact that the amount of the genome not coding for anything varies broadly. In bacteria, for example, the noncoding fraction is negligible, but in eukaryotes it ranges from less than 30%

of the DNA to almost all of it (33). Humans with 3,400,000,000 nucleotides lag far behind the protist *Amoeba dubia* with 670,000,000,000 (107)! However, vertebrates are believed to have the largest number of genes among organisms— about 80,000 (4, 21), which contrasts strongly with the 470 found by complete sequencing of the genome of the bacterium *Mycoplasma genitalium* (70).

Notwithstanding the impressive variation in DNA amounts, using the currency of genetic divergence has a number of actual and potential advantages if attention is restricted to coding sequences. Gene and protein sequences much more closely approximate the ideal characters for inferring degree of phylogenetic divergence than do morphological ones (166), and sequences are much less prone to the problem of saturation. Saturation begins when changes start to occur in characters that have already absorbed a change, as when more than one substitution occurs in a DNA sequence during the divergence of two species. Paradoxically, even though substitutions still occur at a constant rate, the two sequences will eventually diverge more slowly than they did initially, and finally not at all. The inference of the amount of evolution depends on measures to convert observed differences between sequences to estimates of the numbers of substitutions per position, a task made easiest when the rate of divergence is uniform over the time period being studied. Evolutionary rates vary broadly between genes, and genes can be chosen that suit the degree of divergence required (49, 53, 91, 144). Quantitative molecular but not morphological comparisons are possible between flies, humans, bacteria, and nematodes because at the morphological level these organisms present few homologous structures [even within a subclass, efforts to compare morphological divergence levels among hominids with those among frogs (34) were not universally accepted (203).

Gene number—not total DNA content—is then a reasonable measure of the total potential information content of organisms, their "complexity," and sequences of the genes constitute the information content. The notion that complexity increases during evolution has tended to make biologists uneasy, perhaps because "progress" is thought to imply purposive mechanisms; some therefore deny that complexity has increased, at least overall (79). Whether or not complexity has increased "overall," it has undoubtedly increased in many lineages. The information-content approach has popularly conceptualized this trend with the notion of an upper limit in information content for a particular genome (180). The number of cell types and the number of genes may be highly correlated (99), although the strength of this correlation at the upper end of the scale has been questioned (121). Gene number suggests itself as a natural measure of complexity (119, 176), indicating as it does the expected diversity of gene products. For example, the complex systems of interacting loci of hemoglobins and of lactate dehydrogenases in humans owe their origins to the duplication and then divergence of genes stemming from a single ancestral locus in each case. The evolution of mammals from unicellular eukaryotes is readily visualized as a series of events during which gene products are added as new functions and features evolve (110). Many of these additional genes form gene families such as the globins in which the various members perform similar functions or form polymers to yield molecules more efficient than the ancestral monomers (107). Not all genes arise from pre-existing ones, but they can occasionally arise de novo, as in the fortuitous functionality of alternate reading frames (1, 140), and new genes can be produced as the reshuffled subunits of existing genes (107).

According to Bird (20, 21), increases in gene number have been mediated by the need for developmental noise suppression. The noise to be suppressed consists of the expression of genes in tissues or developmental stages in which they would be deleterious, such as perhaps those for crystallins (the transparent proteins of the eye lens) in a blastula or in the pancreas. Bird distinguishes two major thresholds which, when overcome, allowed increased gene number: the formation of the nuclear membrane (allowing the filtering of mRNAs before export to tthe cytoplasm) and the massive increase in methylation of vertebrate genomes (allowing finer control). Whether these particular mechanisms are responsible remains uncertain, especially given the large range of gene numbers of species in the same category [e.g., *Haemophilus influenzae* with 1743 predicted genes (68) as against only 470 in another bacterium, *Mycoplasma genitalium* (70)], but the picture is compelling that additional gene products facilitate complexity and therefore increasing gene number is favored by selection.

Does the vertebrate genome hold $80,000/475 = 168$ times as much information as that of *Mycoplasma genitalium*? No, because members of a gene family are descended from a single ancestral gene and they remain alike. By analogy, the information content of a library of 1000 volumes is not 1000 times that of a single volume if they are all identical; in that case the information content would be that of a single volume plus that needed to record that there are 1000 copies. Information accumulates as copies of the volumes diverge in content, such as by students marking them or ripping pages out, and similarly, the information content of a family of genes diverges as substitutions and deletions (and insertions) accumulate. Eventually it becomes hard to detect that genes are related. Also, genes that evolved under one context become used for another, such as the cooption of a lactate dehydrogenase and a heat shock protein as crystallons (110). A set of ancient conserved sequences has been detected (80, 169) that are present in many genes of all organisms, e.g., in 40% of *Escherichia coli* genes (102). Given such widespread genetic similarity among all living things, it is not as surprising as it would have been 20 years ago that there are similarities in the control of vertebrate and insect development in dorsoventral determination (89), body-plan establishment (32), and even eye placement (78).

In more biological terms, the information content of a set of genes related through homology (the same gene in related species) or in being paralogous (derived via duplication from the same ancestral locus) can perhaps be approximated by the information contained in one sequence of the appropriate length and composition (proportions of nucleotides or amino acids) plus the number of polymorphic positions (preferably weighted by number of alternative alleles) multiplied by the information content per position, recalling that information is additive between messages (66). Given the number of sites, the number of positions expected to be polymorphic can be determined using phylogenies.

MOLECULES AND PHENOTYPE

Do molecular differences reflect differences in other aspects of the organism? The generally close correspondence of molecular phylogenies to those resting on the wisdom of generations of morphological systematists (149) indicates that this is so for morphology, and the similarity of trees derived from metabolic (36) and behavioral (148) characters indicates it is also true for unseen characters. Because molecular data are potentially so numerous and the mechanisms of evolutionary change in them now much better understood, it is not surprising that some well-established morphological phylogenies and inferences on timing have been overturned by molecular results—for example, the place of *Sivapithecus* in the hominid lineage (155, 168), the relationships of cyanobacteria to plants and bacteria (92, 144), whether hut-building bowerbirds are monophyletic (105), the supposed divergence of animal phyla in the Vendian (37, 48, 79, 209), and the relationship among animal phyla (38, 82, 111, 177).

Kadereit (98) suggested that inconsistencies between morphological and molecular phylogenies in plants are due to major gene effects on morphology, and so he argued that such inconsistencies are likely to be particularly illuminating in studies of evolutionary mechanisms. Such changes would include cases in which a suite of characters as defined by the morphologist change as a single entity because of one underlying genetic control mechanism.

"TAXONOMY AS DESTINY"

Interest in phylogenetic aspects of preserving diversity was greatly increased by the discovery that "the" tuatara, *Sphenodon punctatus*, actually comprises two species (46). Further, the various persisting island populations of tuatara are heterogeneous, according to analysis of allozymes. Even if no additional species was recognized, under the viewpoint expressed here the conservation worth of the populations had increased. Naturally, the recognition of two species gives the findings high visibility, truly a case of "taxonomy as destiny" (116).

Although the principle of using phylogeny to determine conservation priorities was widely recognized (2, 90, 129, 175), May's (116) note introduced the first efforts to develop a phylogenetic framework for determining the relative conservation worth in a set of species. These efforts were made by the innovative efforts of a group centered at the British Museum of Natural History, and the efforts hinged exclusively on the topology of the inferred phylogenetic tree (93). The worth of a species was inversely related to the number of nodes between it and the root of its tree, thus tending to give highest priority to species branching off first irrespective of degree of difference (186). Nixon & Wheeler (135) developed this idea a little further, quantifying general ideas on the use of systematics and phylogenies (175) by weighting species inversely to the number of species in their group relative to that of their sister group. Recent developments of the method have eliminated the need for estimating the place of the root and have introduced a method of resolving ties (198), with apparently general agreement now that unrooted trees are the appropriate vehicle for phylogeny-based measures of conservation worth. The more recently developed method counts the numbers of nodes traversed between each candidate species for preservation and the others, according to the dendrogram for the full set of species; the set of species out of those possible (say on economic grounds) that maximizes the sum of the interspecies node distances is the one chosen. This cladistic diversity measure can be given, slightly simplified, as:

$$CD = \sum_{i=1}^{n} \sum_{j=1}^{n} U_{ij}, \qquad\qquad 1.$$

where n = the number of species, and U_{ij} = the number of nodes in the path between species i and species j in the tree.

Responding to May's (116) note, Altschul & Lipman (3) proposed a branch-length measure, but one with some undesirable features (58). Proposals for the use of branch lengths for assessing conservation worth followed rapidly. Pamilo (146) also proposed a weighting scheme for assessing species worth, describing the worth of a subset of species as the length of tree required to represent it. Pamilo's method was designed for rooted trees, using UPGMA or WPGMA, and defined weights for species or groups depending on the "independent evolutionary information" they contain. It was, however, aimed more at tree estimation than at conservation worth.

For Faith phylogenetic diversity (PD) (57, 60, 61, 190) is the sum of the branch lengths in a phylogeny:

$$PD = \sum_{k=1}^{2n-3} d_k, \qquad\qquad 2.$$

where d_k = the length of branch k in the tree.

PD assesses worth in terms of the amount of evolution preserved, i.e., the set among those possible that preserves the longest length of the original tree is favored.

Genetic diversity (GD) (42, 44) estimates the probability that the set of taxa preserves more than one allele per site:

$$GD = 1 - \prod_{k=1}^{2n-3} (1 - p_k), \qquad\qquad 3.$$

where p_k = the proportion of sites with different states at either end of branch k, and hence $0 \leq p \leq 1$.

Genetic diversity was conceived for use with differences rather than evolutionary distances, although when the values are small, the results will be very similar. If applied to measures meeting the requirement of the values being between 0 and 1, GD and PD give results that can be simply interconverted by a procedure using logarithmic transforms (DP Faith, personal communication). Contrary to one statement (59), the rank order of different sets of species is the same whether GD or PD is used, as noted by Krajewski (104). Because PD and GD yield the same rank orders with genetic data, either can be used in exploratory studies; the chief difference between them will be in the relative conservation weights given to sets of taxa. If the different rationales for using distance and difference data are adhered to, then substitution distances should be reconverted to difference estimates for GD estimates. Restriction site polymorphism and DNA-hybridization distances are measures of difference, not distance, and hence are examples of data for which GD is appropriate. Whatever method is used to handle the data, difference measures are preferable because they better reflect the biodiversity being retained, even though they are not desirable for phylogeny inference because they underestimate early divergences.

May (117) recognized the desirability of using the branch length approaches and suggested that, where the branch lengths have not been objectively estimated, it is appropriate to insert the most reasonable estimates available. PH Williams (in Litt) advises that when appropriate branch length information is available, the NMH group now recommends the use of distance-based methods (197).

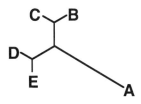

Figure 2   A tree of a group of five species. Branch-length measures (GD and PD) would give the set of taxa A and B higher priority for conservation than the set B and E, but node counting (CD) would do the reverse.

PD and GD would each assign high conservation priority to the extremely divergent tuataras, as does the original root-based node-counting scheme. However, the unrooted node-counting scheme assigns minimal conservation worth to the tuataras because they join the reptile tree in an interior position, even though the branch involved is very long; a simple example is given in Figure 2.

All three measures—CD, GD, and PD—are intended to maximize the preservation of as much of the phylogeny of the complete set of species as possible. In that sense, they are all intended to maximize "phylogenetic diversity," and the particular names associated with them should not obscure that fact.

An alternative and recursive distance measure uses the smallest distance from a threatened species to a species of the existing set (173, 193, 194). With a perfect molecular clock, this measure yields the same answer as PD and GD, but otherwise it may give a species on a short branch a higher worth than one on a long branch (61), which does not seem justified in terms of the preservation of information.

An important issue is that taxon loss is not an all-or-none issue with regard to the provision of protection. All populations face a probability of extinction, and protection reduces this probability. The incorporation of extinction probabilities, with and without protection, has been treated by several authors (173, 193, 194, 207, 208). The general approach would be to weight the trees by the probability of their occurrence in say 50 years time under various possibilities for protection.

STATISTICAL SUFFICIENCY

Tree topology and branch lengths are both estimated quantities. Especially in the case of highly divergent forms, branch lengths are more robustly estimated than topologies, which is fortunate since they provide the more important information for conservation. A preliminary study using a set of bowerbird cytochrome *b* sequences (44) found that conservation worth estimated using genetic distances was highly robust when used to examine bootstrapped data sets. This fact is encouraging but does not eliminate the need for robust statistical advice from scientists concerning the likely effects of management decisions on biodiversity maintenance. Given that it is now routine to determine the statistical sufficiency of phylogenies, users of phylogenetic methods for biodiversity assessment must similarly determine whether the effects of different reserve systems, for example, are really significantly different from each other. One way to do this would be to use only those trees significantly better than any alternatives, but given the expected data set sizes in the near future, this might be difficult.

An approach now available (P-M Agapow, RH Crozier, unpublished) is to prepare bootstrap samples of the data using a package such as PHYLIP (65) and a fast method such as Neighbor Joining (167), and then to use the bootstrap trees to determine bootstrap confidence limits for the conservation worth of the possible sets of species to be preserved. These limits would then enable determination of whether one set preserves significantly more diversity than another.[2] It is hard to see how recommendations without statistical sufficiency would be credible, either for phylogenetic methods or for species richness (for which, methods used for behavioral cataloging (56) might be appropriate).

SURROGACY AND INDICATOR GROUPS

Although technological advances, as discussed below, are rendering the direct estimation of genetic diversity not just increasingly practical but the preferred measure, still for most groups of organisms forms of data other than genetic are more easily obtained. Of these, the most appealing to many researchers is simple species richness—the number of species in a habitat. This measure is preferable to species diversity, determined also by the evenness of species numbers (17), because we are interested in maximizing the potential for the future. Species richness of course takes no account of the degree of divergence between species and is bedevilled by the taxonomic impediment, the growing and serious lack of alpha taxonomists (13, 87, 134)—an impediment the uneven impact of which across taxonomic groups further confounds efforts at total assessment. Rapid procedures for estimating species richness (72, 141, 142) and databanks for biodiversity (71, 96, 163) resembling those for gene sequences (18, 170) will enhance the effectiveness of species richness as a measure of biodiversity.

Williams & Humphries (199) found for several groups a strong relationship between species richness and "character richness," the latter defined as total tree

---

[2]Statistical sufficiency for GD and PD can be estimated using Macintosh program Conserve3 (http://www.gen.latrobe.edu.au/staff/rhc). PD is also implemented in the Diversity package (D.faith@dwe.csiro.au) and node-counting measures in the WorldMap package (195).

length covered, setting each branch to the same unit length (i.e., resembling GD and PD, but with branches equal in length). While this gives some grounds for optimism that species richness may provide an adequate estimator of biodiversity, it is unfortunately true that the method rigidly produces an equal character change per node, thus forcing the result. A valid test would use a group with a well-founded molecular phylogeny, of which many cases are now available. The plots (199) show smaller increases in character richness with increasing species richness, although whether this results from actual saturation or is an effect of the unit-branch method is uncertain.

Topology-based measures appear a natural progression from species richness as a surrogate for direct estimates of biodiversity, but the poor performance of topology measures, alluded to above, in reflecting evolutionary information content impels caution in accepting these as useful approaches.

Higher-taxon richness (52, 73, 75, 196, 198–202), in which higher taxa (e.g., families or genera) are surveyed for comparisons between areas, although originally typified as a root-based method, is essentially a reflection in nomenclature of systematists' impressions of the evolutionary distances between species. Especially given the greater ease of identifying specimens to higher taxa rather than to species, this method is thus promising as a surrogate measure for biodiversity. But the appropriate level will differ between groups because of differences in the degree to which systematists have split species between higher categories (52, 73, 196): families for showy organisms such as some plants, bats, birds, and butterflies, but genera for ants. Tests of how higher taxon richness correlates with biodiversity (as defined here, in terms of genetic information content) should be carried out, and they are possible now for bacteria (see below). For a South African plant community, higher taxon richness correlated strongly with "functional" attributes defined in terms of "pollination systems, seed-dispersal properties, fire-survival mechanisms and nutrient-scavenging systems" (109), but if such characters entered into the definition of the higher categories, then the result may not be as clearcut as it first appears.

Gaston & Blackburn (74), using DNA-hybridization data (171), found that the age of New World bird tribes declines with distance from the equator. While this result does caution against too-ready acceptance of the higher-taxon richness measure, it particularly indicates that species-richness may mean different things in different regions in terms of the average distinctiveness of species. Higher-taxon richness is only a moderately good indicator of species-richness (9, 10). It would be worthwhile to study directly the predictivity of higher-taxon richness to biodiversity as information content.

All studies in which the complete biota are not surveyed are perforce studies of indicator groups: Estimates of biodiversity in one group of organisms are used to estimate the biodiversity of unstudied groups. As alluded to above in connection with higher-taxon richness, the scale of the study is likely to be crucial, and differences in scale (45) may explain contrasting conclusions reached on the success of the indicator group approach (156, 159). Suggested criteria (134, 151) for the choice of indicator groups are sometimes contradictory. As an example of a carefully limited study, the numbers of species in one butterfly subfamily, the Ithomiinae, were found to correlate well with the species numbers of all other butterflies (16). In contrast, a simulation study in which species distributions were placed at random in environmental space showed poor prediction by one group of species of the presence of others (62), suggesting that the frequent success of indicator groups in practice rests on additional evolutionary processes occurring in nature. Null model tests, involving competitive processes affecting the probability of close relatives coexisting, and also preferably habitat heterogeneity and historical effects, would be valuable in the conservation context.

APPLICATIONS FROM THE SPECIES TO THE ECOSYSTEM

Animal and plant breeders have long been concerned about the value of germplasm in the ancestral populations from which agriculturally important breeds are derived. The importance of molecular genetics in assessing germplasm is increasingly recognized (27, 67), as are connections between agricultural and general conservation biology (158). The role of phylogeny has also become clearer, with breeders recognizing that the most phylogenetically divergent breeds are liable to present distinctive characteristics (12, 28, 83), although without application yet of methods such as those described above. Barker (12) considered many of the potential problems and needs in this area, e.g. that because movement of germplasm is often by means of males, the use of mtDNA to determine genetic distances between breeds may not be appropriate. Because population size reduction tends to increase genetic distances, marginal multipliers reflecting the amount of genetic variation in breeds, and hence their selectability, should perhaps be applied to the analysis of conservation worth of breeds in a manner analogous to the use of extinction probabilities (see below).

The stock concept has been important in fisheries for some time. A stock is a set of populations whose exploitation does not endanger that of other stocks of the same species. It is natural for this concept to be imported into conservation biology, especially with genetical interpretations of what constitute stocks (5, 7, 47). The most obvious way to approach this question is simply to determine if samples from different sites are statistically separable on genetic grounds, but it is then uncertain that the populations are evolutionarily distinct.

A more stringent criterion is to determine if the populations are fixed for different alleles. Avise (6) noted particular importance in the characteristics of mtDNA: Not only are mtDNA haplotypes more rapidly fixed than are nuclear ones because the effective population size is smaller for mtDNA due to its occurrence in effectively single doses per individual and its usual matrilinear transmission. In addition, the maternal mode of inheritance means that mtDNA traces female gene flow, and female movement is required before a population can be established.

A more general distinction is that between management units (MUs; groups of populations differing in allele frequencies) and evolutionarily significant units (ESUs; groups of populations derived from different common ancestors). Moritz (125, 126) specifically defined ESUs as groups of populations characterized by "being reciprocally monophyletic for mtDNA alleles and also differ[ing] significantly for the frequency of alleles at nuclear loci." ESU is therefore a narrower category than MU (Figure 3). Moritz (128) recommended that MUs and ESUs be accorded the conservation status of species under the US Endangered Species Act. There are now several important case histories of application of

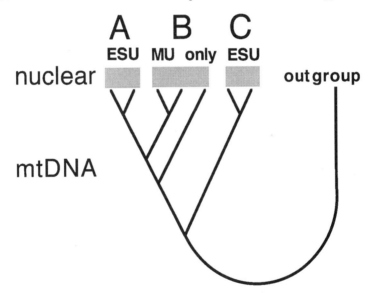

Figure 3 Management units and evolutionarily significant units (125). A tree showing a group of three distinct populations defined by an outgroup; populations are defined as distinct by differing in allele frequencies either for nuclear loci or mitochondrial DNA (mtDNA) haplotypes. Distinct populations are termed "management units" (MUs) because their continuity of allele frequencies indicates interchange between them. MUs that are also monophyletic for mtDNA haplotypes are evolutionarily significant units (ESUs).

the ESU approach, such as the endangered tiger beetles *Cicindella dorsalis* and *C. puritana*, of the US east coast (187–189). For both species, groups of populations could be recognized as ESUs on the basis of monophyly for mtDNA haplotypes. Marine turtles such as *Chelonia mydas*, which often show extreme fidelity to nesting sites, provide another major case study (24).

Rather than preserve the biodiversity of just one group, the goal should be to preserve those areas with high biodiversity overall. This goal implies the use of phylogenies for all included species in the areas in question. A partial approach is to use phylogenies from several species, which, when done at the population level, helps to guard against complications from the potential differences between gene trees and species trees (147), although such concerns should not be overplayed (124). This approach has been taken for Australian rainforest remnants (97), showing strong congruence in four of six vertebrate species surveyed. Moritz & Faith (127) suggested that biodiversity as determined by branch lengths be summed across separate phylogenies for the reserve areas being considered, which may be done for three areas using Venn diagrams. As long as the same kind of data (e.g., same gene region) were used for all the phylogenies, the results should approximate the more rigorous analysis of a complete phylogeny, which might not always be possible.

The ESU concept rests fundamentally not on the biological species concept but on the view that species must be "diagnosable," i.e., distinguishable from each other (40, 188). This dependence on a species concept other than the biological can be seen as concurring with the view frequently expressed that populations and not species are the ultimate currency of biodiversity (14, 42, 44, 113, 114). Possible discordance of the mtDNA tree with the population tree is also a concern (133, 147), but the rarity of statistically significant discordance between nuclear and mtDNA trees (124) suggests that ESUs will usually be correctly determined.

The limits of a genetic clustering approach applicable to rare plant populations were shown in a study of the Australian woody perennial *Grevillea scapigera*, which persists as only 27 individuals. RAPD data were used to derive a UPGMA dendrogram of the plants, and a subset was chosen for seed collection with the intent of maximizing retained genetic variation (165). The measure used was not one of the information measures but the mean distance to all other plants. Given that the dendrogram of members of a sexual population could not be interpreted as a phylogeny, use of the information measures would not be appropriate, but perhaps genetic distance to the most similar plant being preserved might be more appropriate.

Probably in most organisms, populations can be distinguished in terms of allele frequencies on a scale of meters rather than kilometers. Hence the MU and

ESU concepts probably lack operational generality for most groups. However, that the concept has proven applicable for some insects (106, 187, 189) indicates that its use is not restricted to vertebrates.

Species phylogenies have been used to estimate conservation worth for species or sets of species using both node-based (13, 174, 191) and distance-based (173, 193, 194) measures. Bumblebees of the *Bombus sibiricus* group have been extensively used to illustrate application of a range of node-based measures to the selection of general reserves (199, 201, 202), and bowerbirds have been used similarly to illustrate the use of distance-based measures to maximize the retention of genetic diversity in these birds (42, 44). The use of a phylogeny of one group to infer the optimum reserves for others encounters the problem alluded to above for indicator groups but may be preferable to simple species richness measures.

No complete enumeration of any biota, let alone in terms of the phylogeny of the included species, has yet been done, but the potential has emerged to do just this with bacteria, a major and little-studied portion of the total. A problem with even determining bacterial species diversity is that most species cannot be cultured using traditional methods. The magnitude of the uncul-turable world was shown by Torsvik et al (184), who used DNA reannealing analyses to show that a gram of Norwegian forest soil contained 4000 bacte-rial species. An alternative approach, the detection of rRNA genes, had been demonstrated earlier using reverse transcriptase and sequencing primers based on highly conserved parts of the gene sequence (145), but the relative difficulty of the method precluded large-scale application. The advent of the polymerase chain reaction (PCR), which allows specific sequencing of very small amounts of target DNA, has enabled in principle the complete enumeration of bacterial biotas (19, 29, 53, 76, 86, 138, 139, 152, 153, 157, 164, 192). As for the earlier direct-sequencing method using reverse transcriptase, the present approaches use highly conserved regions of the rRNA gene, which flank more variable ones. The studies so far, as might be expected, have revealed that many mi-crobial groups remain to be discovered: The observed diversity is not a simple multiplication of species close to those already known (53). Applications have ranged from surveying bacterial fauna of termite guts (19, 137, 138) to surveys of whole communities (29, 76, 152, 153, 164, 192). Phylogenetic analysis is the best means to place the newly discovered organisms according to gene se-quences. A different approach used simple $(G + C)\%$ of recovered DNA (86), which depends on the very wide variation in $G + C$ content between bacte-rial groups (107). The $G + C$ content approach (86) appears to be very close to higher-taxon richness, in that it returns neither an enumeration of bacterial species nor the true phylogenetic structure, but it does yield an estimate of the bacterial genera present.

DECISION-MAKING IN A PHYLOGENETIC FRAMEWORK

Biological conservation is an economics problem and hence a political one. In the phylogenetic context, the preservation of crane species biodiversity has led to the application of standard economic theory in terms of the divergence between crane species, and their extinction rates without conservation effort (173, 193, 194). Weitzman (190) noted an apparent paradox: It would be desirable to support a less-endangered species rather than a more endangered one when such support increases the probability of at least one surviving (194). The approach is readily extended to potential reserve systems; Faith & Walker (63) analyzed costs and benefits of alternative reserve systems in terms of "biodiversity units." As for ecological economics generally, both cost and benefit have eventually to be expressed in monetary terms for national decision-makers (11, 84, 115, 150, 160, 195). A chief problem is the setting of the discount rate (150), which can logically be seen as different for individuals than for the nations (and world) in which they live (195).

Differences in the information content of genomes led to the realization that, other things being equal, some organisms have intrinsically higher conservation worth than others. For example, vertebrates, with some 80,000 genes have intrinsically higher conservation worth than an insect with 20,000 genes. But this calculation yields an upper value to a weighting factor because many of the vertebrate genes, such as in the *Hox* gene cluster (32, 88), are present in higher copy number than those in insects. Hence, vertebrate species, other things being equal, are worth less than four times as much as an insect species in conservation terms. Other things, of course, are not equal, and under the phylogenetic framework, vertebrates may still average very high conservation value compared with insects if, for example, the branch lengths of insect phylogenies tend to be short compared with those of vertebrates, as might be expected if more species are accommodated in the same time frame. Such issues are open to determination, but in any case it would be highly premature to attempt an overall tree of life approach, although the phylogenetic approach for subgroups is practical now (in particular, complete enumeration of bacterial communities is technically practical as a part of conservation surveys).

Conservation worth in terms of biodiversity is only one factor in decisions, although I have argued that it is the soundest factor scientifically. Many habitats may be marked for preservation not because of any overall higher worth but because they contain species with particular cultural significance, such as birds of paradise or a rhinoceros (42, 44, 101), species known to have potential economic significance, or species of particular scientific value. The choice between habitats will also be governed not only by conservation worth, determined as information content, but also by the costs of maintaining them, and such factors as the interactions with other habitats (e.g., maintenance of an estuary might require somee upstream attention). Although information content as conservation worth is an attainable goal of measurement, implementation will not always be simple. Although it is being applied now, research and development questions abound, such as:

* How much micro-scale variation is there in bacterial communities? If grids of closely spaced samples are compared from various localities, how similar are grid samples from the same locality compared with samples from other localities?

* Are any bacterial species at risk? Are any free-living bacteria of sufficiently restricted distribution to be liable to extinction?

* How well can molecular methods survey groups other than bacteria (e.g., protists and soil cryptozoa)?

* Statistical sufficiency is as important in biodiversity assessment as in phylogenetics: Is one reserve system significantly more biodiverse than another, or does the difference disappear when the underlying phylogeny is bootstrapped (51, 64, 85), for example? Environmental computer packages should provide estimates of statistical confidence for the biodiversity consequences of various policies.

* As the number of species included becomes very large, how much contribution comes from phylogenetically isolated species as against those from a mass of closely related forms? Will species-richness, albeit estimated increasingly using molecular methods, re-emerge as a close approximation to true biodiversity?

* What are the genetical characteristics of the great bulk of species that are not "model organisms"? Conservation biology is still biology, but too much of our attention is focused on a small number of species. What is the spread of gene numbers across taxa, and how different are genes from each other in groups with higher numbers?

ACKNOWLEDGMENTS

I thank the Australian Research Council for supporting my research in evolutionary genetics; A Balmford, MA Burgman, DP Faith, DG Green, DL Hawksworth, C Moritz, O Savolainen, ML Weitzman, and PF Williams for reprints, manuscripts, and citations; and P-M Agapow, JSF Barker, MA

Burgman, DP Faith, J Macy, BP Oldroyd, and PH Williams for comments on versions of the manuscript.

Literature Cited

1. Adelman JP, Bond CT, Douglass J, Herbert E. 1987. Two mammalian genes are transcribed from opposite strands of the same DNA locus. *Science* 235:1514–17
2. Ahern LD, Brown PR, Robertson P, Seebeck JH. 1985. Application of a taxon priority system to some Victorian vertebrate fauna. *Tech. Rep. 42. Arthur Rylah Inst. Environ. Res.,* Victoria
3. Altschul SF, Lipman DJ. 1990. Equal animals. *Nature* 348:493–94
4. Antequera F, Bird A. 1993. Number of CpG islands and genes in human and mouse. *Proc. Natl. Acad. Sci. USA* 90:11995–99
5. Avise JC. 1994. *Molecular Markers, Natural History and Evolution.* New York: Chapman & Hall
6. Avise JC. 1995. Mitochondrial DNA polymorphism and a connection between genetics and demography of relevance to conservation. *Conserv. Biol.* 9:686–90
7. Avise JC, Hamrick JL, eds. 1996. *Conservation Genetics: Case Histories from Nature.* New York: Chapman & Hall
8. Baker JT, Bell JD, Murphy PT. 1996. Australian deliberations on access to its terrestrial and marine biodiversity. *J. Ethnopharmacol.* 51:229–35
9. Balmford A, Green MJB, Murray MG. 1996. Using higher-taxon richness as a surrogate for species richness: 1. Regional tests. *Proc. R. Soc. London Ser. B* 263:1267–74
10. Balmford A, Jayasuriya AHM, Green MJB. 1996. Using higher-taxon richness as a surrogate for species richness. II. Local applications. *Proc. R. Soc. London Ser. B* 263:1571–75
11. Barbier EB, Brown G, Dalmazzone S, Folke C, Gadgil M, et al. 1995. The economic value of biodiversity. In *Global Biodiversity Assessment,* ed. VH Heywood, RT Watson, pp. 822–914. Cambridge, UK: Cambridge Univ. Press
12. Barker JSF. 1994. Animal breeding and conservation genetics. See Ref. 109a, pp. 381–95
13. Barrowclough GF. 1992. Systematics, biodiversity, and conservation biology. In *Systematics, Ecology, and the Biodiversity Crisis,* ed. N Eldredge, pp. 121–43. New York: Columbia Univ. Press
14. Baverstock PR, Joseph L, Degnan S. 1993. Units of management in biological conservation. In *Conservation Biology in Australia and Oceania,* ed. C Moritz, J Kikkawa, pp. 287–93. Chipping Norton, NSW, Aust: Surrey Beatty & Sons
15. Beattie AJ. 1995. Why conserve biodiversity? In *Conserving Biodiversity. Threats and Solutions,* ed. RA Bradstock, TD Auld, DA Keith, RT Kingsford, D Lunney, DP Silversten, pp. 3–10. Chipping Norton, NSW, Aust: Surrey Beatty & Sons
16. Beccaloni GW, Gaston KJ. 1995. Predicting the species richness of neotropical forest butterflies: Ithomiinae (Lepidoptera, Nymphalidae) as indicators. *Biol. Conserv.* 71:77–86
17. Begon M, Harper JL, Townsend CR. 1990. *Ecology. Individuals, Populations and Communities.* Boston: Blackwell. 2nd ed.
18. Benson DA, Boguski M, Lipman DJ, Ostell J. 1996. GenBank. *Nucleic Acids Res.* 24:1–5
19. Berchtold M, Konig H. 1996. Phylogenetic analysis and in situ identification of uncultivated spirochetes from the hindgut of the termite *Mastotermes darwiniensis. Syst. Appl. Microbiol.* 19:66–73
20. Bird A, Tweedie S. 1995. Transcriptional noise and the evolution of gene number. *Philos. Trans. R. Soc. London Ser. B* 349:249–53
21. Bird AP. 1995. Gene number, noise reduction and biological complexity. *Trends Genet.* 11:94–100
22. Booth DE. 1992. The economics and ethics of old-growth forests. *Environ. Ethics* 14:43–62
23. Borris RP. 1996. Natural products research: perspectives from a major

pharmaceutical company. *J. Ethnophar-macol.* 51:29–38

24. Bowen BW, Meylan AB, Ross JP, Limpus CJ, Balazs GH, Avise JC. 1992. Global population structure and natural history of the green turtle (*Chelonia mydas*) in terms of matriarchal phylogeny. *Evolution* 46:865–81

25. Bradshaw HD, Wilbert SM, Otto KG, Schemske DW. 1995. Genetic mapping of floral traits associated with reproductive isolation in monkeyflowers (*Mimulus*). *Nature* 376:762–65

26. Brookfield JFY. 1996. Forced and natural molecular evolution. *Trends Ecol. Evol.* 11:353–54

27. Brown AHD. 1978. Isozymes, plant population genetic structure and genetic conservation. *Theor. Appl. Genet.* 52:145–57

28. Brown AHD, Schoen DJ. 1994. Optimal sampling strategies for core collections of plant genetic resources. See Ref. 109a, pp. 358–70

29. Bruce KD, Osborn AM, Pearson AJ, Strike P, Ritchie DA. 1995. Genetic diversity within *mer* genes directly amplified from communities of noncultivated soil and sediment bacteria. *Mol. Ecol.* 4:605–12

30. Burgman MA, Ferson S, Akakaya HR. 1993. *Risk Assessment in Conservation Biology.* London: Chapman & Hall

31. Callicott JB. 1994. Conservation values and ethics. In *Principles of Conservation Biology,* ed. GK Meff, CR Carroll, pp. 24–49. Sunderland, MA: Sinauer

32. Carroll SB. 1995. Homeotic genes and the evolution of arthropods and chordates. *Nature* 376:479–85

33. Cavalier-Smith T. 1985. *The Evolution of Genome Size.* New York: Wiley

34. Cherry LM, Case SM, Kunkel JG, Wyles JS, Wilson AC. 1982. Body shape metrics and organismal evolution. *Evolution* 36:914–33

35. Chiej R. 1984. *The Macdonald Encyclopedia of Medicinal Plants.* London: Macdonald

36. Clark AG, Wang L. 1994. Comparative evolutionary analysis of metabolism in nine *Drosophila* species. *Evolution* 48:1230–43

37. Conway MS. 1994. Why molecular biology needs palaeontology. *Development* Suppl. S:1–13

38. Conway MS. 1995. Nailing the lophoporates. *Nature* 375:365–66

39. Cox PA, Balick MJ. 1994. The ethnobotanical approach to drug discovery. *Sci. Am.* 270(6):60–65

40. Cracraft J. 1983. Species concepts and speciation analysis. *Curr. Ornithol.* 1:159–87

41. Crosland MWJ, Crozier RH. 1986. *Myrmecia pilosula*: an ant with one pair of chromosomes. *Science* 231:1278

42. Crozier RH. 1992. Genetic diversity and the agony of choice. *Biol. Conserv.* 61:11–15

43. Crozier RH, Dobric N, Imai HT, Graur D, Cornuet J-M, Taylor RW. 1995. Mitochondrial-DNA sequence evidence on the phylogeny of Australian Jackjumper ants of the *Myrmecia pilosula* complex. *Mol. Phylogeny Evol.* 4:20–30

44. Crozier RH, Kusmierski RM. 1994. Genetic distances and the setting of conservation priorities. See Ref. 109a, pp. 227–37

45. Curnutt J, Lockwood J, Lu H-K, Nott P, Russell G. 1994. Hotspots and species diversity. *Nature* 367:326–27

46. Daugherty CH, Cree A, Hay JM, Thompson MB. 1990. Neglected taxonomy and continuing extinctions of tuatara (*Sphenodon*). *Nature* 347:177–79

47. Dixon PI, ed. 1992. *Population Genetics and Its Application to Fisheries Management.* Kensington: Cent. Mar. Sci., Univ. NSW

48. Doolittle RF, Feng DF, Tsang S, Cho G, Little E. 1996. Determining divergence times of the major kingdoms of living organisms with a protein clock. *Science* 271:470–77

49. Doolittle WF, Brown JR. 1994. Tempo, mode, the progenote, and the universal root. *Proc. Natl. Acad. Sci. USA* 91: 6721–28

50. Ecker DJ, Crooke ST. 1995. Combinatorial drug discovery: which methods will produce the greatest value? *Bio-Technology* 13:351–60

51. Efron B, Halloran E, Holmes S. 1996. Bootstrap confidence levels for phylogenetic trees. *Proc. Natl. Acad. Sci. USA* 93:7085–90

52. Eggleton P, Williams PH, Gaston KJ. 1994. Explaining global termite diversity: productivity or history. *Biodivers. Conserv.* 3:318–30

53. Embley TM, Hirt RP, Williams DM. 1994. Biodiversity at the molecular level: the domains, kingdoms and phyla of life. *Philos. Trans. R. Soc. London Ser. B* 345:21–33

54. Erwin TL. 1991. An evolutionary basis for conservation strategies. *Science* 253:750–52

55. Evans LT. 1993. *Crop Evolution, Adaptation and Yield*. Cambridge, UK: Cambridge Univ. Press
56. Fagen RM. 1978. Behavioural catalogue acquisition and completeness: validation of logarithmic regression and curve-fitting methods. *Anim. Behav.* 26:961–62
57. Faith DP. 1992. Conservation evaluation and phylogenetic diversity. *Biol. Conserv.* 61:1–10
58. Faith DP. 1993. Biodiversity and systematics: the use and misuse of divergence information in assessing taxonomic diversity. *Pac. Conserv. Biol.* 1:53–57
59. Faith DP. 1994. Genetic diversity and taxonomic priorities for conservation. *Biol. Conserv.* 68:69–74
60. Faith DP. 1994. Phylogenetic diversity: a general framework for the prediction of feature diversity. See Ref. 68a, pp. 251–68
61. Faith DP. 1994. Phylogenetic pattern and the quantification of organismal biodiversity. *Philos. Trans. R. Soc. London Ser. B* 345:45–58
62. Faith DP, Walker PA. 1996. How do indicator groups provide information about the relative biodiversity of different sets of areas?: on hotspots, complementarity and pattern-based approaches. *Biodivers. Lett.* 3:18–25
63. Faith DP, Walker PA. 1996. Integrating conservation and development: effective trade-offs between biodiversity and cost in the selection of protected areas. *Biodivers. Conserv.* 5:431–46
64. Felsenstein J. 1985. Confidence limits on phylogenies: an approach using the bootstrap. *Evolution* 39:783–91
65. Felsenstein J. 1990. *PHYLIP Manual. Version 3.3*. Berkeley: Univ. Herbarium
66. Feynman RP. 1996. *Feynman Lectures on Computation*. Reading, MA: Addison-Wesley
67. Flavell RB. 1991. Molecular biology and genetic conservation programmes. *Biol. J. Linn. Soc.* 43:73–80
68. Fleischmann RD, Adams MD, White 0, Clayton RA, Kirkness EF, et al. 1995. Whole-genome random sequencing and assembly of *Haemophilus influenzae* Rd. *Science* 269:496–512
68a. Forey PI, Humphries CJ, Vane-Wright RI, eds. 1994. *Systematics and Conservation Evaluation*. Oxford, UK: Clarendon
69. Frankham R. 1995. Conservation genetics. *Annu. Rev. Genet.* 29:305–27
70. Fraser CM, Gocayne JD, White 0, Adams MD, Clayton RA, et al. 1995. The minimal gene complement of *Mycoplasma genitalium*. *Science* 270:397–403
71. Frey AH. 1995. The internet biologist. *FASEB J.* 9:1245–46
72. Gaston KJ. 1996. Species richness: measure and measurement. In *Biodiversity. A Biology of Numbers and Difference*, ed. KJ Gaston, pp. 77–113. Oxford, UK: Blackwell
73. Gaston KJ, Blackburn TM. 1995. Mapping biodiversity using surrogates for species richness: macro-scales and New World birds. *Proc. R. Soc. London Ser. B* 262:335–41
74. Gaston KJ, Blackburn TM. 1996. The tropics as a museum of biological diversity: analysis of the New World avifauna. *Proc. R. Soc. London Ser. B* 263:63–68
75. Gaston KJ, Williams PH. 1993. Mapping the world's species: the higher taxon approach. *Biodivers. Lett.* 1:2–8
76. Giovannoni SJ, Rappé MS, Vergin KL, Adair NL. 1996. 16S rRNA genes reveal stratified open ocean bacterioplankton populations related to the green nonsulfur bacteria. *Proc. Natl. Acad. Sci. USA* 93:7979–84
77. Givnish TJ. 1994. Does diversity beget stability? *Nature* 371:113–14
78. González-Reyes A, Elliott H, St. Johnston D. 1995. Polarization of both major body axes in *Drosophila* by *gurken-torpedo* signalling. *Nature* 375:654–58
79. Gould SJ. 1994. The evolution of life on the earth. *Sci. Am.* 271:85–91
80. Green P. 1994. Ancient conserved regions in gene sequences. *Curr. Opin. Struct. Biol.* 4:404–12
81. Groombridge B, ed. 1992. *Global Biodiversity. Status of the Earth's Living Resources*. London: Chapman & Hall
82. Halanych KM. 1996. Convergence in the feeding apparatuses of lophophorates and pterobranch hemichordates revealed by 18S rDNA: an interpretation. *Biol. Bull.* 190:1–5
83. Hall SRG, Bradley DG. 1995. Conserving livestock breed biodiversity. *Trends Ecol. Evol.* 10:267–70
84. Hardin G. 1993. *Living within Limits: Ecology, Economics, and Population Taboos*. New York: Oxford Univ. Press
85. Hillis DM, Moritz C, eds. 1990. *Molecular Systematics*. Sunderland, MA: Sinauer
86. Holben WE, Harris D. 1995. DNA-based monitoring of total bacterial community structure in environmental samples. *Mol. Ecol.* 4:627–31

87. Holden C. 1989. Entomologists wane as insects wax. *Science* 246:754–56

88. Holland PWH, GarciaFernandez J, Williams NA, Sidow A. 1994. Gene duplications and the origins of vertebrate development. *Development* Suppl. S:125–33

89. Holley SA, Jackson PD, Sasai Y, Lu B, De Robertis EM, et al. 1995. A conserved system for dorsal-ventral patterning in insects and vertebrates involving *sog* and *chordin*. *Nature* 376:249–53

90. Hopper SD, Coates DJ. 1990. Conservation of genetic resources in Australia's flora and fauna. *Proc. Ecol. Soc. Aust.* 16:567–77

91. Hori H, Osawa S. 1987. Origin and evolution of organisms as deduced from 5S ribosomal RNA sequences. *Mol. Biol. Evol.* 4:445–72

92. Humphries CJ, Richardson PM. 1980. Hennig's method and phytochemistry. In *Chemosystematics: Principles and Practice,* ed. FA Bisby, JG Vaughan, CA Wright, pp. 353–78. London: Academic

93. Humphries CJ, Williams PH, Vane-Wright RI. 1995. Measuring biodiversity value for conservation. *Annu. Rev. Ecol. Syst.* 26:93–111

94. Iwu MM. 1996. Biodiversity prospecting in Nigeria: seeking equity and reciprocity in intellectual property rights through partnership arrangements and capacity building. *J. Ethnopharmacol.* 51:209–19

95. Iwu MM. 1996. Implementing the biodiversity treaty: how to make international cooperative agreements work. *Trends Biotechnol.* 14:78–83

96. Jones T. 1994. A biosystematic crisis and a global response to resolve it. *Bull. Entomol. Res.* 84:443–46

97. Joseph L, Moritz C, Hugall A. 1995. Molecular support for vicariance as a source of diversity in rainforest. *Proc. R. Soc. London Ser. B* 260:177–82

98. Kadereit JW. 1994. Molecules and morphology, phylogenetics and genetics. *Bot. Acta* 107:369–73

99. Kauffman SA. 1991. Antichaos and adaptation. *Sci. Am.* 265:64–70

100. Kellert SR. 1993. The biological basis for human values of nature. In *The Biophilia Hypothesis,* ed. SR Kellert, EO Wilson, pp. 42–69. Washington, DC: Island Books

101. Kitching RL. 1994. Biodiversity: political responsibilities and agendas for research and conservation. *Pac. Conserv. Biol.* 1:279–83

102. Koonin EV, Tatusov RL, Rudd KE.

1995. Sequence similarity analysis of *Escherichia coli* proteins: functional and evolutionary implications. *Proc. Natl. Acad. Sci. USA* 92:11921–25

103. Krajewski C. 1991. Phylogeny and diversity. *Science* 254:918–19

104. Krajewski C. 1994. Phylogenetic measures of biodiversity: a comparison and critique. *Biol. Conserv.* 69:33–39

105. Kusmierski RM, Borgia G, Uy A, Crozier RH. 1997. Labile evolution of display traits in bowerbirds indicate reduced effects of phylogenetic constraint. *Proc. R. Soc. London Ser. B.* 264:307–13

106. Legge JT, Roush R, DeSalle R, Vogler AP, May B. 1996. Genetic criteria for establishing evolutionarily significant units in Cryan's Buckmoth. *Conserv. Biol.* 10:85–98

107. Li W-H, Graur D. 1991. *Fundamentals of Molecular Evolution.* Sunderland, MA: Sinauer

108. Linder HP. 1995. Setting conservation priorities: the importance of endemism and phylogeny in the southern African orchid genus *Herschelia. Conserv. Biol.* 9:585–95

109. Linder HP, Midgley JJ. 1994. Taxonomy, compositional biodiversity and functional biodiversity of fynbos. *S. Afr. J. Sci.* 90:329–33

109a. Loeschcke V, Tomiuk J, Jain SK, eds. 1994. *Conservation Genetics.* Basel: Birkhäser Verlag

110. Loomis WH. 1988. *Four Billion Years. An Essay on the Evolution of Genes and Organisms.* Sunderland, MA: Sinauer

111. Mackey LY, Winnepenninckx B, Dewachter R, Backeljau T, Emschermann P, Garey JR. 1996. 18S rRNA suggests that Entoprocta are protostomes, unrelated to Ectoprocta. *J. Mol. Evol.* 42:552–59

112. Mahunnah RLA, Mshigeni KE. 1996. Tanzania's policy on biodiversity prospecting and drug discovery programs. *J. Ethnopharmacol.* 51:221–28

113. Mallet J. 1995. A species definition for the modern synthesis. *Trends Ecol. Evol.* 10:294–99

114. Mallet J. 1996. The genetics of biological diversity: from varieties to species. See Ref. 72, pp. 13–53

115. Mangel M, Talbot LM, Meffe GK, Agardy MT, Alverson DL, et al. 1996. Principles for the conservation of wild living resources. *Ecol. Appl.* 6:338–62

116. May RM. 1990. Taxonomy as destiny. *Nature* 347:129–30

117. May RM. 1994. Conceptual aspects of the quantification of the extent of

biological diversity. *Philos. Trans. R. Soc. London Ser. B* 345:13–20

118. May RM, Nee S. 1995. The species alias problem. *Nature* 378:447–48

119. Maynard Smith J, Szathmáry E. 1995. *The Major Transitions in Evolution.* New York: Freeman

120. Mayr E. 1963. *Animal Species and Evolution.* Cambridge, MA: Harvard Univ. Press

121. McShea DW. 1996. Metazoan complexity and evolution: Is there a trend? *Evolution* 50:477–92

122. Metrick A, Weitzman ML. 1996. Patterns of behavior in endangered species preservation. *Land Econ.* 72:1–16

123. Metsäntutkimuslaitos (Finn. For. Res. Inst.). 1996. *Metsätilastollinen Vuosikirja (Statistical Yearbook of Forestry).* Helsinki: Metsäntutkimuslaitos

124. Moore WS. 1995. Inferring phylogenies from mtDNA variation: mitochondrial-gene trees versus nuclear-gene trees. *Evolution* 49:718–26

125. Moritz C. 1994. Defining 'evolutionarily significant units' for conservation. *Trends Ecol. Evol.* 9:373–75

126. Moritz C. 1995. Uses of molecular phylogenies for conservation. *Philos. Trans. R. Soc. London Ser. B* 349:113–18

127. Moritz C, Faith DP. 1997. Comparative phylogeography and the identification of genetically divergent areas for conservation. *Mol. Ecol.* Submitted

128. Moritz C, Lavery S, Slade R. 1995. Using allele frequency and phylogeny to define units for conservation and management. *Am. Fish. Soc. Symp.* 17:249–62

129. Morowitz HJ. 1991. Balancing species preservation and economic considerations. *Science* 253:752–54

130. Myers N. 1996. Environmental services of biodiversity. *Proc. Natl. Acad. Sci. USA* 93:2764–69

131. Naeem S, Thompson LJ, Lawler SP, Lawton JH, Woodfin RM. 1994. Declining biodiversity can alter the performance of ecosystems. *Nature* 368:734–37

132. Naeem S, Thompson LJ, Lawler SP, Lawton JH, Woodfin RM. 1995. Empirical evidence that declining species diversity may alter the performance of terrestrial ecosystems. *Philos. Trans. R. Soc. London Ser. B* 347:249–62

133. Nei M. 1987. *Molecular Evolutionary Genetics.* New York: Columbia Univ. Press

134. New TR. 1993. Angels on a pin: dimen-

sions of the crisis in invertebrate conservation. *Am. Zool.* 33:623–30

135. Nixon KC, Wheeler QD. 1992. Measures of phylogenetic diversity. In *Extinction and Phylogeny,* ed. MJ Novacek, QD Wheeler, pp. 216–34. New York: Columbia Univ. Press

136. Norton BG. 1987. *Why Preserve Natural Variety?* Princeton, NJ: Princeton Univ. Press

137. Nozick R. 1984. *Anarchy, State, and Utopia.* New York: Basic Books

138. Ohkuma M, Kudo T. 1996. Phylogenetic diversity of the intestinal bacterial community in the termite *Reticulitermes speratus. Appl. Environ. Microbiol.* 62:461–68

139. Ohkuma M, Noda S, Usami R, Horikoshi K, Kudo T. 1996. Diversity of nitrogen fixation genes in the symbiotic intestinal microflora of the termite *Reticulitermes speratus. Appl. Environ. Microbiol.* 62:2747–52

140. Ohno S. 1984. Birth of a unique enzyme from an alternative reading frame of the preexisted, internally repetitious coding sequence. *Proc. Natl. Acad. Sci. USA* 81:2421–25

141. Oliver I, Beattie AJ. 1993. A possible method of rapid assessment of biodiversity. *Conserv. Biol.* 7:562–68

142. Oliver I, Beattie AJ. 1994. A possible method for the rapid assessment of biodiversity. See Ref. 68a, pp. 133–36

143. Orr HA. 1995. The population genetics of speciation: the evolution of hybrid incompatibilities. *Genetics* 139:1805–13

144. Pace NR, Olsen GJ, Woese CR. 1986. Ribosomal RNA phylogeny and the primary lines of evolutionary descent. *Cell* 45:325–26

145. Pace NR, Stahl DA, Lane DJ, Olsen GJ. 1985. Analyzing natural microbial populations by rRNA sequences. *Am. Soc. Microbiol. News* 51:4–12

146. Pamilo P. 1990. Statistical tests of phenograms based on genetic distances. *Evolution* 44:689–97

147. Pamilo P, Nei M. 1988. Relationships between gene trees and species trees. *Mol. Biol. Evol.* 5:568–83

148. Paterson AM, Wallis GP, Gray RD. 1995. Penguins, petrels, and parsimony: Does cladistic analysis of behavior reflect seabird phylogeny? *Evolution* 49:974–89

149. Patterson C, Williams DM, Humphries CJ. 1993. Congruence between molecular and morphological phylogenies. *Annu. Rev. Ecol. Syst.* 24:153–88

150. Pearce D, Moran D. 1994. *The Economic Value of Biodiversity.* London, UK: Earthscan

151. Pearson DL. 1994. Selecting indicator taxa for the quantitative assessment of biodiversity. *Philos. Trans. R. Soc. London Ser. B* 345:75–79

152. Pedersen K, Arlinger J, Hallbeck L, Pettersson C. 1996. Diversity and distribution of subterranean bacteria in groundwater at Oklo in Gabon, Africa, as determined by 16S rRNA gene sequencing. *Mol. Ecol.* 5:427–36

153. Pedrós-Alió C. 1993. Diversity of bacterioplankton. *Trends Ecol. Evol.* 8:86–90

154. Perrot-Minnot MJ, Guo LR, Werren JH. 1996. Single and double infections with *Wolbachia* in the parasitic wasp *Nasonia vitripennis:* effects on compatibility. *Genetics* 143:961–72

155. Pilbeam D. 1996. Genetic and morphological records of the Hominoidea and Hominid origins: a synthesis. *Mol. Phylogeny Evol.* 5:155–68

156. Pomeroy D. 1993. Centers of high biodiversity in Africa. *Conserv. Biol.* 7:901–7

157. Porteous LA, Armstrong JL, Seidler RJ, Watrud LS. 1994. An effective method to extract DNA from environmental samples for polymerase chain reaction amplification and DNA fingerprint analysis. *Curr. Microbiol.* 29:301–7

158. Potthast T. 1996. Inventing biodiversity: genetics, evolution, and environmental ethics. *Biol. Zentralbl.* 115:177–88

159. Prendergast JR, Quinn RM, Lawton JH, Eversham BC, Gibbons DW. 1993. Rare species, the coincidence of diversity hotspots and conservation strategies. *Nature* 365:335–37

160. Principe PP. 1995. Monetizing the pharmacological benefits of plants. In *Medicinal Resources of the Tropical Forest: Biodiversity and Its Importance to Human Health,* ed. MJ Balick, E Elisabetsky, S Laird, pp. 191–218. New York: Columbia Univ. Press

161. Rawls J. 1971. *A Theory of Justice.* Cambridge, MA: Harvard Univ. Press

162. Reid WV. 1996. Gene co-ops and the biotrade: translating genetic resource rights into sustainable development. *J. Ethnopharmacol.* 51:75–92

163. Richardson BJ. 1994. The industrialization of scientific information. See Ref. 68a, pp. 123–31

164. Risatti JB, Capman WC, Stahl DA. 1994. Community structure of a microbial mat: the phylogenetic dimension. *Proc. Natl. Acad. Sci. USA* 91:10173–77

165. Rossetto M, Weaver PK, Dixon KW. 1995. Use of RAPD analysis in devising conservation strategies for the rare and endangered *Grevillea scapigera* (Proteaceae). *Mol. Ecol.* 4:321–29

166. Russo CAM, Takezaki N, Nei M. 1995. Molecular phylogeny and divergence times of drosophilid species. *Mol. Biol. Evol.* 12:391–404

167. Saitou N, Nei M. 1987. The neighborjoining method: a new method for reconstructing phylogenies. *Mol. Biol. Evol.* 4:406–25

168. Sarich VM, Wilson AC. 1967. Immunological time scale for hominoid evolution. *Science* 158:1200–3

169. Shenk MA, Steele RE. 1993. A molecular snapshot of the metazoan Eve. *Trends Biochem. Sci.* 18:459–63

170. Shomer B, Harper RAL, Cameron GN. 1996. Information services of the European Bioinformatics Institute. In *Computer Methods for Macromolecular Sequence Analysis,* ed. RF Doolittle, pp. 3–27. San Diego, CA: Academic

171. Sibley CG, Ahlquist JE. 1990. *Phylogeny and Classification of Birds: A Study of Molecular Evolution.* New Haven, CT: Yale Univ. Press

172. Soejarto DD. 1996. Biodiversity prospecting and benefit-sharing: perspectives from the field. *J. Ethnopharmacol.* 51:1–15

173. Solow A, Polasky S, Broadus J. 1993. On the measurement of biological diversity. *J. Environ. Econ. Manage.* 24:60–68

174. Stiassny MLJ. 1992. Phylogenetic analysis and the role of systematics in the biodiversity crisis. In *Systematics, Ecology, and the Biodiversity Crisis,* ed. N Eldredge, pp. 109–20. New York: Columbia Univ. Press

175. Strahan R. 1989. Conservation priorities: a simple system. In *The Conservation of Threatened Species and Their Habitats,* ed. M Hicks, P Elser, pp. 101–5. Canberra: Aust. Comm. IUCN

176. Szathmáry E, Maynard Smith J. 1995. The major evolutionary transitions. *Nature* 374:227–32

177. Telford MJ, Thomas RH. 1995. Demise of the Atelocerata? *Nature* 376:123–24

178. Templeton AR. 1977. Analysis of head shape differences between two interfertile species of Hawaiian *Drosophila. Evolution* 31:630–41

179. Thompson JN. 1987. Symbiont-induced speciation. *Biol. J. Linn. Soc.* 32:385–93

180. Tiefenbrunner W. 1995. Selektion und Mutation aus informatistheoretischer

Sicht-Grenzen der Erhaltbarkeit von genetischer Information. *Z. Naturforsch. Teil C* 50:883–94

181. Tilman D. 1996. Biodiversity: population versus ecosystem stability. *Ecology* 77:350–63

182. Tilman D, Downing JA. 1994. Biodiversity and stability in grasslands. *Nature* 367:363–65

183. Tilman D, Wedin D, Knops J. 1996. Productivity and sustainability influenced by biodiversity in grassland ecosystems. *Nature* 379:718–20

184. Torsvik V, Goksøyr J, Daae FL. 1990. High density in DNA of soil bacteria. *Appl. Environ. Microbiol.* 56:782–87

185. Turner DM. 1996. Natural product source material use in the pharmaceutical industry: the Glaxo experience. *J. Ethnopharmacol.* 51:39–43

186. Vane-Wright RI, Humphries CJ, Williams PH. 1991. What to protect? Systematics and the agony of choice. *Biol. Conserv.* 55:235–54

187. Vogler AP. 1994. Extinction and the formation of phylogenetic lineages: diagnosing units of conservation management in the tiger beetle *Cicindela dorsalis*. In *Molecular Ecology and Evolution: Approaches and Applications,* ed. B Schierwater, B Streit, GP Wagner, R DeSalle, pp. 261–73. Basel: Birkhäuser Verlag

188. Vogler AP, DeSalle R. 1994. Diagnosing units of conservation management. *Conserv. Biol.* 8:354–63

189. Vogler AP, Knisley CB, Glueck SBJ, Hill JM, DeSalle R. 1993. Using molecular and ecological data to diagnose endangered populations of the puritan tiger beetle *Cicindela puritana*. *Mol. Ecol.* 2:375–83

190. Walker PA, Faith DP. 1995. DIVERSITY-PD: Procedures for conservation evaluation based on phylogenetic diversity. *Biodivers. Lett.* 2:132–39

191. Wallis GP. 1994. Population genetics and conservation in New Zealand: a hierarchical synthesis and recommendations for the 1990s. *J. R. Soc. NZ* 24:143–60

192. Weidner S, Arnold W, Puhler A. 1996. Diversity of uncultured microorganisms associated with the seagrass *Halophila stipulacea* estimated by restriction fragment length polymorphism analysis of PCR-amplified 16S rRNA genes. *Appl. Environ. Microbiol.* 62:766–71

193. Weitzman ML. 1992. On diversity. *Q. J. Econ.* 107:363–405

194. Weitzman ML. 1993. What to preserve? An application of diversity theory to crane preservation. *Q. J. Econ.* 108:157–83

195. Weitzman ML. 1994. On the environmental discount rate. *J. Environ. Econ. Manage.* 26:200–9

196. Williams PH, Gaston KJ. 1994. Measuring more of biodiversity: Can higher-taxon richness predict wholesale species richness? *Biol. Conserv.* 67:211–17

197. Williams PH, Gaston KJ, Humphries CJ. 1994. Do conservationists and molecular biologists value differences between organisms in the same way? *Biodivers. Lett.* 2:67–78

198. Williams PH, Humphries C. 1994. Biodiversity, taxonomic relatedness and endemism in conservation. See Ref. 68a, pp. 269–87

199. Williams PH, Humphries CJ. 1996. Comparing character diversity among biotas. See Ref. 72, pp. 54–76

200. Williams PH, Humphries CJ, Gaston KJ. 1994. Centres of seed-plant diversity: the family way. *Proc. R. Soc. London Ser. B* 256:67–70

201. Williams PH, Humphries CJ, Vane-Wright RI. 1991. Measuring biodiversity: taxonomic relatedness for conservation priorities. *Aust. Syst. Bot.* 4:665–79

202. Williams PH, Vane-Wright RI, Humphries CJ. 1993. Measuring biodiversity for choosing conservation areas. In *Hymenoptera and Biodiversity,* ed. J LeSalle, ID Gauld, pp. 309–28. Wallingford, UK: CAB Int.

203. Wilson A, Kunkel J, Wyles J. 1984. Morphological distance: an encounter between two perspectives in evolutionary biology. *Evolution* 38:1156–59

204. Wilson EO. 1984. *Biophilia: The Human Bond with Other Species.* Cambridge, MA: Harvard Univ. Press

205. Wilson EO. 1992. *The Diversity of Life.* Cambridge, MA: Harvard Univ. Press

206. Wilson EO, Eisner T, Briggs WR, Dickerson RE, Metzenberg RL, et al. 1978. *Life on Earth.* Sunderland, MA: Sinauer. 2nd ed.

207. Witting L, Loeschcke V. 1995. The optimization of biodiversity conservation. *Biol. Conserv.* 71:205–7

208. Witting L, McCarthy MA, Loeschcke V. 1994. Multi-species risk analysis, species evaluation and biodiversity conservation. See Ref. 109a, pp. 239–49

209. Wray GA, Levinton JS, Shapiro LH. 1996. Molecular evidence for deep Precambrian divergences among metazoan phyla. *Science* 274:568–73

Annu. Rev. Ecol. Syst. 1997. 28:269–88

THEORETICAL AND EMPIRICAL EXAMINATION OF DENSITY-DEPENDENT SELECTION

Laurence D. Mueller

Department of Ecology and Evolutionary Biology, University of California, Irvine, California 92697-2525; e-mail: LDMUELLE@UCI.EDU

KEY WORDS: life history, population dynamics, population stability, r- and K-selection

ABSTRACT

The development of theory on density-dependent natural selection has seen a transition from very general, logistic growth-based models to theories that incorporate details of specific life histories. This transition has been justified by the need to make predictions that can then be tested experimentally with specific model systems like bacteria or *Drosophila*. The most general models predict that natural selection should increase density-dependent rates of population growth. When trade-offs exist, those genotypes favored in low-density environments will show reduced per capita growth rates under crowded conditions and vice versa for evolution in crowded environments. This central prediction has been verified twice in carefully controlled experiments with *Drosophila*. Empirical research in this field has also witnessed a major transition from field-based observations and conjecture to carefully controlled laboratory selection experiments. This change in approach has permitted crucial tests of theories of density-dependent natural selection and a deeper understanding of the mechanisms of adaptation to different levels of population crowding. Experimental research with *Drosophila* has identified several phenotypes important to adaptation, especially at high larval densities. This same research revealed that an important trade-off occurs between competitive ability and energetic efficiency.

INTRODUCTION

The idea that the natural environment plays an important and distinctive role in shaping the process of evolution has been understood since Darwin. Indeed,

269

Charles Darwin declared that "The slightest advantage in one being, at any age or during any season, over those with which it comes into competition, or better adaptation in however slight a degree to the surrounding physical conditions, will turn the balance" (23, pg. 442). However, until the 1960s there was no formal theory of evolution and ecology that attempted to describe the consequences of this idea. Great advances had certainly been made in the separate fields of ecology and microevolution or population genetics. One stumbling block to the development of a united theory was the general impression that the ecological factors that mattered to the survival and reproduction of a species were many and complex. This idea was fostered in the late 1950s by concepts like Hutchinson's n-dimensional niche, which developed an abstract description of the ecological requirements of a species and suggested these requirements could not be simply enumerated (41).

Despite this attitude, in the early 1960s MacArthur (52) and later MacArthur & Wilson (53) initiated the development of the theory of density-dependent natural selection, which suggested that one aspect of the environment—density—could be isolated and studied. Further, they suggested that many aspects of an organism's life-history could be due to the population density it had experienced historically. These ideas prompted development of a body of theory (reviewed in the next sections) that would serve as a focus for continuing tension between theoreticians and field scientists both in ecology and evolutionary biology.

At the heart of this tension was the desire of the theoretician or the theoretically inclined to show that elegant theories could describe natural systems, with the hope that the many ecological details not considered by the theory would be unimportant. The field biologist, having accepted the Hutchinsonian reality of ecology, could not believe nature was so simple, and indeed, with little effort, many examples of natural populations could be found that apparently contradicted the expectations of this theory (56). What had been missed through much of the early work in the 1970s was the more sober conclusion that these simple theories might never or only rarely provide complete explanations of real populations. Thus, the only proper way to test these ideas was in controlled settings congruent with the assumptions of the simple models. In the meantime, it was unrealistic to expect the birth, de novo, of a theory in evolutionary ecology that could account for all aspects of the natural environment and their impact on evolutionary processes. Here the lessons from many other more-developed branches of science could be used to suggest that one should try first to understand how simple systems with few variables work and then build upon that to reach the complexity of the real world. Physics has frictionless hockey pucks, thermodynamics has Carnot engines, and evolutionary ecology has r- and K-selection.

THEORY

Verbal Theories

In the next two sections I separate the verbal theory of density-dependent natural selection from the mathematical theory for several reasons. First, the two types of theories have yielded very different predictions about the expected outcome of natural selection. I don't suggest that verbal and mathematical theories will never agree, but agreement is generally lacking in the theories considered here. Second, the assumptions and logic behind the mathematical theories are usually more transparent and easier to evaluate. This doesn't mean the assumptions made in mathematical theories will be better or more reasonable, but in general they are easier to identify.

One of the early discussions of evolution in environments with different levels of competition is given by Dobzhansky (29). Dobzhansky's ideas were known to MacArthur & Wilson (53). Many of the features of r-selected genotypes were outlined in Lewontin's discussion of colonizing species (50). Using an age-structured model, Lewontin provided examples emphasizing that exponential rates of increase are affected most by rapid development and early reproduction. These ideas were developed in a more general setting by Demetrius (27).

The first extensive elaboration of a verbal theory of density-dependent natural selection is found in MacArthur & Wilson's (53) book, *The Theory of Island Biogeography*. MacArthur & Wilson wrote about the differences expected in populations that live at very low and high densities; they were interested in understanding the phenotypes most successful at colonizing new habitats. MacArthur & Wilson used the logistic equation as a backdrop for their discussion, although their predictions—such as evolution favoring productivity or efficiency—are not explicitly derived from any models based on the logistic equation. MacArthur & Wilson originated the terms r- and K-selection, to which the theory of density-dependent selection often refers. MacArthur & Wilson used them to describe the types of selection expected in either uncrowded environments or very crowded environments, respectively. They asserted that r was the appropriate measure of fitness at very low densities and K at very high densities. In fact, the formal theories of density-dependent selection equate per-capita rates of population growth to fitness. Only in the special case when population growth rates follow the logistic equation will fitness be approximated by r and K. Most of the ensuing verbal theories consistently fail to recognize this point. It is also important to recognize that r and K are just two of many phenotypes one could assume are surrogates of fitness, like foraging efficiency, predator avoidance, or clutch size. Later I discuss the theoretical limitations of this idea and the extent to which experimental evidence supports predictions from this theory.

The notion of trade-offs in life-history evolution was a prominent feature of the MacArthur-Wilson theory. The importance of trade-offs was emphasized by Cody (21) in his discussion of the evolution of clutch size. Cody suggested that birds must make the most effective allocation of limited energy to three major competing needs—predator avoidance, interspecific competition, and reproduction. These ideas were further developed by Gadgil & Bossert (30) in an age-structured model. They assumed that age-specific survival and fecundity are affected by reproductive effort and environmental quality. Reproductive effort was defined as the fraction of total energy devoted to reproduction. Gadgil & Bossert showed that in favorable environments reproductive effort increased at all ages. This result also became associated with other predictions about r-selection. Much subsequent work focused on examining energy budgets and the proportion devoted to reproduction vs other activities (104).

The major articulations of the verbal theory of r- and K-selection are found in two papers by Pianka (77, 78). These theories suggest that traits like long life-span, iteroparity, large size, and prolonged development will result from K-selection, whereas a suite of opposing traits would be expected under r-selection. Some of the logical fallacies of these claims have been discussed in previous reviews (11, 96, 97). Many problems stem from the overinterpretation of the parameters r and K. For instance, no logical relationship exists between the logistic equation and age-structured populations, but verbal theory makes specific predictions about phenotypes of age-structured populations. Despite these problems, the verbal theory of r- and K-selection is still influential and finds its way into many ecology textbooks (7).

Population Genetic Models Without Age-Structure

Shortly after the appearance of MacArthur & Wilson's book, several more formal models of density-dependent natural selection appeared (1–3, 16, 20, 90). These models showed explicitly which assumptions were required for density-dependent selection to result in phenotypic differentiation. The discussions in Roughgarden (90, 91) are particularly lucid and are followed here. The primary assumption in these models is that genotypic fitness is assumed equivalent to per-capita rates of population growth. In particular, if fitness varies with density according to the logistic equation, then at a single locus, with multiple alleles, $A_1, A_2, \ldots$ etc, the fitness of genotype $A_i A_j$ is given by

$$W_{ij} = 1 + r_{ij} - r_{ij} N K_{ij}^{-1},$$

where N is the total population size, and r_{ij} and K_{ij} are genotypic-specific measures of sensitivity to density. It is clear that, as a general proposition, this formulation of fitness will not be correct. For instance, consider a species like *Drosophila melanogaster* in which females can store sperm and exert choice over the males they mate with. The rate at which a population of *Drosophila* grows may be limited by how many eggs a female can lay but is almost certainly not limited by male fertility. Nevertheless, in *Drosophila*, male mating success or virility can be an important component of fitness (14, 48) but would not be expected to affect rates of population growth (67). However, in some circumstances genetically based differences in fitness may be reflected in differences in per-capita rates of population growth. This assumption clearly must be tested.

The other major component of these theories is the assumption of some trade-off between the ability to do well under uncrowded conditions and under crowded conditions. Without this assumption, the same genotype would be favored in all environments, and there would be no phenotypic differentiation of populations evolving in environments of different densities. Only with some sort of trade-off does the interesting prediction of different evolutionary outcomes depending on the environment appear. The assumption of trade-offs is a ubiquitous and important one in the theory of life-history evolution (21; Ch. 4 of 98). The trade-off assumption appears most naturally in quantitative genetic models in which some phenotype jointly affects r and K in opposing directions in the logistic (100).

An important question is, then, does the theory of density-dependent natural selection make predictions about the evolution of any trait other than population growth rates? I contend the answer is "no." Any particular organism may have many physiological, behavioral, and even morphological characters that may change and have an effect on rates of population growth. However, the types of characters that are important are likely to vary between organisms. Thus, evolution of population growth rates in an insect may be accomplished by altering the number of eggs laid per unit time by females. We would not expect to see this kind of change in bacteria (certainly the rate of division may change in bacteria, but typically bacteria do not have the option to divide into five daughter cells or 100 daughter cells).

Models with Age-Structure

The major development of this theory of density-dependent selection in age-structured populations is found in Charlesworth (17). Consequent to this theory, some of the predictions of the verbal theory could be formalized. For instance, it is often suggested that low population density would favor semelparous life histories, whereas high density would favor iteroparous life histories. The simplest result from density-independent models of selection in age-structured populations is that improvements in survival or fertility will be most strongly favored by natural selection early in life. The addition of density-dependence to either survival or fertility functions does not alter this conclusion (17). Thus, there is no support for the idea that density-dependent selection will explain the evolution of iteroparity vs semelparity.

Charlesworth (17) showed that natural selection will maximize the equilibrium size of the age-class that is subject to density-dependence. This result extends the earlier work by Roughgarden (91) on populations without age-structure and is consistent with the results of Prout (82) and Iwasa & Teramoto (42) on stage-structured populations.

General vs. Specific Models

The most common life history assumed in both population genetics and population ecology is one that posits discrete generations and populations without stage or age structure. Problems arise when these theories are tested with organisms that depart from these assumptions. For instance, organisms like *Drosophila* can be made to reproduce on a discrete schedule and adult age-classes can be essentially eliminated, but the prereproductive stages of *Drosophila* cannot be removed. In the late 1960s and early 1970s, it was recognized that attempts to estimate fitness coefficients from simple population genetic models with organisms like *Drosophila* could be thwarted by selection acting on the various components of the life cycle (79–81). This, coupled with the necessity to assay adults rather than eggs, meant that the most general models of selection were inappropriate for providing a framework for observations in the simplest of *Drosophila* populations.

Prout (82, 84, 85) recognized that similar problems will occur in simple models of population dynamics. For instance, the simplified life cycle of *Drosophila* in the laboratory can have three census stages (Figure 1), each of which could be the population size in standard models of density-dependent population growth. If selection acts in a density-independent fashion, it is possible for evolution to increase, decrease, or have no effect on equilibrium numbers of particular census stages. In some numerical examples (82) all three results can be observed in the three different census stages (see Figure 1). Consequently, a general claim that selection will maximize population size is not true, just as the claim that mean fitness is always maximized by natural selection is false.

Utilizing the life cycle shown in Figure 1, Prout (83–85) noted that in many organisms fertility depends on pre-adult density. Crowding during these stages often has lasting effects on adult size that in turn affect fertility. This biological phenomenon poses some difficult problems for estimating the underlying population dynamics from data on adult numbers only.

The issues discussed above raise the general question of the most appropriate type of model to use when developing theory in life-history evolution in general. Christiansen (19) distinguishes between phenomenological and explanatory models. The phenomenological models are simple and attempt to

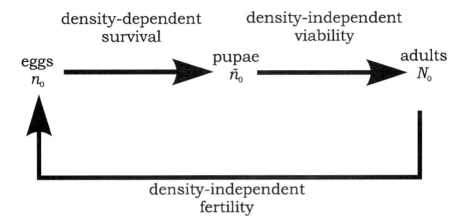

Figure 1 The life cycle of an organism with discrete generations but several pre-reproductive life stages. Selection may act at several places along the life cycle. The interaction of density-dependent survival (egg-to-pupa) and density-independent selection can have complicated outcomes on the equilibrium number of adults (82).

summarize the totality of density-dependence with a single simple function (e.g. the logistic). For these reasons the phenomenological models are thought to have greater generality (49). Explanatory models explicitly take into account specific components of the life cycle of some organism or group of organisms and try to model the response of these life-history components to density. Christiansen argued that this is the more appropriate way to develop theory for the study of life-history evolution in variable environments. Certainly, if theory is being used to make specific predictions about the evolution of a particular population, one cannot use a model that ignores crucial life-history details. Due to Prout's argument, this lesson cannot be overemphasized.

Accordingly, some recent models of density-dependent selection have focused on more explanatory or specific models for organisms that are the current focus of evolutionary research. One model (61, 68) has been employed to develop ecological recursions and single-locus genetic models for *Drosophila* in food-limited environments. This model permits specific examination of the evolution of competitive ability, adult size, efficiency, and equilibrium population size. An interesting theoretical prediction is that density-dependent selection can lead to increased competitive ability, but this will have no lasting impact on the adult equilibrium population size. Thus, verbal theories that attempt to link K of the logistic equation and competitive ability are not sound, at least for the class of organisms modeled by this theory. Second, although adult size may be affected by density-dependent selection, it may either increase or decrease depending on the assumptions made about larval growth and adult size. Thus, this theory yields no single or simple prediction about the evolution of body size due to density-dependent selection. Finally, *Drosophila* populations that follow the life history described by this model are stable only if there is some form of adult density-dependence on female fecundity. This interesting prediction is discussed in more detail later.

Population Stability

A property of discrete time equations of population growth is their ability to exhibit deterministically generated chaos (39, 54, 55). In several surveys of natural (38) and laboratory populations (58, 66, 102, but see 73, 74), most populations appear to have asymptotically stable carrying capacities. However, recent evidence (22, 28, 71, 98a, 105) suggests that unstable dynamics may be more common than previously thought.

These results have stimulated several theoretical investigations of the evolution of population stability. Some (8, 102) have suggested that population stability may evolve by group selection. Those populations that become unstable also become more likely to go extinct, leaving only populations with stable dynamics. There is no empirical evidence to suggest the widespread existence of a population structure that would permit this type of evolution, and thus group selection can hardly be considered a general explanation for the existence of stable population dynamics.

Others (32, 37, 40, 66, 99, 106) have considered individual selection models with some form of density dependence. Details of the particular model (e.g. logistic vs. hyperbolic equations, environmental noise in r vs K) have qualitatively different effects on the outcome of evolution (40, 106). In some cases selection favors increases in population stability, in others stability decreases. Some models that produce the evolution of stable dynamics could do so only by assuming some sort of trade-off in life-history traits (29a, 32, 66, 99). However, the belief in the ubiquity of trade-offs has led some (33a) to suggest that natural selection should typically tend to stabilize populations dynamics.

Several specific theories have identified the biologically important characters that affect population stability. For instance, theoretical predictions that the rate of adult cannibalism of immature stages affects population stability have been verified in laboratory populations of *Tribolium* (22). In *Drosophila*, it appears that the relative amounts of food supplied to adults and larvae determine population stability (61, 71). The theoretical prediction that high levels of food to adults and low amounts to larvae tend to destabilize the population have been verified in experimental populations of *Drosophila* (71). This may be a fairly general mechanism for organisms with distinct life stages in which female fecundity varies with nutritional level of the adult. The characteristic cycles of blowflies (73) may be generated in part by the differing levels of food provided to adults and larvae (71).

A recent analysis of data from Nicholson's classic work (73) has suggested that evolution in the course of a 2-yr experiment may have resulted in the attenuation of population cycles (99). This conclusion must be tempered with the following observations. The experiment involved only a single population, and thus it is unclear if it is a replicable phenomenon. Since the adult population of blowflies went through severe fluctuations and thus bottlenecks, random genetic drift and inbreeding may have caused a reduction in female fecundity, which in turn could have attenuated the cycles. Obviously, these problems could be easily rectified in an experimental system with *Tribolium* or *Drosophila*.

EMPIRICAL RESEARCH

Translation of Theory to Experimental Predictions

The integration of theory and experiments is not easy when dealing with complex biological systems. Generally, scientific research guided by a body of theory will ultimately reach a point at which some theoretical prediction is compared to a set of observations. It is important to realize that agreement between the observations and theoretical expectations does not necessarily mean the theory is correct, nor does a lack of agreement mean the theory is wrong. In an unusually lucid and informative discussion of this issue, Royama (92) notes that the components and structure of a model may be a correct description of a biological system, but because the test systems depart, unintentionally, in substantial ways from the model components, observations and model predictions will disagree.

This point is especially important for models in evolutionary biology, including density-dependent natural selection, because many of our models are quite simple, ignoring many ecological details and thus needing to be tested under conditions that match these simple assumptions as well as possible. Consequently, natural populations are unlikely to be good systems to test theories that make very simple assumptions. Laboratory systems will be more likely to conform to the assumptions of simple models (89), and thus they are more appropriate for testing these models. To say theories that don't explain events in a "real" environment are useless misses an important intellectual thread about how theories of the real world are constructed. These theories are constructed in gradual steps, usually from very simple beginnings. Empirical experiments aid in the elimination of those components that fail to explain even those simple constructs. Thus, physics starts logically with laws of motion that ignore friction. This is not because the physicist believes there is no friction in the real world but rather because systems without friction are useful starting points from which to develop the more complicated theoretical machinery that includes friction and turbulence.

Because even laboratory experiments will often utilize a single species of plant or animal, models that include important aspects of the experimental organism's life history are useful (19). At least two important reasons exist for constructing models that account for specific qualities of an experimental organism. First, model predictions may be affected by life-history details. Second, the appropriate parameters to measure during experimental research may be suggested by the organism-specific modeling effort. Some examples follow.

As an aid for experimental research conducted with the species *Drosophila melanogaster*, I (61) elaborated upon a model developed by others (4, 26, 75) to describe the effects of density-dependent selection in food-limited environments. This model has yielded some important insights into experimental systems utilizing *Drosophila*. The theory demonstrated that competitive ability would not affect the equilibrium population size (61). The same theory provided the appropriate methods for estimating competitive ability in a way congruent with the theory (60, 61, 75).

Another interesting example is Vasi et al (107). In this study, populations of *E. coli* were subjected to a seasonal environment. Periods of exponential growth with adequate resources were followed by periods of growth at saturation densities. The results of these experiments, interpreted by estimating parameters from a model of bacterial population dynamics, showed that these populations had evolved traits that would be most important during the exponential growth phase of the environment, while parameters that would be most important during the periods of saturation density had not changed. An earlier study, by Luckinbill (51), of r- and K-selection in bacterial populations utilized a K-environment similar to the seasonal environment used by Vasi et al (107). Luckinbill observed that K-selected bacteria grew faster than r-selected bacteria at all test densities. The results of Vasi et al suggest Luckinbill's outcome could be due to selection in a seasonal environment favoring most strongly characters that affect growth at low densities. The important conclusion is that the interpretation and design of experimental research can be greatly facilitated by the development of specific models of the experimental organism's life history and population dynamics.

Field Studies and the Comparative Approach

Many early studies of r- and K-selection relied on comparisons of different species or different populations of the same species that occupied different environments. In principle, if the different environments differed only by the density experienced by members of the populations, any genetically based differences ought to be due to the results of density-dependent selection. In practice this almost never happens, and there are always multiple differences between populations, in addition to density, that confound simple interpretations. Many of the problems with these studies have been reviewed previously (11, 97). A short list of the most severe problems would include the following.

Often the exact density history of a particular population was not known with any certainty but was inferred from anecdotal or recent observations (31). In many cases no attempt was made to remove any lingering effects of environmentally induced differences (103). Thus, the genetic basis of any observed differences was uncertain. In essentially all cases of studies in natural populations, no control existed for other environmental factors that might affect life-history evolution, e.g. predation, density-independent sources of mortality, resource variation. Consequently, any result, whether positive or negative, may have been due to one of the uncontrolled factors rather than differences in density per se. Many studies of life-history evolution do not include real replication; the observation of any differences between compared populations is rendered almost completely uninterpretable (36). Thus a difference between a pair of populations could be due to initial sampling events, to genetic drift, or to natural selection. Without replication there is no simple means to distinguish among these alternatives.

These types of problems beset almost all studies of natural populations except the most carefully planned (86, 87). For the reasons outlined above, it had become obvious by the late 1970s (77) that the only powerful way to investigate the theory of density-dependent natural selection would be with the use of populations in controlled laboratory environments.

A recent study (56a) has looked at the relationship between survival and density of Soay sheep in Scotland. This study shows that survival declines with increasing density but is also dependent on the individuals' coat color phenotype, controlled by a single locus, and horn type, controlled by one or two loci. Thus, density-dependent natural selection appears to be an important force in the maintenance of these two genetic polymorphisms. It is remarkable that this is perhaps one of the only studies to document the action of density-dependent selection in a natural population.

Experimental Studies

Pitcher-plant mosquitoes are typically found in more crowded conditions in southern latitudes than in northern latitudes (12). Although estimates of population growth rates show no consistent differences between northern and southern populations, the southern populations are the better competitors, a result consistent with observations from *Drosophila*. Laboratory populations of mosquitoes selected for rapid development also showed no correlated changes in sensitivity to density although their generation time decreased (13).

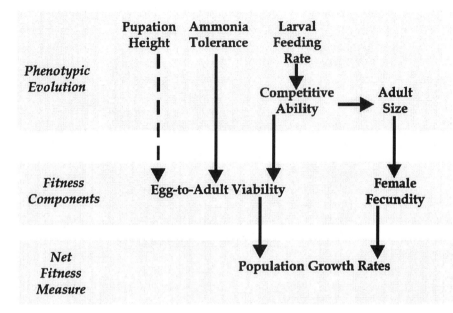

Figure 2 A summary of the effects of density-dependent natural selection in *Drosophila melanogaster*. The dashed line shows an effect that varies with the environment (in this case the type of food used).

Perhaps the most studied experimental organism in this field has been *Drosophila*. Taylor & Condra (101) found that *r*-selected *D. pseudoobscura* developed more rapidly than *K*-selected populations. However, no systematic measurements of population growth rates were made. Barclay & Gregory (5, 6) did selection experiments with *D. melanogaster*, but this study was flawed by a number of technical problems (57) that made the results difficult to interpret.

In Figure 2, I summarize the effects of density-dependent selection that have been inferred from the work of my laboratory over the last 17 years. These results have come from two experimental systems. My early work utilized three populations kept at low adult and larval densities, called *r*-populations, and three kept at high larval and adult densities, called *K*-populations (9, 63–65). These populations had two shortcomings: The adult *K*-population had overlapping generations, whereas *r*-populations were reproduced with young adults only; the *r*-populations were maintained for about 150 generations at a low effective population size (50 or less, compared to 1000 in the *K*-populations). Accordingly, any differences between the *r*- and *K*-populations, especially in the later generations, may have been due to inbreeding effects rather than natural selection. This latter problem was addressed in two ways. Hybrids of the r-populations were made to determine if overdominance occurred, and the phenotypic differences that arose due to inbreeding were eliminated (59, 60). For the phenotypes described in Figure 2, there was no evidence that the phenotypic difference between the r- and K-populations was due to inbreeding effects. After about 180 generations, the r-populations were placed in crowded environments, and then they differentiated from their parental populations in the same manner as K-populations originally had done (34, 70).

Nevertheless, in 1989 in my laboratory, three new types of populations were initiated to investigate the consequences of density-dependent selection. These populations all derived from large, outbred populations cultured in a laboratory for more than 10 years. Each population was replicated five times; the three population-types are called UU, CU, and UC, where the first letter of the name refers to the density at which larvae are raised (U, uncrowded; C, crowded) and the second letter, to the density at which the adults are raised.

A basic prediction from the general theories discussed previously is that rates of population growth should increase in a manner that depends on the environment the population experienced. Two tests of this prediction have yielded consistent results (65, 70): Populations of D. melanogaster raised at high densities experience an increase in population growth rates at high densities (and thus an increase in carrying capacity), but a decrease in population growth rates at low density. These results are supported by earlier studies (3a, 3b, 15a) which show that populations of Drosophila brought into the laboratory and allowed to reach their carrying capacity show an increase in equilibrium population size over many generations.

Three larval behavioral traits have become differentiated in the r- and K-populations. Feeding rates have increased in the K-populations relative to the r-populations (43, 70), and they have also increased in the CU populations relative to the UU populations (69, 93). This trait is highly correlated with larval competitive ability for food (15, 43). Competitive ability in Drosophila larvae is a frequency-dependent process (26, 61, 75). Thus, when food is limited and the superior competitors are rare, they will experience increased viability and adult size. Both of these changes will affect population growth rates (Figure 2). However, these changes will only be short-lived because as the fast feeding types become common, their fitness advantage vanishes.

There is also a demonstrable cost to high feeding-rates. CU populations that have been moved to environments with reduced larval densities show a decline in feeding rates over time relative to similar populations with high larval densities (A Joshi, LD Mueller, unpublished). Part of this cost may be due to reduced efficiency of the larvae, which have evolved high feeding rates. When individual larvae from high-density and low-density populations are given small amounts of food, the larvae from the crowded cultures require more food to complete development (45, 62). The fast-feeding CU larvae also gain more weight during the third instar than do the slow-feeding UU controls (93). However, this additional weight is lost either during the late third instar or during the pupal stage so that adult CU and UU flies are the same size. Together these results suggest that a trade-off between competitive ability (fast feeding rates) and energetic efficiency and that only under extreme larval crowding does the fitness gain due to increased competitive ability outweigh the fitness loss from this trait.

One recent study of density-dependent selection in *D. melanogaster* (88) reports no change in competitive ability after 45 to 50 generations of selection. In this experiment the uncrowded treatment consisted of 50 larvae per vial, whereas the crowded treatment had 150. A concern exists that the strength of selection at the higher density was insufficient to cause changes in feeding rates in this period of time or to overcome the effects of drift (total population sizes were kept at 500 or greater). Support for this view comes from the work of Mueller et al (69). In another experiment (69), no changes in feeding rates were observed for the first 12 generations when larval densities were kept at about 500 per vial. However, within a short time of increasing the larval densities to greater than 1000 per vial, feeding rates in the high-density populations increased.

The position or height above the food at which larvae pupate also differed in the *r*- and *K*-populations (34, 72). Larvae in the crowded cultures tend to pupate higher on the side of the vial and less frequently on the surface than do larvae from the uncrowded populations. This behavior has a dramatic impact on pupal survival; selection for pupation height in crowded environments is either directional or stabilizing, depending on the genotype tested (44). The CU and UU populations have shown inconsistent changes in pupation height; this most likely occurs because the CU and UU populations are raised on softer food than that used by the *r*- and *K*-populations. The soft food probably results in selection for increased pupation height even at low larval densities.

Foraging path length is a larval behavior perhaps largely determined by alleles at a single locus, *for* (24, 25). Two phenotypes have been described, rover and sitter, which differ in the distance traveled on a two-dimensional surface by a larva in a specified period of time (76, 94). Recently, Sokolowski and her colleagues (95) tested the UU and CU populations, the *r*- and *K*-populations and a new set of *D. melanogaster* populations that were handled like the *r*- and *K*-populations. In each of these three independent test systems, the populations kept at high larval densities evolved the more active rover phenotype, whereas the low-density population became predominantly the sitter phenotype. The fitness consequences of the rover phenotype in crowded culture have not been studied explicitly although they may be related to general competitive ability for food.

Adaptation to adult crowding has also been examined in the UU, CU, and UC populations (46, 47). Adults subjected to a 3–5-day episode of adult crowding showed elevated mortality rates; the CU population had the highest mortality rate, and the UC population the lowest (47). This suggests that the UC population adapted to periods of adult crowding and that the CU population was sensitive to such crowding perhaps as a by-product of their adaptation to larval crowding. Adults subjected to these short episodes of crowding suffered a reduction in later rates of survival and fecundity (46). However, the reduction was much greater in the UU and CU populations than it is in the UC population.

Limited food is just one way in which crowded and uncrowded environments are expected to differ; it is also likely that crowded cultures will suffer from increased levels of waste products. One of the few studies of *Drosophila* nitrogenous waste had suggested that urea is a major waste product (10). However, more recent work has shown that ammonia is the primary nitrogen waste product that appears in crowded *Drosophila* cultures and that urea, if present, is at undetectable levels (DJ Borash, AG Gibbs, LD Mueller, unpublished). Concentrations of ammonia in crowded *Drosophila* cultures steadily increase during the entire period of larval foraging and development (about 20 days in the CU cultures, for instance). During this period, other aspects of the environment are also changing; food is being consumed by voracious larvae; sugars and carbohydrates are also being consumed by the growing yeast populations, which produce increasing levels of fermentation products like acetic acid (35). Since development time in crowded cultures increases and becomes more variable (18), the larvae that develop most rapidly and emerge first from crowded cultures are likely to have experienced a very different environment than the larvae that emerge 10 days later. The more slowly developing adults will have experienced the toxic effects of food laced with ammonia for a greater length of time, and the concentrations that they ultimately experience will be much greater than those experienced by the earliest emerging flies.

This description of the environment suggests that our theoretical description of density-dependent selection, even in simple laboratory environments, may be oversimplified. In reality, a crowded environment cannot be characterized by a single number N, which describes the number of larvae or adults that are placed together. A crowded environment is actually one undergoing a profound temporal sequence of degradation that is not equally experienced by all members of the population. This type of environmental heterogeneity may complicate the dynamics of natural selection as much as do other types of environmental heterogeneity (33).

My laboratory has recently documented a polymorphism with the CU populations that is demarcated along an axis of larval development time (DJ Borash, AG Gibbs, LD Mueller, unpublished). Flies that emerge early in the CU populations have very high feeding rates (and thus should be good competitors for limited food), but their absolute viability, especially in food laced with ammonia, is significantly less than that of flies that emerge much later in the CU cultures. These observations suggest that there may be genotypes that specialize on the early part of the crowded environment by developing rapidly and emerging before food levels become very low and waste concentrations very high. However, these early specialists "pay" for this by sacrificing their ability to survive well and take advantage of the resources (albeit less than ideal) that remain in the old crowded cultures. The late specialists develop more slowly but can survive and successfully emerge as viable adults in the low-nutrition and high-waste environment that remains toward the end of the developmental profile.

These observations should be music to the ears of the field ecologist for they suggest one way in which environments become complex. This type of complexity is amenable to study under laboratory conditions and should pave the way to an understanding and appreciation of natural selection in complex environments.

ACKNOWLEDGMENTS

I thank T. Prout for useful comments on the manuscript and the National Science Foundation (DEB-9410281) and the National Institutes of Health (AG09970) for financial support.

Visit the *Annual Reviews home page* at
http://www.annurev.org.

Literature Cited

1. Anderson WW. 1971. Genetic equilibrium and population growth under density-regulation selection. *Am. Nat.* 105:489–98

2. Anderson WW, Arnold J. 1983. Density-regulated selection with genotypic interactions. *Am. Nat.* 121:649–55

3. Asmussen MA. 1983. Density-dependent selection incorporating intraspecific competition. II. A diploid model. *Genetics* 103:335–50

3a. Ayala FJ. 1965. Evolution of fitness in experimental populations of *Drosophila serrata*. *Science* 150:903–5

3b. Ayala FJ. 1968. Genotype, environment, and population numbers. *Science* 162:1453–59

4. Bakker K. 1961. An analysis of factors which determine success in competition for food among larvae of *Drosophila melanogaster. Arch. Neerl. Zool.* 14:200–81

5. Barclay HJ, Gregory PT. 1981. An experimental test of models predicting life-history characteristics. *Am. Nat.* 117:944–61

6. Barclay HJ, Gregory PT. 1982. An experimental test of life history evolution using *Drosophila melanogaster* and *Hyla regilla. Am. Nat.* 120:26–40

7. Begon M, Harper JL, Townsend CR. 1990. *Ecology, Individuals, Populations and Communities.* Cambridge, UK: Blackwell

8. Berryman AA, Millstein JA. 1989. Are ecological systems chaotic—and if not, why not? *Trends Ecol. Evol.* 4:26–28

9. Bierbaum TJ, Mueller LD, Ayala FJ. 1989. Density-dependent evolution of life history characteristics in *Drosophila melanogaster. Evolution* 43:382–92

10. Botella LM, Moya A, Gonzalez MC, Mensua JL. 1985. Larval stop, delayed development and survival in overcrowded cultures of *Drosophila melanogaster:* effect of urea and uric acid. *J. Insect Physiol.* 31:179–85

11. Boyce MS. 1984. Restitution of r- and K-selection as a model of density-dependent natural selection. *Annu. Rev. Ecol. Syst.* 15:427–47

12. Bradshaw WE, Holzapfel CM. 1989. Life-historical consequences of density-dependent selection in the pitcher-plant mosquito, *Wyeomyia smithii. Am. Nat.* 133:869–87

13. Bradshaw WE, Holzapfel CM. 1996. Genetic constraints to life-history evolution in the pitcher-plant mosquito, *Wyeomyia smithii. Evolution* 50:1176–81

14. Britnacher JG. 1981. Genetic variation and genetic load due to the male reproductive component of fitness in *Drosophila. Genetics* 97:719–30

15. Burnet B, Sewell D, Bos M. 1977. Genetic analysis of larval feeding behavior in *Drosophila melanogaster.* II. Growth relations and competition between selected lines. *Genet. Res.* 30:149–61

15a. Buzzati-Traverso AA. 1955. Evolutionary changes in components of fitness and other polygenic traits in *Drosophila melanogaster* populations. *Heredity* 9:153–86

16. Charlesworth B. 1971. Selection in density-regulated populations. *Ecology* 52:469–74

17. Charlesworth B. 1994. *Evolution in Age-Structured Populations.* London: Cambridge Univ. Press. 2nd ed.

18. Chiang HC, Hodson AG. 1950. An analytical study of population growth in *Drosophila melanogaster. Ecol. Monogr.* 20:173–206

19. Christiansen FB. 1984. Evolution in temporally varying environments: density and composition dependent genotypic fitness. In *Population Biology and Evolution,* ed. K Wöhrmann, V Loeschcke, pp. 115–24. Berlin: Springer-Verlag

20. Clarke B. 1972. Density-dependent selection. *Am. Nat.* 106:1–13

21. Cody M. 1966. A general theory of clutch size. *Evolution* 20:174–84

22. Costantino RF, Cushing JM, Dennis B, Desharnais RA. 1995. Experimentally induced transitions in the dynamic behaviour of insect populations. *Nature* 375:227–30

23. Darwin C. 1859. *The Origin of Species.* London: Penguin

24. De Belle JS, Hilliker AJ, Sokolowski MB. 1989. Genetic localization of foraging (for): a major gene for larval behavior in *Drosophila melanogaster. Genetics* 123:157–64

25. De Belle JS, Sokolowski MB. 1987. Heredity of rover/sitter: alternative foraging strategies of *Drosophila melanogaster. Heredity* 59:73–83

26. de Jong G. 1976. A model of competition for food. I. Frequency dependent viabilities. *Am. Nat.* 110:1013–27

27. Demetrius L. 1969. The sensitivity of population growth rate to perturbations in the life cycle components. *Math. Biosci.* 4:129–36

28. Dennis B, Desharnais RA, Cushing JM, Costantino RF. 1995. Nonlinear demographic dynamics: mathematical models, statistical methods, and biological experiments. *Ecol. Monogr.* 65:261–81

29. Dobzhansky Th. 1950. Evolution in the tropics. *Am. Sci.* 38:209–21

29a. Ebenman B, Johanssan A, Jonsson T, Wennergren U. 1996. Evolution of stable population dynamics through natural selection. *Proc. R. Soc. Lond. B.* 263:1145–51

30. Gadgil M, Bossert WH. 1970. Life historical consequences of natural selection. *Am. Nat.* 104:1–24

31. Gadgil M, Solbrig OT. 1972. The concept of r and K selection: evidence from wild flowers and some theoretical considerations. *Am. Nat.* 106:14–31

32. Gatto M. 1993. The evolutionary optimality of oscillatory and chaotic dynamics in simple population models. *Theor. Popul. Biol.* 43:310–36

33. Gillespie J. 1991. *The Causes of Molecular Evolution.* Oxford, UK: Oxford Univ. Press

33a. Godfray HCJ, Cook LM, Hassell MP. 1991. Population dynamics, natural selection and chaos. In *Genes in Ecology,* ed. RJ Berry, JJ Crawford, and GM Hewitt, pp. 55-85. Oxford: Blackwell Sci.

34. Guo PZ, Mueller LD, Ayala FJ. 1991. Evolution of behavior by density-

dependent natural selection. *Proc. Natl. Acad. Sci. USA* 88:10905–6

35. Hageman J, Eisses KT, Jacobs PJM, Scharloo W. 1990. Ethanol in *Drosophila* cultures as a selective factor. *Evolution* 44:447–54

36. Hansen TF. 1992. Evolutionary stability parameters in single-species population models: stability or chaos? *Theor. Popul. Biol.* 42:199–217

37. Hairston NG Jr, Walton WE. 1986. Rapid evolution of a life history trait. *Proc. Natl. Acad. Sci. USA* 83:4831–33

38. Hastings A, Hom CL, Ellner S, Turchin P, Godfray HCJ. 1993. Chaos in ecology: Is mother nature a strange attractor? *Annu. Rev. Ecol. Syst.* 24:1–33

39. Hassell M, Lawton J, May RM. 1976. Pattern of dynamical behavior in single species populations. *J. Anim. Ecol.* 45:471–86

40. Heckel DG, Roughgarden J. 1980. A species near equilibrium size in a fluctuating environment can evolve a lower intrinsic rate of increase. *Proc. Natl. Acad. Sci. USA* 77:7497–500

41. Hutchinson GE. 1957. Concluding remarks. *Cold Spring Harbor Symp. Quant. Biol.* 22:415–27

42. Iwasa Y, Teramoto E. 1980. A criterion of life history evolution based on density dependent selection. *J. Theor. Biol.* 84:545–66

43. Joshi A, Mueller LD. 1988. Evolution of higher feeding rate in *Drosophila* due to density-dependent natural selection. *Evolution* 42:1090–93

44. Joshi A, Mueller LD. 1993. Directional and stabilizing density-dependent natural selection for pupation height in *Drosophila melanogaster. Evolution* 47:176–84

45. Joshi A, Mueller LD. 1996. Density-dependent natural selection in *Drosophila*: trade-offs between larval food acquisition and utilization. *Evol. Ecol.* 10:463–74

46. Joshi A, Mueller LD. 1997. Adult crowding effects on longevity in *Drosophila melanogaster*: increase in age-independent mortality. *Curr. Sci.* In press

47. Joshi A, Wu WP, Mueller LD. 1997. Density-dependent natural selection in *Drosophila*: adaptation to adult crowding. *Evol. Ecol.* In press

48. Kosuda K. 1985. The aging effect on male mating activity in *Drosophila melanogaster. Behav. Genet.* 15:297–303

49. Lewontin RC. 1965. Selection for colonizing ability. In *The Genetics of Coloniz-*

ing Species, ed. HG Baker, GL Stebbins, pp. 77–91. New York: Academic

50. Levins R. 1968. *Evolution in Changing Environments.* Princeton, NJ: Princeton Univ. Press

51. Luckinbill LS. 1978. r- and K-selection in experimental populations of *Escherichia coli. Science* 202:1201–3

52. MacArthur RH. 1962. Some generalized theorems of natural selection. *Proc. Natl. Acad. Sci. USA* 48:1893–97

53. MacArthur RH, Wilson EO. 1967. *The Theory of Island Biogeography.* Princeton, NJ: Princeton Univ. Press

54. May RM. 1974. Biological populations with non-overlapping generations: stable points, stable cycles and chaos. *Science* 186:645–47

55. May RM, Oster G. 1976. Bifurcation and dynamic complexity in simple ecological systems. *Am. Nat.* 110:573–99

56. Menge BA. 1974. Effect of wave action and competition on brooding and reproductive effort in the seastar *Leptasterias hexactis. Ecology* 55:84–102

56a. Moorcroft PR, Albon SD, Pemberton JM, Stevenson IR, Clutton-Brock TH. 1996. Density-dependent selection in a fluctuating ungulate population. *Proc. Roy. Soc. Lond. B* 263:31–38

57. Mueller LD. 1985. The evolutionary ecology of *Drosophila. Evol. Biol.* 19:37–98

58. Mueller LD. 1986. Density-dependent rates of population growth: estimation in laboratory populations. *Am. Nat.* 128:282–93

59. Mueller LD. 1987. Evolution of accelerated senescence in laboratory populations of *Drosophila. Proc. Natl. Acad. Sci. USA* 84:1974–77

60. Mueller LD. 1988. Evolution of competitive ability in *Drosophila* due to density-dependent natural selection. *Proc. Natl. Acad. Sci. USA* 85:4383–86

61. Mueller LD. 1988. Density-dependent population growth and natural selection in food limited environments: the *Drosophila* model. *Am. Nat.* 132:786–809

62. Mueller LD. 1990. Density-dependent natural selection does not increase efficiency. *Evol. Nat.* 4:290–97

63. Mueller LD. 1991. Ecological determinants of life history evolution. *Philos. Trans. R. Soc. London Ser. B* 332:25–30

64. Mueller LD. 1995. Adaptation and density-dependent natural selection. In *Genetics of Natural Populations: The Continuing Importance of Theodosius*

Dobzhansky, ed. L Levine, pp. 222–38. New York: Columbia Univ. Press

65. Mueller LD, Ayala FJ. 1981. Trade-off between r-selection and K-selection in *Drosophila* populations. *Proc. Natl. Acad. Sci. USA* 78:1303–5

66. Mueller LD, Ayala FJ. 1981. Dynamics of single species population growth: stability or chaos? *Ecology* 62:1148–54

67. Mueller LD, Ayala FJ. 1981. Fitness and density-dependent population growth in *Drosophila melanogaster. Genetics* 97:667–77

68. Mueller LD, González-Candelas F, Sweet VF. 1991. Components of density-dependent population dynamics: models and tests with *Drosophila. Am. Nat.* 137:457–75

69. Mueller LD, Graves JL Jr, Rose MR. 1993. Interactions between density-dependent and age-specific selection in *Drosophila melanogaster. Funct. Ecol.* 7:469–79

70. Mueller LD, Guo PZ, Ayala FJ. 1991. Density-dependent natural selection and trade-offs in life history traits. *Science* 253:433–35

71. Mueller LD, Huynh PT. 1994. Ecological determinants of stability in model populations. *Ecology* 75:430–37

72. Mueller LD, Sweet VF. 1986. Density-dependent natural selection in *Drosophila*: evolution of pupation height. *Evolution* 40:1354–56

73. Nicholson AJ. 1957. The self adjustment of populations to change. *Cold Spring Harbor Symp. Quant. Biol.* 2:153–73

74. Nogúes RM. 1977. Population size fluctuations in the evolution of experimental cultures of *Drosophila subobscura. Evolution* 31:200–13

75. Nunney L. 1983. Sex differences in larval competition in *Drosophila melanogaster*: the testing of a competition model and its relevance to frequency dependent selection. *Am. Nat.* 121:67–93

76. Pereira HS, Sokolowski MB. 1993. Mutations in the larval foraging gene affect adult locomotory behavior after feeding in *Drosophila melanogaster. Proc. Natl. Acad. Sci. USA* 90:5044–46

77. Pianka ER. 1970. On r and K selection. *Am. Nat.* 104:592–97

78. Pianka ER. 1972. r and K selection or b and d selection?. *Am. Nat.* 106:581–88

79. Prout T. 1965. The estimation of fitness from population data. *Evolution* 19:546–51

80. Prout T. 1971. The relation between fitness components and population prediction in *Drosophila.* I. The estimation of

fitness components. *Genetics* 68:127–49

81. Prout T. 1971. The relation between fitness components and population prediction in *Drosophila.* II. Population prediction. *Genetics* 68:151–67

82. Prout T. 1980. Some relationships between density-independent selection and density-dependent population growth. *Evol. Biol.* 13:1–68

83. Prout T. 1984. The delayed effect on adult fertility of immature crowding: population dynamics. In *Population Biology and Evolution,* ed. K Wöhrmann, V Loeschcke, pp. 83–86. Berlin: Springer-Verlag

84. Prout T. 1986. The delayed effect on fertility of preadult competition: two species population dynamics. *Am. Nat.* 127:809–18

85. Prout T, McChesney F. 1985. Competition among immatures affects their adult fertility: population dynamics. *Am. Nat.* 126:521–58

86. Reznick DN. 1982. The impact of predation on life history evolution in Trinidadian guppies: the genetic components of observed life history differences. *Evolution* 36:1236–50

87. Reznick DN, Endler JA. 1982. The impact of predation on life history evolution in Trinidadian guppies (*Poecilia reticulata*). *Evolution* 36:160–77

88. Roper C, Pignatelli P, Partridge L. 1996. Evolutionary responses of *Drosophila melanogaster* life history to differences in larval density. *J. Evol. Biol.* 9:609–22

89. Rose MR, Nusbaum TJ, Chippindale AK. 1996. Laboratory evolution: the experimental wonderland and the Cheshire cat syndrome. In *Adaptation,* ed. MR Rose, GV Lauder, pp. 221–41. San Diego: Academic

90. Roughgarden J. 1971. Density-dependent natural selection. *Ecology* 52:453–68

91. Roughgarden J. 1979. *Theory of Population Genetics and Evolutionary Ecology: An Introduction.* New York: Macmillan

92. Royama T. 1971. A comparative study of models for predation and parasitism. *Res. Popul. Ecol. Suppl.* 1:1–91

93. Santos M, Borash DJ, Joshi A, Bounlutay N, Mueller LD. 1997. Density-dependent natural selection in *Drosophila*: evolution of growth rate and body size. *Evolution.* 51:420–32

94. Sokolowski MB. 1980. Foraging strategies of *Drosophila melanogaster*: a chromosomal analysis. *Behav. Genet.* 10:291–302

95. Sokolowski MB, Pereira HS, Hughes K. 1997. Evolution of foraging behavior in *Drosophila* by density dependent selection. *Proc. Natl. Acad. Sci. USA* In press

96. Stearns SC. 1976. Life-history tactics: a review of the ideas. *Q. Rev. Biol.* 51:3–47

97. Stearns SC. 1977. The evolution of life history traits: a critique of the theory and a review of the data. *Annu. Rev. Ecol. Syst.* 8:145–71

98. Stearns SC. 1992. *The Evolution of Life Histories.* Oxford, UK: Oxford Univ. Press

98a. Stenseth NC, Björnstad ON, Saitoh T. 1996. A gradient from stable to cyclic populations of *Clethrionomys rufocanus* in Hakkaido, Japan. *Proc. Roy. Soc. Lond. B* 263:1117–26

99. Stokes TK, Gurney WSC, Nisbet RM, Blythe SP. 1988. Parameter evolution in a laboratory insect population. *Theor. Popul. Biol.* 34:248–65

100. Tanaka Y. 1996. Density-dependent selection on continuous characters: a quantitative genetic model. *Evolution* 50:1775–85

101. Taylor CE, Condra C. 1980. r- and K-selection in *Drosophila pseudoobscura. Evolution* 34:1183–93

102. Thomas WR, Pomerantz MJ, Gilpin ME. 1980. Chaos, asymmetric growth and group selection for dynamical stability. *Ecology* 61:1312–20

103. Tilley SG. 1973. Life histories and natural selection in populations of the salamander *Desmognathus ochropaeus. Ecology* 54:3–17

104. Tinkle DW, Hadley N. 1975. Lizard reproductive effort: caloric estimates and comments on its evolution. *Ecology* 56:427–34

105. Turchin P, Taylor AD. 1992. Complex dynamics in ecological time series. *Ecology* 73:289–305

106. Turelli M, Petry D. 1980. Density-dependent selection in a random environment: an evolutionary process that can maintain stable population dynamics. *Proc. Natl. Acad. Sci. USA* 77:7501–5

107. Vasi F, Travisano M, Lenski RE. 1994. Long term experimental evolution in *Escherichia coli.* II. Changes in life-history traits during adaptation to a seasonal environment. *Am. Nat.* 144:432–56

Annu. Rev. Ecol. Syst. 1997. 28:289–316

TOWARD AN INTEGRATION OF LANDSCAPE AND FOOD WEB ECOLOGY: The Dynamics of Spatially Subsidized Food Webs

Gary A. Polis[1], Wendy B. Anderson[1], and Robert D. Holt[2]

[1]Department of Biology, Vanderbilt University, Nashville, Tennesse 37235;
e-mail: Polisga@ctrvax.Vanderbilt.edu; Anderswb@Ctrvax.Vanderbilt.edu;
[2]Museum of Natural History, University of Kansas, Lawrence, Kansas 66045;
e-mail: Predator@kuhub.cc.ukan.edu

KEY WORDS: food webs, spatial subsidy, trophic dynamics, consumer-resource dynamics, landscape ecology

---

ABSTRACT

We focus on the implications of movement, landscape variables, and spatial heterogeneity for food web dynamics. Movements of nutrients, detritus, prey, and consumers among habitats are ubiquitous in diverse biomes and can strongly influence population, consumer-resource, food web, and community dynamics. Nutrient and detrital subsidies usually increase primary and secondary productivity, both directly and indirectly. Prey subsidies, by movement of either prey or predators, usually enhance predator abundance beyond what local resources can support. Top-down effects occur when spatially subsidized consumers affect local resources by suppressing key resources and occasionally by initiating trophic cascades. Effects on community dynamics vary with the relative amount of input, the trophic roles of the mobile and recipient entities, and the local food web structure. Landscape variables such as the perimeter/area ratio of the focal habitat, permeability of habitat boundaries, and relative productivity of trophically connected habitats affect the degree and importance of spatial subsidization.

INTRODUCTION

Food webs are a central organizing theme in ecology. The organisms that comprise food webs live in a spatially heterogeneous world where habitats vary

289

0066-4162/97/1120-0289$08.00

greatly in productivity, resource abundance, and consumer behavior and demography. Even local communities that appear discrete are open and connected in myriad ways to outside influences (75, 105, 135). The basic components of food webs—nutrients, detritus, and organisms—all cross spatial boundaries. Yet until recently, ecologists neglected to ask how spatial patterns and processes affect web structure and dynamics (135). The core themes of landscape ecology—spatial variation in habitat quality, boundary and ecotonal effects, landscape connections, scaling, and spatial context (187–189)—carry significant implications for food web ecology.

We focus on spatial flows among habitats as a key force in local web dynamics. We first synthesize a large literature documenting the ubiquitous movement of material and organisms among habitats. We then show how spatial subsidies influence consumer-resource and web dynamics, and we propose a preliminary framework to integrate landscape and food web ecology. By spatial subsidy, we mean a donor-controlled resource (prey, detritus, nutrients) from one habitat to a recipient (plant or consumer) from a second habitat which increases population productivity of the recipient, potentially altering consumer-resource dynamics in the recipient system.

LANDSCAPE CONSIDERATIONS

Connectivity varies enormously among real systems, from near total isolation to strong mixing. Factors that influence exchange rate among spatial units are a central focus of landscape ecology (50, 66, 170, 186). "Flow rate" depends on a suite of environmental and organismal attributes (e.g. habitat geometry and area; similarity of, distance between, and relative productivity of interacting habitats; boundary permeability; and organism mobility).

The ratio of "edge" to "interior" (i.e. perimeter-to-area, P/A) is a major determinant of input to a habitat (137), e.g. watershed, riparian, and shoreline to and from streams and lakes (62, 132, 184, 185); the ocean to coastal areas and islands (137); and forest edge to interior (4, 6, 189). P/A is a function of size (larger units have less edge per unit area), shape (e.g. compact vs elongated), and fractal irregularity or folding of the edge (62, 137, 184). Such edge effects are likely universal, governing input, productivity, and dynamics among juxtaposed habitats (135, 137).

The "river continuum concept" (172) illustrates landscape influences on flow rates at several spatial scales. P/A declines from headwater streams to large rivers, with a corresponding decline in the relative importance of local allochthonous inputs compared to in situ productivity. Local production is reduced downstream because of increased depth and turbidity, such that large rivers are net sinks for energy and material derived from smaller order streams; upstream subsidies drive downriver dynamics. Finally, the contribution of allochthonous stream material to lakes or oceans is governed by landscape factors (e.g. small source streams or short rivers should contribute relatively more than longer or larger rivers).

FLOW AND SPATIAL SUBSIDIES: DIRECT EFFECTS

Allochthonous input can influence greatly the energy, carbon (C), and nutrient budget of many habitats. In general, nutrient inputs (nitrogen [N], phosphorus [P], trace elements) increase primary productivity; detrital and prey inputs produce numerical responses in their consumers. Transport across boundaries occurs via either physical or biotic vectors. Wind and water are the primary physical vectors; they transport subsidies either by advection or diffusion (44). Mobile consumers transport nutrients and detritus when they forage in one habitat and defecate in another. We organize trophic flows by origin and destination using two comprehensive categories, water and land.

Movement of Nutrients and Detritus

WATER TO WATER   Water masses often differ substantially in productivity and organic biomass. Transport, both vertical (upwelling, pelagic detrital fallout to benthos) and horizontal (currents, tidal movement, eddy-diffusion), is generally a key determinant of local marine productivity and consequent food webs. In particular, pelagic-benthic coupling is a major route for energy and nutrient flow (7, 13, 14, 16, 90, 98, 140). Much shallow water benthos consists of sessile particle or detritus feeders that rely on settlement of food from the coastal fringe and production from overlying waters. In situ benthic productivity is relatively unimportant (most areas) or totally absent (aphotic zones). Worldwide, the biomass of benthic fauna reflects the productivity of overlying waters (13, 98, 140). Conversely, infusion of nutrients from bottom via both mixing and upwelling controls primary productivity of surface waters (13, 14, 98).

Benthic and pelagic lake habitats are connected via turnover, a process similar to upwelling whereby bottom nutrients, reinfused into photic waters, stimulate productivity. Lakes also receive many nutrients from streams, springs, precipitation, watershed soil and fertilizer runoff, shore vegetation, and litter fall (132, see below). In beaver ponds, biomass input from streams is three times greater than local production (116). One implication of the river continuum concept is that downstream communities are subsidized by upstream "inefficiencies" in C retention and processing (117, 179). In some systems, such input is quite important, e.g. in a Washington estuary, rivers contributed four to eight times more organic material than all local producers combined (165).

Plant detritus produced in one habitat and transported to a second can subsidize detritivores. Productive kelp and seagrass beds fuel dense detritivore populations in the supralittoral (137), littoral (49), intertidal (32), and deep benthic zones (174).

Large consumers transport material among aquatic habitats. Seasonally and daily migrating fish are particularly important conduits. Anadromous fish (e.g. salmon) deposit great amounts of energy and nutrients of marine origin to lakes and nutrient-deficient headwater streams via reproductive products, excretion, and death (51). For example, dead salmon contribute 20–40% of total lake P (51); abundant (9×10^8 to 4.4×10^{10}/year) marine alewives leave up to 146 g/m^2 in freshwater when they die. Nutrients from dead anadromous fish appear critical to sustain productivity of many freshwater and riparian ecosystems (15a, 51, 192).

Daily movement by fish and zooplankton facilitates rapid nutrient translocation across boundaries in freshwater (35, 89, 152, 171) and marine systems (7, 108, 124, 145). Such movement transports great quantities of fecal matter rich in fertilizing nutrients within the water column (the "diel ladder"; 89), between benthic and pelagic waters ("nutrient pump mechanism"; 171), between onshore and offshore waters (22), and to refuge areas (108, 124). In lakes, P input via fish excretion can exceed all other inputs, greatly increasing primary productivity, altering the outcome of phytoplankton competition, and stimulating trophic cascades (34, 171). Detritivorous benthic fish facilitate energy flow through lake webs by infusing DOM and P into the water in forms useful to phytoplankton (171). Marine fish transport nutrients and energy from feeding to resting areas, e.g. N and P from seagrass beds into nutrient-poor waters over corals, which thus increases coral growth rates. These effects are likely general in many marine habitats (26, 109).

Seabirds feeding on fish and invertebrates concentrate and transport great quantities of nutrients in their guano. Guano, a powerful fertilizer, enhances nutrient status and primary production in the intertidal and nearshore marine and estuarine waters (19, 82, 195).

LAND TO WATER   Aquatic and terrestrial systems are often linked functionally by flows of nutrients and organic matter via wind or water moving in the hydrologic cycle (67). In general, food webs in rivers, lakes, and estuaries are fueled by both local primary productivity and allochthonous detritus. Terriginous input is a major factor (along with upwelling and an enhanced light regime) that promotes high primary and secondary productivity in coastal waters, both marine (13, 98) and freshwater (171, 185). Three major conduits shunt material from land to freshwater (67, 179, 185): detritus from leaf and litter fall; dissolved and particulate organic matter (DOM, POM) from soil runoff (107, 110, 113, 117, 170); and detritus, POM, and DOM from floods (64, 114, 178, 179).

The impact of such input on energy budgets and community structure depends on many landscape variables: location in a drainage, nature of the terrestrial surroundings, watershed size, amount of terrestrial runoff, and shoreline to water P/A ratio (40, 93, 109, 132, 152, 165, 172, 185). Often, input greatly exceeds in situ productivity (18, 58, 165). For example, primary production in ponds (2.4% of C budget), stream riffles (10%), and streams (4.2%, 16%) is substantially less than allochthonous input (respectively, 80%, 76%, 91%, 74%; 116, 117). Plants usually benefit greatly from nutrient and DOM input (58, 109, 113).

In floodplain ecosystems, great amounts of detritus, nutrients, and sediments rich in organics are exchanged reciprocally between the river channel and the riparian land via flooding (178, 179, 182). Materials produced on land during dry phases increase productivity of aquatic plants and are a rich food supply for detritivores that move onto the floodplain from the channel during floods (193). Annual floodplain input of total C to a river channel in Georgia was seven times greater than in situ production (55). Blackwater river productivity is powered by inputs from terrestrial systems (15, 64, 106). In Amazonian rivers, "... primary productivity is so low that a food chain could not be built up from endogenous sources alone to support a large biomass of animals" (63: p. 252). The web is based strongly on allochthonous input of organisms and detritus. "The rainforest, in its floodplain manifestation, has come to the trophic rescue of these aquatic ecosystems" (63, p. 252). An estimated 75% of market fish receive substantial input (50–90% of diet) from terrestrial origin (fruit, seeds, insects, small vertebrates).

Birds and mammals foraging on land can transport great quantities of detritus and nutrients to water, e.g. geese, gulls, and hippopotamuses defecate rich feces into water. Well-studied birds bring terrestrial nutrients to lakes via guano (27, 82); e.g. birds bring 36% of the annual P input into some ponds, increasing plant abundance (102). Beaver-transported trees add nutrients and much organic matter (1 ton/beaver/year) to ponds (85), establishing an entire food chain based on wood decomposition (111, 115, 116). In the Amazon Basin, many fish import great amounts of energy and nutrients from terrestrial habitats (riparian, flood forest, and floodplains) to rivers (63, 64).

WATER TO LAND   Conversely, terrestrial organisms benefit from periodic nutrient enhancement from aquatic habitats. This water-land linkage is well known to humans and is exemplified by the agriculturally based cultures along the fertile bottom lands of major rivers (e.g. Nile, Mississippi). The area affected can be great: the Amazon floods 2% (70,000 km^2) of its adjacent forest annually (63). Lake material is an important source of organic matter in bordering land habitats (132). Flooding and winds transport lake plants in quantities

(7.4 kg/m^2/year of shoreline) 4.5 times greater than in situ terrestrial productivity. Such input produces an "edge effect," with greater diversity and densities in the riparian than surrounding habitats (36, 83, 132).

Coastal areas fringing oceans worldwide receive great amounts ($10 \longrightarrow 2000$ kg/m shoreline/year) of organic matter from the sea via shore wrack (algae and carrion) (32, 68, 69, 96, 137) and possibly from N-rich seafoam (78, 162). Seabirds transport substantial nutrients and organic material to land via guano, food scraps, eggs, feathers, and bodies of dead chicks and adults (82, 114, 137). Birds that feed on massive schools of Peruvian anchovetta deposit guano to mean depths of 5.4–28.5 m, with three offshore islands each containing 2.3–5.2×10^6 metric tons (76). Worldwide, seabirds annually transfer 10^4–10^5 tons of P to land (114).

LAND TO LAND   Surprisingly great amounts and variety of windborne detritus and nutrients arrive from near and far and may totally sustain or partially subsidize local webs (10, 53, 78, 97, 154, 162, 164). "Aeolian ecosystems" (162) fueled by windborne input (53, 78) include caves, mountaintops, snowfields, polar regions, new volcanic areas, phytotelmata, and barren deserts and islands. In these systems, local plant productivity is low or absent; yet diverse, detrital-based webs exist with abundant consumers at several trophic positions.

Such webs characterize high-altitude and snow-covered habitats worldwide (10, 53, 97). "Truly immense quantities of pollen grains of many different plants, spores of fungi and of Protozoa, seeds of a great variety of plants, . . . and nearly every conceivable group of winged and apterous insects, spiders, etc brought by the upper air currents, are frozen and entombed in the snow and glacier ice" (97, p. 70). Melted water carries detritus and carcasses to streams to provide a rich food for insects.

Worldwide, nutrient budgets of many terrestrial ecosystems depend on nutrients transported aerially (94). Such subsidies may compensate for low-nutrient soils in temperate forest communities (9). In much of the Amazon Basin, where soils are nutrient poor due to limited river deposition, airborne soils apparently are needed to achieve a nutrient balance (86, 176). Most P, a critical element that limits net primary production, is intercontinental: 13 million tons (13–190 kg/ha/year) is carried by dust blown from the African deserts 5000 km away (163)! Such input doubles the standing stock of P over 4.7–22 ky. Thus, the productivity of Amazon rainforests depends critically upon fertilization from another large ecosystem, separated by an ocean yet atmospherically coupled (163).

Consumers that redistribute large quantities of biomass include mammalian herbivores (e.g. grazers in the Serengeti; 103) and roosting or nesting birds (181). Bats, rats, birds, and crickets are major conduits of energy via guano into caves (42, 79, 82). Great quantities of fecal fruit and seeds from Peruvian oil birds form the base of a diverse cave food web: bacteria, fungus, and >50 species of arthropods (47). Caves worldwide are similar; they receive all energy allochthonously via animals, root exudates, and water flow that deposits surface detritus; such input supports many detritivores, scavengers, fungivores, and a rich predator guild (42, 79, 99).

Movement of Prey

Species produced in one habitat frequently end up as food elsewhere. Movement may be accidental (e.g. by winds), or a product of life history (e.g. migration, ontogenetic habitat switches) or interactions (e.g. interference).

WATER TO WATER   The ubiquitous horizontal and vertical movement of water transports nutrients (see above) and prey. The prey of filter feeders may be produced locally, in adjacent habitats, or far away. Downstream movement of prey characterizes streams and rivers; generally, most productivity is fixed in riffles, yet most consumption occurs in ponds, often subsidizing resident predators (39, 116, 117).

Members of the "deep scattering layer" move 300–1500 m to feed at night in the photic zone and return to deeper waters during the day. These diel migrants carry much primary productivity to depths where they form the prey of large populations of fish and invertebrate predators (7). On deep seamounts in areas of very low in situ productivity, many fish species eat great amounts of prey carried to them by currents and vertical migrations; the population biomass of these fish is an order of magnitude higher than populations that do not receive allochthonous prey (90). Life history migrations of species at all trophic positions at several temporal scales connect food webs of marine pelagic and benthic habitats (the "jellyfish paradigm"; 16; but see 61). Diadromous migration of fish transport large numbers of potential prey between marine and freshwater habitats.

LAND TO WATER   Some aquatic consumers eat terrestrial prey. Fish and aquatic insects eat insects and spiders that drop to streams (63, 193), often in great numbers (100). These amounts may surpass production of in situ aquatic insects. Many salmonids eat "an astonishing diversity" of terrestrial prey (80), at least seasonally; such prey can form >50% of annual energy uptake (80, 100). Land insects from 70 families provide ≈10% of fish diet in a Swedish lake (121).

An abundance of insects blow onto the ocean, both near and far from shore (21, 37, 38). On a typical summer day, an estimated 4.5 billion insects drift over the North Sea from a 30-km coastal strip (38). Greater numbers can occur. Off Nova Scotia, an estimated 800 billion budworm moths formed a floating slick 100×66 km (21). An estimated mean of 2–17 million insects annually drop onto each km^2 of ocean surface worldwide; this equals 2–17 $kg/m^2/year$, or 0.01% of phytoplankton productivity (38). Although insects contribute little to the energetics of ocean ecosystems, they form 27%–60% of the prey volume of some fish and may have large impacts in low-productivity regions.

WATER TO LAND   Many land consumers eat prey of aquatic origin. Emerging aquatic insects are eaten by terrestrial insects, arachnids, amphibians, reptiles, and birds; such consumers often occur in large populations at the margins of water (36, 70, 83, 118, 136, 137, 138, 146). Only ≈3% of emerging insect biomass remained in a desert stream, with 23 g/m²/year (400–155,000 individuals/m²/year) exported to the adjacent riparian zone (83). Such great export can be crucial to terrestrial predators, affecting abundance, territoriality, feeding behavior, and reproductive success (83). Seabirds and their associated ectoparasites are a major food of many consumers (48, 137).

Anadromous fish are important prey of diverse consumers in many terrestrial habitats, e.g. ≈50 species of birds and mammals eat salmon in Alaska (192). These fish "appear to be a keystone food resource for vertebrate predators and scavengers, forging an ecologically significant link between aquatic and terrestrial ecosystems" (192, p. 489) and appear critical to the success of many consumers. This interaction carries important conservation implications; the loss of such fish could exert major effects on these species and their community.

LAND TO LAND   Winds frequently transport prey great distances. Windborne arthropods allow stable populations of predators to persist on barren volcanic fields and new volcanic islands (52, 54, 78, 166, 168). Insects also waft to mountain tops and snowfields (52, 53, 97, 162) where they are eaten by diverse and abundant predators. Over 130 species of arthropods, most from lowland habitats, can occur on barren snowfields (10). "Surprisingly large numbers" of diverse consumers in every trophic category "gorge" with foods "refrigerated in the snows" (97, p. 70). Webs with "every class of feeder" are driven by allochthonous foods (97). In these examples, airborne prey sustained predators (e.g. spiders, insects, lizards, birds, and small mammals) in a system lacking local primary productivity.

Prey movement among habitats often characterizes insect life cycles. Brown & Gange (28) give many examples of the generalized life-cycle of such insects: Females deposit eggs into the soil; larvae, which feed underground on roots or detritus, pupate there; and adults emerge to mate aboveground. Such a life cycle characterizes some of the most successful insects: Almost all termites, ants, cicadas, and many beetles, moths, and flies transport belowground organic material to aboveground consumers (112, 161). In the case of periodic cicadas (the land animal with the greatest biomass/area), predators feed to satiation on emergent adults (28).

Life-history migrations transport prey great distances. For example, monarch butterflies eat temperate milkweed but overwinter in Mexico where they are eaten by tropical birds (33). Similar transport occurs via other migratory animals, e.g. songbirds (92), mammalian grazers (150), and locusts.

Movement of Consumers

Consumers' movement ranges from fine (e.g. local foraging paths) to broad scales (e.g. long-distance migration). Many mobile consumers, migrants, and age classes choose habitats based on relative profitability of forage intake (36, 155, 183). However, some consumers move into habitats with relatively low productivity to avoid interference or predation (73, 127). A vast number of species change feeding habitats during their lifetime (see 135 for examples and food web implications of ontogenetic habitat shifts by consumers).

WATER TO WATER   Aquatic species exhibit a continuum of horizontal and vertical, short and long migrations. The "food availability hypothesis" for the evolution of diadromy in 128 species of fish (65) posits that relative productivity of marine and riverine environments at a given latitude determines if fish feed and grow in the ocean and move to freshwater to reproduce (anadromy) or vice versa (catadromy). Anadromy is more frequent when ocean productivity exceeds neighboring freshwater productivity (temperate, arctic). Many diadromous fish also feed in the less productive habitats and can exert great effects on prey in these places, e.g. anadromous steelhead in California rivers (141). Stable isotope analyses show the relative importance of coupled habitats on anadromous and estuarine fish (15a, 60).

Shorter migrations occur. Krill move annually from the Antarctic ice shelf, where they graze on algae, to pelagic zones, where they eat phytoplankton (157). This cycle may be central to southern oceans ecosystem dynamics. Feeding on ice algae (along with fat storage and cannibalism) allows krill to overwinter and be present in great numbers in the spring. Such seasonal and regional switches of food resources hypothetically explain two mysteries: the maintenance of very large populations and biomass of krill in this oligotrophic community, and the rapid suppression of phytoplankton during their exponential growth phase in spring.

In general, marine zooplankton, fish, birds, and mammals aggregate near regions of high productivity and food density (e.g. upwelling, frontal regions) (3, 14, 173). Crustacean densities inside frontal regions are 74.5 times greater than outside (14). Large populations of consumers migrate from deep water to feed on near-surface resources (7). Both marine and freshwater fish forage across habitats that vary in prey availability, e.g. pelagic and littoral zones (95, 111, 130), coral reef and sand flats (142), river and floodplain (63, 182).

BETWEEN WATER AND LAND   Many species of land and seabirds eat both aquatic and land prey (29, 30, 156, 159). Sixteen Crozet Island species forage in the ocean, in freshwater, and on land (81). Predation by coastal seabirds significantly influences land invertebrates by consuming 24 tons/100 km^2/year; land prey form 12%–25% of their total annual energy intake (156). Many land predators forage along shores (137, 146, 192). Along coasts worldwide, mammals and land birds eat living marine species and carrion (30, 126, 146). Such subsidies allow these species to maintain relatively large coastal populations that also forage on "typical" terrestrial prey (146).

LAND TO LAND   Many taxa move at varying temporal and spatial scales (patches to continents) to use distinct habitats. The dependency on spatially and temporally variable resources (fruit, nectar, insects) necessitates that birds be highly mobile to track changes in resource abundance across geographic scales ranging from within trees to between altitudinal zones, and from intrahabitat shifts to intercontinental migrations (92). Many consumers move 200–500 m from the edge of adjacent fields to exploit forest birds (4, 6, 143, 189). Landscape considerations are important: Nest predation is greater in areas closer to the edge (6, 189) and is a function of P/A ratio of the forest (greater nest predation in smaller fragments; 4, 143). Predators entering habitat islands significantly affect the composition, abundance, and dynamics of avian communities.

Movement by parasitic and pathogenic consumers among "habitats" (hosts) is a key to parasite-host (17) and pathogen-resource dynamics (144, 177). Alternative hosts frequently support parasites (e.g. of humans; 17) or pathogens; infection occurs via movement of either the alternative host or infective stages of pathogens. A typical control measure is to eliminate "reservoir" species of alternative hosts. However, such measures are often only partially successful—e.g. local infection rates are influenced by spores from distant outbreaks (144, 177). Further, most parasites with complex life histories and many plant pathogens require "landscape complementation" (50) of resources. For example, some rust fungi must move among alternative hosts to develop, e.g. cedar-apple rust alternates between apples and eastern red cedar (144).

FLOW AND SPATIAL SUBSIDIES: INDIRECT EFFECTS ON TROPHIC DYNAMICS

"Tropho-spatial" linkage is often a key factor in local dynamics. Recipient species almost always benefit from inputs. Food web effects depend on web configuration and the trophic roles of the mobile entity (e.g. nutrients, basal species, predators) and recipient (basal, intermediate, or top species) (1, 139).

Donor Control

Interactions between consumers and allochthonous resources are typically donor controlled, i.e. consumers benefit from but do not affect resource renewal rate (i.e. no recipient control; 131, 135, 139). Donor control occurs whenever a resource population is spatially partitioned into subpopulations that occupy different compartments, only one of which is accessible to consumers (36, 139). For mobile consumers, some feedback is likely, i.e. recipient control of resources by consumers may occur in either habitat.

Movement of Nutrients

We suggest that nutrient input is a major factor in open systems (and we argue that most systems are open); by contrast in closed systems, in situ herbivores and decomposers regulate nutrient recycling rates and availability to plants (171). If allochthonous nutrients enrich plants, primary productivity will increase, often dramatically (see earlier). Nutrient enrichment also increases plant quality. Herbivore survival and reproduction often depends more on host N concentration than on C availability (101, 139). Consequently, nutrient-subsidized systems often exhibit elevated densities of herbivores and higher-level consumers.

Mixing and upwelling of nutrients in the ocean stimulate phytoplankton blooms, followed by numerical responses through the web: Zooplankton increase, and nekton and vertebrates move to plankton concentrations (13, 14, 98). Upwelling drives oceanic production: Upwelling ecosystems account for just 0.1% of the ocean's surface, but 50% of the world's fish catch (13, 14). Cessation of upwelling sharply depresses pelagic productivity at all levels; e.g. El Niño events produce population collapses of seabirds, mammals, marine iguanas, and invertebrates off the Americas (14).

Exchange of nutrients between pelagic/littoral zones and intertidal zones can be quite important. Nutrient enhancement from coastal upwelling allows intertidal algae and higher-level consumers to increase productivity and standing stock (19). Nutrient input from oceanic waters may be a key "bottom up" factor in intertidal community structure (104). Fertilization effects of seabird guano on intertidal algae also can propagate up the web (25, 194): Invertebrate consumers grow faster and larger, reproduce more, and increase in density. The density of birds eating these invertebrates is 2–3.8 times greater than in unaffected areas (194). Conversely, nearshore productivity can increase secondary productivity of adjacent marine waters (98, 119).

In freshwater systems, large populations of herbivores/detritivores are fueled by detrital input from land and by aquatic plants using nutrients ultimately derived from terrestrial systems (40, 57, 147). Areas receiving P and N in runoff develop productive, "nutrient subsidized" phytobenthic assemblages that then support a rich zoobenthic community (179). Anthropogenic nutrients (e.g. from sewage, fertilizer, phosphate detergents) increase productivity, cause eutrophication, and significantly alter lake and estuarine communities (34, 44). A striking "dead zone" in the Gulf of Mexico is linked to fertilizer input from North America's agricultural regions (87).

Nutrient subsidies likewise affect terrestrial systems. Fixed N from atmospheric pollution arrives in sufficiently large quantity via rain to change plant species composition and productivity, alter the outcome of plant competition, and disrupt entire communities worldwide (56, 84, 175). Guano, rich in N and P, enhances the quality and quantity of land plants and underlies entire food webs in coastal and insular ecosystems worldwide (19, 31, 82, 91, 102, 137; WA Anderson, GA Polis, unpublished data). In inland Antarctica, allochthonous guano provides the only nutrients to sustain lichens, which then support microorganisms and arthropods (149). These communities occur only around bird colonies.

Nutrient enrichment subsidizes plants and also indirectly influences herbivores and predators (34, 35, 139, 171). Thus, although nutrient input increases productivity, effects on system stability are unclear (44, 45, 160). The "paradox of enrichment" (148) suggests that reducing nutrient limitation can destabilize plant-herbivore interactions—plant populations become susceptible to overgrazing by indirectly subsidized herbivores (44, 148). For example, bird guano increases N concentration and growth rates of mangroves; fertilized plants lose four times more tissue to abundant herbivorous insects than do unfertilized plants (129).

Movement of Detritus

Cross-habitat flows of detrital subsidies (122, 164) often produce bottom-up effects in marine, freshwater, and terrestrial systems: Detritivores and their consumers increase throughout the web (see earlier). Diverse webs form even where local productivity is largely (or totally) absent: caves (42, 47, 79, 99), barren oceanic islands (74, 137, 138, 166, 167, 168) and deserts (154), light-limited zones of oceans and lakes (13, 174), blackwater rivers (15, 63, 64), and aeolian environments (162) such as lava flows (78, 168), mountaintops (53, 97), and polar areas (149, 164). The most biomass-rich community on earth is supported 100% by allochthonous detritus: 15–300 m mats of detrital surfgrass and kelp are converted into >1 kg/m^3 of benthic crustaceans (up to 3×10^6 individuals/m^3); large numbers of trophically distinct fish feed in these "hotspots" (174).

In other marine systems, detrital input allows species throughout the web to increase productivity and standing stock (20, 23, 32, 49, 104). Faunal biomass on beaches receiving various energy subsidies, from either upwelling or plankton blooms, is one to three orders of magnitude greater than on beaches without subsidies (12). Islands receiving detrital shore wrack often support diverse and abundant consumers (68, 137, 138). Abundant beach detritus from a successful kelp restoration project allowed several seabird species that eat kelp detritivores to recover (23). Inputs from extremely productive marshes, estuaries, and sea-grass and mangrove areas contribute substantially to secondary productivity in adjacent coastal waters; e.g. an estimated 3.5–8 metric tons/ha of mangrove detritus is exported offshore annually (123). Many stream consumers directly and indirectly rely heavily on terrestrially produced detritus as a major energy source (40, 41, 184, 185)

Numerical responses of subsidized detritivores can depress in situ resources. Intertidal grazers can occur at very high densities if they receive kelp detritus that originates sublittorally (32, 49, 105). These dense intertidal herbivores then graze noncoralline algae to low cover. Leaffall is the major energy source producing great numbers of herbivorous stream snails; snails, so subsidized, depress in situ algae (147).

Movement of Prey

Prey input allows predators to increase locally, as observed for diverse consumers in many habitats (see earlier). Top down effects occur when subsidized consumers increase densities and depress local resources. The dynamics of such donor-controlled interactions exhibit several features. First, because subsidized consumers cannot depress the renewal rate of imported prey, they are assured of a food supply that they cannot overexploit. Second, consumer success is decoupled at least partially from the constraints of local productivity and prey dynamics. Third, subsidized consumers can depress local resources below levels possible from isolated in situ consumer-resource dynamics in an interaction parallel to apparent competition (72, 77); however, instead of an alternative prey, an alternative habitat furnishes resources to consumers (or provides food for mobile consumers; see below). Thus, imported foods permit consumers to overexploit resident prey, even to extinction, without endangering the predator itself. Note that donor-controlled input decouples resource suppression from in situ productivity (73, 126, 127, 138, 139). Consequently, spatial subsidies can lead to dynamics and abundance patterns inconsistent with consumer-resource models (e.g. 8, 125) based solely on in situ productivity (135, 139). Fourth, effects are generally asymmetric: Prey in less productive habitats are affected more adversely than are those in more productive habitats.

We illustrate these dynamics. In a California vinyard, two-spotted mites move from grass to less productive grapevines. This prey input allows populations of predaceous mites on the vines to increase and suppress populations of an in situ pest, the Willamette mite, to lower densities than without spatial subsidy (59). In the Serengeti, nomadic herds track rainfall to forage in relatively productive habitats (155). Migratory prey (e.g. wildebeest) are thought to allow resident lions to increase to the point that they depress resident species (e.g. warthogs, impala; 150). Heavy poaching of mobile Cape buffalo outside the Serengeti lowered their numbers in the game park, causing lions to decrease substantially, with increases in several alternative prey species (AR Sinclair, personal communication).

Subsidized consumers can influence entire communities if they suppress key species. Communities should be more stable if subsidies allow consumers to suppress species capable of explosive reproduction (e.g. sea otter example below). Subsidized predators can increase so much that they depress herbivores, thus allowing plants to be more successful—an "apparent trophic cascade"— apparent because energy sustaining high consumer densities is not from in situ productivity (as usually modeled) but arises outside the focal habitat (138).

We cite four examples of trophic cascades subsidized by allochthonous prey. Large numbers of lumpsucker fish that migrate periodically from deeper waters are eaten by sea otters; such prey help maintain equilibrium densities of otters and allow control of sea urchins, thus releasing kelp from intense herbivory (180). Abundant coastal spiders eat many marine *Diptera* and suppress insect herbivores; plant damage is significantly less than on plants unprotected by subsidized spiders (136). Spiders along German rivers, subsidized by abundant aquatic insect prey, suppress herbivores, thus lessening damage to plants (J Henschel, personal communication). Detritivorous soil insects from gaps (herb/grass layer) in tropical forests are the major food of canopy anoles (46). Landscape considerations are important: Canopies immediately downwind from a gap support twice the flying insects and lizards as do closed canopies or those adjacent to, but upwind of, a gap. In subsidized areas, anoles depress resident herbivores, and plants show significantly less damage. Subsidized anoles also depress arboreal spiders, thus indirectly allowing small insects to increase. In each example above, prey import is donor controlled, spatially subsidized consumers exert recipient control on local prey, and such prey depression indirectly makes resources of these prey more successful.

Finally, prey movement can homogenize patches of differing productivity by linking predators with prey produced elsewhere. Prey flow from riffles to pools can be great: Drifting insects ($50–1300/100 \text{ m}^3$) subsidize resident pool predators and overwhelm predator effects on local prey populations (39). This process is a function of the number of prey moving, habitat isolation, and rates of immigration versus local depletion (41, 72). Thus, although predators may remove many prey individuals, prey input may make mask predator effects (39).

Movement of Consumers

Consumer movement produces effects generally similar to those of prey movement (but see below): Consumers usually persist at densities higher than possible in isolated habitats. We expect that cross-habitat foraging greatly affects resource dynamics at several spatial scales ranging from long-distance migrations of birds, marine and terrestrial mammals, and fish to short-distance foraging behaviors of predators among patches. Although such effects are un-doubtedly widespread, few examples document how such movement facilitates resource depression or even how consumers benefit by cross-habitat foraging. This situation exists because the process is difficult to study and the question has not been well focused theoretically.

We focus on how consumer movement affects resources directly via con-sumption, and community structure indirectly via food web effects. If mobile consumers feed only in one area (e.g. migrating gray whales, predaceous stone-flies), they exert no top-down effects in alternative habitats (although they may fall prey to resident predators, e.g. tropical birds feeding on monarchs; 33). Feeding in two or more areas (e.g. songbirds, ungulates, diadromous fishes, metamorphic insects or amphibians) can sustain consumers in less productive habitats (e.g. summer breeding vs. winter feeding grounds). Movement may even maintain consumers in a habitat too small or unproductive to sustain the population solely on in situ resources.

Mobile consumers can depress prey: e.g. spiders and insects depressed by baboons traveling from productive riparian areas to adjacent desert dunes (24), halos in seagrass beds caused by intense herbivory from fish resident in adjacent reefs (142), coastal birds eating inland invertebrates (156, 159), pathogens among plant hosts (144, 177). Agricultural changes in southern wintering grounds favorable to lesser snow geese may have caused destruction of littoral vegetation on the shores of Hudson Bay; geese, subsidized to very high densities by crops, overgraze lawns of reeds and grass to near zero cover (88).

Consumer movement may influence the stability and structure of entire com-munities if subsidized consumers suppress key species or movement facilitates trophic cascades. In two well-studied freshwater cascades, fish predators are subsidized by noncascade, allochthonous prey to population levels that can suppress local prey. Adult and juvenile bass derive much food from littoral prey; bass predation on planktivorous fish tops the cascade in the pelagic zone of Wisconsin lakes (34, 152). Steelhead grow most in the ocean and migrate to California rivers, where they initiate strong cascades if conditions are suitable (141).

CASE STUDY: ISLANDS AND THE OCEAN-LAND INTERFACE

We have shown how single inputs to specific trophic positions directly and indirectly influence dynamics. In nature, however, a variety of inputs are used by many species within a community. The significance of such inputs on entire communities is well studied in two systems. Information on fresh-water streams and lakes is presented throughout the paper. Here we describe the other system—island and coastal habitats affected greatly by multiple input from the sea. We suggest that the processes we describe occur worldwide along the ocean-land interface. This "coastal ecotone" forms a major ecosystem that occupies about 8% of the earth's surface along 594,000 km of coastline (137).

We describe systems that, without marine input, would be fairly simple; in reality, myriad allochthonous inputs create a complex system. Primary productivity on Gulf of California islands off Baja California is low, yet material from the very productive ocean supports high densities of many consumers (135–138). Two features allow smaller islands in the Gulf (and elsewhere) to receive more marine subsidies than large islands and the adjacent mainland. First, seabirds frequently nest and rest on small islands (their predators are usually absent). Second, small islands exhibit a greater P/A ratio and are thus influenced relatively more by marine detritus along the shore.

Seabirds are one conduit by which nutrients, detritus, and prey enter islands. Seabirds deposit large quantities of N- and P-rich guano, which indirectly affects consumers (31, 82, 114, 137, 156, 158, 191). When coupled with adequate precipitation, these nutrients make terrestrial primary productivity on Gulf islands 13.6 times greater than on islands unaffected by seabirds. Plant quality also increases; plant tissue has three to four times more N and P (WA Anderson, GA Polis, unpublished data). The quantity and quality of plant detritus is likewise higher, thus indirectly allowing larger herbivore and detritivore populations (134). Seabirds also directly facilitate large populations of ectoparasitic and scavenger arthropods that eat bird tissue (48, 137). Overall, insects are 2.8 times more abundant on islands with seabirds (137). These prey stimulate large populations of higher-level consumers, e.g. on average, spiders are 4.1 times and lizards, 4.9 times more abundant on Gulf islands with seabirds. Ants, when present, appear to limit tick populations on islands worldwide; this cascades to produce more successful seabird breeding (48).

Algal wrack and carrion deposited on the shore form the second conduit. About 28 kg/year enter each meter of shore on Gulf islands (much lower than in many areas where seagrass and kelp contribute 1000 $\longrightarrow$ 2000 kg/year/m; 137). The ratio of biomass from marine input (MI) to terrestrial productivity (TP) by plants is 0.5–22 on most islands; overall, 42 of 68 Gulf islands are predicted to have more MI than TP; five others have MI/TP ratios of 0.5 to 1.0. Many islands worldwide receive more energy from the sea than from land plants (30, 68, 69, 96, 137).

Marine input supports abundant detritivore and scavenger populations on the coast. Some of these consumers fall prey to local and mobile terrestrial predators. In the Baja system, insects, spiders, scorpions, lizards, rodents, and coyotes are 3–24 times more abundant on the coast and small islands compared to inland areas and large islands (136, 137, 138, 146). (Coastal carnivores worldwide are often dense; 146.) In Baja, coastal spiders are six times more abundant than inland spiders; ^{13}C and ^{15}N stable isotope analyses confirm that their diet is significantly more marine based than is that of inland counterparts (5). Such analyses show that marine matter contributes significantly to the diet of many coastal taxa worldwide (5). On the Baja mainland, coastal coyotes eat ≈50% mammals and ≈50% marine prey and carcasses (146). Here, coastal rodent populations are significantly less dense than on islands lacking coyotes, suggesting that marine-subsidized coyotes depress local rodent populations.

Complex webs based on multiple allochthonous inputs are well studied in two other systems. On Marion Island (31, 156, 158, 191), manuring by penguins, seabirds, and seals significantly influences terrestrial processes. Guano and other material are deposited at 0.4 tons/ha/year, contributing 87% of all N to terrestrial plants; almost 1 ton of carcasses/km^2/year are also deposited. Most carcasses are eaten by predatory and scavenger birds and mammals. Feral cats eat great numbers of seabirds: 2100 cats ate 400,000 petrels (35.4 tons; 1.7 kg/ha/year/cat; 30, 190).

On the Mercury Islands off New Zealand, guano from dense seabird poulations adds K, N, and P (11, 43, 169). Enriched soils support luxuriant plant growth; abundant detritus is used by a trophically diverse and dense fauna. Three groups that eat detritivores and/or seabirds occur at very high densities and biomass: centipedes, lizards (10/m^2 supralittorally), and tuatara (densities as high as 2000/ha, with an immense population biomass, as individuals average 450 g). Tuatara eat many seabirds and terrestrially based prey indirectly supported by seabird nutrients (43).

In these systems, it is impossible to explain consumer dynamics solely by local productivity, and it is short-sighted to focus on one conduit of energy flow. The relative importance of each conduit varies due to the structure of the land-water interface (e.g. P/A ratios), temporal variability in climate (e.g. rain-stimulated plant growth decreases MI/TP on Gulf islands, 137, 138), and changes in marine productivity (e.g. depressed productivity due to El Niño events reduces kelp production, seabird populations, and marine input to islands; 14, 137, 138).

DIRECT AND INDIRECT EFFECTS OF SPATIAL SUBSIDIES: THEORY

Spatial subsidies, in theory, influence all aspects of food web structure and dynamics (135). Theory of consumer-resource dynamics in spatially heterogeneous landscapes suggests key effects expected in natural systems. Some predictions match empirical patterns; others need assessment. We focus on how spatial subsidies affect stability and abundance in stable systems. In unstable systems, effects of flows can be counterintuitive and are poorly explored.

For simplicity, consider a landscape where focal habitat A is coupled to a much larger or more productive habitat, with little or no reciprocal impact of A on the larger habitat. The influence of the larger habitat on A is represented by splicing input and emigration terms into standard predator-prey and food web models, such as this model for a food-limited predator eating a local prey: $dN/dt = NF(N) - aPN + I_N - e_N N$, $dP/dt = P[g(aN) - m] + I_P - e_P P$, where N and P are prey and predator abundance in habitat A; $F(N)$, local prey growth rate; aN, predator functional response (a is the per-predator attack rate, per prey); g, local predator birth rate (an increasing function of aN); m, predator mortality. The input terms I_N, I_P are spatial subsidies for prey and predator (in simple cases, constant input rates); e_P and e_N scale losses to the larger habitat (e.g. emigration, wash-out). Without immigration, predator persistence requires local prey abundance to exceed a threshold. In stable, isolated systems, increasing prey production sustains more predators, with large-amplitude oscillations possible at high productivities (195). This model is useful to examine a range of scenarios, e.g. direct density-dependence, or additional prey species.

Prey Flow, Specialist Predator ($I_P = 0$)

Prey input, mimicking enhanced prey productivity (e.g. Schoener's [e.g. 153] models of competition in donor-controlled systems) can enhance predator numbers, but with little effect on equilibrial prey numbers. By contrast, prey emigrating in response to predator abundance (i.e. $e = e'P$) can strongly reduce prey abundance and net productivity, indirectly depressing predator abundance. The openness of prey dynamics puts a floor of I_N/e on prey numbers and is analogous to incorporating a fixed-number refuge (74). Generally, refuges tend to stabilize unstable predator-prey interactions (e.g. 2, 75, 151). Even a low-productivity refuge with little effect on mean abundance can exert strong effects by damping the destabilizing impact of prey fluctuations in productive habitats (2, 151). This partially donor-controlled interaction involving allochthonous input makes less likely the classic "paradox of enrichment" (148).

Flow of Alternative Prey, Generalist Predator

To model input of a second prey not recruited locally, we add an equation for the local dynamics of this second prey and express predator growth as a function of the abundance of both prey. This model is one of apparent competition between alternative prey of a generalist, food-limited predator (71, 72, 74, 77). The interaction between the second prey and predator is donor controlled. This allochthonous prey indirectly increases predation pressure on the local prey.

If input or quality of the second prey is sufficiently great, overconsumption can drive the local prey extinct. We suggest this model describes many of our empirical cases and illuminates how allochthonous input often depresses in situ prey.

The effects of consumer movement differ from those of nutrients, detritus, or prey: Consumers using one habitat can affect resource dynamics in another upon their return. General models suggest that predator dispersal in heterogeneous environments can stabilize otherwise unstable predator-prey systems (72).

Predator Flow, Resident Prey ($I_N = 0$)

The rules a predator uses to select among habitats can greatly influence local dynamics. With passive immigration, as in the above model, the predator persists (at $P* = I_P/e_P$), even without local prey, and so can depress unproductive local prey populations. Predator immigration likewise tends to depress local prey; if r is prey intrinsic growth rate, when $I/e > r/a$, prey are eliminated. Sufficiently high predator flows destabilize at the community level, as resident prey in low-productivity environments can be eliminated; lower predator flows tend to stabilize, with depressed prey numbers. If the resident prey shows inverse density-dependence at low abundance, the system can also exhibit alternative stable states, with or without the prey (76).

Alternatively, predators may use optimal foraging rules to move among habitats. If consumers exhibit ideal free habitat selection, at equilibrium abundances of both, consumers and resources are those expected from local dynamics (72, 128). This prediction sometimes holds (armored catfish-algae; 128). However, the multiple examples of resource depression caused by predator subsidies (see above) suggest this is not the norm in nature.

Landscape Variables Influencing Immigration and Input Rates (I_X)

Our models assume constant rates for I_N and I_P. However, input rates may vary both spatially and temporally. Immigration or input rates are undoubtedly a function of landscape variables such as perimeter and permeability of focal habitat A, and the distance between trophically connected habitats A and B. The probability that habitat A will intercept a subsidy or consumer moving from B is related directly to the perimeter (p) of A and inversely to the distance (d) between A and B. Furthermore, the probability that a moving entity will enter A once A is intercepted is a function of the boundary permeability (M) of A. Thus, $I_X = (p_A M_A/d_{AB})$ and is variable in time and among habitats. Such variability in input rates, although not incorporated into our models, likely exerts substantial impact on the dynamics in the recipient habitat.

OVERVIEW

Ecologists are now aware that dynamics are rarely confined within a focal area and that factors outside a system may substantially affect (and even dominate) local patterns and dynamics. Local populations are linked closely with other populations through such spatially mediated interactions as source-sink and metapopulation dynamics, supply-side ecology, and source pool-dispersal effects (75, 135). The identification of landscape ecology as a specific discipline is recognition of multihabitat dynamics (50, 66, 170, 195). Here we dramatize the need to integrate landscape and food web ecology. This requires consideration of issues not in this paper: e.g. spatial scaling (186), landscape influences on food web assembly (75), and reciprocal effects of web dynamics on landscapes (e.g. large herbivores creating patchiness). However, the themes explored here will be central to an integrated discipline of landscape and food web ecology.

Our synthesis suggests several general principles: the movement of nutrients, detritus, prey, and consumers among habitats is ubiquitous in diverse biomes and is often a central feature of population, consumer-resource, food web and community dynamics. Bottom-up effects that increase secondary productivity are initiated frequently by inputs of nutrient to producers in the herbivore channel or detritus to decomposers in the saprovore channel. Top-down effects occur when spatially subsidized consumers affect in situ resources. These effects then can propagate indirectly throughout the entire web to affect species abundance and stability properties, often in complex ways (120). The strength of these effects depends on the flow rates of resources and consumers, each a function of landscape variables. A natural avenue of future work is to examine the relative impact of input when added to a landscape perspective and particular food web configurations.

One strong insight for applied ecology is that the dynamics of seemingly distinct systems are intimately linked by spatial flow of matter and organisms. Land management of local areas (e.g. agricultural and forestry practices, fragmentation, desertification) affects not only other terrestrial habitats (e.g. 6, 84, 88, 92, 163, 175, 188) but the productivity, food webs, and community structure of streams, rivers, lakes, estuaries, and oceans (67, 87, 165, 181, 183). Conversely, processes and policies in aquatic systems (e.g. eutrophication, fisheries) affect both aquatic (51, 165, 172) and terrestrial (23, 105, 137, 191) systems. The message is clear: Ecosystems are closely bound to one another, be they stream and lake, pelagic and intertidal zones, farms and the sea, forest and river, or ocean and desert.

We end by noting that "tropho-spatial" phenomena (movement of nutrients, food, and consumers; subsidized consumers; resource suppression in low-productivity habitats; altered stability properties) exert their influence at all scales throughout ecology. Although most of our examples used distinct habitats, such dynamics can occur "sympatrically" among microhabitat patches (e.g. 46, 59) or at immense distances (e.g. 92, 163, 176). An integration of landscape perspectives with consumer-resource and food web interactions will enrich models, complement our understanding of the dynamics of populations and communities, help design better protocols for biological control of pest species, and improve techniques for the protection of critical habitats and endangered species.

ACKNOWLEDGMENTS

We thank the many people who alerted us to relevant and new studies. This work was supported by grants from the National Science Foundation to GAP and RDH, and the Natural Science Committee and University Research Council of Vanderbilt University.

> Visit the *Annual Reviews home page* at
> http://www.annurev.org.

Literature Cited

1. Abrams PA. 1993. Effects of increased productivity on the abundance of trophic levels. *Am. Nat.* 141:351–71
2. Abrams PA, Walters CJ. 1996. Invulnerable prey and the paradox of enrichment. *Ecology* 77:1125–33
3. Ainley DG, Demaster DP. 1990. The upper trophic levels in polar ecosystems. In *Polar Oceanography,* ed. WO Smith, pp. 599–630. Orlando, FL: Academic
4. Ambuell B, Temple SA. 1983. Area-dependent changes in the bird communities and vegetation of southern Wisconsin forests. *Ecology* 64:1057–68
5. Anderson WA, Polis GA. 1997. Evidence from stable carbon and nitrogen isotopes showing marine subsidies of island communities in the Gulf of California. *Oikos.* In press
6. Andren H, Anglestam P. 1988. Elevated predation rates as an edge effect in habitat islands: experimental evidence. *Ecology* 69:544–47
7. Angel MV. 1984. Detrital organic fluxes through pelagic ecosystems. In *Flows of Energy and Materials in Marine Ecosystems,* ed. MJR Fasham, pp. 475–516. New York: Plenum
8. Arditi R, Ginzburg LR. 1989. Coupling in predator-prey dynamics: ratio dependence. *J. Theor. Biol.* 139:311–26

9. Art HW, Bormann FH, Voigt G, Woodwell G. 1974. Barrier island forest ecosystem: role of meterologic nutrient inputs. *Science* 184:60–62
10. Ashmole NP, Nelson JM, Shaw MR, Garside MR. 1983. Insects and spiders on snow fields in the Cairngorms, Scotland. *J. Nat. Hist.* 17:599–613
11. Atkinson IAE. 1964. The flora, vegetation and soils of Middle and Green Islands Group. *NZ J. Bot.* 6:385–402
12. Bally R. 1987. The ecology of sandy beaches of the Benguela ecosystem. *S. Afr. J. Mar. Sci.* 5:759–70
13. Barnes RSK, Hughes RN. 1988. *An Introduction to Marine Ecology.* Oxford, UK: Blackwell Sci.
14. Barry JP, Dayton PK. 1991. Physical heterogeneity and the organization of marine communities. In *Ecological Heterogenity,* ed. J Kolasa, ST Pickett, pp. 270–320. New York: Springer Verlag
15. Benke AC, Wallace JB. 1997. Trophic basis of production among riverine caddisflies: implications for food web analysis. *Ecology.* 78:1132–45
15a. Bilby RE, Fransen BR, Bisson PA. 1996. Incorporation of nitrogen and carbon from spawning coho salmon into the trophic system of small streams: evi-

dence from stable isotopes. *Can. J. Fish. Aquat. Sci.* 53:164–73

16. Boero F, Belmonte G, Fanelli G, Piraino S, and Rubino F. 1996. The continuity of living matter and the discontinuities of its constituents: Do plankton and benthos really exist? *Trends Ecol. Evol.* 11:177–80

17. Bogitsh, BJ, Cheng TC. 1990. *Human Parasitology.* Philadelphia, PA: Saunders Coll.

18. Boling RH, Goodman E, VanSickle J, Zimmer J, Cummins K, et al. 1975. Towards a model of detritus processing in a woodland stream. *Ecology* 56:141–51

19. Bosman AL, Hockey PAR. 1986. Seabird guano as a determinant of rocky intertidal community structure. *Mar. Ecol. Prog. Ser.* 32:247–57

20. Bosman AL, Hockey PAR, Seigfried WR. 1987. The influence of coastal upwelling on the functional structure of rocky intertidal communities. *Oecologia* 72:226–32

21. Bowden J, Johnson CG. 1976. Migrating and other terrestrial insects at sea. In *Marine Insects,* ed. L Cheng, pp. 99–117. Amsterdam: North Holland

22. Braband Å, Faafeng A, Nilssen JPM. 1990. *Can. J. Fish. Aquat. Sci.* 47:364–72

23. Bradley RA, Bradley DW. 1993. Wintering shorebirds increase after kelp (*Macrocystis*) recovery. *Condor* 95:372–76

24. Brain C. 1990. Spatial usage of desert environment by baboons. *J. Arid Environ.* 18:67–73

25. Branch GM, Barkai A, Hockey PA, Hutchings L. 1987. Biological interactions: causes or effects of variability in the Benguela ecosystem. *S. Afr. J. Mar. Sci.* 5:425–46

26. Bray RN, Miller AC, Geesey GG. 1981. The fish connection: a trophic link between planktonic and rocky reef communities. *Science* 214:204–05

27. Brinkhurst RO, Walsh B. 1967. Rostherne Mere, England, a further instance of guanotrophy. *J. Fish. Res. Board Can.* 24:1299–309

28. Brown VK, Gange AC. 1990. Insect herbivory below ground. *Adv. Ecol. Res.* 20:1–59

29. Burger AE. 1982. Foraging behaviours of Lesser Sheathbills Chionis minor exploiting invertebrates on a sub-antarctic island. *Oecologia* 52:236–45

30. Burger AE. 1985. Terrestrial food webs in the sub-Antarctic: island effects. In *Antarctic Nutrient Cycles and Food Webs,* ed. WR Siegfried, PR Condy, RM Laws, pp. 582–91. New York: Springer Verlag

31. Burger AE, Lindebloom HJ, Williams AJ. 1978. The mineral and energy contributions of guano of selected species of birds to the Marion Island terrestrial ecosystem. *S. Afr. J. Antarctic Res.* 8:59–70

32. Bustamante RH, Branch GM, Eekhout S. 1995. Maintenance of an exceptional intertidal grazer biomass in South Africa: subsidy by subtidal kelps. *Ecology* 76:2314–29

33. Calvert WH, Hedrick LE, Brower LP. 1979. Mortality of the monarch butterfly (*Danaus plexippus* L.): avian predation at five overwintering sites in Mexico. *Science* 204:847–51

34. Carpenter SR, Kitchell JF. 1993. *The Trophic Cascade in Lakes.* Cambridge, UK: Cambridge Univ. Press

35. Carpenter SR, Kraft CE, Wright R, He X, Soranno PA, Hodgson JR. 1992. Resilience and resistence of a lake phosphorus cycle before and after a food web manipulation. *Am. Nat.* 140:781–98

36. Charnov EL, Orians GH, Hyatt K. 1976. Ecological implications of resource depression. *Am. Nat.* 110:247–59

37. Cheng L, Birch MC. 1977. Terrestrial insects at sea. *J. Mar. Biol. Assoc. UK* 57:995–97

38. Cheng L, Birch MC. 1978. Insect flotsam: an unstudied marine resource. *Ecol. Entomol.* 3:87–97

39. Cooper SD, Walde SJ, Peckarsky BL. 1990. Prey exchange rates and the impact of predators on prey populations. *Ecology* 71:1503–14

40. Covich AP. 1988. Geographical and historical comparisons of neotropical streams: biotic diversity and detrital processing in highly variable habitats. *J. N. Am. Benth. Soc.* 7:361–86

41. Covich AP, Crowm T, Johnson S, Varza D, Certain D. 1991. Post-hurricane Hugo increases in atyidshrimp abundances in a Puerto Rican montane stream. *Biotropica* 23:448–54

42. Culver D. 1982. *Cave Life.* Cambridge, MA: Harvard Univ. Press

43. Daugherty CH, Towns DR, Atkinson IAE, Gibbs GW. 1990. The significance of the biological resources of New Zealand islands for ecological restoration. In *Ecological Restoration of the New Zealand Islands,* ed. DR Towns, CH Daugherty, IAE Atchinson, pp. 9–21. *Conserv. Sci. Publ., No. 2*

44. DeAngelis DL. 1992. *Dynamics of Nutrient Cycling and Food Webs.* London: Chapman & Hall

45. DeAngelis DL, Loreau M, Neergaard D, Mulholland PJ, Marzolf ER. 1995. Modelling nutrient-periphyton dynamics in streams: the importance of transient storage zones. *Ecol. Model* 80:149–60

46. Dial R. 1992. *A food web for a tropical rain forest: the canopy view from Anolis.* PhD diss. Stanford, CA: Stanford Univ.

47. Dourojeanni M, Tovar A. 1974. Notas sobre el ecosistema y la conservacion de la cueva de las Lechuzas (Pargue Nacional de Tingo Maria, Peru). *Rev. For. Peru* 5:28–45

48. Duffy D. 1991. Ants, ticks, and nesting seabirds: dynamic interactions. In *Bird-Parasite Interactions,* ed. J Loye, M Zuk, pp. 242–57. New York: Oxford Univ. Press

49. Duggins DO, Simenstad CA, Estes JA. 1989. Magnification of secondary production by kelp detritus in coastal marine ecosystems. *Science* 245:170–73

50. Dunning JB, Danielson BJ, Pulliam HR. 1992. Ecological processes that affect populations in complex landscapes. *Oikos* 65:169–75

51. Durbin AS, Nixon SW, Oviatt CA. 1979. Effects of the spawning migration of the alewife, *Alosa pseudoharengus,* on freshwater ecosystems. *Ecology* 60:8–17

52. Edwards JS. 1986. Derelicts of dispersal: arthropod fallout on Pacific Northwest volcanoes. In *Insect Flight, Dispersal and Migration*, ed. W Danthanaryana, pp. 186–203. Berlin: Springer-Verlag

53. Edwards JS. 1987. Arthropods of alpine aeolian ecosystems. *Annu. Rev. Entomol.* 32:163–79

54. Edwards JS, Crawford RL, Sugg PM, Peterson M. 1986. Arthropod colonization in the blast zone of Mt. St. Helens: five years of progress. In *Mt. St. Helens; Five Years Later*, ed. S Keller. Spokane, WA: E. Wash. Univ. Press

55. Edwards RT, Meyer JL. 1987. Metabolism of a sub-tropical low gradient blackwater river. *Freshwater Biol.* 17:251–63

56. Field CB, Chapin FS III, Matson PA, Mooney HA. 1992. Responses of terrestrial ecosystems to the changing atmosphere: a resource-based approach. *Annu. Rev. Ecol. Syst.* 23:201–35

57. Fisher SG, Likens GE. 1972. Stream ecosystem: organic energy processes. *BioScience* 22:33–37

58. Fisher SG, Likens GE. 1973. Energy flow in Bear Brook, New Hampshire: an integrative approach to stream ecosystem metabolism. *Ecol. Monogr.* 43:421–39

59. Flaherty DL. 1969. Ecosystem trophic complexity with Willamette mite, *Eotetranychus willamettei* Ewing (Acarina: Tetranychidae), densities. *Ecology* 50:911–15

60. France R. 1995. Stable nitrogen isotopes in fish: literature synthesis on the influence of ecotonal coupling. *Estuar. Coast. Shelf Sci.* 41:737–42

61. France RL. 1996. Benthic-pelagic uncoupling of carbon flow. *Trends Ecol. Evol.* 11:471

62. Gasith A, Hasler AD. 1976. Airborne litterfall as a source of organic matter in lakes. *Limnol. Oceanogr.* 21:253–58

63. Goulding M. 1980. *The Fishes and the Forest.* Berkeley, CA: Univ. Calif. Press

64. Goulding M, Carvalho ML, Ferreira EG. 1988. *Rio Negro: Rich Life in Poor Water.* The Hague, The Netherlands: SPB Academic

65. Gross MR, Coleman RM, McDowall RM. 1988. Aquatic productivity and the evolution of diadromous fish migration. *Science* 239:1291–93

66. Hansen AJ, di Castri F, eds. 1992. *Landscape Boundaries: Consequences for Biotic Diversity and Ecological Flows.* New York: Springer Verlag

67. Hasler AD, ed. 1975. *Coupling of Land and Water Systems.* New York: Springer-Verlag

68. Heatwole H. 1971. Marine-dependent terrestrial biotic communities on some cays in the coral sea. *Ecology* 52:363–66

69. Heatwole H, Levins R, Byer M. 1981. Biogeography of the Puerto Rican Bank. *Atoll Res. Bull.* 251:1–62

70. Henschel J, Stumpf H, Mahsberg D. 1996. Increase in arachnid abundance and biomass at water shores. *Rev. Suisse Zool* Vol. hors serie:269–78

71. Holt RD. 1977. Predation, apparent competition, and the structure of prey communities. *Theor. Popul. Biol.* 12:197–229

72. Holt RD. 1984. Spatial heterogeneity, indirect interactions, and the coexistence of prey species. *Am. Nat.* 124:377–406

73. Holt RD. 1985. Population dynamics of two patch environments: some anomalous consequences of an optimal habitat

distribution. *Theor. Popul. Biol.* 28:181–208

74. Holt RD. 1987. Prey communities in patchy environments. *Oikos* 50:276–90

75. Holt RD. 1993. Ecology at the mesoscale: the influence of regional processes on local communities. In *Community Diversity: Historical and Biogeographical Perspective,* ed. R Ricklefs, D Schulter, pp. 77–88. Chicago, IL: Univ. Chicago Press

76. Holt RD. 1996. Food webs in space: an island biogeographic perspective. See Ref. 139a, pp. 313–23

77. Holt RD, Lawton JH. 1994. The ecological consequences of shared natural enemies. *Annu. Rev. Ecol. Syst.* 25:495–520

78. Howarth FG. 1979. Neogeoaeolian habitats on new lava flows on Hawaii island: an ecosystem supported by windborne debris. *Pac. Insects* 20:133–44

79. Howarth FG. 1983. Ecology of cave arthropods. *Annu. Rev. Entomol.* 28:365–89

80. Hunt RL. 1975. Use of terrestrial invertebrates as food by salmonids. See Ref. 67, pp. 137–52

81. Hureau JC. 1985. Interactions between antarctic and sub-antarctic marine, freshwater and terrestrial organisms. In *Antarctic Nutrient Cycles and Food Webs,* ed. WR Siegfried, PR Condy, RM Laws, pp. 626–29. New York: Springer-Verlag

82. Hutchinson GE. 1950. Survey of existing knowledge of biogeochemistry: 3. The biogeochemistry of vertebrate excretion. *Bull. Am. Mus. Nat. Hist.* 96:554p

83. Jackson JK, Fisher SG. 1986. Secondary production, emergence and export of aquatic insects of a Sonoran Desert stream. *Ecology* 67:629–38

84. Jefferies RL, Maron JL. 1997. The embarrassment of riches: atmospheric deposition of nitrogen and community and ecosystem processes. *Trends Ecol. Evol.* 12:74–78

85. Johnston CA, Naiman RJ. 1987. Boundary dynamics at the aquatic-terrestrial interface: the influence of beaver and geomorphology. *Landsc. Ecol.* 1:47–57

86. Jordan CF. 1985. *Nutrient Cycling in Tropical Forest Ecosystems: Principles and Their Application in Management and Conservation.* New York: Wiley & Sons

87. Justic D, Rabalais NN, Turner RE, Wiseman WJ Jr. 1993. Seasonal coupling between riverborne nutrients, net productivity, and hypoxia. *Mar. Pollut. Bull.* 26:184–89

88. Kerbes RH, Kotanen PM, Jefferies RL. 1990. Destruction of wetland habitats by lesser snow geese: a keystone species on the west coast of Hudson Bay. *J. Anim. Ecol.* 27:242–58

89. Kitchell JF, O'Neill RV, Webb D, Gallepp GW, Bartell SM, et al. 1979. Consumer regulation of nutrient cycling. *BioScience* 29:28–34

90. Koslow JA. 1997. Seamounts and the ecology of deep-sea fisheries. *Am. Sci.* 85:168–76

91. Leevantaar P. 1967. Observations in guanotrophic environments. *Hydrobiologia* 29:441–89

92. Levey DJ, Stiles FG. 1992. Evolutionary precursors of long-distance migration: resource availability and movement patterns in neotropical landbirds. *Am. Nat.* 140:447–76

93. Likens GE. 1984. Beyond the shoreline: a watershed-ecosystem approach. *Int. Ver. Theor. Angew. Limnol., Ver.* 22:1–22

94. Likens GE, Bormann FH. 1975. An experimental approach in New England landscapes. See Ref. 67, pp. 7–29

95. Lodge DM, Barko JW, Strayer D, Melack JM, Mittlebach GG, et al. 1988. Spatial heterogeneity and habitat interactions in lake communities. In *Complex Interactions in Lake Communities,* ed. SR Carpenter, pp. 181–208. New York: Springer Verlag

96. Lord WD, Burger JF. 1984. Arthropods associated with Herring Gull (*Larus argentatus*) and Great Black-backed Gull (*Larus marinus*) carrion on islands in the Gulf of Maine. *Environ. Entomol.* 13:1261–68

97. Mani MS. 1968. *Ecology and Biogeography of High Altitude Insects.* The Hague, The Netherlands: Junk

98. Mann KH, Lazier JRN. 1991. *Dynamics of Marine Ecosystems.* London: Blackwells

99. Martin JL, Oromi P. 1986. An ecological study of Cueva de los Roques lava tube (Tenerife, Canary Islands). *J. Nat. Hist.* 20:375–88

100. Mason CF, MacDonald SM. 1982. The input of terrestrial invertebrates from tree canopies to a stream. *Freshw. Biol.* 12:305–11

101. Mattson WJ. 1980. Herbivory in relation to plant nitrogen content. *Annu. Rev. Ecol. Syst.* 11:119–61

102. McColl JG, Burger J. 1976. Chemical input by a colony of Franklin Gulls nesting in cattails. *Am. Mid. Nat.* 96:270–80

103. McNaughton SJ. 1985. Ecology of a grazing ecosytem: the Serengeti. *Ecol. Monogr.* 55:259–94

104. Menge B. 1992. Community regulation: Under what conditions are bottom-up factors important on rocky shores? *Ecology* 73:755–65

105. Menge BA. 1995. Joint 'bottom-up' and 'top-down' regulation of rocky intertidal algal beds in South Africa. *Trends Ecol. Evol.* 10:431–32

106. Meyer JL. 1990. A blackwater perspective on riverine ecosystems. *Bioscience* 40:643–51

107. Meyer JL, Schultz ET. 1985. Migrating haemulid fishes as a source of nutrients and organic matter on coral reefs. *Limnol. Oceanogr.* 30:146–56

108. Meyer JL, Schultz ET, Helfman GS. 1983. Fish schools: an asset to corals. *Science* 220:1047–49

109. Meyer JL, Tate CM. 1983. The effects of watershed disturbance on dissolved organic carbon dynamics of a stream. *Ecology* 64:33–44

110. Minshall GW. 1967. Role of allochthonous detritus in the trophic structure of a woodland stream. *Ecology* 48:139–49

111. Mittlebach GG, Osenberg CW. 1993. Stage structured interactions in bluegill: consequences of adult resource variation. *Ecology* 74:2381–94

112. Moran VC, Southwood TRE. 1982. The guild composition of arthropod communities in trees. *J. Anim. Ecol.* 51:289–306

113. Mulholland PJ. Organic carbon flow in a swamp-stream ecosystem. *Ecol. Monogr.* 51:307–22

114. Murphy GI. 1981. Guano and the anchovetta fishery. *Res. Man. Environ. Uncert.* 11:81–106

115. Naiman RJ, Melillo JM. 1984. Nitrogen budget of a subarctic stream altered by beaver. *Oecologia* 62:150–55

116. Naiman RJ, Melillo JM, Hobbie JE. 1986. Ecosystem alteration of boreal forest streams by beaver. *Ecology* 67:1254–69

117. Naiman RJ, Melillo JM, Lock MA, Ford TE, Reice SR. 1987. Longitudinal patterns of ecosystem processes and community structure in a subarctic river continuum. *Ecology* 68:1139–56

118. Nelson JM. 1965. A seasonal study of aerial insects close to a moorland stream. *J. Anim. Ecol.* 34:573–79

119. Newell RC. 1984. The biological role of detritus in the marine environment. In

120. Nisbet RM, Briggs CJ, Gurney WSC, Murdoch WW, Stewart-Oaten A. 1993. Two-patch metapopulation dynamics. In *Patch Dynamics*, ed. SA Levin, TM Powell, JH Steele, pp. 125–35. Berlin: Springer-Verlag

121. Norlin A. 1967. Terrestrial insects on lake surfaces, their availability and importance as fish food. *Rep. Inst. Freshwater Res. Drottningholm* 47:39–55

122. Odum EP. 1971. *Fundamentals in Ecology.* Philadelphia, PA: Saunders. 3rd ed.

123. Odum WE, Heald EJ. 1975. Mangrove forests and aquatic productivity. See Ref. 67, pp. 129–36

124. Ogden JC, Gladfelter EA. 1983. Coral reefs, seagrass beds and mangroves: their interaction in the coastal zones of the Caribbean. *UNESCO Rep. Mar. Sci.* 23:1–130

125. Oksanen L, Fretwell S, Arruda J, Niemela P. 1981. Exploitation ecosystems in gradients of primary productivity. *Am. Nat.* 118:240–61

126. Oksanen L, Oksanen T, Ekerholm P, Moen J, Lundberg P, et al. 1996. Structure and dynamics of arctic-subarctic grazing webs in relation to primary productivity. See Ref. 139a, pp. 231–42

127. Oksanen T. 1990. Exploitative ecosystems in heterogeneous habitat complexes. *Evol. Ecol.* 4:220–34

128. Oksanen T, Power ME, Oksanen L. 1995. Ideal free habitat selection and consumer-resource dynamics. *Am. Nat.* 146:565–85

129. Onuf CP, Teal JM, Valiela I. 1977. Interactions of nutrients, plant growth and herbivory in a mangrove ecosystem. *Ecology* 58:514–26

130. Osenburg CW, Mittelbach GG, Wainwright PC. 1992. Two-stage life histories in fish: the interaction between juvenile competition and adult performance. *Ecology* 73:255–67

131. Persson L, Bengtsson J, Menge BA, Power ME. 1996. Productivity and consumer regulation: concepts, patterns,, and mechanisms. See Ref. 139a, pp. 396–434

132. Pieczynska E. 1975. Ecological interactions between land and the littoral zones of lakes. See Ref. 67, pp. 263–76

133. Deleted in proof

134. Piñero FS, Polis GA. 1997. Marine donor controlled dynamics on islands in the Gulf of California: subsidy of de-

tritivore tenebrionid beetles by seabirds. *Ecology.* Submitted

135. Polis GA, Holt RD, Menge BA, Winemiller K. 1996. Time, space and life history: influences on food webs. See Ref. 139a, pp. 435–60

136. Polis GA, Hurd SD. 1995. Extraordinarily high spider densities on islands: flow of energy from the marine to terrestrial food webs and the absence of predation. *Proc. Natl. Acad. Sci. USA* 92:4382–86

137. Polis GA, Hurd SD. 1996. Linking marine and terrestrial food webs: allochthonous input from the ocean supports high secondary productivity on small islands and coastal land communities. *Am. Nat.* 147:396–423

138. Polis GA, Hurd SD. 1996. Allochthonous input across habitats, subsidized consumers, and apparent trophic cascades: examples from the ocean-land interface. See Ref. 139a, pp. 275–85

139. Polis GA, Strong D. 1996. Food web complexity and community dynamics. *Am. Nat.* 147:813–46

139a. GA Polis, KO Winemiller, eds. 1996. *Food Webs: Integration of Patterns and Dynamics.* London: Chapman & Hall

140. Pomeroy LR. 1979. Secondary production mechanisms of continental shelf communities. In *Ecological Processes in Coastal and Marine Systems,* ed. RJ Livingston, pp. 163–86. New York: Plenum

141. Power ME. 1990. Effects of fish in river food webs. *Science* 250:811–14

142. Randall JE. 1965. Grazing effect on sea grasses by herbivorous reef fishes in the West Indies. *Ecology* 46:255–60

143. Robbins CS. 1980. Effect of forest fragmentation on breeding bird populations in the piedmont of the Mid-Atlantic region. *Am. Nat.* 33:31–36

144. Roberts DA, Boothroyd CW. 1972. *Fundamentals of Plant Pathology.* San Francisco: WH Freeman

145. Roger C, Grandperrin R. 1976. Pelagic food webs in the tropical Pacific. *Limnol. Oceanogr.* 21:731–35

146. Rose M, Polis GA. 1997. The distribution and abundance of coyotes: the importance of subsidy by allochthonous foods coming from the sea. *Ecology.* In press

147. Rosemond AD, Mulholland PJ, Elwood JW. 1993. Top down and bottom up control of stream periphyton: effects of nutrients and herbivores. *Ecology* 74:1264–80

148. Rosenzweig MLK. 1971. Paradox of enrichment: destabilization of exploitation ecosystems in ecological time. *Science* 171:385–87

149. Ryan PG, Watkins BP. 1989. The influence of physical factors and ornithogenic products on plant and arthropod abundance at an island group in Antarctica. *Polar Biol.* 10:152–60

150. Schaller GB. 1972. *The Serengeti Lion.* Chicago, IL: Univ. Chicago Press

151. Scheffer M, de Boer RJ. 1995. Implications of spatial heterogeneity for the paradox of enrichment. *Ecology* 76: 2270–77

152. Schindler DE, Carpenter SR, Cottingham KL, He X, Hodgson JR, et al. 1996. Food-web structure and littoral zone coupling to pelagic trtophic cascades. See Ref. 139a, pp. 96–105

153. Schoener TW. 1976. Alternatives to Lotka-Volterra competition: models of intermediate complexity. *Theor. Popul. Biol.* 10:309–33

154. Seely MK. 1991. Sand dune communities. In *The Ecology of Desert Communities,* ed. GA Polis, pp. 348–82. Tucson, AZ: Univ. Ariz. Press

155. Senft RL, Coughenour MB, Bailey DW, Rittenhouse LR, et al. 1987. Large herbivore foraging and ecological hierarchies. *BioScience* 37:789–99

156. Siegfried WR. 1981. The role of birds in ecological processes affecting the functioning of the terrestrial ecosystem at sub-antarctic Marion Island. *Com. Natl. Francais Rech. Antarct.* 51:493–99

157. Smetacek V, Scharek R, Nothig EM. 1990. Seasonal and regional variation in the pelagial and its relationship to the life history cycle of krill. In *Antarctic Ecosystems: Ecological Change and Conservation,* ed. KR Kerry, G Hempel, pp. 103–14. New York: Springer Verlag

158. Smith VR. 1979. The influence on seabird manuring on the phosphorus status of Marion island (sub-Antarctic) soils. *Oecologia* 41:123–26

159. Stahl JC, Weimerskirch H. 1982. Le ségrégation écologique des deux especes de sternes des Iles Crozet. *Com. Natl. Francais Rech. Antarct.* 51:449–56

160. Stone L, Gabric A, Berman T. 1996. Ecosystem resilience, stability, and productivity: seeking a relationship. *Am. Nat.* 148:892–903

161. Stork NE. 1991. The composition of the arthropod fauna a Bornean lowland rain forest trees. *J. Trop. Ecol.* 7:161–80

162. Swan LW. 1963. Aeolian zone. *Science* 140:77–78

163. Swap R, Garstang M, Greco S, Talbot R, Kållberg P. 1992. Saharan dust in the Amazon Basin. *Tellus B* 44:133–49

164. Teeri JA, Barrett PE. 1975. Detritus transport by wind in a high arctic terrestrial system. *Arctic Alpine Res.* 7:387–91

165. Thom RM. 1981. *Primary productivity and carbon input to Grays Harbor estuary, Washington.* US Army Corps Eng., Seattle Dist., Seattle, WA

166. Thornton IWB,New TR. 1988. Krakatau invertebrates: the 1980's fauna in the context of a century of recolonization. *Philos. Trans. R. Soc. London Ser. B* 322:493–522

167. Thornton IWB, New TR, McLaren DA, Sudarman HK, Vaughan PJ. 1988. Air-borne arthropod fall-out on Anak Krakatau and a possible pre-vegetation pioneer community. *Philos. Trans. R. Soc. Lond. B* 322:471–79

168. Thornton IWB, New TR, Zann RA, Rawlinson PA. 1990. Colonization of the Krakatau islands by animals: a perspective from the 1980s. *Philos. Trans. R. Soc. Lond. B* 328:131–65

169. Towns DR. 1975. Ecology of the black shore skink, *Leiolopisma suteri* (Lacertilia: Scincidae), in boulder beach habitats. *NZ J. Zool.* 2:389–408

170. Turner MG. 1989. Landscape ecology: the effect of pattern on processes. *Annu. Rev. Ecol. Syst.* 20:171–97

171. Vanni MJ. 1996. Nutrient transport and recycling by consumers in lake food webs: implications for algal communities. See Ref. 139a , pp. 81–95

172. Vannote RL, Minshall GW, Cummins KW, Sedell JR, Cushing CE. 1980. The river continuum concept. *Can. J. Fish. Aquat. Sci.* 37:130–37

173. Veit RR, Silverman ED, Everson I. 1993. Aggregation patterns of pelagic predators and their principal prey, Antarctic krill, near South Georgia. *Ecology* 62:551–64

174. Vetter E. 1994. Hotspots of benthic production. *Nature* 372:47

175. Vitousek PM. 1994. Beyond global warming: ecology and global change. *Ecology* 75:1861–76

176. Vitousek PM, Sanford RL. 1986. Nutrient cycling in moist tropical forest. *Annu. Rev. Ecol. Syst.* 17:137–67

177. Walker JC. 1969. *Plant Pathology.* New York: McGraw-Hill. 3rd ed.

178. Ward JV. 1988. Riverine-wetland interactions. In *Freshwater Wetlands and Wildlife,* ed. RR Sharitz, JW Gibbons. Oak Ridge, TN: Off. Sci. Tech. Info., US Dep. Energy

179. Ward JV. 1989. The four dimensional nature of lotic ecosystems. *J. N. Am. Benth. Soc.* 8:2–8

180. Watt J, Siniff DB, Estes JA. 1997. Diet and foraging behavior of an "equilibrium density" sea otter population: the influence of episodic oceanic subsidies. *J. Anim. Ecol.* Submitted

181. Weir JS. 1969. Importation of nutrients into woodlands by rooks. *Nature* 221:487–88

182. Welcomme R. 1979. *Fisheries Ecology of Floodplain Rivers.* London: Longmans

183. Werner E, Gilliam J. 1984. The ontogenetic niche and species interactions in size-structured populations. *Annu. Rev. Ecol. System.* 15:393–426

184. Wetzel RG. 1983. *Limnology.* Philadelphia, PA: WB Saunders. 2nd ed.

185. Wetzel RG. 1990. Land-water interfaces: metabolic and limnological regulators. *Int. Ver. Theor. Angew. Limnologie, Ver.* 24:6–24

186. Wiens JA. 1992. Ecological flows across landscape boundaries: a conceptual overview. In *Landscape Boundaries: Consequences for Biotic Diversity and Ecological Flows,* ed. AJ Hansen, F di Castri, pp. 217–35. New York: Springer Verlag

187. Wiens JA. 1995. Landscape mosaics and ecological theory. In *Mosaic Landscapes and Ecological Processes,* ed. L Hansson, L Fahrig, G Merriam, pp. 1–26. London: Chapman & Hall

188. Wiens JA. 1997. Metapopulation dynamics and landscape ecology. In *Metapopulation Biology: Ecology, Genetics and Evolution,* ed. I Hanski, M Gilpin, pp. 43–62. San Diego, CA: Academic

189. Wilcove DS, McLellan CH, Dodson AP. 1986. Habitat fragmentation in the temperate zone. In *Conservation Biology,* ed. ME Soule, pp. 237–56. Sunderland, MA: Sinauer Assoc.

190. Williams AJ. 1978. Mineral and energy contributions of petrels (Procellariiformes) killed by cats, to the Marion Island terrestrial ecosystem. *S. Afr. J. Antarct. Res* 8:49–53

191. Williams AJ, Burger AE, Berruti A. 1978. Mineral and energy contributions of carcasses of selected species of seabirds to the Marion Island terrestrial ecosystem. *S. Afr. J. Antarct. Res.* 8:53–59

192. Willson MF, Halupka KC. 1995. Anadromous fish as keystone species in vertebrate communities. *Conserv. Biol.* 9:489–97

193. Winemiller KO. 1990. Spatial and temporal variation in tropical fish trophic networks. *Ecol. Monogr.* 60:331–67

194. Wootton JT. 1991. Direct and indirect effects of nutrients on intertidal community structure: variable consequences of seabird guano. *J. Exp. Mar. Biol. Ecol.* 151:139–53

195. Yodzis P. 1989. *Introduction to Theoretical Ecology.* New York: Harper & Row

Annu. Rev. Ecol. Syst. 1997. 28:317–39

SETTLEMENT OF MARINE ORGANISMS IN FLOW

Avigdor Abelson
Institute for Nature Conservation Research, Tel Aviv University, Tel Aviv, Israel
69978; e-mail: avigdor@ccsg.tau.ac.il

Mark Denny
Hopkins Marine Station, Biological Sciences Department, Stanford University,
Pacific Grove, California 93950-3094

KEY WORDS: settlement, flow, propagules, larvae, spores

ABSTRACT

A feature common to many benthic marine plants and animals is the release of propagules that serve as the organism's only mechanism of dispersal. Successful dispersal depends to a large extent on the process of settlement—the transient phase between the pelagic life of the propagule and the benthic existence of the adult. The flow of water may affect settlement on three levels: 1. Flow can act by exerting hydrodynamic forces on settling propagules. These forces may affect the propagule's encounter with the substratum, its behavior following encounter, or both. 2. Flow may provide a settlement cue that induces active behavior of motile propagules. 3. Flow may act to mediate various settlement cues (e.g. sediment load and the concentration of attractants). We discuss these three levels of flow effects as a means of examining the potential importance of flow in the settlement process, and then we explore the ecological consequences of settlement in different flow-regimes in light of the direct effects of flow and flow-derived factors.

INTRODUCTION

Evidence is accumulating for the significant involvement of water motion in diverse biological and ecological processes in the sea (e.g. 13, 24, 39, 62). These effects include the flux of nutrients, food, and wastes to and from organisms, external fertilization of gametes, disturbance by hydrodynamic forces of benthic organisms, and the ability of organisms to detect scents. In addition,

317

0066-4162/97/1120-0317$08.00

flow may control or strongly regulate the dispersal of benthic species. A feature common to many benthic plants and animals is the release of propagules (e.g. spores and larvae), which are, in most cases, the organism's only mechanism of dispersal. Several possible advantages accrue to a life history that includes dispersing propagules: wide distribution of offspring, gene flow, and coexistence with disturbance (39, 73). Fulfilment of these tasks depends to a large extent on the process of settlement—the transient phase between the pelagic dispersal phase and the sessile existence of the adult. During this phase propagules must first encounter the substratum. Following its arrival at the seabed, a propagule must determine the adequacy of the substratum for adult requirements by using a variety of settlement cues, which may include surface contour (17, 95), substratum type (15), chemistry (55, 65), the presence of a microbial film (82), and flow conditions (e.g. 58, 100). Finally, the propagule must effectively attach, thereby allowing metamorphosis and development into a reproductive adult.

Flow may affect settlement of marine organisms on three levels. First, flow may act as an agent exerting hydrodynamic forces on settling propagules, which may affect either the propagule's encounter with the substratum, the propagule's behavior (passive or active) following encounter, or both. Second, flow may act as a settlement cue that induces active behavior of motile propagules. Only a few studies (most of which deal with invertebrate larvae) have examined the role of flow regime in determining site selection by propagules (e.g. 4, 58). Finally, flow may act as a mediating factor affecting various settlement cues (e.g. sedimentation, chemical distribution, and light intensity).

There are three approaches to the study of flow effects on settlement. The first, and probably the most common, is to monitor settlement on substrata in the field (e.g. 27, 56, 57). By characterizing the flow pattern induced by substrata, and comparing this to the settlement distribution of larvae, one can indirectly relate the settlement pattern to flow. The primary problem with this approach, however, is a practical one; it is difficult to observe the behavior of propagules under natural flow conditions. The second approach circumvents this observational problem by directly observing settlement behavior in the laboratory under controlled flow conditions (e.g. 4, 15, 21, 58). The disadvantage of this approach is that it cannot simulate the exact flow conditions experienced by propagules in the field. The third method employs numerical modeling of flow conditions prevailing in the field and of propagules' behavior during dispersal and settlement (e.g. 25, 28, 34). This last approach depends largely on data obtained by the other two to supply realistic values for parameters and boundary conditions. To be mathematically tractable, models must often be simplistic. Despite this limitation, models may reveal neglected aspects of empirical studies and can provide new directions of research. A synthesis of all three approaches is likely best to augment understanding of the effects of flow on the settlement of propagules.

In the present review we examine the three levels of flow effects and the various structural and behavioral features of propagules that may interact with flow during settlement. We then explore the ecological consequences of settlement as they are affected by the flow characteristics of differing substratum types.

SUBSTRATUM TYPES AND FLOW PATTERNS

The vast majority of studies that deal with settlement, whether in the field, in the lab, or on a computer, examine behavior and settlement patterns over planar substrata. In nature, however, the surfaces available for settlement (e.g. coral reefs, kelp forests, and artificial reefs and other engineered structures) include a diversity of shapes. To understand flow effects on settlement and to assess the flow-generated forces that a propagule must withstand, we first must characterize the flow patterns associated with common substrata in the sea. To this end we briefly describe the flow patterns induced by three major substratum types: planar surfaces, protruding bodies, and depressions. Our qualitative description is based on the simplifying assumption that the undisturbed, mainstream flow is steady. In the field, however, it is not uncommon to find flow that is unsteady due to tides and/or wave action (for a more thorough description of benthic flow environments, see 62).

Planar Substrata

Examples of planar surfaces are numerous in the marine environment and include substrata of limited area, such as rock faces, as well as more spacious substrata, such as soft bottoms. If we assume that the flow is steady over a planar substratum oriented parallel to the direction of water motion, the flow near the substratum (in the benthic boundary layer) is similarly steady. The boundary layer over a planar substratum develops from a very thin, laminar flow just downstream of the leading edge of the surface, to a turbulent, thick boundary layer further along the surface at a certain distance downstream from the edge (Figure 1). If the planar substratum is spacious and hydrodynamically smooth, a fully developed boundary layer is characterized by three zones: a thin viscous sublayer adjacent to the seafloor, a turbulence-dominated layer adjacent to the mainstream flow, and a buffer layer between the two (Figure 1). Although the ocean floor is seldom smooth, viscous sublayers that may exist there (16) can be up to 6 mm thick. For purposes of calculating the forces acting on small propagules during settlement, the relevant zone is this thin sublayer.

Protruding Bodies

The surfaces of marine substrata are frequently rough, and the flow over rough substrata can be divided into various flow regimes (54, 97, 98). Some of these involve "protruding bodies," those elements or bodies that project above their

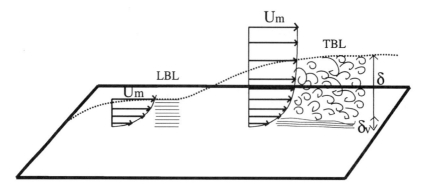

Figure 1   Schematic representation of the flow regimes in the vicinity of planar substratum. The laminar boundary layer (LBL) is indicated by solid lines. The fully turbulent boundary layer (TBL) is characterized by eddies and sweeps (indicated by curliques). The buffer layer is indicated by small eddies over the viscous sublayer. Um = free-stream flow; the broken line indicates the outer edge of the boundary layer, the thickness of which is δ; arrows indicate the flow, and their length proportional to flow velocity. δ_v = thickness of the viscous sublayer.

neighbors and the presence of which alters the flow pattern: kelp stipes, coral branches, rocks, and manmade structures such as pilings and artificial reefs. Although protruding bodies offer less surface area than do planar ones, they nonetheless form an important component of the substratum in many marine environments. In our analysis of flow pattern in the vicinity of protruding bodies, we use a circular cylinder as a simple model (Figure 2).

The presence of a cylinder causes flow to accelerate as water moves toward its widest cross-section (at $\theta = 90°$ for the circular cylinder in Figure 2), and flow near the cylinder can reach velocities twice that of mainstream. On the downstream side of the cylinder, flow decelerates. Due to the adverse pressure gradient associated with the deceleration, a separation zone is induced resulting in a downstream wake (Figure 2).

Depressions

Depressions (e.g. pits and crevices) are abundant on the surfaces of submerged substrata including organisms, manmade structures, and both soft bottoms and hard bottoms. Generally speaking, depressions have the opposite effect from protruding bodies—they increase the cross-sectional area through which water flows, and water consequently slows when flowing over and into a depression (62). The decreasing velocity leads to an adverse pressure gradient that results in a separation of the flow at the upstream edge of the depression and a vortical flow within the cavity itself (Figure 3). Flow often reattaches to the substratum at the downstream rim (96; Figure 3C).

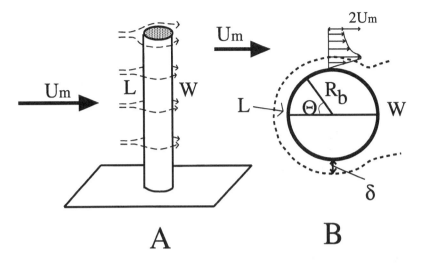

Figure 2   Schematic representation of a protruding body (*A*), its cross section (*B*), and the flow regime in its vicinity. The flow pattern induced by protruding bodies (in high Reynolds numbers) is of two zone types: accelerated laminar, thin boundary layer flow (L) in upstream turns, and retarded turbulent flow in downstream wakes (W). *Um* = free-stream flow (note that the velocity at the boundary-layer edge can reach 2 Um); *broken lines* indicate the outer edge of the boundary layers; *broken arrows* indicate streamlines, and *full arrows* flow velocity where length is proportional to velocity. Rb = body radius, θ = angle between a given point and the upstream stagnation point; δ = thickness of the boundary layer.

The maximum tangential velocities of vortices within depressions are approximately one tenth the mainstream flow velocity and do not exceed two tenths mainstream velocity (e.g. 40, 77, 91). Periodically, the vortex within the cavity becomes unstable and is shed from the cavity. This vortex-shedding leads to bulk replacement of fluid within the depression and, thereby, potentially to the bulk delivery of propagules to the cavity. The shedding of vortices is strongly enhanced by the presence of a distinct downstream edge to the cavity (69).

Most studies dealing with cavity flow have examined depressions below a planar substratum, in which case sharply defined vortices are the dominant patterns of cavity flow. Depressions in hard substrata may be formed, however, from walls raised above the local topography, and they may produce a different pattern of cavity flow (A Abelson, personal observation). This flow pattern includes a relatively slow entrainment of fluid out of the cavity by the "mainstream flow" (an outward current that occupies more than three quarters of the cavity cross section), and a rapid inward current near the downstream lip (Figure 3*C*). Delivery of propagules to the cavity under such circumstances

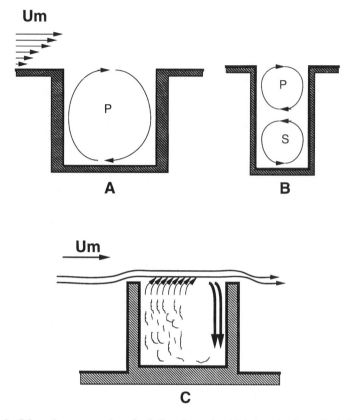

Figure 3 Schematic representation of a shallow depression (*A*) (where the depression's depth is equal to or less than its length in the direction of flow) and deep depressions (*B*, *C*) (where depth is greater than length) and the flow regimes in their vicinity. Separation and reattachment points are illustrated by the streamlines (in *C*). *Solid arrows* indicate flow direction; *arrow thickness* indicates relative flow velocity; *wavy lines* indicate the zone of slow outward current; Um = free-stream flow; P = primary vortex; S = secondary vortex.

may be enhanced by the inward, rapid current, while passive escape of propagules is unlikely because the outward current is slower than the propagules sink. Unfortunately, no quantitative study has yet addressed the transport rates to cavities under these circumstances.

FLOW AS A FORCING AGENT

Settlement can be divided into two critical stages: the delivery of propagules to the substratum, and the subsequent establishment of the larvae or spores, a stage that includes both attachment and metamorphosis. Here, we describe possible effects of flow as a physical forcing mechanism that acts on settling propagules during these two stages. Transport mechanisms of propagules to each substratum type are listed, along with the relevant problems each poses for propagules. We then examine problems that propagules might encounter in exploring the substratum and achieving a residence time sufficient for the creation of a firm attachment—a process that is a function of time (e.g. 29).

Mechanisms of Propagule Transport To Substrata

Delivery of propagules to the substratum can be achieved through active swimming, passive transport by the ambient flow, or both. Passive, flow-mediated delivery may be divided into two major categories depending on whether the flow is laminar or turbulent.

DELIVERY BY ACTIVE SWIMMING   Marine propagules can be classified according to their ability to transport themselves. Nonmotile propagules include various bacteria, bacterial capsules, fungal and algal spores, larvae of some invertebrate species, dispersive fragments of both plants and animals, and the slow-moving juveniles and adults of tiny species. Motile propagules include some algal spores and (primarily) invertebrate larvae. These motile propagules can reach 5 mm in length (10) but are frequently much smaller, whereas the nonmotile fragments can be much larger—up to several centimeters long (e.g. 35).

Among propagules that are nominally motile, most are weak swimmers, capable of maintaining speeds of less than 1 cm/s (19, 51). This speed is less than the flow speeds common near ocean substrata (e.g. 13), suggesting that the flow may limit the maneuverability of most self-propelled propagules. Accordingly, we assume that under most prevailing turbulent flow conditions, transport of propagules to the substratum is due to passive motion.

DELIVERY BY LAMINAR FLOW   The basic mechanisms by which particles can be delivered to the substratum in laminar flow were introduced to the biological literature by Rubenstein & Koehl (72) in the context of aerosol filtration theory. Subsequently, a number of studies (e.g. 8, 42, 75) have contributed to our understanding of the mechanisms controlling particle entrapment by marine organisms. Four of these mechanisms may also operate to transport propagules to substrata: 1. Direct interception, in which neutrally buoyant particles follow streamlines; if particles are carried within one particle radius of the substratum, they encounter it. 2. Inertial impaction, in which particles that are denser than the fluid diverge from streamlines as flow is accelerated while passing around a protruding body; under appropriate circumstances, dense particles move in the direction of the substratum. 3. Gravitational deposition, a mechanism similar to inertial impaction, but one in which deviation from a streamline is caused by a gravity. 4. Diffusional deposition, in which very tiny propagules may exhibit random, Brownian motion, thereby causing them to encounter the substratum.

DELIVERY BY TURBULENT TRANSPORT   The nature of turbulent flows near the bed is characterized by an alternating sequence of "ejections" and "sweeps," which are random in space and time (84). The ejection phase is a slow movement of vortices that creates horseshoe-shaped structures in the flow, which move away from the substratum. The sweep phase is characterized by high-speed flow that penetrates the viscous sublayer as fluid moves toward the substratum. Sweeps have been suggested to be a mechanism by which propagules are carried through the viscous sublayer to the substratum (18). To date, information on the efficiency of sweeps as a transport mechanism, and on the encounter rates of particles transported by sweeps, is lacking, and we must rely on the traditional theory of turbulent diffusion for a qualitative estimation of the efficiency by which turbulence transports propagules (e.g. 25, 28).

TRANSPORT TO DIFFERENT SUBSTRATUM TYPES   The transport of passive propagules to horizontal, planar substrata is dominated by turbulent transport, gravitational deposition, and downward swimming (e.g. 24, 30, 34). Turbulence intensity (an index of the efficacy of turbulent transport) is a function of both flow velocity (24) and whether the flow is accelerating (50). At high mainstream velocities in decelerating or steady flows, turbulence intensity is typically high, and the dominant transport mechanism to a planar substratum is likely to be turbulent transport regardless of propagule size. However, under conditions of very low turbulence intensities (e.g. slow and accelerating flows), transport may instead be dominated by gravitational deposition (especially for large propagules that consequently sink fast) and downward swimming. Given the vast area of planar substrata in the sea and the variety of effective transport mechanisms that can combine to maintain high encounter rates over a wide range of flow velocities and propagule sizes, it can be argued that, among the three substratum types, the chances of a propagule encountering planar substrata are the highest.

A preliminary model of transport to protruding bodies suggests that direct interception is the dominant mechanism (A Abelson, unpublished data). The efficiency of direct interception is proportional to both flow velocity and propagule size (42, 72). Therefore, the transport rates of propagules (which are typically small) in the low-flow velocities common in the marine environment are expected to be low.

The separated flow in a depression inhibits most of the transport mechanisms that prevail outside cavities. The sole exception, as suggested by Yager et al (96), is gravitational deposition, which, in this case, is coupled to the process of vortex-shedding. Particles can effectively settle only in the interim between shedding events. The frequency with which vortices are shed is determined by the flow velocity, cavity dimension, and a dimensionless index (the Strouhal number), which is a function of the cavity shape and Reynolds number (92).

Gravitational deposition is likely to be the dominant mechanism transporting propagules into depressions. One should bear in mind, however, that particle dynamics in vortices can be complicated and are not well understood. For example, submicron particles can segregate from the flow due to the rotational motion of the flow (33), a process for which a physical explanation is lacking. In deep depressions with low flow-velocities, diffusional deposition and directed swimming may also be important.

The Limits to Exploration and Attachment

Once a propagule encounters the substratum, the next step in settlement is either exploration of the surface or immediate attachment. In either case, the propagule must be able to control its position in space, whether to stay in one spot for time sufficient to establish a permanent attachment or to move in search of an optimal site.

The ability of a propagule to control its position depends on whether it can resist the hydrodynamic forces imposed on it. Under steady flow conditions, a propagule on the substratum is exposed to drag and lift forces, and if the flow accelerates either spatially or temporally, an acceleration reaction is also imposed. In addition, the velocity gradient in the benthic boundary layer exerts a torque on the propagule, a "rotary force" that rolls the propagule along the substratum (A Abelson, P Sanjines, S Monismith, submitted).

In the absence of adhesion, even the slightest forces are sufficient to dislodge passive particles from the substratum, with the result that nonadhesive propagules are likely to settle only where the flow is exceedingly slow (e.g. in depressions). Under such benign flow conditions, directed swimming may also be effective in allowing a propagule to maintain its position.

Assuming that the surfaces of substrata are equally rough, the relative forces on given propagule sizes at the substratum surface may depend on the substratum type. Protruding bodies induce the largest hydrodynamic forces among the three substratum types. These relatively strong forces are due primarily to the steep velocity gradients over the protruding-body surfaces, as well as to the increased flow velocity along the sides of the body (which, in cases of circular cylinders, may reach values twice the mainstream flow velocity; Figure 2). An additional force, the acceleration reaction, may play an important role in determining the resultant force on relatively large propagules. Planar substrata present intermediate values of mean forces exerted on settling propagules. These forces are lower than those exerted by the flow in the vicinity of protruding bodies but higher than these in depressions. Propagules settling on planar substrata, however, can experience instantaneous shear stresses due to sweeps that may exceed 30 times the average (26).

Proposed Solutions To Force-Related Problems

In the discussion above, we identified two factors that can act as potential barriers to the settlement of marine propagules—the inefficient manner in which they are delivered to protruding bodies and the need to resist hydrodynamic forces while maintaining a position on the substratum. The solutions to these problems are of two kinds: mechanisms that increase the probability of encounter with the substratum, and mechanisms of temporary, instantaneous attachment (or adhesion), which enable the propagule to achieve the residence time required for a permanent attachment.

As has been noted, the efficiency of transport to protruding bodies is positively correlated with propagule size; hence, an obvious mechanism to increase the probability of a propagule encountering a protruding body is to increase the propagule's size. However, large propagules experience large flow-related forces that can detach them from the substratum. Likewise, large propagules are costly to produce. Therefore, the optimal solution should be one that increases the propagule's effective size prior to encounter but does not increase the force exerted after encounter.

The evolved solutions to this problem are based on the use of flexible appendages or other temporary devices that increase the propagule's volume. For example, mucus sheaths envelope the propagules of red algae (12) and diatoms (E Gaiser, personal communication); the sheath of a red-algal spore may increase its actual volume by nearly an order of magnitude (Figure 4A). Another strategy is to extend projections from the propagule's body (Figure 4B), a solution found in bacterial capsules (20) and pili (60), fungal spores (38), and the dispersive fragments of red algae (47) and soft corals (Y Benayahu, personal communication). These fragments produce sticky crowns of rhizoids.

Attachment devices that cover the whole surface area of the propagule body or a large portion of it might be impractical for motile propagules, which in many cases utilize their body surface for locomotion (by cilia, for instance). Motile propagules such as larvae and dispersive juveniles or adults of invertebrates are more likely to use a single, projecting appendage which, due to the continuous reorientation of the propagule in flow, may nonetheless occupy a much larger effective space than a propagule without an appendage (4) (Figure 4C). Projecting appendages include the mucous threads secreted by larvae of hydrozoans (37; A Abelson, unpublished information), sea anemones (76), corals (4, 88), and polychaetes (32), and by juveniles and small dispersive adults of snails (49, 90) and bivalves (23, 87).

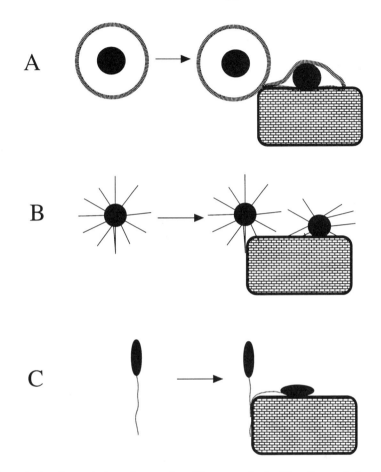

Figure 4 Attachment devices of propagules. (*A*) mucous sheath; (*B*) surface projections (e.g. bacterial pili); (*C*) threads (e.g. mucous and byssus).

Single projecting appendages have been observed in a flow tank to give rise to enhanced encounter rates of larvae with protruding bodies (4). However, the increased probability of encounter is largely dependent on the orientation of the propagule and its appendage relative to the protruding body. The orientation of appendage-bearing propagules may be affected, among other factors, by their aspect ratio and the turbulent dissipation energy (e.g. 41).

After the propagule has reached the substratum, it has two potential remaining problems. If it is to settle there permanently, it must maintain its position until a permanent attachment is made. Alternatively, it may be advantageous to explore the substratum in search of a more appropriate site. In both cases, some form of adhesion is required.

Often an adhesive is secreted prior to the organism's encounter with the substratum, examples being the mucous threads and sheaths described above. These allow the propagule to adhere instantly upon contact with the substratum. If the point of contact is inappropriate, some propagules can cut themselves off the substratum and return to the flow (4). This trial-and-error strategy may be an inefficient means of searching for a favorable site to settle. Alternatively, adhesion can be effected by devices that enable the propagule to move while keeping its attachment. The most prominent example is the barnacle cypris larva, which uses its antennular attachment organs to hold itself on the substratum while it explores (e.g. 21, 99). The exact adhesive mechanism of these antennules is not clear, but it seems that the larva can control the adhesiveness of its antennules. A similar role is played by the pyriform organ of the cyphonautes larva of bryozoans. This organ is reported to secrete a thin mucus sheet that aids locomotion and helps to fasten the larva temporarily to the substratum (83); larvae of the bryozoan *Membranipora membranacea* have been observed to explore substrata in flow (1), although the exact mechanism of locomotion in cyphonautes larvae is not yet clear.

The morphology of a larva can affect its exploratory locomotion when the organism is exposed to flow. This effect has been examined using three morphological models (spheres, symmetrical elongated models, and asymmetrical elongated models) in a flow tank (A Abelson, P Sanjines, S Monismith, submitted). The three morphologies behave differently when exposed to the shear flow of a benthic boundary layer. Specifically, spherical models were subjected to torque levels stronger than the forces most larvae can generate; they rolled along the substratum. Thus, spherical larvae may have trouble controlling their orientation. The symmetrical elongated models autorotated in a pattern that included two weak equilibrium positions, so larvae of this shape may also have problems controlling their orientation. The asymmetrically elongated models kept a stable position parallel to the flow direction and seemed to have the morphology best suited for locomotion.

FLOW AS A SETTLEMENT CUE

It is reasonable to assume that the various effects of flow on settling organisms may have led to the utilization of the local flow regime as a settlement cue. Indeed, some field studies show active site selection by larvae that may be attributed to larval response to different flows (e.g. 57, 93). The role of flow may be indirect, acting through its effect on other environmental parameters (see the following section). Nevertheless, the few studies on active behavior of settling propagules in response to flow show that larvae can select settlement sites by somehow sensing the local flow regime (e.g. 4, 43, 58).

Parameters that may be used by larvae to characterize the flow regime in their vicinity include: flow direction, shear stress, pressure gradient (adverse or favorable gradients), turbulence intensity, boundary-layer or viscous sublayer thickness, and acceleration. These flow parameters are often interrelated (e.g. shear stress is a function of the boundary layer thickness, and turbulence intensity may be linked to acceleration). As a consequence, it can be difficult to determine which parameter is affecting behavior. For example, Crisp (21) studied the behavior of barnacle larvae in tubes and on rotating plates. He demonstrated that larvae tend to move actively against the flow both by swimming and by using their adhesive antennules. Crisp argued that the velocity gradient close to the substratum is the flow parameter that affects the attachment of larvae of sessile species, and the nominal speed of the water is important only for its influence on the velocity gradient. In the same study, however, when referring to the exploratory behavior of larvae before attachment, Crisp related larval migration to flow direction rather than to the velocity gradient along the substratum. Unfortunately, the experimental design in Crisp's study did not enable discrimination between larval response to flow direction and velocity gradient.

In another study of settlement, coral larvae were shown to actively select settlement sites that differ from other sites only in their flow regime (4). Although the coral larvae select sites that can be characterized by different accelerations (acceleration, steady flow, deceleration), the precise flow parameters used to select preferred sites were not clear.

The studies cited above failed to isolate the specific flow parameters responsible for the observed larval behavior. It is possible, however, to isolate the effects of flow parameters by using substrata that induce distinct flow regimes. For example, it is possible to discriminate between larval responses to flow direction and shear stress using circular cylinders and flat plates in a flow tank. A cylinder affects the flow such that shear stress increases in the direction of flow because the flow accelerates. In contrast, on a flat plate the shear decreases in the direction of flow because the boundary layer grows. It was found that cyphonautes larvae respond to flow direction rather than to changing trends in shear stress (1).

FLOW AS A MEDIATING FACTOR AFFECTING SETTLEMENT CUES

Flow affects various environmental parameters, some of which are used as settlement cues by propagules of benthic marine organisms. Despite the likelihood of such indirect effects of flow on settlement, this is probably the least studied aspect of the role of flow in settlement. The most prominent environmental factors that are affected by flow and used as settlement cues are the concentration of chemical attractants or settlement inducers, the magnitude of sediment load, the grain size distribution of soft-substrata, and the characteristics of the microfilm.

Flow and Chemical Cue Distribution

Numerous studies (for reviews, see 55, 65) describe the induction and/or enhancement of settlement in invertebrate larvae in response to chemical factors produced by conspecific adults, host and prey species, and microbial films. On the other hand, very few studies have examined the effect of chemicals on settling propagules under flow conditions (67, 89), and, to the best of our knowledge, only one study has examined the combined effects of water-soluble chemical cues and flow on settling propagules (89).

Despite our poor knowledge of larval response to waterborne chemicals under flow conditions, it is generally accepted that chemotaxis toward a preferred substratum in flow is unlikely to play a role in settlement (e.g. 22, 65). This conclusion is based on three lines of reasoning. First, turbulent flow over a body releasing a waterborne substance is expected to dilute the substance to negligible concentrations within a short distance from the source (65). Second, the small sizes of larvae would seem to make orientation and navigation in a concentration gradient unlikely (22). Finally, given the limited locomotory capabilities of larvae, it has been assumed that they do not move upstream (see, for example, 7). Because waterborne chemicals cannot be tracked by downstream exploration, the possibility of tracking settlement sites along gradients of waterborne chemicals would appear to be impractical.

It is always dangerous to underestimate the abilities of organisms, however, and the advantage of being able to track waterborne chemicals would seem to be large. For example, many organisms inhabit spatially limited, specific substrata (e.g. the bodies of host organisms), and for them the probability of encountering the appropriate substratum without any remote indicator is likely to be extremely low. The ability of dispersed chemicals to indicate at a distance the presence of the required substratum could be used by a capable larva to move toward its host. Indeed, we believe that larvae that settle in association with other organisms may (while on the substratum) possess the capacity to track waterborne settlement inducers toward their preferred substratum under flow conditions. We base our view on the following considerations.

First, while the expectation is that turbulence rapidly mixes chemical plumes, it must be remembered that mixing near the wall is relatively weak, so that plumes might remain coherent for substantial distances (52). Additionally, while low concentrations might be measured in the mean in turbulent flow, these are generally the result of intermittent pulses of high concentration, rather than of a steady supply of mixed fluid (64). Relatively strong chemical cues may thus be available occasionally to guide settlement. Crabs track waterborne odors from distances of 0.5 m and 1 m in a turbulent flow (94).

Second, the arguments against the ability of larvae to orient and navigate in a concentration gradient are not grounded in experimental results. Moore et al (53) suggested that if significant spatial differences in chemical signal structure consistently occur at spatial scales of a few hundred mm, as they found in their study, then directional and distance information from odor plumes could be derived via differential sensory input between paired sensors in small organisms. Furthermore, the case against chemotaxis in larvae is at odds with a large body of evidence of chemotaxis in other, much smaller organisms (e.g. 5, 8). For example, neutrophil cells (a type of phagocytic white blood cells) are able to detect very low concentrations (ca 10^{-10} M) of a specific chemoattractant and can detect a 1% difference in the concentration across the cell, allowing them to migrate up a chemical gradient (5).

Finally, some larvae can explore substrata in the upstream direction. Crisp (21) found that barnacle larvae can actively move against the flow by swimming extremely fast (4–5 cm/s) and using their adhesive antennules. Crisp's results have been criticized for the experimental design, which did not simulate the natural conditions of larvae (58). However, it is reasonable to assume that barnacle larvae are capable of locomoting against the stream. The cyphonautes larvae of *Membranipora membranacea* can explore substrata against currents that are much faster than their locomotion speeds by crawling over mucus sheets that attach them to the substratum (1).

In summary, we suggest that larvae of benthic marine organisms may be tracking waterborne settlement cues under flow conditions. However, a future study will have to resolve this issue by examining whether plumes carrying chemical settlement cues can induce upstream substratum exploration by propagules that utilize these cues to track their preferred habitats.

Flow and Sediment Load

Sedimentation exerts harmful effects on hard-bottom dwellers in diverse manners (e.g. 70). Sediment grains in flow may abrade the organism's tissue; suspended grains can interfere with filter feeding; and deposited particles can cover organisms, preventing them from carrying out various vital processes. In light of the deleterious effects of sediment on organisms inhabiting hard substrata, sediment load is likely to have an influence on the habitat choice of settling larvae. This issue has not been explored in depth, but in certain cases settlement rate has been shown to be reduced in areas with high sediment loads (e.g. 36, 86).

Sediment load is the outcome of two counteracting processes, sediment erosion (or entrainment) and deposition. Flow is a dominant agent governing these sedimentological processes and therefore is likely to be an important indirect factor affecting propagule site selection and settlement.

Flow and Soft-Substratum Type

Size of sediment grains is an important parameter in the biology and ecology of soft-bottom dwellers (e.g. 79). Grain size may affect particle selection by deposit feeders (74), and grain size may determine the amount of organic matter in the sediment and consequently the feeding efficiency of deposit feeders (e.g. 85). The effects of grain size on soft-bottom dwellers explains what has been known for decades—that sediment grain size can determine site selection, settlement, and metamorphosis of soft-bottom dwellers (reviews in 13, 79).

Flow characteristics (e.g. turbulence intensity and shear stress) are crucial in determining the grade and structure of deposited sediments at a given site (within the limiting frame of the sediment-grain coomposition of the potential sources; e.g. 6). Combining the role of flow in sediment distribution and the role of grain size in site selection, we can argue that flow probably plays an important role mediating larval settlement in response to grain size.

Flow and Biofilm Type

There is mounting evidence for the role of flow in bacterial attachment, removal, growth, and spatial density, as well as for its effect on metabolic activity of biofilm communities (11, 45, 100). Flow causing different shear stresses may induce bacterial films of different morphology and structure. Biofilm communities function as pioneer communities developed during the first stages of succession on a newly cleared substratum, and they act both as an attractant and as the actual substrate upon which algae and animals subsequently settle. For example, barnacle larvae of two species respond differently to each of two films (61). Flow can thus indirectly affect late settlers by its effect on pioneer communities of microorganisms.

CONCLUSIONS AND SPECULATIONS

Active Versus Passive Behavior in Settlement

Most models of propagular settlement assume, for simplicity, that when a larva encounters the substratum, it immediately adheres and settles (e.g. 25, 78). Similarly, models of benthic invertebrate recruitment treat larvae as passive particles (e.g. 71). Butman and coworkers (58, 67) pointed out that such studies disregard the importance of behavior in settlement, supporting their argument with results clearly showing that active behavior may greatly affect recruitment patterns. Currently, the accepted paradigm is that settlement is a combined result of both flow effects and larval behavior (e.g. 30). It seems that the delivery of larvae to potential habitats is governed mainly by flow-related processes, whereas exploratory larval behavior may be important over small spatial scales after the initial encounter with the substratum (e.g. 13, 67, 93).

Nonmotile and motile propagules exhibit major difference in behavior during settlement. Nonmotile propagules, by definition, do not actively explore the substratum and can select an optimal habitat only by initiating permanent adhesion when they come across a suitable site. Although motile propagules are able to explore and select an appropriate settlement site, a wide range of flow regimes exist under which these organisms are unable to explore the surface. Support for this assumption is given, for example, by Pawlik & Butman (66), who have observed "erosion" of larvae from the bed at shear velocities (shear velocity is a measure of the magnitude and correlation of turbulent fluctuations in velocity near the substratum) higher than 1.03 cm s^{-1}. In such flow conditions, swimming of motile propagules can be considered ineffective.

We therefore suggest that differential settlement (i.e. deviation from the settlement distribution predicted from the distribution of initial contacts of propagules with the substratum) under conditions of hydrodynamic forces higher than the propagule swimming forces is due to desertion of unfavorable sites rather than to exploration and active selection of an appropriate site. Invertebrate larvae have been observed to desert substrates following encounter in flow tanks (e.g. 4, 14, 58). Moreover, larvae reportedly have deserted their settlement site after settlement and metamorphosis due to unfavorable conditions (e.g. 68).

An alternative explanation for differential settlement in flow is provided by the "adhere and explore" mechanisms of barnacles and bryozoans. These mechanisms may explain deviations of settlement distribution from encounter distributions of barnacles and bryozoans under flow conditions (as was found by Walters, 93). The potential importance of "adhere and explore" mechanisms is that they save the propagule the need to return to the flow if it encounters unfavorable sites.

Ecological Significance of Settlement in Environments of Different Flow Conditions

Flow causes food particles to be distributed nonhomogeneously in sites that differ in height, or over morphologically different substrata of benthic habitats (3, 59), and organisms may be distributed in accordance with their potential food distribution (2). Larvae of sessile particle-feeders may settle in flow patterns that enhance the concentration of their "preferred" food.

In this sense, larvae of suspension-, bedload- and deposit-feeding species (for definitions, see 3, 42, 48) are expected to select distinct microhabitats that differ in their flow patterns and, as a consequence, their prevailing particle compositions. Specifically, larvae that feed on fine, suspended particles may have evolved mechanisms (e.g. adhesive appendages) that allow them to settle on protruding bodies where the fluxes of fine, suspended particles are the highest and where the potentially disturbing coarse bedload particles are in relatively low quantities. On the other hand, propagules of bedload- and deposit-feeders would be expected to have evolved without these mechanisms because, for example, adhesive appendages might stick them to sites unfavorable in terms of food supply. The larvae of bedload-feeders are expected to favor sites on planar substrata in which the bedload fluxes are the highest, whereas deposit-feeders are more likely to prefer depressions.

Larvae that tend to settle in depressions may recognize differences in flow conditions that result from different pit sizes (80). Likewise, larvae of different species, or of the same species in different seas, may differ in their responses to the same pit under the same flow conditions (17, 80). Alternatively, depressions may be chosen not only for the flow pattern they induce. For instance, small depressions, equal in size to or smaller than many propagules and with little effect on flow, may be chosen as preferred settlement sites, possibly because they provide firm attachment (46).

Postsettlement strategies also exist that enable organisms to persist and grow on a wide range of substratum types. For example, the flow pattern in the organisms' vicinity can be altered. This can be accomplished either by changing the substratum morphology (an option open mainly to soft-bottom dwellers, e.g. the ampharetid polychaete *Amphicteis scaphobranchiata*; 63), or by building a body that changes the flow pattern experienced by the organism's feeding devices (e.g. diverse coral species; 2).

CONCLUDING REMARKS

We have presented several themes regarding the effects of flow on settling propagules that should receive more attention in studies of benthic ecology. First, not all substrata are planar. The morphology of the substratum is of great importance in determining the flow pattern in its vicinity and consequently may play a pivotal role in affecting the species composition of settling propagules. The studies of Mullineaux & Butman (57, 58), for instance, clearly show the distinctive effects of induced flow patterns over substrata on the distribution of larvae at settlement. These studies have concentrated, however, on planar substrata and do not deal with other substrata of great importance (but see 80). The neglect of substratum morphology and its resultant flow regime is exemplified by field studies that attempt to explain the distribution of organisms on cylindrical bodies (such as mangrove prop roots) by examining differential settlement on planar substrata (e.g. ceramic tiles; 9). Studies with different substratum morphologies are necessary to assess the exact role of morphology on site selection of propagules and survivorship of the developing recruits.

Second, flow can simultaneously play a role as a settlement cue (which may influence active behavior of motile propagules) and as a factor preventing or disturbing the settlement process. Distinguishing between these two roles is important for understanding the settlement process. No significant attempt has been made to characterize environments in which active exploration is possible and those in which active locomotion is prevented by excessive hydrodynamic forces.

Third, where differential settlement is correlated with flow conditions, flow is often considered to be the sole factor directly affecting the distribution of settling propagules. Flow, however, may affect diverse settlement cues. Parameters other than flow but influenced by flow may determine the differential settlement. These factors include biofilm structure, sediment load, substratum type, distribution of chemicals, and light intensity (due to turbidity).

Finally, further work should be conducted to examine the hypothesis that projecting appendages or other adhesive devices are associated with propagules of species inhabiting substrata characterized by high velocities of flow (e.g. protruding bodies). The corollary here is that species lacking these appendages are likely to settle in sites with low flow velocity (e.g. depressions). Some support for this hypothesis is provided by settlement patterns of several coral species that were found to be correlated to the adult distribution that match their food-particle distribution (4). The differences in settlement patterns of these species can be related to mechanisms affecting larval encounter with and attachment to the substratum. Most mechanisms that facilitate settlement on protruding bodies have been previously described in contexts other than settlement in flow and have elicited various explanations as to their role in the biology and ecology of propagules. The most widely accepted explanations for the role of projecting appendages on larvae concern their role in drifting and dispersion (e.g. 44, 90), in feeding (e.g. 31), and in creating aggregations (20). Strathmann (81) argues that there is no evidence that projecting appendages are used to maintain larvae in suspension, as is often suggested. Likewise, the presence of projecting appendages among tiny propagules that exhibit very low settling velocities is not likely to affect dramatically their dispersal range, especially for motile propagules that can maintain their vertical position by swimming. We do not deny alternative explanations for the roles of projecting appendages. However, many organisms possessing them are apparently able to settle on protruding bodies and in other environments of strong, accelerating flow.

In summary, the information of previous studies suggests that flow may act on settling propagules through one or more of three levels: as an agent exerting hydrodynamic forces, as a settlement cue, and as a mediating factor affecting various other settlement cues. A limitation to interpreting the exact role of each level is that discrimination among the three levels requires a complicated experimental setup that in many cases may pose "unresolved" technical problems. Of the three levels, flow as a mediating factor affecting various settlement cues seems to be the least studied. The role of flow as an agent exerting hydrodynamic forces on settling propagules has been clearly shown in various studies. Likewise, there is increasing evidence for the use of flow by propagules (mainly larvae) in selecting preferred sites that fit the adult requirements. To use the flow as a settlement cue, propagules should possess mechanisms to cope with force-related problems. Here we have explored such mechanisms whose effectiveness remains to be examined. Taking into account the significant involvement of flow in diverse biological and ecological processes in the sea, it is reasonable to argue that whatever the relative weight of each level, the overall effect of flow may be crucial for many benthic species.

ACKNOWLEDGMENTS

We are thankful to J Eckman, D Fautin, P Jumars, and R Strathmann for their constructive comments on previous versions of this paper. This research was supported by Fulbright scholarship of the USIEF and the Ben-Gurion Fellowship of the Israel Ministry of Science to A.A.

Visit the *Annual Reviews home page* at
http://www.annurev.org.

Literature Cited

1. Abelson A. 1997. Settlement in flow: up-stream exploration of substrata by weakly swimming larvae. *Ecology* 78:160–66
2. Abelson A, Loya Y. 1995. Cross-scale patterns of particulate-food acquisition in marine benthic environments. *Am. Nat.* 145:848–54
3. Abelson A, Miloh T, Loya Y. 1993. Flow patterns induced by substratum and body morphology of benthic organisms, and their role in determining food particle availability. *Limnol. Oceanogr.* 38:1116–24
4. Abelson A, Weihs D, Loya Y. 1994. Hydrodynamic impedance to settlement of marine propagules, and trailing-filament solutions. *Limnol. Oceanogr.* 39:164–69
5. Alberts B, Bray D, Lewis J, Raff M, Roberts K, et al. 1989. *Molecular Biology of the Cell.* New York: Garland. 1219 pp.
6. Allen JRL. 1985. *Physical Sedimentol-*

ogy. London: Allen & Unwin. 272 pp.
7. Andre C, Jonsson PR, Lindegarth M. 1993. Predation on settling bivalve larvae by benthic suspension feeders: the role of hydrodynamics and larval behavior. *Mar. Ecol. Prog. Ser.* 97:183–92
8. Berg S. 1983. *Random Walks in Biology.* Princeton, NJ: Princeton Univ. Press. 142 pp.
9. Bingham BL. 1992. Life histories in an epifaunal community: coupling of adult and larval processes. *Ecology* 73:2244–59
10. Bingham BL, Young CM. 1991. Larval behavior of the ascidian *Ecteinascidia turbinata* (Herdman): an in situ experimental study of the effects of swimming on dispersal. *J. Exp. Mar. Biol. Ecol.* 145:189–204
11. Blenkinsopp SA, Lock MA. 1994. The impact of storm-flow on river biofilm architecture. *J. Phycol.* 30:807–18

12. Boney AD. 1975. Mucilage sheaths of spores of red algae. *J. Mar. Biol. Assoc. UK* 55:511–18

13. Butman CA. 1987. Larval settlement of soft-sediment invertebrates: the spatial scales of pattern explained by active habitat selection and the emerging role of hydrodynamical processes. *Oceanogr. Mar. Biol. Annu. Rev.* 25:113–65

14. Butman CA, Grassle JP. 1992. Active habitat selection by *Capitella* sp. I larvae. I. Two-choice experiments in still water and in flume flows. *J. Mar. Res.* 50:669–715

15. Butman CA, Grassle JP, Webb CM. 1988. Substrate choices made by marine larvae settling in still water and in a flume flow. *Nature* 333:771–73

16. Caldwell DR, Chriss TM. 1979. The viscous sublayer at the sea floor. *Science* 205:1131–32

17. Chabot R, Bourget E. 1988. Influence of substratum heterogeneity and settled barnacle density on the settlement of cypris larvae. *Mar. Biol.* 97:45–56

18. Characklis WG. 1981. Fouling biofilm development: a process analysis. *Biotech. Bioeng.* 23:1923–60

19. Chia FS, Buckland-Nicks J, Young CM. 1984. Locomotion of marine invertebrate larvae: a review. *Can J. Zool.* 62:1205–22

20. Cowen JP. 1992. Morphological study of marine bacterial capsules: implications for marine aggregates. *Mar. Biol.* 114:85–95

21. Crisp DJ. 1955. The behaviour of barnacle cyprids in relation to water movement over surfaces. *J. Exp. Biol.* 32:569–90

22. Crisp DJ. 1965. Surface chemistry, a factor in the settlement of marine invertebrate larvae. *Bot. Gothob. Acta Univ. Gothob.* 3:51–65

23. DeBlok JW, Tan-Maas M. 1977. Function of byssus threads in young postlarval *Mytilus. Nature* 267:558

24. Denny MW. 1988. *Biology and the Mechanics of the Wave-Swept Environment.* Princeton, NJ: Princeton Univ. Press. 329 pp.

25. Denny MW, Shibeta MF. 1989. Consequences of surf-zone turbulence for settlement and external fertilization. *Am. Nat.* 134:859–89

26. Dyer KR, Soulsby RL. 1988. Sand transport on the continental shelf. *Annu. Rev. Fluid Mech.* 20:295–324

27. Eckman JE. 1983. Hydrodynamic processes affecting benthic recruitment. *Limnol. Oceanogr.* 28:241–57

28. Eckman JE. 1990. A model of passive settlement by planktonic larvae onto bottoms of differing roughness. *Limnol. Oceanogr.* 35:887–901

29. Eckman JE, Savidge WB, Gross TF. 1990. Relationship between duration of cyprid attachment and drag forces associated with detachment of *Balanus amphitrite* cyprids. *Mar. Biol.* 107:111–18

30. Eckman JE, Werner FE, Gross TF. 1994. Modelling some effects of behavior on larval settlement in a turbulent boundary layer. *Deep-Sea Res.* 41:185–208

31. Emlet RB, Strathmann RR. 1985. Gravity, drag, and the feeding currents of small zooplankton. *Science* 228:1016–17

32. Gee JM, Knight-Jones EW. 1962. The morphology and larval behaviour of a new species of *Spirorbis* (serpulidae). *J. Mar. Biol. Assoc. UK* 42:641–54

33. Goldshtik M, Husain HS, Hussain F. 1992. Loss of homogeneity in a suspension by kinematic action. *Nature* 357:141–42

34. Gross TF, Werner FE, Eckman JE. 1992. Numerical modeling of larval settlement in turbulent bottom boundary layers. *J. Mar. Res.* 50:611–42

35. Highsmith RC. 1982. Reproduction by fragmentation in corals. *Mar. Ecol. Prog. Ser.* 7:207–26

36. Hodgson G. 1990. Sediment and the settlement of larvae of the reef coral *Pocillopora damicornis. Coral Reefs* 9:41–43

37. Hughes RG. 1977. Aspects of the biology and life-history of *Nemertesia antennina*. *J. Mar. Biol. Assoc. UK* 57:641–57

38. Jones EBG, Moss ST. 1978. Ascospore appendages of marine ascomycetes: an evaluation of appendages as taxonomic criteria. *Mar. Biol.* 49:11–26

39. Jumars PA. 1993. *Concepts in Biological Oceanography: An Interdisciplinary Primer.* New York: Oxford Univ. Press. 347 pp.

40. Koseff JR. 1983. *Momentum transfer in a complex recirculating flow.* PhD thesis. Stanford Univ., CA

41. Krushkal EM. Gallily I. 1988. On the orientation distribution function of nonspherical aerosol particles in a general shear flow. II. The turbulent case. *J. Aerosol Sci.* 19:197–211

42. LaBarbera M. 1984. Feeding currents and particle capture mechanisms in suspension-feeding animals. *Am. Zool.* 24:71–84

43. Lacoursiere JO. 1992. A laboratory study of fluid flow and microhabitat selection by larvae of *Simulium vittatum* (Diptera: Simuliidae). *Can. J. Zool.* 70:582–96

44. Lane DJW, Beaumont AR, Hunter JR. 1985. Byssus drifting and the threads

of the young post-larval mussel *Mytilus edulis*. *Mar. Biol.* 84:301–8

45. Lau YL, Liu D. 1993. Effects of flow rate on biofilm accumulation in open channels. *Water Res.* 27:355–60

46. LeTourneux F, Bourget E. 1988. Importance of physical and biological settlement cues used at different spatial scales by the larvae of *Semibalanus balanoides*. *Mar. Biol.* 97:57–66

47. Lipkin Y. 1977. Centroceras, the 'missile'-launching marine red alga. *Nature* 270:48–49

48. Lopez GR, Levinton JS. 1987. Ecology of deposit-feeding animals in marine sediments. *Q. Rev. Biol.* 62:235–60

49. Martel A, Diefenbach T. 1993. Effects of body size, water current and microhabitat on mucous-thread drifting in postmetamorphic gastropods *Lacuna* spp. *Mar. Ecol. Prog. Ser.* 99:215–20

50. Middleton GV, Southard JB. 1984. *Mechanics of Sediment Movement*. Tulsa: SEPM. 2nd ed.

51. Mileikovsky SA. 1973. Speed of active movement of pelagic larvae of marine bottom invertebrates and their ability to regulate their vertical position. *Mar. Biol.* 23:11–17

52. Monismith SG, Koseff JR, Thompson J, O'Riordan CA, Nepf H. 1990. A study of model bivalve siphonal currents. *Limnol. Oceanogr.* 35:680–96

53. Moore PA, Zimmer-Faust RK, Bement SL, Weissburg MJ, Parrish JM, et al. 1992. Measurement of microscale patchiness in a turbulent aquatic odor plume using a semiconductor-based microprobe. *Biol. Bull.* 183:138–42

54. Morris HM. 1955. Flow in rough conduits. *Trans. Am. Soc. Civ. Eng.* 120:373–98

55. Morse DE. 1990. Recent progress in larval settlement and metamorphosis: closing the gaps between molecular biology and ecology. *Bull. Mar. Sci.* 46:465–83

56. Mullineaux LS. 1988. The role of settlement in structuring a hard-substratum community in the deep sea. *J. Exp. Mar. Biol. Ecol.* 120:247–61

57. Mullineaux LS, Butman CA. 1990. Recruitment of encrusting benthic invertebrates in boundary-layer flows: a deepwater experiment on Cross Seamount. *Limnol. Oceanogr.* 35:409–23

58. Mullineaux LS, Butman CA. 1991. Initial contact, exploration and attachment of barnacle (*Balanus amphitrite*) cyprids settling in flow. *Mar. Biol.* 110:93–103

59. Muschenheim DK. 1987. The dynamics of near-bed seston flux and suspension-feeding benthos. *J. Mar. Res.* 45:473–96

60. Nakasone N, Iwanaga M. 1990. Pili of *Vibrio cholerae* non-01. *Infect. Immun.* 58:1640–46

61. Neal AL, Yule AB. 1994. The tenacity of *Elminius modestus* and *Balanus cyprids* to bacterial films grown under different shear regimes. *J. Mar. Biol. Assoc. UK* 74:251–57

62. Nowell ARM, Jumars PA. 1984. Flow environments of aquatic benthos. *Annu. Rev. Ecol. Syst.* 15:303–28

63. Nowell ARM, Jumars PA, Fauchald K. 1984. The foraging strategy of a subtidal and deep-sea feeder. *Limnol. Oceanogr.* 23:645–49

64. O'Riordan CA, Monismith SG, Koseff JR. 1993. A study of concentration boundary-layer formation over a bed of model bivalves. *Limnol. Oceanogr.* 38:1712–29

65. Pawlik JR. 1992. Chemical ecology of the settlement of benthic marine invertebrates. *Oceanogr. Mar. Biol. Annu. Rev.* 30:273–335

66. Pawlik JR, Butman CA. 1993. Settlement of marine tube worm as a function of current velocity: interacting effects of hydrodynamics and behavior. *Limnol. Oceanogr.* 38:1730–40

67. Pawlik JR, Butman CA, Starczak VR. 1991. Hydrodynamic facilitation of gregarious settlement of a reef-building tube worm. *Science* 251:421–24

68. Richmond RH. 1985. Reversible metamorphosis in coral planula larvae. *Mar. Ecol. Prog. Ser.* 22:181–85

69. Rockwell D, Naudascher E. 1978. Review: self-sustaining oscillations of flow past cavities. *ASME J. Fluids Eng.* 100:152–65

70. Rogers CR. 1989. Responses of coral reefs and reef organisms to sedimentation. *Mar. Ecol. Prog. Ser.* 62:185–202

71. Roughgarden J, Gaines S, Possingham H. 1988. Recruitment dynamics in complex life cycles. *Science* 241:1460–66

72. Rubenstein DI, Koehl MAR. 1977. The mechanisms of filter feeding: some theoretical considerations. *Am. Nat.* 111:981–94

73. Scheltema RS. 1974. Biological interactions determining larval settlement of marine invertebrates. *Thalassia Jugosl.* 10:263–69

74. Self RFL, Jumars PA. 1988. Crossphyletic patterns of particle selection by deposit feeders. *J. Mar. Res.* 46:119–43

75. Shimeta J, Jumars PA. 1991. Physical mechanisms and rates of particle capture

by suspension-feeders. *Oceanogr. Mar. Biol. Annu. Rev.* 29:191–257

76. Siebert AE. 1974. A description of the embryology, larval development, and feeding of the sea anemones *Anthopleura elegantissima* and *A. xanthogrammica. Can. J. Zool.* 52:1383–88

77. Sinha SN, Gupta AK, Oberai MM. 1982. Laminar separating flow over backsteps and cavities. Part II: Cavities. *AIAAJ* 20:370–75

78. Smith IR. 1982. A simple theory of algal deposition. *Freshw. Biol.* 12:445–49

79. Snelgrove PVR, Butman CA. 1994. Animal-sediment relationships revisited: cause versus effect. *Oceanogr. Mar. Biol. Annu. Rev.* 32:111–77

80. Snelgrove PVR, Butman CA, Grassle JP. 1993. Hydrodynamic enhancement of larval settlement in the bivalve *Mulinia lateralis* and the polychaete *Capitella* sp. I in microdepositional environments. *J. Exp. Mar. Biol. Ecol.* 168:71–109

81. Strathmann RR. 1993. Hypotheses on the origin of marine larvae. *Annu. Rev. Ecol. Syst.* 24:89–117

82. Strathmann RR, Branscomb ES, Vedder K. 1981. Fatal errors in set as a cost of dispersal and the influence of intertidal flora on set of barnacles. *Oecologia* 48:13–18

83. Stricker SA. 1988. Metamorphosis of the marine bryozoan *Membranipora membranacea:* an ultrastructural study of rapid morphogenetic movements. *J. Morphol.* 196:53–72

84. Sumer BM, Oguz B. 1978. Particle motions near the bottom in turbulent flow in an open channel. *J. Fluid Mech.* 86:109–27

85. Taghon GL, Self RFL, Jumars PA. 1978. Predicting particle selection by deposit feeders: a model and its implications. *Limnol. Oceanogr.* 23:752–59

86. Te FT. 1992. Response to higher sediment loads by *Pocillopora damicornis* planulae. *Coral Reefs* 11:131–34

87. Titman CW, Davis PA. 1976. The dispersal of young post-larval bivalve molluscs by byssus threads. *Nature* 262:386–87

88. Tranter PRG, Nicholson DN, Kinchington D. 1982. A description of spawning and post-gastrula development of the cool temperate coral *Caryophyllia smithi. J. Mar. Biol. Assoc. UK* 62:845–54

89. Turner EJ, Zimmer-Faust RK, Palmer MA, Luckenbach M, Pentcheff ND. 1994. Settlement of oyster (*Crassostrea virginica*) larvae: effects of water flow and water-soluble chemical cue. *Limnol. Oceanogr.* 39:1579–93

90. Vahl O. 1983. Mucus drifting in the limpet *Helicion pellucidus. Sarsia* 68:209–11

91. Varzaly AM. 1978. *Some features of low-speed flow over a rectangular cavity.* Engineer thesis. Stanford Univ., CA. 189 pp.

92. Vogel S. 1981. *Life in Moving Fluids.* Boston: Willard Grant. 352 pp.

93. Walters LJ. 1992. Field settlement locations on subtidal marine hard substrata: Is active larval exploration involved. *Limnol. Oceanogr.* 37:1101–7

94. Weissburg MJ, Zimmer-Faust RK. 1993. Life and death in moving fluids: hydrodynamic effects on chemosensory-mediated predation. *Ecology* 74:1428–43

95. Wethey DS. 1986. Ranking of settlement cues by barnacle larvae: influence of surface contour. *Bull. Mar. Sci.* 39:393–400

96. Yager PL, Nowell ARM, Jumars PA. 1993. Enhanced deposition to pits: a local food source for benthos. *J. Mar. Sci.* 51:209–36

97. Young WJ. 1992. Clarification of the criteria used to identify near-bed flow regimes. *Freshw. Biol.* 28:383–91

98. Young WJ. 1993. Field techniques for the classification of near-bed flow regimes. *Freshw. Biol.* 29:377–83

99. Yule AB, Walker G. 1987. Adhesion in barnacles. In *Crustacean Issues 5: Barnacle Biology,* ed. AJ Southard. Rotterdam: Balkema

100. Zheng D, Taylor GT, Gyananath G. 1994. Influence of laminar flow velocity and nutrient concentration on attachment of marine bacterioplankton. *Biofouling* 8:107–20

Annu. Rev. Ecol. Syst. 1997. 28:341–58

SPECIES RICHNESS OF PARASITE ASSEMBLAGES: Evolution and Patterns

Robert Poulin

Department of Zoology, University of Otago, P.O. Box 56, Dunedin, New Zealand;
e-mail: robert.poulin@stonebow.otago.ac.nz

KEY WORDS: community structure, helminths, nestedness, phylogenetic influences, sampling
effort

ABSTRACT

Parasite communities are arranged into hierarchical levels of organization, covering various spatial and temporal scales. These range from all parasites within an individual host to all parasites exploiting a host species across its geographic range. This arrangement provides an opportunity for the study of patterns and structuring processes operating at different scales. Across the parasite faunas of various host species, several species-area relationships have been published, emphasizing the key role of factors such as host size or host geographical range in determining parasite species richness. When corrections are made for unequal sampling effort or phylogenetic influences, however, the strength of these relationships is greatly reduced, casting a doubt over their validity. Component parasite communities, or the parasites found in a host population, are subsets of the parasite fauna of the host species. They often form saturated communities, such that their richness is not always a reflection of that of the entire parasite fauna. The species richness of component communities is instead influenced by the local availability of parasite species and their probability of colonization. At the lowest level, infracommunities in individual hosts are subsets of the species occurring in the component community. Generally, their structure does not differ from that expected from a random assembly of available species, although comparisons with precise null models are still few. Overall studies of parasite communities suggest that the action of processes determining species richness of parasite assemblages becomes less detectable as focus shifts from parasite faunas to infracommunities.

Introduction

The two fundamental questions at the core of community ecology are: What determines the number of species in an assemblage, and do these species form

341

0066-4162/97/1120-0341$08.00

a truly structured community or a random assemblage (97)? Structure implies the presence of statistical associations among species, creating predictable patterns of species co-occurrence that depart from null models of random species assemblages. Both the availability of species and the forces structuring communities combine to determine how many species occur in an assemblage. Patterns in community structure and species richness, however, are dependent on the scale of the investigation. Mechanisms operating at one scale may generate patterns that can be detected only at a different scale (62). The necessity to study communities at various temporal and spatial scales places constraints on investigations of the determinants of community richness and structure.

The hierarchical arrangement of parasite communities makes them ideal models to tackle these issues (26, 46, 94). Parasite assemblages range from those inside individual hosts, formed through local ecological processes acting over short periods of time, all the way to assemblages of parasite species exploiting entire host species, formed over long periods of time by evolutionary events acting across the host's entire geographic range. Parasite communities thus provide good opportunities to investigate the determinants of richness or structure at various scales. Communities of larval trematodes in snail hosts are one of two types of parasite communities that have received much attention from ecologists (61, 95). The other type are communities of metazoan parasites in vertebrate hosts. This review summarizes recent developments in the study of this latter type of parasite community. In particular, I emphasize the determinants of parasite species richness at all hierarchical levels of the community and the statistical artifacts and biases that may complicate the detection of underlying patterns of species richness.

Hierarchy of Parasite Assemblages

A common problem in community ecology is the need to specify boundaries for a species assemblage when no obvious physical barriers delimit the area occupied by the assemblage. This obstacle is easily overcome in parasite assemblages if we focus on adult stages in hosts and disregard larval stages in the environment. Parasites live in or on discrete habitats, or hosts, so whatever the level of study chosen it is possible to define the physical boundaries of the assemblage.

Parasite assemblages occur in host individuals. All parasites of different species within the same host form an infracommunity (46). They consist of species that may interact, either positively or negatively, and therefore ecological interactions may be important in determining their composition. The possibility of examining many host individuals from the same host population facilitates statistical tests of species co-occurrence and of the predictability of infracommunity composition. Infracommunities are often short lived, their maximum lifespan being that of the host, and in constant turnover because of new parasites being recruited and old ones dying out. Therefore, they represent communities assembled in ecological time from a pool of currently and locally available species.

This pool consists of all parasite species exploiting the host population at one point in time. These species form what is known as the component community (46). Each infracommunity is a subset of the species present in the component community, and thus the maximum richness of an infracommunity equals that of the component community. The component community is longer-lived than any of its infracommunities; it lasts at least a few host generations. Its composition changes as parasite species become locally extinct and as others arrive when new hosts migrate into the population from other host populations. These newcomers may bring parasite species adapted to the host species but not present until that moment in the host population because of historical events.

No single host population is likely to include all species of parasites known to exploit the host species. Instead, each component community is a subset of a larger collection of species that I refer to as the parasite fauna of the host species. The term *community* does not apply to this larger set because in many host species some parasite species in the fauna may never co-occur in the same host population. Parasite faunas are artificial rather than biological entities, worth discussing because they have been the subject of many studies. The theoretical maximum number of species in a component community is set by the size of the parasite fauna. In contrast with component communities and especially infracommunities, parasite faunas are much longer-lived and are assembled over evolutionary time. New species join the fauna as they switch host and add host species to the range that they can exploit. This occurs initially in one host population, i.e. in one parasite component community, and the new parasite species can subsequently spread to other host populations. Species are lost from parasite faunas when they become extinct from all host populations.

Thus, parasite assemblages can be studied at various hierarchical levels and at various spatial and temporal scales. At one extreme are infracommunities, formed within one host generation by ecological processes such as infection and death. At the other extreme are parasite faunas, assembled over the entire phylogenetic history of a host lineage by evolutionary processes such as host switching and extinction. The next sections review what is known about the determinants of species richness in parasite faunas, in their component communities, and in the infracommunities of which the latter consist.

Evolution and Richness of Parasite Faunas

Parasite faunas can gain new species in two ways. First, parasite species from sympatric host species can switch host, or colonize a new host species, and join the parasite fauna of that host species. This acquisition of parasite species may happen when different host species coexist and one (or more) provide suitable living conditions for another's parasites. Second, parasite lineages may undergo speciation within a host lineage without the host speciating. This would require gene flow to be interrupted between parts of a parasite population but not among its host population. The result of this could be one or more related parasite species exploiting the same host species. This phenomenon may be common, as the richness of parasite faunas is often inflated by the presence of many congeneric species (51). Parasite faunas can lose species through extinction, when a parasite species disappears from all populations of a host species.

Because colonization and extinction are important processes in the evolution of parasite faunas, as they are in other assemblages, and because of the insular nature of hosts as habitats, MacArthur & Wilson's (67) theory of island biogeography has proven popular among parasite ecologists (60, 94). That theory and others like it predict that variables associated with rates of extinction and colonization should explain the variability in species richness among communities. Several host life-history or ecological characteristics proposed as determinants of parasite faunal richness explain some of the variability in parasite species richness among host species. Host-species body size and geographic range, in particular, have proven relatively good predictors of faunal richness. Larger-bodied host species may provide more space and a greater diversity of niches to parasites. They also are more likely to be colonized by parasites because they consume more prey that may harbor larval parasites, and they live longer and are thus less ephemeral habitats than small-bodied, short-lived hosts. Similarly, a greater geographic range can result in encounters with and colonization by a greater number of parasite species. Other host characteristics have also been implicated as determinants of the richness of parasite faunas because of their potential influence on colonization or extinction rates. These include host density, diet, behavior, and various life-history traits (5, 33, 71, 85, 88). Also, because parasites are more likely to colonize a new host species related to their current host (81), the phyletic diversity of a host taxon may help to shape its parasite fauna.

Many species-area relationships involving either host body size or geographic range have been published (Tables 1, 2). In some of these studies, the number of species in either the richest or the typical component community was used instead of the richness of the entire parasite fauna, but this is unlikely to matter in such comparative analyses. The relationships obtained are almost invariably positive and often statistically significant. Some general trends are apparent; for instance, the richness of the parasite fauna correlates more strongly with body size among fish hosts and more strongly with geographic range among amphibian and reptile hosts than among hosts of other taxa (Tables 1, 2). However, because these have been obtained using widely different methods and

Table 1 Published relationships among host species between the species richness of the parasite fauna and either host body size or host geographic range

Host taxon[a]	Assemblage type[b]	Parasite group	Body size[c]	Geographic range[c]	References
Fish					
British freshwater fish (34)	F	Helminths	0.28	0.82	85
British freshwater fish (27)	F	Helminths		0.64	32
British freshwater fish (27)	F*	Helminths		0.17	32
British freshwater fish (32)	C	Helminths		0.68	36
Canadian freshwater fish (87)	C*	Helminths	0.55	0.32	5
North American freshwater fish (68)	F	All parasites	0.33	0.50	15
African freshwater fish (19)	F*	Monogenea	0.86	0.30	38
Marine fish (102)	C	Ectoparasites	0.31		86
Fish in general (72)	C*	Helminths	0.64		82
Fish in general (40)	C*	Ectoparasites	0.19		82
Amphibians					
Salamandridae (?)	F	Helminths		0.90	1
Plethodontidae (?)	F	Helminths		0.69	1
Ambystomatidae (?)	F	Helminths		0.94	1
Ranidae (?)	F	Helminths		0.77	1
Bufonidae (?)	F	Helminths		0.88	1
Reptiles					
Iguanidae (?)	F	Helminths		0.44	1
Scincidae (?)	F	Helminths		0.95	1
Emydidae (?)	F	Helminths		0.66	1
Birds					
Holarctic waterfowl (37)	F	Helminths	0.35	0.86	32
Holarctic waterfowl (35)	F*	Helminths	0.19	0.45	32
Birds in general (54)	C*	Helminths	0.36		82
Mammals					
North American rodents (40)	F	Mites		0.61	59
North American rodents (40)	F*	Mites		0.43	59
Desert rodents from Israel (12)	C*	Fleas	0.85	0.10	58
Indian mammals (12)	C	Protozoans and helminths	0.15	−0.10	101
Mammals in general (77)	C*	Helminths	0.43		82

[a]Number of host species included in the analyses is shown in parentheses.
[b]Patterns were observed using either the richness of the known parasite fauna of a host species (F) or the richness of the component communities in one or more populations of a host species (C). An asterisk denotes that the analysis controlled for sampling effort.
[c]Values are correlation coefficients, computed in different ways.

because many analyses have been performed on similar data sets, a rigorous meta-analysis is impossible and overall trends cannot be confirmed.

Two important statistical flaws weaken several of the results presented in Table 1. First, unequal sampling effort of the different host species in a study may either cause or mask a relationship between richness of the parasite fauna and either host body size or geographic range (32, 100). As more individual hosts are examined and as more host populations are sampled, the number of known parasite species in the parasite fauna increases asymptotically. A host species with few known parasites may have a species-poor fauna; it may also

Table 2 Published relationships among host species between the species richness of the parasite fauna and either host body size or host geographic range, in which corrections have been made for phylogenetic influences

Host taxon[a]	Assemblage type[b]	Parasite group	Body size[c]	Geographic range[c]	References
Fish					
Marine fish (67)	C	Ectoparasites	0.22		84
Fish in general (49)	C	Helminths	0.52		82
Fish in general (32)	C	Ectoparasites	0.28		82
Fish in general (51)	C	Helminths	0.32		34
Birds					
Holarctic waterfowl (12)	F	Helminths		0.60	32
Soviet birds (84)	F	Nematoda	0.23	0.05	33
Soviet birds (84)	F	Trematoda	0.30	0.10	33
Soviet birds (83)	F	Cestoda	0.13	0.10	33
Birds in general (41)	C	Helminths	0.19		82
Birds in general (48)	C	Helminths	0.08		34
Mammals					
Mammals in general (56)	C	Helminths	0.06		82
Mammals in general (70)	C	Helminths	0.18		34

[a]Number of phylogenetic contrasts included in the analyses is shown in parentheses.

[b]Patterns were observed using either the richness of the known parasite fauna of a host species (F) or the richness of the component communities in one or more populations of a host species (C). Richness measures were corrected for sampling effort in all studies.

[c]Values are correlation coefficients, computed through the origin.

have a rich fauna that is yet to be recorded. When corrections for unequal sampling effort are made in comparative analyses of parasite faunas, relationships with either host body size or geographic range become weaker (32, 59; see Table 1). The reason for this is that larger vertebrates with broader geographic ranges are the subject of more investigations and there are more complete lists of their parasites. Distinguishing between the respective effects of sampling effort and host size or range may prove complicated (37), but this needs to be done to properly assess the role of host characteristics.

The second flaw of many recent analyses is that they have overlooked the potential influence of host phylogeny in determining the richness of parasite faunas. When a host lineage splits into two new daughter lineages, the parasite species harbored by the ancestral host are likely to be inherited by the daughter host species. These sister lineages may then diverge with respect to body size or other ecological traits, and experience different rates of parasite colonization and extinction. However, since they start out with almost identical parasite faunas, they will have faunas more similar to one another than to those of other host species. Computer simulations generating host phylogenies in which parasite species are gained or lost in relation to host characteristics suggest that unless rates of parasite colonization and/or extinction are very high and strongly linked with host traits, phylogenetic inheritance of parasites is likely to obscure any effects of host traits (WL Vickery, R Poulin, in preparation).

Methods exist to control for phylogenetic influences in comparative analyses (40), and these should always be used when investigating patterns of variation in parasite faunal richness among host species. Controlling for phylogenetic effects can completely change the results of comparative analyses. As seen by comparing the results in Table 1 with those in Table 2, the strength of the relationships between faunal richness and host characteristics can be decreased following adequate corrections for host phylogeny (82). Also, different results are obtained because of a loss of statistical power when degrees of freedom are derived from independent phylogenetic contrasts rather than from species numbers. For example, when host species are treated as independent observations, aquatic vertebrates appear to have richer parasite faunas than do their terrestrial relatives (10). When phylogenetic contrasts are used to address the same question with the same data, there is no significant difference between the faunal richness of aquatic and terrestrial vertebrates (34, 82). Other trends, such as the greater richness of helminth communities in endotherms than in ectotherms, possibly resulting from differences in vagility and food consumption (30, 53), cannot be tested statistically using phylogenetically independent contrasts. If corrections for sampling effort and phylogenetic influences have such a big impact on correlations between characteristics of host species and the richness of their parasite fauna, much of the available evidence (i.e. most of the results presented in Table 1) may be unreliable.

Richness of Component Communities

Component communities consist of subsets of the species forming the parasite fauna of the host species. Different component communities from the same parasite fauna (i.e. the same host species) often share parasite species but only occasionaly have identical species composition. Component communities are at the interface between different parasite faunas. A new parasite species acquired from another host species during a host-switching event is at first present in a single-component community. Over time it may spread to other component communities following exchanges of host individuals among host populations. Therefore, host vagility and migrations of individual hosts across populations will influence the similarity in richness and composition among parasite component communities.

Comparisons among component communities from the same host species usually suffer from the same sampling effort bias as do comparisons among

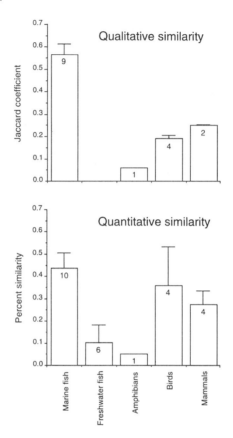

Figure 1 Qualitative and quantitative similarity (mean ± standard error) among component communities of helminth parasites in different groups of vertebrate hosts. Jaccard coefficients indicate the average proportion of parasite species shared by different component communities, whereas percentage similarity quantifies the similarity of different component communities based on the proportional abundance of their various parasite species; both measures range from zero to one. Numbers of host species on which the averages are based are shown in the columns. Data are from Refs. 1, 9, 21, 25, 27, 28, 43–45, 50, 68, 76, 106.

parasite faunas from different host species (100). Although corrections for un-equal sampling effort are never made, many studies have looked at the variation in richness and composition among component communities. In host species with populations isolated from one another, such as freshwater fish and am-phibians in which movements of individuals among populations are very lim-ited, typically, the similarity among component communities is much less than among component communities of vagile hosts such as marine fish and birds (Figure 1). Exchanges between component communities can take other routes, however. For instance, in parasite communities of freshwater fish, helminths using fish as intermediate hosts but maturing in birds have a greater likelihood of colonizing other component communities in other lakes than do helminths maturing in fish. A parasite fauna comprising many such parasites can consist of highly similar and predictable component communities (27).

What determines the richness of a component community? Although the variability among component communities within a parasite fauna often suggest they are merely stochastic assemblages of available species (27, 49), some characteristics of host populations or their habitat may in fact determine their composition and richness (39). For instance, the richness of the component communities in fish hosts often correlates with selected physicochemical characteristics of the lakes they inhabit, such as surface area, altitude, or depth (48, 68). Particular species of parasites may be excluded from certain component communities, i.e. certain lakes, because of unfavorable physicochemical conditions or because of the absence of suitable intermediate hosts (20).

When host populations are fragmented and displaced by natural events, or introduced to new locations by humans, their parasite component communities are likely to be species-poor in the period immediately following the displacement. This results from essential intermediate hosts being left behind or from founder effects, such as too few individuals of a parasite species to maintain itself in the new host population. Over time, the component community may reacquire these lost parasites when new host individuals migrate into the host population, as well as acquire new parasite species through host switching. There is evidence suggesting that time since the establishment of a host population can explain much of the variation in species richness among component communities (36). Not only the time since the displacement but also its distance may affect the richness and composition of parasite component communities. In freshwater salmonid fishes, host populations in their heartland, i.e. the geographical area in which the host taxon originated, harbor rich component communities consisting of salmonid specialist helminths (52). Salmonids have been introduced to many areas outside of their heartland; as the distance from the heartland increases, component communities become increasingly species-poor and increasingly formed of nonsalmonid specialist parasites. The distance from the heartland affects the likelihood of new parasite species joining a given component community as well as the nature of these new species.

Ultimately the richness of a component community is limited by the richness of the parasite fauna to which it belongs. Recent evidence, however, suggests that component communities reach a saturation level of species richness well below that of the parasite fauna (1, 54). Thus, the richness of component communities belonging to rich parasite faunas is not significantly influenced by the size of the species pool available in the fauna but rather by the finite number of niches available for parasites and/or by the local availability of parasite species.

Richness of Infracommunities

Just as component communities are subsets of species present in the parasite fauna, infracommunities are subsets of species present in the component community. Again, the upper limit on richness in infracommunities should be the richness of the component community to which they belong. However, saturation of infracommunities makes component community richness a poor predictor of infracommunity richness, at least in some fish (55) and mammal (83) hosts. In avian hosts, richer component communities consist of richer infracommunities, and no obvious saturation is apparent (83). Nevertheless, the richest infracommunity in a component community rarely includes all species found in the component community (see Table 3).

If individual hosts are random samplers of the parasites available in their environment, then infracommunities should be random assemblages of species found in the component community. One way of testing whether infracommunities are random assemblages is to compare the frequency distribution of their richnesses to that predicted by a null model (83). The Poisson distribution has been used as a null model (30), but it assumes equal probability of infection for all parasite species present in the component community. In reality, different species occur with different prevalences. A better null model can be obtained by computing the expected frequency of all possible combinations of species based on their respective prevalences (47). When the results of published studies are compared to the null model, it is clear that the observed distribution of infracommunity richnesses usually agrees with that expected from a random assemblage of species (Table 3).

What is the cause of the apparent statistical randomness of infracommunity species richness? There are at least three possibilities. First, the null models may not be entirely appropriate and could lead to some patterns escaping detection. This may be the case here: Prevalence is a good measure of parasite occurrence only if it is independent of species interactions, i.e. if the absence of a species from a host reflects a lack of infection rather than the elimination of that species by others following infection. Second, the statistical power of the test used may be too low to detect real but small departures from null expectations. Again, this may have been a problem in the analyses because of varying host sample sizes (Table 3). Third, the observed randomness may simply be real. All three explanations may apply here as well as to other uses of null models, and it is impossible to assess their respective contribution to the results. In general, though, the analyses in Table 3 suggest that at least some if not most component communities in vertebrate hosts consist of isolationist

Table 3 Distribution of infracommunity richnesses in the parasite component communities of vertebrates

Type (host sample size)[a]	Component community richness	Infracommunity richness, mean (range)	Prevalence[b] <0.1	Prevalence[b] >0.8	0 or 1 sp./host[c] Obs.	0 or 1 sp./host[c] Exp.	p[d]	References
Fish hosts								
Ecto (16)	20	8.4 (5–11)	.25	.30	0	0		87
Ecto (11)	16	8.2 (5–11)	.25	.38	0	0		87
Ecto (11)	12	7.3 (5–9)	.17	.42	0	0		87
Ecto (52)	4	0.3 (0–2)	.75	.00	49	51.3		57
Ecto (35)	13	3.4 (0–13)	.31	.00	11	2.7	<.001	35
Endo (295)	2	1.0 (0–2)	.00	.50	263	265.6		56
Endo (24)	4	3.6 (1–4)	.00	1.00	1	0.1		19
Amphibian hosts								
Endo (50)	4	0.7 (0–3)	.25	.00	42	44.7		30
Endo (115)	10	2.0 (0–7)	.60	.10	48	41.6	<.01	30
Endo (125)	10	1.9 (0–6)	.50	.00	54	56.4	<.05	30
Endo (107)	8	1.1 (0–4)	.50	.00	75	77.4		30
Bird hosts								
Endo (19)	2	0.1 (0–1)	1.00	.00	19	18.9		17
Endo (78)	4	0.1 (0–1)	1.00	.00	78	77.6		80
Endo (41)	4	1.6 (1–3)	.00	.00	22	14.6	<.05	78
Endo (104)	3	0.9 (0–2)	.33	.00	94	93.3		22
Endo (215)	42	5.1 (1–14)	.64	.00	3	3.6		98
Endo (45)	26	8.1 (3–14)	.27	.00	0	0	<.05	31
Endo (40)	11	2.8 (0–5)	.45	.09	3	3.3		90
Endo (1265)	5	0.1 (0–1)	1.00	.00	1265	1261.3		103
Endo (26)	4	1.2 (0–2)	.25	.00	18	17.5		7
Mammal hosts								
Proto (496)	5	1.5 (0–5)	.20	.00	236	257.7	<.001	92
Endo (215)	6	1.4 (0–4)	.17	.17	120	30.3	<.001	4
Endo (38)	6	1.8 (0–4)	.17	.00	18	15.3		91
Endo (400)	12	2.0 (0–7)	.33	.00	167	189.8	<.001	13
Endo (114)	8	2.6 (0–5)	.63	.13	9	7.3		29
Endo (56)	3	1.3 (0–3)	.33	.00	32	51.9	<.001	2
Endo (55)	11	2.2 (0–4)	.45	.00	9	10.3		3
Endo (14)	3	0.9 (0–3)	.00	.00	11	10.9		24
Endo (7)	10	6.3 (4–10)	.00	.40	0	0		6
Endo (21)	6	1.5 (0–3)	.00	.17	11	2.7	<.01	16
Endo (104)	23	3.5 (1–10)	.48	.00	17	5.9	<.01	77
Endo (46)	14	4.1 (1–8)	.43	.14	6	0.7	<.05	18
Endo (31)	6	0.9 (0–2)	.50	.00	23	21.7		42
Endo (36)	3	0.9 (0–2)	.67	.00	30	33.9		69
Endo (177)	20	6.5 (2–10)	.35	.10	0	0.1		79
Endo (100)	8	0.7 (0–3)	.88	.00	89	90.7		73
Endo (154)	12	2.4 (0–5)	.33	.00	37	36.9		23
Endo (53)	11	4.0 (1–8)	.27	.09	4	2.3	<.01	99
Endo (96)	10	3.0 (0–7)	.20	.10	14	14.7		14
Endo (150)	31	6.0 (0–19)	.52	.00	22	0.6	<.001	8
Endo (13)	7	3.5 (1–6)	.29	.14	1	0.3		72
Endo (84)	21	4.3 (0–9)	.43	.00	7	2.3		64
Endo (47)	13	2.8 (1–7)	.54	.08	6	5.0		64

[a]Type of parasite communities: Ecto = ectoparasitic metazoans, Endo = internal helminths, Proto = intestinal protozoans.

[b] Proportion of rare (prevalence < 0.1) and common (prevalence > 0.8) parasite species in the component community.

[c]Observed and expected frequencies of hosts harbouring either no parasites or a single species of parasite.

[d]Chi-square test comparing observed and expected distributions of infracommunity richnesses.

infracommunities, in which species interactions play no discernable role in structuring the assemblages.

The cases in which the observed distribution of infracommunity richnesses departs significantly from the null model may have several explanations. First, extreme heterogeneity among host individuals in susceptibility to infection may cause the discrepancies. Second, nonrepresentative samples of infracommunities may deviate from the expectations of the null model, whereas the whole component community does not. This explanation does not seem to apply to the list of examples in Table 3, as there were no differences in the number of hosts examined between component communities fitting the null model and those departing from it. Third, these examples can represent interactive communities, where the assemblage is structured to some extent by interspecific interactions among parasites. Positive interactions, such as the presence of one parasite species facilitating infection by other species, can generate more rich infracommunities than expected by chance. In Table 3, these could explain cases in which fewer hosts harbor one or no parasite species than is expected from the null model. On the other hand, negative interactions such as competitive exclusion can lead to fewer rich infracommunities than expected by chance and could account for instances in Table 3 where there were more uninfected hosts or hosts harboring a single parasite species than predicted by the null model.

There are two common methods used to test for interspecific interactions among parasites within infracommunities. The first consists of comparing the realized spatial niches of parasites in both the absence and presence of competing species (12, 75, 96). Observing a niche shift by one species in the presence of other parasites may indicate that interspecific interactions have selected for reductions in niche overlap through spatial segregation. Such functional responses, however, may lead to few if any changes in parasite numbers and in infracommunity composition and are of limited importance in the context of this review.

The second way of testing for interactions among parasite species is to contrast the number of positive pairwise associations with the number of negative associations among parasite species, across infracommunities within a component community (e.g. 41, 65, 70, 87). In parasite ecology (93, 94) as well as in ecology in general (102), there have been calls for observed patterns of species associations to be tested against truly adequate null models. Variance tests on binary presence-absence data for parasite species in infracommunities often assume that the number of positive covariances should equal the number of negative ones if infracommunities are random assemblages (89). Recently, Lotz & Font (66) used randomization procedures to refine this null expectation. They showed that a high proportion of rare parasite species, with low prevalence in the component community, can lead to an excess of negative associations, and that a high proportion of common species with high prevalence can produce an excess of positive associations. Among the results listed in Table 3, component communities in which the distribution of infracommunity richnesses departs from that expected from random assembly of species do not show unusual proportions of rare or common species. Still, such departures from random patterns should be examined individually using precise null models such as those obtained by Lotz & Font (66).

If the results of Table 3 are representative, most component communities of parasites may consist of infracommunities that are nothing more than unstructured, random assemblages of available species. Another approach to the detection of structuring forces or assembly rules involves testing for a nested subset pattern among infracommunities. This pattern in species composition is commonly reported from insular communities of free-living organisms (74, 104). In a parasite community, it would imply that the species forming a species-poor infracommunity are nonrandom subsets of progressively richer infracommunities. In other words, common parasite species would be found in infracommunities of various richnesses, but rare species would occur only in species-rich ones. This pattern was first investigated and indeed observed in an ectoparasite component community of a freshwater fish (35). However, further analyses of 38 component communities of fish ectoparasites (105) and 2 component communities of helminths of mammals (83) indicate that nested patterns may be very rare in parasite communities. These latest results also suggest that infracommunities are typically not predictable and structured but instead are assembled at random.

With the exception of trematode communities in their snail intermediate host (61, 95), most studies of parasite communities have focused on helminths in their vertebrate definitive hosts. Searching for patterns of species richness among infracommunities in definitive hosts may be fruitless, however, because of the way in which infracommunities recruit new members. Several helminths are acquired by vertebrates when the latter consume infected prey that serve as intermediate hosts for parasites. Clearly, because intermediate hosts often harbor many larval parasites of one or more species, new recruits to infracommunities in definitive hosts arrive as "packets" rather than as individuals (11, 63). Infracommunities in definitive hosts are simply the sum of randomly selected infracommunities in intermediate hosts minus the larval parasites that fail to establish in the definitive host. This phenomenon could explain many positive associations between pairs of species in definitive hosts; these could merely reflect associations within the intermediate host transmitted to the definitive host, and they would imply no actual species interactions (63, 66). In fact, if assemblages in intermediate hosts have a structure, then this structure gets passed on to the infracommunity in the definitive host as intermediate hosts are eaten by definitive hosts. Searching for structure in infracommunities of vertebrates may be pointless unless we expect species interactions in the definitive host such as competition or facilitation to be very influential and capable of reshaping the structure acquired from intermediate hosts. In general though, infracommunities may often be truly random assemblages, as suggested by the results summarized above.

Conclusions

Parasite communities in vertebrates are neatly arranged in hierarchical levels of organization and allow patterns of community richness or structure to be studied at various scales. At both the component community and infracommunity levels, easy replication is possible and allows for robust statistical testing. Despite these apparent advantages, the search for determinants of richness and structure is proving just as frustrating for parasite communities as it has in communities of free-living organisms. At the parasite fauna and component community levels, corrections for sampling effort and/or phylogenetic influences are not always made; when they are, patterns are typically less likely to be found. Future studies will need to include these corrections and to look at a broader range of host species. At the infracommunity level, comparisons of observed patterns with appropriate null models are only now becoming common practice, and these often reveal that infracommunity structure does not depart significantly from that of random assemblages. A shift of emphasis from infracommunities themselves toward recruitment processes is required, as much of the structure of infracommunities may originate from the arrival of new recruits in discrete packets. Based on the evidence currently available and until more results come in, it would seem that species richness in helminth communities in vertebrates rarely departs from random patterns at the infracommunity level but follows general trends at higher levels only. This is in sharp contrast with the evidence from larval trematode communities in snails (61, 95) and points to the importance of studying different types of parasite communities before drawing any general conclusions.

Literature Cited

1. Aho JM. 1990. Helminth communities of amphibians and reptiles: comparative approaches to understanding patterns and processes. See Ref. 26, pp. 157–95
2. Babero BB. 1953. Some helminth par-
asites of Alaskan beavers. *J. Parasitol.* 39:674–75
3. Babero BB, Lee JW. 1961. Studies on the helminths of nutria, *Myocastor coypus* (Molina), in Louisiana with check-list

of other worm parasites from this host. *J. Parasitol.* 47:378–90

4. Beck C, Forrester DJ. 1988. Helminths of the Florida manatee, *Trichecus manatus latirostris*, with a discussion and summary of the parasites of sirenians. *J. Parasitol.* 74:628–37

5. Bell G, Burt A. 1991. The comparative biology of parasite species diversity: internal helminths of freshwater fish. *J. Anim. Ecol.* 60:1047–64

6. Beverley-Burton M. 1971. Helminths from the Weddell seal, *Leptonychotes weddelli* (Lesson, 1826), in the Antarctic. *Can. J. Zool.* 49:75–83

7. Brooks DR, Hoberg EP, Houtman A. 1993. Some platyhelminths inhabiting white-throated sparrows, *Zonotrichia albicollis* (Aves: Emberizidae: Emberizinae), from Algonquin Park, Ontario, Canada. *J. Parasitol.* 79:610–12

8. Bucknell D, Hoste H, Gasser RB, Beveridge I. 1996. The structure of the community of strongyloid nematodes of domestic equids. *J. Helminthol.* 70:185–92

9. Bush AO. 1990. Helminth communities in avian hosts: determinants of pattern. See Ref. 26, pp. 197–232

10. Bush AO, Aho JM, Kennedy CR. 1990. Ecological versus phylogenetic determinants of helminth parasite community richness. *Evol. Ecol.* 4:1–20

11. Bush AO, Heard RW, Overstreet RM. 1993. Intermediate hosts as source communities. *Can. J. Zool.* 71:1358–63

12. Bush AO, Holmes JC. 1986. Intestinal helminths of lesser scaup ducks: an interactive community. *Can. J. Zool.* 64:142–52

13. Calero C, Ortiz P, De Souza L. 1950. Helminths in rats from Panama City and suburbs. *J. Parasitol.* 36:426

14. Calero C, Ortiz P, De Souza L. 1951. Helminths in cats from Panama City and Balboa, C.Z. *J. Parasitol.* 37:326

15. Chandler M, Cabana G. 1991. Sexual dichromatism in North American freshwater fish: Do parasites play a role? *Oikos* 60:322–28

16. Choquette LPE, Gibson GG, Pearson AM. 1969. Helminths of the grizzly bear, *Ursus arctos* L., in northern Canada. *Can. J. Zool.* 47:167–70

17. Christensen ZD, Pence DB. 1977. Helminths of the plain chachalaca, *Ortalis vetula maccalli*, from the South Rio Grande Valley. *J. Parasitol.* 63:830

18. Conti JA, Forrester DJ, Brady JR. 1983. Helminths of black bears in Florida. *Proc. Helminthol. Soc. Wash.* 50:252–56

19. Curran S, Caira JN. 1995. Attachment site specificity and the tapeworm assemblage in the spiral intestine of the blue shark (*Prionace glauca*). *J. Parasitol.* 81:149–57

20. Curtis MA, Rau ME. 1980. The geographical distribution of diplostomiasis (Trematoda: Strigeidae) in fishes from northern Quebec, Canada, in relation to the calcium ion concentrations of lakes. *Can. J. Zool.* 58:1390–94

21. Custer JW, Pence DB. 1981. Ecological analyses of helminth populations of wild canids from the Gulf coastal prairies of Texas and Louisiana. *J. Parasitol.* 67:289–307

22. Dancak K, Pence DB, Stormer FA, Beasom SL. 1982. Helminths of the scaled quail, *Callipepla squamata*, from northwest Texas. *Proc. Helminthol. Soc. Wash.* 49:144–46

23. Deblock S, Pétavy AF, Gilot B. 1988. Helminthes intestinaux du renard commun (*Vulpes vulpes* L.) dans le Massif Central (France). *Can. J. Zool.* 66:1562–69

24. DeBruin D, Pfaffenberger GS. 1984. Helminths of desert cottontail rabbits (*Sylvilagus auduboni* (Baird)) inhabiting prairie dog towns in eastern New Mexico. *Proc. Helminthol. Soc. Washington* 51:369–70

25. Edwards DD, Bush AO. 1989. Helminth communities in avocets: importance of the compound community. *J. Parasitol.* 75:225–38

26. Esch GW, Bush AO, Aho JM. 1990. *Parasite Communities: Patterns and Processes.* London: Chapman & Hall

27. Esch GW, Kennedy CR, Bush AO, Aho JM. 1988. Patterns in helminth communities in freshwater fish in Great Britain: alternative strategies for colonization. *Parasitology* 96:519–32

28. Fedynich AM, Pence DB, Gray PN, Bergan JF. 1996. Helminth community structure and pattern in two allopatric populations of a nonmigratory waterfowl species (*Anas fulvigula*). *Can. J. Zool.* 74:1253–59

29. Forrester DJ, Pence DB, Bush AO, Lee DM, Holler NR. 1987. Ecological analysis of the helminths of round-tailed muskrats (*Neofiber alleni* True) in southern Florida. *Can. J. Zool.* 65:2976–79

30. Goater TM, Esch GW, Bush AO. 1987. Helminth parasites of sympatric salamanders: ecological concepts at infracommunity, component and compound community levels. *Am. Midl. Nat.* 118:289–300

31. Gray CA, Gray PN, Pence DB. 1989. Influence of social status on the helminth

community of late-winter mallards. *Can. J. Zool.* 67:1937–44

32. Gregory RD. 1990. Parasites and host geographic range as illustrated by waterfowl. *Funct. Ecol.* 4:645–54

33. Gregory RD, Keymer AE, Harvey PH. 1991. Life history, ecology and parasite community structure in Soviet birds. *Biol. J. Linn. Soc.* 43:249–62

34. Gregory RD, Keymer AE, Harvey PH. 1996. Helminth parasite richness among vertebrates. *Biodiv. Cons.* 5:985–97

35. Guégan J-F, Hugueny B. 1994. A nested parasite species subset pattern in tropical fish: host as major determinant of parasite infracommunity structure. *Oecologia* 100:184–89

36. Guégan J-F, Kennedy CR. 1993. Maximum local helminth parasite community richness in British freshwater fish: a test of the colonization time hypothesis. *Parasitology* 106:91–100

37. Guégan J-F, Kennedy CR. 1996. Parasite richness/sampling effort/host range: the fancy three-piece jigsaw puzzle. *Parasitol. Today* 12:367–69

38. Guégan J-F, Lambert A, Lévêque C, Combes C, Euzet L. 1992. Can host body size explain the parasite species richness in tropical freshwater fishes? *Oecologia* 90:197–204

39. Hartvigsen R, Halvorsen O. 1994. Spatial patterns in the abundance and distribution of parasites of freshwater fish. *Parasitol. Today* 10:28–31

40. Harvey PH, Pagel MD. 1991. *The Comparative Method in Evolutionary Biology* Oxford, UK: Oxford Univ. Press

41. Haukisalmi V, Henttonen H. 1993. Coexistence in helminths of the bank vole *Clethrionomys glareolus*. I. Patterns of co-occurrence. *J. Anim. Ecol.* 62:221–29

42. Hoberg EP, McGee SG. 1982. Helminth parasitism in raccoons, *Procyon lotor hirtus* Nelson and Goldman, in Saskatchewan. *Can. J. Zool.* 60:53–57

43. Hoberg EP, Ryan PG. 1989. Ecology of helminth parasitism in *Puffinus gravis* (Procellariiformes) on the breeding grounds at Gough Island. *Can. J. Zool.* 67:220–25

44. Hogans WE, Dadswell MJ, Uhazy LS, Appy RG. 1993. Parasites of American shad, *Alosa sapidissima* (Osteichthyes: Clupeidae), from rivers of the North American Atlantic coast and the Bay of Fundy, Canada. *Can. J. Zool.* 71:941–46

45. Holmes JC. 1990. Helminth communities in marine fishes. See Ref. 26, pp. 101–30

46. Holmes JC, Price PW. 1986. Communities of parasites. In *Community Ecology: Patterns and Processes,* ed. J Kikkawa, DJ Anderson, pp. 187–213. Oxford: Blackwell

47. Janovy J Jr, Clopton RE, Clopton DA, Snyder SD, Efting A, Krebs L. 1995. Species density distributions as null models for ecologically significant interactions of parasite species in an assemblage. *Ecol. Model.* 77:189–96

48. Kennedy CR. 1978. An analysis of the metazoan parasitocoenoses of brown trout *Salmo trutta* from British lakes. *J. Fish Biol.* 13:255–63

49. Kennedy CR. 1990. Helminth communities in freshwater fish: structured communities or stochastic assemblages? See Ref. 26, pp. 131–56

50. Kennedy CR. 1995. Richness and diversity of macroparasite communities in tropical eels *Anguilla reinhardtii* in Queensland, Australia. *Parasitology* 111:233–45

51. Kennedy CR, Bush AO. 1992. Species richness in helminth communities: the importance of multiple congeners. *Parasitology* 104:189–97

52. Kennedy CR, Bush AO. 1994. The relationship between pattern and scale in parasite communities: a stranger in a strange land. *Parasitology* 109:187–96

53. Kennedy CR, Bush AO, Aho JM. 1986. Patterns in helminth communities: Why are birds and fish different? *Parasitology* 93:205–15

54. Kennedy CR, Guégan J-F. 1994. Regional versus local helminth parasite richness in British freshwater fish: saturated or unsaturated parasite communities? *Parasitology* 109:175–85

55. Kennedy CR, Guégan J-F. 1996. The number of niches in intestinal helminth communities of *Anguilla anguilla*: Are there enough spaces for parasites? *Parasitology* 113:293–302

56. Kennedy CR, Moriarty C. 1987. Coexistence of congeneric species of Acanthocephala: *Acanthocephalus lucii* and *A. anguillae* in eels *Anguilla anguilla* in Ireland. *Parasitology* 95:301–10

57. Kozel TR, Whittaker FH. 1985. Ectoparasites of the blackstripe topminnow, *Fundulus notatus*, from Harrods Creek, Oldham County, Kentucky. *Proc. Helminthol. Soc. Wash.* 52:314–15

58. Krasnov BR, Shenbrot GI, Medvedev SG, Vatschenok VS, Khokhlova IS. 1997. Host-habitat relations as an important determinant of spatial distribution of flea assemblages (Siphonaptera) on rodents in

the Negev Desert. *Parasitology* 114:159–73

59. Kuris AM, Blaustein AR. 1977. Ectoparasitic mites on rodents: application of the island biogeography theory? *Science* 195:596–98

60. Kuris AM, Blaustein AR, Alió JJ. 1980. Hosts as islands. *Am. Nat.* 116:570–86

61. Kuris AM, Lafferty, KD. 1994. Community structure: larval trematodes in snail hosts. *Annu. Rev. Ecol. Syst.* 25:189–217

62. Levin SA. 1992. The problem of pattern and scale in ecology. *Ecology* 73:1943–67.

63. Lotz JM, Bush AO, Font WF. 1995. Recruitment-driven, spatially discontinuous communities: a null model for transferred patterns in target communities of intestinal helminths. *J. Parasitol.* 81:12–24

64. Lotz JM, Font WF. 1985. Structure of enteric helminth communities in two populations of *Eptesicus fuscus* (Chiroptera). *Can. J. Zool.* 63:2969–78

65. Lotz JM, Font WF. 1991. The role of positive and negative interspecific associations in the organization of communities of intestinal helminths of bats. *Parasitology* 103:127–38

66. Lotz JM, Font WF. 1994. Excess positive associations in communities of intestinal helminths of bats: a refined null hypothesis and a test of the facilitation hypothesis. *J. Parasitol.* 80:398–413

67. MacArthur RH, Wilson EO. 1967. *The Theory of Island Biogeography* . Princeton, NJ: Princeton Univ. Press

68. Marcogliese DJ, Cone DK. 1991. Importance of lake characteristics in structuring parasite communities of salmonids from insular Newfoundland. *Can. J. Zool.* 69:2962–67

69. Meyer MC, Chitwood BG. 1951. Helminths from fisher (*Martes p. pennanti*) in Maine. *J. Parasitol.* 37:320–21

70. Moore J, Simberloff D. 1990. Gastrointestinal helminth communities of bobwhite quail. *Ecology* 71:344–59

71. Morand S, Poulin R. 1997. Density, body mass and parasite species richness of terrestrial mammals. *Evol. Ecol.* In press

72. Nickel PA, Hansen MF. 1967. Helminths of bats collected in Kansas, Nebraska and Oklahoma. *Am. Midl. Nat.* 78:481–86

73. Palmieri JR, Thurman JB, Andersen FL. 1978. Helminth parasites of dogs in Utah. *J. Parasitol.* 64:1149–50

74. Patterson BD, Atmar W. 1986. Nested subsets and the structure of insular mammalian faunas and archipelagos. *Biol. J. Linn. Soc.* 28:65–82

75. Patrick MJ. 1991. Distribution of enteric helminths in *Glaucomys volans* L. (Sciuridae): a test for competition. *Ecology* 72:755–58

76. Pence DB. 1990. Helminth community of mammalian hosts: concepts at the infracommunity, component and compound community levels. See Ref. 26, pp. 233–60

77. Pence DB, Crum JM, Conti JA. 1983. Ecological analyses of helminth populations in the black bear, *Ursus americanus*, from North America. *J. Parasitol.* 69:933–50

78. Pence DB, Murphy JT, Guthery FS, Doerr TB. 1983. Indications of seasonal variation in the helminth fauna of the lesser prairie chicken, *Tympanuchus pallidicinctus* (Ridgway) (Tetraonidae), from northwestern Texas. *Proc. Helminthol. Soc. Wash.* 50:345–47

79. Pence DB, Windberg LA. 1984. Population dynamics across selected habitat variables of the helminth community in coyotes, *Canis latrans*, from south Texas. *J. Parasitol.* 70:735–46

80. Pence DB, Young VE, Guthery FS. 1980. Helminths of the ring-necked pheasant, *Phasianus colchicus* (Gmelin) (Phasianidae), from the Texas Panhandle. *Proc. Helminthol. Soc. Wash.* 47:144–47

81. Poulin R. 1992. Determinants of host-specificity in parasites of freshwater fishes. *Int. J. Parasitol.* 22:753–58

82. Poulin R. 1995. Phylogeny, ecology, and the richness of parasite communities in vertebrates. *Ecol. Monogr.* 65:283–302

83. Poulin R. 1996. Richness, nestedness, and randomness in parasite infracommunity structure. *Oecologia* 105:545–51

84. Poulin R, Rohde K. 1997. Comparing the richness of metazoan ectoparasite communities of marine fishes: controlling for host phylogeny. *Oecologia* 110:278–83

85. Price PW, Clancy KM. 1983. Patterns in number of helminth parasite species in freshwater fishes. *J. Parasitol.* 69:449–54

86. Rohde K, Hayward C, Heap M. 1995. Aspects of the ecology of metazoan ectoparasites of marine fishes. *Int. J. Parasitol.* 25:945–70

87. Rohde K, Hayward C, Heap M, Gosper D. 1994. A tropical assemblage of ectoparasites: gill and head parasites of *Lethrinus miniatus* (Teleostei, Lethrinidae). *Int. J. Parasitol.* 24:1031–53

88. Sasal P, Morand S, Guégan J-F. 1997. Determinants of parasite species richness in Mediterranean marine fishes. *Mar. Ecol. Prog. Ser.* In press

89. Schluter D. 1984. A variance test for detecting species associations, with some example applications. *Ecology* 65:998–1005
90. Scott ME. 1984. Helminth community in the belted kingfisher, *Ceryle alcyon* (L.), in southern Quebec. *Can. J. Zool.* 62:2679–81
91. Seidenberg AJ, Kelly PC, Lubin ER, Buffington JD. 1974. Helminths of the cotton rat in southern Virginia, with comments on the sex ratios of parasitic nematode populations. *Am. Midl. Nat.* 92:320–26
92. Seville RS, Stanton NL, Gerow K. 1996. Stable parasite guilds: coccidia in spermophiline rodents. *Oikos* 75:365–72
93. Simberloff D. 1990. Free-living communities and alimentary tract helminths: hypotheses and pattern analyses. See Ref. 26, pp. 289–319
94. Simberloff D, Moore J. 1996. Community ecology of parasites and free-living animals. In *Host-Parasite Evolution: General Principles and Avian Models,* ed. DH Clayton, J Moore, pp. 174–97. Oxford, UK: Oxford Univ. Press
95. Sousa WP. 1992. Interspecific interactions among larval trematode parasites of freshwater and marine snails. *Am. Zool.* 32:583–92
96. Stock TM, Holmes JC. 1988. Functional relationships and microhabitat distributions of enteric helminths of grebes (Podicipedidae): the evidence for interactive communities. *J. Parasitol.* 74:214–27
97. Strong DR, Simberloff D, Abele LG, Thistle AB. 1984. *Ecological Communities: Conceptual Issues and the Evidence.* Princeton, NJ: Princeton Univ. Press
98. Thul JE, Forrester DJ, Abercrombie CL. 1985. Ecology of parasitic helminths of wood ducks, *Aix sponsa,* in the Atlantic flyway. *Proc. Helminthol. Soc. Wash.* 52:297–310
99. Waid DD, Pence DB. 1988. Helminths of mountain lions (*Felis concolor*) from southwestern Texas, with a redescription of *Cylicospirura subaequalis* (Molin, 1860) Vevers, 1922. *Can. J. Zool.* 66:2110–17
100. Walther BA, Cotgreave P, Price RD, Gregory RD, Clayton DH. 1995. Sampling effort and parasite species richness. *Parasitol. Today* 11:306–10
101. Watve MG, Sukumar R. 1995. Parasite abundance and diversity in mammals: correlates with host ecology. *Proc. Natl. Acad. Sci. USA* 92:8945–49
102. Weiher E, Keddy PA. 1995. Assembly rules, null models, and trait dispersion: new questions from old patterns? *Oikos* 74:159–64
103. Williams IC, Newton I. 1969. Intestinal helminths of the bullfinch, *Pyrrhula pyrrhula* (L.), in southern England. *Proc. Helminthol. Soc. Wash.* 36:76–82
104. Worthen WB. 1996. Community composition and nested-subset analyses: basic descriptors for community ecology. *Oikos* 76:417–26
105. Worthen WB, Rohde K. 1996. Nested subset analyses of colonization-dominated communities: metazoan ectoparasites of marine fishes. *Oikos* 75:471–78
106. Yanez DM, Canaris AG. 1988. Metazoan parasite community composition and structure of migrating Wilson's phalarope, *Steganopus tricolor* Viellot, 1819 (Aves), from El Paso County, Texas. *J. Parasitol.* 74:754–62

Annu. Rev. Ecol. Syst. 1997. 28:359–89

HYBRID ORIGINS OF PLANT SPECIES

Loren H. Rieseberg

Biology Department, Indiana University, Bloomington, Indiana 47405;
e-mail: lriesebe@bio.indiana.edu

KEY WORDS: plants, hybridization, introgression, reproductive isolation, speciation

ABSTRACT

The origin of new homoploid species via hybridization is theoretically difficult because it requires the development of reproductive isolation in sympatry. Nonetheless, this mode is often and carelessly used by botanists to account for the formation of species that are morphologically intermediate with respect to related congeners. Here, I review experimental, theoretical, and empirical studies of homoploid hybrid speciation to evaluate the feasibility, tempo, and frequency of this mode. Theoretical models, simulation studies, and experimental syntheses of stabilized hybrid neospecies indicate that it is feasible, although evolutionary conditions are stringent. Hybrid speciation appears to be promoted by rapid chromosomal evolution and the availability of a suitable hybrid habitat. A selfing breeding system may enhance establishment of hybrid species, but this advantage appears to be counterbalanced by lower rates of natural hybridization among selfing taxa. Simulation studies and crossing experiments also suggest that hybrid speciation can be rapid—a prediction confirmed by the congruence observed between the genomes of early generation hybrids and ancient hybrid species. The frequency of this mode is less clear. Only eight natural examples in plants have been rigorously documented, suggesting that it may be rare. However, hybridization rates are highest in small or peripheral populations, and hybridization may be important as a stimulus for the genetic or chromosomal reorganization envisioned in founder effect and saltational models of speciation.

INTRODUCTION

Hybridization may have several evolutionary consequences, including increased intraspecific genetic diversity (2), the origin and transfer of genetic adaptations

359

(2, 93), the origin of new ecotypes or species (42, 102), and the reinforcement or breakdown of reproductive barriers (27, 55, 77). Although the frequency and importance of these outcomes are not yet clear in either plants or animals, a critical body of data is now available for assessing the mechanistic basis and frequency of one of these—the origin of new species. The last comprehensive review of this topic in relation to plants was Grant's (42) monograph "Plant Speciation." Grant listed six mechanisms by which the breeding behavior of hybrids could be stabilized, thus providing the potential for speciation:

1. asexual reproduction;

2. permanent translocation heterozygosity;

3. permanent odd polyploidy;

4. allopolyploidy;

5. the stabilization of a rare hybrid segregate isolated by postmating barriers;

6. the stabilization of a rare hybrid segregate isolated by premating barriers.

The first three of these mechanisms generate flocks of clonal or uniparental microspecies that span the range of morphological variability between the parental species. Sexual reproduction among microspecies is limited or absent, making it difficult to discuss their origin and evolution in the context of sexual isolation and speciation. By contrast, the latter three mechanisms generate sexual derivatives and therefore have the potential to give rise to new biological species.

This review focuses on the origin of sexual, homoploid hybrid species (mechanisms 5 and 6), (but see 50, 89 for reviews of polyploidy in plants). After clarification of concepts and terminology, the historical basis of our current understanding of hybrid speciation is reviewed. This is followed by examination of the frequency of natural hybridization and an exploration of experimental and theoretical studies that test the feasibility of homoploid hybrid speciation. Once the feasibility of this mode of speciation has been established, I briefly critique the methods used for identifying homoploid hybrid species in nature and then focus on those examples of hybrid speciation that are well established. Finally, I discuss promising areas for future research and possible approaches that may facilitate studies of this mode.

WHAT IS A HYBRID SPECIES?

Both "hybrid" and "species" can have several meanings for evolutionary biologists. The term hybrid can be restricted to organisms formed by cross-fertilization between individuals of different species, or it can be defined more broadly as the offspring between individuals from populations "which are distinguishable on the basis of one or more heritable characters" (44). I prefer this broader definition of hybrids, as it provides greater flexibility in usage. Nonetheless, in this review, I focus on hybrids formed by crosses between species.

The term species has a much wider variety of definitions, ranging from concepts based on the ability to interbreed to those based on common descent. Mayr's (59) biological species concept—"species are groups of interbreeding natural populations which are reproductively isolated from all other such groups"—is perhaps the most widely accepted of these. Although I have previously expressed concern about the limitations of this concept (73), its emphasis on reproductive isolation does offer a straightforward approach to the study of speciation (20). Moreover, the evolution of reproductive barriers is particularly crucial to the successful origin of new hybrid species; otherwise, the new hybrid lineage will be swamped by gene flow with its parents. Thus, the focus of this review is on the evolution of reproductive isolation between new hybrid lineages and their parents.

HISTORICAL PERSPECTIVE

The hypothesis that new species may arise via hybridization appears to have originated with Linnaeus (58; cited in 84), who wrote "it is impossible to doubt that there are new species produced by hybrid generation. ... For thence it appears to follow, that the many species of plants in the same genus in the beginning could not have been otherwise than one plant, and have arisen from this hybrid generation." This represents a modification of the orthodox view of special creation, which asserted that all existing species were created by the hand of God and which denied the existence of constant hybrids (15). However, Linnaeus' observations were limited to F_1 hybrids, and he was unaware of potential difficulties with his hypothesis such as segregation and sterility.

Rigorous experimental study of plant hybridization was initiated by Joseph Kölreuter in 1760 and led to two critical discoveries (84). First, Kölreuter found that a hybrid from *Nicotiana paniculata* $\times$ *N. rustica* produced no seeds—the first "botanical mule." As a result, Kölreuter concluded that hybrid plants are produced only with difficulty and are unlikely to occur in nature in the absence of human intervention or disturbance to the habitat. Second, Kölreuter and his successor, Carl Gartner, discovered that later generation hybrids tended to revert back to the parental forms, thus refuting the existence of constant hybrids and supporting the orthodox view of special creation (84). The views of Kölreuter and Gartner on the lack of constancy of hybrids (although not necessarily on creation) were held by most other prominent botanical hybridizers during the eighteenth and nineteenth centuries, including Charles Darwin, John

Goss, Thomas Laxton, Patrick Shireff, Gregor Mendel, Charles Naudin, and WO Focke (84).

Nonetheless, reports of constant hybrids continued to arise. For instance, Herbert (49) noted that hybrid varieties of plants sometimes preserve themselves almost as distinctly as species. These reports, although controversial, were taken seriously by prominent botanists such as Mendel (61), who emphasized in his *Pisum* paper: "This feature is of particular importance to the evolutionary history of plants, because constant hybrids attain the status of *new species*" (emphasis in the original). Naudin (66) also recognized the possibility that hybrid characters may become fixed in later generations and that this may facilitate species formation. This represented the first explicit recognition of the possibility that later-generation hybrids may become stabilized and thus foreshadowed modern models of hybrid speciation.

The potential role of hybridization in species formation was taken a step further by Anton Kerner (51), who recognized the important role of habitat in governing hybrid species establishment. Kerner realized that although hybrids were frequently formed in nature, their successful establishment required suitable open habitat that was not occupied by the parental species. This was a significant contribution, as ecological divergence plays an important role in current models of hybrid speciation. However, Kerner restricted his discussion to fertile hybrids, thus ignoring the sterility problems associated with many hybrids. Furthermore, like Linnaeus, Kerner failed to recognize the potential problem of segregation, and it was not until the rediscovery of Mendel's work in the early portion of the twentieth century that the old problem of hybrid speciation could be stated correctly—what is the mechanism by which a new fertile and constant hybrid lineage could arise and become reproductively isolated from its parents?

The first major contribution to this problem was made by Winge (103), who postulated that a fertile and constant hybrid species could be derived instantaneously by the duplication of a hybrid's chromosome complement (i.e. allopolyploidy). This hypothesis was quickly confirmed experimentally in a variety of plant species, and allopolyploidy is now recognized to be a prominent mode of speciation in flowering plants and ferns (89, 93).

By contrast, the feasibility of hybrid speciation in the absence of ploidal change remained unsolved until it was addressed by Müntzing (65). He postulated that the sorting of chromosomal rearrangements in later-generation hybrids could, by chance, lead to the formation of new population systems that were homozygous for a unique combination of chromosomal sterility factors (Figure 1). The new hybrid population would be fertile, stable, and at the same ploidal level as its parents, yet partially reproductively isolated from both parental species due to a chromosomal sterility barrier. Although early authors

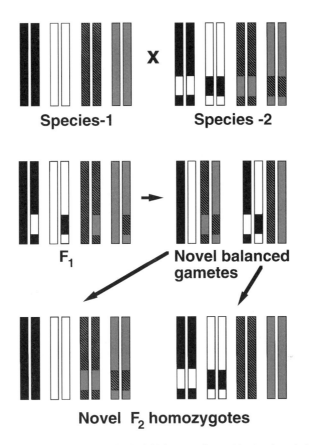

Figure 1   Simple chromosomal model for the initial stages of recombinational speciation. The two parental species have the same diploid chromosome number ($2n = 8$), but differ by two reciprocal translocations. The first generation hybrid will be heterozygous for the parental chromosomal rearrangements and will generate 16 different classes of gametes with respect to chromosome structure (not shown). Twelve of these will be unbalanced and presumably inviable due to deletions and/or insertions. The remaining four will be balanced and viable. Two of the four will recover parental chromosome structures, whereas the final two will have recombinant karyotypes. If selfed, a small fraction of F_2 individuals will be recovered that possess a novel homokaryotype. These F_2s will be fertile and stable but will be at least partially intersterile with the parental species.

focused on chromosomal rearrangements (33, 65), it is clear that the sorting of genic sterility factors should generate similar results. Thus, current models incorporate both genic and chromosomal sterility factors. Modern contributions to the study of this process, termed recombinational speciation (39), include rigorous experimental and theoretical tests of the model as well as the gradual accumulation of well-documented case studies from nature.

Concomitant with the development of the recombinational speciation model was the growing recognition that new hybrid species might become isolated from their parents by premating barriers rather than hybrid sterility or inviability. In fact, this view was implicit in Kerner's (51) account of the hybrid origin of *Rhodendron intermedium*, which appears to be partially isolated from its parents due to soil preferences and the behavior of pollinators. More recently, Grant (38) presented a model for sympatric speciation in flowering plants based on flower constancy of pollinators and then suggested that hybridization might be as likely as mutation in generating the new floral structures required for speciation. Unfortunately, rigorous experimental or theoretical studies of this hypothesis have not been conducted. Nonetheless, a number of empirical studies have identified hybrid taxa that are isolated by premating barriers alone, and new recombinational species are typically isolated from the parental species by both premating and postmating barriers.

THE FREQUENCY OF INTERSPECIFIC HYBRIDIZATION

Data on naturally occurring plant hybrids are quite extensive, and several compilations illustrate the extent of hybridization in nature. Perhaps the most comprehensive listing is Knobloch's (52) compilation of 23,675 putative examples of interspecific or intergeneric hybrids, but this figure must be interpreted with caution. Some of the hybrids appear to be fanciful, and many known hybrids have been omitted (92). Moreover, Knobloch included both natural and artificial hybrids, and it is not clear what fraction of this listing comprises natural hybrids.

A more reliable indicator of the frequency of hybridization comes from a recent review of five biosystematic floras (28). The frequency of natural hybrids when compared to the total number of species in the flora ranged from approximately 22% for the British flora to 5.8% for the intermountain flora of North America, with an average of 11% over all five floras. Assuming a similar frequency of natural hybrids worldwide, this would suggest a worldwide total of 27,500 hybrid combinations among the 250,000 described plant species. Although this is a sizable number, it may represent a substantial underestimate; many hybrids have gone undetected due to inadequate systematic attention to certain groups in certain floras.

Natural hybrids also were found to be unevenly distributed taxonomically (28). Only 16%–34% of plant families and 6%–16% of genera have one or more reported hybrids. Thus, contemporary hybridization may not be as common or ubiquitous as believed, but appears to be concentrated in a small fraction of families and genera. Notably, the life-history characteristics significantly associated with these hybridizing genera include perennial habit, outcrossing breeding systems, and asexual reproductive modes that allow stabilization of hybrid reproduction.

The estimates provided above indicate that the rate of natural hybrid formation in plants is sufficiently high to provide ample opportunity for homoploid hybrid speciation. Although hybrid speciation seems most likely to be important in family or genera with high rates of contemporary hybridization, even rare hybridization events can be evolutionarily important, as a single, partially fertile, hybrid individual can suffice as the progenitor of a new evolutionary lineage (28).

THEORY

Models

Homoploid hybrid speciation is unusual because not only does it involve hybridization between taxa at the same ploidal level, but it also represents a type of sympatric speciation, as the parental species must co-occur geographically to produce hybrids. Chromosomal models for this process were proposed by Stebbins (94) and Grant (39) and can be summarized as follows (Figure 1):

1. Two parental species are distinguished by two or more separable chromosomal rearrangements.

2. Their partially sterile hybrid gives rise via segregation and recombination to new homozygous recombinant types for the rearrangements.

3. The recombinant types are fertile within the line but at least partially sterile with both parents.

Grant (39) also noted that the formation of new structural homozygotes in the progeny of a hybrid is more likely under conditions of inbreeding than of outbreeding. This leads to the prediction that recombinational speciation should be more common in selfing species than outcrossers.

A more general model has recently been proposed by Templeton (98) that incorporates both chromosomal and genic incompatibility and recognizes the important roles of selection and ecological divergence. Moreover, this model can be applied to the stabilization of hybrid segregates isolated by either premating or postmating barriers. There are four critical steps in Templeton's model:

1. Hybridization is followed by inbreeding and hybrid breakdown due to chromosomal or genic incompatibilities.

2. Hybrid segregates with the highest fertility or viability are favored by selection. (This differs from the model of Grant & Stebbins, who emphasized the role of chance in generating novel homokaryotypes).

3. A new hybrid genotype may become stabilized if it becomes reproductively isolated from the parental species. Otherwise, it will be overcome by gene flow with its parents. Presumably, reproductive isolation will evolve as a by-product of selection for increased viability or fertility, rather than by selection for particular chromosomal rearrangements (although see 86).

4. Once a hybrid genotype becomes stabilized, it must co-exist with one or both parents or occupy a new ecological niche. Either outcome requires ecological divergence.

Templeton emphasized two factors that appear to facilitate this mode. First, he argues that the evolution of reproductive barriers between the stabilized hybrid genotype and its parents could be facilitated by rapid chromosomal evolution. The presence of chromosomal rearrangements in a segregating population can lead to further chromosomal breakage (81, 86), particularly if accompanied by inbreeding (56). Moreover, there is substantial evidence that genic mutation rates also increase in hybrid populations (13). The elevated chromosomal and genic mutation rates in hybrid populations are often referred to as hybrid dysgenesis, and it now appears to be the rule rather than the exception in hybrid populations (13). The rate of chromosomal evolution should also be enhanced by inbreeding and/or population subdivision because both will reduce effective population sizes and increase the fixation of novel chromosomal rearrangements through drift.

A second factor that Templeton considered even more critical to rates of speciation via this mode is the availability of suitable habitat. The importance of habitat for the establishment and success of hybrids was previously recognized by Kerner (51) and Anderson (2). Kerner commented on the critical role of "open habitats" for the establishment of hybrids, whereas Anderson emphasized the importance of habitat disturbance for facilitating breakdown of premating reproductive barriers between previously isolated parental species, and for providing suitable habitat for hybrid segregates, which often diverge ecologically from both parents (2). Moreover, as is discussed in detail later in this review, most bona fide hybrid species are found in habitats that are extreme relative to the parental species (71, 100), implying a critical role for habitat availability and ecological divergence in homoploid hybrid speciation.

Simulation Studies

The only detailed quantitative study of the feasibility and dynamics of hybrid speciation was by McCarthy et al (60) who focused on the strict chromosomal or recombinational model and used computer simulations to test how various parameters affect rates of establishment and spread of recombinational species in a spatially structured environment. The model simulated breeding in a hybrid zone between two hermaphroditic plant species with nonoverlapping discreet generations and no seed dormancy. The effects of five parameters were tested: (a) fertility of F_1s, (b) relative fitnesses of stabilized derivatives, (c) selfing rate, (d) number of chromosomal differences between the parental species, and (e) hybrid zone interface length (this parameter increases the number of hybrid matings per generation).

Most of the findings were intuitive and consistent with the simple genetic models of Grant (39) and Stebbins (94). In general, the process was facilitated by increased F_1 fertility, a great selective advantage for recombinant types, high selfing rates, a small number of chromosomal differences between the parental species, and a long hybrid zone interface. However, recombinational speciation was possible even with F_1 fertilities as low as 0.018, although the number of generations required for the new species to become established was high. Likewise, outcrossing retarded the speciation process, but given a sufficiently high selective advantage for the stabilized derivatives ($\alpha = 2.0$), the process became feasible even under conditions of obligate outcrossing. Similarly, the process was slowed but not ruled out by increased numbers of chromosomal rearrangements differentiating the parental species or a short hybrid zone interface.

In addition to establishing the feasibility of the recombinational speciation model, the simulations also revealed several new insights into the dynamics of the process. In particular, hybrid speciation was "punctuated": long periods of hybrid zone stasis were followed by abrupt transitions in which the selected type became established and rapidly displaced the parental species. Apparently, the critical factor in this process is the number of individuals of the optimal recombinant type, as mates of their own kind are scarce when numbers are low. This leads to a feedback effect as increased numbers of the optimal type leads to an increased chance of finding a similar mate, which in turn increases the number of optimal types in the next generation, and so forth.

Given the critical effect of the number of optimal type individuals, it is perhaps not surprising that the spatial distribution of these individuals is also important. Optimal recombinant–type individuals that are clumped are more likely to mate than those that are evenly dispersed. Thus, it appears that high selfing but low dispersal rates should facilitate this mode.

An important parameter in this model is the relative fitnesses of the new recombinant types. To implement selection in their model, the homoploid hybrid derivative was assumed to have a new, better-adapted combination of genes, resulting in a selective advantage over all other types (both hybrids and parentals). Although an increasing number of studies suggest that a subset of hybrid genotypes may be more fit than the parental genotypes in certain habitats (reviewed in 7), none has measured lifetime fitness. Thus, the validity of this assumption is untested. Also, it seems unlikely that the hybrid genotypes would be more fit than the parentals in all habitats in the hybrid zone. However, the occurrence of new hybrid species in habitats in which neither parent can survive (103) does suggest that particular hybrid genotypes may be more fit than their parents in novel habitats (see below).

Other Considerations

All of these discussions have assumed that the establishment of the novel hybrid type will occur in sympatry with both parents. Although hybrid speciation must be initiated in sympatry, Charlesworth (18) argued that this mode is most likely when "a group of hybrid plants colonize a new locality and are by chance spatially or ecologically isolated from the parental species." Thus, hybrid founder events might be viewed as foci of speciation. The possibility that a hybrid derivative might be stabilized in parapatry or allopatry should not be seen as minimizing the importance of the development of reproductive barriers. As the hybrid derivative becomes established and expands its geographic distribution, it most likely will come back in contact with its parents. Presumably, the existence of reproductive barriers will allow it to survive the challenge of sympatry.

The similarity of homoploid hybrid speciation to saltational (56) and founder effect models of speciation (16, 59) has previously been recognized (37, 71). Rieseberg (71) suggested that intra- or interspecific hybridization might be a stimulus for the chromosomal repatterning envisioned by Lewis's saltational model. Similarly, Grant & Grant (37) argued that hybridization is more likely than founder events to generate the genetic reorganization proposed in founder effect models. Theoretical studies indicate that for founder events to provide significant isolation, founding populations must contain high levels of genetic load, which is subject to strong epistatic selection (12, 98). Populations with these characteristics are instantaneously created by hybridization (79).

The small, peripheral populations emphasized in the saltational and founder models also are those most prone to hybridization. In general, the smaller the population the larger the relative proportion of foreign pollen, seeds, or spores (27). Moreover, because pollinators spend more time periods foraging in large than small populations, the proportion of interpopulational and presumably interspecific matings increases in the latter (27). Peripheral populations also are likely to experience higher levels of foreign reproductive propagules because the proportion of potential interspecific mates is likely to be greater at the boundary of a species range. These observations provide additional support for an important role for hybrid founder events in speciation.

ESTIMATES OF HYBRID FITNESS

If homoploid hybrid speciation occurs in sympatry, the fitness of the new hybrid lineage must play a critical role. If the new hybrid lineage is more fit than either parent in all habitats, as assumed in the McCarthy et al (60) model, it quickly replaces the parental species. However, if the fitness advantage of the hybrid lineage is restricted to a divergent habitat, as assumed in Templeton's (98) model, then it must co-exist with the parental species. Finally, if the new hybrid lineage is less fit in both parental and divergent habitats, it cannot be maintained in sympatry.

Hybrid fitness should be less immediately critical in hybrid founder events, as the potential for gene flow or competition with the parental species is reduced. However, once established, the new hybrid lineage is likely to expand its range and come back in contact with the parental species. At this point, both reproductive isolation and fitness become critical as the hybrid lineage could be eliminated by either gene flow or competition.

Note that the primary concern is with the average fitnesses of fertile, stabilized hybrid lineages, not F_1s or early segregating hybrid classes. Nonetheless, fitness estimates of early generation hybrids can provide insights into the likelihood of homoploid hybrid speciation and can suggest possible fitness expectations for stabilized hybrid derivatives. It also is useful to distinguish between the average fitness of a genealogical class of hybrids and the fitnesses of particular genotypes (V Grant, personal communication). Finally, as has been stressed by Arnold (5), the interaction between habitat and fitness must be recognized.

The average viability and fertility of early hybrid generations (e.g. F_2s, F_3s, etc) is predicted to be lower than that of the parental species due to the break-up of adaptive gene combinations (24). This is generally what is found, particularly for species with strong postmating reproductive barriers. Well-characterized examples include *Zauschneria* (19), *Layia* (19), *Gilia* (40), and *Helianthus* (45). This makes sense because hybridizing species would merge if the average fitness of the hybrids were greater than that of the parents. The fact that most plant hybrid zones are limited in extent also implies that hybrids are on average less fit than their parents (21), at least in parental habitats. Although a number of recent studies have described the replacement of populations of rare taxa by hybrid swarms (14, 55, 77), this appears to be due to genetic swamping by a numerically larger congener rather than high average hybrid fitness.

On the other hand, there are several examples in which the average fitness of a particular class or classes of hybrids appears to be equivalent to or to exceed that of their parents, at least for those fitness parameters measured (reviewed in 7). For example, *Artemisia* hybrids were more developmentally stable and had higher seed germination rates than either parent (31, 35), and *Iris* (29) and *Oryza* (53) hybrids had higher vegetative growth rates than their parents in parental or hybrid habitats. Unfortunately, none of these studies measured lifetime fitness, so whether they represent valid exceptions to the general rule of reduced average hybrid fitness is unclear. Nonetheless, enhanced hybrid fitness does seem plausible if postmating isolating barriers are weak and the hybrids occupy a novel habitat (30, 63), or if environmental conditions change (2, 36). These conditions are common in hybridizing plant species, so rank-order fitness estimates that occasionally favor hybrids should not be unexpected.

Even if the average fitness of a hybrid generation is lower than that of the parental species, this does not rule out the possibility that a particular hybrid genotype might be more fit than either parent, particularly in novel or "hybrid" habitats. Although lifetime fitness estimates are not yet available for individual hybrids, indirect evidence is accumulating that particular hybrid genotypes may be more fit than their parents in hybrid habitats. For example, significant genotype-habitat associations are often reported for hybrid swarms (2, 23, 68, 95). Presumably, this indicates that a selective advantage accrues for certain hybrid genotypes when found in favorable habitats, although these correlations could also result from historical factors (16). Likewise, studies that describe fertility, viability, or other fitness parameters in hybrids almost invariably report the presence of a small fraction of hybrid genotypes that are more fit than parental individuals, even if the hybrids on average exhibit reduced fitness (40, 45). These observations are supported by genetic studies which suggest that a small fraction of gene combinations may be favorable in interspecific hybrids, perhaps allowing them to colonize previously unoccupied adaptive peaks. For instance, 5% of the interspecific gene combinations tested in sunflower interspecific hybrids were favorable (79). Finally, the occurrence of hybrid species in habitats in which neither parent can survive (71) does suggest that assumptions of greater fitness of certain hybrid genotypes in novel habitats are realistic.

EXPERIMENTAL STUDIES

Experimental Verification of the Recombinational Speciation Model

Most experimental studies of homoploid hybrid speciation have focused on the recovery of fertile recombinant types following hybridization between chromosomally differentiated parents. A relevant early study in the genus *Crepis* was conducted by Gerassimova (33), who crossed two *Crepis tectorum* lines that differed by two reciprocal translocations. Selfing of a semisterile F_1 hybrid resulted in the recovery of a fertile F_2 plant that was homozygous for both translocations and, as a result, semisterile with the parental lines. Although this study demonstrated that fertile, recombinant individuals can be derived by hybridization between chromosomally divergent parents, conclusions that can be drawn from this experiment are limited due to the small number of generations analyzed and the weakness of the sterility barriers among the parental taxa and their hybrid derivative. These limitations were corrected in a series of comprehensive experiments involving *Elymus* (94), *Nicotiana* (87), and *Gilia* (40, 41).

Snyder (88) suggested that some of the morphologically diverse microspecies assigned to the selfing grass species *Elymus glaucus* might in fact result from introgression between ancestral *E. glaucus* and two related species known to hybridize in nature, *Sitanion jubatum* and *S. hystrix*. To test this hypothesis, Stebbins (94) generated several F_1 hybrids between microspecies of *E. glaucus* and either *S. jubatum* or *S. hystrix*. Although the vast majority of F_1 florets did not produce seeds (>99.99%), a small number of seeds were generated by four F_1s. In three cases, the F_1 seeds were not useful because the offspring had either recovered the morphology of their maternal parent, had undergone polyploidization, or were sterile. However, a single seed from the fourth F_1 appeared to result from a backcross toward *E. glaucus*. This plant had a seed fertility of 30% and was selfed for two generations. The resulting progeny were vigorous and had normal seed fertility (88%–100%). Moreover, crosses with the original *E. glaucus* parent indicated almost complete reproductive isolation; pollen fertility in the progeny of these crosses ranged from 0% to 3%.

These experiments not only verified the plausibility of Snyder's hypothesis, they also indicated that the origin of homoploid hybrid species need not be restricted to a selfing reproductive mode, particularly when the F_1 hybrids are highly sterile. This is important because backcross progeny are typically more easily generated and more fertile than self- or sib-crosses in early hybrid generations. In addition, the parental species used in these crosses have no apparent chromosomal differences, suggesting that it was the assortment of genic rather than chromosomal sterility factors that resulted in the isolation of the artificial neospecies.

Although the *Elymus* experiment appears to have involved unintentional selection for fertility, the first direct test of the effects of selection on the genetic isolation and morphological divergence of interspecific hybrids was conducted by Smith & Daly (87) in *Nicotiana*. They generated interspecific hybrids between the large-flowered *N. sanderae* and small-flowered *N. langsdorffii*.

Individual F_1 plants with large, intermediate, and small flowers were used to initiate three self-pollinating hybrid lineages, which were then subjected to selection for flower size (large, intermediate, and small, respectively) over 10 generations. By the tenth generation, all lines bred true for both floral (selected) and vegetative (unselected) morphological characters, and statistical analyses revealed that the three lines were separable from each other and from the parental species on the basis of either type of character. Investigation of reproductive barriers also revealed that each selected line was isolated from its parents by one or more genetic barriers such as crossability, meiotic aberration frequency, and pollen abortion. These results indicated that morphological divergence and genetic isolation can arise following strong selection in self-fertilizing hybrid populations.

The most convincing experimental validation of the recombinational speciation model comes from a series of elegant studies (40, 41) involving hybrids of *Gilia malior* × *G. modocensis*. The two species are selfing annual tetraploids with a relatively high chromosome number ($2n = 36$). First generation hybrids are highly sterile, with pollen and seed fertility of 2% and 0.007%, respectively; abnormal meiotic pairing suggests that this reduction in fertility is due to structural chromosomal differences between the parental genomes. To generate fertile and meiotically normal hybrid lines, the most fertile and viable plants were artificially selected from each generation, thus augmenting natural selection on the same traits. Although early generation plants were weak and partially sterile (hybrid breakdown), vigor and fertility improved rapidly. By the F_8 and F_9, full vigor, normal chromosomal pairing, and full fertility had been recovered in three hybrid lineages or branches. Branch I and branch III each possessed a new combination of morphological and cytogenetic features (40), whereas branch II reverted largely to the *G. modocensis* parent both morphologically and in terms of crossability (41). As in the case of *Elymus*, the two recombinant *Gilia* lineages were strongly isolated from their parents (4%–18% pollen fertility). This is concordant with theoretical expectations that the strength of genetic isolation between hybrid derivatives and their parents should be strongly correlated with barrier strength between the parents themselves (39).

The Role of Gene Interactions in Hybrid Speciation

The crossing experiments discussed above demonstrate that fertile, stable hybrid lines can arise via crosses between both weakly and strongly isolated species and that these new lines may be morphologically divergent relative to the parents. However, these studies tell us little about the forces governing the merger of differentiated parental species genomes or whether the results from these experimental studies are readily extrapolated to speciation in nature. To address these questions, Rieseberg et al (79) compared the genomic composition of experimentally synthesized hybrid lineages (*Helianthus annuus* × *H. petiolaris*) with that of an ancient hybrid species, *H. anomalus* (71). Interactions among genes that affect hybrid fitness and, indirectly, hybrid genomic composition were detected by analyzing parental marker segregation in the synthesized lineages.

Three hybrid lineages were synthesized: lineage I, $P-F_1-BC_1-BC_2-F_2-F_3$; lineage II, $P-F_1-F_2-BC_1-BC_2-F_3$; and lineage III, $P-F_1-F_2-F_3-BC_1-BC_2$. Crosses were performed by applying pooled pollen from all plants from a given generation to stigmas of the same individuals—a strategy that facilitates natural selection for increased fertility. Fifty-six or 58 progeny from the final generation of each hybrid lineage were then surveyed for 197 mapped *H. petiolaris* markers. The marker surveys were used to estimate the genomic composition of each of the 170 hybrid progeny individually and each of the three hybrid lineages cumulatively.

Comparison of the genomic composition of the ancient hybrid and synthetic hybrid lineages revealed that although generated independently, all three synthesized hybrid lineages converged to nearly identical gene combinations, and this set of gene combinations was recognizably similar to that found in *H. anomalus*. Concordance in genomic composition between the synthetic and ancient hybrids suggests that selection rather than chance largely governs hybrid species formation. Because the synthetic hybrid lineages were generated in the greenhouse rather than under natural conditions, congruence in genomic composition appears to result from fertility selection rather than selection for adaptation to a xeric habitat. This conclusion is supported by the rapid increase in fertility observed in the three hybrid lineages; average pollen fertility increased from 4% in F_1s to over 90% in the fifth generation hybrids. Congruence in genomic composition also implies that the genomic structure and composition of hybrid species may be essentially fixed within a few generations after the initial hybridization event and remain relatively static thereafter. This observation is concordant with the experimental studies described above and simulation studies (60) that suggest a rapid tempo for hybrid speciation.

Analysis of patterns of parental marker distributions in the experimental hybrids also allowed insights into the genetic processes governing hybrid genomic composition. The two parental species differ by a minimum of ten interchromosomal translocations and inversions (Figure 2), and hybrid plants heterozygous for one or more of these rearrangements exhibit reduced fertility (Figure 2). Because all backcrosses in the synthetic hybrids were in the direction of *H. annuus*, selection against chromosomally heterozygous individuals appears to have greatly reduced the frequency of *H. petiolaris* chromosomal fragments in the rearranged portion of the genome—an observation that holds for both the synthetic and ancient hybrid lineages. However, chromosomal

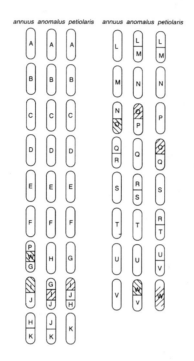

Figure 2   Linkage relationships between *Helianthus annuus, H. petiolaris,* and their putative hybrid derivative, *H. anomalus,* as inferred from comparative linkage mapping (81). Lines of shading within linkage groups indicate inversions.

rearrangements alone cannot explain the concordance in genomic composition between the synthesized and ancient hybrids as significant congruence is observed for both collinear and rearranged linkage groups.

Gene interactions also appear to play an important role in controlling hybrid genomic composition. Evidence for gene interactions comes from two sources: marker frequencies and associations. Most *H. petiolaris* markers (71% to 85%) introgressed at significantly lower than expected frequencies in the synthesized hybrids, suggestive of unfavorable interactions between loci tightly linked to these markers and *H. annuus* genes. By contrast, favorable interspecific gene interactions are implied by the significantly higher than expected rates of introgression observed for 5% to 6% of *H. petiolaris* markers. Concordance of marker frequency across the three synthesized hybrid lineages further suggests that these interactions remain largely constant regardless of hybrid genealogy.

Analyses of associations among segregating parental markers allow detection of specific interactions among chromosome segments that affect hybrid fertility, rather than the general interactions inferred from the frequency data. The rationale for this approach is that selection will favor the retention of genes that interact favorably. The signature of this epistatic selection should be detectable by positive associations or correlations among markers linked to these interacting genes. Likewise, genes that interact negatively should be detectable by negative associations among markers linked to these genes.

To test for these interactions, Rieseberg et al (79) analyzed all unlinked introgressed markers for significant two-way and three-way associations. The results from these analyses were compelling. Ten or more significant two-way associations were observed in each of the three synthetic hybrid lineages. In the more powerful three-way analysis, 21, 29, and 15 significant three-way associations were observed, generating complex epistatic webs. It is noteworthy that even though very stringent significance levels were employed in this analysis ($\alpha \leq 0.0001$), many of the same two- and three-way associations were observed in multiple hybrid lineages. Because the hybrid lineages were generated independently, selection rather than drift must account for these shared associations. Moreover, markers with epistatic interactions were more often found in all three lineages than markers lacking epistasis, suggesting that these interactions influence hybrid genomic composition.

The convergence of the synthetic and ancient hybrid sunflower lineages toward a similar set of gene combinations also suggests that hybrid speciation may be more repeatable than previously believed (21). However, if this is the case, one might ask why, in contrast to the sunflower results, the experimentally synthesized *Nicotiana* (87) and *Gilia* hybrid lineages (40, 41) diverged from each other in terms of morphology and cytogenetics. There are several possible reasons for this. In *Nicotiana*, diversifying selection was employed, essentially guaranteeing the generation of divergent lineages. Moreover, in both *Nicotiana* and *Gilia*, lineages were initiated and maintained by self-pollination of single, selected individuals—thus ensuring a major role for drift. By contrast, population sizes of 20 or more were maintained in *Helianthus*, plants were outcrossed, and natural fertility selection was allowed to proceed via pollen pooling. Another critical factor is that two generations of backcrossing toward *H. annuus* were employed in the generation of all three synthetic sunflower hybrid lineages. This appears to have resulted in the retention of that subset of *H. petiolaris* chromosomal fragments that interacted in a neutral or positive manner with the *H. annuus* genetic background. Backcrossing toward *H. annuus* also is likely to have occurred during the formation of the ancient hybrid species, *H. anomalus,* as backcrosses in this direction are more fertile and easily produced than other genotypic classes (48).

NATURAL HYBRID SPECIES

The Difficulty of Unambiguous Documentation

The experimental and theoretical studies discussed above indicate that homoploid hybrid speciation is a workable process under artificial conditions and may therefore occur in nature. Nevertheless, the actual extent of this mode of speciation in nature is unclear. This is not due to a lack of proposed examples in the taxonomic literature. In fact, there are few monographs that do not invoke hybridization to account for the origin of at least one or two morphologically intermediate or mosaic taxa (48, 67, 90). However, it is well known that morphological intermediacy can arise from forces other than hybridization. Dobzhansky (24), for example, recognized that intermediacy could arise by convergent morphological evolution. He also noted that remnants of the ancestral population from which two species differentiated might have the appearance of hybrids—an early and explicit recognition of plesiomorphy (the retention of primitive characteristics). Other authors have expressed skepticism concerning the use of quantitative phenotypic data to identify hybrids in the absence of information regarding the genetic basis of the characters being scored (10, 34, 47).

In many studies, morphological information has been augmented with evidence from secondary chemistry, ecological, and geographic data, and/or the production of synthetic hybrids that resemble the natural hybrids. Each of these approaches has its strengths, but as with morphological data, it is often difficult to determine whether intermediacy for chemical characters, ecological distribution, and/or geographic range actually result from hybridity. And putative examples of hybrid species based on these traditional biosystematic data sets often cannot be verified with molecular evidence (64, 74, 80, 91, 104).

Molecular markers represent a more powerful tool for identifying hybrid taxa (82), but even this approach can generate ambiguous results. As with morphological characters, a taxon can share molecular markers of related taxa due to the joint retention of alleles following speciation in a polymorphic ancestor (symplesiomorphy). This phenomenon has also been referred to as lineage sorting when discussed in the context of gene lineage data (8). As a result, it is much easier to reject the hypothesis of hybrid origin than to confirm it with molecular data sets. The use of multiple loci (32), linked markers (25), and gene lineage data (9, 69, 74) greatly enhances the probability of distinguishing between symplesiomorphy and hybridization. For example, if a putative hybrid species possessed multiple, linked markers of potential parents, and/or additivity for diagnostic parental markers at multiple loci, the probability that this situation could be attributed to symplesiomorphy or convergence is minimized. Likewise, hybridization becomes an increasingly probable explanation for shared alleles of identical sequence as species divergence times increase (43).

Even if evidence in favor of hybridity is unambiguous, this does not mean it had anything to do with speciation. For example, discordant organellar and nuclear phylogenies apparently due to hybridization are being reported with increasing frequency (83), but the evolutionary outcome of most of these ancient cases of hybridization appears to be introgression rather than hybrid speciation. The distinction between introgression and hybrid speciation can be difficult as well (101, 104). Incongruence between cytoplasmic- and nuclear-based phylogenetic trees, for example, suggest that hybridization played a role in the evolution of a wild species of cotton, *Gossypium bickii* (101). However, it is not clear whether this ancient hybridization event was important in its origin, as *G. bickii* does not appear to have a biparental nuclear genome typical of hybrid species.

Case Studies

A survey of the botanical literature identified more than 50 putative examples of homoploid hybrid species representing over 20 families of seed plants. However, only 17 examples have been rigorously tested with molecular markers. In my judgment, homoploid hybrid speciation has been convincingly documented in eight of these cases (Table 1), whereas in the remaining nine cases, hybrid speciation was disproved (78, 80, 91, 104) or the molecular marker data were ambiguous with regard to hybrid origin (1, 22, 64, 101).

HYBRID SPECIES ISOLATED BY POSTMATING BARRIERS   The classification of hybrid species by their mode of reproductive isolation is somewhat arbitrary as most probably have both postmating and premating barriers. Premating barriers are especially ubiquitous as all hybrid species appear to have diverged ecologically from their parents. Nonetheless, the examples discussed in this section do differ from their parents by chromosomal or genic sterility factors, whereas those in the next section appear to be isolated primarily due to habitat differences (Table 1).

The first application of molecular methods to the study of homoploid hybrid speciation was in *Stephanomeria* (Compositae) (32). The hybrid species, *S. diegensis*, and its parents, *S. exigua* and *S. virgata*, are self-incompatible annuals with the same diploid chromosome number ($2n = 16$). The parental species are widespread and largely allopatric, but they do co-occur and hybridize in Southern California. First generation hybrids are semisterile (14% pollen viability), apparently due to chromosomal structural differences between the parental species.

Stephanomeria diegensis is an abundant native of coastal southern California and morphologically is an "amalgam" of its parents. Analyses of 20 isozyme

Table 2 Confirmed examples of homoploid hybrid species in flowering plants

Taxon	Evidence[a]	Isolating mechanisms[b]	Growth form	Breeding system	Pollination syndrome	Reference
Helianthus anomalus	C,I,L,M,N,P	E,G	annual herb	outcrossing	animal	71
Helianthus deserticola	C,I,M,N,P	E,G	annual herb	outcrossing	animal	71
Helianthus paradoxus	C,I,M,N,P	E,G	annual herb	outcrossing	animal	71, 74
Iris nelsonii	C,I,M,N,P	E,G	annual herb	outcrossing	animal	4, 6
Peaonia emodi	N	E	perennial herb	outcrossing	animal	85
Peaonia species group	N	E	perennial herb	outcrossing	animal	85
Pinus densata	I,M,P	E	tree	outcrossing	wind	99, 100
Stephanomeria diagensis	C,I,M	E,G	annual herb	outcrossing	animal	32

[a]C = cytological or crossing studies; I, isozymes; L, linkage mapping; M, morphology; N, nuclear markers; P, plastid markers.
[b]E = ecogeographic isolation (premating); G, isolation due to genic or chromosomal sterility factors (postmating).

loci revealed that *S. diegensis* was a composite of the genes of *S. exigua* and *S. virgata* and had only one very rare unique allele. Furthermore, artificial F_1 hybrids between *S. diegensis* and its parents averaged 1% to 2% pollen viability, and thus they are significantly less fertile than those between the parental species. Rapid chromosomal evolution may have facilitated the speciation process as has been shown for *Helianthus* (below).

Molecular studies of the annual *Helianthus* species of section *Helianthus* have identified three species that appear to be derived via this mode: *H. anomalus, H. deserticola*, and *H. paradoxus* (71, 72, 74, 75, 79, 81). Although morphologically distinctive and allopatric, all three species appear to be derived from the same two parents (*H. annuus* and *H. petiolaris*) as they combine parental allozymes and nuclear ribosomal repeat units and share the chloroplast DNA (cpDNA) haplotype of one or both parental species. Like their parents, the three hybrids are self-incompatible and have a haploid chromosome number of 17. They differ from the parental species, however, in terms of geographic distribution and habitat preferences. *Helianthus paradoxus* is endemic to saline brackish marshes in west Texas, whereas *H. anomalus* and *H. deserticola* are xeric species restricted to the Great Basin desert of the southwestern United States. By contrast, the parental species are widespread throughout the central and western portion of the United States, with *H. annuus* found primarily in mesic soils and *H. petiolaris* in dry, sandy soils. Artificial crossing experiments indicate that all three hybrid species are semisterile with their parents, apparently due to chromosomal sterility barriers (below).

To test the hypothesis that rapid karyotypic evolution (98) can facilitate the development of reproductive isolation between a new hybrid lineage and its parents, genetic linkage maps were generated for *H. annuus, H. petiolaris*, and one of their hybrid derivatives, *H. anomalus*. Gene order comparisons revealed that 6 of the 17 linkage groups were co-linear among all three species, whereas the remaining 11 linkages were not conserved in terms of gene order (Figure 2). The two parental species, *H. annuus* and *H. petiolaris*, differed by at least 10 separate structural rearrangements, including three inversions, and a minimum of seven interchromosomal translocations. The genome of the hybrid species, *H. anomalus*, was extensively rearranged relative to its parents (Figure 2). For 4 of the 11 rearranged linkages, *H. anomalus* shared the linkage arrangement of one parent or the other. For the remaining seven linkages, however, unique linkage arrangements were displayed. In fact, a minimum of three chromosomal breakages, three fusions, and one duplication are required to achieve the *H. anomalus* genome from its parents. It is noteworthy that all seven novel rearrangements in *H. anomalus* involve linkage groups that are structurally divergent in the parental species, suggesting that structural differences may induce additional chromosomal rearrangements upon recombination. Similar increases in chromosomal mutation rates following hybridization have previously been reported in grasshoppers (86).

To reduce gene flow, chromosomal structural differences must enhance reproductive isolation. This does appear to be the case in *H. anomalus*, in which first generation hybrids with its parents are partially sterile, with pollen stainabilities of 1.8%–4.1% (*H. annuus*) and 2%–58.4% (*H. petiolaris*) (17, 46). Meiotic analyses revealed multivalent formations and bridges and fragments suggesting that chromosomal structural differences are largely responsible for hybrid semisterility (17). Thus, the rapid karyotypic evolution inferred from these mapping data does satisfy genetic models for speciation through hybrid recombination (98).

Thus far, all of the examples of hybrid speciation have involved two parental species, but there is no particular reason why additional species cannot be involved. An example comes from the Louisiana irises where *Iris nelsonii* contains genetic markers from *I. fulva*, *I. hexagona*, and *I. brevicaulis* (4, 6). Speciation appears to have occurred very recently, as individual plants cannot be unequivocally distinguished from either *I. fulva* or certain hybrid genotypes from contemporary hybrid zones. However, taken as a whole, the genetic make-up of *I. nelsonii* does differ significantly from parental and contemporary hybrid populations, and the species is stable and distinct in terms of morphology, ecological preference, and karyotype.

HYBRID SPECIES ISOLATED BY PREMATING BARRIERS   Good examples of homoploid hybrid species that are isolated by premating barriers only are difficult to find. One possible example is *Rhodendron intermedium*, which appears to be partially isolated from its parents due to soil preferences and the behavior of pollinators (51), but Kerner's hypothesis has not been tested using modern methods. More recently, Grant (38) presented a model for sympatric speciation in flowering plants based on flower constancy of pollinators and suggested that hybridization might be as likely as mutation in generating the new floral structures. Straw (96) postulated that two species of *Penstemon* might represent homoploid hybrid species that were derived via this manner. However, molecular studies do not support this hypothesis (104).

This mode is well-documented in Asian pines. Closely related species of *Pinus* hybridize frequently, and interspecific hybrids often are fertile and vigorous. Nonetheless, species remain distinct, apparently due to habitat isolation. Although several Asian pine species are thought to be of hybrid origin (62), the only example that has been analyzed rigorously is the putative origin of *P. densata* from *P. tabulaeformis* and *P. yunnanensis*. The three species have different ecological requirements, with *P. densata* endemic to high mountain elevations where neither of its putative parents is found. The geographic distribution of *P. densata* overlaps slightly with both parents, but the parents themselves are allopatric. Isozyme studies indicate that *P. densata* does combine the allozymes of its putative parents (100). However, analyses of cpDNA variation revealed the presence of three cpDNA haplotypes in *P. densata* (99). Two were identical to those found in the putative parents, but the third could not be found in extant Asian pine species. As a result, Wang et al (99) concluded that *P. densata* may have been derived via hybridization between two extant and one extinct Asian pine species.

A final study that may provide insights into the evolutionary potential of homoploid hybrid speciation concerns the genus *Paeonia* (85). Additivity of individual nucleotide positions in the internal transcribed spacer region of nuclear ribosomal genes implied a hybrid origin of a single species, *P. emodi*, as well as that of an entire lineage of ten species. The latter discovery is important because it suggests that homoploid hybrid species are not evolutionary deadends, but can found dynamic and speciose lineages.

THE BIOLOGY OF HOMOPLOID HYBRID SPECIES

The eight confirmed examples of hybrid speciation include one tree, three perennial herbs, and four annual herbs (Table 1). Although this sample is too small to make valid generalizations about homoploid hybrid speciation, one surprising result is that all have an outcrossing breeding system. The presence of outbreeding is unexpected, because hybrid speciation is predicted to occur more readily in highly inbreeding populations (42, 60, 98). Likewise, the high proportion of annual species of confirmed hybrid origin is unusual as hybridization appears to be more frequent in perennials.

All of the proposed and confirmed hybrid species differ from their parental species in habitat preference. This is expected, of course, as two species cannot occupy the same niche. What is unexpected, however, is that the habitat occupied by the hybrid taxa is often novel or extreme rather than intermediate relative to that of the parental species. Examples include the high altitude habitat of *Pinus densata* (100) and the xeric or marshy habitats of the three *Helianthus* hybrid species (71).

Morphologically, confirmed hybrid species exhibit a large proportion of extreme or novel characteristics when compared with their parents (34, 70, 76, 78). At least some of this morphological divergence appears to arise as a direct result of hybridization, as studies of synthetic hybrids from 33 plant genera reveal that over 10% of morphological characteristics are extreme in first generation hybrids and greater than 30% are extreme in later generation hybrids (76). This phenomenon is often referred to as transgressive segregation (97), and it appears to be the rule rather than the exception in segregating progenies from

Table 2   Isozyme variability in confirmed hybrid species (Hd) and their parents (Pt)

Taxon	Percentage of loci polymorphic[a]	Mean no. of alleles per locus	Mean heterozygosity[b]	Reference
Helianthus:				
annuus (Pt)	23.5	1.3	0.065	72
anomalus (Hd)	17.6	1.2	0.069	72
deserticola (Hd)	5.0	1.1	0.022	72
paradoxus (Hd)	5.9	1.1	0.027	72
petiolaris (Pt)	35.3	1.5	0.123	72
Iris:				
brevicaulis (Pt)	54.0	1.7	0.167	6
fulva (Pt)	41.5	1.4	0.122	6
hexagona (Pt)	43.6	1.6	0.152	6
nelsonii (Hd)	38.0	1.5	0.134	6
Pinus:				
densata (Hd)	61.5	2.5	0.210	100
tabulaeformis (Pt)	53.8	2.8	0.195	100
yunnanensis (Pt)	46.2	2.2	0.169	100
Stephanomeria:				
diegensis (Hd)	36.3	2.3	0.082	32
exigua (Pt)	33.6	2.6	0.098	32
virgata (Pt)	34.0	2.6	0.109	32

[a]For *Helianthus*, *Iris*, and *Stephanomeria*, a locus was considered polymorphic if the frequency of the most common allele did not exceed 0.99, whereas for *Pinus*, a frequency of 0.95 was used to define a polymorphic locus.

[b]For *Helianthus*, *Iris*, and *Pinus*, values are for mean expected heterozygosity, whereas for *Stephanomeria*, values represent observed mean heterozygosity per individual.

interspecific crosses (76). Botanists have speculated that the morphological and ecological novelty created by hybridization might allow hybrid populations to spread onto previously unoccupied adaptive peaks (3, 5, 54, 93).

Earlier workers predicted that hybrid taxa would be more variable genetically and have greater evolutionary potential than their parental species because they would combine the alleles of both parents (3, 39, 93). Although this is a reasonable argument, it is not supported by data from confirmed hybrid species (Table 2). The three *Helianthus* hybrid species exhibited lower levels of genetic diversity than either parent as measured by estimates of percentage polymorphic loci, mean number of alleles per locus, and mean heterozygosity (Table 2). *Pinus densata* was slightly more variable genetically than either parent, whereas *Stephanomeria diegensis* and *Iris nelsonii* were roughly equivalent to their parents in terms of variability (Table 2). The lower-than-predicted levels of diversity in the *Helianthus* hybrids may indicate that a small number of parental individuals was involved in their origin, possibly via hybrid founder events as discussed earlier.

CONCLUSIONS AND FUTURE DIRECTIONS

Satisfactory understanding of any mode of speciation requires answers to the following questions: Is the mode theoretically possible? Is there evidence for it in nature? How does it occur? Under what evolutionary conditions is it most likely? How quickly does it occur? And how frequent is it?

Most of these questions can be answered adequately for homoploid hybrid speciation. As discussed above, experimental and theoretical data indicate that this mode is feasible, and molecular marker data provide convincing evidence for its operation in nature. Likewise, the evolutionary processes accompanying or facilitating this mode of speciation are well understood. Important components such as the sorting of genic and chromosomal sterility factors, rapid chromosomal evolution, strong fertility and viability selection, and ecological divergence have been verified by both theoretical and experimental studies. Many of the critical ecological parameters that promote this mode of speciation have also been identified: the availability of suitable hybrid habitat, a selective advantage for hybrids in a hybrid habitat, and a long hybrid zone interface, which enhances the number of hybrid matings per generation. However, the observation that confirmed hybrid species are outbreeding (Table 1) contradicts theoretical studies indicating that rates of hybrid speciation should increase with selfing (60). Perhaps the advantage of selfing for hybrid species establishment is counterbalanced by lower rates of natural hybridization among selfing lineages.

Less evidence is available concerning the tempo and frequency of this mode. However, both experimental and theoretical data point to a rapid tempo of speciation. For example, fertile and stable hybrid segregants can often be obtained after only a few generations of hybridization, and simulation studies suggest that hybrid speciation is punctuated: Long periods of hybrid zone stasis are followed by the rapid establishment and growth of the new hybrid lineage (62). Congruence in genomic composition of synthetic and ancient hybrid species also suggests that hybrid genomes are likely to be stabilized quickly, with little change thereafter (79). Possibly, the tempo of speciation in natural hybrid species could be tested by analyzing the sizes of parental chromosomal fragments. Due to recombination, fragment sizes should decline in a predictable manner over time (11), perhaps making it feasible to estimate the number of generations of hybridization required to stabilize a hybrid species genome.

Estimating the frequency of hybrid speciation in nature is more speculative. Only eight examples in plants have been rigorously documented (Table 1), and even fewer in animals (26), suggesting that this mode may be rare. However, these low numbers may be an artifact of the difficulty of detecting and rigorously documenting homoploid hybrid species, particularly if the hybridization events are ancient. A much larger number of hybrid species have been proposed, and phylogenetic studies continually uncover unexpected cases of ancient hybridization in many evolutionary lineages (83). Although attempts to estimate the frequency of homoploid hybrid speciation are probably premature, hybridization may play a major role as the creative stimulus for speciation in small or peripheral populations. Hybridization rates appear to be highest in populations with these characteristics (28), and hybridization may be a more plausible mechanism than population bottlenecks for generating the genetic or chromosomal reorganization proposed in founder effect or saltational models of speciation (37, 71).

Although substantial progress has been made in studying this mode, much remains to be understood. A major gap in our knowledge relates to the origin of homoploid hybrid species that are isolated from their parents by premating barriers only. Empirical data indicate that species have arisen in this manner, but experimental and theoretical studies have focused on the strict recombinational model that involves the sorting of genic and chromosomal sterility factors. Because hybrid speciation is both reticulate and rapid, it is particularly amenable to experimental manipulation and replication. Thus, it should be feasible experimentally to synthesize new homoploid hybrid species isolated by premating barriers only. The design of these experimental studies could be informed by theoretical studies that identify parameters critical to this mode.

Another important issue relates to the fitness of hybrid genotypes. The ecological divergence required by the hybrid speciation model implies that the average fitness of the new hybrid lineage must exceed that of the parental species in hybrid habitat. This does not mean that early generation hybrids are more fit on average than their parents, but it does imply the existence of interspecific gene combinations that convey a fitness advantage in hybrid habitats. This hypothesis could be tested by comparing the lifetime fitnesses of individual hybrid and parental genotypes in hybrid habitats. Presumably, later generation hybrid segregants resulting from several generations of habitat and fertility selection would be most likely to exhibit a fitness advantage. The value of these experiments would be enhanced if the genomic composition of the hybrids was known (79), so that the effects of particular gene combinations on fitness could be determined. An alternative approach would be selection experiments that compare the responses of hybrid and parental populations to selective regimes approximating those expected in hybrid habitats (57).

Questions also remain concerning the genetic processes by which new hybrid species arise: 1. How do hybrid populations move to a new adaptive peak? Sunflower genetic mapping experiments suggest that this is accomplished primarily by selection rather than drift, and that selection can act directly on gene combinations as well as all individual genes. However, it is not clear whether this result is generalizable to hybrid lineages in which selection is less intense. Detailed mapping studies of synthetic hybrid lineages from less divergent species crosses may be required to address this question satisfactorily. 2. Do postmating reproductive barriers in new hybrid species arise primarily by the sorting of preexisting parental sterility factors or via the high genic and chromosomal mutation rates characteristic of hybrid populations? This question could be addressed by comparing the locations of quantitative trait loci (QTL) contributing to postmating reproductive barriers between the parental species with those isolating the hybrid species from its parents. If the "sterility QTL" in the hybrid taxon are a subset of those found in its parents, then the sorting hypothesis would be accepted. The presence of unique sterility QTL in the hybrid taxon would be more difficult to interpret, as this could be attributed either to rapid evolution during speciation or to divergent evolution following speciation. 3. What fraction of the morphological and ecological novelty observed in hybrid species is created by hybridization versus divergent evolution following speciation? As with the previous question, the critical issue here is to elucidate the direct role of hybridization in the speciation process, in this case with respect to the origin of morphological and ecological novelty. Although this question could be answered by comparing the morphological and ecological characteristics of synthetic and natural hybrid species, these comparisons have yet to be made.

Finally, I want to emphasize the continuing importance of molecular phylogenetic studies that identify and document natural hybrid species, as these studies are critical to reliable generalizations about the frequency and evolutionary significance of this mode. If designed well, these studies not only provide a means for "cleansing" the literature of oft-cited but incorrect examples of hybrid speciation, but they also provide an efficient strategy for testing large numbers of plant and animal groups for the existence of hybrid species (71). Clearly, studies that generate trees for multiple, unlinked loci and sample several populations per species will be most successful. The use of multiple loci is of particular importance as five independent gene trees are required to achieve 95% confidence that a given reticulation event will be detected (78). Greater phylogenetic resolution also will increase the chance of detecting ancient and possibly speciose hybrid lineages.

ACKNOWLEDGMENTS

I thank Shanna Carney, Keith Gardner, Rick Noyes, Jean Pan, Rhonda Rieseberg, Mark Ungerer, Jeannette Whitton, and Diana Wolf for criticism of the manuscript. The author's research on hybrid speciation has been supported by the NSF and the USDA.

Literature Cited

1. Allan GJ, Clark C, Rieseberg LH. 1997. Distribution of parental DNA markers in *Encelia virginensis* (Asteraceae): a diploid species of putative hybrid origin. *Plant Syst. Evol.* In press
2. Anderson E. 1948. Hybridization of the habitat. *Evolution* 2:1–9
3. Anderson E. 1949. *Introgressive Hybridization.* New York: Wiley
4. Arnold ML. 1993. *Iris nelsonii* (Iridaceae): origin and genetic composition of a homoploid hybrid species. *Am. J. Bot.* 80:577–83
5. Arnold ML. 1997. *Natural Hybridization and Evolution.* Oxford, UK: Oxford Univ. Press
6. Arnold ML, Hamrick JL, Bennett BD. 1990. Allozyme variation in Louisiana irises: a test for introgression and hybrid speciation. *Heredity* 84:297–306
7. Arnold ML, Hodges SA. 1995. Are natural hybrids fit or unfit relative to their parents? *Trends Ecol. Evol.* 10:67–71
8. Avise JC. 1994. *Molecular Markers, Natural History, and Evolution.* New York: Chapman & Hall
9. Avise JC, Saunders NC. 1984. Hybridization and introgression among species of sunfish (*Lepomis*): analysis of mitochondrial DNA and allozyme markers. *Genetics* 108:237–55
10. Baker HG. 1947. Criteria of hybridity. *Nature* 159:221–23
11. Baird SJE. 1995. A simulation study of multilocus clines. *Evolution* 49:1038–45
12. Barton NH. 1989. Founder effect speciation. In *Speciation and Its Consequences,* ed. D Otte, JA Endler, 10:229–54. Sunderland, MA: Sinauer
13. Barton NH, Hewitt GM. 1985. Analysis of hybrid zones. *Annu. Rev. Ecol. Syst.* 16:113–48
14. Brochmann C. 1984. Hybridization and distribution of *Argranthemum coronopifolium* (Asteraceae-Anthemideae) in the Canary Islands. *Nord. J. Bot.* 4:729–36
15. Callender LA. 1988. Gregor Mendel: an opponent of descent with modification. *Hist. Sci.* 26:41–75
16. Carson HL, Templeton AR. 1984. Genetic revolutions in relation to speciation phenomena: the founding of new populations. *Annu. Rev. Ecol. Syst.* 15:97–131
17. Chandler JM, Jan C, Beard BH. 1986. Chromosomal differentiation among the annual *Helianthus* species. *Syst. Bot.* 11:353–71
18. Charlesworth D. 1995. Evolution under the microscope. *Curr. Biol.* 5:835–36
19. Clausen J. 1951. *Stages in the Evolution of Plant Species.* Ithaca, NY: Cornell Univ. Press
20. Coyne JA. 1992. Genetics and speciation. *Nature* 355:511–15
21. Coyne JA. 1996. Speciation in action. *Science* 272:700–1
22. Crawford DJ, Ornduff R. 1989. Enzyme electrophoresis and evolutionary relationships among three species of *Lasthenia* (Asteraceae: Heliantheae). *Am. J. Bot.* 76:289–96
23. Cruzan MB, Arnold ML. 1993. Ecological and genetic associations in an Iris hybrid zone. *Evolution* 47:1432–45.
24. Dobzhansky TH. 1941. *Genetics and the Origin of Species.* New York: Columbia Univ. Press
25. Doebley J, Goodman MM, Stuber CW. 1987. Patterns of isozyme variation between maize and Mexican annual teosinte. *Econ. Bot.* 41:234–46
26. Dowling TE, Secor S. 1997. The role of hybridization in the evolutionary diversification of animals. *Annu. Rev. Ecol. Syst.* 28:XXX–XX
27. Ellstrand NC, Elam DR. 1993. Population genetic consequences of small population size: implications for plant conservation. *Annu. Rev. Ecol. Syst.* 24:217–42
28. Ellstrand NC, Whitkus R, Rieseberg LH. 1996. Distribution of spontaneous plant hybrids. *Proc. Natl. Acad. Sci. USA* 93:5090–93
29. Emms SK, Arnold ML. 1997. The effect of habitat on parental and hybrid fitness: reciprocal transplant experiments with Louisiana irises. *Evolution.* In press
30. Endler JA. 1977. *Geographic Variation, Speciation, and Clines.* Princeton, NJ: Princeton Univ. Press
31. Freeman DC, Graham JH, Byrd DW, McArthur ED, Turner WA. 1995. Narrow hybrid zone between two subspecies of big sagebrush (*Artemisia tridentata*: Asteraceae). III. Developmental

instability. *Am. J. Bot.* 82:1144–52

32. Gallez GP, Gottlieb, LD. 1982. Genetic evidence for the hybrid origin of the diploid plant *Stephanomeria diegensis.* *Evolution* 36:1158–67

33. Gerassimova H. 1939. Chromosome alterations as a factor of divergence of forms. I. New experimentally produced strains of *C. tectorum* which are physiologically isolated from the original forms owing to reciprocal translocation. *C. R. Acad. Sci. URSS* 25:148–54

34. Gottlieb LD. 1972. Levels of confidence in the analysis of hybridization in plants. *Ann. Mo. Bot. Gard.* 59:435–46

35. Graham JH, Freeman DC, McArthur ED. 1995. Narrow hybrid zone between two subspecies of big sagebrush (*Artemisia tridentata:* Asteraceae). II. Selection gradients and hybrid fitness. *Am. J. Bot.* 82: 709–16

36. Grant BR, Grant PR. 1993. Evolution of Darwin's finch hybrids caused by a rare climatic event. *Proc. R. Soc. London B Biol. Sci.* 251:111–17

37. Grant PR, Grant BR. 1994. Phenotypic and genetic effects of hybridization in Darwin's finches. *Evolution* 48:297–316

38. Grant V. 1949. Pollination systems as isolating mechanisms in flowering plants. *Evolution* 3:82–97

39. Grant V. 1958. The regulation of recombination in plants. *Cold Spring Harbor Symp. Quant. Biol.* 23:337–63

40. Grant V. 1966. Selection for vigor and fertility in the progeny of a highly sterile species hybrid in *Gilia. Genetics* 53:757–75

41. Grant V. 1966. The origin of a new species of *Gilia* in a hybridization experiment. *Genetics* 54:1189–99

42. Grant V. 1981. *Plant Speciation.* New York: Columbia Univ. Press

43. Hanson MA, Gaut BS, Stec AO, Fuerstenberg SI, Goodman MM, et al. 1996. Evolution of anthocyanin biosynthesis in maize kernels: the role of regulatory and enzymatic loci. *Genetics* 143:1395–407

44. Harrison RG. 1990. Hybrid zones: windows on evolutionary process. *Oxford Surv. Evol. Biol.* 7:69–128

45. Heiser CB. 1947. Hybridization between the sunflower species *Helianthus annuus* and *H. petiolaris. Evolution* 1:249–62

46. Heiser CB. 1958. Three new annual sunflowers (*Helianthus*) from the southwestern United States. *Rhodora* 60:271–83

47. Heiser CB. 1973. Introgression reexamined. *Bot. Rev.* 39:347–66

48. Heiser CB, Smith DM, Clevenger S, Martin WC. 1969. The North American sunflowers (*Helianthus*). *Mem. Torrey Bot. Club* 22:1–218

49. Herbert W. 1847. On hybridization amongst vegetables. *J. Hortic. Soc.* 2:1–107

50. Hilu KW. 1993. Polyploidy and the evolution of domesticated plants. *Am. J. Bot.* 80:1494–99

51. Kerner A. 1894–1895. *The Natural History of Plants.* Vols. 1, 2. London: Blackie & Son

52. Knobloch IW. 1971. Intergeneric hybridization in flowering plants. *Taxon* 21:97–103

53. Langevin SA, Clay K, Grace JB. 1990. The incidence and effects of hybridization between cultivated rice and its related weed red rice (*Oryza sativa* L.). *Evolution* 44:1000–8

54. Levin DA, ed. 1979. *Hybridization: An Evolutionary Perspective.* Stroudsberg, PA: Dowden, Hutchinson & Ross

55. Levin DA, Francisco-Ortega J, Jansen RK. 1996. Hybridization and the extinction of rare species. *Conserv. Biol.* 10:10–16

56. Lewis H. 1966. Speciation in flowering plants. *Science* 152:167–72

57. Lewontin RC, Birch LC. 1966. Hybridization as a source of variation for adaptation to new environments. *Evolution* 20:315–36

58. Linné C. 1760. Disquisitio de sexu plantarum, ab Academia Imperiali Scientiarum Petropolitana praemio ornata. *Amoenitates Academicae* 10:100–31

59. Mayr E. 1963. *Animal Species and Evolution.* Cambridge, MA: Harvard Univ. Press

60. McCarthy EM, Asmussen MA, Anderson WW. 1995. A theoretical assessment of recombinational speciation. *Heredity* 74:502–9

61. Mendel G. 1866. Experiments on plant hybrids (English translation). In *The Origin of Genetics: A Mendel Source Book,* ed. C Stern, ER Sherwood, 1:1–48. San Francisco: WH Freeman

62. Mirov NT. 1967. *The Genus Pinus.* New York: Ronald Press

63. Moore WS. 1977. An evaluation of narrow hybrid zones in vertebrates. *Q. Rev. Biol.* 52:263–67

64. Morrell P. 1996. Molecular tests of the proposed diploid hybrid origin of *Gilia achilleifolia* (Polemoniaceae). *Am. J. Bot.* Suppl. 83:180

65. Müntzing A. 1930. Outlines to a genetic monograph of the genus *Galeopsis. Hereditas* 13:185–341

66. Naudin C. 1863. Nouvelles recherches sur l'hybridité dans les végétaux. *Ann. Sci. Nat. Bot. Biol. Veg.* 19:180–203

67. Ornduff R. 1966. A biosystematic survey of the foldfield genus *Lasthenia. Univ. Calif. Publ. Bot.* 40:1–92

68. Potts BM, Reid JB. 1988. Hybridization as a dispersal mechanism. *Evolution* 42:1245–55

69. Powell JR. 1983. Interspecific cytoplasmic gene flow: evidence from *Drosophila. Proc. Natl. Acad. Sci. USA* 80:492–95

70. Randolph LF. 1966. *Iris nelsonii*, a new species of Louisiana iris of hybrid origin. *Baileya* 14:143–69

71. Rieseberg LH. 1991. Homoploid reticulate evolution in *Helianthus*: evidence from ribosomal genes. *Am. J. Bot.* 78:1218–37

72. Rieseberg LH, Beckstrom-Sternberg S, Liston A, Arias DM. 1991. Phylogenetic and systematic inferences from chloroplast DNA and isozyme variation in *Helianthus* sect. *Helianthus (Asteraceae). Syst. Bot.* 16:50–76

73. Rieseberg LH, Broulliet L. 1994. Are many plant species paraphyletic? *Taxon* 43:21–32

74. Rieseberg LH, Carter R, Zona S. 1990. Molecular tests of the hypothesized hybrid origin of two diploid *Helianthus* species (Asteraceae). *Evolution* 44:1498–511

75. Rieseberg LH, Choi H, Chan R, Spore C. 1993. Genomic map of a diploid hybrid species. *Heredity* 70:285–93

76. Rieseberg LH, Ellstrand NC. 1993. What can morphological and molecular markers tell us about plant hybridization? *Crit. Rev. Plant Sci.* 12:213–41

77. Rieseberg LH, Gerber D. 1995. Hybridization in the Catalina mahogany: RAPD evidence. *Conserv. Biol.* 9:199–203

78. Rieseberg LH, Morefield JD. 1995. Character expression, phylogenetic reconstruction, and the detection of reticulate evolution. *Monog. Syst. Bot. Mo. Bot. Gard.* 53:333–54

79. Rieseberg LH, Sinervo B, Linder CR, Ungerer M, Arias DM. 1996. Role of gene interactions in hybrid speciation: evidence from ancient and experimental hybrids. *Science* 272:741–45

80. Rieseberg LH, Soltis DE, Palmer JD. 1988. A molecular re-examination of introgression between *Helianthus annuus* and *H. bolanderi* (Compositae). *Evolution* 42:227–38

81. Rieseberg LH, Van Fossen C, Desrochers A. 1995. Hybrid speciation accompanied by genomic reorganization in wild sunflowers. *Nature* 375:313–16.

82. Rieseberg LH, Wendel J. 1993. Introgression and its consequences in plants. In *Hybrid Zones and the Evolutionary Process*, ed. R. Harrison, 4:70–109. New York: Oxford Univ. Press

83. Rieseberg LH, Whitton J, Linder R. 1996. Molecular marker discordance in plant hybrid zones and phylogenetic trees. *Acta Bot. Neerl.* 45:243–62

84. Roberts HF. 1929. *Plant Hybridization before Mendel*. Princeton, NJ: Princeton Univ. Press

85. Sang T, Crawford DJ, Stuessy TF. 1995. Documentation of reticulate evolution in peonies (*Paeonia*) using ITS sequences of nrDNA: implications for biogeography and concerted evolution. *Proc. Natl. Acad. Sci. USA* 92:6813–17

86. Shaw DD, Wilkinson P, Coates DJ. 1983. Increased chromosomal mutation rate after hybridization between two subspecies of grasshoppers. *Science* 220:1165–67

87. Smith HH, Daly K. 1959. Discrete populations derived by interspecific hybridization and selection. *Evolution* 13:476–87

88. Snyder LA. 1951. Cytology of inter-strain hybrids and the probable origin of variability in *Elymus glaucus. Am. J. Bot.* 38:195–202

89. Soltis DE, Soltis PS. 1993. Molecular data and the dynamic nature of polyploidy. *Crit. Rev. Plant Sci.* 12:243–75

90. Spooner DM. 1990. Systematics of *Simsia* (Compositae-Heliantheae). *Syst. Bot. Monogr.* 30:1–90

91. Spooner DM, Sytsma KJ, Smith JF. 1991. A molecular reexamination of diploid hybrid speciation of *Solanum raphanifolium. Evolution* 45:757–64

92. Stace CA, ed. 1975. *Hybridization and the Flora of the British Isles*. London: Academic Press

93. Stebbins GL. 1950. *Variation and Evolution in Plants*. New York: Columbia Univ. Press

94. Stebbins GL. 1957. The hybrid origin of microspecies in the *Elymus glaucus* complex. *Cytologia Suppl.* 36:336–40

95. Stebbins GL, Daly GK. 1961. Changes in the variation of a hybrid population of *Helianthus* over an eight-year period. *Evolution* 15:60–71

96. Straw RM. 1955. Hybridization, homogamy and sympatric speciation. *Evolution* 9:441–44

97. Tanksley SD. 1993. Mapping polygenes. *Annu. Rev. Genet.* 27:205–33

98. Templeton AR. 1981. Mechanisms of speciation—a population genetic approach. *Annu. Rev. Ecol. Syst.* 12:23–48

99. Wang X-R, Szmidt AE. 1994. Hybridization and chloroplast DNA variation in a *Pinus* complex from Asia. *Evolution* 48:1020–31

100. Wang X-R, Szmidt AE, Lewandowski A, Wang Z-R. 1990. Evolutionary analysis of *Pinus densata* (Masters) a putative Tertiary hybrid. 1. Allozyme variation. *Theor. Appl. Genet.* 80:635–40

101. Wendel JF, Stewart JM, Rettig JH. 1991. Molecular evidence for homoploid reticulate evolution among Australian species of *Gossypium*. *Evolution* 45:694–711

102. Whitham TG, Morrow PA, Potts BM. 1994. Plant hybrid zones as center for biodiversity: the herbivore community of two endemic Tasmanian eucalypts. *Oecologia* 97:481–90

103. Winge Ø. 1917. The chromosomes: their number and general importance. *C. R. Trav. Lab. Carlsberg* 13:131–275.

104. Wolfe AD, Elisens WJ. 1995. Evidence of chloroplast capture and pollen-mediated gene flow in *Penstemon* section *Peltanthera* (Scrophulariaceae). *Syst. Bot.* 20:395–412

Annu. Rev. Ecol. Syst. 1997. 28:391–435

EVOLUTIONARY GENETICS OF LIFE CYCLES

Alexey S. Kondrashov

Section of Ecology and Systematics, Cornell University, Ithaca, New York 14853;
e-mail: ask3@cornell.edu

KEY WORDS: life cycle, ploidy, apomixis, amphimixis, evolution

ABSTRACT

The life cycles of cellular species are reviewed from the genetic perspective. Almost all life cycles include stages during which only one genome is transmitted from a parent to its offspring. This, together with interorganismal gene exchange, which occurs regularly in at least some prokaryotes and in the majority of eukaryotes, allows selection to evaluate different alleles more or less independently. Regular genetic changes due to intraorganismal ploidy cycles or recombination may also be important in life cycles of many unicellular forms. Eukaryotic amphimixis is generally similar in all taxa, but the current data on the phylogeny and reproduction of unicellular eukaryotes are insufficient to determine whether it evolved several times or just once. Theoretically, gradual origin of amphimixis from apomixis, with each step favored by natural selection, is feasible. However, we still do not know how this process occurred nor what selection caused it. For reasons not entirely clear, some properties of amphimictic life cycles are much less variable and more conservative than the others. Evolution of many aspects of reproduction requires more theoretical studies, while the existing data are insufficient to choose among the currently competing hypotheses.

INTRODUCTION

The life cycle of a species is the succession of organisms connected by reproductive processes that repeats itself more or less regularly. I consider the life cycles of cellular forms from the point of view of transmission genetics and evolution. Fascinating viral life cycles (167, 297) require separate treatment. First, the parameters of reproduction are reviewed, and consistent terminology introduced. Second, the data on the life cycles of various taxa is presented systematically. Third, the evolution of life cycles is discussed.

391

PARAMETERS OF LIFE CYCLES

Description of a life cycle requires several parameters. In nature, different parameters often vary more or less independently, making any general classification of life cycles (6, 10, 12) not very useful. Thus, I treat these parameters separately. Only typical phenomena are considered here, while some peculiar processes of secondary origin are mentioned later.

Statics

ORGANISMS   More or less continuously existing entities that deal independently with their environments are called organisms or individuals. Separation of organisms is to some extent arbitrary (a fly remains the same organism from egg to imago, while a strawberry runner eventually produces a new organism). An organism may consist of one or many cells. Cells are usually clearly distinct, although large volumes of protoplasm with many nuclei may be viewed both as cells (e.g. in many amoebae) and as plasmodia (e.g. in acellular slime molds). In unicellulars each cell division creates a new organism, whereas in multicellular organisms some cell divisions (usually mitotic) increase only the number of cells within an organism.

GENOMES   The minimal complete complement of genetic material of a form of life is called its genome. A genome consists of DNA molecules, chromosomes, which are usually circular in prokaryotes (exceptions: *Agrobacterium tumefaciens, Borrellia burgdorferi, Streptomyces griseus*; 146, 184, 203) and in eukaryotic organelles (exception: mitochondria of most cnidarians and some other forms; 30, 199), but linear in eukaryotic nuclei. Prokaryotes usually have just one chromosome (exceptions: *Rhodobacter sphaeroides, Brucella melitensis, Agrobacterium tumefaciens,* and *Burkholderia cepacia*; 54, 146, 177). Eukaryotes usually have 5–25 chromosomes (245), but sometimes hundreds, e.g. in homosporous ferns (282), or just one (an ant *Myrmecia croslandi* and a nematode *Parascaris univalens*; 65, 100).

The genome of a form of life is to some extent variable. Nonspecialized variability of genomes usually involves small differences, with some exceptions (long insertions, deletions, inversions, etc). Occasionally, the whole chromosome may be present or absent (B-chromosomes, facultative plasmids, etc; 23, 145, 254). Generally, nonspecialized genetic variability is not important in the context of life cycles. In contrast, some forms of specialized variability, in particular, those related to sex or mating-type determination, are directly relevant to gene transmission (see below).

PLOIDY   A euploid cell contains only complete copies of its genome, while an aneuploid cell contains different numbers of copies of different parts of its genome. Prokaryotes are euploid, except those having multiple obligate plasmids. Possessing genetic material of different origin, which is normal in eukaryotes, usually leads to aneuploidy, because organelle DNA is present in more copies than nuclear DNA. Nuclear parts of eukaryotic genomes are normally euploid, although aneuploidy is common in some unicellulars.

A euploid cell can possess just one (haploidy, $x = 1$) or many (polyploidy, $x = 2, 3, \ldots$) copies of its genome. Within a cell, multiple genomes can reside in the same or separate compartments (nuclei or organelles). Polyploidy affects the life cycle more when multiple genomes are similar (autopolyploidy, intraspecific polyploidy). In contrast, with allopolyploidy (amphiploidy, interspecific polyploidy), multiple genomes often behave as the parts of one supergenome (amphidiploidy) (181, 283).

Dynamics

REPRODUCTION OF ORGANISMS   The genome is the unit of gene function. Remarkably, it is also the unit of gene transmission because usually whole genomes are transmitted from parents to offspring. The only process that creates new genomes, DNA replication, is necessary for reproduction.

After cell division, ploidy remains unchanged only if every DNA molecule was replicated exactly once. Such division, mitosis, produces two daughter cells genetically identical to the maternal one. After the division of a polyploid cell, multiple genomes may be distributed unevenly between the daughter cells, especially when they are located in different compartments (nuclei or organelles), leading to irregular mitosis.

When a new organism appears from the single cell, each process that produces this cell (mitosis, meiosis, or syngamy) is also a mode of reproduction. Reproductive mitosis occurs in a mitocyte and produces two mitospores (101, 256), whereas reproductive meiosis occurs in meiocytes and usually produces four meiospores. Syngamy, which normally involves the complete merger of two gametes, results in a zygote.

With vegetative reproduction, an offspring develops from multiple initial cells and, thus, from multiple genomes (164). Genetically, vegetative reproduction can lead to the same consequences as reproduction through irregular mitosis in a polyploid cell.

Substantial genetic changes can occur within an organism. This happens if such changes occur without cell division or merger, or if meiosis or syngamy is not followed by reproduction, leading to formation of multicellular organisms with genetically distinct parts.

INCREASE OF PLOIDY   One way to increase ploidy is by combining different DNAs in the same cell. In eukaryotes this usually occurs through syngamy.

Karyogamy, the fusion of gamete nuclei, may occur almost immediately after syngamy, or after a long delay, or even after many mitoses. Sometimes, karyogamy never happens. Fertilization as a synonym of syngamy or karyogamy is ambiguous because it also means insemination at the organismal level. Alternatively, only parts of some genomes are included in the genetic material of the cell (transformation, conjugation), leading to at least transitory aneuploidy. A possibility that union of incomplete but complementary parts of several genomes producing the complete genome is apparently never realized in cellular forms of life. The alternative way of ploidy increase is endomitosis—DNA replication not followed by cell division (165). This process may involve either full genomes or only parts, thus causing aneuploidy.

REDUCTION OF PLOIDY   Ploidy will be reduced after a cell division not compensated by previous replication. Surprisingly, the usual option in eukaryotes, two-step meiosis, is actually a succession of two cell divisions preceded by only one replication. Meiosis requires polyploidy ($x = 2, 4, \ldots$) and leads to both reduction and segregation (131, 244). Usually homologous chromosomes conjugate in meiosis, although regular segregation may also occur without conjugation (142, 228). The data on one-step meiosis, one cell division not preceded by replication, are conflicting (see below). Normally segregation is Mendelian, i.e. every genome has the same probability of being transmitted to a viable meiospore, but occasionally some genomes are transmitted more frequently than others (3, 191, 255, 298). The ploidy of a cell also can be reduced by genome destruction, occurring by genome expulsion from a cell, its intracellular destruction, or by removal of chromosomes one by one (parasexual process) (26, 202, 263, 303, 306, 314, 315). This leads to various degrees of reduction and aneuploidy. In polyploid cells, processes intermediate between mitosis and meiosis may be possible, if only some genomes replicate before cell division.

GENOME ALTERATIONS   Mutation sensu stricto alters a DNA sequence without using information from other sequences. Mutations can occur at any moment owing to errors in DNA replication or repair, and they are not known to play any specific role in life cycles, although they are involved in somatic genome differentiation (e.g. of immunoglobulin genes).

In contrast, recombination—creation of new genomes from the parts of preexisting ones—is an essential part of many life cycles. In eukaryotes, homologous reciprocal recombination (in which no sequences are either lost or created) between different genomes occurs regularly in meiosis and involves both reassortment of nonhomologous chromosomes and crossing-over between homologous chromosomes. Usually, 1–3 cross-overs occur per chromosome (10), despite enormous diversity of the amount of DNA per chromosome. Mitotic recombination also may be important in some life cycles (87). In prokaryotes, nonreciprocal homologous intergenomic recombination occurs after transformation or conjugation and results in only one viable recombinant genome.

Many other forms of recombination are possible. They include reciprocal recombination between nonhomologous (either paralogous or unrelated) sequences, causing unequal crossing-over, translocations, and inversions, as well as nonreciprocal recombination, leading to gene conversion and replicative transpositions. Conversion involves either homologous (alleles of the same locus in different genomes) or nonhomologous sequences from either the same or different genomes (119).

Events of this kind resemble mutations if they occur rarely. However, such recombination, different from homologous intergenomic recombination, may also occur regularly in some life cycles (see below). It may also be involved in changing of the genomes that are never transmitted (macronuclei of ciliates, chromatin diminution in nematodes and copepods, immune cell differentiation of vertebrates).

ASSORTMENT OF THE UNITING GENOMES   With panmixia, every two genomes in a population have the same chance to form a zygote. Two kinds of departures from panmixia are possible. First, genomes may be subdivided into exogamous classes, with union possible only between those from different classes. If genomes from different classes are similar, their random union is genetically close to true panmixia. Second, there may be preferential union of similar or dissimilar (assortative mating) or of related or unrelated (inbreeding or its avoidance) genomes (183).

Exogamous classes are called sexes or mating types. The distinction between them is not sharp. The word sexes is used when there are only two morphologically different classes (anisogamy or oogamy, but not isogamy) and/or some profound genetic differences are involved (32, 47). In contrast, mating types may be numerous, and their differences are controlled by only one or few loci, although the alleles (idiomorphs; 11) at them are often very different (46, 49). Sexes and mating types often coexist in angiosperms. An organism can produce gametes of only one sex (dioecy, bisexuality, gonochorism) or of both sexes (monoecy, hermaphroditism, unisexuality, cosexuality). Mixed situations (e.g. gynodioecy or androdioecy) are possible.

Selfing (autogamy, homomixis, self-fertilization), as opposed to outcrossing (outbreeding, allogamy), is the closest form of inbreeding in which uniting genomes are produced by a single individual. Still, selfing involves union of genomes produced via different mitoses (gamoid selfing) or via different meioses (zygoid selfing, see below) (259). Sexes and mating types may be viewed as an extreme form of negative assortative mating.

Structure of the Life Cycle

APOMIXIS   Apomictic life cycles involve no genetic exchange between different organisms. Typical apomixis involves reproduction only by single mitospores (itself called apomixis, as well as parthenogenesis or asexual, agamous, uniparental, or unisexual reproduction). Apomixis of haploids is the only strictly unigenomic mode of reproduction. Vegetative reproduction may also be a part of apomictic life cycles. Genetic changes during apomictic life cycles may occur due either to ploidy cycles (that is, alternations of endomitoses and reductions leading to changes of ploidy; 165) or to intraorganismal recombination (see below).

AMPHIMIXIS   Sexual reproduction as a synonym for amphimixis is ambiguous because sexes may be absent; thus we prefer the term amphimixis. Such cycles involve genetic exchange between different cells and organisms. Prokaryotic amphimixis involves the exchange of only parts of the genome. Typical eukaryotic amphimixis involves regular alternation of ploidy increase by syngamy and ploidy reduction by meiosis. This divides a life cycle into two phases with a two-fold difference in ploidy—gamophase (n, from reduction to syngamy) and zygophase (2n, from syngamy to reduction). The corresponding cells are gamoid or zygoid, while the organisms may be called gamobionts and zygobionts. The terms haploid and diploid are misleading in this context because gamoid polyploid cells are not haploid.

A phase is unreduced if mitoses occur in it that lead to formation of multicellular organisms and/or to reproduction. When the gamophase is unreduced, gametes are produced via haploid mitoses, while otherwise meiospores serve as gametes. When the zygophase is unreduced, meiocytes are produced via diploid mitoses, while otherwise zygotes act as meiocytes (often after a long delay). The two phases may be represented by either morphologically similar (isomorphic) or different (heteromorphic) organisms. Rarely, life cycles of individuals of different sexes are different, as with arrhenotoky, when males are gamoid and females are zygoid. Life cycles with both phases reduced are unknown, although with two-step meiosis they are feasible because such a cycle would double the number of cells.

Thus, mitosis is always present in amphimictic cycles. Apomictic reproduction via mitosis occurs in all unreduced phases of unicellular forms and is a regular part of amphimixis (heterogony) in some multicellular forms (rhodophytes and cyclical apomicts; 1, 10, 12). In such cases the complete amphimictic cycle lasts not just one generation ("from egg to egg") but longer. The same is true for some apomictic ploidy cycles. Multigenerational amphimictic cycles are usually more or less synchronous within a population.

The terms haplontic, diplontic, and diplohaplontic are used for life cycles in which zygoid, gamoid, or neither phase is reduced. Meiosis in these cases is sometimes called zygotic, gametic, or sporic (or somatic). I prefer to state explicitly which phase, if any, is reduced. Calling the gamophase and zygophase "sexual and asexual generations" is misleading: a generation includes the complete life cycle, both phases are "sexual" in the sense that they are parts of an amphimictic cycle, and zygobionts are often sexually dimorphic, e.g. in animals. Sporophyte and gametophyte are sometimes used as synonyms of zygobiont and gamobiont. If, however, meiospores function as gametes, zygobionts may be called gametophytes, while with apomixis in the gamophase, gamobionts may be called sporophytes. Parthenogenesis as a term for development of a gamoid organism from a meiospore, as well as reproduction by automixis (see below), is also confusing.

Several forms of atypical amphimixis are known. One of them involves exchange of whole genomes between multinuclear cells, apparently without any intragenomic recombination (273, 274). If the gamophase is haploid, this would lead to no genetic consequence, but polyploid cells after such exchange may have new sets of genomes.

Facultative modes of reproduction, such that alternative modes of reproduction are possible simultaneously, are very common. Sometimes one mode is much more frequent, e.g. in many almost-obligate amphimicts in which, nevertheless, accidental apomixis (tychoparthenogenesis) may occur. Quite often, however, two or more processes (e.g. amphimixis and apomixis or vegetative reproduction) have comparable frequencies.

LIFE CYCLES IN VARIOUS TAXA

The life cycles in different taxa are known to very different extents. While the data on multicellulars are based on direct cytological and morphological evidence, indirect evidence from genetic structure of a population may by crucial in unicellulars (301).

Prokaryotes

We know only a fraction of prokaryotic diversity (8, 81), and even for the forms already described, the current data on life cycles are insufficient for general conclusions.

Polyploidy is apparently rare in bacteria, although it does occur and may play an important role in, e.g. *Deinococcus radiodurans* (219) and *Azotobacter vinelandii* (80). Ploidy cycles are possible, with reduced ploidy in cysts (198). Intragenomic recombination may be a regular part of the life cycle, e.g. in *Neisseria gonorrhoeae* (295).

Until recently, prokaryotes were believed to be mostly apomictic. This view was supported by the early data on the genetic structure of natural populations of *Escherichia coli* (271), despite the known possibility of conjugation in this species. Now it is clear, however, that genetic exchange due to transformation and/or conjugation (212) can be quite intensive in some prokaryotes (68, 123, 213, 300). While *Salmonella* and *Haemophilus* are essentially clonal, amphimixis significantly affects the populations of *Neisseria*, *Rhizobium*, *Burkholderia*, *Helicobacter* (97, 213, 330), and, perhaps, even *E. coli* (108). Thus, amphimixis occurs in several distant taxa.

Eukaryotes

PHYLOGENY   Despite recent progress, our understanding of the phylogeny of eukaryotes remains imprecise because (*a*) different traits, both molecular and ultrastructural, often suggest contradictory phylogenies; (*b*) some taxa are probably artificial, containing rather distantly related forms; and (*c*) many poorly studied "flagellates," "amoebae," "heliozoans," etc, have unclear affinities, while others may await discovery (36, 234). I briefly consider the phylogeny of nuclear genes, which are essential for life cycles, using classifications of Cavalier-Smith (37) and Corliss (61) and ignoring taxa with obscure affinities or modes of reproduction. Phylogeny of organelles is different due to symbioses (17).

The phylogenetic tree of eukaryotes consists of several clades that diverged early, followed by the large group of "higher" clades that apparently diverged during a relatively short period around 1500 Mya (42, 70, 235, 280, 307). Within this group, 80S ribosomes, Golgi dictyosomes, and mitochondria with flat or tubular cristae are typical. Some earlier clades have the same set of ultrastructural traits, while others are different (70S ribosomes, no mitochondria or mitochondria with discoid cristae, etc), although it is not always clear which trait states are primitive (43, 127a). Below I list possible early clades in the order that might coincide with that of their divergence from the trunk of the eukaryotic tree (171, 237).

1. Metamonada, amitochondriate flagellates, including diplomonads (e.g. *Giardia*, *Trepomonas*, and *Hexamita*), retortomonads, and oxymonads (*Pyrsonympha*) (43, 115, 158).

2. Microsporidia, amitochondriate intracellular parasites (43, 334). However, microsporidians may also be fungi, modified by parasitism (78, 151).

3. Jakobid flagellates, a plausible sister group to all mitochondriate taxa (172, 229) and possibly also to parabasalians (127a) and archamoebids (55).

4. Parabasalia, flagellates with 70S ribosomes, hydrogenosomes (modified mitochondria; 31), and Golgi dictyosomes, including trichomonads and hypermastigotes (106, 127a, 334).

5. Percolozoa, flagellates, amoeboflagellates, or amoebae, usually having mitochondria but lacking Golgi dictyosomes, including percolomonads, heteroloboseans (*Naegleria, Vahlkampfia*, and "cellular slime mold" *Acrasis*) (124, 262), lyromonads, and pseudociliates.

6. Euglenozoa, flagellates with Golgi dictyosomes and mitochondria with discoid cristae, including euglenoids and kinetoplastids (*Bodo, Trypanosoma*) (158, 226).

7. Foraminifera (237, 317), perhaps the earliest known clade with the overall ultrastructure typical of the higher clades.

8. Archamoebae (*Entamoeba, Pelomyxa, Phreatamoeba*). The monophyly of this "clade" is questionable. Some data indicate their early branching (78, 125), while other data place it much higher (55, 73, 115, 258, 334). *Amoeba proteus* might also belong to this clade or to another early branch (238).

   Below I list clades that diverged later. They are presented in an order that might coincide with that of their divergence from the trunk of the eukaryotic tree (171), although the topology of this group may be more complex. Parentheses indicate plausible sister taxa relationships within clades.

9. Alveolata, including (Apicomplexa and Dinoflagellata), (Ciliophora and Plasmodiophora) (127), and Haplosporidia (90).

10. Heterokonta, including Bicoecia, (Labirinthulidae and Thraustochytridae), (Oomycetidae and Hyphochytridae), followed by chromophyte algae Diatomea (or Bacillariophyta), (Chrysophycea and Eumastigophycea), and (Xanthophycea and Phaeophycea). The positions of several other heterokont taxa are unclear (40, 41, 176, 308).

11. Haptophyta (41).

12. Glaucophyta together with Cryptophyta (15) and some Lobosea (lobose amoebas), e.g. *Acanthamoeba* and *Hartmannella* (323).

13. Chlorarachniophyta together with Filosea (filose amoebae) (16).

14. Mycetozoa, cellular or plasmodial slime molds, including protosteleans (*Protostelium*), myxogastreans (*Physarum*), and dictyosteleans (*Dictyostelium*) (286). The conclusion that this is a clade is preliminary. Some data indicate that myxogastreans (70) and/or dictyostelids (59, 171) represent early branches, while other data place dictyostelids among the higher clades (4, 74, 193, 262).

15. Rhodophyta, which includes some unicellular and many multicellular forms, with Bangiophycea being an early clade, and diverse Florideaphycea branching off later (243).

16. Chlorophyta, with Prasinophycea as the early clade, followed by two independent clades: one consisting of Chlorophyceae, Microthamnales, and Ulvophyceae, and the other consisting of Charophyceae and land plants (216, 288).

17. Choanoflagellata together with the flagellate *Apusomonas* and several parasites (*Dermocystidium, Ichthyophonus, Psorospermium,* and "rosette agent") are closely related to the common ancestor of Fungi and Animalia and may form a clade (39, 152, 243a).

18. Fungi, including Chytridiomycota, Zygomycota, Trichomycota, and (Ascomycota and Basidiomycota) (127, 185). Apparently, Fungi and Animalia are sister taxa (42, 318), but see (107).

19. Animalia, including Myxospora (268, 276) and perhaps Mesozoa.

ANEUPLOIDY   Aneuploidy appeared many times. It apparently plays an important role in long-term evolutionary changes, eventually leading to the formation of modified genomes (e.g. 154). However, the role of aneuploidy in life cycles appears to be limited, especially in multicellular amphimicts. In unicellulars, aneuploidy is probably less detrimental per se, and variable aneuploidy reported in *Leishmania* (66), *Botrytis cinerea* (35), *Trypanosoma brucei* (309), *Entamoeba histolytica* (188), *Candida maltosa* (9), and *Physarum polycephalum* (169a) suggests that processes resulting in aneuploidy may be more or less regular components of their life cycles.

The only pervasive exception is aneuploidy due to cytoplasmic DNA, because organelle chromosomes usually (with some exceptions, e.g. *Chlamydomonas* chloroplast DNA) are present with much higher ploidy than nuclear chromosomes (20). Chromosomes of obligate symbiotic intracellular bacteria, if counted as a part of the genome (223), would also cause aneuploidy.

POLYPLOIDY   Allopolyploidy may lead to the origin of new taxa, but, as with aneuploidy, its role in life cycles appears to be limited, except that it may trigger secondary losses of amphimixis. In contrast, autopolyploidy may affect the life cycles significantly, being a stage of the ploidy cycle (see below).

Among the few unicellular forms studied, many are polyploid, apparently due to autopolyploidy. The examples are *Giardia*, which have two nuclei, each with several genomes (83), many amoebae, perhaps unrelated, which have either one polyploid nucleus (*Amoeba proteus* and several species of *Entamoeba* and

Acanthamoeba; 2, 335) or many nuclei (275, 287, 329), large radiolarians (244) and foraminiferans (261), and *Physarum* (181, 324).

In alveolates, in addition to the well-known "somatic" polyploidy of macronuclei in ciliates, polyploidy and/or aneuploidy is common, perhaps as a temporary condition, in dinoflagellate nuclei (277). In heterokonts, polyploidy is known in Oomycota (181). In Fungi, polyploidy is reported for all major groups, including *Allomyces* from Chytridiomycota (181) and ascomycetes (322).

In green algae, polyploid series are reported in several groups. They are common among monoecious Charophyceae (181). Autopolyploids make up about 80% of the most diverse bryophyte group, Bryopsida, but are rare in other bryophytes (242). Polyploidy is common in ferns (319), although many species with very high chromosome numbers are genetically diploids (282). In gymnosperms, polyploidy is rare (181). In angiosperms, allopolyploidy, often associated with apomixis, is common. There are many autopolyploid forms (usually, with $x = 2$ in the gamophase), e.g. within genera *Saxifraga* (101), *Luzula* (139), *Crepis* (144), and *Plantago* (310), which may have originated via gametes with zygoid ploidy (28, 85). Many plant polyploid lineages survived for a long time to give rise to taxa of ancient polyploids, e.g. Sphagnopsida (279), homosporous ferns (93), and several families of flowering plants. Approximately 70% of flowering plant species went through polyploidization (perhaps, mostly allopolyploidization) in their evolutionary history (206, 283).

In Animalia, polyploidy is generally uncommon but may be frequent in some groups (e.g. 57, 181). The evolution of amphimictic polyploids in nature probably involves 2x-gametes and 3x-apomicts as intermediate steps (5). Several taxa of fishes, including the families Salmonidae and Catostomidae, consist of ancient polyploids (24). In amphibians, polyploidy (3x) may be associated with apomixis (*Ambystoma*), but autopolyploidy also exists (with $x = 2$, 3, or 4 in the gamophase) in amphimictic frogs from several genera (e.g. 109, 205). In reptiles, polyploidy (3x) is exclusively associated with apomixis (181). Polyploidy is lethal in birds and animals.

NONAMPHIMICTIC GENETIC CHANGES   Amphimixis is by far the best-studied mechanism for regular genetic changes. However, such changes can also occur due to other mechanisms, in both apomicts and amphimicts. The importance of nonamphimictic genetic changes is still not fully appreciated by evolutionary biologists.

Ploidy cycles   In some life cycles, autopolyploidy with a high x alternates with haploidy or with autopolyploidy with a smaller x. Such ploidy cycles occur in some unicellulars, which may be primitively apomictic, including *Entamoeba, Pelomyxa*, and *Phreatamoeba* (2, 165). It also exists in some forms that lost amphimixis recently (170). Finally, the ploidy cycle can coexist with an amphimictic n-2n cycle. So far, the only definite example of this is a brown alga *Ectocarpus siliculosus*, in which endomitosis may occur after syngamy, followed by two successive meioses, which causes an n-2n-4n-2n-n cycle (224). Other examples may be Radiolaria, which undergo ploidy cycle and which are strongly suspected to be amphimictic (244), foraminiferans (261), *Physarum polycephalum* (169a), dinoflagellates (277), raphidophytes (333), and some red algae (111).

Nonamphimictic recombination   So far, there are few well-described cases where recombination, different from homologous intergenomic recombination, is the regular component of a life cycle. These cases are mating type switching in *Saccharomyces cerevisiae* (304) and generation of antigen-coding gene variation in several parasites, including *Trypanosoma* (25), *Pneumocystis carinii* (316), and *Plasmodium falciparum* (122). Either a ploidy cycle or some nonreciprocal intergenomic recombination (conversion) is suggested by the polyploidy of *Giardia*. Many more cases perhaps await discovery (272, 299).

DISTRIBUTION OF AMPHIMIXIS   Here I consider the presence of amphimictic life cycles, together with their variability due to different duration of its phases and formation of complex multicellular organisms. Deviations from typical syngamy and meiosis, as well as combinations of amphimixis and apomixis, are addressed later.

1. Metamonada. Amphimixis has never been observed in diplomonads and retortamonads (204) but was reported for oxymonads (56, 244, 246). Either phase may be reduced in Pyrsonymphida. In species with a reduced gamophase, diplobionts copulate and each produces two haploid nuclei. Nuclear exchange produces a "double zygote" consisting of two diploid nuclei, which divide into two single zygotes. Therefore, a haploid nucleus never occupies a separate cell, which may be considered the deepest reduction of the gamophase (compare with conjugation of ciliates and diatoms). Because Metamonada are extremely important phylogenetically, more data on their reproduction, in particular on the cytology of meiosis, are necessary.

2. Microsporidia. Amphimixis is apparently common, although difficult to observe (71, 120, 173, 175, 204, 296). It was even proposed that amphimixis is polyphyletic within Microsporidia (7). The nature of meiosis is controversial. Contrary to the common view that it has an unusual nature (see below), it has been recently proposed that it is the typical two-step meiosis (89).

3. Jakobid flagellates. Nothing is known about amphimixis in this group.

4. Parabasalia. All the data on amphimixis on this phylogenetically critical group are due to Cleveland (56). Both gamo- and zygophase reduction were described. In one species with a reduced zygophase, a large and a small gamete are produced mitotically by the same haploid cell.

5. Percolozoa. Amphimixis was never observed in percolozoan amoebae or acrasid slime molds (204). However, the genetic structure of populations of the free-living amoeba *Naegleria lovaniensis* strongly suggests some form of genetic exchange (239).

6. Euglenozoa. There are some unconfirmed reports of amphimixis in euglenoids (204, pp. 270; 278, 231; 244, pp. 202–204), and at least some euglenozoans are diploid (53, 121, 241, 252). Amphimixis occurs at least in some kinetoplastids, as implied by the data on genetic recombination in crosses (95) and on the genetic structure of some populations (289). In *Leishmania*, syngamy and meiosis occur within host cells (169), while in *Herpetomonas,* syngamy of individuals occurred in culture (285).

7. Foraminifera. Amphimixis is well known. Usually both phases are unreduced (102; 204), but in some forms gamophase is reduced (236). Both gamobionts and zygobionts, although unicellular and initially uninuclear, become multinucleate (or polyploid; 261) because of endomitoses.

8. Archamoebae. Amphimixis has never been reported, although some forms are relatively well studied.

9. Alveolata. Amphimixis is the basic form of life cycles in this clade. Several dinoflagellates have a reduced zygophase, although zygotes may grow and persist for a long time, being the resting stage (18, 91, 156). Unreduced zygophase in *Noctiluca* is controversial (269). Apicomplexans apparently always have a reduced zygophase (e.g. 195, 196).

In contrast, the gamophase is overreduced in ciliates so that they have no haploid cells in the life cycle (11). Syngamy is replaced with conjugation of two zygobionts, each of which undergoes meiosis and a mitotic division of one of the four meiotic nuclei (the other three disintegrate), after which one nucleus from each cell moves to the other cell and nuclear fusions produce two genetically identical cells. Thus, in contrast to Pyrsonymphida, conjugation of Ciliophora involves a two-step meiosis and one haploid mitosis (253). Zygobionts also have an additional vegetative polyploid macronucleus, developing via endomitoses and partial genome destruction.

In plasmodiophorids, the amphimixis and meiosis were detected by observations of the synaptonemal complex (27), and the zygophase is apparently reduced (246). Life cycles of haplosporidians are unknown.

10. Heterokonta. In bicoecids, amphimixis is unknown. Amphimixis with reduced gamophase is known for labyrinthulids, but it was never observed in the related thraustochytrids and hyphochytrids. In contrast, amphimixis with reduced gamophase is well known in oomycetes and may strongly affect the genetic structure of their populations (327).

Amphimixis is normal in diatoms. Gamophase is reduced to the same extent as in pyrsonymphids and ciliates, and zygobionts conjugate (200, 225). Raphidophytes apparently have a reduced gamophase (333). Insufficiently studied chrysophytes have a reduced zygophase (264), but it is unclear how common amphimixis is among them. Xanthophytes are also poorly studied, but amphimixis with either both phases unreduced or with a reduced zygophase is known (103, 230).

All phaeophytes are multicellular and most are amphimictic. Both phases are unreduced in all orders except Fucales, in which the gamophase is reduced and sometimes new zygobionts start developing on a maternal zygobiont (compare with angiosperms and animals) (12, 13, 192). In some orders (e.g. Dictyotales), both phases are equally developed; in one order (Scytosiphonales) the gamophase is predominant, while in many others (e.g. Laminariales), zygobionts are much larger than gamobionts. Facultative apomixis is common in both phases, and a cell may act as either a gamete or mitospore, depending on external conditions.

11. Haptophyta. Amphimixis is apparently common. Both phases are unreduced with gamoid and zygoid cells morphologically distinct and multicellularity (colony formation) in the zygophase (79, 204, 312). Apparently, this is the only case in which one phase is multicellular while the other is unicellular but unreduced.

12. Glaucophyta, Cryptophyta, and some Lobosea. Amphimixis is poorly studied in this possible clade. Syngamy was observed in cryptomonads and apparently the zygophase is reduced (169b). Amphimixis, perhaps with reduced gamophase, was found in *Arcella vulgaris* (218), a lobose amoeba (92).

13. Chlorarachniophyta and Filosea. Amphimixis was reported in two chlorarachniophytes, apparently, with both phases unreduced (14a, 104) and in a vampyrellid filose amoeba *Lateromyxa gallica*, apparently, with a reduced gamophase (260). More data on this clade are needed.

14. Mycetozoa. There are no data on amphimixis in protostelids. In myxogastrean *Physarum* amphimixis is well known, with both phases represented by polynuclear plasmodia (204, 281). The gamophase consists of a succession of unicellular organisms reproducing apomictically. After syngamy, a diploid plasmodium grows and nuclei multiply via endomitoses. Different plasmodia may merge, producing chimeric diplobionts. Meiosis occurs in spores, starting a new gamophase. Apomixis, although common in the gamophase, does not occur in the zygophase.

Amphimixis is common in dictyosteleans. The zygophase is reduced, although zygotes live long and feed (including cannibalistically) actively (182). Three out of four meiospores die before germination.

15. Rhodophyta. Amphimixis is unknown in unicellular rhodophytes. There is one report of amphimixis, apparently with both phases unreduced, in Bangiophycidae (118). In Florideophycidae, both phases are unreduced, and the zygophase usually consists of two successive multicellular organisms: The zygote develops in the carposporophyte, which apomictically produces the tetrasporophyte, in which meiosis occurs. These three stages may form various complex organisms (12, 192, 204, 251). Gene expression may be substantially different in different phases (189).

16. Chlorophyta. The data on Prasinophyceae are insufficient, but amphimixis was reported (129; 227, pp. 129–130; 291). Apparently, both phases may be unreduced.

In Chlorophyceae, amphimixis is common, and the zygophase is always reduced. In Ulvophyceae, both phases may be unreduced, and they may be similar or one of them may dominate (12). In Dasicladales, the zygophase is reduced, but the single diploid cell becomes large (*Acetabularia*), and after meiosis many haploid mitoses occur inside it. In Caulerpales, the gamophase is usually reduced (143). A zygote may remain attached to a maternal gamobiont, and new gamobionts (of both sexes) may develop on a somatic body of a zygobiont (Prasiolales).

Charophyceae have a reduced zygophase. Bryophytes have both phases unreduced, and zygobionts live on gamobionts. In vascular plants, both phases are unreduced with the zygoid phase predominant. Gamobionts are either independent (pteridophytes) or live on diplobionts (seed plants, gymnosperms, and angiosperms). In flowering plants, gamophase reduction is the deepest but is not complete, as two mitoses in male haplobionts and several mitoses in female haplobionts still occur. Sometimes a female haplobiont is chimeric, when more than one meiospore from the same meiosis take part in its formation (96).

In angiosperms, syngamy leading to production of a zygote is accompanied by another syngamy ("double fertilization"), which leads to development of endosperm with $x = 2, 3$, or higher (96, Figure 20-15). Thus, the seed contains diploid tissue of the parental zygobiont (seed coat), endosperm, and a newly formed diplobiont tissue (embryo).

Diploids that function as gamobionts can be artificially obtained in bryophytes. This does not necessarily involve apomixis, as diploid gamobionts may produce diploid gametes, causing polyploidization (181, 319). The opposite phenomenon, development of haploids according to the zygobiont program (apogamy), occurs in ferns and flowering plants. To be fertile, such organisms have to undergo diploidization by endomitosis (149).

17. Choanoflagellata. No direct data exist on this group.

18. Fungi. Chytridiomycota usually have a reduced zygophase, but sometimes both phases are unreduced. The same cell may serve as a gamete or a gamoid mitospore producing a new gamobiont. The zygophase is reduced in Zygomycota. In Ascomycota and Basidiomycota, both phases are unreduced. In multicellular Ascomycota, the gamophase is predominant, while in yeasts either phase may be predominant. In Basidiomycota, usually the zygophase is predominant, but in Uredinales complex mycelia develop in both phases. In Ascomycota and, especially, Basidiomycota, after syngamy, karyogamy usually is delayed for many cell divisions so that two haploid nuclei occur in a single zygoid cell (dikaryon) (326).

19. Animalia. In amphimictic Animalia, the gamophase is reduced. The only exception is arrhenotoky with gamoid males and zygoid females, which evolved independently in some Nematoda, Rotifera, Acarina, and Insecta (species of orders Homoptera, Thysanoptera, and Coleoptera and the whole order Hymenoptera). Haploid females may occur sporadically (292). Haploids are probably inviable in all vertebrates, although completely homozygous fertile individuals can be produced by artificial automixis (290).

Finally, amphimixis is known in several protozoans with uncertain phylogenetic positions that may belong to advanced clades. In Opalinata, mitoses occur in both phases, although the zygophase predominates, and fusing gametes differ in size (204). In the only class of Actinopoda in which amphimixis has been observed, Heliozoa, the gamophase is reduced. Amphimixis has never been observed in Radiolaria, although it may well be present. In the obligate parasites Paramyxea, probably both phases are unreduced and represented by multinuclear plasmodia (204).

ATYPICAL AMPHIMICTIC INCREASE OF PLOIDY   Syngamy followed by complete unification of all the genetic material of both gametes is by far the most common way of amphimictic ploidy increase, and it appears to be the primitive one. Still, there are several other possibilities.

Incomplete syngamy   Apart from organelle chromosomes (see below), there may be other cases in eukaryotic amphimixis in which only a part of one of the two uniting genomes is incorporated into the genetic material of the zygote (which is normal in prokaryotes). Recently, such a case was described in a fish, *Poecilia formosa* (266), and more cases probably exist (14).

Multiple syngamies within a cell   Merging of two polynuclear gamoid cells, instead of typical gametes, can lead to multiple karyogamies and to formation of many zygotes together. This process (gamontogamy) occurs in some Dasicladales (Acetabularia), in many foraminiferans, and in gregarines (113).

Overreduction of the gamophase   In several taxa (Pyrsonympha, ciliates, and diatoms), reduction of the gamophase is so deep that no independent gamoid cells are produced. This leads to various modifications of syngamy involving, instead of free gametes, direct interactions of zygobionts that undergo meiosis (gametangiogamy, conjugation).

Automixis   The union of two genomes that originate in the same meiosis can be viewed as modified syngamy, modified meiosis, and/or an extreme form of inbreeding. Its consequences depend on how the zygoid level of ploidy is assured in the product of automixis, the autozygote, and the results vary from immediate complete homozygosity (suppression of the first haploid mitosis in the products of meiosis), through less rapid loss of heterozygosity (fusion of various products of meiosis), to preservation of the genotype of the parental zygobiont (additional replication before meiosis with synapsis only between sister chromosomes or intrameiotic restitution if chromosomes pair in the second meiotic division) (75, 292).

Automixis as the obligate mode of amphimixis was described in Heliozoa (this needs verification) and in several plant and animal species (in particular, insects). Facultative automixis is more widespread, both in unicellulars (diatoms and ciliates) and multicellulars (220). In one species of ciliates, polymorphism in the ability to undergo autoconjugation was found to depend on a single locus with two alleles (72). Even in birds, occasional automixis can lead to development of completely homozygous males in turkeys (292) and chickens (265).

ATYPICAL AMPHIMICTIC REDUCTION OF PLOIDY   Although typical two-step meiosis is prevalent, there are exceptions that may be more common and important than currently thought.

Multiple meioses within a cell   In a multinuclear zygoid cell, many reduction divisions can occur simultaneously. This happens in foraminiferans, where the whole body of a zygobiont is converted into gametes (schizogony).

Meiotic vegetative reproduction   Instead of being independent, related meiospores may contribute to the same organism, producing one chimeric gamobiont, e.g. in red algae (204). In several fungi, specialized meiosis produces two-nuclear spores bearing different incompatibility alleles, also leading to a chimeric gamobiont.

One-step meiosis   This process, apparently more logical than the usual two-step meiosis (131), remains controversial (246). It was reported in Pyrsonymphida (Oxymonadida), Parabasalia (Hypermastogotes), Microsporidia (but see 89), some Dinoflagellata (in which two-step meiosis has also been reported, 240), some Apicomplexa (56, 244, 246), and a fungus *Pneumocystis carinii* (197). For all these forms, however, more data are necessary. In all the alleged cases of one-step meiosis, no crossing-over was reported, although a priori diads of chromosomes, formed without premeiotic replication, can be as capable of recombination as tetrads.

Nonmeiotic reduction   Reduction due to genome expulsion has evolved several times. In some cases, separate chromosomes are lost independently, causing aneuploidy (at least temporarily, until further losses restore euploidy). Some Deiteromycetes have paraamphimixis involving syngamy, formation of chimeric mycelia (heterokaryosis), karyogamy, crossing-over and diploid mitosis, and haploidization via loss of chromosomes (26, 87).

Alternatively, whole genomes can be expelled at various moments of the life cycle. With hybridogenesis, the paternal genome is eliminated during meiosis, and as a result the genotype of every meiospore is identical to that of the maternal gamete (69). This extreme form of non-Mendelian segregation can be viewed as modified meiosis or as a mode of genome destruction. Hybridogenesis suppresses recombination and effectively leads to apomixis (semiclonality, see below). It is known in several fishes (*Poeciliopsis*) and amphibians (*Rana*) (306, 315). In other cases, genome expulsion occurs after syngamy, during early stages of the zygobiont formation. Again, it affects the paternal genome and leads to formation of gamoid males (263).

ATYPICAL AMPHIMICTIC RECOMBINATION   Crossing-over may be confined to one sex, usually the homogametic one (34). No unambiguous cases of complete genome-wide loss of crossing-over are known, even though segregation can proceed normally without it. Segregation of different chromosomes is usually independent: By the time of meiosis, they "forget" from which genome they came. However, several cases of weak linkage between genes on different chromosomes are known (pseudolinkage) (337). Reciprocal homologous recombination occasionally occurs in mitosis and plays an important role in some life cycles (26, 147).

ORGANELLE INHERITANCE   Organellar parts of the genome are usually transmitted polygenomically and via irregular mitosis because organelle DNA replication is asynchronized. However, the effective number of human mitochondrial chromosomes transmitted to an offspring may also be quite low due to developmental bottlenecks (128). Syngamy is frequently incomplete with respect to uniparentally transmitted organelle chromosomes, effectively leading to their apomixis. Mitochondria are maternally transmitted in mycetozoans, green algae, and fungi (exception: *Neurospora*; 58), vascular plants (94, 321) and animals (exception: *Mytilus*; 336). Chloroplasts are inherited maternally in ferns (94), paternally in conifers (321), and either maternally or biparentally in flowering plants (153, 302). Uniparental organelle inheritance has evolved many times and occurs owing to different mechanisms, while recombination between organelle genomes appears to be uncommon (20, 150).

ASSORTMENT OF THE UNITING GENOMES   Assortment of genomes depends on the assortment of gametes and/or of zygoid organisms. Both gamoid and zygoid influences either can be in the form of exogamous classes of genomes or can lead to deviations from panmixia due to inbreeding, assortative mating, and mate choice.

Gamoid vs. zygoid control   In unicellulars, the zygophase directly influences the genome assortment only if the gamophase is overreduced. In multicellulars, direct zygophase influence may occur by two mechanisms. First, mating can involve zygobionts, after which gametes are assorted within the pairs, e.g. in oomycetes and animals, especially those with internal fertilization. Second, gametes may have their sexes dictated by the zygobionts that produced them or their parental gamobionts (all animals and vascular plants). Mating between zygobionts also leads to zygoid control of inbreeding, assortative mating, and mate choice.

Absence of exogamous classes   This situation is sometimes referred to as homothally. However, this word is also used to denote production of different classes of gametes by the same organism (e.g. due to dioecy) or frequent switches between mating types (e.g. in many yeast forms) because in these cases amphimixis is possible within a population originating from one organism. Real absence of exogamous classes is known in cellular and plasmodial mycetozoans (60, 204); zygomycetes, ascomycetes, and basidiomycetes (58, 87, 222, 248); and charophytes (204).

Sexes and mating types   Mechanisms of sex determination are extremely variable. In monoecious forms, gametes of both sexes may always be produced. Sex may be determined environmentally (51,52), e.g. in oomycetes, the sex of a gamete may depend on the mating partner of the zygobiont that produces it (247,248). Genetic sex determination may involve sex chromosomes, acting either in gamoid (some phaeophytes and bryophytes) or zygoid (vascular plants and animals) phases, as well as more complex mechanisms (46,47).

Multiple exogamous classes are more common in taxa with a predominant zygophase (ciliates, basidiomycetes, angiosperms, ascidians) but are also possible in some taxa with a predominant gamophase, dictyostelids and Zygomycetes (*Mucor*) (33,46,49,132,180,328). Both gamoid and zygoid genotypes can affect restrictions on panmixia imposed by mating types.

Differentiation of gametes   Isogamy is the only known state in slime molds and chrysophytes, and it also occurs in dinoflagellates, diatoms, xanthophytes, phaeophytes, chlorophytes (12), and, perhaps, in other taxa. Isogamy may be the primitive state for eukaryotes as a whole (132), although oogamy might be primitive within major taxa (diatoms) (200). Anisogamy is common in dinoflagellates, apicomplexans, diatoms, xanthophytes, phaeophytes, chlorophytes, and Fungi (many Ascomycota and Uredinales). Oogamy exists in all oomycetes, some phaeophytes (Dictyotales and Fucales), and xanthophytes (*Vaucheria*), rhodophytes, charophytes, bryophytes, vascular plants, and animals.

Monoecy and dioecy   These terms can be applicable to independent gamobionts, to zygobionts (if gamophase is reduced), or to complex zygoid-gamoid organisms on which gametes are produced. Gamoid dioecy is known in phaeophytes (Laminariales), charophytes, chlorophytes, bryophytes (332), and heterosporous ferns. Dioecy that involves the zygophase occurs in fungi (in which it is rare), many flowering plants, and most animals. Monoecy exists in all major taxa. Even in animals, it is common in invertebrates (e.g. Ctenophora, Turbellaria, Annelida, Mollusca, Tunicata, Briozoa, and Chaetognatha; 10) and occurs in fishes (but not in other Vertebrata). Sexual dimorphism occurs to various extents in most dioecious forms, including Oedogoniales (chlorophytes).

Selfing   Monoecy or the absence of sexes is an obvious precondition for selfing. Selfing of gamobionts produces totally homozygous zygotes. It occurs in foraminiferans, charophytes, and homosporous ferns (282). Mating between closely related gametes of unicellulars, which is possible in the absence of sexes and mating types, has the same consequences and occurs in slime molds, dinoflagellates, and others.

Zygoid selfing also increases homozygosity, but only by 50% per generation. It is known in heliozoans in which a zygoid cell divides mitotically producing two cells, each of which undergoes meiosis producing one gamete, and then syngamy occurs (244), oomycetes, and many other monoecious zygobionts (190, 267). The only known example of selfing in Vertebrata is in the fish *Rivulus marmoratus* (114).

COMBINATIONS OF AMPHIMIXIS AND APOMIXIS   Such combinations are almost universal and lead to the most significant diversity of amphimictic life cycles.

Regular apomixis   Apomixis is a necessary part of the life cycles of all unicellulars. In fact, an unreduced phase in such cycles is the phase in which apomictic reproduction occurs. The number of apomictic generations during an amphimictic phase is usually relatively large and uncertain. Amphimixis (meiosis followed by syngamy in zygobionts, or syngamy in gamobionts) is frequently triggered by bad conditions (starvation, deficit of nitrogen, etc) and/or by the introduction of potential partners. However, in pennate diatoms, the successive apomicts are different (they become progressively smaller), so amphimixis must happen regularly.

Among multicellulars, a similar phenomenon occurs in floridean rhodophytes, in which the diploid phase consists of a succession of two different organisms, carposporophyte and tetrasporophyte, connected by apomictic reproduction. More often, all apomictic generations are identical. Regular apomixis of this form is common in phaeophytes and chlorophytes, but not in vascular plants. Again, switching to amphimixis (which begins from production of males in dioecious forms) is triggered by deteriorating conditions. In animals, regular apomixis, "cyclical parthenogenesis," evolved independently in trematodes, rotifers, cladocerans, aphids, cecidomyids (gall midges), and cynipid wasps (1, 22, 292).

Facultative apomixis   In many cases, apomixis is not a necessary part of the amphimictic life cycle but still occurs often. It is common in green algae, except Caulerpales and Charophyceae. Gamoid apomixis is possible in bryophytes via mitospores called gemmae. Zygoid apomixis occurs in ferns and flowering plants but is unknown in gymnosperms. In ferns it always involves formation (via a modified meiosis without recombination) of a peculiar organism, which is morphologically a gamobiont but has the same 2n-genotype as the preceding zygobiont (apospory). This "2n-gamobiont" produces zygobionts apomictically (76, 319). In apomictic flowering plants, meiosis may be absent, and a 2n-embryo may form directly from diplobiontic tissue (probably most of these cases are, in fact, vegetative reproduction) (256). Haploid apomixis is absent in all vascular plants, including pteridophytes, with free-living haplobionts, although false vegetative reproduction is possible (164). Facultative apomixis via various types of spores is very common in the gamophases of Zygomycota and Ascomycota. No facultative apomixis is known in vertebrates.

Occasional apomixis   In most taxa an offspring may occasionally develop apo-mictically, due to a variety of mechanisms (tychoparthenogenesis) (292). Mam-malia is the only taxon in which this appears to be impossible because maternal and paternal genomes are required because of genome imprinting (293).

OBLIGATE APOMIXIS   Amphimixis is unknown in several high-level taxa of unicellulars. Some of them may be primitively apomictic, but this is far from certain (see above). In all other cases, obligate apomixis is a secondary phe-nomenon, although a quite common one. Many cases are of very recent origin (210), with clear traces of amphimictic ancestry, as with pseudogamy (gyno-genesis), in which a male gamete is needed to activate an ovum but does not contribute any genes to it. However, some obligate multicellular apomicts may be over 1 My old (44, 145a).

   Obligately apomictic forms apparently exist among Oomycota, Phaeophyta, and Rhodophyta. Fungi in which amphimixis has never been observed are called Deiteromycetes. Some of them probably are obligate apomicts, mostly of ascomycetic or basidiomycetic origin (222). In flowering plants, obligate apomixis is usually associated with polyploidy (101, 221, 313).

   Obligate apomixis is reported in 10 phyla of invertebrate animals, especially in Arthropoda (292), and may have evolved many times in the same species (62, 331). In vertebrates, obligate apomixis is described in seven genera of fishes, in the amphibian genus *Ambystoma*, in five lizard families, and in single species of chameleons and snakes (77, 315). In fishes and amphibians, pseu-dogamy is required, with related species serving as sperm donors. Apomixis in vertebrates has always evolved via modified meiosis. With the exception of *Poe-cilia formosa*, chromosome synapsis occurs, but in all the cases studied, both maternal genomes are transmitted unchanged. Most vertebrate apomicts are polyploids [usually 3x of presumably recent hybrid origin (67, 292)]. Apomixis is unknown in birds and mammals.

VEGETATIVE REPRODUCTION   Only exceptionally is vegetative reproduction a necessary part of amphimictic life cycles (e.g. chimeric gamobionts in red algae described above and monozygotic super-twinning in armadillos). In con-trast, facultative vegetative reproduction is common in all amphimictic taxa, excluding the most advanced animals such as Insecta, Mollusca, and Vertebrata (excepting monozygotic twins in *Homo sapiens*) (10, 130, 138, 256).

   Obligate vegetative reproduction is always a secondary phenomenon, be-cause multicellulars are primitively amphimictic. A small number of Fungi ("*Mycelia sterilia*") are known to reproduce only vegetatively by sclerotia. A sclerotium usually develops from a single cell, so that genetically this form of vegetative reproduction is close to apomixis. There are also some cases of obligate vegetative reproduction in plants and animals (164, 292).

General Trends

I outline three trends that are clearly visible in the huge diversity of life cycles in nature. The reasons for their existence are discussed later.

VARIABLE VS. FIXED PARAMETERS   Some parameters are much less variable than others. Euploidy appears to be the general rule in all cellular forms, except for cytoplasmic inheritance. Almost all life cycles regularly pass through unicellular haploid stages, when a new organism originates from a single cell with one genome. This can happen because of apomixis of haploids, ploidy cycles, and/or amphimixis.

Amphimixis always involves strict alternation of gamo- and zygophases. Although superimposition of the ploidy cycle can occasionally lead to two successive meioses, two successive syngamies (not separated by a meiosis) are unknown. Ploidy of gamo- and zygophases always differs by the factor of two. The absence of, say, n-3n cycles has sometimes been erroneously referred to as the existence of only two sexes (88, 250). Mendelian segregation and two-step meiosis are also remarkably uniform.

In contrast, all possible relative durations of gamoid and zygoid phases can be found. Various bizarre mixed organisms and modes of combining amphimixis with apomixis exist in some taxa. The number and genetic determination of exogamous classes of gametes, the degree of difference between sexes, and systems of inbreeding also show a great deal of variability.

CONSERVATIVE VS. FLEXIBLE PARAMETERS   Some variable parameters are taxonomically conservative, being usually invariant within high-order taxa, while others can differ between closely related species and even within populations. Surprisingly, ploidy is rather flexible, with apomictic polyploids or amphimictic autotetraploids evolving repeatedly in, for example, flowering plants. Mating types (58) and levels of inbreeding (137) are also flexible. Selfing and automixis evolved many times in different ways (292). Transitions to obligate secondary apomixis also may be quite fast.

In contrast, phase lengths are rather conservative and often uniform even within a phylum (12, 159). Differentiation between sexes and modes of organelle inheritance is less conservative but still not very flexible.

DIRECTION OF CHANGES   Many parameters are more variable within primitive, unicellular taxa than within advanced, multicellular taxa. This becomes evident if we compare mammals, which have practically no aneuploidy, no polyploidy, obligate amphimixis with meiotic recombination only, two sexes, and oogamy, with unicellular eukaryotes. The mammalian traits are generally typical of multicellular forms, although there are many exceptions. Evolution of the amphimictic cycle usually proceeded in the direction of an expanded zygophase (12).

EVOLUTION OF LIFE CYCLES

Parameters of life cycles affect both the physiology of reproducing organisms and gene transmission. The fundamental problem is to find which influences are crucial to the evolution of a particular parameter (immediate benefit vs. variation and selection explanations; 163). The answers may differ in different cases.

There is an inherent contradiction between gene functioning and gene transmission. For efficient functioning, all genes of an organism must act together. In contrast, efficient selection requires that genes are regularly reshuffled during transmission, allowing their independent evaluation. Two features of a life cycle determine efficiency of selection: One is the number of genomes transmitted to an offspring, and the second is the degree of genetic exchange among different, independently selected organisms. Unigenomic reproduction allows selection to test an allele regardless of the other alleles at the same locus. Genetic exchange, together with recombination, allows selection to test an allele regardless of the alleles at other loci.

Therefore, selection is most efficient under obligate amphimixis with a haploid gamophase, reduced zygophase, strict alternation of full-genome syngamy and reduction with Mendelian segregation, and free recombination. Strikingly, life cycles of most multicellular eukaryotes are close to this ideal, except that selection during zygophase is usually very important. Small deviations probably do not cause a profound decline in the efficiency of selection because even occasional separation of genes may be enough to allow their almost independent evaluation.

Analysis of the evolutionary choice between two alternative states of the life cycle must be done in terms of group selection if there is no amphimixis in at least one state or if there is amphimixis in both states, but the differences between them are so profound (e.g. diploid vs. autotetraploid zygophase) that these states are reproductively isolated (209). In both cases evolution involves only competition between genetically independent lineages; eventually the life cycle that lead to the most efficient selection will evolve, as long as functional constraints do not preclude this. In contrast, the evolution of reproduction within an amphimictic population must be considered in terms of individual selection, and the efficiency of selection is not necessarily maximized in this case.

Origin of Genomes and Euploidy

In primitive life, genetic information probably was not organized into well-defined genomes. However, genomes appeared very early, so the common ancestor of all extant cellular forms almost certainly had one (214, 305). Factors of evolution of the physical organization of the genomes are obscure. Why are some chromosomes circular while others are linear, and what determines

their number? Even if linearity and moderate numbers of chromosomes in eukaryotic genomes appeared due to purely physiological reasons, or by accident,
these features may be necessary preconditions for meiosis to evolve (see below). Too many chromosomes with amphimixis lead to freer recombination but
diminish the genetic impact of intrachromosomal crossing-over, which might
eventually cause its loss (161).

Euploidy may also be common mostly because of its immediate benefits. If
so, it is unclear why aneuploidy caused by numerous organelle chromosomes
is so pervasive in eukaryotes and why these, despite losing most of their genes,
do not disappear completely (20), with only one known exception (31). In
most cases, however, a higher copy number of some genes is achieved by
the evolution of multigene families, instead of aneuploidy. Drastic changes
in genome organization, e.g. due to polyploidy (178), may lead to abnormal
intrapopulation variability and eventually to aneuploidy (105). Nevertheless,
sometimes such changes are successful, as indicated by numerous cases of
ancient polyploidy.

Evolution of Ploidy

For a cell to be bigger than a certain size, it must be polyploid. This is probably
the reason for polyploidy of many unicellular organisms. It is not clear whether
haploidy is primitive in cellular organisms. However, originating a new organism from just one genome leads to the most efficient selection, reducing, for
example, mutation load. Although polyploidy may have a temporal advantage
by masking partially recessive mutant alleles, eventually it leads to a higher
load because the overall mutation rate per polyploid genome is higher (165).

Physiological advantages of polyploidy and transmissional advantages of
haploidy can be reconciled by cyclical polyploidy, or the ploidy cycle. If
occasionally, preferably before the conditions get worse, a new organism is
haploid (after which polyploidy is quickly restored by endomitoses), the efficiency of selection is almost as high as with permanent haploidy. Such
cycles, to some extent similar to those consisting of amphimixis and regular
apomixis (see above), may be quite common in unicellular apomicts (165). Extra apomictic ploidy cycles superimposed on the amphimictic ploidy cycle may
also be a mutation-load–reducing mechanism used by the phase with the highest
ploidy.

Irregular polyploidy in the organelle parts of eukaryotic genomes also increases the mutation load (164). Low effective numbers of organelle chromosomes (128) improve the efficiency of selection and may evolve as a load-
reducing mechanism.

In amphimicts, haploidy of the gamophase leads to the most efficient selection. Autopolyploidy makes segregation less efficient because Hardy-Weinberg

equilibrium is approached only asymptotically (64, 278). This causes substantial deviations from Hardy-Weinberg equilibrium only under very strong selection (or very high x) because there are no combinations of genomes that are never broken. In addition, autopolyploidy increases the mutation rate per organism. However, because under amphimixis the mutation load depends on gene interactions, autopolyploidy can in fact reduce the load (84). Allopolyploidy does not affect segregation at individual loci. However, as long as it has the form of amphidiploidy, it effectively increases the genome size and, thus, the mutation rate. This might explain the common association between allopolyploidy and apomixis because both conditions can be successful when selection is relaxed.

Still, the long-term success of so many polyploids is puzzling. Perhaps it has nothing to do with the impact of polyploidy on gene transmission and may instead be caused either by immediate benefits of polyploidy even in multicellulars (29) or by the fact that polyploidy, especially allopolyploidy, leads to instant significant changes at the genomic level that cannot be produced by any other means (284). Gradual reversion to haploidy of the gamophase, due to divergence of the different genomes (diploidization), often makes it impossible to say whether the ancestral polyploid was originally auto- or allopolyploid.

A phenomenon similar to diploidization must occur in ancient apomictic diploids, originated by transition of zygobionts to obligate apomixis, because gradual accumulation of differences between the two genomes will finally make each of them insufficient to start an offspring. Such haploidization may be used to identify ancient apomicts with diploid zygophase (21).

Vegetative Reproduction

Vegetative reproduction is similar to polyploidy because in both cases an offspring receives many genomes. The genetic consequences of any obligate polygenomic reproduction depend on the relatedness of the genomes transmitted to an offspring. If all these genomes had a common ancestor only recently, the efficiency of selection is reduced, relative to unigenomic reproduction, only slightly (164), analogous to the case of the ploidy cycle. In contrast, obligate vegetative reproduction with no coalescence of genealogies of the cells and genomes transmitted to an offspring leads to the least efficient selection and to the highest mutation load. Also, origination of an organism from many initial cells facilitates intercellular selection (217) in the same way as biparental inheritance of organelles may facilitate selection among them (58). The winning cells (or organelles) may be selfish and nonoptimal from the organismal perspective. Not surprisingly, obligate vegetative reproduction is rare (10).

Intraorganismal Recombination

Nonreciprocal, nonhomologous intragenomic recombination appears to be an attractive way to change the genome in a regular and potentially reversible way. Currently, we know of only a few forms, mostly parasites (174), in which such recombination is an essential part of the life cycle. More data are needed, for both apomicts and amphimicts. If this phenomenon is, indeed, uncommon, this may imply the rarity of regular and drastic fluctuations of selection when every genotype becomes maladapted soon after becoming frequent (as may be the case in parasites, due to the host's immune response).

Intense nonreciprocal homologous recombination (e.g. conversion) between genomes in a polyploid cell will reduce its effective ploidy, making selection more efficient. This may be equivalent to the ploidy cycle and may prevent divergence among the genomes without amphimixis (21). No formal studies of this have been done, and I do not know of any data on whether this process actually occurs.

In contrast, reciprocal homologous recombination among genomes, which are always selected together (residing in the same organism from the time of their origin from their common ancestral genome), cannot lead to any genetic consequences because it will not reduce linkage disequilibrium. Thus, I do not agree with Cavalier-Smith (38) that such recombination may exist in apomicts and be the basis for the later evolution of amphimixis. The discovery of regular reciprocal homologous recombination between the genomes of a polyploid obligate apomict would be a very interesting surprise.

Origin of Interorganismal Gene Exchange

Amphimixis hardly confers any significant physiological benefits (but see 19), while in terms of gene transmission it may be beneficial if (a) the population is genetically variable, (b) there are linkage disequilibria that it can destroy, and (c) this destruction is beneficial (135a, 163). Prokaryotic amphimixis is mechanistically relatively simple and may evolve under the appropriate conditions, e.g. if disequilibria appear due to epistatic selection against mutations (252a). We do not know enough about prokaryote biology to determine whether this mechanism was important or why some bacteria are amphimictic, whereas others are clonal.

Eukaryotic amphimixis is more complex. The main problem is the origin of meiosis, because syngamy can appear easily. Despite early skepticism, gradual origin of meiosis, with each step favored by natural selection, is possible (38, 166, 214). Reduction becomes necessary after the origin of syngamy in order to prevent the unlimited increase of ploidy, although strict alternation of syngamy and reduction can hardly be the primitive condition. With a multichromosomal genome (common in protists), reduction from the very beginning may be accompanied by recombination due to their independent assortment. Such primitive amphimixis (without crossing-over) may already confer a substantial advantage. Even if the primitive reduction occurred owing to random chromosome loss, leading to a high cost due to aneuploidy, its benefit may outweigh this cost, provided that the number of chromosomes was not too high (below 5–7) and the intensity of selection fluctuates, occasionally becoming high (166). During times of weak selection, apomixis can be expected, as it is the case in all modern unicellulars.

Crossing-over probably originated later, when regular reduction and chromosome conjugation were already in place. The only detailed analysis of the origin of amphimixis from the population genetic perspective performed so far (166) was based on mutational deterministic hypothesis (163). Other possibilities, e.g. selection with fluctuating direction, should also be studied.

Some features of the ancestral apomicts could preadapt them to the origin of amphimixis. An apomictic ploidy cycle may have provided the primitive amphimicts with a ready mechanism for reduction (135, 165, 214). Proteins responsible for crossing-over have probably evolved from those involved in DNA repair. Even intragenomic or nonhomologous intergenomic recombination (but probably not homologous intergenomic recombination, see above) may predate amphimixis and provide the basis for the evolution of crossing-over.

Two-step meiosis is a paradoxical process because it starts reduction from DNA replication (131). This may be the vestige of its origin from mitosis, in which case two-step meiosis is primitive. Alternatively, one-step meiosis can be primitive and represent an intermediate step in the evolution of two-step meiosis (246).

A crucial question is whether amphimixis is mono- or polyphyletic. Clearly, monophyly of such a complex and basically uniform (246) adaptation as meiosis is the most parsimonious hypothesis (38). However, we cannot reject a priori the polyphyletic origin of meiosis because traits that evolved many times under the same selective pressure on the basis of the same preadaptations may be very similar.

Knowledge about the phylogeny of eukaryotes and their modes of reproduction is still insufficient for any definite conclusions. Monophyly of amphimixis would imply its very early origin, as long as amphimictic microsporidians are not modified Fungi, or parabasalians really have amphimixis, or, at least, kinetoplastids and foraminiferans are, as claimed, two early clades. In addition, repeated secondary losses of amphimixis must occur in the ancestors of several major apomictic protozoan taxa, including those that are perhaps close to the ancestors of several amphimictic clades, i.e. jakobid flagellates, bicoecids, unicellular rhodophytes, and choanoflagellates and their allies. This seems unlikely because in multicellulars, major taxa never appear after the loss of amphimixis. Of course, amphimixis can also be overlooked in some or all of these taxa. However, if they are, in fact, primitively apomictic, amphimixis must be polyphyletic, having originated independently in many clades (kinetoplastids, foraminiferans, heterokonts, rhodophytes, etc). Neither of these alternatives looks very attractive. Thus, the origin of amphimixis remains a very interesting puzzle and more data are needed.

Evolution of Amphimictic Cycles

The enormous diversity of amphimixis is the result of its extensive evolution. This evolution may be fast because modifications of meiosis and gamete assortment leading to automixis and even apomixis appear readily and may have a very simple genetic basis (137, 325). While some evolutionary processes considered below can only alter the character of amphimixis, evolution of facultative apomixis, linkage, or inbreeding can effectively lead to its gradual destruction.

DURATION OF PHASES   With strict alternation of pairwise syngamy and twofold reduction, the duration of phases is the only flexible parameter of the structure of the amphimictic cycle. Selection during gamophase evaluates genomes more independently and, thus, may be more efficient, thereby promoting the reduction of the zygophase. Still, in an amphimictic population, selection with a higher ploidy (zygophase) can also be more efficient if it allows the population to exploit better the benefits of synergistic epistasis, leading, in particular, to a lower mutation load (166a).

However, group selection arguments are insufficient to study evolution within an amphimictic population, and a modifier allele that shifts selection to the zygophase can win even if it eventually causes a higher load (231, 239a), with outcrossing and recombination favoring such modifiers (141, 233, 253).

It seems, however, that analysis only in terms of transmission genetics cannot explain the evolution of life cycles in which both phases are unreduced (140), and immediate benefits of this may be more important (13, 194, 249, 294). Still, the taxonomic distribution of phase reductions can hardly be explained exclusively by such benefits.

MULTICELLULARITY AND MIXED ORGANISMS   Because multicellularity has appeared in amphimicts, its origin may be viewed as the substitution of reproductive mitoses by somatic ones within the amphimictic cycle. This has profound consequences for selection because selection among cells is very different from that among organisms (116, 217, 232). The formation of genetically mixed multicellular organisms has never been studied from the transmission genetics perspective. Of course, immediate benefits of multicellularity and of different phases remaining together may be the leading factor here, and the approach of life-history theory may be the most useful one (311).

EXOGAMOUS CLASSES OF GENOMES   Sexes and mating types evolved independently many times, which resulted in their great heterogeneity (32, 46, 47, 49, 126, 132, 133, 148, 186, 270). Complex systems of incompatibility are mostly known from taxa with selection in the zygophase (136). Because such selection may lead to inbreeding depression, this suggests that avoidance of inbreeding was the factor of evolution of these systems. Dioecy may evolve due to similar reasons (208). Gamoid selfing and selection is identical to facultative apomixis, so that the conditions favoring sexes and mating types in this case may be identical to those favoring obligate amphimixis (162, 207). The existence of just two sexes can follow from the evolution of cytoplasmic inheritance (133).

REALLOCATION OF SELECTION AND GENETIC CONFLICTS   Natural selection is unavoidable, but organisms can exercise some control over when and how to be selected. Particularly interesting situations may appear when entities of some kind can regulate intensity of selection among the entities of the other kind. Zygobionts can influence selection among pollen (63), among non–free living gametes (201), among the cells within themselves (116, 157), or among different genomes within a cell (117, 132). Females can encourage competition among males and among their offspring.

In these and similar cases the analysis of conflicts between different stages of the life cycles, different sexes, or nuclear vs. organelle chromosomes may be essential (20, 110, 134, 253). Apparently, such conflicts, which may lead to evolution which decreases the mean population fitness, are possible only in amphimicts because in apomicts the genotype remains essentially the same within a lineage, and evolution must proceed in the direction increasing the group fitness. This extra disadvantage of amphimixis might lead to the transition to apomixis in some cases but is perhaps not too large in the forms that remain amphimictic.

ASYMMETRIC PATTERNS   One of the most striking features of amphimixis is that symmetry during reduction and syngamy is often evolutionarily unstable, so that various asymmetric patterns evolve. Currently, there is no general theory for the evolution of such patterns, which may actually be driven by different forces in different cases.

Non-Mendelian segregation   When not all the meiospores survive, preferential transmission to a viable meiospore is obviously advantageous for a gamoid genotype. Several mechanisms may explain why meiotic drive is rare (82, 86, 112), but it is unclear what actually happens in nature.

Anisogamy and sexual dimorphism   Repeated origins of anisogamy and oogamy may be caused by their immediate benefits (10) or by the conflict between nuclear chromosomes and cytoplasmic hereditary entities, organelle chromosomes or parasites (117). Such a conflict can favor the restriction of cytoplasm transmission to only one of the two uniting gametes. Once present, anisogamy may evolve due to competition among gametes produced by the same (201) or different (179) organisms. Many explanations were proposed for the evolution of sexual dimorphism of zygobionts (255a, 262a).

Haploid males   Independent origin of male haploidy in seven animal taxa appears to be caused by transmission genetics factors. An attractive possibility is that mothers, which control the ploidy of their offspring, use this mechanism to increase the efficiency of selection (e.g. selection against mutations) among their offspring (99).

Degeneration of heterozygous sex chromosomes   Slow degeneration of permanently nonrecombining Y or W chromosomes occurred many times and is probably caused by stochastic population effects, which facilitate fixation of deleterious mutations and prevent fixation of beneficial mutations in such chromosomes (47).

Uniparental inheritance of organelles   It may evolve because selection favors nuclear genes, which prevents competition between cytoplasmic chromosomes of different origin (20, 132, 133) or due to some other form of conflict between nuclear and cytoplasmic chromosomes (98).

ABRUPT TRANSITION TO OBLIGATE APOMIXIS   Despite its gradual origin, amphimixis in some taxa may be abruptly converted into obligate apomixis. Immediate benefit arguiments seem insufficient to explain why this does not happen more often because, if anything, they favor obligate apomixis (209). There are ≈20 hypotheses on what makes amphimixis advantageous at the population level due to genetic changes it causes (135a, 163, 209). However, the lack of data, in particular on deleterious mutation rates and on the amplitudes of the fluctuations of selection, limits our understanding.

REGULAR APOMIXIS   If the intensity of selection fluctuates widely, apomixis during times of weak selection is the best strategy, with transitions to amphimixis before episodes of strong selection (160). Remarkably, this is the case with all unicellulars, in which average selection is weak because each organism produces just two offspring, making a genetic load over 0.5 impossible. In multicellulars, such regular apomixis appears when the organisms can predict episodes of tough selection.

   If episodes of tough selection disappear, regular apomixis may become obligate, eliminating amphimixis (62). Gradual increase in the number of apomictic generations between successive amphimictic ones provides the mechanism for smooth transition to obligate apomixis. These processes still await investigation in terms of individual selection.

FACULTATIVE APOMIXIS   Gradual reallocation of resources to the production of apomictic offspring in the case of facultative apomixis provides another mechanism for smooth transition to obligate apomixis. Obligate amphimixis represents a large evolutionary problem because only very strong selection makes disadvantageous a modifier causing a low or moderate rate of facultative apomixis (135a, 162). Occasional incomplete syngamy may be enough for almost independent selection in some prokaryotes in which linkage disequilibria are absent (211). Difficulty in producing offspring in two ways may lead to the immediate disadvantage of facultative apomixis.

EVOLUTION OF RECOMBINATION RATES   Even in a multichromosomal genome, selection for crossing-over may be significant. However, if there is already about one crossover per chromosome, selection for further increase of recombination rate is very weak (48, 161). Individual selection usually favors increased recombination under conditions more stringent than those under which group selection favors amphimixis (45, 166b).

Repeated losses of crossing-over in the heterogametic sex show that it is not necessary for proper segregation (34). It is not clear why these losses evolved. Their immediate benefit may involve avoidance of nonreciprocal exchanges between different sex chromosomes (47). Inversion polymorphisms reduce recombination within individual chromosomes. They may be protected by the induced overdominance due to accumulation of different partially dominant mutations in different inversions (320).

EVOLUTION OF OUTCROSSING   Without outcrossing, the uniting genomes would be identical, and amphimixis would be genetically equivalent to endomitosis. Paradoxically, exogamous classes of genomes, while restricting their free assortment, may, in fact, make this assortment more random by preventing inbreeding (187, 207). Random union of gametes (or gamobionts) and random mating of zygobionts (see below) followed by random assortment of gametes within each pair are genetically equivalent only without selection (168). Selection against deleterious mutations can promote the evolution of obligate or facultative outcrossing (50, 162).

CONCLUSIONS AND PERSPECTIVES

We have good knowledge of the life cycles in plants and animals and, to a lesser extent, in other multicellular eukaryotes. In contrast, life cycles of prokaryotes and unicellular eukaryotes, critical from the evolutionary perspective, have been studied insufficiently. Amphimixis as well as regular genetic changes of nonamphimictic nature may be overlooked in many forms. One possibility for detecting amphimixis in protists is to look for meiosis-specific genes. Many such genes have already been identified in *S. cerevisiae*, the complete genome of which is now known. If these genes are conservative enough, as at least some genes involved in mitosis are (257), their homologs can be sought in various eukaryotes. However, a mei-218 gene product of *D. melanogaster* (215) does not have any close similarities in the *S. cerevisiae* genome (SA Shabalina, personal communication), which might indicate polyphyly of meiosis. We need more data on molecular architecture of meiosis in various taxa as well as on genetic structures of protozoan populations and on their phylogeny.

Our ability to draw conclusions from such data is limited for two reasons. First, much theoretical work remains to be done. Evolution of many features of the life cycles either has never been modeled or has been addressed only from the perspective of selection against mutations. Thus, it is still unclear whether any transmission genetic factors can explain obligate amphimixis (135a). Second, we need to know more about the physiological effects of various life cycle features and about the parameters relevant to population genetic theory, such as mutation rates and modes of fluctuating selection. Comprehensive understanding of the evolutionary genetics of life cycles will be an enormous achievement, but we are still far from it.

ACKNOWLEDGMENTS

I have taken into account several very useful comments by Laurence Hurst and two members of *ARES* editorial board. This work was supported by NSF grant DEB-9417753.

Visit the *Annual Reviews home page* at
http://www.annurev.org.

Literature Cited

1. Adler H.   1894. *Alternating Generations. A Biological Study of Oak Galls and Gall Flies.* Oxford: Clarendon

2. Afon'kin SYu. 1986. Spontaneous 'depolyploidization' of cells in *Amoeba* clones with increased nuclear DNA content. *Arch. Protist.* 131:101–12

3. Agulnik SI, Sabantsev ID, Orlova GV, Ruvinsky AO. 1993. Meiotic drive on aberrant chromosome 1 in the mouse is determined by a linked distorter. *Genet. Res.* 61:91–96

4. Angata K, Kuroe K, Yanagisawa K, Tanaka Y. 1995. Codon usage, genetic code and phylogeny of *Dictyostelium discoideum* mitochondrial DNA as deduced from a 7.3 kb region. *Curr. Genet.* 27:249–56

5. Astaurov BL. 1969. Experimental polyploidy in animals. *Annu. Rev. Genet.* 3:99–126

6. Bacci G. 1965. *Sex Determination.* Oxford, UK: Pergamon

7. Baker MD, Vossbrinck CR, Didier ES, Maddox JV, Shadduck JA. 1995. Small subunit ribosomal DNA phylogeny of various microsporidia with emphasis on AIDS-related forms. *J. Euk. Microbiol.* 42:564–70

8. Barns SM, Delwiche CF, Palmer JD, Pace NR. 1996. Perspectives on archeal diversity, thermophily and monophyly from environmental rRNA sequences. *Proc. Natl. Acad. Sci. USA* 93:9188–93

9. Becher D, Schulze S, Kasusk A, Stoll R, Wedler H, Oliver SG. 1995. Chromosome polymorphisms close to the cm-ADE1 locus of *Candida maltosa. Mol.*

Gen. Genet. 247:591–602

10. Bell G. 1982. *The Masterpiece of Nature.* San Francisco: Univ. Calif. Press

11. Bell G. 1988. *Sex and Death in Protozoa.* Cambridge, UK: Cambridge Univ. Press

12. Bell G. 1994. The comparative biology of the alternation of generations. In *The Evolution of Haploid-Diploid Life Cycles,* ed. M Kirkpatrick, *Lectures on Math. in Life Sci.* 25:1–26. Providence: Am. Math. Soc.

13. Bell G. 1997. The evolution of the life cycle of brown seaweeds. *Biol. J. Linn. Soc.* 60:21–38

14. Beukeboom LW, Seif M, Mettenmeyer T, Plowman AB, Michiels NK. 1996. Paternal inheritance of B chromosomes in a parthenogenetic hermaphrodite. *Heredity* 77:646–54

14a. Beutlich A, Schnetter R. 1993. The life cycle of *Cryptochlora perforans* (Chlorarachniophyta). *Bot. Acta* 106:441–47

15. Bhattacharya D, Helmchen T, Bibeau C, Melkonian M. 1995. Comparisons of nuclear-encoded small-subunit ribosomal RNAs reveal the evolutionary position of the Glaucocystophyta. *Mol. Biol. Evol.* 12:415–20

16. Bhattacharya D, Helmchen T, Melkonian M. 1995. Molecular evolutionary analyses of nuclear-encoded small subunit ribosomal RNA identify an independent rhizopod lineage containing the Euglyphina and the Chlorarachniophyta. *J. Euk. Microbiol.* 42:65–69

17. Bhattacharya D, Medlin L. 1995. The phylogeny of plastids: a review based on comparisons of small-subunit ribosomal RNA coding regions. *J. Phycol.* 31:489–98

18. Bhaud Y, Soyer-Gobillard MO, Salmon JM. 1988. Transmission of gametic nuclei through a fertilization tube during mating in a primitive dinoflagellate *Prorocentrum micans.* Ehr. *J. Cell Sci.* 89:197–206

19. Birdsell J, Wills C. 1996. Significant competitive advantage conferred by meiosis and syngamy in the yeast *Saccharomyces cerevisiae. Proc. Natl. Acad. Sci. USA* 93:908–12

20. Birky CW Jr. 1995. Uniparental inheritance of mitochondrial and chloroplast genes: mechanisms and evolution. *Proc. Natl. Acad. Sci. USA* 92:11331–38

21. Birky CW Jr. 1996. Heterozygosity, heteromorphy, and phylogenetic trees in asexual eukaryotes. *Genetics* 144:427–37

22. Blackwelder RE, Shepherd BA. 1981. *The Diversity of Animal Reproduction.* Boca Raton: CRC Press

23. Boeskorov GG, Kartavtseva IV, Zagorodnyuk IV, Belyanin AN, Lyapunova EA. 1995. Nucleolus organizer regions and B-chromosomes of wood mice (Mammalia, Rodentia, *Apodemus*). *Genetika* 31:185–92 (In Russian)

24. Boron A. 1992. Natural polyploidy in fish. *Przeglad Zoologiczny* 36:67–76

25. Borst P, Rudenko G. 1993. Antigenic variation in African trypanosomes. *Science* 264:1873–74

26. Bos CJ. 1996. Somatic recombination. In *Fungal Genetics: Principles and Practice,* ed. CJ Bos. Mycology Ser. 13:73–95. New York: Marcel Dekker

27. Braselton JP. 1989. Karyotypic analysis of *Ligniera verrucosa* Plasmodiophoromycetes. *Can. J. Bot.* 67:1216–18

28. Bretagnolle F, Thompson JD. 1995. Gametes with the somatic chromosome number: mechanisms of their formation and role in the evolution of autopolyploid plants. *New Phytol.* 129:1–22

29. Bretagnolle F, Thompson JD. 1996. An experimental study of ecological differences in winter growth. *J. Ecol.* 84:343–51

30. Bridge D, Cunningham CW, Schierwater B, DeSalle R, Buss LW. 1992. Class-level relationships in the phylum Cnidaria: evidence from mitochondrial genome structure. *Proc. Natl. Acad. Sci. USA* 89:8750–53

31. Bui ETN, Bradley PJ, Johnson PJ. 1996. A common evolutionary origin of mitochondria and hydrogenosomes. *Proc. Natl. Acad. Sci. USA* 93:9651–56

32. Bull JJ. 1983. *Evolution of Sex-Determining Mechanisms.* Menlo Park, CA: Benjamin-Cummings

33. Bull JJ, Pease CM. 1989. Combinatorics and variety of mating-type systems. *Evolution* 43:667–71

34. Burt A, Bell G, Harvey PH. 1991. Sex differences in recombination. *J. Evol. Biol.* 4:259–77

35. Buttner P, Koch F, Voigt K, Quidde T, Risch S, et al. 1994. Variations of ploidy among isolates of *Botrytis cinerea*: implications for genetic and molecular analyses. *Curr. Genet.* 25:445–50

36. Cavalier-Smith T. 1993. Evolution and diversity of zooflagellates. *J. Euk. Microbiol.* 40:603–05

37. Cavalier-Smith T. 1993. Kingdom Protozoa and its 18 phyla. *Microbiol. Rev.* 57:953–94

38. Cavalier-Smith T. 1995. Cell cycles, diplokaryosis and the Archezoan origin

of sex. *Arch. Protist.* 145:189–207

39. Cavalier-Smith T, Allsopp MTEP. 1996. Corallochytrium, an enigmatic non-flagellate protozoan related to choanoflagellates. *Eur. J. Protistol.* 32: 306–10

40. Cavalier-Smith T, Allsopp MTEP, Chao EE. 1994. Thraustochytrids are chromists, not Fungi: 18S rRNA signatures of Heterokonta. *Philos. Trans. R. Soc. London B* 346:387–97

41. Cavalier-Smith T, Allsopp MTEP, Haueber MM, Gothe G, Chao EE, et al. 1996. Chromobiote phylogeny: The enigmatic alga *Reticulosphaera japonensis* is an aberrant haptophyte, not a heterokont. *Eur. J. Phycol.* 31:255–63

42. Cavalier-Smith T, Chao EE. 1995. The opalozoan *Apusomonas* is related to the common ancestor of animals, fungi, and choanoflagellates. *Proc. R. Soc. Lond. B* 261:1–6

43. Cavalier-Smith T, Chao EE. 1996. Molecular phylogeny of the free-living archezoan *Trepomonas agilis* and the nature of the first eukaryote. *J. Mol. Evol.* 43:551–62

44. Chaplin JA, Hebert PDN, 1997. *Cyprinotus incongruens* (Ostracoda): an ancient asexual? *Mol. Ecol.* 6:155–68

45. Charlesworth B. 1993. Directional selection and the evolution of sex and recombination. *Genet. Res.* 61:205–24

46. Charlesworth B. 1994. The nature and origin of mating types. *Curr. Biol.* 4: 739–41

47. Charlesworth B. 1996. The evolution of chromosomal sex determination and dosage compensation. *Curr. Biol.* 6: 149–62

48. Charlesworth B, Barton NH. 1996. Recombination load associated with selection for increased recombination. *Genet. Res.* 67:27–41

49. Charlesworth D. 1995. Multi-allelic self-incompatibility polymorphisms in plants. *BioEssays* 17:31–38

50. Charlesworth D, Morgan MT, Charlesworth B. 1993. Mutation accumulation in finite outbreeding and inbreeding populations. *Genet. Res.* 61:39–56

51. Charnov EL. 1982. *The Theory of Sex Allocation.* Princeton, NJ: Princeton Univ. Press

52. Charnov EL. 1989. Evolution of the breeding sex ratio under partial sex change. *Evolution* 43:1559–61

53. Chaudhary BR, Prasad RN. 1986. Contributions to the karyology of euglenoid flagellates. IV. *Thrachelomonas*

Ehrenberg emend. Deflandre. *Cytologia* 51:723–29

54. Choudhary M, MacKenzie C, Nereng KS, Sodergren E, Weinstock GM, Kaplan S. 1994. Multiple chromosomes in bacteria: structure and function of chromosome II of *Rhodobacter sphaeroides* 2.4.1-T. *J. Bacteriol.* 176:7694–702

55. Clark CG, Roger AJ. 1995. Direct evidence for secondary loss of mitochondria in *Entamoeba histolytica.* *Proc. Natl. Acad. Sci. USA* 92:6518–21

56. Cleveland LR. 1947. The origin and evolution of meiosis. *Science* 105:287–89

57. Coates KA. 1995. Widespread polyploid forms of *Lumbricillus lineatus* (Mueller) (Enchytraeidae: Oligochaeta): comments on polyploidism in the enchytraeids. *Can. J. Zool.* 73:1727–34

58. Coenen A, Croft JH, Slakhorst M, Debets F, Hoekstra R, 1996. Mitochondrial inheritance in *Aspergillus nidulans.* *Genet. Res.* 67:93–100

59. Cole RA, Slade MB, Williams KL, 1995. *Dictyostelium discoideum* mitochondrial DNA encodes a NADH:ubiquinone oxidoreductase subunit which is nuclear encoded in other eukaryotes. *J. Mol. Evol.* 40:616–21

60. Collins OR. 1979. Myxomycete biosystematics: some recent developments and future research opportunities. *Bot. Rev.* 45:145–201

61. Corliss JO. 1994. An interim utilitarian ("user-friendly") hierarchical classification and characterization of the protists. *Acta Protozool.* 33:1–51

62. Crease TJ, Stanton DJ, Hebert PDN. 1989. Polyphyletic origins of asexuality in *Daphnia pulex* II. Mitochondrial DNA variation. *Evolution* 43:1016–26

63. Cresti M, Gori P, Pacini E, eds. 1988. *Sexual Reproduction in Higher Plants.* Berlin: Springer

64. Crow JF, Kimura M. 1970. *An Introduction to Population Genetics Theory.* New York: Harper & Row

65. Crozier RH, Dobric N, Imai HT, Graur D, Cornuet JM, Taylor RW. 1995. Mitochondrial-DNA sequence evidence on the phylogeny of Australian Jack-Jumper ants of the *Myrmecia pilosula* complex. *Mol. Phyl. Evol.* 4:20–30

66. Cruz AK, Titus R, Beverley SM. 1993. Plasticity in chromosome number and testing of essential genes in *Leishmania* by targeting. *Proc. Natl. Acad. Sci. USA* 90:1599–603

67. Darevsky IS, Kupriyanova LA, Uzzel T. 1985. Parthenogenesis in reptiles. In

Biology of the Reptilia, ed. C. Gaus, F. Billett 15B:411–526. New York: Wiley

68. Davies J. 1994. Inactivation of antibiotics and the dissemination of resistance genes. *Science* 264:375–82

69. Dawley RM, Bogart JP. 1989. *Evolution and Ecology of Unisexual Vertebrates.* Albany: New York State Mus.

70. De Rijk P, Van De Peer Y, Van Den Broeck I, De Wacher R. 1995. Evolution according to large ribosomal subunit RNA. *J. Mol. Evol.* 41:366–75

71. Diarra K, Toguebaye BS, 1994. Light and electron microscope study of the octosporus phase of the life cycle of *Amblyospora senegalensis* n. sp. (Microspora, Amblyospora), parasite of *Culex thalassius* (Diptera, Culicidae). *Arch. Protist.* 144:212–20

72. Dini F, Luporini P. 1980. Genetic determination of the autogamy trait in the hypotrich ciliate *Euplotes crassus. Genet. Res.* 35:107–19

73. Doolittle RF, Feng DF, Tsang S, Cho G, Little E. 1996. Determining divergence times of the major kingdoms of living organisms with a protein clock. *Science* 271:470–77

74. Drouin G, Moniz-De-Sa M, Zuker M. 1995. The *Giardia lamblia* actin gene and the phylogeny of eukaryotes. *J. Mol. Evol.* 41:841–49

75. Dubois A. 1991. Nomenclature of parthenogenetic, gynogenetic, and hybridogenetic vertebrate taxons: new proposals. *Alytes* 8:61–74

76. Dyer AF, Page CN, eds. 1985. Biology of Pteridophytes. *Proc. R. Soc. Edinburgh B,* Vol. 86 entire

77. Echelle AA, Dowling TE, Moritz CC, Brown WM. 1989. Mitochondrial-DNA diversity and the origin of the *Menidia clarkhubbsi* complex of unisexual fishes (Atherinidae). *Evolution* 43:984–93

78. Edlind TD, Li J, Visvesvara GS, Vodkin MH, McLaughlin GL, Katiyar SK. 1996. Phylogenetic analysis of β-tubulin sequences from amitochondrial protozoa. *Mol. Phyl. Evol.* 5:359–67

79. Edvardsen B, Vaulot D. 1996. Ploidy analysis of the two motile forms of *Chrysochromulina polylepis* (Prymnesiophyceae). *J. Phycol.* 32:94–102

80. Efuet ET, Pulakat L, Gavini N. 1996. Investigations on the cell volumes of *Azotobacter vinelandii* by scanning electron microscopy. *J. Basic Microbiol.* 36:229–34

81. Embley TM, Hirt RP, Williams DM. 1994. Biodiversity at the molecular level: the domains, kingdoms and phyla

of life. *Philos. Trans. R. Soc. London B* 345:21–33

82. Eshel I. 1985. Evolutionary genetic stability of Mendelian segregation and the role of free recombination in the chromosomal system. *Am. Nat.* 125:412–20

83. Fan J-B, Korman SH, Cantor CR, Smith CL. 1991. *Giardia lamblia* haploid genome size determined by pulsed field gel electrophoresis is less than 12 Mb. *Nucl. Acids Res.* 19:1905–8

84. Felber F. 1991. Establishment of a tetraploid cytotype in a diploid population: effect of relative fitness of two cytotypes. *J. Evol. Biol.* 4:195–207

85. Felber F, Bever JD. 1997. Effect of triploid fitness on the coexistence of diploids and tetraploids. *Biol. J. Linn. Soc.* 60:95–106

86. Feldman MW, Otto SP. 1991. A comparative approach to the population genetics theory of segregation distortion. *Am. Nat.* 137:443–56

87. Fincham JRS, Day PR, Radford A. 1979. *Fungal Genetics.* Oxford: Blackwell. 4th ed.

88. Fisher RA. 1930. *The Genetical Theory of Natural Selection.* Oxford, UK: Clarendon

89. Flegel TW, Pasharawipas T. 1995. A proposal for typical eukaryotic meiosis in microsporidians. *Can. J. Microbiol.* 41:1–11

90. Flores BS, Siddall ME, Burreson E. 1996. Phylogeny of the Haplosporidia (Eukaryota: Alveolata) based on small subunit ribosomal RNA gene sequence. *J. Parasitol.* 82:616–23

91. Fritz L, Anderson DM, Triemer RE. 1989. Ultrastructural aspects of sexual reproduction in the Red Tide dinoflagellate *Gonyaulax tamarensis. J. Phycol.* 25:95–107

92. Friz CT. 1994. An analysis of the phylogenetic relationships of fifteen taxa of family Amoebidae. *Arch. Protist.* 144:31–46

93. Gastony GJ. 1991. Gene silencing in a polyploid homosporous fern: paleopolyploidy revisited. *Proc. Natl. Acad. Sci. USA* 88:1602–05

94. Gastony GJ, Yatskievych G. 1992. Maternal inheritance of the chloroplast and mitochondrial genomes in cheilanthoid ferns. *Am. J. Bot.* 79:716–22

95. Gibson W, Bailey M. 1994. Genetic exchange in *Trypanosoma brucei:* evidence for meiosis from analysis of a cross between drug-resistant transformants. *Mol. Biochem. Parasitol.* 64:241–52

96. Gifford EM, Foster AS. 1989. *Morphology and Evolution of Vascular Plants.* New York: Freeman

97. Go MF, Kapur V, Graham DY, Musser JM. 1996. Population genetic analysis of *Helicobacter pylori* by multilocus enzyme electrophoresis: extensive allelic diversity and recombinational population structure. *J. Bacteriol.* 178:3934–38

98. Godelle B, Reboud X. 1995. Why are organelles uniparentally inherited? *Proc. R. Soc. London B* 259:27–33

99. Goldstein DB. 1994. Deleterious mutations and the evolution of male haploidy. *Am. Nat.* 144:176–83

100. Gonzalez-Garcia JM, Rufas JS, Antonio C, Suja JA. 1995. Nucleolar cycle and localization of NORs in early embryos of *Parascaris univalens. Chromosoma* 104:287–97

101. Grant V. 1975. *Genetics of Flowering Plants.* New York: Columbia Univ. Press

102. Grell KG. 1967. Sexual reproduction in Protozoa. In *Research in Protozoology,* ed. TT Chen. 2:147–213

103. Grell KG. 1990. *Reticulosphaera japonensis*—new species Heterokontophyta from tide pools of the Japanese coast. *Arch. Protist.* 138:257–69

104. Grell KG. 1990. Some light microscopic observations on *Chlorarachnion reptans* Geitler. *Arch. Protist.* 138:271–90

105. Green DM. 1990. Muller's ratchet and the evolution of supernumerary chromosomes. *Genome* 33:818–24

106. Gunderson J, Hinkle G, Leipe D, Morrsion HG, Stickel SK, et al. 1995. Phylogeny of trichomonads inferred from small-subunit rRNA sequences. *J. Euk. Microbiol.* 42:411–15

107. Gupta RS. 1995. Phylogenetic analysis of the 90 kD heat shock family of protein sequences and an examination of the relatioship among animals, plants, and fungi species. *Mol. Biol. Evol.* 12:1063–73

108. Guttman DS, 1997. Recombination and clonality in natural populations of *Escherichia coli. Trends Ecol. Evol.* 12:16–22

109. Haddad CFB, Pombal JP Jr, Batistic RF. 1994. Natural hybridization between diploid and tetraploid species of leaffrogs, genus *Phyllomedusa* (Amphibia). *J. Herpetol.* 28:425–30

110. Haig D. 1986. Conflicts among megaspores. *J. Theor. Biol.* 123:471–80

111. Haig D. 1993. Alternatives to meiosis: the unusual genetics of red algae, Microsporidia, and others. *J. Theor. Biol.* 163:15–31

112. Haig D, Grafen A. 1991. Genetic scrambling as a defence against meiotic drive. *J. Theor. Biol.* 153:531–58

113. Hall DW, Hostettler N. 1993. Septate gregarines from *Reticulitermes flavipes* and *Reticulitermes virginicus* Isoptera Rhinotermitidae. *J. Euk. Microbiol.* 40: 29–33

114. Harrington RW. 1961. Oviparous hermaphroditic fish with internal self-fertilization. *Science* 134:1749–50

115. Hashimoto T, Nakamura Y, Kamaishi T, Nakamura F, Adachi J, et al. 1995. Phylogenetic place of mitochondrion-lacking protozoan, *Giardia lamblia,* inferred from amino acid sequences of elongation factor 2. *Mol. Biol. Evol.* 12:782–93

116. Hastings IM. 1991. Germline selection: population genetic aspects of the sexual/asexual lifecycle. *Genetics* 129:1167–76

117. Hastings IM. 1992. Population genetic aspects of deleterious cytoplasmic genomes and their effect on the evolution of sexual reproduction. *Genet. Res.* 59:215–25

118. Hawkes MW. 1988. Evidence of sexual reproduction in *Smithora naiadum* Erythropeltidales Rhodophyta and its evolutionary significance. *Br. Phycol. J.* 23:327–36

119. Hayashida H, Kikuno R, Yasunaga T, Miyata T. 1985. A classification of gene conversion and its evolutionary implications. *Proc. Jpn. Acad. B* 61:149–52

120. Hazard EI, Brookbank JW. 1984. Karyogamy and meiosis in an *Amblyospora* sp. (*Microspora*) in the mosquito *Culex salinarius. J. Invert. Pathol.* 44:3–11

121. Henriksson J, Aslund L, Marcina RA, De-Cazzulo BMF, Cazzulo JJ, et al. 1990. Chromosomal localization of seven cloned antigen genes provides evidence of diploidy and further demonstration of karyotype variability in *Trypanosoma cruzi. Mol. Biochem. Parasitol.* 42:213–24

122. Hernandez-Rivas R, Hinterberg K, Scherf A. 1996. Compartmentalization of genes coding for immunodominant antigens to fragile chromosome ends leads to dispersed subtelomeric gene families and rapid gene evolution in *Plasmodium falciparum. Mol. Biochem. Parasitol.* 78:137–48

123. Hill SA. 1996. Limited variation and maintenance of tight genetic linkage

characterize heteroallelic pilE recombination following DNA transformation of
Neisseria gonorrhoeae. Mol. Microbiol.
20:507–18

124. Hinkle G, Sogin ML. 1993. The evolution of the Vahlkampfiidae as deduced
from 16S-like ribosomal RNA analysis.
J. Euk. Microbiol. 40:599–603

125. Hinkle G, Leipe DD, Narad TA, Sogin ML. 1994. The unusually long small
subunit ribosomal RNA of *Phreatamoeba balamuthi. Nucleic Acids Res.*
22:465–69

126. Hoekstra RF. 1987. The evolution of
sexes. In *The Evolution of Sex and Its
Consequences,* ed. SC Stearns, pp. 59–
91. Basel: Birkhauser

127. Hoiland K. 1995. The origin and early evolution of fungi. *Blyttia* 53:27–
42

127a. Horner DS, Hirt RP, Llyod SKD, Embley TM. 1996. Molecular data suggest an
early acquisition of the mitochondrion
endosymbiont. *Proc. R. Soc. London B*
263:1053–59

128. Howell N, Kubacka I, Mackey DA.
1996. How rapidly does the human mitochondrial genome evolve? *Am. J. Hum.
Gen.* 59:501–9

129. Huber ME, Lewin RA. 1986. An
electrophoretic survey of the genus
Tetraselmis Chlorophyta Prasinophyceae. *Phycologia* 25:205–9

130. Huges RN. 1989. *A Functional Biology
of Clonal Animals.* London: Chapman
& Hall

131. Hurst LD. 1993. Evolutionary genetics: drunken walk of the diploid. *Nature*
365:206–07

132. Hurst LD. 1995. Selfish genetic elements and their role in evolution: the
evolution of sex and some of what that
entails. *Philos. Trans. R. Soc. London B*
349:321–32

133. Hurst LD. 1996. Why are there only two
sexes? *Proc. R. Soc. Lond. B* 263:415–
22

134. Hurst LD, Atlan A, Bengtsson BO. 1996.
Genetic conflicts. *Q. Rev. Biol.* 71:317–
64

135. Hurst LD, Nurse P. 1991. A note on
the evolution of meiosis. *J. Theor. Biol.*
150:561–63

135a. Hurst LD, Peck JR. 1996. Recent advances in understanding of the evolution
and maintenance of sex. *Trend. Ecol.
Evol.* 11:46–52

136. Iwasa Y, Sasaki A. 1987. Evolution of
the number of sexes. *Evolution* 41:49–
65

137. Jain SK. 1985. Probability of selfing in

Amaranthus is easily selected. *Genetica*
66:21–121

138. Jackson JBC, Buss LW, Cook RE. 1985.
*Population Biology and Evolution of
Clonal Organisms.* New Haven, CT:
Yale Univ. Press

139. Jarolimova V, Kirschner J. 1995. Tetraploids in *Luzula multiflora* (Juncaceae)
in Ireland: karyology and meiotic behaviour. *Folia Geobotanica et Phytotax-
onomica* 30:389–96

140. Jenkins CD. 1993. Selection and the evolution of genetic life cycles. *Genetics*
133:401–10

141. Jenkins CD, Kirkpatrick M. 1995. Deleterious mutations and the evolution of
genetic life cycles. *Evolution* 49:512–20

142. John B. 1990. *Meiosis.* Cambridge, UK:
Cambridge Univ. Press.

143. John DM. 1994. Alternation of generations in algae: its complexity, maintenance and evolution. *Biol. Rev. Camb.
Philos. Soc.* 69:275–91

144. Jones GH. 1994. Meiosis in autopolyploid *Crepis capillaris*: III. Comparison
of triploids and tetraploids; evidence for
nonindependence of autonomous pairing sites. *Heredity* 73:215–19

145. Jones RN. 1995. B chromosomes in
plants. *New Phytol.* 131:411–34

145a. Judson OP, Normark BB. 1996. Ancient
asexual scandals. *Trends Ecol. Evol.*
11:41–46

146. Jumas-Bilak E, Maugard C, Michaux-
Charachon S, Allardet-Servent A, Per-
rin A, et al. 1995. Study of the organization of the genomes of *Escherichia coli,
Brucella melitensis* and *Agrobacterium
tumefaciens* by insertion of a unique restriction site. *Microbiology* 141:2425–
32

147. Kain JM, Destombe C. 1995. A review
of the life history, reproduction and phenology of *Gracilaria. J. Appl. Phycol.*
7:269–81

148. Karlin S, Lessard S. 1986. *Theoretical
Studies on Sex Ratio Evolution.* Princeton, NJ: Princeton Univ. Press

149. Kasha KJ, ed. 1974. *Haploids in Higher
Plants.* Guelph, Ontario: Univ. Guelph
(Huddleston & Barney)

150. Kawano S, Takano H, Kuroiwa T. 1995.
Sexuality of mitochondria: fusion, recombination, and plasmids. *Int. Rev. Cy-
tol.* 161:49–110

151. Keeling PJ, Doolittle WF. 1996. Alpha-
tubulin from early-diverging eukaryotic
lineages and the evolution of the tubulin family. *Mol. Biol. Evol.* 13:1297–
305

152. Kerk D, Gee A, Standish M, Wain-

wright PO, Drum AS, et al. 1995. The rosette agent of Chinook salmon (*Oncorhynchus tshawytscha*) is closely related to choanoflagellates, as determined by the phylogenetic analyses of its small ribosomal subunit RNA. *Mar. Biol.* 122:187–92

153. Keys RN, Smith SE, Mogensen HL. 1995. Variation in generative cell plastid nucleoids and male fertility in *Medicago sativa*. *Sex. Plant Repr.* 8:308–12

154. Kim I. 1994. Aneuploidy in flowering plants: Asteraceae and Onagraceae. *Korean J. Plant Taxon.* 24:265–78

155. Deleted in proof

156. Kita T, Fukuyo Y. 1995. Life history of *Peridinium volzii* (Dinophyceae) in a natural pond. *Bull. Plankt. Soc. Japan* 42:63–73

157. Klekowski EJ. 1988. *Mutation, Developmental Selection, and Plant Evolution.* New York: Columbia Univ. Press

158. Klenk HP, Zillig W, Lanzendorfer M, Grampp B, Palm P. 1995. Location of protist lineages in a phylogenetic tree inferred from sequences of DNA-dependent RNA polymerases. *Arch. Protistenkunde* 145:221–30

159. Klinger T. 1993. The persistence of haplodiploidy in algae. *Trends Ecol. Evol.* 8:256–58

160. Kondrashov AS. 1984. A possible explanation of cyclical parthenogenesis. *Heredity* 52:307–08

161. Kondrashov AS. 1984. Deleterious mutations as an evolutionary factor. I. The advantage of recombination. *Genet. Res.* 44:199–217

162. Kondrashov AS. 1985. Deleterious mutations as an evolutionary factor. II. Facultative apomixis and selfing. *Genetics* 111:635–53

163. Kondrashov AS. 1993. Classification of hypotheses on the advantage of amphimixis. *J. Hered.* 84:372–87

164. Kondrashov AS. 1994. Mutation load under vegetative reproduction and cytoplasmic inheritance. *Genetics* 137:311–18

165. Kondrashov AS. 1994. The asexual ploidy cycle and the origin of sex. *Nature* 370:213–16

166. Kondrashov AS. 1994. Gradual origin of amphimixis by natural selection. In *The Evolution of Haploid-Diploid Life Cycles,* ed. M Kirkpatrick, *Lectures on Math. in the Life Sci.* 25:27–51. Providence, RI: Am. Math. Soc.

166a. Kondrashov AS, Crow JF. 1991. Haploidy or diploidy: Which is better? *Nature* 351:314–15

166b. Kondrashov AS, Yampolsky LYu. 1996. Evolution of amphimixis and recombination under fluctuating selection in one and many traits. *Genet. Res.* 68:165–73

167. Koonin EV, Dolja VV. 1993. Evolution and taxonomy of positive-strand RNA viruses—implications of comparative analysis of amino acid sequences. *Crit. Rev. Biochem. Mol. Biol.* 28:375–430

168. Korzuhin MD, Kaganova OZ. 1982. Genotype dynamics in populations with breeding pairs. *J. Theor. Biol.* 96:281–94

169. Kreutzer RD, Yemma JJ, Grocl M, Tesh RB, Martin TI. 1994. Evidence of sexual reproduction in the protozoan parasite *Leishmania* (Kinetoplastida: Trypanosomatidae). *Am. J. Trop. Med. Hyg.* 51:301–07

169a. Kubbies M, Wick R, Hildebrandt A, Sauer HW. 1986. Flow cytometry reveals a high degree of genomic size variation and mixoploidy in various strains of the acellular slime mold *Physarum polycephalum. Cytometry* 7:481–85

169b. Kugrens P, Lee RE. 1988. Ultrastructure of fertilization in a cryptomonad. *J. Phycol.* 24:385–93

170. Kuhlenkamp R, Muller DG, Whittick A. 1993. Genotypic variation and alternating DNA levels at constant chromosome numbers in the life history of the brown alga *Haplospora globosa* Tilopteridales. *J. Phycol.* 29:377–80

171. Kumar S, Rzhetsky A. 1996. Evolutionary relationships of eukaryotic kingdoms. *J. Mol. Evol.* 42:183–93

172. Lang BF, Goff LJ, Gray MW. 1996. A 5S rRNA gene is present in the mitochondrial genome of the protist *Reclinomonas americana* but is absent from red algal mitochondrial DNA. *J. Mol. Biol.* 261:607–13

173. Lange CE. 1994. Sexuality in *Perezia dichroplusae* Lange (Microspora: Pereziidae). *Neotropica* 40:29–33

174. Lanzer M, Fischer K, LeBlanc SM. 1995. Parasitism and chromosome dynamics in protozoan parasites: Is there a connection? *Mol. Biochem. Parasitol.* 70:1–8

175. Larsson JIR. 1994. *Trichoctospora pygopellita* gen. et sp. nov. (Microspora, Thelohaniidae), a microsporidian parasite of the mosquito *Aedes vexans* (Diptera, Culicidae). *Arch. Protist.* 144:147–61

176. Leipe DD, Wainright PO, Gunderson JH, Porter D, Patterson DJ, et al. 1994. The stramenopiles from a molecular perspective: 16S-like rRNA sequences

from *Labyrinthuloides minuta* and *Cafeteria roenbergensis*. *Phycologia* 33: 369–77

177. Lessie TG, Hendrickson W, Manning BD, Devereux R. 1996. Genomic complexity and plasticity of *Burkholderia cepacia*. *FEMS Microb. Let.* 144:117–28

178. Levin DA. 1983. Polyploidy and novelty in flowering plants. *Am. Nat.* 122:1–25

179. Levitan DR. 1996. Effects of gamete traits on fertilization in the sea and the evolution of sexual dimorphism. *Nature* 382:153–55

180. Lewis D. 1979. *Sexual Incompatibility in Plants*. London: Edward Arnold

181. Lewis WH, ed. 1980. *Polyploidy: Biological Relevance*. New York: Plenum

182. Lewis KE, O'Day DH. 1996. Phagocytosis in *Dictyostelium*: nibbling, eating and cannibalism. *J. Euk. Microbiol.* 43:65–69

183. Lewontin RC, Kirk D, Crow JF. 1966. Selective mating, assortative mating and inbreeding: definitions and implications. *Eugen. Q.* 15:141–43

184. Lezhava A, Kameoka D, Sugino H, Goshi K, Shinkawa H, et al. 1997. Chromosomal deletions in *Streptomyces griseus* that remove the *afs*A locus. *Mol. Gen. Genet.* 253:478–83

185. Li J, Heath IB, Packer L. 1993. The phylogenetic relationships of the anaerobic chytridiomycetous gut Fungi Neocallimasticaceae and the Chytridiomycota II. Cladistic analysis of structural data and description of Neocallimasticales ord.nov. *Can. J. Bot.* 71:393–407

186. Liberman U, Feldman MW. 1980. On the evolutionary significance of Mendel's ratios. *Theor. Pop. Biol.* 17:1–15

187. Liberman U, Feldman MW, Eshel I, Otto SP. 1990. Two-locus autosomal sex determination on the evolutionary genetic stability of the even sex ratio. *Proc. Natl. Acad. Sci. USA* 87:2013–17

188. Lioutas C, Schmetz C, Tannich E. 1995. Identification of various circular DNA molecules in *Entamoeba histolytica*. *Exp. Parasitol.* 80:349–52

189. Liu QY, Baldauf SL, Reith ME. 1996. Elongation factor 1-α genes of the red alga *Porphyra purpurea* include a novel, developmentally specialized variant. *Plant. Mol. Biol.* 31:77–85

190. Lloyd DG. 1980. Demographic factors and mating patterns in angiosperms. In *Demography and Evolution in Plant Populations,* ed. OT Solbrig. Oxford: Blackwell

191. Lloyd DG. 1984. Gene selection of Mendel's rules. *Heredity* 53:613–24

192. Lobban CS, Wynne MJ, eds. 1981. *The Biology of Seaweeds*. Berkeley, CA: Univ. Calif. Press

193. Loomis WF, Smith DW. 1995. Consensus phylogeny of *Dictyostelium*. *Experientia* 51:1110–15

194. Loreau M, Ebenhoen W. 1994. Competitive exclusion and coexistence of species with complex life cycles. *Theor. Pop. Biol.* 46:58–77

195. Mackenstedt U, Wagner D, Heydorn AO, Melhorn H. 1990. DNA measurements and ploidy determination of different stages in the life cycle of *Sarcocystis muris*. *Parasitol. Res.* 76:662–68

196. Mackenstedt U, Gauer M, Fuchs P, Zapf F, Schein E, Mehlhorn H. 1995. DNA measurements reveal differences in the life cycles of *Babesia bigemina* and *B. canis*, two typical members of the genus *Babesia*. *Parasitol. Res.* 81:595–604

197. Mackenstedt U, Ungar C, Sahm M, Seitz HM, Mehlhorn H. 1995. New aspect in the life cycle of *Pneumocystis carinii*, revealed by DNA measurements. *Eur. J. Protistol.* 31:127–36

198. Maldonado R, Jimenez J, Casadesus J. 1994. Changes of ploidy during the *Azotobacter vinelandii* growth cycle. *J. Bact.* 176:3911–19

199. Maleszka R. 1993. Electrophoretic analysis of the nuclear and organellar genomes in the ultra-small alga *Cyanidioschyzon merolae*. *Curr. Genet.* 24:548–50

200. Mann DG. 1993. Patterns of sexual reproduction in diatoms. *Hydrobiologia* 269–270:11–20

201. Manning JT, Chamberlain AT. 1994. Sib-competition and sperm competitiveness: an answer to 'why so many sperms?' and the recombination/sperm number correlation. *Proc. R. Soc. London B* 256:177–82

202. Mantovani B, Scali V. 1992. Hybridogenesis and androgenesis in the stick-insect *Bacillus rossius* × *Bacillus grandii-benazzii* (Insecta, Phasmatodea). *Evolution* 46:783–96

203. Marconi RT, Casjens S, Munderloh UG, Samuels DS. 1996. Analysis of linear plasmid dimers in *Borrelia burgdorferi* sensu lato isolates: implications concerning the potential mechanisms of linear plasmid replication. *J. Bact.* 178:3357–61

204. Margulis L, Corliss JO, Melkonian M, Chapman DJ, eds. 1990. *Handbook of Protoctista*. Boston: Jones & Bartlett

205. Marsden JE, Schwager SJ, May B. 1987. Single-locus inheritance in the tetraploid treefrog *Hyla versicolor* with an analysis of expected progeny ratios in tetraploid organisms. *Genetics* 116:299–312

206. Masterson J. 1994. Stomatal size in fossil plants: evidence for polyploidy in majority of angiosperms. *Science* 264:421–24

207. Mather K. 1942. Heterothally as an outbreeding mechanism in fungi. *Nature* 142:54–56

208. Mayer SS, Charlesworth D. 1992. Genetic evidence for multiple origins of dioecy in the Hawaiian shrub *Wikstroemia* Thymelaeaceae. *Evolution* 46:207–15

209. Maynard Smith J. 1978. *The Evolution of Sex.* Cambridge, UK: Cambridge Univ. Press.

210. Maynard Smith J. 1986. Contemplating life without sex. *Nature* 324:300–01

211. Maynard Smith J. 1994. Estimating the minimum rate of genetic transformation in bacteria. *J. Evol. Biol.* 7:525–34

212. Maynard Smith J, Dowson CG, Spratt BG. 1991. Localized sex in bacteria. *Nature* 349:29–31

213. Maynard Smith J, Smith NH, O'Rourke M, Spratt BG. 1993. How clonal are bacteria? *Proc. Natl. Acad. Sci. USA* 90:4384–88

214. Maynard Smith J, Szathmary E. 1995. *The Major Transitions in Evolution.* Oxford, UK: Freeman

215. McKim KS, Dahmus JB, Hawley RS. 1996. Cloning of the *Drosophila melanogaster* meiotic recombination gene mei-218: a genetic and molecular analysis of interval 15E. *Genetics* 144:215–28

216. Melkonian M, Surek B. 1995. Phylogeny of the chlorophyta: congruence between ultrastructural and molecular evidence. *Bull. Soc. Zool. France* 120:191–208

217. Michod RE. 1996. Cooperation and conflict in the evolution of individuality. II. Conflict mediation. *Proc. R. Soc. London B* 263:813–22

218. Mignot J-P, Raikov IB. 1992. Evidence for meiosis in the testate amoeba *Arcella*. *J. Protozool.* 39:287–89

219. Minton KW. 1994. DNA repair in the extremely radioresistant bacterium *Deinococcus radiodurans. Mol. Microbiol.* 13:9–15

220. Mogie M. 1986. Automixis: its distribution and status. 1986. *Biol. J. Linn. Soc.* 28:321–29

221. Mogie M. 1988. A model for the evolution and control of generative apomixis. *Biol. J. Linn. Soc.* 35:127–54

222. Moore-Landecker E. 1990. *Fundamentals of the Fungi.* Englewood Cliffs, NJ : Prentice Hall. 3rd ed.

223. Moran N, Baumann P. 1994. Phylogenetics of cytoplasmically inherited microorganisms of arthropods. *Trends Ecol. Evol.* 9:15–20

224. Muller DG. 1967. Generationswechel, Kernphasenwechesel and Sexualitat der Braunalge *Ectocarpus siliculosus. Planta* 75:39–54

225. Nakahara H, Ichimura T. 1992. Convergent evolution of gametangiogamy both in the zygnematalean green algae and in the pennate diatoms. *Jpn. J. Phycol.* 40:161–66

226. Nakamura Y, Hashimoto T, Kamaishi T, Adachi J, Nakamura F, et al. 1996. Phylogenetic position of kinetoplastid protozoa inferred from the protein phylogenies of elongation factor 1-alpha and 2. *J. Biochem.* 119:70–79

227. Norris RE. 1980. Prasinophytes. In *Phytoflagellates,* ed. ER Cox, pp. 85–145. New York: Elsevier

228. Oakley HA. 1985. Meiosis in *Mesostoma ehrenbergii ehrenbergii* (Turbellaria, Rhabdocoela) III. Univalent chromosome segregation during the first meiotic division in spermatocytes. *Chromosoma* 91:95–100

229. O'Kelly CJ. 1993. The Jakobid flagellates: structural features of Jakoba, Reclinomonas and Histiona and implications for the early diversification of eukaryotes. *J. Euk. Microbiol.* 40:627–36

230. Ott DW, Oldham-Ott CK, 1996. Oogenesis in *Vaucheria litorea* Hofman ex C. agardh. *J. Phycol.* 32(3 Suppl.):36.

231. Otto SP, Goldstein DB. 1992. Recombination and the evolution of diploidy. *Genetics* 131:745–51

232. Otto SP, Orive ME. 1995. Evolutionary consequences of mutation and selection within an individual. *Genetics* 141:1173–87

233. Otto SP, Marks JC. 1996. Mating systems and the evolutionary transition between haploidy and diploidy. *Biol. J. Linn. Soc.* 57:197–218

234. Patterson DJ, Zolffel M. 1991. Heterotrophic flagellates of uncertain taxonomic position. In *The Biology of Free-Living Heterotrophic Flagellates,* ed. DJ Patterson, J Larson, pp. 427–36. Oxford, UK: Clarendon

235. Patterson DJ, Sogin ML. 1993. Eukaryote origins and protistan diversity. In *The Origin and Evolution of Prokaryotic*

and Eukaryotic Cells, ed. H Hartman, K Matsuno, pp. 13–46. New Jersey: World Sci. Publ. Co.

236. Pawlowski J, Lee JJ. 1992. The life cycle of *Rotaliella elatiana* new species a tiny macroalgavorous foraminifer from the gulf of Elat. *J. Protozool.* 39:131–43

237. Pawlowski J, Bolivar I, Fahrni JF, Cavalier-Smith T, Gouy M. 1996. Early orgin of foraminifera suggested by SSU rRNA gene sequences. *Mol. Biol. Evol.* 13:445–50

238. Perasso R, Baroin-Tourancheau A. 1992 (1993). The eukaryogenesis: a model derived from rRNA molecular phylogenies. *Compt. Rend. Sean. Soc. Biol. Fil.* 186:656–65

239. Pernin P, Ataya A, Cariou ML. 1992. Genetic structure of natural populations of the free-living amoeba, *Naegleria lovaniensis.* Evidence for sexual reproduction. *Heredity* 68:173–81

239a. Perrot V, Richerd S, Valero M. 1991. Transition from haploidy to diploidy. *Nature* 351:315–17

240. Pfiester LA, Anderson DM. 1987. Dinoflagellate reproduction. In *The Biology of Dinoflagellates,* ed. FJR Taylor, pp. 611–48. Oxford, UK: Blackwell

241. Prasad RN, Chaudhary BRL. 1987. Contributions to the karyology of euglenoid flagellates. II. Lepocinclis Petry. *Cytologia* 52:357–60

242. Przywara L, Kuta E. 1995. Karyology of bryophytes. *Polish Bot. Stud.* 0:1–83.

243. Ragan MA, Gutell RR. 1995. Are red algae plants? *Bot. J. Linn. Soc.* 118:81–105

243a. Ragan MA, Goggin CL, Cawthorn RJ, Cerenius L, Jamieson AVC, et al. 1996. A novel clade of protistan parasites near the animal-fungal divergence. *Proc. Natl. Acad. Sci. USA* 93:11907–12

244. Raikov IB. 1982. *The Protozoan Nucleus.* Wien: Springer

245. Raikov IB. 1989. Nuclear genome of the Protozoa. In *Progress in Protozoology,* ed. DJ Patterson, JO Corliss, 3:21–86. Bristol: Biopress

246. Raikov IB. 1995. Meiosis in protists: recent advances and persisting problems. *Eur. J. Protistol.* 31:1–7

247. Raper JR. 1966. Life cycles, basic patterns of sexuality, and sexual mechanisms. In *The Fungi,* ed. GC Ainsworth, AS Sussman, II:473–511. New York: Academic Press

248. Raper JR. 1966. *Genetics of Sexuality in Higher Fungi.* New York: Ronald Press

249. Raper JR, Flexer AS. 1970. The road to diploidy with emphasis on a detour.

Symp. Soc. Gen. Microbiol. 20:401–32

250. Rashevsky N. 1970. Contributions to the theory of organismic sets: Why are there only two sexes? *Bull. Math. Biophys.* 22:293–301

251. Raven JA. 1993. The roles of the Chantransia phase of *Lemanea* (Lemaneaceae, Batrachospermales, Rhodophyta) and of the 'Mushroom' phase of Himanthalia (Himanthaliaceae, Fucales, Phaeophyta). *Bot. J. Scot.* 46:477–85

252. Rawson JRY. 1975. The characterization of *Euglena gracilis* DNA by its reassociation kinetics, BBA 402:171–78

252a. Redfield RJ. 1988. Evolution of bacterial transformation: Is sex with dead cells ever better than no sex at all? *Genetics* 119:213–21

253. Reed JN, Hurst LD. 1996. Dynamic analysis of the evolution of a novel genetic system: the evolution of ciliate meiosis. *J. Theor. Biol.* 178:355–68

254. Reed KM, Beukeboom LW, Eickbush DG, Werren JH. 1994. Junctions between repetitive DNAs on the PSR chromosome of *Nasonia vitripennis*: association of palindromes with recombination. *J. Mol. Evol.* 38:352–62

255. Rhoades MM. 1952. Preferential segregation in maize. In *Heterosis,* ed. JW Gowen, pp. 66–80. Ames: Iowa State Coll. Press

255a. Rice WR. 1996. Sexually antagonistic male adaptation triggered by experimental arrest of female evolution. *Nature* 381:232–34

256. Richards AJ. 1986. *Plant Breeding Systems.* London: Allen & Unwin

257. Riley DE, Campbell LA, Puolakkainen M, Krieger JN. 1993. *Trichomonas vaginalis* and early evolving DNA and protein sequences of the CDC2–28 protein kinase family. *Mol. Microbiol.* 8:517–19

258. Riley DE, Krieger JN. 1995. Molecular and phylogenetic analysis of PCR-amplified cyclin-dependent kinase (CDK) family sequences from representatives of the earliest available lineages of eukaryotes. *J. Mol. Evol.* 41:407–13

259. Ritland K, Soltis DE, Soltis PS. 1990. A two-locus model for the joint estimation of intergametophytic and intragametophytic selfing rates. *Heredity* 65:289–96

260. Roepstorf P, Huelsmann N, Hausmann K, 1993. Karyological investigations on the vampyrellid filose amoeba

Lateromyxa gallica Huelsmann 1993. *Eur. J. Protistol.* 29:302–10

261. Roettger R, Schmaljohann R, Zacharias H, 1989. Endoreplication of zygotic nuclei in the larger foraminifer *Heterostegina depressa* Nummultidae. *Eur. J. Protistol.* 25:60–66

262. Roger AJ, Smith MW, Doolittle RF, Doolittle WF, 1996. Evidence for the Heterolobosea from phylogenetic analysis of genes encoding glyceraldehyde-3-phosphate dehydrogenase. *J. Euk. Microbiol.* 43:475–85

262a. Rowe L, Houle D. 1996. The lek paradox and the capture of genetic variance by condition dependent traits. *Proc. R. Soc. London B* 263:1415–21

263. Sabelis MW, Nagelkerke CJ. 1988. Evolution of pseudo-arrhenotoky. *Exp. Appl. Acarology* 4:301–18

264. Sandgren CD, Flanagin J. 1986. Heterothallic sexuality and density dependent encystment in the chrysophycean alga *Synura petersenii. J. Phycol.* 22:206–16

265. Sarvella P. 1973. Adult parthenogenetic chickens. *Nature* 243:171

266. Schartl M, Nanda I, Schlupp I, Wilde B, Epplen JT, et al. 1995. Incorporation of subgenomic amounts of DNA as compensation for mutational load in a gynogenetic fish. *Nature* 373:68–71

267. Schemske DW, Lande R. 1985. The evolution of self-fertilization and inbreeding depression in plants. II. Empirical observations. *Evolution* 39:41–52

268. Schlegel M, Lom J, Stechmann A, Bernhard D, Leipe D, et al. 1996. Phylogenetic analysis of complete small subunit ribosomal RNA coding region of *Myxidium lieberkuehni*: evidence that Myxozoa are Metazoa and related to the bilateria. *Arch. Protist.* 147:1–9

269. Schnepf E, Drebes G. 1993. Anisogamy in the Dinoflagellate Noctiluca? *Helgolaender Meeresuntersuchungen* 47:265–73

270. Schuster P, Sigmund K. 1982. A note on the evolution of sexual dimorphism. *J. Theor. Biol.* 94:107–10

271. Selander RK, Levin BR. 1980. Genetic diversity and structure in *Escherichia coli* populations. *Science* 210:545–47

272. Selker EU. 1990. Premeiotic instability in *Neurospora. Annu. Rev. Genet.* 24:579–619.

273. Seravin LN, Gudkov AV. 1983. Spontaneous cell fusion of marine amoeba *Hyperamoeba fallax, Tsitologiya* 25:194–99 (in Russian)

274. Seravin LN, Gudkov AV. 1984. The main types and forms of agamic cell fusion in *Protozoa. Tsitologiya* 26:123–31 (in Russian)

275. Seravin LN, Gudkov AV. 1985. Similarity and difference between two marine limax amoebae, *Gruberella flavescens* and *Euhyperamoeba fallax* (Lobosea, Gymnamoebia), *Zoologicheskij Zhurnal* 64:1090–93 (in Russian)

276. Siddall ME, Martin DS, Bridge D, Desser SS, Cone DK. 1995. The demise of a phylum of protists: phylogeny of myxozoa and other parasitic cnidaria. *J. Parasitol.* 81:961–67

277. Silva ES, Faust MA. 1995. Small cells in the life history of dinoflagellates (Dinophyceae): a review. *Phycologia* 34:396–408

278. Singh MK, Kumar RA. 1996. On the unique equilibrium state of tetraploid under selection. *Biometrics* 52:717–20

279. Smith AJE. 1983. Chromosomes in the evolution of the Bryophyta. In *Chromosomes in Evolution of Eukaryotic Groups,* ed. AK Sharma, A Sharma, 1:225–44. Boca Raton: CRC Press

280. Sogin ML, Morrison HG, Hinkle G, Silberman JD. 1996. Ancestral relationships of the major eukaryotic lineages. *Microbiologia* 12:17–28

281. Solnica-Krezel L, Bailey J, Gruer DP, Price JM, Dove WF, et al. 1995. Characterization of npf mutants identifying developmental genes in Physarum. *Microbiology* 141:799–816

282. Soltis DE, Soltis PS. 1987. Polyploidy and breeding systems in homosporous Pteridophyta: a reevaluation. *Am. Nat.* 130:219–32

283. Soltis DE, Soltis PS. 1993. Molecular data and the dynamic nature of polyploidy. *Crit. Rev. Plant Sci.* 12:243–73

284. Song K, Lu P, Tang K, Osborn TC. 1995. Rapid genome change in synthetic polyploids of *Brassica* and its implications for polyploid evolution. *Proc. Natl. Acad. Sci. USA* 92:7719–23

285. Sousa MA. 1994. Cell-to-cell interactions suggesting a sexual process in *Herpetomonas megaseliae* (Kinetoplastida: Trypanosomatidae). *Parasitol. Res.* 80:112–16

286. Spiegel FW, Lee SB, Rusk SA. 1995. Eumycetozoans and molecular systematics. *Can. J. Bot.* 73 (SUPPL. 1 SECT. E-H): S738-S746

287. Spoon DM, Hogan CJ, Chapman GB. 1995. Ultrastructure of a primitive, multinucleate, marine, cyanobacteriophagous amoeba (*Euhyperamoeba biospherica* n. sp.) and its possible

significance in the evolution of eukaryotes. *Invert. Biol.* 114:189–201

288. Steinkotter J, Bhattacharya D, Semmelroth I, Bibeau C, Melkonian M. 1994. Prasinophytes form independent lineages within the Chlorophyta: evidence from ribosomal RNA sequence comparisons. *J. Phycol.* 30:340–45

289. Stevens JR, Tibayrenc M. 1996. *Trypanosoma brucei* s.l.: evolution, linkage and the clonality debate. *Parasitology* 112:481–88

290. Streisinger G, Walker C, Dower N, Knauber D, Singer F. 1981. Production of clones of homozygous diploid zebra fish (*Brachydanio rerio*). *Nature* 291:293–96

291. Suda S, Watanabe MM, 1989. Evidence for sexual reproduction in the primitive green alga *Nephroselmis olivacea* Prasinophyceae. *J. Phycol.* 25:596–600

292. Suomalainen E, Saura A, Lokki J. 1987. *Cytology and Evolution in Parthenogenesis.* Boca Raton, FL: CRC

293. Surani MA. 1991. Influence of genome imprinting on gene expression, phenotypic variations and development. *Hum. Reprod.* 6:45–57

294. Svedelius N. 1927. Alternation of generations in relation to reduction division. *Bot. Gaz.* 83:362–84

295. Swanson J, Morrison S, Barrera O, Hill S. 1990. Piliation changes in transformation-defective gonococci. *J. Exp. Med.* 171:2131–40

296. Sweeny AW, Doggett SL, Piper RG. 1993. Life cycle of a new species of *Duboscqia* (Microspora, Thelohaniidae) infecting the mosquito *Anopheles hilli* and an intermediate copepod host *Apocyclops dengizicus. J. Invert. Pathol.* 62:137–46

297. Temin HM. 1993. Retrovirus variation and reverse transcription: abnormal strand transfers result in retrovirus genetic variation. *Proc. Natl. Acad. Sci. USA* 90:6900–03

298. Temin RG, Ganetzky B, Powers PA, Lyttle TW, Pimpinelli S, et al. 1991. Segregation-distortion in *Drosophila melanogaster*: genetic and molecular analyses. *Am. Nat.* 137:287–331

299. Thaler DS. 1994. Sex is for sisters: intragenomic recombination and homology-dependent mutation as sources of evolutionary variation. *Trends Ecol. Evol.* 9:108–10

300. Tibayrenc M. 1996. Towards a unified evolutionary genetics of microorganisms. *Annu. Rev. Microbiol.* 50:401–29

301. Tibayrenc M, Kjellberg F, Arnaud J, Oury B, Breniere SF, et al. 1991. Are eukaryotic microorganisms clonal or sexual? A population genetics vantage. *Proc. Natl. Acad. Sci. USA* 88:5129–33

302. Tinley-Bassett RAE. 1994. Nuclear controls of chloroplast inheritance in higher plants. *J. Hered.* 85:347–54

303. Tinti F, Scali V. 1992. Genome exclusion and gametic DAPI-DNA content in the hybridogenetic *Bacillus rossius-grandii-benazzii* complex (Insecta, Phasmatodea). *Mol. Rep. Dev.* 33:235–42

304. Thon G, Klar AJS. 1993. Directionality of fission yeast mating type interconversion is controlled by the location of the donor loci. *Genetics* 134:1045–54

305. Trifonov EN. 1995. Segmented structure of protein sequences and early evolution of genome by combinatorial fusion of DNA elements. *J. Mol. Evol.* 40:337–42

306. Tunner HG, Heppich-Tunner S. 1991. Genome exclusion and two strategies of chromosome duplication in oogenesis of a hybrid frog. *Naturwissenschaften* 78:32–34

307. Van de Peer Y, Neefs JM, De Rijk P, De Wachter R. 1993. Evolution of eukaryotes as deduced from small ribosomal subunit RNA sequences. *Biochem. Syst. Ecol.* 21:43–55

308. Van de Peer Y, Van der Auwera G, De Wachter R. 1996. The evolution of stamenopiles and alveolates as derived by "substitution rate calibration" of small ribosomal subunit RNA. *J. Mol. Evol.* 42:201–10

309. Van der Ploeg LHT, Smith CL, Polvere RI, Gottesdiener KM. 1989. Improved separation of chromosome-sized DNA from *Trypanosoma brucei*, stock 427–60. *Nucleic Acids Res.* 17:3217–27

310. Van Dijk P, Hartog M, Van Delden W. 1992. Single cytotype areas in autopolyploid *Plantago media* L. *Biol. J. Linn. Soc.* 46:315–31

311. Vance RR. 1992. Optimal somatic growth and reproduction in a limited constant environment: the general case. *J. Theor. Biol.* 157:51–70

312. Vaulot D, Birrien JL, Marie D, Casotti R, Veldhuis MJW, et al. 1994. Morphology, ploidy, pigment composition, and genome size of cultured strains of *Phaeocystis* (Prymnesiophyceae). *J. Phycol.* 30:1022–35

313. Vielle-Calzada JP, Crane CF, Stelly DM. 1996. Apomixis: the asexual revolution. *Science* 274:1322–23

314. Vinogradov AE, Borkin LJ, Gunther R, Rozanov JM. 1990. Genome elimination in diploid and triploid *Rana esculenta* males: cytological evidence from DNA flow cytometry. *Genome* 33:619–27

315. Vrijenhoek RC. 1993. The origin and evolution of clones versus the maintenance of sex in *Poeciliopsis. J. Hered.* 84:388–95

316. Wada M, Sunkin SM, Stringer JR, Nakamura Y. 1995. Antigenic variation by positional control of major surface glycoprotein gene expression in *Pneumocystis carinii. J. Infect. Dis.* 171:1563–68

317. Wade CM, Darling KF, Kroon D, Brown AJL. 1996. Early evolutionary origin of the planktonic Foraminifera inferred from small subunit rDNA sequence comparisons. *J. Mol. Evol.* 43:672–77

318. Wainright PO, Hinkle G, Sogin ML, Stickel SK. 1993. Monophyletic origins of the Metazoa: an evolutionary link with Fungi. *Science* 260:340–42

319. Walker TG. 1984. Chromosomes and evolution in pterydophytes. In *Chromosomes in Evolution of Eukaryotic Groups*, ed. AK Sharma, A Sharma, 2:104–41. Boca Raton, FL: CRC

320. Wasserman M. 1972. Factors influencing fitness in chromosomal strains in *Drosophila subobscura. Genetics* 72:691–708

321. Watano Y, Imazu M, Shimizu T. 1996. Spatial distribution of CpDNA and MtDNA haplotypes in a hybrid zone between *Pinus pumila* and *P. parviflora*var. *pentaphylla* (Pinaceae). *J. Plant Res.* 109:403–08

322. Weber E. 1992. *Studies of the Reproduction and Ploidy of Various Ascomycetes. Bibliotheca Mycologica* v. 140

323. Weekers PHH, Gast RJ, Fuerst PA, Byers TJ. 1994. Sequence variations in smallsubunit ribosomal RNAs of *Hartmannella vermiformis* and their phylogenetic implications. *Mol. Biol. Evol.* 11:684–90

324. Werenskiold AK, Schreckenbach T, Valet G. 1988. Specific nuclear elimination in polyploid plasmodia of the slime mold *Physarum polycephalum. Cytometry* 9:261–65

325. Werner JE, Peloquin SJ. 1990. Inheritance and two mechanisms of 2n egg formation in 2x potatoes. *J. Hered.* 81:371–74

326. Wessels JGH, Meinhardt F, eds. 1994.

The Mycota, I: Growth, Differentiation and Sexuality. Berlin: Springer.

327. Whisson SC, Drenth A, MacLean DJ, Irwin JAG. 1994. Evidence for outcrossing in *Phytophthora sojae* and linkage of a DNA marker to two avirulence genes. *Curr. Genet.* 27:77–82

328. Wichterman R. 1986. *The Biology of Paramecium.* New York: Plenum

329. Willumsen NBS, Siemensma F, Suhr-Jessen P. 1987. A multinucleate amoeba, *Parachaos zoochlorellae* (Willumsen 1982) comb. nov., and a proposed division of the genus *Chaos* into the genera *Chaos* and *Parachaos* (Gymnamoebia, Amoebidae). *Arch. Protist.* 134:303–13

330. Wise MG, Shimkets LJ, McArthur JV. 1995. Genetic structure of a lotic population of *Burkholderia* (*Pseudomonas*) *cepacia. Appl. Env. Microbiol.* 61:1791–98

331. Woehrmann K, Tomiuk J. 1988. Life cycle strategies and genotypic variability in populations of aphids. *J. Genet.* 67:43–52

332. Wyatt R, Anderson LE. 1984. Breeding systems in bryophytes. In *The Experimental Biology of Bryophytes*, ed. AF Dryer, JG Duckett, pp. 39–64. London: Academic Press

333. Yamaguchi M, Imai I. 1994. A microfluorometric analysis of nuclear DNA at different stages in the life history of *Chattonella antiqua* and *Chattonella marina* (Raphidophyceae). *Phycologia* 33:163–70

334. Yamamoto A, Hashimoto T, Asaga E, Hasegawa M, Goto N. 1994. Plylogenetic position of the mitochondrionlacking protozoan *Trichomonas tenax*, based on amino acid sequences of elongation factors 1α and 2. *J. Mol. Evol.* 44:98–105

335. Yang Q, Zwick MG, Paule MR. 1994. Sequence organization of the *Acanthamoeba* rRNA intergenic spacer: Identification of transcriptional enhancer. *Nucleic Acids Res.* 22:4798–805

336. Zouros E, Ball AO, Saavedra C, Freeman KR. 1994. An unusual type of mitochondrial DNA inheritance in the blue mussel *Mytilus. Proc. Natl. Acad. Sci. USA* 91:7463–67

337. Zhuchenko AA, Korol AB. 1985. *Recombination in Evolution and Selection.* Moscow: Nauka (in Russian).

Annu. Rev. Ecol. Syst. 1997. 28:437–66

PHYLOGENY ESTIMATION AND HYPOTHESIS TESTING USING MAXIMUM LIKELIHOOD

John P. Huelsenbeck
Department of Biology, University of Rochester, Rochester, NY 14627;
e-mail: johnh@onyx.si.edu

Keith A. Crandall
Department of Zoology and M. L. Bean Museum, Brigham Young University, Provo, UT 84602

KEY WORDS: maximum likelihood, phylogeny, hypothesis test, evolutionary model, molecular evolution

---

Abstract

One of the strengths of the maximum likelihood method of phylogenetic estimation is the ease with which hypotheses can be formulated and tested. Maximum likelihood analysis of DNA and amino acid sequence data has been made practical with recent advances in models of DNA substitution, computer programs, and computational speed. Here, we describe the maximum likelihood method and the recent improvements in models of substitution. We also describe how likelihood ratio tests of a variety of biological hypotheses can be formulated and tested using computer simulation to generate the null distribution of the likelihood ratio test statistic.

INTRODUCTION

Only recently has phylogenetics been recognized as a field that has basic relevance to many questions in biology. Phylogenies have proven to be important tools of research in fields such as human epidemiology (42, 86), ecology (7), and evolutionary biology (43). In fact, for any question in which history may be a confounding factor, phylogenies have a central role (25). To the outsider interested in using a phylogeny, one of the most frustrating aspects of the field

437

of systematics is the lack of agreement as to which of the many methods of analysis to use (54). Which method of analysis is best suited for a particular data set is a question of current, and often vehement, debate. The evolutionary biologist is often left asking not only, "Which method of analysis do I use?", but, if different methods lead to conflicting genealogies, "Which tree do I believe?". As researchers become drawn to the opinion that phylogeny reconstruction is a problem of statistical inference, it is important to examine phylogenetic methods for their statistical properties and assumptions.

Several criteria can be used as a basis to choose among methods. One criterion is accuracy; that is, how well do different methods estimate the correct tree? This criterion has captured most of the attention of systematists, and there is a veritable scientific cottage industry producing papers that examine the performance of different phylogenetic methods for simulated data sets (29, 44, 49, 50, 54, 62, 97, 108), well-supported phylogenies (2, 17, 52), and experimental phylogenies (16, 48). However, criteria besides accuracy are also important. For example, a phylogenetic method should provide some means of falsifying the assumptions made during the analysis (88). All phylogenetic methods, by necessity, must make specific assumptions about the evolutionary process. Typical assumptions include a bifurcating tree as a model to describe the genealogy of a group and a model of character change. Yet, many methods do not provide a means of testing these assumptions. Furthermore, some provision for choosing among different models of evolution should be available. In choosing between a simple model of character change and a more complex model, for example, how can one justify using the complex model in a phylogenetic analysis? Finally, the methods should be able to estimate the confidence in the phylogeny and provide a framework for testing phylogenetic hypotheses.

Inasmuch as phylogenies are important for many evolutionary questions, the criteria posed above are important. In this review, we concentrate on one method of phylogenetic estimation—maximum likelihood. Recent advances have made maximum likelihood practical for analysis of DNA and amino acid sequence data. Many of the advances consist of improvements in the models of DNA substitution implemented by maximum likelihood. However, increased computer speed and improved computer programs have also played an important role. In this paper, we review the recent advances made in maximum likelihood estimation of phylogenetic trees. Specifically, we examine how maximum likelihood has been used for phylogeny estimation and hypothesis testing.

THE LIKELIHOOD PRINCIPLE

The method of maximum likelihood is usually credited to the English statistician RA Fisher, who described the method in 1922 and first investigated its properties (28). The method of maximum likelihood depends on the complete specification of the data and a probability model to describe the data. The probability of observing the data under the assumed model will change depending on the parameter values of the model. The maximum likelihood method chooses the value of a parameter that maximizes the probability of observing the data.

A Coin Tossing Experiment

Consider the simple experiment of tossing a coin with the goal of estimating the probability of heads for the coin. The probability of heads for a fair coin is 0.5. For this example, however, we assume that the probability of heads is unknown (perhaps the fairness of the coin is being tested) and must be estimated. The three main components of the maximum likelihood approach are (a) the data, (b) a model describing the probability of observing the data, and (c) the maximum likelihood criterion.

Assume that we performed the coin flip experiment, tossing a coin n times. An appropriate model that describes the probability of observing h heads out of n tosses of a coin is the binomial distribution. The binomial distribution has the following form

$$\Pr[h \mid p, n] = \binom{n}{h} p^h (1 - p)^{n-h}, \qquad\qquad 1.$$

where p is the probability of heads, the binomial coefficient $\binom{n}{h}$ gives the number of ways to order h successes out of n trials, and the vertical line means "given."

Assuming independence of the individual and discrete outcomes, the likelihood function is simply the joint probability of observing the data under the model. For the coin toss experiment in which a binomial distribution is assumed, the likelihood function becomes

$$L(p \mid h, n) = \binom{n}{h} p^h (1 - p)^{n-h}. \qquad\qquad 2.$$

Often the log likelihood is used instead of the likelihood for strictly computational purposes. Taking the natural log of the function does not change the value of p that maximizes the likelihood.

Figure 1 shows a plot of likelihood, L, as a function of p for one possible outcome of $n = 10$ tosses of a coin (six heads and four tails). The likelihood appears to be maximized when p is the proportion of the time that heads appeared in our experiment. This illustrates a computational way to find the maximum likelihood estimate of p (change p in small increments until a maximum is found). Alternatively, the maximum likelihood estimate of p can be found analytically by taking the derivative of the likelihood function with respect to p and finding where the slope is zero. If this is done, we find that the estimate

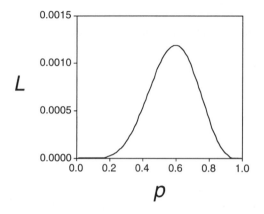

Figure 1   The likelihood surface for one possible realization of the coin tossing experiment. Here, six heads and four tails are observed. The likelihood appears to be maximized when $p = 0.6$.

of p is $\hat{p} = {}^h/_n$. The estimate of p is simply the proportion of heads that we observed in our experiment.

Maximum likelihood estimates perform well according to several criteria. Statistical estimators are evaluated according to their consistency, efficiency, and bias. These criteria provide an idea of how concentrated an estimate of a parameter is around the true value of that parameter. A method is consistent if the estimate converges to the true value of the parameter as the number of data increases. An efficient method provides a close estimate of the true value of the parameter (i.e., the variance of the estimate is small), and an unbiased estimator does not consistently under- or over-estimate the true value of the parameter. Mathematically, an estimate $\hat{\theta}_n$ of a parameter θ based on a sample of size n is consistent if

$$P(|\hat{\theta}_n - \theta| > \varepsilon) \rightarrow 0, \quad \text{as } n \rightarrow \infty \qquad 3.$$

for any $\varepsilon > 0$ where θ is the true value of the parameter. The mean square error (MSE) measures the variance and bias of an estimate. The MSE of an estimate $\hat{\theta}$ is

$$MSE(\hat{\theta}) = E(\hat{\theta} - \theta)^2$$
$$= \text{Var}(\hat{\theta}) + (E(\hat{\theta}) - \theta)^2. \qquad 4.$$

The first term of the MSE is the variance of the estimate, and the second term is the bias. A method is unbiased if the expectation of the estimate equals the true value of the parameter [$E(\hat{\theta}) = \theta$]. When two competing estimators of a parameter are considered, the estimate with the smaller MSE is said to be more efficient. Maximum likelihood estimates are typically consistent under the model. Furthermore, they are asymptotically efficient, meaning that the variance of a maximum likelihood estimate is equal to the variance of any unbiased estimate as the sample size increases. However, maximum likelihood estimates are often biased (e.g., the maximum likelihood estimate of the parameter σ^2 of a normal distribution is biased).

MAXIMUM LIKELIHOOD IN PHYLOGENETICS

The application of maximum likelihood estimation to the phylogeny problem was first suggested by Edwards & Cavalli-Sforza (20). However, they found the problem too computationally difficult at the time and attempted approximate methods instead (thereby introducing the minimum evolution—later to be known as the parsimony—and the least squares methods of phylogeny estimation). The subsequent history of maximum likelihood can be read as a steady progress in which computational barriers were broken and the models used were made more biologically realistic. For example, Neyman (83) applied maximum likelihood estimation to molecular sequences (amino acids or nucleotides) using a simple model of symmetric change that assumed substitutions were random and independent among sites. It was not until Felsenstein's implementation (24), however, that a general maximum likelihood approach was fully developed for nucleotide sequence data. Below, we outline the basic strategy for obtaining maximum likelihood estimates of phylogeny, given a set of aligned nucleotide or amino acid sequences. We then show how the basic strategy can be complicated by biological reality.

Conceptually, maximum likelihood in phylogenetics is as simple as the example given above for estimating the probability of heads in a coin toss experiment. The data for molecular phylogenetic problems are the individual site patterns. We assume that the sequences have been aligned, though the alignment procedure can be explicitly incorporated into the estimation of phylogeny (111, 112). For example, for the following aligned DNA sequences of $s = 4$ taxa,

Taxon 1 ACCAGC
Taxon 2 AACAGC
Taxon 3 AACATT
Taxon 4 AACATC,

the observations are $x_1 = \{A, A, A, A\}'$, $x_2 = \{C, A, A, A\}'$, $x_3 = \{C, C, C, C\}'$, $x_4 = \{A, A, A, A\}'$, $x_5 = \{G, G, T, T\}'$, and $x_6 = \{C, C, T, C\}'$. If one were interested in coding the data as amino acids, the above sequences, if in frame, would be represented as $x_1 = \{Thr, Asn, Asn, Asn\}'$ and $x_2 = \{Ser, Ser, Ile, Ile\}'$. The sample consists of n vectors (as many vectors as there are sites in the sequence, with the elements of each vector denoting the nucleotide state for each taxon for site i.

Note that two of the sites exhibit the same site pattern ($\mathbf{x}_1$ and $\mathbf{x}_4$). There are a total of $r = 4^s$ site patterns possible for s species. The number of sites exhibiting different site patterns can also be considered as the data in a phylogenetic analysis. For example, the above data matrix can also be described as

$$
\begin{array}{ll}
\text{Taxon 1} & \text{AAAAAAAAA}\ldots\text{C}\ldots\text{C}\ldots\text{C}\ldots\text{G}\ldots\text{TTT} \\
\text{Taxon 2} & \text{AAAAAAAAA}\ldots\text{A}\ldots\text{C}\ldots\text{C}\ldots\text{G}\ldots\text{TTT} \\
\text{Taxon 3} & \text{AAAACCCCG}\ldots\text{A}\ldots\text{C}\ldots\text{T}\ldots\text{T}\ldots\text{TTT} \\
\text{Taxon 4} & \text{ACGTACGTA}\ldots\text{A}\ldots\text{C}\ldots\text{C}\ldots\text{T}\ldots\text{CGT}
\end{array}
$$

$$
\text{Number}\quad 2\,0\,0\,0\,0\,0\,0\,0\,0\ldots1\ldots1\ldots\ 1\ldots1\ldots0\,0\,0
$$

where the matrix is now a 4×256 matrix of all $r = 4^4 = 256$ site patterns possible for four species. The site patterns are labeled $1, 2, \ldots, r$. Most of the possible site patterns for our sample data set are not observed. However, 5 site patterns are observed (now labeled $\mathbf{y}_1 = \{A, A, A, A\}'$, $\mathbf{y}_{65} = \{C, A, A, A\}'$, $\mathbf{y}_{86} = \{C, C, C, C\}'$, $\mathbf{y}_{94} = \{C, C, T, C\}'$, and $\mathbf{y}_{176} = \{G, G, T, T\}'$). The numbers of sites exhibiting each site pattern are contained in a vector $\mathbf{n}$ ($n_1 = 2, n_{65} = 1$, $n_{86} = 1, n_{94} = 1, n_{176} = 1$, with all other $n_i = 0$ for the example data matrix).

Maximum likelihood assumes an explicit model for the data. Just as with the coin tossing experiment, the data are considered as random variables. However, instead of two possible outcomes, there are $r = 4^s$ possible outcomes for DNA sequences. Hence, the data can be described using a multinomial distribution. The multinomial distribution is a generalization of the binomial distribution and has the following form:

$$
\Pr[n_1, n_2, \ldots, n_r \mid p_1, p_2, \ldots, p_r] = \binom{n}{n_1, n_2, \ldots, n_r} \prod_{i=1}^{r} p_i^{n_i}, \qquad 5.
$$

where $\binom{n}{n_1, n_2, \ldots, n_r}$ is the number of ways that n objects can be grouped into r classes, n_i is the number of observations of the ith site pattern, and p_i is the probability that site pattern i occurs. A maximum likelihood estimate of p_i is $\hat{p}_i = {}^{n_i}/n$ (that is, the probability of the ith class is the proportion of the time it was observed). The likelihood, then, can be calculated assuming a multinomial distribution by setting the likelihood equal to Equation 5. However, by using a multinomial distribution, one cannot estimate topology or other biologically interesting parameters. Hence, models that incorporate phylogeny are assumed. The difference in the log likelihood of the data under multinomial and phylogenetic models, however, represents the cost associated with assuming

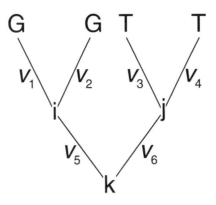

Figure 2 The likelihood method assumes that observed sequences (here, the nucleotides for site pattern 176 from the example in the text; $\mathbf{y}_{176} = \{G, G, T, T\}'$) are related by a phylogenetic tree (τ) with branch lengths ($v_1, \ldots, v_6$) specified in terms of expected number of substitutions per site. The probability of observing the data ($\mathbf{y}_{176}$) is a sum over the possible assignments of nucleotides to the internal nodes, $i, j,$ and k.

a phylogenetic tree and substitution model (32, 80, 93) and has also been used to show that maximum likelihood in phylogenetics is consistent (117).

Just like the multinomial probability model, phylogenetic models specify the probability of observing different site patterns. At a minimum, a phylogenetic model for molecular data includes a tree (τ) relating the sequences with branch lengths of the tree (**v**) specified in terms of expected number of changes per site and a model of sequence change. Consider just one of the nucleotide site patterns, above ($\mathbf{y}_{176}$), for the tree of Figure 2. Because the identities of the nucleotides at the internal nodes $i, j,$ and k are unknown, the probability of observing site pattern 176 is a sum of 64 terms (the $4^3 = 64$ possible assignments of nucleotides to nodes $i, j,$ and k),

$$\Pr[\mathbf{y}_{176} = \{G, G, T, T\} \mid \tau, v_1, \ldots, v_6, \Theta]$$

$$= \sum_{i=1}^{4} \sum_{j=1}^{4} \sum_{k=1}^{4} \pi_k \, p_{Gi}(v_1, \Theta) \, p_{Gi}(v_2, \Theta) \, p_{Tj}(v_3, \Theta) \qquad \text{6.}$$

$$\times \, p_{Tj}(v_4, \Theta) \, p_{ik}(v_5, \Theta) \, p_{jk}(v_6, \Theta),$$

where nucleotides, A, C, G, or T are assumed at the internal nodes if $i, j,$ or k are equal to 1, 2, 3 or 4, respectively, π_i is the equilibrium frequency of nucleotide i, and $p_{xy}(v_i, \Theta)$ is the probability of observing nucleotides x and y at the tips of a branch given the branch length and other parameters, Θ, of the substitution model. Felsenstein (24) pointed out that the above calculation can be performed much more quickly by taking advantage of the tree structure when performing the summations over nucleotides at interior nodes. If instead of DNA sequences, amino acid sequences are used, the above equation involves a summation over $20^3 = 8000$ possible assignments of amino acids to the internal nodes, and $p_{xy}(v_i, \Theta)$ represents the probability of observing amino acids x and y at the tips of a branch given the branch length and other parameters of the substitution model.

Assuming independence among sites, the likelihood of a tree (τ) is

$$L(\tau, \mathbf{v}, \Theta \mid \mathbf{y}_1, \ldots, \mathbf{y}_r) = \binom{n}{n_1, n_2, \ldots, n_r} \prod_{i=1}^{r} \Pr[\mathbf{y}_i \mid \tau, \mathbf{v}, \Theta]^{n_i}, \qquad 7.$$

where $\mathbf{v}$ is a vector containing the lengths of the branches and is either $\mathbf{v} = (v_1, \ldots, v_{2s-2})$ for rooted trees or $\mathbf{v} = (v_1, \ldots, v_{2s-3})$ for unrooted trees (s is the number of sequences), and r is the total number of site patterns possible for s sequences. The multinomial coefficient $\binom{n}{n_1, n_2, \ldots, n_r}$ is a constant and is usually disregarded when calculating the likelihood of a tree. Also, to speed computation of the likelihood, the product is taken only over observed site patterns. The likelihood as formulated in Equation 7 does not consider the order of the site patterns. However, for several models of DNA substitution, the order of the sites is of interest and cannot be disregarded (27, 120). For such problems, the likelihood function is the product over all sites

$$L(\tau, \mathbf{v}, \Theta \mid \mathbf{x}_1, \ldots, \mathbf{x}_n) = \prod_{i=1}^{n} \Pr[\mathbf{x}_i \mid \tau, \mathbf{v}, \Theta]. \qquad 8.$$

The method of Felsenstein (24) is to choose the tree that maximizes the likelihood as the best estimate of phylogeny. This application of likelihood is unusual because the likelihood function changes depending on the tree (81, 124). In principle, to find the maximum likelihood tree, one must visit each of the $\frac{(2s-5)!}{2^{s-3}(s-3)!}$ possible unrooted bifurcating trees in turn (23). For each tree, one finds the combination of branch lengths and other parameters that maximizes the likelihood of the tree (that maximizes the likelihood function, above). The maximum likelihood estimate of phylogeny is the tree with the greatest likelihood.

This procedure of visiting all possible trees and calculating the likelihood for each is computationally expensive. Fortunately, there are many short cuts that can substantially speed up the procedure. As mentioned above, Felsenstein described an efficient method to calculate the likelihood by taking advantage of the tree topology when summing over all possible assignments of nucleotides to internal nodes (24). There are also efficient ways of optimizing branch lengths that involve taking the first and second derivatives of the likelihood function with respect to the branch of interest (see 67). Finally, rather than visiting each possible tree, search algorithms concentrate on only those trees that have a good chance of maximizing the likelihood function (see 105).

POISSON PROCESS MODELS

To calculate the probability of observing a given site pattern, the transition probabilities $[P_{xy}(v_i, \Theta)]$ need to be specified. All current implementations of likelihood assume a time-homogeneous Poisson process to describe DNA or amino acid substitutions.

An Example with Two-Character States

As an example of how transition probabilities are calculated, consider a very simple case for which only two character states exist (0 or 1). The rate of change from 0 to 1 or from 1 to 0 in an infinitesimal amount of time, δt, is specified by the rate matrix, $\mathbf{Q}$

$$\mathbf{Q} = \{q_{ij}\} = \begin{pmatrix} -\lambda\pi_1 & \lambda\pi_1 \\ \lambda\pi_0 & -\lambda\pi_0 \end{pmatrix}. \qquad 9.$$

The states are ordered 0, 1 along the rows and diagonals; λ is the rate of change from 0 to 1 or from 1 to 0; and π_0 and π_1 are the equilibrium frequencies of states 0 and 1, respectively. The diagonals of the rate matrix are negative to satisfy the mathematical requirement that the row sums are zero. The matrix may be multiplied by a constant such that the average rate of substitution is one, and time (t) is then measured by the expected number of substitutions per site (v). The matrix $\mathbf{Q}$, above, is reversible because it satisfies the requirement that $\pi_i q_{ij} = \pi_j q_{ji}$.

To calculate the probability of observing a change over an arbitrary interval of time, t, the following matrix calculation is performed: $\mathbf{P}(t, \Theta) = p_{ij}(t, \Theta) = e^{\mathbf{Q}t}$ (15). The vector Θ contains the parameters of the substitution model (in this case $\Theta = \{\pi_0, \pi_1\}$). For many substitution models, the transition probability matrix $\mathbf{P}(t, \Theta)$ can be calculated analytically. For example, the transition probabilities for the two-state case are

$$\mathbf{P}(t, \Theta) = \{p_{ij}(t, \Theta)\} = \begin{pmatrix} \pi_0 + (1 - \pi_0)e^{-\lambda t} & \pi_1 - \pi_1 e^{-\lambda t} \\ \pi_0 - \pi_0 e^{-\lambda t} & \pi_1 + (1 - \pi_1)e^{-\lambda t} \end{pmatrix}. \qquad 10.$$

However, for complicated rate matrices, the probability matrix can be calculated numerically (e.g., 118).

Models of DNA Substitution

Many of the advances in maximum likelihood analysis in phylogenetics have come through improvements in the models of substitution assumed. One of the simplest models of DNA substitution—the Jukes-Cantor (JC69) model—assumes that the base frequencies are equal ($\pi_A = \pi_C = \pi_G = \pi_T$) and that the rate of change from one nucleotide to another is the same for all possible changes (58). However, the JC69 model, like several other models, is simply a special case of a general model of DNA substitution for which the instantaneous rate matrix $\mathbf{Q}$ has the following form:

$$\mathbf{Q} = \{q_{ij}\} = \begin{pmatrix} \cdot & r_2\pi_C & r_4\pi_G & r_6\pi_T \\ r_1\pi_A & \cdot & r_8\pi_G & r_{10}\pi_T \\ r_3\pi_A & r_7\pi_C & \cdot & r_{12}\pi_T \\ r_5\pi_A & r_9\pi_C & r_{11}\pi_G & \cdot \end{pmatrix} \qquad 11.$$

(3, 94, 109, 118). The rows and columns are ordered A, C, G, and T. The matrix gives the rate of change from nucleotide i (arranged along the rows) to nucleotide j (along the columns). For example, $r_2\pi_C$ gives the rate of change from "A" to "C". Let $\mathbf{P}(v, \Theta) = \{p_{ij}(v, \Theta)\}$ be the transition probability matrix, where $p_{ij}(v, \Theta)$ is the probability that nucleotide i changes into j over branch length v. The vector Θ contains the parameters of the substitution model (e.g., π_A, π_C, π_G, π_T, r_1, r_2, r_3, ...). As for the two-state case, to calculate the probability of observing a change over a branch of length v, the following matrix calculation is performed: $\mathbf{P}(v, \Theta) = e^{\mathbf{Q}v}$.

A model based on the matrix $\mathbf{Q}$ (Equation 11) represents the most general 4×4 model of DNA substitution currently available (Figure 3). The model is nonreversible, meaning that the rate matrix $\mathbf{Q}$ does not satisfy the reversibility condition ($\pi_i q_{ij} = \pi_j q_{ji}$). Because the model is nonreversible, the likelihood of an unrooted tree changes depending on the root position. Therefore, the likelihood must be maximized over the $\frac{(2s-3)!}{2^{s-2}(s-2)!}$ rooted trees. For reversible models of DNA substitution, on the other hand, the likelihood is maximized over the $\frac{(2s-5)!}{2^{s-3}(s-3)!}$ unrooted trees because, for a given unrooted topology, the likelihood is the same regardless of where the tree is rooted. Many commonly used models of DNA substitution are subsets of this general model. Table 1 shows the parameter settings of the general model that give a variety of models of DNA substitution. Maximum likelihood explicitly incorporates a model of substitution into the estimation procedure, as do distance methods; parsimony methods, on the other hand, incorporate variations of these models implicitly. Because other models of DNA substitution are subsets of this general model, and because they are often subsets of one another as well, likelihood ratio tests of the model of DNA substitution can be easily performed testing whether a particular parameter provides a significant improvement in the likelihood (as is discussed later).

The four-state character models describe the substitution process at a single site. The assumption of independence among sites is necessary in phylogenetic

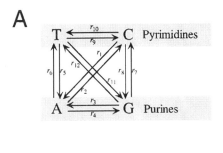

A

T $\rightleftarrows$ C Pyrimidines

A $\rightleftarrows$ G Purines

B

$$\mathbf{Q} = \{q_{ij}\} = \begin{pmatrix} \cdot & r_2\pi_C & r_4\pi_G & r_6\pi_T \\ r_1\pi_A & \cdot & r_8\pi_G & r_{10}\pi_T \\ r_3\pi_A & r_7\pi_C & \cdot & r_{12}\pi_T \\ r_5\pi_A & r_9\pi_C & r_{11}\pi_G & \cdot \end{pmatrix}$$

Figure 3   The likelihood method assumes that substitutions follow a Poisson process, with the rate of nucleotide substitution specified by a rate matrix, $\mathbf{Q}$. The rates of change from one nucleotide to another are specified by the parameters $r_1, \ldots, r_{12}$ (A). The rate matrix, $\mathbf{Q}$, corresponding to the diagram (A) is shown in B. Different models of substitution are just special cases of the general matrix shown here. For example, to obtain the Jukes-Cantor (JC69; 58) model, the rates of change among all nucleotides are equal ($r_1 = r_2 = \cdots = r_{12}$) and base frequencies are equal ($\pi_A = \pi_C = \pi_G = \pi_T$).

analyses and is assumed when calculating the likelihood of a tree (the joint probability of observing all site patterns is the product of the individual site patterns). However, biologists know that the substitution processes at different sites in a sequence often are not independent. For example, hydrogen-bonded sites in the stem regions of ribosomal genes are not independent because a substitution in one nucleotide changes the probability that a compensatory substitution will occur in its partner.

Several authors have examined the effect of non-independent substitution among pair-bonded stem nucleotides. For example, Dixon & Hillis (18) examined the appropriate weights that stem sites should receive to correctly accommodate non-independence in a parsimony analysis. Others have devised time-homogeneous Poisson process models to describe substitutions in stem regions. Instead of a 4×4 matrix, these models assume a 16×16 rate matrix ($\mathbf{Q}$) of all possible nucleotide doublets possible in a stem (76, 95, 98, 99). The instantaneous rate of change is set to zero if more than one substitution is required to change from doublet i to doublet j (e.g. $q_{ij} = 0$ for AC $\rightarrow$ CG). The other rate parameters of the $\mathbf{Q}$ matrix are specified in different ways depending

Table 1 Parameter settings for a variety of evolutionary models employed in maximum likelihood analyses. Parameters of the substitution matrix, **Q**, are shown in Figure 3.

Model	Nucleotide frequencies	Rates of change	Reference
JC69	$\pi_A = \pi_C = \pi_G = \pi_T$	$r_1 = r_2 = r_3 = r_4 = r_5 = r_6$ $= r_7 = r_8 = r_9 = r_{10} = r_{11} = r_{12}$	58
K80	$\pi_A = \pi_C = \pi_G = \pi_T$	$r_3 = r_4 = r_9 = r_{10}; r_1 = r_2 = r_5$ $= r_6 = r_7 = r_8 = r_{11} = r_{12}$	60
K3ST	$\pi_A = \pi_C = \pi_G = \pi_T$	$r_3 = r_4 = r_9 = r_{10}; r_5 = r_6 = r_7$ $= r_8; r_1 = r_2 = r_{11} = r_{12}$	61
F81	$\pi_A; \pi_C; \pi_G; \pi_T$	$r_1 = r_2 = r_3 = r_4 = r_5 = r_6$ $= r_7 = r_8 = r_9 = r_{10} = r_{11} = r_{12}$	24
HKY85	$\pi_A; \pi_C; \pi_G; \pi_T$	$r_3 = r_4 = r_9 = r_{10}; r_1 = r_2 = r_5$ $= r_6 = r_7 = r_8 = r_{11} = r_{12}$	45
TrN	$\pi_A; \pi_C; \pi_G; \pi_T$	$r_3 = r_4; r_9 = r_{10}; r_1 = r_2 = r_5$ $= r_6 = r_7 = r_8 = r_{11} = r_{12}$	107
SYM	$\pi_A = \pi_C = \pi_G = \pi_T$	$r_1 = r_2; r_3 = r_4; r_5 = r_6; r_7 = r_8; r_9$ $= r_{10}; r_{11} = r_{12}$	126
GTR	$\pi_A; \pi_C; \pi_G; \pi_T$	$r_1 = r_2; r_3 = r_4; r_5 = r_6; r_7 = r_8; r_9$ $= r_{10}; r_{11} = r_{12}$	64

on the assumptions the biologist is willing to make. For example, Schöniger & von Haeseler (98) considered the simplest case by setting $q_{ij} = \pi_j$ if doublets i and j differ at one nucleotide (e.g., $q_{AC,AG} = \pi_{AG}$).

Other models consider the substitution process at the level of the codon. Fundamental to interpreting changes in substitution rates is an accurate assessment of how these changes influence the resulting protein; that is, does a substitution produce a change in the amino acid (a nonsynonymous change), or does the substitution not alter the protein (a synonymous change) (100). An increase in the relative amount of nonsynonymous change can be strong evidence for adaptive evolution (72). Many methods exist for estimating synonymous and nonsynonymous substitution rates (68–70, 75, 82, 87, 89). Recently, maximum likelihood has been used to estimate synonymous and nonsynonymous rates of change (33, 77, 78). These authors modeled the substitution process at the level of the codon. They used a 61×61 matrix to describe the instantaneous rate of change from one codon to another (the three stop codons are not considered). For the model of Muse & Gaut (78) the instantaneous rate of change from codon i to j is zero ($q_{ij} = 0$) if the change requires more than one substitution, $q_{ij} = \beta \pi_{n_{ij}}$ if the substitution causes a change in the amino acid, and $q_{ij} = \alpha \pi_{n_{ij}}$ if the substitution is synonymous (where n_{ij} is the equilibrium frequency of the substituted nucleotide). The model of Goldman & Yang (33) is similar except that they allow a transition/transversion rate bias and consider the physicochemical properties of the 20 amino acids by using Grantham's distances (34). The parameters of both models can be estimated using maximum likelihood. Unlike many of the standard methods, these approaches do not rely on the assumption that the number and location of silent/replacement sites do not change over time. The codon-based approach has been used successfully to explain the rate heterogeneity found in the chloroplast genome as being due primarily to differences in nonsynonymous substitution rates (78). As an alternative to modeling the substitution process at the level of the codon, the substitution process has also been modeled at the amino acid level (1, 9, 110).

The assumption of equal rates at different sites can also be relaxed. Several models of among-site rate heterogeneity have been developed (e.g., assume that the sites are log normally distributed—39, 85; assume an invariant rate class—12, 45, 91; estimate the rates in different data partitions separately, see 105; or use a combination of different rate distributions—38, 114). Probably the most widely used model is one that assumes that rates are gamma distributed (57, 116). The gamma distribution is a continuous probability density function that has wide application in probability and statistics (the well-known χ^2 and exponential distributions are special cases of the gamma distribution). Systematists have co-opted this distribution for their own use because the shape of the gamma distribution changes dramatically depending on the value of the shape parameter, α, and the scale parameter, λ. Systematists set $\alpha = 1/\lambda = a$ so that the mean of the gamma distribution is 1.0 and the variance is $1/a$. Rates at different sites, then, are thought of as random variables drawn from a gamma distribution with shape parameter a. When a is equal to infinity, the gamma model of among-site rate heterogeneity collapses to the equal rates case. However, most empirical estimates of the shape parameter a fall in the range of 0.1 to 0.5 (121), indicating substantial rate variation among sites. Yang (116, 119) provides details on how to calculate the likelihood under a gamma model of rate heterogeneity.

An advantage of the likelihood approach is that the models can be made complicated to incorporate other biologically important processes. For example, the models of substitution can be modified to account for insertion and deletion events (5, 111, 112), secondary structure of proteins (76a, 110), and correlated rates among sites (11, 27, 120). In the course of estimating phylogeny, the maximum likelihood method provides estimates of model parameters that may be of interest to the biologist. If the biologist is only interested in phylogeny, then these additional parameters are considered nuisance parameters (i.e., parameters not of direct interest to the biologist but which must be accommodated in the analysis by either integrated likelihood or maximum relative likelihood methods; see 31). However, maximum likelihood estimates of parameters such as the variance in the rate of substitution among sites or the bias in the substitution process are of interest to students of molecular evolution. Interestingly, many of the models that are currently implemented in likelihood were algebraically intractable and therefore unusable for other methods of phylogenetic inference. However, now that computer speed is affordable, these models and estimates of their parameters are feasible. Furthermore, Monte Carlo methods promise to make tractable models that are currently difficult or impossible to implement (37, 63, 84, 125).

LIKELIHOOD RATIO TESTS IN PHYLOGENETICS

Currently, one of the most debated subjects in the field of phylogenetics concerns the role that assumptions play in a phylogenetic analysis (8, 21, 74). All phylogenetic methods make assumptions about the process of evolution. An assumption common to many phylogenetic methods, for example, is a bifurcating tree to describe the phylogeny of species. However, additional assumptions are made in a phylogenetic analysis. For example, the assumptions of a maximum likelihood analysis are mathematically explicit and, besides the assumption of independence among sites, include parameters that describe the substitution process, the lengths of the branches on a phylogenetic tree, and among-site rate heterogeneity. The assumptions made in a parsimony analysis include independence and a specific model of character transformation (often called a step-matrix or weighting scheme; a commonly used weighting scheme is to give every character transformation equal weight). Phylogenetic methods can estimate the correct tree with high probability despite the fact that many of the assumptions made in any given analysis are incorrect. In fact, the maximum likelihood, parsimony, and several distance methods appear to be robust to violation of many assumptions, including making incorrect assumptions about the substitution process, among-site rate variation, and independence among sites (50, 55, 99). The advantage of making explicit assumptions about the evolutionary process is that one can compare alternative models of evolution in a statistical context. Instead of being viewed as a disadvantage, the use of explicit models of evolution in a phylogenetic analysis allows the systematist not only to estimate phylogeny, but to learn about processes of evolution through hypothesis testing.

One measure of the relative tenability of two competing hypotheses is the ratio of their likelihoods. Consider the case in which L_0 specifies the likelihood under the null hypothesis, whereas L_1 specifies the likelihood of the same data under the alternative hypothesis. The likelihood ratio is

$$\Lambda = \frac{\max[L_0(\text{Null Model} \mid \text{Data})]}{\max[L_1(\text{Alternative Model} \mid \text{Data})]} \qquad 12.$$

(19, 59, 71, 101). Here, the maximum likelihood calculated under the null hypothesis (H_0) is in the numerator, and the maximum likelihood calculated under the alternative hypothesis (H_1) is in the denominator. When Λ is less than one, H_0 is discredited and when Λ is greater than one, H_1 is discredited. Λ greater than one is only possible for non-nested models. When nested models are considered (i.e., the null hypothesis is a subset or special case of the alternative hypothesis), $\Lambda < 1$ and $-2 \log \Lambda$ is asymptotically χ^2 distributed under the null hypothesis with q degrees of freedom, where q is the difference in the number of free parameters between the general and restricted hypotheses.

An alternative means of generating the null distribution of $-2 \log \Lambda$ is through Monte Carlo simulation (parametric bootstrapping; 13, 14). Felsenstein (26) first suggested the use of the parametric bootstrap procedure in phylogenetics. Goldman (32) was among the first to apply the method in phylogenetics and to demonstrate that the usual χ^2 approximation of the null distribution is not appropriate for some tests involving phylogeny. In parametric bootstrapping, replicate data sets are generated using simulation under the assumption that the null hypothesis is correct. The maximum likelihood estimates of the model parameters under the null hypothesis are used to parameterize the simulations. For the phylogeny problem, these parameters would include the tree topology, branch lengths, and substitution parameters (e.g., transition:transversion rate ratio or the shape parameter of the gamma distribution). For each simulated data set, $-2 \log \Lambda$ is calculated anew by maximizing the likelihood under the null and alternative hypotheses. The proportion of the time that the observed value of $-2 \log \Lambda$ exceeds the values observed in the simulations represents the significance level of the test. Typically, the rejection level is set to 5%; if the observed value for the likelihood ratio test statistic is exceeded in less than 5% of the simulations, then the null hypothesis is rejected. Although there are good statistical reasons for test statistics based on the probability density of the data, the parametric bootstrap procedure may also be used to determine the null distribution of other test statistics (47).

Maximum likelihood allows the easy formulation and testing of phylogenetic hypotheses through the use of likelihood ratio tests (though also see 96). Furthermore, likelihood ratio tests are known to have desirable statistical properties. For example, they are known to be uniformly most powerful when simple hypotheses are considered and often outperform other hypothesis tests for composite hypotheses (92). Over the past two decades, numerous likelihood ratio tests have been suggested. These include tests of the null hypotheses that (a) a model of DNA substitution adequately explains the data (32, 80, 93), (b) rates of nucleotide substitution are biased (32, 80, 93), (c) rates of substitution are constant among lineages (24, 65, 79, 115), (d) rates are equal among sites (123), (e) rates of substitution are the same in different data partitions (30, 66, 122),

(f) substitution parameters are the same among data partitions (122), (g) the same topology underlies different data partitions (51), (h) a prespecified group is monophyletic (53), (i) hosts and associated parasites have corresponding phylogenies (56), (j) hosts and parasites have identical speciation times (56), and (k) rates of synonymous and nonsynonymous substitution are the same (77). Here, we describe a few of these tests. Our goal is not to provide an exhaustive list and description of all the likelihood ratio tests that have been proposed in phylogenetics but rather to illustrate how biological questions can be addressed in a simple way using likelihood.

Testing the Model of DNA Substitution

Current models implemented by maximum likelihood and distance methods assume that DNA substitutions follow a time-homogeneous Poisson process. As mentioned above, these models have been made complex to incorporate biological reality by the addition of parameters that allow for biased substitution and among-site rate variation. Likelihood provides the systematist with a rationale for choosing among different possible models through the use of likelihood ratio tests.

Typically, the question asked is "Does the addition of a substitution parameter provide a significant increase in the likelihood"? For example, one possible null hypothesis to consider is that transitions and transversions occur at the same rate. The Felsenstein (24; designated F81) model assumes equal rates for all substitutions and could be used to calculate the likelihood under the null hypothesis. The likelihood under the null hypothesis is compared to the likelihood under an alternative hypothesis that assumes a model of DNA substitution that allows a different rate of substitution for transitions and transversions. In this case, the HKY85 (45) model would be an appropriate model of DNA substitution to assume for the alternative hypothesis; the F81 and HKY85 models are identical except that the HKY85 model includes a parameter that allows for a different rate for transitions and transversions. The likelihood ratio test statistic ($-2 \log \Lambda$) is calculated, and the significance level is approximated by comparing $-2 \log \Lambda$ to χ^2 distribution with 1 degree of freedom.

As an example, consider data from five species of vertebrates. The data consist of aligned DNA sequences of 1383 sites from the albumin gene (Fish: *Salmo salar*, X5297; Frog: *Xenopus laevis*, M18350; Bird: *Gallus gallus*, X60688; Rat: *Rattus norvegicus*, J00698; Human: *Homo sapiens*, L00132). For this set of taxa, the phylogeny is almost certainly (Fish,(Frog,(Bird,(Rat,Human)))) (52). As mentioned above, these data could be analyzed using any one of several models. Here, we consider a hierarchy of hypotheses (Figure 4). Ideally, the hierarchy of hypotheses to be considered should be formulated before analysis begins. The null hypotheses considered are (a) base frequencies are equal,

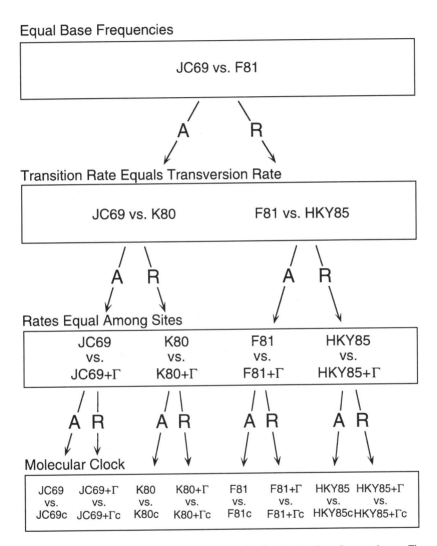

Figure 4   The hierarchy of hypotheses examined for the albumin data from five vertebrates. The parameters of the models are explained in Table 1. At each level, the null hypothesis is either accepted, "A," or rejected, "R."

(b) transitions and transversions occur at the same rate, (c) the rate of substitution is equal among sites, and (d) rates among lineages are constant through time (i.e., the molecular clock holds).

Figure 5 shows the trees and log likelihoods of the albumin data under several different models of DNA substitution. For the first null hypothesis considered—that base frequencies are equal—the likelihood under the JC69 model is compared to the likelihood under the alternative hypothesis calculated assuming the F81 model. These two models are identical except that the F81 model allows for unequal base frequencies and the JC69 model assumes equal base frequencies. Furthermore, the models are nested because the JC69 model is simply a special case of the F81 model. Therefore, the likelihood ratio test statistic, $-2 \log \Lambda$, can be compared to a χ^2 distribution with 3 degrees of freedom. For the test of equal base frequencies, $-2 \log \Lambda = 17.56$, a value much greater than 7.82, which represents the 95% critical value from a χ^2 distribution with 3 degrees of freedom. Therefore, the null hypothesis that base frequencies are equal is rejected, and the F81 model is preferred to the JC69 model of DNA substitution. The other three null hypotheses considered can also be tested using likelihood ratio tests. Table 2 shows the results of these tests. For the albumin data set, the most parameter-rich model considered (HKY85+Γ) is the best fitting model (i.e., provides a statistically significant increase in likelihood over the other models considered). Although the HKY85+Γ model was found to best fit the data, a general test of model adequacy indicates that the model is inadequate to explain the data (H_0: HKY85+Γ, H_1: unconstrained model, H_0 rejected at $P < 0.01$; 32). As expected, our models of DNA substitution do not fully describe the process of evolution leading to the observed sequences.

Although this conclusion sounds dire, it should be taken with a grain of salt because we know a priori that our models are inadequate to explain all the features of the evolutionary process. However, although the model is in some sense false, this does not detract from the utility of the model for estimating parameters such as topology and branch lengths, especially given the observation that phylogenetic methods in general, and maximum likelihood in particular, can be robust to violation of assumptions (50). Note that although a hierarchy of hypotheses was considered in this example, an alternative means of specifying the tests would be to treat the most general model as the alternative against which the other models are compared. Also note that the same tree was estimated for each of these models and that this tree is the one generally acknowledged as the best based on other sources of evidence (e.g., fossil and morphological data; 4). This is true even though the assumptions of all of the models are violated to some degree and some of the models considered (e.g., the JC69 model) poorly describe the data. Hence, the contention that methods using wrong models are

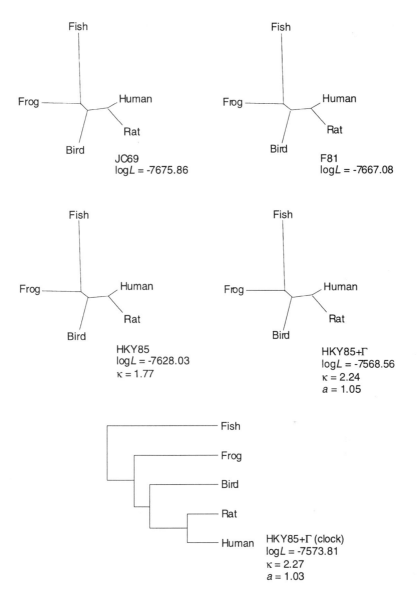

Figure 5 The phylogenies estimated for the albumin data under five different models of DNA substitution. The same phylogeny is estimated in each case, and this phylogeny is consistent with the traditional phylogeny of vertebrates [(Fish,(Frog,(Bird,(Rat,Human))))]. Note that the estimate of the transition:transversion rate ratio (κ) changes depending on whether or not among-site rate variation is accounted for; κ is underestimated when rate variation is not accounted for. Models that assume gamma distributed rates are denoted with a "+Γ."

Table 2 The results of likelihood ratio tests performed on the albumin DNA data from five vertebrates

Null hypothesis	Models compared	$\log L_0$	$\log L_1$	$-2\log \Lambda$	d.f.	P
Equal base frequencies	H_0: JC69 H_1: F81	-7675.86	-7667.08	17.56	3	2.78×10^{-5}
Transition rate equals transversion rate	H_0: F81 H_1: HKY85	-7667.08	-7628.03	78.10	1	9.75×10^{-19}
Equal rates among sites	H_0: HKY85 H_1: HKY85+Γ	-7628.03	-7568.56	118.94	1	0
Molecular clock	H_0: HKY85+Γc H_1: HKY85+Γ	-7573.81	-7568.56	10.5	3	1.47×10^{-2}

L_0 and L_1 denote the likelihoods under the null (H_0) and alternative (H_1) hypotheses, respectively. P represents the probability of obtaining the observed value of the likelihood ratio test statistic ($-2\log \Lambda$) if the null hypothesis were true. Because multiple tests are performed, the significance value for rejection of the null hypothesis should be adjusted using a Bonferroni correction (hence, the significance level for rejection of the null hypothesis is set to 1.25×10^{-2}).

"poor estimators of hierarchical pattern when the assumptions of the models are violated" (8, p. 426) appears overstated. Just as with other methods of phylogenetic estimation, the maximum likelihood method can be robust to a variety of model violations (50).

Tests of Topology

IS A PRESPECIFIED GROUP MONOPHYLETIC?   A taxonomic group is monophyletic if all its members share a most recent common ancestor. Many phylogenetic studies are aimed at determining the monophyly, or lack thereof, of some group of organisms. The most controversial of these studies have questioned the monophyly of groups, such as the rodents, bats, and toothed whales (35, 36, 73, 90), long defined as having a common evolutionary history based on morphological similarities. Often, the monophyly of a group has important evolutionary implications, particularly with respect to selection and adaptation. Pettigrew, for example, argued on the basis of neurological characters that megachiropterans (flying foxes) are more closely related to primates than they are to microchiropterans (90). This hypothesis of relationship implies that bats are not a monophyletic group and that either flight evolved twice (independently) in mammals or that flight evolved once in mammals but was subsequently lost in the primates.

How can the Pettigrew hypothesis that bats do not form a monophyletic group be tested? An analysis of the interphotoreceptor retinoid binding protein (IRBP) gene, that has been sequenced in primates, bats, and other mammals, provides an example of a likelihood ratio test that can be generalized to other questions of relationship (53, 103, 113). The maximum likelihood tree for the IRBP gene is shown in Figure 6A (113). The tree is consistent with the monophyly of bats. The best tree based on the assumption that bats are not monophyletic has a log likelihood 40.16 less than the best tree (Figure 6B). There are two ways to explain this result: (a) the Pettigrew hypothesis that bats are not monophyletic is correct, but estimation error has resulted in a tree consistent with bat monophyly; or (b) bats are a monophyletic group.

The ratio of the likelihoods calculated under the null model (bats constrained to be nonmonophyletic) and under the alternative hypothesis (no constraints placed on relationships) provides a measure of the relative support of the two hypotheses. In this case, the ratio of the likelihoods is $-\log \Lambda = 40.16$ (113). How damaging is this likelihood ratio to the Pettigrew hypothesis? Figure 6C shows the distribution of $-\log \Lambda$ that would be expected if the null hypothesis that bats are not monophyletic is true. The observed value of $-\log \Lambda$ is much greater than would be expected under the null hypothesis. Hence, the Pettigrew hypothesis can be rejected for the IRBP gene with a significance level of less than 1% (113).

What are the statistical properties of the likelihood ratio test of monophyly? Simulation study suggests that the method can be powerful (i.e., frequently rejects the null hypothesis when, in fact, the null hypothesis is false). Figure 7 shows the results of a study in which one of two trees was simulated (53). Simulating data for Tree 1 generates the distribution under the null hypothesis, whereas simulating data for Tree 2 generates data under the alternative hypothesis. The graph shows that the likelihood ratio test of monophyly can be powerful; the power of the test increases, as expected, when the number of sites in the analysis is increased. Although promising, the simulations presented in Figure 7 represent an ideal situation for the likelihood ratio test of monophyly. Additional simulations suggest that the test also performs well when the overall rate of substitution is low and an incorrect model is implemented in the likelihood analysis. However, when an incorrect model is used and the overall rate of substitution is high, the test rejects the null hypothesis too often. Hence, the likelihood ratio test of monophyly should be implemented with as biologically realistic a model as possible to prevent false rejection of the null hypothesis.

DO DIFFERENT DATA SETS CONVERGE TO SIGNIFICANTLY DIFFERENT TREES? The comparison of phylogenetic trees estimated from different data partitions has been used to address a variety of biological questions. Here, a data partition is defined as a division of characters into two or more subsets. The characters in each subset are either suspected to or have been demonstrated to have evolved according to different processes (e.g., different rates of substitution, different levels of selection, or different underlying phylogenies). Examples of potential

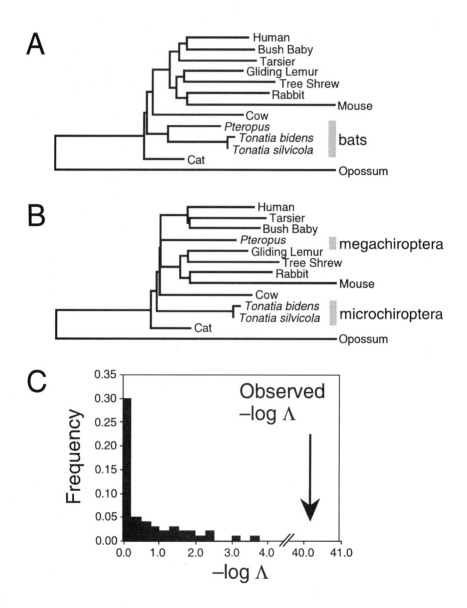

partitions of DNA characters include first- and second- versus third-codon positions, different genes, different coding regions within genes, or different genomic regions (e.g., nuclear versus mitochondrial, or different viral segments). The incongruence of the trees estimated from different genes in bacteria has been used to demonstrate horizontal gene transfer and recombination in bacteria (40). Similarly, the incongruence of trees estimated from different viral DNA segments has been used to show that reassortment of the segments has frequently occurred in the hanta virus (46). Such an approach can also be used to indicate method failure; if the partitioned data have evolved on the same underlying phylogenetic tree, but different trees have been estimated from each data partition, then either sampling error or systematic bias by the phylogenetic method are at fault (10). Finally, the congruence of trees estimated from host and associated parasites can be used to infer cospeciation (6, 41, 102).

How can incongruence of phylogenetic trees estimated from different data partitions be tested? A simple likelihood ratio test can be used to test whether the same phylogeny underlies all data partitions (51). The likelihood under the null hypothesis (L_0) is calculated by assuming that the same phylogenetic tree underlies all of the data partitions. However, branch lengths and other parameters of the substitution model are estimated independently in each. The likelihood under the alternative hypothesis (L_1) is calculated by relaxing the constraint that the same tree underlies each data partition. The alternative hypothesis allows the possibility that the histories of all data partitions are different. The likelihood ratio test statistic [$-2 \log \Lambda = -2 (\log L_0 - \log L_1)$] is compared to a null distribution generated using parametric bootstrapping.

Rejection of the null hypothesis of homogeneity (i.e., the same phylogeny for all data partitions) can indicate one of several different processes. One possibility is that a different history underlies different data partitions; incongruence of this sort could be caused by recombination, horizontal gene transfer, or ancestral polymorphism. Another possibility is that the phylogenetic methods have failed for one or more data partitions. All phylogenetic methods can

←───

Figure 6   The phylogenetic relationship of bats and other mammals. Trees were estimated using maximum likelihood under a model that allows an unequal transition:transversion rate, unequal base frequencies, and among-site rate heterogeneity (the HKY85+Γ model). Maximum likelihood trees were obtained using a tester version of the program PAUP* 4.0 (104). Bats form a monophyletic group in the maximum likelihood tree (log $L = -5936.52$) (*A*). The best tree under the Pettigrew hypothesis (that megachiroptera are constrained to be a sister group with the primates) (log $L = -5976.69$) (*B*). The distribution of the likelihood ratio test statistic under the assumption that the null hypothesis (the Pettigrew hypothesis) of relationship is correct (*C*). The observed likelihood ratio test statistic ($-\log \Lambda = 40.16$) is significant at $P < 0.01$. Hence, the Pettigrew hypothesis is rejected.

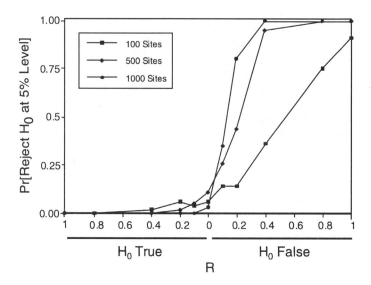

Figure 7   The power of the likelihood ratio test of monophyly. The graph represents the results of a computer simulation of two unrooted four taxon trees. The null hypothesis is true when the tree ((A, B), C, D) is simulated and false when ((A, C), B, D) is simulated. The external branches were constrained to be equal in length. R represents the ratio of the length of the internal branch to the external branches; when $R = 1.0$, all branches of the tree are equal in length. The null hypothesis is that the taxa A and B are a group (that the taxon bipartition {A, B} {C, D} exists). The null hypothesis is rarely rejected when it is true and frequently rejected when false.

produce biased estimates of phyolgeny when their assumptions are severely violated. For example, the parsimony method is known to produce inconsistent estimates of phylogeny for some very simple four-species trees (22). A method is inconsistent when it converges to an incorrect phylogenetic tree as more nucleotide or amino acid sites are included in the analysis. Regardless of the cause of the incongruence, combining the data is unwarranted because one runs the risk of either obscuring an interesting evolutionary phenomenon (e.g., different histories for different data partitions) or providing a poor estimate of phylogeny when method failure for one or more data partitions is at fault (10).

Null Distributions: χ^2 or Parametric Bootstrapping?

For many of the tests discussed here, nested hypotheses are considered (i.e., the null hypothesis is a special case of the alternative hypothesis). For most statistical problems involving nested hypotheses, the χ^2 distribution with q degrees of freedom (where q is the difference in the number of free parameters between the alternative and null hypotheses) can be used to test the significance of the likelihood ratio test statistic ($-2 \log \Lambda$). However, the phylogeny problem is an unusual statistical problem, and for many nested phylogenetic hypotheses, it is known that the χ^2 distribution is not appropriate (32). For example, for the tests of monophyly and topological incongruence among different data partitions, the χ^2 distribution cannot be used to determine the significance level of the likelihood ratio test statistic. On the other hand, the χ^2 distribution appears to be appropriate when testing whether the addition of a substitution parameter provides a significant improvement in the likelihood. The reason that the χ^2 distribution is appropriate for some phylogenetic probelms but not for others appears to be related to the fact that a tree topology is not a standard statistical parameter (124). In fact, for tests involving maximization over trees, it is difficult to determine the appropriate degrees of freedom because it is not clear how many parameters a topology represents. Probably the safest course for many likelihood ratio tests is to generate the null distribution using computer simulation (parametric bootstrapping). For null hypotheses that are composite, the maximum likelihood values under the null hypothesis can be used to parameterize the simulation (13, 14). This procedure has the advantage that it does not rely on asymptotic results and can be applied to non-nested as well as to nested hypotheses. The sensitivity of the paramertic bootstrap procedure to incorrect assumptions, however, has not been widely tested (though see 53).

CONCLUSIONS

Phylogenies are being applied to a wider variety of biological questions than ever before. One of the challenges for systematists is to develop appropriate tests to address the questions posed by evolutionary biologists. Statistical tests of phylogenetic questions can be formulated in many different ways. However, for many of the tests posed to date, the underlying assumptions are not clear. In fact, for some tests the null hypothesis is unclear (106). The testing of phylogenetic hypotheses in a likelihood framework should prove useful in the future. Likelihood provides a unified framework for the evaluation of alternative hypotheses. With the use of parametric bootstrapping, likelihood ratio tests can be applied to questions for which the null distribution is difficult to determine analytically.

The application of likelihood ratio tests in phylogenetics is a recent phenomenon, with most of the research activity occurring in the past five years. However, in that time the approach has proven powerful. Likelihood ratio tests have already provided information on the pattern of DNA substitution. Furthermore, the approach has been applied to questions involving topology and has even allowed the examination of whether hosts and parasites cospeciated (56). Future research can investigate the statistical properties of likelihood ratio tests with the objective of determining the power and robustness of the tests.

Another avenue of research involves the development of likelihood ratio tests to address additional null hypotheses. The likelihood approach should be particularly useful because the development of a likelihood ratio test is straightforward as long as the null and alternative hypotheses can be precisely described.

ACKNOWLEDGMENTS

We thank Rasmus Nielsen, Jack Sites, and Spencer Muse for comments on earlier versions of this manuscript. Spencer Muse, especially, made many helpful comments pointing out inadequacies of an earlier manuscript. This work was supported by a Miller postdoctoral fellowship awarded to JPH and an Alfred P. Sloan Young Investigator Award to KAC.

> **Visit the *Annual Reviews* home page at**
> **http://www.annurev.org.**

Literature Cited

1. Adachi J, Hasegawa M. 1992. Amino acid substitution of proteins coded for in mitochondrial DNA during mammalian evolution. *Jpn. J. Genet.* 67:187–97
2. Allard MW, Miyamoto MM. 1992. Perspective: testing phylogenetic approaches with empirical data, as illustrated with the parsimony method. *Mol. Biol. Evol.* 9:778–86
3. Barry D, Hartigan JA. 1987. Statistical analysis of hominoid molecular evolution. *Stat. Sci.* 2:191–210
4. Benton MJ. 1990. Phylogeny of the major tetrapod groups: morphological data and divergence dates. *J. Mol. Biol.* 30:409–24
5. Bishop MJ, Thompson EA. 1986. Maximum likelihood alignment of DNA sequences. *J. Mol. Biol.* 190:159–65
6. Brooks DR. 1981. Hennig's parasitological method: a proposed solution. *Syst. Zool.* 30:229–49
7. Brooks DR, McLennan DA. 1991. *Phylogeny, Ecology, and Behavior.* Chicago, IL: Univ. Chicago Press. 434 pp.
8. Brower AVZ, DeSalle R, Vogler A. 1996. Gene trees, species trees, and systematics: a cladistic perspective. *Annu. Rev. Ecol. Syst.* 27:423–50
9. Bruno WJ. 1996. Modeling residue usage in aligned protein sequences via maximum likelihood. *Mol. Biol. Evol.* 13:1368–74
10. Bull JJ, Huelsenbeck JP, Cunningham CW, Swofford DL, Waddell PJ. 1993. Partitioning and combining data in phylogenetic analysis. *Syst. Biol.* 42:384–97
11. Churchill GA. 1989. Stochastic models for heterogeous DNA sequences. *Bull. Math. Biol.* 51:79–94
12. Churchill GA, von Haeseler A, Navidi WC. 1992. Sample size for a phylogenetic inference. *Mol. Biol. Evol.* 9:753–69
13. Cox DR. 1961. Tests of separate families of hypotheses. *Proc. 4th Berkeley Symp. Math. Stat. Prob.* 1:105–23
14. Cox DR. 1962. Further results on tests of separate families of hypotheses. *J. R. Stat. Soc. B* 24:406–24
15. Cox DR, Miller HD. 1977. *The Theory of Stochastic Processes.* London: Chapman & Hall
16. Crandall KA. 1994. Intraspecific cladogram estimation: accuracy at higher levels of divergence. *Syst. Biol.* 43:222–35
17. Cummings MP, Otto SP, Wakeley J. 1995. Sampling properties of DNA sequence data in phylogenetic analysis. *Mol. Biol. Evol.* 12:814–22
18. Dixon MT, Hillis DM. 1993. Ribosomal RNA secondary structure: compensatory mutations and implications for phylogenetic analysis. *Mol. Biol. Evol.* 10:256–67
19. Edwards AWF. 1972. *Likelihood.* Cambridge: Cambridge Univ. Press
20. Edwards AWF, Cavalli-Sforza LL. 1964. Reconstruction of evolutionary trees. In *Phenetic and Phylogenetic Classification,* ed. J McNeill, pp. 67–76. London: Syst. Assoc.
21. Farris JS. 1983. The logical basis of phylogenetic analysis. In *Advances in Cladistics,* ed. NI Platnick, VA Funk, 2:7–36.

New York: Columbia Univ. Press

22. Felsenstein J. 1978. Cases in which parsimony or compatibility methods will be positively misleading. *Syst. Zool.* 27:401–10

23. Felsenstein J. 1978. The number of evolutionary trees. *Syst. Zool.* 27:27–33

24. Felsenstein J. 1981. Evolutionary trees from DNA sequences: a maximum likelihood approach. *J. Mol. Evol.* 17:368–76

25. Felsenstein J. 1985. Phylogenies and the comparative method. *Am. Nat.* 125:1–15

26. Felsenstein J. 1988. Phylogenies from molecular sequences. *Annu. Rev. Genet.* 22:521–65

27. Felsenstein J, Churchill GA. 1996. A hidden Markov model approach to variation among sites in rate of evolution. *Mol. Biol. Evol.* 13:93–104

28. Fisher RA. 1922. On the mathematical foundations of theoretical statistics. *Philos. Trans. R. Soc. London Ser. A* 222:309–68

29. Gaut BS, Lewis PO. 1995. Success of maximum likelihood phylogeny inference in the four-taxon case. *Mol. Biol. Evol.* 12:152–62

30. Gaut BS, Weir BS. 1994. Detecting substitution-rate heterogeneity among regions of a nucleotide sequence. *Mol. Biol. Evol.* 11:620–29

31. Goldman N. 1990. Maximum likelihood inference of phylogenetic trees with special reference to a Poisson process model of DNA substitution and to parsimony analyses. *Syst. Zool.* 39:345–61

32. Goldman N. 1993. Statistical tests of models of DNA substitution. *J. Mol. Evol.* 36:182–98

33. Goldman N, Yang Z. 1994. A codon-based model of nucleotide substitution for protein-coding DNA sequences. *Mol. Biol. Evol.* 11:725–36

34. Grantham R. 1974. Amino acid difference formula to help explain protein evolution. *Science* 185:862–64

35. Grauer D, Hide WA, Li W-H. 1991. Is the guinea-pig a rodent? *Nature* 351:649–52

36. Grauer D, Hide WA, Zharkikh A, Li W-H. 1992. The biochemical phylogeny of guinea-pigs and gundis, and the paraphyly of the order rodentia. *Comp. Biochem. Physiol. B* 101:495–98

37. Griffiths RC, Tavaré S. 1994. Simulating probability distributions. *Theor. Popul. Biol.* 46:131–59

38. Gu X, Fu Y-X, Li W-H. 1995. Maximum likelihood estimation of the heterogeneity of substitution rate among nucleotide sites. *Mol. Biol. Evol.* 12:546–57

39. Guoy M, Li W-H. 1989. Phylogenetic analysis based on rRNA sequences supports the archaebacterial rather than the eocyte tree. *Nature* 339:145–47

40. Guttman DS. 1997. Recombination and clonality in natural populations of *Escherichia coli. Trends Ecol. Evol.* 12:16–22

41. Hafner MS, Nadler SA. 1988. Phylogenetic trees support the coevolution of parasites and their hosts. *Nature* 332:258–59

42. Harvey PH, Nee S. 1994. Phylogenetic epidemiology lives. *Trends Ecol. Evol.* 9:361–63

43. Harvey PH, Pagel MD. 1991. *The Comparative Method in Evolutionary Biology.* Oxford: Oxford Univ. Press. 239 pp.

44. Hasegawa M, Kishino H, Saitou N. 1991. On the maximum likelihood method in molecular phylogenetics. *J. Mol. Evol.* 32:443–45

45. Hasegawa M, Kishino K, Yano T. 1985. Dating the human-ape splitting by a molecular clock of mitochondrial DNA. *J. Mol. Evol.* 22:160–74

46. Henderson WW, Monroe MC, St. Jeor SC, Thayer WP, Rowe JE, et al. 1995. Naturally occurring sin nombre virus genetic reassortants. *Virology* 214:602–10

47. Hillis DM. 1997. Biology recapitulates phylogeny. *Science* 276:218–19

48. Hillis DM, Bull JJ, White ME, Badgett MR, Molineux IJ. 1992. Experimental phylogenetics: generation of a known phylogeny. *Science* 255:589–91

49. Huelsenbeck JP. 1995. Performance of phylogenetic methods in simulation. *Syst. Biol.* 44:17–48

50. Huelsenbeck JP. 1995. The robustness of two phylogenetic methods: four-taxon simulations reveal a slight superiority of maximum likelihood over neighbor joining. *Mol. Biol. Evol.* 12:843–49

51. Huelsenbeck JP, Bull JJ. 1996. A likelihood ratio test to detect conflicting phylogenetic signal. *Syst. Biol.* 45:92–98

52. Huelsenbeck JP, Cunningham CW, Graybeal A. 1997. The performance of phylogenetic methods for a well supported phylogeny. *Syst. Biol.* In press

53. Huelsenbeck JP, Hillis DM, Nielsen R. 1996. A likelihood-ratio test of monophyly. *Syst. Biol.* 45:546–58

54. Huelsenbeck JP, Hillis DM. 1993. Success of phylogenetic methods in the four-taxon case. *Syst. Biol.* 42:247–64

55. Huelsenbeck JP, Nielsen R. 1997. The effect of non-independent substitution on phylogenetic accuracy. *Syst. Biol.* Submitted

56. Huelsenbeck JP, Rannala B, Yang Z. 1997. Statistical tests of host parasite

cospeciation. *Evolution* 51:410–19
57. Jin L, Nei M. 1990. Limitations of the evolutionary parsimony method of phylogenetic inference. *Mol. Biol. Evol.* 7:82–102
58. Jukes TH, Cantor CR. 1969. Evolution of protein molecules. In *Mammalian Protein Metabolism*, ed. HM Munro, pp. 21–132. New York: Academic
59. Kendall M, Stuart A. 1979. *The Advanced Theory of Statistics*, 2:240–80. London: Charles Griffin
60. Kimura M. 1980. A simple method for estimating evolutionary rate of base substitutions through comparative studies of nucleotide sequences. *J. Mol. Evol.* 16:111–20
61. Kimura M. 1981. Estimation of evolutionary distances between homologous nucleotide sequences. *Proc. Natl. Acad. Sci. USA* 78:454–58
62. Kuhner MK, Felsenstein J. 1994. A simulation comparison of phylogeny algorithms under equal and unequal evolutionary rates. *Mol. Biol. Evol.* 11:459–68
63. Kuhner MK, Yamato J, Felsenstein J. 1995. Estimating effective population size and mutation rate from sequence data using Metropolis-Hastings sampling. *Genetics* 140:1421–30
64. Lanave C, Preparata G, Saccone C, Serio G. 1984. A new method for calculating evolutionary substitution rates. *J. Mol. Evol.* 20:86–93
65. Langley CH, Fitch WM. 1974. An examination of the constancy of the rate of molecular evolution. *J. Mol. Evol.* 3:161–77
66. Learn GH, Shore JS, Furnier GR, Zurawski G, Clegg MT. 1992. Constraints on the evolution of plastid introns: the group II intron in the gene encoding tRNA-Val(UAC). *Mol. Biol. Evol.* 9:856–71
67. Lewis PO, Huelsenbeck JP, Swofford DL. 1996. Maximum likelihood. In *PAUP*: Phylogenetic Analysis Using Parsimony (and Other Methods,) Version 4.0*. Sunderland, MA: Sinauer
68. Lewontin RC. 1989. Inferring the number of evolutionary events from DNA coding sequence differences. *Mol. Biol. Evol.* 6:15–32
69. Li W-H. 1993. Unbiased estimation of the rates of synonymous and nonsynonymous substitution. *J. Mol. Evol.* 36:96–99
70. Li W-H, Wu C-I, Luo C-C. 1985. A new method for estimating synonymous and nonsynonymous rates of nucleotide substitution considering the relative likelihood of nucleotide and codon changes. *Mol. Biol. Evol.* 2:150–74
71. Lindgren BW. 1976. *Statistical Theory*.

New York: Macmillan. 3rd ed.
72. Messier W, Stewart C-B. 1997. Episodic adaptive evolution of primate lysozymes. *Nature* 385:151–54
73. Milinkovitch MC, Orti G, Meyer A. 1993. Revised phylogeny of whales suggested by mitochondrial DNA. *Nature* 361:346
74. Mindell DP, Thacker CE. 1996. Rates of molecular evolution: phylogenetic issues and applications. *Annu. Rev. Ecol. Syst.* 27:279–303
75. Miyata T, Yasunaga T. 1980. Molecular evolution of mRNA: a method for estimating rates of synonymous and amino acid substitution from homologous sequences and its application. *J. Mol. Evol.* 16:23–26
76. Muse SV. 1995. Evolutionary analyses when nucleotides do not evolve independently. In *Current Topics on Molecular Evolution*, ed. M Nei, N Takahata, pp. 115–24. University Park, PA: Penn. State Univ., Inst. Mol. Evol. Genet.
76a. Muse SV. 1995. Evolutionary analysis of DNA sequences subject to constraints on secondary structure. *Genetics* 139:1429–39
77. Muse SV. 1996. Estimating synonymous and nonsynonymous substitution rates. *Mol. Biol. Evol.* 13:105–14
78. Muse SV, Gaut BS. 1994. A likelihood approach for comparing synonymous and nonsynonymous nucleotide substitution rates with application to the chloroplast genome. *Mol. Biol. Evol.* 11:715–24
79. Muse SV, Weir BS. 1992. Testing for equality of evolutionary rates. *Genetics* 132:269–76
80. Navidi WC, Churchill GA, von Haeseler A. 1991. Methods for inferring phylogenies from nucleic acid sequence data by using maximum likelihood and linear invariants. *Mol. Biol. Evol.* 8:128–43
81. Nei M. 1987. *Molecular Evolutionary Genetics*. New York: Columbia Univ. Press
82. Nei M, Gojobori T. 1986. Simple methods for estimating the numbers of synonymous and nonsynonymous nucleotide substitutions. *Mol. Biol. Evol.* 3:418–26
83. Neyman J. 1971. Molecular studies of evolution: a source of novel staistical problems. In *Statistical Decision Theory and Related Topics*, ed. SS Gupta, J Yackel, pp. 1–27. New York: Academic
84. Nielsen R. 1997. A likelihood approach to populations samples of microsatellite alleles. *Genetics* 146:711–16
85. Olsen GJ. 1987. Earliest phylogenetic branchings: comparing rRNA-based evolutionary trees inferred with various

techniques. *Cold Spring Harbor Symp. Quant. Biol.* 52:825–37

86. Ou C-Y, Ciesielski CA, Myers G, Bandea CI, Luo C-C, et al. 1992. Molecular epidemiology of HIV transmission in a dental practice. *Science* 256:1165–71

87. Pamilo P, Bianchi NO. 1993. Evolution of the *Zfx* and *Zfy* genes: rates and interdependence between the genes. *Mol. Biol. Evol.* 10:271–81

88. Penny D, Hendy MD, Steel MA. 1992. Progress with methods for constructing evolutionary trees. *Trends Ecol. Evol.* 7:73–79

89. Perler R, Efstratiadis A, Lomedico P, Gilbert W, Klodner R, Dodgson J. 1980. The evolution of genes: the chicken preproinsulin gene. *Cell* 20:555–66

90. Pettigrew JD. 1986. Flying primates? Megabats have the advanced pathway from eye to midbrain. *Science* 231:1304–6

91. Reeves JH. 1992. Heterogeneity in the substitution process of amino acid sites of proteins coded for by mitochondrial DNA. *J. Mol. Evol.* 35:17–31

92. Rice JA. 1995. *Mathematical Statistics and Data Analysis.* Belmont, CA: Duxbury

93. Ritland K, Clegg MT. 1987. Evolutionary analysis of plant DNA sequences. *Am. Nat.* 130:S74–100

94. Rodriguez F, Oliver JF, Marin A, Medina JR. 1990. The general stochastic model of nucleotide substitutions. *J. Theor. Biol.* 142:485–501

95. Rzhetsky A. 1995. Estimating substitution rates in ribosomal RNA genes. *Genetics* 141:771–83

96. Rzhetsky A, Nei M. 1995. Tests of applicability of several substitution models for DNA sequence data. *Mol. Biol. Evol.* 12:131–51

97. Saitou N, Imanishi T. 1989. Relative efficiencies of the Fitch-Margoliash, maximum-parsimony, maximum-likelihood, minimum-evolution, and neighbor-joining methods of phylogenetic tree construction in obtaining the correct tree. *Mol. Biol. Evol.* 6:514–25

98. Schöniger M, von Haeseler A. 1994. A stochastic model for the evolution of autocorrelated DNA sequences. *Mol. Phylogeny Evol.* 3:240–47

99. Schöniger M, von Haeseler A. 1995. Performance of the maximum likelihood, neighbor joining, and maximum parsimony methods when sequence sites are not independent. *Syst. Biol.* 44:533–47

100. Sharp PM. 1997. In search of molecular darwinism. *Nature* 385:111–12

101. Silvey SD. 1975. *Statistical Inference.* London: Chapman & Hall

102. Simberloff D. 1987. Calculating probabilities that cladograms match: a method of biogeographic inference. *Syst. Zool.* 36:175–95

103. Stanhope MJ, Czelusniak J, Si J-S, Nickerson J, Goodman M. 1992. A molecular perspective on mammalian evolution from the gene encoding Interphotoreceptor Retinoid Binding Protein, with convincing evidence for bat monophyly. *Mol. Phylogeny Evol.* 1:148–60

104. Swofford DL. 1996. *PAUP*: Phylogenetic Analysis Using Parsimony (and Other Methods), Version 4.0.* Sunderland, MA: Sinauer Assoc.

105. Swofford DL, Olsen GJ, Waddell PJ, Hillis DM. 1996. Phylogenetic inference. In *Molecular Systematics,* ed. DM Hillis, C Moritz, BK Mable, pp. 407–514. Sunderland, MA: Sinauer. 2nd ed.

106. Swofford DL, Thorne JS, Felsenstein J, Wiegmann BM. 1996. The topology-dependent permutation test for monophyly does not test for monophyly. *Syst. Biol.* 45:575–79

107. Tamura K, Nei M. 1993. Estimation of the number of nucleotide substitutions in the control region of mitochondrial DNA in humans and chimpanzees. *Mol. Biol. Evol.* 10:512–26

108. Tateno Y, Takezaki N, Nei M. 1994. Relative efficiencies of the maximum-likelihood, neighbor-joining, and maximum-parsimony methods when substitution rate varies with site. *Mol. Biol. Evol.* 11:261–77

109. Tavaré S. 1986. Some probabilistic and statistical aspects of the primary structure of nucleotide sequences. In *Lectures on Mathematics in the Life Sciences,* ed. RM Miura, pp. 57–86. Providence, RI: Am. Math. Soc.

110. Thorne JL, Goldman N, Jones DT. 1996. Combining protein evolution and secondary structure. *Mol. Biol. Evol.* 13:666–73

111. Thorne JL, Kishino H, Felsenstein J. 1991. An evolutionary model for maximum likelihood alignment of DNA sequences. *J. Mol. Evol.* 33:114–24

112. Thorne JL, Kishino H, Felsenstein J. 1992. Inching toward reality: an improved likelihood model of sequence evolution. *J. Mol. Evol.* 34:3–16

113. Van Den Bussche RA, Baker RJ, Huelsenbeck JP, Hillis DM. 1997. Base compositional bias and phylogenetic analysis: a test of the "flying DNA" hypothesis. *Mol. Phyl. Evol.* Submitted

114. Waddell PJ, Penny D. 1996. Extending hadamard conjugations to model sequence evolution with variable rates across sites. Available by anonymous ftp from onyx.si.edu.
115. Weir BS. 1990. *Genetic Data Analysis.* Sunderland, MA: Sinauer
116. Yang Z. 1993. Maximum likelihood estimation of phylogeny from DNA sequences when substitution rates differ over sites. *Mol. Biol. Evol.* 10:1396–401
117. Yang Z. 1994. Statistical properties of the maximum likelihood method of phylogenetic estimation and comparison with distance methods. *Syst. Biol.* 43:329–42
118. Yang Z. 1994. Estimating the pattern of nucleotide substitution. *J. Mol. Evol.* 39:105–11
119. Yang Z. 1994. Maximum likelihood phylogenetic estimation from DNA sequences with variable rates over sites: approximate methods. *J. Mol. Evol.* 39:306–14
120. Yang Z. 1995. A space-time process model for the evolution of DNA sequences. *Genetics* 139:993–1005
121. Yang Z. 1996. Among-site rate variation and its impact on phylogenetic analyses. *Trends Ecol. Evol.* 11:367–72
122. Yang Z. 1996. Maximum likelihood models for combined analyses of multiple sequence data. *J. Mol. Evol.* 42:587–96
123. Yang Z, Goldman N, Friday AE. 1994. Comparison of models for nucleotide substitution used in maximum-likelihood phylogenetic estimation. *Mol. Biol. Evol.* 11:316–24
124. Yang Z, Goldman N, Friday AE. 1995. Maximum likelihood trees from DNA sequences: a peculiar statistical estimation problem. *Syst. Biol.* 44:384–99
125. Yang Z, Rannala B. 1997. Bayesian phylogenetic inference using DNA sequences: a Markov chain Monte Carlo method. *Mol. Biol. Evol.* 14:In press
126. Zharkikh A. 1994. Estimation of evolutionary distances between nucleotide sequences. *J. Mol. Evol.* 39:315–29

Annu. Rev. Ecol. Syst. 1997. 28:467–94

SPECIES TURNOVER AND THE REGULATION OF TROPHIC STRUCTURE

Mathew A. Leibold, Jonathan M. Chase, Jonathan B. Shurin, and Amy L. Downing

Department of Ecology and Evolution, University of Chicago, Chicago, Illinois 60637; e-mail: mleibold@pondside.uchicago.edu

KEY WORDS: top-down vs bottom-up, biomanipulation, productivity, trophic cascade, compositional change

---

ABSTRACT

Trophic structure, the partitioning of biomass among trophic levels, is a major characteristic of ecosystems. Most studies of the forces that shape trophic structure emphasize either "bottom-up" or "top-down" regulation of populations and communities. Recent work has shown that these two forces are not mutually exclusive alternatives, but efforts to model their interaction still often yield unrealistic predictions. We focus on the problems involved with modeling situations in which community composition, including both the number of trophic levels and the species composition within a trophic level, can change. We review the development of these ideas, emphasizing in particular how compositional change can alter theoretical expectations about the regulation of trophic structure. A comparison of studies on the effects of predators and resource productivity in limnetic ecosystems reveals an intriguing disparity between the results of manipulative experiments and those of correlational studies. We suggest that this contrast is a result of the difference in the temporal scales operating in the two types of studies. Ecosystem-level variables may appear to approach an equilibrium in short-term press experiments; however, processes such as invasion and extinction of species will not have time to play out in most such experiments. We found that the responses of ecosystems to short-term experimental treatments involve less change in species composition than is found in natural communities that have diverged in response to local conditions over longer periods. We argue that the results of short-term experiments support the predictions of models in which

467

the species pool does not change, whereas correlational studies among systems support theories that incorporate compositional change.

Introduction

The interplay of forces that shape community structure has fascinated ecologists at least since Darwin (31). Elton (41), in particular, developed the notion that feeding relations in combination with energetic constraints and the diversification of organisms into functional roles could elucidate many aspects of community structure, including the partitioning of organisms into trophic levels (i.e. Elton's "pyramid of numbers"). Since then, many have attempted to explain how feeding relations among species influence major aspects of community structure. Trophic structure, defined here as the partitioning of biomass into trophic levels and "guilds," is perhaps the most obvious aspect of community structure related to the transfer of energy (in the form of fixed carbon and materials) implied by feeding relations (41, 83, 104).

Currently two artificially distinct perspectives attempt to explain variation in the trophic structure of ecosystems. The first, influenced by the perspectives of Elton (41) and Lindeman (83), assumes that major features of ecosystems are regulated primarily from the "bottom up" (99, 174, 175). This view suggests that the biomass of organisms at any trophic level is a function of the productivity of their resource base. Two predictions emerge from this approach (99): that more productive ecosystems will have more trophic levels, and that the biomass of organisms at all trophic levels will increase with the basal productivity of the ecosystem. Though these arguments have intuitive appeal, they are at odds with the predictions of the simplest mathematical formulations of predator-prey interactions that include any dynamic feedback from consumers to their resources.

Alternatively, a "top-down" approach focuses on how the number of trophic levels in a system influences partitioning of biomass among all the trophic levels. This view was most forcefully argued by Hairston, Smith, & Slobodkin (54), hereafter referred to as HSS (and further elaborated by Fretwell—44). Top-down forces have received renewed attention in aquatic systems, especially due to the work of Carpenter, Kitchell, and their collaborators (23, 25), and because of the potential for biomanipulation (purposeful management of fish populations) to alleviate the symptoms of lake eutrophication (65, 147). Based on a dualistic assumption that a given trophic level is regulated either by resource competition or by predation, HSS argued that the number of trophic levels functioning in an ecosystem determines its trophic structure. Plants are expected to dominate in ecosystems with odd numbers of trophic levels, whereas herbivores will dominate in ecosystems with an even number of levels. Based in part on previous work in lake eutrophication (142), Carpenter & Kitchell (23) acknowledged a large role for nutrient loading in lake ecosystems. They used the same fundamental approach as HSS to argue that the abundance of secondary carnivores accounts for much of the variation in plant and herbivore biomass in lakes that is not explained by nutrient levels.

These two contrasting views (bottom-up vs. top-down) make very different predictions about patterns of covariation in biomass at adjacent trophic levels; empirical evidence lends support for both perspectives. A bottom-up approach argues that all trophic levels should increase with productivity. Numerous studies in aquatic systems and some evidence in terrestrial systems show patterns of positive covariation between plant and herbivore biomass, supporting the "bottom-up" perspective. In contrast, much experimental evidence for trophic cascades in enclosure and biomanipulation studies in aquatic systems, and an increasing number of similar studies in terrestrial systems, argue for the "top-down" perspective. Clearly these two forces are not mutually exclusive, and it is evident that the dualism between them is artificial and uninformative. Recent conceptual work has tried to synthesize the two views into one that examines how productivity and predation jointly affect trophic structure (23, 51, 52, 64, 67, 78, 79, 99, 106, 117, 118, 128). Below we summarize the results of these approaches. We then review evidence from both correlational and experimental studies of trophic structure in lake ecosystems, and we identify an important contrast between the results of the two types of studies. We argue that this contrast cannot easily be explained by most current models, and we suggest that a resolution can be found in models that allow for compositional species replacement along environmental gradients. We further argue for the inclusion of species turnover in models by comparing the predictions of such models with other documented patterns in community and ecosystem structure.

Simple Theories of Trophic Interactions and Trophic Structure

Most recent models of trophic structure fall into one of two broad categories. Models that emphasize "vertical structuring" focus on predator-prey interactions and examine the effects of varying the number of trophic levels. This type of model considers communities to be organized as food chains. The second type incorporates "horizontal structuring" in which multiple species at a trophic level compete for resources and share predators.

HSS (54), for example, focused almost exclusively on vertical structuring. They argued that the number of trophic levels present under different conditions influences the pattern of biomass partitioning among trophic levels. They viewed resource limitation and predator limitation as having relatively exclusive roles, predicting that biomass partitioning into trophic levels would depend on whether there were an even or odd number of trophic levels. Since then (44, 106, 149), these ideas have been modified to account for the joint regulation of organisms by a dynamic balance between predation (and other sources of mortality) and resource competition. This has led to the conclusion that the number of trophic levels (and therefore the importance of trophic cascades) depends on the productivity of the ecosystem. Within the range of conditions in which the number of trophic levels is fixed, adjacent trophic levels respond differently to increasing productivity (i.e. one increases while the other remains constant; see Figure 1A). Despite the conceptual appeal of this

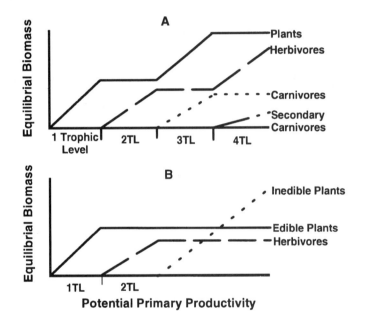

Figure 1 Biomass accrual of various trophic levels by enhanced ecosystem productivity under different simple models: A: Accrual in the simple food chain model developed by Oksanen et al (106). The number of trophic levels present is predicted to increase as shown by four zones on the x-axis. Biomass accrual among organisms at adjacent trophic levels is predicted to be uncorrelated unless the number of trophic levels also changes. Figure modified from Leibold (78). Though nonlinearities may modify these predictions slightly (e.g. 106), most cases result in asymmetric patterns of biomass accrual of organisms at adjacent trophic levels. B: Biomass accrual of edible and inedible plants and of herbivores with enhanced ecosystem productivity as developed by Phillips (122). The pattern is similar to that found in the Oksanen model (106) until productivity is high enough that inedible plants invade. Thereafter, enhanced productivity is completely shunted to inedible plants, and edible plants and herbivores are unaffected. Predictions can be modified somewhat by having only partial plant resistance (64, 79), but such cases also result in a strong shunt toward increasingly resistant plants without commensurately large increases in herbivores.

theory, the empirical evidence that the number of trophic levels correlates with productivity, and that biomass responses to productivity depend on the number of trophic levels, remains mixed (10, 20, 35, 94, 100, 123).

In response to HSS (54), several ecologists (39, 102, 169) have emphasized ways in which horizontal structuring that allows compensation among competing species within a trophic level could reduce the importance of top-down regulation of trophic level biomass. For example, when species differ in their vulnerability to consumers, more resistant species can compensate by increasing in abundance when more susceptible species are consumed. Such compensation is still the most common objection to trophic cascades (67, 78, 153). The role of compensatory effects is particularly evident in models in which the bottom-most trophic level is heterogeneous in its edibility and edible plants compete with inedible ones (21, 52, 64, 78, 79, 122). In such cases (Figure 1*B*), enhanced productivity is less likely to lead to a greater number of trophic levels or to enhanced grazer densities, because the enhanced production primarily benefits the inedible plants.

Both horizontal and vertical factors are critical in determining how trophic structure, and different functional groups within trophic levels (e.g. edible vs inedible plants) respond differently to variation in productivity. Abrams (1) modeled all possible food web configurations involving up to three trophic levels and up to two species per trophic level. He found that different food web configurations resulted in very heterogenous responses to productivity. The responses were further complicated by nonlinearities in the interactions among species such that very few testable generalizations could be made. Leibold & Wilbur (82) documented qualitatively distinct effects of nutrients on biomass partitioning into trophic levels under different food-web architectures in ponds. More recent, but untested, models have investigated situations in which populations are not at equilibrium and may oscillate or be "chaotic" (4) and when adaptive foraging behavior affects food web interactions (3).

The work on food chains without horizontal structure illustrates one of the important ways that novel large-scale predictions arise when there are vertical compositional changes as well as population or biomass changes in response to bottom-up regulation (106). In models of food chains, correlations between herbivore and plant biomass in response to nutrient levels are explained by variation in the number of trophic levels. However, more subtle but comparable effects can also occur when compositional change occurs within a trophic level (i.e. with horizontal structuring; 79, 122, 156). We argue that these effects are particularly important when there is extensive compositional turnover of species. However, before discussing these models in more detail, we review the experimental and observational literature to highlight discrepancies in our current understanding of top-down and bottom-up regulation of food webs.

Incongruences Between Experimental
Results and Correlational Studies

Some limnologists have felt uncomfortable with the dualistic alternation of resource and consumer limitation between adjacent trophic levels suggested by the HSS model, and they have argued more for the bottom-up perspective (99, 173). Support for the importance of resource supply, in contrast with the simple HSS formulation, has arisen from surveys that reveal that plant and herbivore biomass are both positively correlated with eutrophication (increases in nutrient levels in lakes that enhance their productivity) (49, 56, 92, 178). However, the evidence has been largely phenomenological rather than deductive, and correlational data on top-down effects are more difficult to collect (57, 101, 117, 119). Arguments against the importance of trophic cascades and the utility of biomanipulation generally invoke the role of functional heterogeneity among organisms, particularly emphasizing plant compensation (14, 22, 34, 59, 60, 134, 163) but also compensation among animals (25, 40, 72). Changes in phytoplankton composition along a gradient of trophic status are one line of evidence against top-down control (6, 80, 113, 133, 171, 172). Though there is some debate, resistant or toxic algae often increase with eutrophication (134, 111, 172). Similar patterns have been used by terrestrial ecologists to suggest that trophic cascades might be uncommon in terrestrial systems (39, 102, 126, 153). Some researchers have attempted to circumvent this problem at the population level by focusing their analyses on pairwise interactions between herbivores and "edible" plants (93), ignoring the role of "inedible" plants. An important implication of models that include inedible plants is that herbivore populations should not respond strongly to nutrient levels in lakes, but instead, excess nutrients should be sequestered in inedible algae (as shown in Figure 1*B*). However, experiments reviewed by Leibold (78) and Brett & Goldman (18, 19), and correlational studies such as those of McCauley et al (93), show that herbivores do respond positively to nutrient levels despite the presence of "inedible" algae.

Heterogeneity among herbivores may also limit top-down control. For instance, it has been argued that trophic cascades and successful biomanipulations are more likely in situations in which grazers are dominated by members of the genus *Daphnia*. Several experiments support this claim (78, 144). *Daphnia*'s importance is often explained by its diet breadth and strong population responses. To evaluate the effects on algae, Sarnelle (139) analyzed experiments in which *Daphnia* grazers have been directly or indirectly (via the use of fish predators) manipulated. Using the difference in algal biomass in the two grazer treatments divided by the biomass in the grazer-free treatment (which he named the algal response factor or ARF), he showed that grazers have larger proportional effects on phytoplankton in more eutrophic lakes. Similarly, Mazumder

(88) showed that the regression between algal biomass and total phosphorus levels is different in lakes where *Daphnia* are abundant (which he inferred to indicate food webs with even numbers of "functional" trophic levels) from those where *Daphnia* are rare (which he associated with an odd number of trophic levels). Both of these authors used their evidence to support top-down models of interactions in plankton, especially in eutrophic situations where Carney (22) has argued that "inedible" algae would prevent its occurrence.

Few studies have simultaneously discussed both experimental and correlational evidence for trophic regulation (25, 99), and there has been remarkably little discussion what a lack of congruence between experimental and correlational patterns might imply. McQueen et al (99) compared regressions of trophic level biomasses in surveys related to eutrophication with data from experiments manipulating fish predation, concluding that both top-down and bottom-up processes were important. They further argued that bottom-up effects were likely to be more important because top-down effects tended to "dissipate" as they proceeded to lower and lower trophic levels, particularly at the plant-herbivore interface. This analysis is, however, mostly phenomenological, and it suffers because the contrast between bottom-up and top-down effects is confounded with the type of evidence used (experiments vs surveys) (141).

Carpenter & Kitchell (25) also discussed the relative merits of short-term experiments (usually replicated in relatively small enclosures) with the longer studies involving whole-lake manipulations; they concluded that important additional processes occurring in whole lakes make extrapolation from most experiments difficult because such extrapolations ignore processes that occur on longer and larger scales. They have particularly focused on the roles of horizontal migrations by fishes and the coupling of benthic and pelagic processes.

We argue that differences between experimental and correlational patterns have the potential to be just as informative as similarities in distinguishing among models. Here we explore the contrasts between studies of lake communities that have responded to variation in nutrient levels and predation over long time scales (surveys of lakes that have diverged over tens to hundreds of years), whole-lake experiments (lasting from one to five years), and short-term enclosure experiments (lasting less than one year). Differences between them might inform us about the role of additional long-term, large-scale processes that are absent in the small-scale studies. A direct comparison of three approaches to the same questions allows us to qualitatively evaluate the effects of temporal and spatial scales.

We start by considering large-scale patterns in relation to nutrient levels and predation in lakes. To evaluate the patterns associated with nutrient levels, we focus on survey data. Data such as these have previously been examined by others (32, 56, 92, 110). We pool data from these and additional studies and present them as a standard by which to evaluate the experimental evidence we discuss below. Because virtually all proponents of trophic cascades in lakes acknowledge the role of nutrients, we also want to evaluate the long-term, whole-lake effects of fish predation on plankton communities in a way that controls (as much as possible) for nutrient levels. The best sources of data to investigate whole-lake responses to predator effects were biomanipulation studies and studies of historical changes in fish predation. We analyzed all studies (starting with those listed in Refs. 13 and 146 and supplemented by a search for more recent studies) that included phytoplankton and herbivore biomass data and, if possible, nutrient data.

Because models make different predictions about the relative responses of plants and herbivores to nutrients, we focus on the relationship between algal biomass (converted to chlorophyll-a concentration) and zooplankton biomass (usually restricted to crustaceans but sometimes including rotifers, converted to dry weight concentration). The data are shown in Figures 2A and 2B. As in previous studies, our compilation shows a strong correlation between algal and zooplankton biomass. These data disagree with the predictions of both of the simplest models of food chains, and of the models incorporating inedible plants described above (Figures 1A and 1B). The pattern is clearly related to variation in nutrient levels and thus substantiates the bottom-up view, suggesting that interactions between resources and their consumers might be "donor-controlled" without feedback from herbivores on plants. At first glance, this pattern tends to be incompatible with the evidence cited above about the role of grazers in lake ecosystems, especially since many of the lakes in the survey contain *Daphnia*, which are often able to control algae (88, 139).

---→

Figure 2   Patterns of variation in unmanipulated lakes. *A*: Herbivore biomass (evaluated as log zooplankton biomass) is correlated with plant biomass (evaluated as log chlorophyll-a concentration) in lakes with different total phosphorus levels. The scaling factor (slope of the log-log regression) is 0.404 with a standard error of 0.017; and the correlation coefficient (0.48) is highly significant ($p < 0.001$). Both variables are also strongly correlated with log total phosphorus (TP), indicated by the size of the symbol for each data point. Solid circle size is proportional to logTP, and open triangles denote studies in which data on TP was not available. *B*: Relative partitioning of biomass between zooplankton and algae expressed as the log of the ratio of zooplankton dry weight (ug/L) to the log of chlorophyll-a concentration (ug/L). Assuming that plant dry weight is about 100 times the chlorophyll concentration (6), these data show that over the three orders of magnitude in chlorophyll-a concentration, the ratio of herbivore to plant biomass declines from roughly equal partitioning between plants and herbivores to a trophic pyramid in which plant biomass is roughly 100-fold higher than herbivore biomass. Data are from References (32, 56, 61, 63, 76, 92, 109, 135, 148, 152, 159).

A

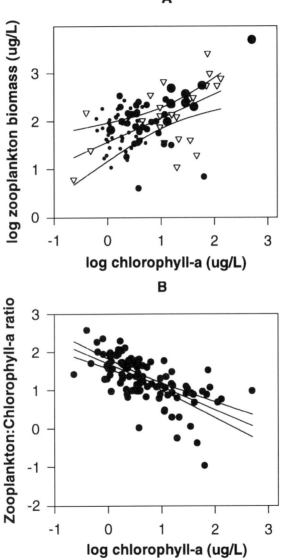

B

These data, however, also indicate just how dramatically trophic structure can vary with nutrient levels. The ratio of herbivore to plant biomass in these lakes varies by over two orders of magnitude (Figure 2*B*) and hints at the huge variability in ecosystem function that must accompany such variation. As in previous studies, the biomass of herbivores declines relative to plants as lakes become more productive, and this again seems superficially inconsistent with claims that grazers (at least *Daphnia*) are more important in eutrophic systems (88, 139).

This positive correlation between plant and herbivore biomass with eutroph-ication is also evident in the biomanipulation studies (Figure 3). When algal and zooplankton biomass are plotted without regard to fish densities, the data show the same relationship as in the survey studies described above. However, with a few notable exceptions, increases in fish populations within lakes tend

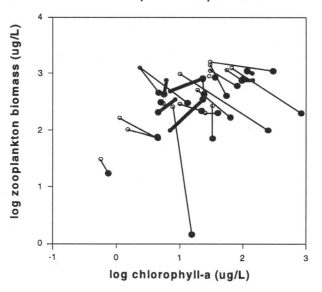

Biomanipulation experiments

Figure 3 Effects of biomanipulation on plants and herbivores as measured by log chlorophyll concentration and log zooplankton dry weight from whole-lake experiments. Each line shows the results of a single study in which planktivore densities have been altered. Lines connect low planktivory data (*open circles*) to the high planktivory data (*solid circles*). Most cases (shown with *thinner lines*) point from the upper left to the lower right, indicating a trophic cascade. The six exceptions are illustrated with *thicker lines* and show patterns inconsistent with trophic cascades. Data are from References (15, 16, 25, 37, 47, 71, 84, 98, 120, 132, 138, 150, 151, 161, 168, 162, 170).

to alter trophic structure by decreasing herbivore abundances and increasing plants (i.e. a trophic cascade). Values for pre- and post-manipulated lakes fall mostly within the range of values found in the survey, supporting the view that much of the variation in trophic structure that is not explained by nutrient levels can be attributed to variation in fish predation (23).

Information from these surveys and biomanipulation studies can be supplemented by a growing body of replicated, controlled experiments usually conducted in artificial pond arrays and enclosures. Such experiments have advantages over the survey and unreplicated biomanipulation experiments described above. In particular, the biomanipulation studies have the advantage (and the flaw) that they allow an enormous array of additional processes to influence their outcome, which cannot occur in the smaller enclosures (e.g. whole lake processes, unique historical phenomena, uncontrolled climatic effects). Replicated experiments control for many of these possibilities and allow a more focused understanding of mechanisms behind the results. Here we explore the outcome of food-web interactions when local populations interact on the time scale of weeks to months and are not confounded with some of the longer-time and larger-scale processes, which are not encompassed by the theories described above in which there is no compositional change.

To analyze these data we define a unitless metric, the "zooplankton response factor" (ZRF), analogous to ARF (139). We divide the magnitude of nutrient effects on zooplankton and algae by their densities in the low nutrient treatment. For predator manipulations we standardize zooplankton by their biomass in the low predation treatment and algae by their biomass in the high predation treatment (where the value more likely reflects their "carrying capacity"). The relationship between ARF and ZRF should correspond to the slope of the response on log-log plots of their respective biomasses. This method allows us to compare relative responses without having to account for differences either in the magnitude of the manipulations or in baseline environmental conditions, with expectations from survey patterns.

We analyzed data from published studies that manipulated planktivorous fish or nutrients (nitrogen and/or phosphorus) in mesocosms or artificial experimental ponds, and which reported both phytoplankton and zooplankton standing crops. Multiple comparisons were made both for factorial studies that manipulated both nutrients and predators (for instance, high vs low fish at both high and low nutrients) and for studies with more than two treatment levels. The results of nutrient manipulations show a surprising level of heterogeneity (Figure 4). In some cases, zooplankton responded by increasing much more strongly than algae, whereas in other cases, algae increased much more strongly than zooplankton. Neither extreme responses corresponded with expectations from the survey, which predicted much less asymmetric responses. The 95%

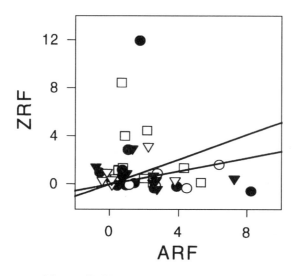

Figure 4   Responses of plants and herbivores to nutrient additions in replicated enclosure and small pond experiments. The plant response is plotted as the unitless algal response factor (ARF) equal to the difference between the two treatments in some measure of plant biomass (usually chlorophyll-a concentration) divided by the biomass in the low nutrient treatment. The zooplankton response is plotted as the identically defined (using zooplankton biomass instead of algal biomass) zooplankton response factor (ZRF). *Shaded symbols* denote cases in which *Daphnia* were present, whereas *open symbols* denote experiments conducted in situations where *Daphnia* were absent. *Triangles* denote experiments conducted in the presence of fish predators, whereas *circles* denote experiments conducted in the absence of fish. *Open squares* denote results of whole-lake manipulations (regardless of *Daphnia* incidence or fish presence). The two lines indicate the joint 95% confidence interval expected if ARF and ZRF resulted from biomass responses observed in the survey data (Figure 2). Responses are significantly more heterogeneous than expected ($p < 0.001$). Data are from References (26, 38, 42, 43, 55, 58, 68, 74, 75, 77, 90, 103, 114, 130, 179).

confidence interval of the slope from the survey relationship of algal and zooplankton biomass, plotted in Figure 4 along with the results of the experiments described above, highlights this dichotomy.

Taken together, these results support the notion that an unknown aspect of each of the local ecosystems (such as the number of trophic levels or some other aspect of food-web architecture) interacts with nutrient enhancement to produce dramatically different results. One possibility is that some of the experiments were conducted in situations with an odd number of functional trophic levels, whereas others were conducted with an even number of trophic levels. Another possibility is that the occurrence of *Daphnia* alters the response to nutrients because it influences the likelihood that some component of the algal community will be inedible. To examine these possibilities, we classified each study as to whether fish were present in the enclosures and whether *Daphnia* were present in the original zooplankton assemblage. No striking patterns of association emerged between the results and the number of trophic levels or occurrence or *Daphnia* (Figure 4).

We also reviewed the results of whole-lake experimental manipulations of nutrients, often conducted to enhance fish production in otherwise oligotrophic conditions. Figure 4 shows ARFs and ZRFs obtained in these larger-scale and longer-term experiments (usually 1–3 years of monitoring after nutrient additions). The results are as variable as are those of the enclosure studies described above. Toxic effects of nutrients on zooplankton have been invoked in some of the cases in which algal blooms have occurred without corresponding increases in zooplankton (86), but no direct evidence for toxicity was presented. Because most of the effects in which zooplankton responded without similarly large increases in algae occurred in the presence of *Daphnia*, these results tended to reinforce the view that trophic cascades are more important, and donor control less important, in such situations.

Finally, we examined experimental manipulations of fish ("top-down" manipulation) in enclosures and replicate artificial ponds. In order to compare experimental fish manipulations to the whole-lake biomanipulation studies, we calculated the ARF and ZRF (Figures 5A and 5B). Generally, the distribution of results from the enclosures are in good agreement with the biomanipulation studies as indicated by the occurrence of most of the data points in the lower right quadrant. This quadrant corresponds to increased fish predation that leads to a decrease in zooplankton and an increase in algae. Thus, trophic cascades commonly occur, but some noteworthy exceptions involve all three other possible outcomes. In situations in which an increase in fish predation is associated with an increase in zooplankton biomass and a decrease in algae (the upper left quadrant), the effects are small and possibly not significant. However, in some cases increased fish predation results in an increase in zooplankton biomass as well as increased algal biomass (the upper right quadrant), and some studies show both effects were negative (the lower left quadrant). These may reflect cases in which effects were largely mediated through changes in nutrient levels (33, 163, 164) or other unknown indirect pathways. Responses in trophic structure in the enclosure experiments as expressed by the ratio of herbivore to plant biomass were also remarkably similar to the biomanipulation studies, suggesting that most trophic structure responses are not often strongly affected by whole-lake processes, though the literature does document particular instances in which such phenomena occur (25).

In summary, these analyses reveal a striking dichotomy between the responses of trophic structure to experimental nutrient and predator manipulations and variation in trophic structure along natural nutrient gradients in lakes.

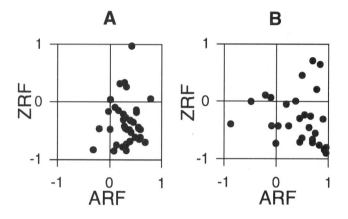

Figure 5 Responses of plants and herbivores to manipulations of fish. *A*: Results of experiments in replicated enclosures and small ponds. ARF and ZRF are defined as in Figure 4 except that the differences between treatments in resulting plant biomass are standardized by the plant biomass observed in the high-planktivory situation (most closely related to maximum algal "carrying capacity"), whereas differences in zooplankton biomass are standardized by the plant biomass observed in the low-planktivore treatment (related to zooplankton "carrying capacity"). Data points in the lower right quadrat (declines in zooplankton associated with increases in algae) are consistent with the trophic cascade. Other points are not, and these indicate mostly either simultaneous increases or decreases in both plants and herbivores.   Data are from References (42, 57, 58, 61, 96, 89, 90, 114, 115, 130, 158, 165 , 166, 167). *B*: Results of whole-lake biomanipulation studies plotted identically to panel *A*. Most data are consistent with a trophic cascade (in the lower right quadrat), but some cases also show simultaneous increases or decreases in both. Data are from References (15, 16, 25, 26, 37, 47, 71, 73, 84, 98, 120, 132, 138, 150, 151, 161, 162, 168, 170).

One possibility is that whole-lake processes are absent in small-scale experiments that make them nonrepresentative of natural conditions. However, the observation of similar dichotomous reactions to whole-lake fertilization experiments (Figure 4) argues against the possibility that restricted spatial scale is responsible for the contrast.   Further, the similar responses of whole-lake biomanipulation and small-scale enclosures to manipulations of fish predators also argue that such whole-lake processes, though they may be present, are not sufficiently large to account for the discrepancy.

The occurrence of parallel strong responses by plants and herbivores to nutrients is now understood to be a major line of evidence against simple food chain theories, and other hypotheses have been forwarded to explain this pattern.   These explanations have emphasized (*a*) compensation by organisms within trophic levels with different edibilities (52, 78, 153), (*b*) ratio-dependent models in which the functional response of consumers is dependent on the ratio of consumers to resources rather than resource density (9, 46, 62), (*c*) direct interference among consumers (91, 108, 149, 176), (d) nonlinear effects associated with adaptively plastic responses of organisms to predation risk and growth (3), and (e) the possibility that non-equilibrium communities will have average patterns that may differ qualitatively from patterns in predicted variation in the equilibrium points (4). Though many of these remain interesting possibilities, compensation (at least among planktonic organisms), ratio-dependence, interference, and nonlinear effects should occur in enclosures as well as in the lakes they supposedly mimic. These models consequently do not adequately explain why patterns among lakes that have diverged over many years differ so much from the results of experiments conducted both in enclosures and on whole-lake ecosystems over shorter periods of time.

Our thesis is that a major shortcoming of the current set of models is the limited extent to which they consider compositional changes within a trophic level. To evaluate whether compositional change along natural gradients was different than that from experiments, we compared the patterns of pairwise similarity in phytoplankton composition among naturally occurring lakes with the average similarity between different treatments in replicated experiments and compositional changes associated with biomanipulation studies. We obtained a matrix of pairwise similarities for an array of 40 lakes along a wide trophic gradient in Florida (155) and related these differences to the pairwise differences in the logarithm of their chlorophyll concentrations. Because our emphasis is on changes associated with the invasion and extinction of species rather than on patterns of relative dominance, we calculated Jaccard coefficients of similarity on the incidence matrix (presence/absence) and ignored changes associated with variation in relative dominance. The data (Figure 6) show that pairwise similarities are generally lower than about 0.5 and that they decrease with increasing differences in levels of eutrophication ($R = -0.21, p < .001$). An analysis of a similar gradient in northern fishless ponds (80) shows a very similar pattern.

We compared this distribution with pairwise values obtained from nutrient addition experiments, fish manipulation experiments, and two estimates from biomanipulation experiments that provided data on algal composition with enough resolution to provide useful contrasts with the survey data. Changes in phytoplankton composition resulting from experimental manipulations were substantially less than in the survey and did not show any strong pattern of association with the magnitude of change in plant biomass. This raises the possibility that whatever limits the development of such compositional turnover, especially colonization by species from outside the system, also constrains the responses in trophic structure that we document above. Additional evidence that compositional change is important comes from a mesocosm study (not included in the analyses described above) manipulating nutrients in which initial

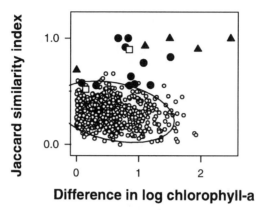

Figure 6   Pairwise similarity (measured using the Jaccard index on dichotomous presence/absence data) in algal composition among lakes (*small open symbols*), enclosure experiments (*solid symbols, circles* denote nutrient addition experiments and *triangles* denote fish manipulation experiments), and whole lake biomanipulation experiments (*open squares*) in relation to pairwise log-scale differences in mean plant biomass (measured as chlorophyll-a concentration). The data from the experiments (from references (8, 12, 85, 116, 154, 163, 165, 166) are significantly higher than the 95% confidence interval derived from the survey of lakes (155).

phytoplankton and zooplankton diversity was strongly enriched by pooling from numerous ponds with various nutrient levels. This experiment created initial conditions that allowed for species composition to sort out along a nutrient gradient. Roughly parallel responses in algal and zooplankton biomass to nutrients were observed (82). Jaccard similarity of algal composition between low- and high-nutrient treatments in this experiment averaged 0.27, a value akin to the range found in the eutrophic survey shown in Figure 6.

Compositional Change and Trophic Structure

The experimental and observational literature highlights several important discrepancies in our understanding of the regulation of trophic structure by nutrients and predators in lakes. In the first place, results of most experimental work support models that do not allow for compositional change, predicting asymmetric responses by plants and herbivores to eutrophication, whereas correlational analyses suggest roughly proportional increases in both. Secondly, demonstrably less compositional change by algae occurs in experimental enclosures and in whole lake manipulations than in surveys across lakes. These observations imply that compositional change may modify predictions about the regulation of trophic structure from those predicted on the basis of the simpler models described above.

How might this occur? We seek models that allow the simultaneous analysis of biomass responses at all trophic levels, that allow compositional change within trophic levels, and that function under joint regulation by both predation from above and resource production from below (simultaneous top-down and bottom-up regulation). Simulation models incorporate some of these features (33). However, we focus on simple analytic models in which some species consume and potentially compete for a single resource and are consumed by a single predator (11, 52, 64, 79, 160). Here we concentrate on the analysis of the "keystone predation" model presented by Leibold (79). The model predicts a series of species replacements as the supply of the resource is increased. Greater productivity favors increasingly resistant species, whereas species that are more vulnerable but more efficient at exploiting resources are lost (Figure 7). In this scenario, the trophic structure undergoes small changes as species are replaced along trophic gradients. As in food chain models, the densities of the bottom-most resource and of the predator are enhanced by enhanced nutrient levels, but in addition the intermediate trophic levels are enhanced as well. The community as a whole tracks equilibrium points along a productivity gradient. Species that are good resource competitors but susceptible to predators will dominate at low resource supply rates, whereas highly defended species that are poor resource competitors will replace them as primary productivity increases. The key feature of the model is the "tracking" of a large number of equilibria associated with taxonomic turnover that results in proportional responses by organisms at all three trophic levels.

Clearly, the "keystone predation" model can be only a caricature of the complex array of factors that affect compositional changes because it analyses compositional and trophic structure variation in a guild of species controlled by a single predator and a single resource. Most lakes have, at any time, dozens of planktonic herbivores and dozens to hundreds of plant species. Furthermore, compositional change in both of these groups can be driven by factors other than variation in productivity or top predators per se (including nutrient supply ratios, disturbance, habitat variation, pH, and temperature). There are two ways to interpret patterns of variation in limnetic communities in the context of this model. First, the model might be an adequate descriptor of one subsystem in the more complex array of food web interactions, but this subsystem is strong enough to provide a detectable pattern despite the occurrence of many other complex patterns of interaction. Alternatively, the qualitative predictions made by the model may adequately predict the cumulative behavior of many subsystems that act in parallel but roughly additive ways. Jager et al (70) conducted simulations of more complex food webs (5 species each in 5 trophic levels) in ways that are consistent with the model and found that most of the predictions made by Leibold (79) are obtained even with these additional complexities, supporting the validity of either or both of these interpretations.

Further, because the model assumes that local assemblages track the equilibria along a productivity gradient, it may be that other predictions would hold

A)

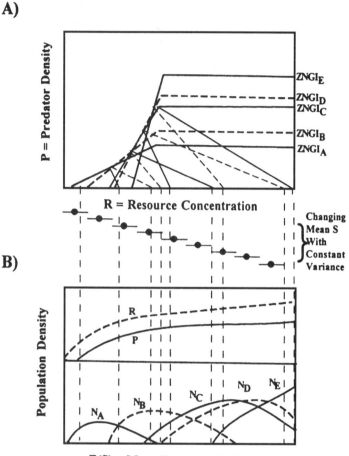

E(S) = Mean Resource Productivity

Figure 7 The keystone predator model (64, 79). *A*: Zero net growth isoclines (labeled "ZNIGI$_i$" for a group of species ($i = A$-E) that share a common predator (whose density is denoted P) and a common resource (with density R). The species show a trade-off between exploitation ability (minimum resource requirements in the absence of predators) and their susceptibility to predators (the slope of the ZNGI). For each of the pairwise equilibria that denote coexistence of uninvasible pairs, extensions of the two species' "impact vectors" are denoted that identify what level of ecosystem productivity will allow each equilibrium. *B*: Predicted qualitative patterns of densities of predators (P), resource levels (R), and densities of species A-E whose ZNGIs are shown in panel A. In a situation with local spatiotemporal variability (denoted by *dots with lines across them* denoting the mean and variance of such variability), both predator and resource densities increase monotonically. Additionally, there is a gradient of species replacements with increasing summed densities of the intermediate group of species. These species consist of good resource exploiters (e.g. species "A") at low productivity and predator-resistant species (species "E") at high productivity. Figure is from Leibold (79) who describes the model in detail.

under non-equilibrium situations. We would seek models that investigate trophic structure dynamics under conditions in which local population dynamics and invasion-colonization dynamics are both non-equilibrial. We know of no such models relevant to this situation but anticipate that they will provide additional insights into problems associated with the responses of ecosystems at different time scales. However, we believe the predictions made with the assumptions of equilibrium can serve as a useful starting point for such investigations.

Associations with other Community and Ecosystem Attributes

Because the keystone predation model includes predictions about both the horizontal and vertical structure of food webs, it can also make other predictions about community structure along productivity gradients and link them to predictions that strictly involve trophic structure. These predictions, developed by Holt et al (64), Grover (51, 52), and Leibold (79) and supported by the simulations of Jager et al (70), include the following:

1. *A unimodal relation between species richness within a trophic level and the rate of supply of resources to that trophic level.* In lakes, both algal diversity (5, 80, 105, 129) and zooplankton diversity (36, 80) are unimodally related to various measures of eutrophication (especially to ambient TP levels).

2. *A series of species replacements along a productivity gradient.* Alternative models of biodiversity can predict that even with a unimodal diversity curve as described above, species distribution patterns should consist of nested subsets rather than a series of species replacements. The intermediate peak occurs via the addition of specialists to otherwise less diverse communities dominated by generalists that occur at low and at high productivity (137, 156). Leibold (80) presented an analysis that supports the conclusion that species show replacements rather than nestedness along a productivity gradient in ponds.

3. *Dominance shifts from good resource-exploiters at low productivity to predation-tolerant species at high productivity.* Such predictions are relatively hard to verify because there are relatively few data to link these traits to the wide array of species found along limnetic productivity gradients. In general, however, it is recognized that resistant (or "inedible") algae dominate at higher productivity (80, 81, 171, 172).

Further work on the model of keystone predation involving the effects of compositional turnover may also generate further predictions. For example,

McPeek (95) has used this model to explain patterns of species distributions involving habitat generalists and specialists. Additionally, Grover (51) has examined how such a model acts to determine the "community assembly" process of diversification in local assemblages.

The predictions generated by the keystone predator model are testable and in many cases contrast with the predictions of other models. If our understanding of the various possibilities is sufficiently well developed, it may even be possible to conduct such comparative studies to identify and test alternative models. Leibold (80) conducted a correlative approach to argue that the keystone predation hypothesis is better able to explain variation in community structure in fishless ponds than the "paradox of enrichment" (136), resource heterogeneity (156), or resource ratio (156, 157) hypotheses. Such comparative approaches based on a method of strong inference (125) can strongly complement experimental approaches.

Conclusions

The relative roles of resources and predators in communities and ecosystems are themes that run deep in both ecology and limnology (29, 31, 41, 45, 50, 83, 104, 112). Many simplifying approaches have been taken to discern patterns in the "entangled bank" that results from the joint action of these two processes. Ideas about the regulation of trophic structure of ecosystems provide a fascinating example of how such ideas develop and become progressively more sophisticated as the demands for generality, rigor, and precision have increased. For example, we can trace ideas about the importance of predators from the HSS (54) hypothesis, which was aimed primarily at explaining why plants were common in terrestrial systems (i.e. why the "world was green" rather than "brown"), to Fretwell's (44) and Oksanen et al's (106) more sophisticated models that recognized that the productivity of the ecosystem might influence the number of trophic levels and thus explain variation in plant biomass across habitat gradients (i.e. why eastern US grasslands are "greener" than western ones). Subsequently, we can identify trophic models to explain residual variation in nutrient biomass relations (23), as well as models that include variation in plant edibility (52, 78, 122). These more refined models helped to account for variation in plant biomass in the absence of information about the number of trophic levels present. They also predict well the visible patterns in compositional change within trophic levels along productivity gradients (e.g. plant communities). Our goal in this paper is to argue that changes in species composition drive ecosystem-level patterns rather than simply functioning as a by-product of such patterns.

Other approaches have tried to circumvent developing such increasingly sophisticated models. For example, "ratio-dependent models" (in which functional responses of consumers are determined by the ratio of resource to consumer densities) have been proposed as alternatives to the prey-dependent models (in which functional responses are determined only by the density of the resource) we have emphasized here. Though there is much disagreement about their overall validity (2, 48, 107, 140), ratio-dependent models might be most useful as simplifications of the more complex mechanisms (7, 17). This simplification, however, cannot facilitate our understanding of the association between community and trophic structure that is the focus of this paper. Alternatively, others have advocated approaches that completely ignore mechanisms of population regulation processes in favor of "brute force" empirically predictive methods (121). This approach embodies a sense that ecosystems are too complex to "un-entangle." We realize that shortcomings in the more realistic, but still somewhat simplistic, models we advocate, such as the keystone predation model, will likely become evident. However, we believe that much of the evidence from the study of trophic structure in lakes and ponds supports many of these predictions, and we suggest that such approaches may be a useful standard for future model development and elaboration.

Trophic cascades have been documented in nonlimnetic ecosystems (e.g. 87, 127, 177). Terrestrial ecosystems also show patterns of biomass accrual similar to those in limnetic systems (30, 94, 100). Trophic interactions in some of these biomes also frequently involve compositional change (e.g. 27, 143). These and other lines of evidence suggest that trophic structure in other ecosystems may be regulated by the same basic processes that seem to work in lakes and ponds (but see 53).

An important aspect of the keystone predator model (and related models) is its ability to synthesize the effects of both "vertical" and "horizontal" structuring as documented in numerous food web studies (28, 123, 124). This synthesis of competitive and predator-prey interrelations is particularly appealing because it begins to provide better links between community and ecosystem approaches (including both the process-oriented, and currently mostly experimental, approaches, as well as the more descriptive "faunistic" or "floristic" approach that describes variation in biotas with environmental gradients). Our recent insights into the community-dependent structure of lake ecosystems, as exemplified by the work surveyed from the experimental, whole-lake biomanipulation and comparative approaches, suggest the importance of striving for better understanding of the links between these currently disparate approaches.

ACKNOWLEDGMENTS

We thank Tom Miller, Jean Tsao, Tim Wootton, and an anonymous reviewer for helpful comments on the manuscript and Paul del Georgio for supplying data. MAL was supported by NSF grants BSR-8817806 and DEB-9509004.

488   LEIBOLD ET AL

JMC, JBS, and ALD were supported by GAANN fellowships from the US Department of Education.

Literature Cited

1. Abrams PA. 1993. Effects of increased productivity on the abundances of trophic levels. *Am. Nat.* 141:351–71
2. Abrams PA. 1994. The fallacies of "ratio-dependent" predation. *Ecology* 75:1842–50
3. Abrams PA. 1996. Dynamics and interactions in food webs with adaptive foragers. In *Food Webs: Integration of Patterns and Dynamics,* ed. GA Polis, KO Winnemiller, pp. 113–21. London/New York: Chapman & Hall
4. Abrams PA, Roth JD. 1994. The effects of enrichment on three-species food chains with nonlinear functional responses. *Ecology* 75:1118–30
5. Agusti S, Duarte CM, Canfield DE. 1991. Biomass partitioning in Florida phytoplankton communities. *J. Plankton Res.* 13:239–45
6. Ahlgren G. 1970. Limnological studies of Lake Norrviken, a eutrophied Swedish lake. *Schwei. Zeitsch. Hydrol.* 32:353–96
7. Akçakaya HR, Arditi R, Ginzburg LR. 1995. Ratio-dependent predation: an abstraction that works. *Ecology* 6:995–1004
8. Andersson G, Berggren H, Cronberg G, Gelin C. 1978. Effects of planktivorous and benthivorous fish on organisms and water chemistry in eutrophic lakes. *Hydrobiologia* 59:9–15
9. Arditi R, Ginzburg LR. 1989. Coupling in predator-prey dynamics: ratio-dependence. *J. Theor. Biol.* 139:311–26
10. Arditi R, Ginzburg LR, Akçakaya HR. 1991. Variation in plankton densities among lakes: a case for ratio-dependent population models. *Am. Nat.* 138:1287–96
11. Armstrong RA. 1979. Prey species replacement along a gradient of nutrient enrichment: a graphical approach. *Ecology* 60:76–84
12. Beklioglu M, Moss B. 1996. Mesocosm experiments on the interaction of sediment influence, fish predation and aquatic plants with the structure of phytoplankton and zooplankton communities. *Freshwater Biol.* 36:315–25
13. Benndorf J. 1990. Conditions of effective biomanipulation; conclusions derived from whole-lake experiments in Europe. *Hydrobiologia* 200/201:187–203
14. Benndorf J. 1995. Possibilities and limits for controlling eutrophication by biomanipulation. *Int. Rev. Gesamten Hydrobiol.* 80:519–34
15. Benndorf J, Kneschke H, Kossatz K, Penz E. 1984. Manipulation of the pelagic food web by stocking with predacious fishes. *Int. Rev. Gesamten Hydrobiol.* 69:407–28
16. Benndorf J, Schultz H, Benndorf A, Unger R, Penz E, et al. 1988. Food-web manipulation by enhancement of piscivorous fish stocks: long-term effects in the hypertrophic Bautzen resevoir. *Limnologica* 19:97–110
17. Berryman AA, Gutierrez AP, Arditi R. 1995. Credible, parsimonius and useful predator-prey models—a reply to Abrams, Gleeson and Sarnelle. *Ecology* 76:1980–85
18. Brett MT, Goldman CR. 1996. A meta-analysis of the freshwater trophic cascade. *Proc. Natl. Acad. Sci. USA* 93:7723–26
19. Brett MT, Goldman CR. 1996. Consumer versus resource control in freshwater pelagic food webs. *Science* 275:384–86
20. Briand F, Cohen JE. 1987. Environmental correlates of food chain length. *Science* 238:956–60
21. Briand F, McCauley E. 1978. Cybernetic mechanisms in lake plankton systems: how to control undesirable algae. *Nature* 273:228–30
22. Carney HJ. 1990. A general hypothesis for the strength of food web interactions in relation to trophic state. *Int. Ver. Theor. Angew. Limnol. Verh.* 24:487–92
23. Carpenter SR, Kitchell JF. 1984. Plankton community structure and limnetic primary production. *Am. Nat.* 124:159–72
24. Deleted in proof
25. Carpenter SR, Kitchell JF. 1993. *The Trophic Cascade in Lakes.* Cambridge, UK: Cambridge Univ. Press. 385 pp.

26. Carpenter SR, Kitchell JF, Cottingham KL, Schindler DE, Christensen DL, et al. 1996. Chlorophyll variability, nutrient input, and grazing: evidence from whole-lake experiments. *Ecology* 77:725–35

27. Chase JM. 1997. Effects of a central place forager on food chain dynamics and spatial pattern in meadows. *Ecology*. In press

28. Cohen JE, Briand F, Newman CM. 1990. *Community Food Webs: Data and Theory*. New York: Springer-Verlag

29. Connell J. 1961. The influence of interspecific competition and other factors on the distribution of the barnacle *Chthamalus stellatus. Ecology* 42:710–23

30. Cyr H, Pace ML. 1993. Magnitude and patterns of herbivory in aquatic and terrestrial ecosystems. *Nature* 361:148–50

31. Darwin CR. 1859. *The Origin of Species*. Reprinted 1976. New York: Macmillian

32. del Giorgio PA, Gasol JM. 1995. Biomass distribution in freshwater plankton communities. *Am. Nat.* 146:135–52

33. DeAngelis DL. 1992. *Dynamics of Nutrient Cycling and Food Webs*. New York/London: Chapman & Hall

34. DeMelo R, France R, McQueen DJ. 1992. Biomanipulation: hit or myth. *Limnol. Oceanogr.* 37:192–207

35. Diehl S, Lundberg PA, Gardfjell H, Oksanen L, Persson L. 1993. Daphnia-phytoplankton interactions in lakes: Is there a need for ratio-dependent consumer-resource models? *Am. Nat.* 142:1052–61

36. Dodson SI. 1991. Species richness of crustacean zooplankton in European lakes of different sizes. *Int. Ver. Theor. Angew. Limnol. Verh.* 24:1223–29

37. Duncan A. 1990. A review: limnological management and biomanipulation in the London reservoirs. *Hydrobiologia* 200/201:541–48

38. Edmonson WT, Litt AH. 1982. Daphnia in Lake Washington. *Limnol. Oceanogr.* 27:272–93

39. Ehrlich PR, Birch LC. 1967. The "balance of nature" and "population control". *Am. Nat.* 101:97–107

40. Elser JJ, Luecke C, Brett MT, Goldman CR. 1995. Effects of food web compensation after manipulation of a rainbow trout in an oligotrophic lake. *Ecology* 76:52–69

41. Elton C. 1927. *Animal Ecology*. London: Sidgwick & Jackson

42. Faafeng BA, Hessen DO, Brabrand A, Nilssen JP. 1990. Biomanipulation and food-web dynamics—the importance of seasonal stability. *Hydrobiologia* 200/201:119–28

43. Findlay DL, Kasian SEM. 1987. Phyto-plankton community responses to nutrient addition in lake 226, Experimental Lakes Area, Northwestern Ontario. *Can J. Fish. Aquat. Sci.* 44:35–46 (Suppl.)

44. Fretwell SD. 1977. The regulation of plant communities by food chains exploiting them. *Perspect. Biol. Med.* 20:169–85

45. Gause GF. 1934. *The Struggle for Existence*. New York: Hafner

46. Getz WM. 1984. Population dynamics: a per capita resource approach. *J. Theor. Biol.* 108:623–43

47. Giussani G, de Bernardi R, Ruffoni T. 1990. Three years experience in biomanipulating a small eutrophic lake: Lago di Candia (Northern Italy). *Hydrobiologia* 200/201:357–66

48. Gleeson SK. 1994. Density dependence is better than ratio dependence. *Ecology* 75:1834–35

49. Gliwicz ZM. 1975. Effect of zooplankton grazing on photosynthetic activity and composition of phytoplankton. *Int. Ver. Theor. Angew. Limnol. Verh.* 19:1490–97

50. Grinnell J. 1917. The niche-relationships of the California Thrasher. *The Auk* 34:427–33

51. Grover JP. 1994. Assembly rules for communities of nutrient limited plants and specialist herbivores. *Am. Nat.* 94:421–25

52. Grover JP. 1995. Competition, herbivory, and enrichment: nutrient-based models for edible and inedible plants. *Am. Nat.* 145:746–74

53. Hairston NG Jr, Hairston NG Sr. 1993. Cause-effect relationships in energy flow, trophic structure, and interspecific interactions. *Am. Nat.* 142:379–411

54. Hairston NG, Smith FE, Slobodkin LB. 1960. Community structure, population control, and competition. *Am. Nat.* 44:421–25

55. Hall DJ, Cooper WE, Werner EE. 1970. An experimental approach to the production dynamics and structure of freshwater animal communities. *Limnol. Oceanogr.* 15:839–928

56. Hanson JM, Peters RH. 1984. Empirical prediction of crustacean zooplankton biomass and profundal macrobenthos biomass in lakes. *Can. J. Fish. Aquat. Sci.* 41:439–45

57. Hansson L-A. 1992. The role of food chain composition and nutrient availability in shaping algal biomass development. *Ecology* 73:241–47

58. Hansson L-A, Carpenter SR. 1993. Relative importance of nutrient availability and food chain for size and community composition in phytoplankton. *Oikos* 67:257–63

59. Harris GP. 1986. *Phytoplankton Ecology: Structure, Function and Fluctutation.* Cambridge, UK: Cambridge Univ. Press

60. Harris GP. 1994. Pattern, process, and prediction in aquatic ecology—a limnological view of some general ecological problems. *Freshwater Biol.* 32:143–60.

61. Havens KE. 1993. Responses to experimental fish manipulations in a shallow, hypereutrophic lake: the relative importance of benthic nutrient recycling and trophic cascade. *Hydrobiologia* 254:73–80

62. Herendeen RA. 1995. A unified quantitative approach to trophic cascade and bottom-up: top-down hypotheses. *J. Theor. Biol.* 176:13–26

63. Hessen DO, Andersen T, Lyche A. 1990. Carbon metabolism in a humic lake: pool sizes and cycling through zooplankton. *Limnol. Oceanogr.* 35:84–99

64. Holt RD, Grover J, Tilman D. 1994. Simple rules for interspecific dominance in systems with exploitation and apparent competition. *Am. Nat.* 144:741–71

65. Hrbacek J, Dvorakova M, Korinek V, Prochazkova L. 1961. Demonstration of the effects of fish stock on the species composition of zooplankton and the intensity of the metabolism of the whole plankton association. *Ver. Theor. Angew. Limnol. Verh.* 14:192–95

66. Huismann J, Weissing FJ. 1995. Competition for nutrients and light in a mixed water column: a theoretical analysis. *Am. Nat.* 146:536–64

67. Hunter MD, Price PW. 1992. Playing chutes and ladders: heterogeneity and the relative roles of bottom-up and top-down forces in natural communities. *Ecology* 73:724–32

68. Hyatt KD, Stockner JG. 1985. Response of Sockeye Salmon (*Oncorhynchus nerka*) to fertilization of British Columbia coastal lakes. *Can J. Fish. Aquat. Sci.* 42:320–31

69. Deleted in proof

70. Jager HI, Gardner RH, DeAngelis DL, Post WM. 1984. *A simulation approach to understanding the processes that structure food webs. ORNL/TM-8904.* Oak Ridge, TN: Oak Ridge National Laboratory 171 pp.

71. Jeppesen E, Sondergaard M, Mortensen E, Kristensen P, Riemann B, et al. 1990. Fish manipulation as a lake restoration tool in shallow, eutrophic temperate lakes 1: cross-analysis of three Danish case-studies. *Hydrobiologia* 200/201: 205–8

72. Kerfoot WC, DeMott WR. 1984. Food web dynamics: dependent chains and vaulting. In *Trophic Interactions within Aquatic Ecosystems,* ed. DG Meyers, JR Strickler, pp. 347–381. Washington, DC: Westview

73. Langeland A. 1990. Biomanipulation in Norway. *Hydrobiologia* 200/201:535–40

74. Langeland A, Reinertsen H. 1982. Interaction between phytoplankton and zooplankton in a fertilized lake. *Holarct. Ecol.* 5:253–72

75. Larocque I, Mazumder A, Proulx M, Lean DRS, Pick FR. 1996. Sedimentation of algae: relationships with biomass and size distribution. *Can. J. Fish. Aquat. Sci.* 53:1133–42

76. Lean DRS, Fricker HJ, Charlton MN, Cuhel RL, Pick FR. 1987. The Lake Ontario life support system. *Can. J. Fish. Aquat. Sci.* 44:2230–40

77. LeBrasseur RJ, McAllister CD, Barraclough WE, Kennedy OD, Manzee J, et al. 1978. Enhancement of Sockeye Salmon (*Oncorhynchus nerka*) by lake fertilization in Great Central Lake: summary report. *J. Fish. Res. Board Can.* 35:1580–96

78. Leibold MA. 1989. Resource edibility and the effects of predators and productivity on the outcome of trophic interactions. *Am. Nat.* 134:922–49

79. Leibold MA. 1996. A graphical model of keystone predators in food webs: trophic regulation of abundance, incidence and diversity patterns in communities. *Am. Nat.* 147:784–812

80. Leibold MA. 1997. Biodiversity and nutrient enrichment in pond plankton communties. *Evol. Ecol.* In press

81. Leibold MA. 1997. Do competition models predict nutrient availabilities in limnetic ecosystems? *Oecologia* 110:132–42

82. Leibold MA, Wilbur HM. 1992. Interactions between food-web structure and nutrients on pond organisms. *Nature* 360:341–43

83. Lindeman RL. 1942. The trophic-dynamic aspect of ecology. *Ecology* 23:399–418

84. Lyche A, Faafeng BJ, Brabrand A. 1990. Predictability and possible mechanisms of plankton reponse to reduction of planktivorous fish. *Hydrobiologia* 200/201:251–61

85. Lynch M, Shapiro J. 1981. Predation, enrichment, and phytoplankton community structure. *Limnol. Oceanogr.* 26:86–102

86. Malley DF, Chang PSS, Findlay DL, Linsey GA. 1988. Extreme perturbation of the zooplankton community of a small precambrian shield lake by the addition of nutrients. *Verh. Int. Verein. Limnol.* 23:2237–47

87. Marquis RJ, Whelan CJ. 1994. Insectivorous birds increase growth of white oak through consumption of leaf-chewing insects. *Ecology* 75:2007–14

88. Mazumder A. 1994. Patterns of algal biomass in dominant odd- vs. even-link lake ecosystems. *Ecology* 75:1141–49

89. Mazumder A, Lean DRS. 1994. Consumer-dependent responses of lake ecosystems to nutrient loading. *J. Plankton Res.* 16:1567–80

90. Mazumder A, Taylor WD, Lean DRS, McQueen DJ. 1992. Partitioning and fluxes of phosphorus: mechanisms regulating the size-distribution and biomass of plankton. *Arch. Hydrobiol. Beih. Ergebn. Limnol.* 35:121–43

91. McCarthy MA, Ginzburg LR, Akakaya HR. 1995. Predator interference across trophic chains. *Ecology* 76:1310–19

92. McCauley E, Kalff J. 1981. Empirical relationships between phytoplankton and zooplankton biomass in lakes. *Can. J. Fish. Aquat. Sci.* 38:458–63

93. McCauley E, Murdoch WW, Watson S. 1988. Simple models and variation in plankton densities among lakes. *Am. Nat.* 132:383–403

94. McNaughton SJ, Oesterheld M, Frank DA, Williams KJ. 1989. Ecosystem-level patterns of primary productivity and herbivory in terrestrial grasslands. *Nature* 341:142–144

95. McPeek MA. 1996. Trade-offs, food web structure, and the coexistence of habitat specialists and generalists. *Am. Nat.* 148:S124–38 (Suppl.)

96. McQueen DJ, Post JR. 1985. Enclosure experiments: the effects of planktivorous fish. *Lake and Reservoir Management, EPA Doc. EP8.2:L14/1985,* pp. 313–318. Washington, DC: USGPO

97. McQueen DJ, France R, Kraft C. 1992. Confounded impacts of planktivorous fish on freshwater biomanipulations. *Arch. Hydrobiol.* 125:1–24

98. McQueen DJ, Johannes MRS, Post JR, Stewart TJ, Lean DRS. 1989. Bottom-up and top-down impacts on freshwater pelagic community structure. *Ecol. Monogr.* 59:289–309

99. McQueen DJ, Post JR, Mills E. 1986. Trophic relationships in freshwater pelagic ecosystems. *Can. J. Fish. Aquat. Sci.* 43:1571–1581

100. Moen J, Oksanen L. 1991. Ecosystem trends. *Nature* 353:510

101. Moss B, McGowan S, Carvalho L. 1994. Determination of phytoplankton crops by top-down and bottom-up mechanisms in a group of English lakes, the West Midland meres. *Limnol. Oceanogr.* 39:1020–29

102. Murdoch WW. 1966. "Community structure, population control, and competition": a critique. *Am. Nat.* 100:219–26

103. O'Brien WJ, deNoyelles F Jr. 1974. Relationship between nutrient concentration, phytoplankton density, and zooplankton density in nutrient enriched experimental ponds. *Hydrobiologia* 44:105–25

104. Odum EP. 1971. *Fundamentals of Ecology.* Philadelphia: Saunders. 3rd ed.

105. Ogawa Y, Ichimura S. 1984. Phytoplankton diversity in inland waters of different trophic status. *Jpn. J. Limnol.* 45:173–77

106. Oksanen L, Fretwell SD, Arrüda J, Miemela P. 1981. Exploitation ecosystems along gradients of primary productivity. *Am. Nat.* 118:240–61

107. Oksanen L, Moen J, Lundberg PA. 1992. The time-scale problem in exploiter-victim models: Does the solution lie in ratio-dependent exploitation? *Am. Nat.* 140:938–960

108. Oksanen T, Power ME, Oksanen L. 1995. Ideal free habitat selection and consumer-resource dynamics. *Am. Nat.* 146:565–85

109. Pace ML. 1984. Zooplankton community structure, but not biomass, influences the phosphorus-chlorophyll a relationship. *Can. J. Fish. Aquat. Sci.* 41:1089–96

110. Pace ML. 1986. An empirical analysis of zooplankton community size structure across lake trophic gradients. *Limnol. Oceangr.* 31:45–55

111. Paerl HW. 1988. Nuisance phytoplankton blooms in coastal, estuarine, and inland waters. *Limnol. Oceanogr.* 33:823–47

112. Paine RT. 1966. Food web complexity and species diversity. *Am. Nat.* 100:65–75

113. Palmer CM. 1959. *Algae in water supplies. An illustrated manual on the identification, significance, and control of algae in water supplies. US Dept. Public Health Service.* Cincinnati, OH: Taft Sanitary Engin. Ctr.

114. Pérez-Fuentetaja A, McQueen DJ, Demers E. 1996. Stability of oligotrophic and eutrophic planktonic communities after disturbance by fish. *Oikos* 75:98–110

115. Pérez-Fuentetaja A, McQueen DJ, Ramcharan CW. 1996. Predator-induced

bottom-up effects in oligotrophic systems. *Hydrobiologia* 317:163–176

116. Pérez-Martinez C, Cruz-Pizarro L. 1995. Species-specific phytoplankton responses to nutrients and zooplankton manipulations in enclosure experiments. *Freshwater Biol.* 33:193–203

117. Persson L, Andersson G, Hamrin SF, Johansson L. 1988. Predator regulation and primary production along the productivity gradient. In *Complex Interactions in Lake Communities*, ed. SR Carpenter, pp. 45–68. New York: Springer-Verlag

118. Persson L, Bengtsson J, Menge BA, Power ME. 1996. Productivity and consumer relations-concepts, patterns and mechanisms. In *Food Webs: Integration of Patterns and Dynamics,* ed. GA Polis, KO Winnemiller, pp. 396–434. London/New York: Chapman and Hall

119. Persson L, Diehl S, Johansson L, Andersson G, Hamrin SF. 1992. Trophic interactions in temperate lake ecosystems: a test of food chain theory. *Am. Nat.* 140:59–84

120. Persson L, Johansson L, Andersson G, Diehl S, Hamrin SF. 1993. Density-dependent interactions in lake ecosystems: whole lake perturbation experiments. *Oikos* 66:193–208

121. Peters RH. 1986. The role of prediction in limnology. *Limnol. Oceanol.* 31:1143–59

122. Phillips OM. 1974. The equilibrium and stability of simple marine systems. II. Herbivores. *Arch. Hydrobiol.* 73:310–33

123. Pimm SL. 1982. *Food Webs.* London: Chapman & Hall

124. Pimm SL. 1991. *The Balance of Nature?* Chicago, IL: Univ. Chicago Press

125. Platt JR. 1964. Strong inference. *Science* 146:347–53

126. Polis GA, Strong DR. 1996. Food web complexity and community dynamics. *Am. Nat.* 147:813–846

127. Power ME. 1990. Effects of fish in river food webs. *Science* 250:811–814

128. Power ME. 1992. Top-down and bottom-up forces in food webs: Do plants have primacy? *Ecology* 73:733–746

129. Proulx M, Pick FR, Mazumder A, Hamilton PB, Lean DRS. 1996. Experimental evidence for interactive impacts of human activities on lake algal species richness. *Oikos* 76:191–95

130. Proulx M, Pick FR, Mazumder A, Hamilton PB, Lean DRS. 1996. Effects of nutrients and planktivorous fish on the phytoplankton of shallow and deep aquatic systems. *Ecology* 77:1556–1572

131. Deleted in proof

132. Reinertsen H, Jensen A, Koksvik JI, Langeland A, Olsen Y. 1990. Effects of fish removal on the limnetic ecosystem of a eutrophic lake. *Can. J. Fish. Aquat. Sci.* 47:166–73

133. Reynolds CS. 1984. The ecological basis for the successful biomanipulation of aquatic communities. *Arch. Hydrobiol.* 130:1–33

134. Reynolds CS. 1984. *The Ecology of Freshwater Phytoplankton.* Cambridge, UK: Cambridge Univ. Press

135. Riemann B, Sondergaard M, Schierup HH, Bossel-Mann S, Christensen G, et al. 1982. Carbon metabolism during a spring diatom bloom in the eutrophic Lake Mosso, Denmark. *Int. Rev. Gesamten Hydrobiol.* 67:145–85

136. Rosenzweig ML. 1971. Paradox of enrichment: destabilization of exploitation ecosystems in ecological time. *Science* 171:385–87

137. Rosenzweig ML. 1995. *Species Diversity in Space and Time.* Cambridge, UK: Cambridge Univ. Press

138. Sanni S, Waervagen SB. 1990. Oligotrophication as a result of planktivorous fish removal with rotenone in the small, eutrophic, Lake Mosvatn, Norway. *Hydrobiologia* 200/201:263–74

139. Sarnelle O. 1992. Nutrient enrichment and grazer effects on phytoplankton in lakes. *Ecology* 73:551–60

140. Sarnelle O. 1994. Inferring process from pattern: trophic level abundances and imbedded interactions. *Ecology* 75:1835–41

141. Sarnelle O. 1996. Predicting the outcome of trophic manipulation in lakes—a comment on Harris (1994). *Freshwater Biol.* 35:339–42

142. Schindler DW, Armstrong FA, Holmgren SK, Brunkill GJ. 1971. Eutrophication of lake 227, Experimental Lakes Area, Northwest Ontario, by addition of phosphorus and nitrates. *J. Fish. Res. Board Can.* 28:1763–82

143. Schmitz OJ. 1994. Resource edibility and trophic exploitation in an old-field food web. *Proc. Natl. Acad. Sci. USA* 91:5364–67

144. Schoenberg SA, Carlson RE. 1984. Direct and indirect effects of zooplankton grazing on phytoplankton in a hypereutrophic lake. *Oikos* 42:291–302

145. Shapiro J. 1990. Biomanipulation: the next phase—making it stable. *Hydrobiologia* 200/201:15–27

146. Shapiro J. 1990. Current beliefs regarding dominance by blue-greens: the case

for the importance of CO_2 and pH. *Ver. fur theor. angew. Limnol., Verhandlungen* 24:

147. Deleted in proof

148. Shortreed KS, Stockner JG. 1986. Trophic status of 19 subarctic lakes in the Yukon Territory. *Can. J. Fish. Aquat. Sci.* 43:797–805

149. Smith FE. 1969. Effects of enrichment in mathematical models. In *Eutrophication: Causes, Consequences, Corrections.* Washington, DC: Natl. Acad. Sci.

150. Stenson JAE, Bohlin T, Henrikson L, Nilsson BI, Nyman HG, et al. 1978. Effects of fish removal from a small lake. *Verh. Int. Verein. Limnol.* 20:794–801

151. Stenson JAE, Svensson JE. 1994. Manipulations of planktivore fauna and development of crustacean zooplankton after restoration of the acidificied Lake Gardsjön. *Arch. Hydrobiol.* 131:1–23

152. Stockner JG, Shortreed KS. 1989. Algal picoplankton production and contribution to food-webs in oligotrophic British Columbia lakes, Canada. *Hydrobiologia* 173:151–66

153. Strong DR. 1992. Are trophic cascades all wet? Differentiation and donor-control in speciose ecosystems. *Ecology* 73:747–54

154. Tátrai IL, Tóth G, Istánovics V, Zlinsky J. 1990. The importance of higher trophic level in the process of eutrophication in enclosure. *Int. Rev. ges. Hydrobiol.* 75:175–88

155. Taylor WD, Hiatt FA, Hern SC, Hilgert JW, Lambou VW, et al. 1978. *Distribution of phytoplankton in Florida lakes. EPA-600/3–78–085.* Las Vegas: USEPA

156. Tilman D. 1982. *Resource Competition and Community Structure.* Princeton, NJ: Princeton Univ. Press

157. Tilman D, Pacala S. 1993. The maintenance of species richness in plant communities. In *Species Diversity in Ecological Communities* ed. RE Ricklefs, D Schluter, pp. 13–25. Chicago: Univ. Chicago Press

158. Turner AM, Mittelbach GG. 1992. Effects of grazer community composition and fish on algal dynamics. *Can. J. Fish. Aquat. Sci.* 49:1908–15

159. Vadstein O, Harkjen BO, Jensen A, Olsen Y, Reinertsen H. 1989. Cycling of organic carbon in the photic zone of a eutrophic lake with special reference to the heterotrophic bacteria. *Limnol. Oceanogr.* 34:840–55

160. Vance RR. 1978. Predation and resource partitioning in one-predator-two prey model communities. *Am. Nat.* 112:797–813

161. van der Molen DT, Boers PCM. 1996. Changes in phosphorus and nitrogen cycling following food web manipulations in a shallow Dutch lake. *Fresh. Biol.* 35:189–202

162. van Donk E, Grimm MP, Gulati RD, Klein Breteler JPG. 1990. Whole-lake food-web manipulation as a means to study community interactions in a small ecosystem. *Hydrobiologia* 200/201:275–89

163. Vanni M.J. 1987. Effects of nutrients and zooplankton size on the structure of a phytoplankton community. *Ecology* 68:624–35

164. Vanni MJ. 1996. Nutrient transport and recycling by consumers in lake food webs: implications for algal communities. In *Food Webs: Integration of Patterns and Dynamics,* ed. GA Polis, KO Winnemiller, pp. 81–95. New York: Chapman & Hall

165. Vanni MJ, Findlay DL. 1990. Trophic cascades and phytoplankton community structure. *Ecology* 71:921–37

166. Vanni MJ, Layne CD. 1996. "Top-down" trophic interactions in lakes: effects of fish on nutrient dynamics. *Ecology* 78:1–20

167. Vanni MJ, Layne CD, Arnott SE. 1996. Nutrient recycling and herbivory as mechanisms in the "top-down" effect of fish on algae in lakes. *Ecology* 78:21–40

168. Vanni MJ, Luecke C, Kitchell JF, Allen Y, Temte J, et al. 1990. Effects on lower trophic levels of massive fish mortality. *Nature* 344:333–35

169. Van Valen L. 1973. Pattern and the balance of nature. *Evol. Theor.* 1:31–49

170. Wagner KJ. 1986. *Biological management of a pond ecosystem to meet water use objectives. Lake and Reservoir Management EPA 1986 EP8.2:1 141985*

171. Watson S, McCauley E. 1988. Comparing patterns of net- and nanoplankton production and biomass among lakes. *Can J. Fish. Aquat. Sci.* 45:915–20

172. Watson S, McCauley E, Downing JA. 1992. Sigmoid relationships between phosphorus, algal biomass, and algal community structure. *Can. J. Fish. Aquat. Sci.* 49:2605–10

173. Wetzel RG. 1983. *Limnology.* Philadelphia: Saunders. 2nd ed.

174. White TCR. 1978. The importance of relative shortage of food in animal ecology. *Oecologia* 33:71–86

175. White TCR. 1993. *The Inadequate Environment: Nitrogen and the Abundance of Animals.* Berlin: Springer-Verlag

176. Wollkind DJ. 1976. Exploitation in three trophic levels: an extension allowing

intraspecies carnivore interaction. *Am. Nat.* 110:431–47

177. Wootton JT. 1995. Effects of birds on sea urchins and algae: A lower-intertidal trophic cascade. *Ecoscience* 2:321–28

178. Yan ND. 1986. Empirical prediction of crustacean zooplankton biomass in nutrient-poor Canadian Shield lakes. *Can. J. Fish. Aquat. Sci.* 43:788–96

179. Yan ND, Lafrance CJ, Hitchin GG. 1982. Plankton fluctuations in a fertilized, acidic lake: the role of invertebrate predators. In *Acid Rain and Fisheries*, ed. TA Haines, RE Johnson. Bethesda, MD: Am. Fisheries Soc.

Annu. Rev. Ecol. Syst. 1997. 28:495–516

EXTINCTION VULNERABILITY AND SELECTIVITY: Combining Ecological and Paleontological Views

Michael L. McKinney

Department of Geological Science and Department of Ecology and Evolutionary Biology, University of Tennessee, Knoxville, Tennessee 37996; e-mail: mmckinne@utk.edu

KEY WORDS: extinction, vulnerability, selectivity, risk proneness, phylogenetic, mass extinction

ABSTRACT

Extinction is rarely random across ecological and geological time scales. Traits that make some species more extinction-prone include individual traits, such as body size, and abundance. Substantial consistency appears across ecological and geological time scales in such traits. Evolutionary branching produces phylogenetic (as often measured by taxonomic) nesting of extinction-biasing traits at many scales. An example is the tendency, seen in both fossil and modern data, for higher taxa living in marine habitats to have generally lower species extinction rates. At lower taxononomic levels, recent bird and mammal extinctions are concentrated in certain genera and families. A fundamental result of such selectivity is that it can accelerate net loss of biodiversity compared to random loss of species among taxa. Replacement of vulnerable taxa by rapidly spreading taxa that thrive in human-altered environments will ultimately produce a spatially more homogenized biosphere with much lower net diversity.

INTRODUCTION

Extinction selectivity, or relative vulnerability, is of growing interest in both ecology (1, 3, 62, 93) and paleontology (49, 74, 75, 105). Despite the great difference in scale of observation in the two fields, much evidence about which groups are more likely to become extinct is consistent. Such evidence indicates a largely untapped potential for understanding why biased extinction is

495

0066-4162/97/1120-0495$08.00

common at many scales. For conservation biology, the importance of understanding extinction proneness is to "provide a basis for proactive conservation," instead of current approaches that are largely "reactive and piecemeal" (1, p. 144). The increasing scale of biodiversity loss will require proactive measures that go beyond the population scales that have characterized most of conservation biology, to examine extinction risk differences among taxa and habitats (46, 71).

Applying fossil data to modern extinction problems is desirable because they document extinctions across large scales of time and space (48, 49). Also, such data provide our only opportunity to study "natural" extinction patterns. Human impacts have been so profound that not a single case of nonanthropogenic species extinction can be documented in the last 8000 years (16). Furthermore, nearly all of today's anthropogenic causes of extinction, including habitat loss and biotic exchanges, were also major causes of extinctions in the geological past (48, 49).

I examine the best-studied biological traits that influence risk, especially those defining "generalist" taxa. I then examine ecological and fossil evidence on how these traits are nonrandomly distributed among three basic categories of biodiversity: 1. populations and species, 2. higher taxa, and 3. geographic biotas. This nonrandomness identifies three major areas of risk assessment that need urgent scrutiny by conservation planners. For example, evolution has nonrandomly nested species risk among higher taxa (62, 74).

BRIEF HISTORY OF THOUGHT ON EXTINCTION VULNERABILITY

That some groups are at greater risk of extinction is an old idea. Lamarck speculated that marine organisms were less prone to extinction than terrestrial organisms because they were more "buffered" against environmental change (15). In *Principles of Geology*, Charles Lyell (63) devoted two chapters to a surprisingly modern discussion of the selective nature of extinction. Examples include extinction of cold-intolerant plants during times of climatic cooling and the higher risk of extinction in species with narrow geographic ranges. Darwin devoted considerable attention to dominant species that resisted extinction because they were both more abundant locally and had a very wide geographic range (10). Another early theme still common in modern discussions of species risk (e.g. 78, 80) is that of generalist species. Simpson (96), for example, argued that broadly adapted species persisted longer in the geological record.

The dawning "extinction crisis" in the 1960s led to a flurry of ecological studies in the 1970s that produced a rapid growth of publications, especially in ecology, about extinction proneness. Ehrenfeld (21) produced an influential list of factors promoting species extinction risk that is still used in textbooks (45, 78, 80): specialized habitat preferences, restricted distribution, intolerance of human presence, reproduction in aggregates, low fecundity (low litter size, slow maturation, long gestation), large size and predatory habits, less adaptable behavior patterns, and excessive hunting. Another commonly cited list is Terborgh's (102): high trophic level, largest guild members, poor dispersal ability, continental endemics, island endemics, colonial nesting, and migratory habits. Terborgh's list (and many others) reflects the influence of bird studies, where many risk data continue to originate (54, 55, 56, 65). Current articles and books discussing traits extinction risk in both ecology (45, 78, 80) and paleontology (48, 88, 89) typically draw heavily, if not entirely, on these early lists.

Nonrandom Extinctions in Ecological and Geological Time

Brown (10) discussed the traditionally dichotomous approach that ecologists have taken toward extinction, regarding it as either deterministic or stochastic. A similar dichotomy has existed in paleontology, illustrated in Raup's (88) subtitle *Bad Genes or Bad Luck*. Hedrick and others (41) argued that this dichotomy is artificial and largely reflects our ignorance of the complex processes that create stochastic patterns.

My present concern is with the role of this dichotomous view in discussions of selectivity, or relative risk between groups. One of the most general deterministic ecological patterns is the similar composition of depauperate subsets found in modern faunas that represent fragments of formerly more species-rich faunas. Global warming since the last glacial episode has isolated once-continuous bird and mammal faunas on mountain ranges and on islands due to sea level rise. A number of studies (e.g. 10, 19) show that "extinction is a highly deterministic process: extinctions occurred in approximately the same sequence throughout the region, despite wide variation in extinction rates" (19, p. 496). The sequence of extinction can often be related to specific traits. Species with large body size, high trophic level, specialized habitat needs, and poor dispersal are among the most consistently extinction-prone, and they disappear from all but the largest "islands" (10, 19). These same nested patterns are also found in studies of current habitat fragmentation of modern ecosystems. Mammals (60), birds (20, 56, 102), and plants (103), for example, show that the depauperate subsets of the smallest remnant habitats tend to consist mainly of small, generalized, and widely dispersing species.

Nonrandom extinction is also very common in the fossil record. Despite persistent claims of evolutionary randomness (34), both background and mass extinctions consistently show nonrandom ecological and taxonomic patterns of extinction, discussed in detail below. A key observation by Jablonski (49, p. 34) is that modern extinctions so far "conform mainly to intensified versions of background expectations."

TRAITS PROMOTING EXTINCTION VULNERABILITY

Population extinction can be fruitfully viewed as "correlated death" of individuals (39). When population size is large enough to avoid the intrinsic extinction causes of genetic and demographic fluctuations, it generally takes one or more extrinsic environmental perturbations to cause such correlated death of a population and species (59). These perturbations can be categorized in a variety of ways such as Diamond's (20) "evil quartet" of habitat loss, species introductions, extinction cascades, or overexploitation, and they can occur over many time scales.

Figure 1 illustrates how the traits cited as promoting population extinction risk, often in both ecology and paleontology, can be separated into categories that influence the risk of population extinction via "correlated death" by external perturbations.

Individual Traits

The most basic category consists of traits of the individual organism. Most extinction-biasing traits discussed in both ecology and paleontology are in this category. Examples listed in Table 1 include specialization and large body size.

Individual traits can increase population extinction risk by their effect on (*a*) probability of death per individual, and (*b*) number of individuals. For example, other things being equal, the risk of correlated death generally increases as probability of individual death increases and mean abundance decreases. Other things are rarely equal, and the interaction among these variables can be complex, as demonstrated by life history theory (101) and population biology

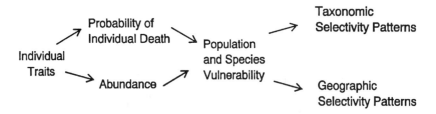

Figure 1   Individual traits influence extinction by their effect on the probability of individual death and on abundance. This translates into population and then species vulnerability. Phylogenetic nesting of species vulnerability among taxa from evolutionary constraints results in nonrandom taxonomic and geographic patterns of selectivity.

Table 1 Biological traits cited in the ecological and paleontological literature as increasing extinction risk. Symbols: $++$ = many citations identified, $+$ = at least one citation identified, $-$ = none identified so far. Citations for modern extinctions are 5, 10, 20, 21, 29, 30, 45, 61, 62, 78, 80, 84, 92, 102, 108. Citations for fossil extinctions are 2, 8, 22, 29, 48, 49, 51, 66, 72, 75, 86, 89, 95, 99, 100, 105

	Modern extinctions	Fossil extinctions
Individual Traits		
Specialization:	$++$	$++$
* Stenothermy	$++$	$++$
* Specialized diet	$++$	$++$
* High trophic level	$++$	$++$
* Symbiotic	$++$	$+$
* Large body size	$++$	$++$
* Low fecundity	$++$	$+$
* Long-lived	$++$	$+$
* Slow growth/development	$++$	$+$
* Complex morphology	$+$	$++$
* Complex behavior	$+$	$+$
* Limited mobility	$++$	$-$
* Migratory	$++$	$-$
Aquatic biotas:		
* Planktic	$+$	$++$
* Epifaunal	$+$	$++$
* Filter-feeder	$+$	$++$
- Coarse-filter feeder	$-$	$+$
* Non-benthic larvae	$+$	$+$
* Non-brooding larvae	$-$	$+$
Abundance Traits		
Low mean abundance (K):	$++$	$++$
* localized range	$++$	$++$
* low density	$++$	$++$
High abundance variation	$++$	$+$
Low intrinsic growth (r)	$++$	$-$
Seasonal aggregations	$++$	$-$
Low genetic variation	$++$	$-$
Aquatic biotas:		
* Small colonies (corals)	$-$	$+$

(84). However, Figure 1 does help differentiate between two major categories of risk traits that are usually lumped together: traits that describe the individual versus traits that describe the population.

Table 1 summarizes the individual traits most commonly found in the ecological and/or paleontological literature. It is far from exhaustive and should be viewed as an initial attempt to compare the relative frequency with which various traits are cited in the literature. General sources for this Table include many excellent reviews (and citations within them) of extinction-biasing traits cited in the caption. Additional references are discussed below.

Table 1 shows that, despite the immense differences in temporal and spatial scales of observation between ecological and geological data, there are many consistencies. Some cases in which these scales do not agree seem to stem from scale differences in sampling. Ecological data on extinction risk are much better for terrestrial than marine biota (71), whereas the reverse is true for the fossil record (49). Because space does not permit a detailed discussion of Table 1, I focus on two major themes that subsume many key practical and theoretical issues—specialization and body size.

Specialization

Many traits in Table 1 are associated with the idea that extinction-prone organisms are "specialized" or narrowly adapted. In fact, specialization is arguably the most fundamental concept in the history of thought on extinction risk. More than 20 years ago, Wilson & Willis (108) referred to "truncation of ecological guilds" as the "well-known but seldom emphasized early loss of specialists and large species." Since then, evidence for this has continued to accumulate, and specialization is found in virtually all conservation textbooks as a major trait promoting extinction (e.g. 45, 78, 80). Similarly, many paleontologists have emphasized the extinction resistance of generalist species in both background and many mass extinctions (8, 72, 75, 99, 105).

Many extinction-promoting traits are associated with being specialized (Table 1). This implies that such traits may covary. Covariation is supported by the growing evidence for Brown's "niche-breadth hypothesis" (10) that niche breadth tends to covary among niche parameters, so that a species broadly adapted in one parameter such as temperature tolerance is also broadly adapted in other parameters such as diet (10, 28, 61, 62). There is also evidence that niche narrowness is related to decreased local abundance and decreased geographic range (10, 61, 62). Specialized species may thus be prone to multiple jeopardies produced by a synergistic combination of fewer individuals with narrower tolerances to change (61, 62). Among living freshwater fish, for example, specialization is correlated with extinction-proneness by limiting geographic range and also by increasing sensitivity to anthropogenic environmental changes

(1). At paleontological scales, this may explain why tropical taxa tend to have higher evolutionary turnover rates (see below).

Explanations of extinction risk in terms of specialization have been plagued by poorly defined terms (28). Not specifying the valid taxonomic and phylogenetic scales of comparison has often confused the issue. A classic illustration is the traditional r-K paradigm, which is widely recognized as invalid at the population level for which it was originally formulated (73, 101). But Stearns (101) noted that many of the original assertions of the model are true at higher taxonomic scales of comparison because clade-level life-history and related traits tend to become embedded relatively early in development. They are thus shared by all species in the clade. Mammals, for example, are more specialized and K-selected than insects, in terms of being longer-lived, larger, and less abundant (14), and having other classic K-selected traits (many of which are listed in Table 1). This is also true of coarse comparisons within the mammals, where rodents are more generalized and r-selected than large mammals (14). Such generalizations may contribute to our understanding of why some clades (however defined) are more at risk than others.

At finer taxonomic and phylogenetic scales of comparison, we thus might not expect broad generalizations of strict covariance among traits to hold as often as generalizations made among coarse scales (101). For example, a phylogenetic analysis (6) showed that, among British birds, species with fast development tend to be abundant, as predicted by r-K. But no life history variables consistently correlated with abundance within taxa (6). Estimates of relative extinction risk should thus specify exactly the taxonomic and phylogenetic levels and the traits. The increasing use of phylogenetic analysis and the comparative method among ecologists (40) promises to do this and may greatly increase the precision of extinction-risk estimates. An excellent example is Gittleman's (31) phylogenetic analysis of the giant panda. He showed how its high extinction risk arises from having an exceptionally slow development, low fecundity, and other high-risk traits compared to related species.

The large majority of risk traits in Table 1 were deduced from data that did not include detailed phylogenetic information. Rather, the data were based on comparisons of extinction rates among related taxa. For example, extinction selectivity within bird guilds is among the best-studied patterns. Studies of tropical forest fragmentation showed that large frugivores and large insectivores are the first groups to go extinct in small fragments (56). Such ecological patterns of selectivity translate into taxonomic patterns because the species most at risk are not randomly distributed among the families studied (see below). Extinction patterns among related taxa after habitat fragmentation are also well documented in mammals (60) and plants (103): Poorly dispersing species of mammals and plants tend to go extinct sooner than widely dispersing relatives. In addition, a few generalized species tend to increase in abundance (108).

Paleontological patterns of selectivity against specialists in Table 1 have usually been, not surprisingly, derived from analyses of coarse taxonomic patterns. However, comparisons of species-level with genus- and higher-level selectivity patterns generally yield the same patterns (4, 27). Foote (26) reviewed evidence that fossil taxonomic patterns are generally valid indicators of underlying phylogenetic patterns.

Body Size

Large body size is one of the most commonly cited traits promoting extinction in both ecology (10, 45, 78, 80) and paleontology (48, 73, 105). In some cases, large body size has itself been seen as a type of specialization (97). In other cases, large size has been viewed as a correlate of specialization, as in the traditional r-K paradigm (14, 73, 101). In still other cases, the most extinction-prone species are described as the largest and most specialized of a feeding guild (55, 56, 102), indicating that large size and specialization can be independent traits.

Even when large body size is separated from other risk-promoting traits, it is not a universal predictor of extinction. Studies can be found reporting positive, negative, and no relationship between body size and probability of extinction (5). A main reason for these discrepancies is that these studies "concern a variety of taxa, in different habitats, at different spatial scales and whose extinction has been caused by different processes, and so are difficult to reconcile" (5, p. 472). These scaling differences produce different results because body size influences population abundance (29, 30, 62) and fluctuations (29, 30, 62, 84) in complex ways that differ among taxa.

We may therefore expect to find that body effects on extinction will vary among taxa. In the aquatic realm, in marine invertebrates and freshwater fish, large size apparently does not increase extinction risk. Jablonski (50) found that, at least for mollusks in the late Cretaceous (K-T) mass extinction, body size was not a factor in determining species survivorship. He suggested that this was because dispersal ability and geographic range—factors known to strongly influence molluscan extinction resistance (49)—are not correlated with body size. Similarly, small body size often promotes extinction in freshwater fish, perhaps because small fish disperse poorly (1).

In contrast, large body size in terrestrial vertebrates is often correlated with increased geographic range and many other traits that affect extinction probability (10, 11, 14, 73). In particular, two general patterns seem to emerge. One is based on the finding that, of closely related species, the larger is often more abundant, perhaps because of competitive advantages (18). We might thus predict that populations of large-bodied species are less extinction-prone than populations of closely related smaller species. Data are scarce, but this is apparently true in shrews (83) and some birds (33).

A second prediction is that, of more distantly related taxa, large body size increases extinction proneness because of the commonly discussed liabilities of large size in the "mouse-to-elephant" curves (14, 73). Much evidence, from both fossil and modern extinctions, seems to support this. In Brown & Nicoletto's study (11), 11 of 13 mammal species that disappeared from 24 local habitats in North America weighed more than 2 kg. Among birds, both globally endangered species and historically extinct species (since 1600) tend to be larger than average (5).

Similar patterns are seen at the coarse taxonomic scales of comparison in fossil vertebrate extinctions. In his overview of the K-T extinctions, Archibald (2) reported that 57% of large (over 10 kg) vertebrates went locally extinct at the Hell Creek site compared to 30% of small vertebrates. Large mammal species had lower survivorship than smaller species during the late Pleistocene megafaunal extinctions (66) and apparently during most background time (98). However, in the late Eocene extinctions, large-bodied mammal species were not selected against (106). This points out the importance of external causal mechanisms; the late Eocene extinctions were generally related to global cooling (86), and large body size is often favored in endotherms (14, 73).

Abundance Traits

Abundance can be seen as a derived trait of more basic individual traits (Figure 1). Abundance at any time is produced by the interaction of intrinsic individual traits, which determine individual survival and reproduction, with the extrinsic abiotic and biotic environment. Rarity is often cited as the single best predictor of extinction likelihood across many scales (29, 65, 100). Abundance is also easier to measure than many individual traits, such as mean body size, complexity, mobility, and developmental rate.

Gaston (29) discussed the many ways that rarity, or low mean abundance, can be described. Two of the most common ways are included in Table 1: localized geographic range and low density. Both of these metrics are strongly correlated with increased extinction rate on ecological (29, 84, 92) and geological (48, 49, 75) scales. There are many exceptions, however. A substantial number of abundant, widespread species have become extinct through human activity (92, 107).

The influence of rarity on extinction depends on taxonomic scale and the types of rarity examined. According to an exemplary study by Mace & Kershaw (65), for South African birds, small population size is a better predictor of extinction risk than small geographic range or habitat specificity. A neglected type of rarity is artificial rarity, i.e. formerly abundant species now decimated by humans. Species displaying this type of rarity may be much more likely to go extinct than species that are evolutionarily adapted to rarity (58).

The two types of rarity in Table 1 are not independent. There is evidence at many scales that geographic range and local abundance are generally correlated so that widespread species tend also to be locally dense (10, 29, 36, 61, 62). Brown's niche-breadth hypothesis is that this correlation occurs because broadly adapted species are able to exploit a wide range of resources, both locally and geographically (10). However, other factors are also likely involved in this correlation. Metapopulation dynamics, for instance, indicates that locally rare species are more likely to have a narrower geographic range because they have poorer dispersal and thus fewer source populations (36). A major implication of both the niche-breadth and metapopulation explanations is that rare species are faced with synergistic forces that may make them much more prone to extinction than abundant species (61, 62). In addition to being more localized, sparse, and relatively more specialized, rare species have more fragmented geographic ranges that can amplify population decline through metapopulation and edge effects (70).

Two other important abundance traits in Table 1 are high abundance variation and low rate of intrinsic growth. Other things being equal, a population with greater temporal variation will have a greater probability of extinction than one with less (84). But other things, such as spatial distribution, life history, and other variables, are rarely equal (61, 62). This is apparently why a consistent relationship between temporal abundance variation and extinction probability was not found in the extensive survey by Gaston & McArdle (30). The role of spatial distribution is especially evident given the recent findings that variation in spatial and temporal abundance are correlated. Species with patchy, highly variable spatial abundance also tend to have highly variable temporal abundance (10, 69). Fossil foraminifera species with higher spatial abundance variation also tend to have higher temporal abundance fluctuations (77).

Low intrinsic growth rates can contribute to extinction by reducing what Pimm (84) has called "resilience," i.e. the ability of a population to rapidly recover from disturbances. Empirical evidence is seen in Amazonian mammals; hunting has caused higher local extinction rates among primates than in artiodactyls and large rodents because of primates' lower intrinsic growth rates (7).

Population Vulnerability Versus Species Vulnerability

Conservation texts often imply that population vulnerability and species vulnerability are correlated (45, 78, 80). Even a current overview on population dynamics of extinctions has "not made a distinction between the local extinction of populations, and the global extinction of species" (62, p. 148). On the other hand, Brown (10, p. 164) warned that "it is hazardous to make sweeping predictions about species extinctions and conservation policy from studies of small populations. . .." For example, one of the primary messages of metapopulation dynamics is that species with high population turnover (extinction and recolonization) can, in theory at least, have relatively low species turnover (36–38, 77).

Three emerging lines of evidence tentatively imply that population risk is generally correlated with species risk. More cautiously, this evidence indicates that species risk is not independent of population risk.

The first line of evidence is the documentation by Harrison (37, 38) that many, perhaps most, species do not have classic metapopulation structures, consisting of population patches in an equilibrium state of local extinction and recolonization. Instead, species persistence often depends on one or a few large mainland or "source" populations that track or migrate along with environmental changes. The vulnerability of these source populations would likely be good indicators of species vulnerability. Alternatively, some mobile species maintain a very patchy population structure, with substantial gene flow within the species range, so it is essentially a single large population. Again, the vulnerability of this population to environmental change would seem to be a good indicator of species risk.

The second line of evidence is the nested, predictable sequence of population extinction discussed above, in which the same species tend to be the first to go locally extinct (10, 19). This sequence indicates that population vulnerability will ultimately translate into species extinction as losses accumulate. Thus, bird species that disappeared earliest on Barro Colorado Island also tended to have lower survival rates on the mainland (54). Similarly, species that are most at risk globally are often the first to disappear from local communities. For example, primate species are among the most imperiled mammals globally (46), and they are also among the first mammal species to disappear locally (7).

The third line of evidence is that the same traits that population extinction risk, such as specialization and rarity, also seem to explain increased extinction rates among species and higher taxa. Low intrinsic rate of population increase, long generation time, and long lifespans have been used to explain high rates of extinction (or threat) in primate populations (46) and in primate species (7).

EXTINCTION VULNERABILITY AMONG TAXA

Phylogenetic constraints can nonrandomly affect population and species extinction vulnerability (risk) in many ways among taxa because body size, abundance, niche-breadth, and many other extinction-biasing traits in Table 1 have a significant genetic (heritable) component (29, 61, 62). Evolutionary branching can therefore concentrate extinction-biasing traits into certain taxa at many taxonomic levels. Specific evidence for this includes the nonrandom taxonomic clustering of rare species in some North American bird families (68) and plant families (94), widespread species in some genera of living echinoids (74) and fossil mollusks (47), and large-bodied birds in some higher taxa (5, 6). The rise of the comparative method, especially as applied to ecological traits (40), should soon provide many crucial insights into such patterns.

Understanding phylogenetic effects on extinction vulnerability is crucial to maximizing biodiversity because, as Foote (24, 25) has shown in many fossil taxa, highly selective extinction can reduce biodiversity much more than random extinction. When species extinction is highly concentrated within certain higher taxa (or areas of morphospace), there is a greater loss of net diversity than when similar amounts of extinction are randomly distributed among species. This has troubling implications for today's extinction crisis which, as discussed next, shows considerable evidence for nonrandom extinction at many taxonomic levels.

Vulnerability Differences Among Higher Taxa

A large literature documents consistent taxonomic differences in fossil (49, 51, 75, 91, 100, 105) and modern (46, 71) species extinction rates among very high taxa such as classes or orders. Fossils indicate that these extinction patterns are often consistent both across "background" geologic time and during mass extinctions (22, 49, 51, 72, 91, 100, 105), indicating that the same groups are generally less susceptible to extinction from all possible causes.

Table 2 summarizes some of the recent data on average extinction rates in fossil taxa, expressed in terms of average species duration. There are many uncertainties in these estimates, and they are only averages of a wide range of species durations in each taxon (49). But the estimates for each taxon have been surprisingly stable since Simpson began his early estimates in the 1940s (98). Also, modeling results indicate that extinction rate differences, such as lower rates for bivalves than mammals, cannot be attributed solely to differences in fossil preservation potential (27). Finally, there is tentative agreement with modern extinction rates. Fossil primates, for example, have a relatively shorter species duration than other mammals (Table 2), in agreement with modern data that primates are more extinction-prone from human activities than are most other mammals (7, 46).

A key implication of Table 2 is that marine taxa are consistently less vulnerable than terrestrial taxa. This can be viewed statistically: The probability that all six land taxa (counting the three mammal entries as only one) would end up

Table 2   Estimated mean duration of fossil species. Except insects, invertebrate data are for marine species. Where estimates involved a range of values, the midrange value is shown

Taxon	Species duration (my)	Reference
Reef corals	25	100
Bivalves	23	100
Benthic forams	21	12,13
Bryozoa	12	43
Gastropods	10	100
Planktic forams	10	81
Echinoids	7	100
Crinoids	6.7	4
Non-marine		
Monocot plants	4	100
Horses	4	44
Dicot plants	3	100
Freshwater fish	3	100
Birds	2.5	100
Mammals	1.7	100
Primates	1	67
Insects	1.5	100

as the six highest out of 14 total entries due to random independent sorting is about 1% (76).

Are the inferences drawn from fossil data valid for modern human-caused extinction dynamics? Figure 2 plots the fossil species duration data from Table 2 against the percentage of threatened species within each taxon using a recent compilation (71). There is a fairly high correlation between the modern and fossil data. As with the fossil extinction data, worrisome biases appear in the estimates of threatened living species. In particular, both fossil (49, 89, 90) and modern (71) extinction data tend to underestimate the number of extinct or threatened species because many rare species are often not recorded. Given these caveats, the pattern in Figure 2 is that terrestrial taxa tend to be more vulnerable to extinction today, as in the past (76). This provides tentative quantitative support to the conventional wisdom that marine species are less prone to extinction (15, 82). Direct evidence indicates only four confirmed modern extinctions of marine invertebrates, all gastropods (15).

As Jablonski (51) noted, it is difficult to relate the differences in vulnerability among such higher taxa to specific biological variables. We can, however, point

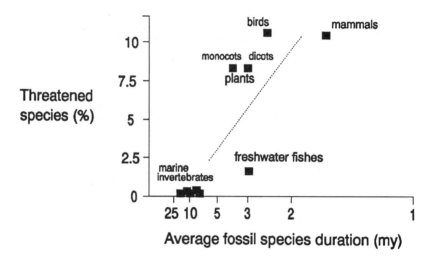

Figure 2 Proportion of species in a group that is currently threatened versus average fossil duration of species in that group. Regression line is least squares.

to direct fossil evidence that persistence of higher taxa is strongly influenced by two variables, the number of species and the resistence of each species to extinction. Increased species-richness has been documented to increase the geological persistence of many higher taxa (48, 49). Such taxa may be at considerable risk during the modern extinction crisis if they rely on high rates of speciation to replace frequently lost species, should anthropogenic reduction of speciation occur.

The other factor, differences in species extinction resistance, is important because without it, very diverse clades would never die out (89). Such differences are apparent in Figure 2, in which some higher taxa, especially marine, seem to have reduced species vulnerability. In this case, a broad biological explanation for the pattern is the common suggestion that marine species tend to have wider geographic ranges than do terrestrial species (82). Rapoport (87) quantified this in a comparison of geographic ranges between marine and terrestrial species. He derived an index of cosmopolitanism four times higher in marine than terrestrial species. This is consistent with findings that geographic range is a primary determinant of fossil species persistence in marine taxa (49, 50).

Vulnerability Among Lower Taxa

Comprehensive comparisons of species extinction rates among lower taxa such as families are often difficult. Major reasons for this include the difficulty of accurately estimating fossil and modern species extinction rates, noted above, and of accurately reconstructing phylogenetic relationships (26, 32, 40). Fossil studies suffer the additional burden that many of the traits influencing extinction-proneness (see Table 1), such as physiological and life-history parameters, are not preserved. This obstructs efforts to explain patterns of differential extinction. For example, McKinney (74) found evidence that extinction has been concentrated in certain genera of echinoids over about the last 75 my. But, aside from the fact that widespread species are clustered within certain modern echinoid genera, it was impossible to determine what caused the nonrandom fossil extinctions. More information on vulnerability among lower taxa is obviously available for living species. Phylogenetic analysis shows that carnivores tend to have higher extinction rates than most other mammals through geologic time (32).

The most useful data on taxonomic vulnerability patterns come from studies of historical anthropogenic extinctions. The findings of Kattan et al (56) show that forest fragmentation strongly impacts large frugivorous and insectivorous birds and disproportionately affects such major taxa as parrots, toucans, and antbirds (55). Kattan (55) cited other studies showing that bird families such as Columbidae were more resistant to Pleistocene extinctions than other families such as Picidae and Cracidae because of phylogenetic nesting of dietary, body-size, and abundance traits. An example among plants is that the cactus family, Cactaceae, contains a disproportionately high number of extinct and endangered species, because of the generally restricted habitat and specialized adaptations of the family (42).

A predictive approach to future vulnerability patterns has been pioneered by Russell and others (93) in a study of recent bird and mammal extinctions. Data on the extinction rate of these two groups are of the highest quality of all major groups because of popular interest in them (46). Using rarefaction techniques, Russell and others (93) found that bird and mammal species' extinctions since 1600 are nonrandomly concentrated within species-poor genera. Much of this selectivity in birds was due to the high extinction rates of endemic island birds (e.g. rails), which are often evolutionarily isolated and tend to be classified into their own genera (93). The causes in mammals seem to be more complex, with multiple types of ecological selectivity including overhunting and effects of large body size.

Russell and others (93) predicted future selectivity by applying a probability formula from Mace (64) to species now listed as threatened on the IUCN Red List (46). Future bird extinctions are predicted to have less taxonomic selectivity than in the past (93). This projected decrease in selectivity may occur because island bird faunas have already been so devastated that future extinctions will largely concentrate on continental bird populations containing more species-rich faunas (93). Mammals may show declining future selectivity (93) because widespread habitat loss on continents is replacing more selective extinction causes such as overhunting of large mammals (46).

EXTINCTION VULNERABILITY AMONG GEOGRAPHIC BIOTAS

The extinction vulnerability literature contains three major categories of habitats that are often contrasted: terrestrial-marine, island-continental, and tropical-nontropical, where the first of each pair is usually considered to contain more extinction-prone species. Such habitat, or geographic, categories are seldom independent of taxonomic extinction patterns. Evolutionary diversification within regions produces phylogenetic proximity that is often correlated with geographic proximity (9).

The terrestrial-marine pair illustrates a relationship between taxonomic and habitat vulnerability. Higher taxa in the marine environment tend to have lower extinction rates than mainly terrestrial taxa (Figure 2). This appears true for both the current extinction crisis and past geological background extinctions (76). In addition to having more widespread species, the marine environment may buffer species from rapid physical changes (82). Within the marine realm, deep-water species seem to have preferentially survived the late Devonian (72) and late Permian (22) mass extinctions. Similarly, the current extinction crisis is most strongly affecting shallow, nearshore habitats (82).

One terrestrial habitat—freshwater—deserves special mention. The very high modern extinction rates in these systems seem to be matched by relatively high turnover rates during geological time (Figure 2). However, some mass extinctions apparently showed the opposite effect—preferential survival of freshwater biota compared to marine biota. The global cooling that caused the late Devonian extinction (72), and the sea level fluctuations and meteorite impact that caused the late Mesozoic extinction (2) are both characterized by nonrandom survival of freshwater biota. This has been attributed to the ability of freshwater biota in the late Devonian to better tolerate seasonal and other physical fluctuations (72). Freshwater biotic survival of the late Cretaceous meteorite impact has been related to a detritus-based food web that buffered the effects (95). The high extinction rate of freshwater species today (46, 78, 80) is apparently because the causes are so multifaceted, including pollution, dams, exotic species, and other disturbances (78, 80), that any advantages in tolerance or buffering effects relative to marine habitats are overwhelmed.

For modern extinctions, island species are well known to have generally higher rates than continental species (45, 46, 78, 80, 84). Thus, the fossil record notably shows that islands have often provided refugia during mass extinctions, particularly for marine taxa; many higher taxa may have survived on such refugia during times of sea level lowering (52). In recent times islands near New Zealand have served as refugia to transplant ground birds threatened by European exotics (45).

Higher extinction rates of tropical species compared to nontropical species are well documented for modern times (46, 62, 71, 78). This disparity also seems to occur at geological time scales. Briggs (8) and especially Flessa & Jablonski (23) reviewed the considerable evidence from marine groups that tropical species are geologically younger and, on average, have higher extinction rates than nontropical species. This could be due to narrower ecological tolerances and/or narrower geographic ranges of tropical species (8, 23). In past mass extinctions, including the late Devonian (72), late Permian (22), late Cretaceous (2), and some smaller Cenozoic extinction events (86, 99), tropical taxa have been preferentially eliminated. As today, tropical reefs have been consistently decimated during such mass extinctions (57).

EXTINCTION FILTERS: PAST SELECTIVITY AND TODAY'S EXTINCTIONS

Ecologists increasingly appreciate that extinction patterns during the Cenozoic Era, especially since the Oligocene Epoch, may explain poorly understood vulnerability differences among living biotas (3, 17, 53). Two general areas are briefly reviewed here.

Climatic Filters

Since the late Eocene, about 40 Mya, the earth's climate has undergone cycles of cooling. While these have caused no global mass extinctions, they have caused many regional extinctions on land and in the sea (86). Stanley (99) and Briggs (8) reviewed evidence that these cycles consistently produced impoverished, broadly adapted biotas by selectively eliminating tropical biotas. The warming cycle that followed these extinctions is characterized by the evolutionary rediversification of more specialized tropical forms that were affected during the subsequent cooling episodes.

That we are currently in an impoverished phase in the oceans is seen in the evidence that the severe cooling and sea level changes of the Pleistocene had little impact on marine biota, at least in the northern hemisphere (49, 53). Coope's review (17) made a similar point about land faunas, showing that ice-age insect faunas of western Europe experienced very little extinction. A basic theme is that these filtered, broadly adapted marine and terrestrial biotas are able to migrate, with shifting geographic ranges, in response to climate change

(17, 104). The implication for impending anthropogenic global warming is that these biotas may respond in the same way.

Human Filters

Extinctions caused by humans before the modern extinction crisis are particularly important filters of current vulnerability. Perhaps the best known example is the role of humans in the geologically recent megafaunal extinctions of the late Pleistocene Epoch (66). Another example is the devastation of island bird faunas by human settlements (85). Previous loss of species that are most sensitive to human activities can, in an ironic way, be viewed as a positive sign in that the remaining species are likely to be less vulnerable (3). Some Pacific islands have experienced bird loss for thousands of years; the islands with the longest periods of human habitation have generally witnessed the fewest recent recorded bird extinctions (85). Such cryptic (unrecorded) extinctions seem also to have occurred in Mediterranean plants. Recorded historic extinction rates are much higher for plants in Australia and California than in the Mediterranean area, where intensive human settlement has existed for much longer (35).

A main implication of past filtering is that current conservation efforts should place greater emphasis on preserving areas where humans have had less impact. As Balmford (3) noted, this is where the greatest amount of relative biodiversity can be saved, especially in terms of effort expended.

SYNTHESIS: BIOTIC HOMOGENIZATION AND THE SPREAD OF WEEDY TAXA

Traits that make populations more prone to extinction include: 1. individual traits, especially those related to narrow niches, and 2. abundance traits, such as low mean abundance and abundance fluctuations. The apparent lack of pure metapopulation dynamics in most species, along with other evidence, implies that these traits promote species extinction as well as population extinction. Evidence for Brown's niche-breadth hypothesis suggests that some species are extremely vulnerable because they have synergistic combinations of extinction-promoting traits. These include specialization and rarity with fragmented ranges. Such generalizations about extinction vulnerability are most useful in specific phylogenetic, or at least taxonomic, contexts. Large body size, for example, seems to increase extinction vulnerability in mammals, but not in many taxa of marine invertebrates.

A phylogenetic view also demonstrates that traits promoting extinction are not randomly distributed among taxa at many scales. This has enormous, largely unexplored, implications for conservation biology. Among very high taxa, marine organisms seem to be much less extinction-prone both now and in the geologic past. Selectivity among lower taxa is illustrated by the biased loss of certain families of forest birds and by historical extinctions concentrated in certain genera and families of birds and mammals. Such selectivity may decrease as more widespread, catastrophic causes of extinction increase. Phylogenetic selectivity often translates into geographic selectivity as illustrated by terrestrial and tropical species, which are generally more prone to extinction now and have been in the past than are marine and nontropical species. Predicting future phylogenetic extinction selectivity allows biologists to detect patterns of vulnerability at many taxonomic scales while simultaneously considering the phylogenetic uniqueness of those taxa in making conservation decisions.

A crucial prediction of the vulnerability patterns reviewed here is that the future biosphere will clearly become progressively less diverse as the more extinction-prone taxa and subtaxa become extinct. This biodiversity loss will be increasingly magnified to the degree that phylogenetic and taxonomic selectivity are concentrated in certain groups; such selectivity produces disproportionate reduction of overall biodiversity (24, 25).

The converse of the patterns considered here is that of positive human selective impacts on some biota. Some taxa, generally categorized as commensals or weeds, not only have preferentially survived in human-altered environments but have thrived (79). The rapid spread of such taxa, in conjunction with the increasing extinction of more vulnerable taxa, will produce a biosphere that is not only less generally diverse, but also more spatially homogenized (10, 92).

ACKNOWLEDGMENTS

I thank John Gittleman and Tom Brooks for their comments. Support by the National Science Foundation and Petroleum Research Fund is gratefully acknowledged.

> Visit the *Annual Reviews home page* at
> http://www.annurev.org.

Literature Cited

1. Angermeier, PL. 1995. Ecological attributes of extinction-prone species: loss of freshwater fishes in Virginia. *Conserv. Biol.* 9:143–58
2. Archibald JD. 1996. *Dinosaur Extinction and the End of an Era.* New York: Columbia Univ. Press. 237 pp.
3. Balmford A. 1996. Extinction filters and current resilience: the significance of past selection pressures for conservation biology. *Trends Ecol. Evol.* 11:193–97
4. Baumiller TK. 1993. Survivorship analysis of Paleozoic Crioidea: effect of filter morphology on evolutionary rates. *Paleobiology* 19:304–21
5. Blackburn TM, Gaston KJ. 1994. Animal body size distributions: patterns, mechanisms, and implications. *Trends Ecol. Evol.* 9:471–74
6. Blackburn TM, Lawton JH, Gregory RD. 1996. Relationships between abundances and life histories of British birds. *J. Anim. Ecol.* 65:52–62
7. Bodmer RE, Eisenberg JF, Redford KH. 1997. Hunting and the likelihood of

extinction of Amazonian mammals. *Conserv. Biol.* 11:460–67

8. Briggs JC. 1995. *Global Biogeography.* Amsterdam: Elsevier. 452 pp.

9. Brooks DR, McLennan DA. 1993. Historical ecology: examining phylogenetic components of community evolution. In *Species Diversity in Ecological Communities,* ed. RE Ricklefs, D Schluter, pp. 267–280. Chicago, IL: Univ. Chicago Press

10. Brown JH. 1995. *Macroecology.* Chicago, IL: Univ. Chicago Press. 270 pp.

11. Brown JH, Nicoletto PF. 1991. Spatial scaling of species composition: body masses of North American land mammals. *Am. Nat.* 138:1478–512

12. Buzas MA, Culver SJ. 1989. Biogeographic and evolutionary patterns of continental margin benthic foraminifera. *Paleobiology* 15:11–19

13. Buzas MA, Culver SJ. 1991. Species diversity and dispersal of benthic foraminifera. *BioScience* 41:483–92

14. Calder WA. 1984. *Size, Function and Life History.* Cambridge, MA: Harvard Univ. Press. 387 pp.

15. Carlton JT. 1993. Neoextinctions in marine invertebrates. *Am. Zool.* 33:499–507

16. Caughley G. 1994. Directions in conservation biology. *J. Anim. Ecol.* 63:215–44

17. Coope GR. 1995. Insect faunas in ice age environments: Why so little extinction? See Ref. 61a, pp. 55–74

18. Cotgreave P. 1993. The relationship between body size and abundance in animals. *Trends Ecol. Evol.* 8:244–48

19. Cutler AH. 1991. Nested faunas and extinction in fragmented habitats. *Conserv. Biol.* 5:496–505

20. Diamond JM. 1984. "Normal" extinctions of isolated populations. In *Extinctions,* ed. MH Nitecki, pp. 191–246. Chicago, IL: Univ. Chicago Press

21. Ehrenfeld DW. 1970. *Biological Conservation.* New York: Holt, Rinehart & Winston. 421 pp.

22. Erwin DH. 1993. *The Great Paleozoic Crisis.* New York: Columbia Univ. Press. 327 pp.

23. Flessa KW, Jablonski D. 1996. The geography of evolutionary turnover: a global analysis of extant bivalves. See Ref. 51a, pp. 376–97

24. Foote M. 1993. Discordance and concordance between morphological and taxonomic diversity. *Paleobiology* 19:185–204

25. Foote M. 1996. Models of morphological diversification. See Ref. 51a, pp. 62–86

26. Foote M. 1996. Perspective: evolutionary patterns in the fossil record. *Evolution* 50:1–11

27. Foote M, Raup DM. 1996. Fossil preservation and the stratigraphic ranges of fossils. *Paleobiology* 22:121–40

28. Futuyma DJ, Moreno G. 1988. The evolution of ecological specialization. *Annu. Rev. Ecol. Syst.* 19:207–33

29. Gaston KJ. 1994. *Rarity.* London: Chapman & Hall. 205 pp.

30. Gaston KJ, McArdle BH. 1994. The temporal variability of animal abundances: measures, methods and patterns. *Philos. Trans. R. Soc. Lond.* 345:335–58

31. Gittleman JL. 1994. Are the pandas successful specialists or evolutionary failures? *BioScience* 44:456–64

32. Gittleman JL, Anderson CG, Cates SE, Luh H., Smith JD. 1997. Detecting ecological pattern in phylogenies. In *Biodiversity Dynamics: Turnover of Populations, Taxa, and Communities,* ed. M McKinney, pp. xx–xx. New York: Columbia Univ. Press. In press

33. Gotelli NJ, Graves GR. 1990. Body size and the occurrence of avian species on land-bridge islands. *J. Biogeogr.* 17:315–25

34. Gould SJ. 1996. *Full House.* New York: Harmony. 243 pp.

35. Greuter W. 1995. Extinctions in Mediterranean areas. See Ref. 61a, pp. 88–97

36. Hanski I, Kouki J, Halkka A. 1993. Three explanations of the positive relationship between distribution and abundance of species. In *Species Diversity in Ecological Communities,* ed. RE Ricklefs, D Schluter, pp. 108–116. Chicago, IL: Univ. Chicago Press

37. Harrison S. 1994. Metapopulations and conservation. In *Large-scale Ecology and Conservation Biology,* ed. PJ Edwards, RM May, N Webb, pp. 111–128. New York: Blackwell

38. Harrison S, Hastings A. 1996. Genetic and evolutionary consequences of metapopulation structure. *Trends Ecol. Evol.* 11:180–85

39. Harrison S, Quinn JF. 1989. Correlated environments and the persistence of metapopulations. *Oikos* 56:293–98

40. Harvey PH. 1996. Phylogenies for ecologists. *J. Anim. Ecol.* 65:255–63

41. Hedrick PW, Lacy RC, Allendorf FW, Soule ME. 1996. Directions in conservation biology: comments on Caughley. *Conserv. Biol.* 10:1312–20

42. Hernandez HM, Barcenas RT. 1996. Endangered cacti in the Chihuahuan Desert. *Conserv. Biol.* 10:1200–9

43. Horowitz AS, Pachut JF. 1994. Lyellian bryozoan percentages and the fossil record of the recent Bryozoan fauna. *Palaios* 9:500–5

44. Hulbert RC. 1993. Taxonomic evolution in North American Neogene horses: the rise and fall of an adaptive radiation. *Paleobiology* 19:216–34

45. Hunter ML. 1996. *Fundamentals of Conservation Biology.* Cambridge, MA: Blackwell. 482 pp.

46. IUCN. 1996. 1996. *IUCN Red List of Threatened Animals.* Gland, Switzerland: IUCN. 368 pp.

47. Jablonski D. 1987. Heritability at the species level: analysis of geographic ranges of Cretaceous mollusks. *Science* 238:360–63

48. Jablonski D. 1991. Extinctions: a paleontological perspective. *Science* 253:754–57

49. Jablonski D. 1995. Extinctions in the fossil record. See Ref. 61a, pp. 25–44

50. Jablonski D. 1996. Body size and macroevolution. See Ref. 51a, pp. 256–289

51. Jablonski D. 1996. Mass extinctions: persistent problems and new directions. *Geol. Soc. Am. Spec. Pap.* 307:1–9

51a. Jablonski D, Erwin DH, Lipps JH, eds. 1996. *Evolutionary Paleobiology.* Chicago, IL: Univ. Chicago Press

52. Jablonski D, Flessa KW. 1986. The taxonomic structure of shallow-water marine faunas: implications for Phanerozoic extinctions. *Malacologia* 27:43–66

53. Jackson JBC. 1995. Constancy and change of life in the sea. See Ref. 61a, pp. 45–54

54. Karr JR. 1990. Avian survival rates and the extinction process on Barro Colorado Island, Panama. *Conserv. Biol.* 4:391–97

55. Kattan GH. 1992. Rarity and vulnerability: the birds of the Cordillera Central of Colombia. *Conserv. Biol.* 6:64–70

56. Kattan GH, Alvarez-Lopez H, Giraldo M. 1994. Forest fragmentation and bird extinctions: San Antonio eighty years later. *Conserv. Biol.* 8:138–46

57. Kauffman EG, Fagerstrom JA. 1993. The Phanerozoic evolution of reef diversity. In *Species Diversity in Ecological Communities,* ed. RE Ricklefs, D Schluter, pp. 315–329. Chicago, IL: Univ. Chicago Press

58. Kunin WE, Gaston KJ, ed. 1997. *The Biology of Rarity.* London: Chapman & Hall. 280 pp.

59. Lande R. 1988. Genetics and demography in biological conservation. *Science* 241:1455–60

60. Laurance WF. 1994. Rainforest fragmentation and the structure of small mammal communities in tropical Queensland. *Biol. Conserv.* 69:23–32

61. Lawton JH. 1995. Population dynamic principles. See Ref. 61a, pp. 147–163

61a. Lawton JH, May RM, eds. 1995. *Extinction Rates.* Oxford, UK: Oxford Univ. Press

62. Lawton JH, Nee S, Letcher A, Harvey, P. 1994. Animal distributions: patterns and processes. In *Large-Scale Ecology and Conservation Biology,* ed. PJ Edwards, RM May, N Webb, pp. 41–58. New York: Blackwell

63. Lyell C. 1832. *Principles of Geology.* London: Murray

64. Mace GM. 1994. Classifying threatened species: means and ends. *Philos. Trans. R. Soc. London* 344:91–97

65. Mace GM, Kershaw M. 1997. Extinction risk and rarity on an ecological timescale. In *The Biology of Rarity,* ed. WE Kunin, K. Gaston, pp. 130–149. London: Chapman & Hall

66. Martin PS. 1984. Prehistoric overkill: the global model. In *Quaternary Extinctions,* ed. PS Martin, R Klein, pp. 354–403. Tucson: Univ. Ariz. Press

67. Martin RD. 1993. Primate origins: plugging the gaps. *Nature* 363:223–34

68. Maurer BA. 1991. Concluding remarks: birds, body size and evolution. *Acta XX Congr. Int. Ornith.* 85:835–37

69. Maurer BA. 1994. *Geographical Population Analysis.* Oxford: Blackwell. 130 pp.

70. Maurer BA, Nott MP. 1997. Geographic range fragmentation and the evolution of biological diversity. In *Biodiversity Dynamics: Turnover of Populations, Taxa, and Communities,* ed. M McKinney, pp. xx-xx. New York: Columbia Univ. Press. In press

71. May RM, Lawton JH, Stork NE. 1995. Assessing extinction rates. See Ref. 61a, pp. 1–24

72. McGhee GR. 1996. *The Late Devonian Mass Extinction.* New York: Columbia Univ. Press. 302 pp.

73. McKinney ML. 1990. Trends in body size evolution. In *Evolutionary Trends,* ed. KJ McNamara, pp. 75–118. Tucson: Univ. Arizona Press

74. McKinney ML. 1995. Extinction selectivity among lower taxa: gradational patterns and rarefaction error in extinction estimates. *Paleobiology* 21:300–13

75. McKinney ML. 1997. How do rare species avoid extinction? A paleontological view. In *The Biology of Rarity,* ed. KJ

Kunin, WE Gaston, pp. 110–129. London: Chapman & Hall

76. McKinney ML. 1997. Is marine biodiversity at less risk? *Biodiversity Letters.* In press

77. McKinney ML, Allmon WD. 1995. Metapopulations and disturbance: from patch dynamics to biodiversity dynamics. In *New Approaches to Speciation in the Fossil Record,* ed. D Erwin, R Anstey, pp. 123–183. New York: Columbia Univ. Press

78. Meffe GK, Carroll CR. 1994. *Principles of Conservation Biology.* Sunderland, MA Sinauer. 600 pp.

79. Morris DW, Heidinga L. 1997. Balancing the books on biodiversity. *Conserv. Biol.* 11:287–89

80. New TR. 1995. *An Introduction to Invertebrate Conservation Biology.* Oxford, UK: Oxford Univ. Press. 194 pp.

81. Norris RD. 1991. Biased extinction and evolutionary trends. *Paleobiology* 17:388–99

82. Norse E. ed. 1993. *Global Marine Biological Diversity.* Washington, DC: Island. 308 pp.

83. Peltonen A, Hanski I. 1991. Patterns of island occupancy explained by colonization and extinction rates in shrews. *Ecology* 72:1698–708

84. Pimm SL. 1991. *The Balance of Nature?* Chicago, IL: Univ. Chicago Press. 434 pp.

85. Pimm SL, Moulton MP, Justice LJ. 1995. Bird extinctions in the Pacific. See Ref. 61a, pp. 75–87

86. Prothero DR. 1994. *The Eocene-Oligocene Transition.* New York: Columbia Univ. Press. 291 pp.

87. Rapoport EH. 1994. Remarks on marine and continental biogeography: an areographical viewpoint. *Philos. Trans. R. Soc. Lond.* 343:71–78

88. Raup DM. 1991. *Extinction: Bad Genes or Bad Luck?* New York: Norton. 210 pp.

89. Raup DM. 1994. The role of extinction in evolution. *Proc. Nat. Acad. Sci.* 91:6758–63

90. Raup DM. 1996. Extinction models. See Ref. 51a, pp. 419–36

91. Raup DM, Boyajian GE. 1988. Patterns of generic extinction in the fossil record. *Paleobiology* 14:109–25

92. Rosenzweig ML. 1995. *Species Diversity in Space and Time.* Cambridge, UK: Cambridge Univ. Press. 436 pp.

93. Russell GJ, Brooks TM, McKinney ML, Anderson CG. 1997. Decreased taxo-

nomic selectivity in the future extinction crisis. *Conserv. Biol.* 11: In press

94. Schwartz MW. 1993. The search for patterns among rare plants: Are primitive plants likely to be rare? *Biol. Conserv.* 64:121–27

95. Sheehan PM, Coorough P, Fastovsky DE. 1996. Biotic selectivity during the K/T and Late Ordovician extinction events. *Geol. Soc. Am. Spec. Pap.* 307:477–89

96. Simpson GG. 1953. *The Major Features of Evolution.* New York: Columbia Univ. Press. 289 pp.

97. Stanley SM. 1973. An explanation for Cope's Rule. *Evolution* 27:1–26

98. Stanley SM. 1979. *Macroevolution.* New York: WH Freeman. 211 pp.

99. Stanley SM. 1990. Delayed recovery and the spacing of major extinctions. *Paleobiology* 16:401–14

100. Stanley SM. 1990. The general correlation between rate of speciation and rate of extinction. In *Causes of Evolution,* ed. RM Ross, WD Allmon, pp. 103–127. Chicago, IL: Univ. Chicago Press

101. Stearns SC. 1992. *The Evolution of Life Histories.* Oxford, UK: Oxford Univ. Press. 234 pp.

102. Terborgh J. 1974. Preservation of natural diversity: the problem of extinction prone species. *BioScience* 24:715–22

103. Turner IM, Chua KS, Ong JY, Soong BC, Tan HT. 1996. A century of plant species loss from an isolated fragment of lowland tropical rainforest. *Conserv. Biol.* 10:1229–44

104. Valentine JW, Jablonski J. 1993. Fossil communities: compositional variation at many time scales. In *Species Diversity in Ecological Communities,* ed. RE Ricklefs, D Schluter, pp. 341–49. Chicago, IL: Univ. Chicago Press

105. Van Valen LM. 1994. Concepts and the nature of natural selection by extinction: Is generalization possible? In *The Mass Extinction Debates,* ed. W Glen, pp. 200–216. Stanford, CA: Stanford Univ. Press

106. Van Valkenburgh B. 1994. Extinction and replacement among predatory mammals in the North American Late Eocene and Oligocene. *Hist. Biol.* 8:129–50

107. Vermeij GJ. 1993. Biogeography of recently extinct marine species: implications for conservation. *Conserv. Biol.* 7:391–97

108. Wilson EO, Willis EO. 1975. Applied biogeography. In *Ecology and Evolution of Communities,* ed. ML Cody, JM Diamond, pp. 522–34. Cambridge, MA: Belknap

Annu. Rev. Ecol. Syst. 1997. 28:517–44

TREE-GRASS INTERACTIONS IN SAVANNAS[1]

R. J. Scholes
Division of Water, Environment and Forest Science and Technology, CSIR, PO Box 395, Pretoria 0001, South Africa; e-mail: bscholes@csir.co.za

S. R. Archer
Department of Rangeland Ecology and Management, Texas A&M University, College Station, Texas 77843-2126, USA; e-mail: sarcher@vms1.tamu.edu

KEY WORDS: competition, facilitation, herbivory, stability, fire

ABSTRACT

Savannas occur where trees and grasses interact to create a biome that is neither grassland nor forest. Woody and gramineous plants interact by many mechanisms, some negative (competition) and some positive (facilitation). The strength and sign of the interaction varies in both time and space, allowing a rich array of possible outcomes but no universal predictive model. Simple models of coexistence of trees and grasses, based on separation in rooting depth, are theoretically and experimentally inadequate. Explanation of the widely observed increase in tree biomass following introduction of commercial ranching into savannas requires inclusion of interactions among browsers, grazers, and fires, and their effects on tree recruitment. Prediction of the consequences of manipulating tree biomass through clearing further requires an understanding of how trees modify light, water, and nutrient environments of grasses. Understanding the nature of coexistence between trees and grass, which under other circumstances are mutually exclusive or unequal partners, yields theoretical insights and has practical implications.

INTRODUCTION

The term savanna has been widely used and variously defined. The prevailing ecological usage denotes communities or landscapes with a continuous grass layer and scattered trees. This mixture of contrasting life-forms, coupled with

[1] Send reprint requests to Dr. S. Archer at above address.

517

strong alternation of wet and dry seasons in tropical and subtropical regions, distinguishes savanna structure and function from that of forest, grassland, and desert biomes (35, 80, 148, 160). Savannas are among the most striking vegetation types where contrasting plant life forms co-dominate. They are geographically extensive and socioeconomically important in tropical (177, 208) and temperate (4, 40, 105) regions. Tropical savannas cover about 1600 M ha (147), an eighth of the global land surface; they cover over half the area of Africa and Australia, 45% of South America, and 10% of India and Southeast Asia (196). Temperate savannas in North America occupy over 50 M ha (105). More importantly, savannas contain a large and rapidly growing proportion of the world's human population and a majority of its rangelands and livestock.

Savannas have been broadly subdivided based on the stature, canopy cover, and arrangement of woody elements (40, 49, 88, 137). "Savanna grasslands" contain widely scattered trees or shrubs, and these may grade into "tree savanna," "shrub savanna," or "savanna woodland." "Savanna parkland" is a two-phase mosaic landscape in which circular clumps, groves, or "mottes" of woody plants (discrete phase) are dispersed throughout a grassy matrix (continuous phase) (13, 111, 136, 203). In some regions, the wooded patches in these two-phase mosaics occur as linear bands arrayed parallel to slope contour lines (114, 176, 202). The phrase "savanna landscape" denotes areas where savanna vegetation is dominant but may be interspersed with riparian or gallery forest, or patches of woodland, swamps, or marshes.

The origin, age, nature, and dynamics of savannas are not well understood. The spatial pattern and relative abundance of grasses and woody plants in savannas are dictated by complex and dynamic interactions among climate, topography, soils, geomorphology, herbivory, and fire (15, 186). These interactions may be synergistic or antagonistic and may reflect stochastic variation or positive feedbacks. In addition, some savanna vegetation has undoubtedly been derived and maintained by prehistoric, historic, or recent human activities. In many areas, "natural" and anthropogenic factors interact, making it difficult to identify, isolate, or quantify the key determinants of savanna structure.

In this review we focus on the ecological processes that regulate the balance between woody plants and herbaceous vegetation. Postulated mechanisms and conceptual models of life-form interactions are evaluated using field observations and experimental evidence. We use "savanna" to refer to mixed tree-grass communities on well-drained soils; and we exclude treeless or nearly treeless grasslands, marshes, bogs, and sites with seasonally flooded soils. The word "tree" is used as shorthand for the more general phrase "woody plant," which includes arborescents, fruticose shrubs, and long-lived lignified forbs. "Grass" includes both the true grasses (Poaceae) and the sedges (Cyperaceae). In both cases the terms are used to describe life-forms rather than species.

HUMANS IN SAVANNAS

Much of the rich history of hominoid evolution has occurred in savannas (72), and activities of humans have influenced the structure and function of savanna ecosystems. Deliberate use of fire by hominids (by 2.5 MYA) likely increased the fire frequency in African savannas (38). Aboriginal fires in Australia began at least 40,000–60,000 years BP, and perhaps as long as 140,000 years BP (16). Fire was undoubtedly used extensively by prehistoric humans in North and South America as well (162). Current patterns of vegetation distribution across savanna landscapes may still reflect activities of prehistoric inhabitants. For example, the distinct communities dominated by *Acacia tortilis* that occur within *Burkea africana* savannas of South Africa appear to have developed on Iron Age settlement sites (32). In Kenya, past distributions of temporary corrals used by nomadic pastoralists markedly influence patterns of *A. tortilis* distribution and recruitment (131).

Human population growth and widespread Anglo-European expansion and settlement in the eighteenth and nineteenth centuries have influenced tree-grass mixtures worldwide. Extensive clearing of trees for fuel, lumber, and cropland have fragmented forests and produced anthropogenic or degraded savannas (63, 154, 177, 208). In other areas, fire suppression, overhunting or eradication of indigenous savanna animals, and the introduction of livestock and exotic trees have caused herbaceous degradation and a progressive increase in woody plant density, known as bush or brush encroachment (1, 8, 65, 113, 119). As a result, areas that were once forest may now be savanna-like, while areas that were once grassland or open savanna may now be shrublands or woodlands with little grass biomass. Documentation of historical changes has been facilitated by spatially explicit reconstructions from ground and aerial photographs, analysis of stable isotopes of soil carbon, dendrochronology, and biogenic opals (11).

STUDIES OF WOODY PLANT-GRASS INTERACTIONS

The Effect of Woody Plants on Grasses

The presence of woody plants can alter the composition, spatial distribution, and productivity of grasses in savannas. The effects of trees on grasses ranges from positive to neutral to negative, and these may depend on (*a*) the ecophysiological or specific characteristics of the tree and grass growthforms (canopy architecture, rooting patterns), photosynthetic pathway (C_3, C_4, CAM), photosynthetic habit (evergreen, deciduous), and resource requirements (light, water, nutrients); (*b*) availability of resources as influenced by interannual variability in the amount and seasonality of precipitation and topoedaphic properties; (*c*) extent of selective grazing, browsing, or granivory; and (*d*) frequency, intensity, and extent of disturbances such as fire. Though not well quantified, the nature of tree-grass interactions can change with time. At decadal time scales, tree aging and factors that influence tree size and density must be considered.

EFFECTS OF ISOLATED TREES ON HERBACEOUS SPECIES COMPOSITION   At the scale of the tree, the species composition of the herbaceous layer may change along gradients extending from the bole to the canopy drip-line and into the adjoining inter-tree zone. In subtropical or temperate systems, C_3 grasses and herbaceous dicots may occur primarily beneath tree canopies, with C_4 grasses dominating patches beyond the canopy (61, 75, 127, 151). Herbaceous species respond individualistically to tree influences. Some grasses may be ubiquitously distributed beneath and between tree canopies, whereas others may congregate at the drip-line or have clear affinities for one microhabitat or another (10, 75, 94, 100, 195). Differences in species composition under and away from savanna trees are more distinct in low than in high rainfall zones (25, 27), suggesting that environmental gradients are stronger in habitats where effects of the radiant energy regime or root competition have a greater influence on species interactions. Grazing and browsing pressure may also alter patterning of herbaceous vegetation in savannas. In heavily grazed savannas, few differences between tree-crown and grassland zones may occur (28). Conversely, browsing of trees with low, dense, evergreen canopies can enhance morning and afternoon light levels, facilitate establishment of unique grasses, and increase total herbaceous biomass beneath the tree canopy (61). In other cases, browsing may stimulate tree growth and thus suppress grasses (164).

EFFECTS OF ISOLATED TREES ON HERBACEOUS PRODUCTION   Savanna trees affect herbaceous phenology, production, and biomass allocation (root/shoot and leaf/stem) as well as species composition. Trees have historically been viewed as competitors with grasses, especially in temperate zones, and are widely regarded as having a negative impact on herbaceous production, particularly where livestock production is a primary land use. However, well-documented exceptions to this generalization indicate that assessments should be made on a case-by-case basis. The productivity of areas under tree canopies may be enhanced by improved water and nutrient status but be suppressed by low irradiance and competition between trees and grasses for belowground resources. Thus, at the scale of the individual tree, the net effect on grass production can be negative, neutral, or positive and can change with tree age or size and density. As with species composition, the relationship is influenced by a variety of factors including grazing, browsing, and rainfall.

Experimental evidence for positive interactions, or facilitation, among plants has increased markedly during the past 10 years, and many of the examples involve trees, shrubs, and herbaceous vegetation (45). The beneficial effects of trees on grasses can occur via amelioration of harsh environmental conditions, alteration of substrate characteristics, or increased resource availability (26, 45). For example, increased herbaceous production beneath tree canopies in a Kenyan savanna was associated with lower soil temperatures, lower plant water stress, and greater soil organic matter concentrations, mineralizable N, and microbial biomass compared to those of adjacent grassland away from tree canopies (25, 195). Experimental manipulations suggest that factors related to soil fertility (23) and amelioration of radiant energy regimes (124) variously interact to influence herbaceous production. The magnitude of tree enhancement of grass growth varies with annual rainfall. In oak (*Quercus douglasii*) savannas in the western United States, herbaceous production is enhanced under tree canopies in drier regions but is reduced under tree canopies where annual rainfall exceeds 500 mm (104). In drier regions of Kenya (450 mm annual rainfall), tree species (*Acacia tortilis* and *Adansonia digitata*) had a similar and positive influence on herbaceous production (95% higher under trees than in the open). However, on a more mesic site (750 mm annual rainfall), tree canopy enhancement of herbaceous productivity was substantially diminished and differed with tree species (52% higher under *A. tortilis* canopies than in the open, but only 18% higher under *A. digitata*) (27). While these results support the hypothesis that facilitation is most likely to occur in stressful environments (45), a comparison of 10 studies showed that enhancement or suppression of grass production by tree canopies was unrelated to annual rainfall (117).

Tree roots may extend well beyond their canopies, the extent depending upon tree species, tree age/size, soil type, and annual rainfall. Soil moisture depletion (42) and trenching studies (5, 6) indicate that shallow, lateral roots can be important in water uptake and maintenance of leaf area and transpiration, even in trees widely regarded as phreatophytic. It is thus possible that elevated herbaceous production beneath tree canopies is partially the result of reductions in herbaceous growth resulting from root competition beyond the canopy, where the benefits of enriched soil nutrients and amelioration of the radiant energy are minimal. In southern Texas savanna parklands, herbaceous biomass was reduced within tree clusters but did not vary with distance from canopy edge nor with size of cluster, suggesting that tree effects did not extend beyond the canopy zone. The evergreen arborescent *Juniperus virginiana* also suppresses herbage biomass markedly beneath its canopy but has little impact on biomass away from its canopy over tree heights ranging from 2 to 6 m (57). In contrast, removal of *J. pinchotii* plants elicited a significant herbaceous response up to 6 m from canopy margins in shallow, rocky soils, but such removal had little influence on herbaceous production beyond the canopied zone on deeper soils (56). Tree-root exclusion on a *Burkea africana* site in South Africa resulted in increased grass height and basal area over a two-year period (93), and *Acacia karroo* trees suppressed grass growth up to 9 m from their canopies (164). Tree root effects on herbaceous production may also be mediated by rainfall amount. Trenching experiments in a Kenyan savanna, where tree roots extended up to 9 m beyond crowns, failed to demonstrate a statistically significant release in herbaceous production under or away from tree canopies at a low rainfall site (23). However, trenching of the same tree species at a high rainfall site stimulated herbaceous production 16–36%.

TREE SIZE AND HERBACEOUS PRODUCTION   The magnitude of differences in herbaceous production under, versus away from, savanna trees can also vary with tree age, size, and density. Time of tree occupation will influence soil properties, and microclimate can change as the tree canopy develops (10). The positive effects noted above may therefore not occur until many years after tree establishment. In some cases, the negative effects of trees on grasses may not be apparent until plants reach a critical age/size (46, 57, 61, 108, 164). In other cases suppression of herbaceous production may be comparable over a wide range of tree sizes (5–70 m^2 canopy area) (73), despite significant differences in microclimate and rainfall interception (74). Thus, tree effects can be over-ridden by other site or environmental effects. When trees are young and small, facilitation may be more important than competition, and grass production is enhanced; as trees and shrubs become larger, competition may overshadow facilitation and adversely affect herbaceous production (2, 10). At the landscape level of resolution, grass composition and production are also affected by tree density as well as tree size/age. As described in the subsequent paragraphs, there is often a strong, negative correlation between tree density or cover and grass cover or biomass. Because of the potentially positive effects of trees on grasses, herbaceous diversity and production may be greater where there are a few trees than where there are no trees, but the trend is reversed at high tree densities (46, 151, 165).

TREE DENSITY AND HERBACEOUS PRODUCTION   Encroachment of trees and shrubs into what were formerly sparsely treed grasslands has resulted from commercial livestock grazing over the past two centuries in semi-arid savanna regions (8, 10, 12, 119). As woody plant cover or density increases, grass production typically declines dramatically. Encroachment of trees may further intensify grazing pressure, as landholders seldom destock in response to decreases in grass production associated with increases in tree density. Because bush encroachment may lead to failure of commercial ranching systems, landowners commonly engage in extensive bush clearing. The bush encroachment problem has thus been the main impetus for research into tree-grass interactions in savannas. Herbaceous production generally increases after the removal of trees, except in some low rainfall zones (20) or where grazing or past land use has limited the availability of perennial grass propagules or favors exotic weeds and shrubs (24). Factors affecting herbage response to tree-clearing include initial tree size or age-class distribution, density, method of tree removal, condition and composition of understory vegetation at the time of treatment, and posttreatment climate, grazing, and browsing (71, 149).

The negative effect of trees on grasses may result from rainfall interception, litter accumulation, shading, root competition, or a combination of these factors; and the effect depends on the leaf area, canopy architecture, and rooting patterns of the tree. Where trees were removed, the typical relationship between grass production and tree biomass, cover, or basal area was a negative exponential, with the steepest decline in grass production resulting from the initial increments of tree cover (41, 129, 190). Similar relationships between grass growth and tree cover were found in response to naturally occurring variations in tree density (33, 85, 115, 127, 151).

The degree of curvature in the relationship between grass growth and tree biomass is mediated by environmental conditions, resource availability, and the understory composition. For example, the relationship can be nearly linear on highly productive sites (or in wet years on less productive sites), becoming progressively more concave on marginal sites and in drier years (141, 191). The relationship may also be linear on ungrazed sites, shifting to nonlinear when preferential grazing shifts the balance in favor of woody plants by reducing grass competition and fire frequency (110). The functional form of the relationship may also differ for C_3 vs. C_4 grasses (positive vs. negative correlation—127) and for grass, forb, shrub, or total understory production (linear vs. nonlinear negative correlation—181). Inconsistencies in overstory-understory relationships may therefore be explained when growth-form, site, and environmental factors are considered (106).

The response of grasses to tree removal is typically greatest by those growing on patches with physicochemical properties that are enhanced by trees (170, 171). Grass growth after clearing may be further augmented by nutrients released from decomposing tree residues. As a result, the magnitude of "release" of herbaceous production after tree removal may be greatest on landscapes with high cover of mature, large woody plants. Relative to those not influenced by trees, soils enriched by trees have the potential to sustain higher levels of grass production long (>13 yr) after tree removal (173, 174). However, this potential may not be realized because trees and shrubs may rapidly (within 3 to 10 years) recolonize the site and depress herbaceous production to pre-clearing levels (71, 75). For sprouting woody plants, stem densities may increase above pretreatment levels and cause a shift in tree composition to species less desirable for livestock and wildlife (62). In other cases, bush clearing has successfully enhanced wildlife diversity (29) and fostered co-existence of livestock and wildlife (150). Past bush clearing activities have too often ignored impacts on nontarget plants, noncommercial or nontarget herbivores, and system-level properties and processes.

The Effect of Grasses on Woody Plants

WOODY PLANT ESTABLISHMENT   For woody plants with potentially long lifespans and low post-establishment mortality rates, seedling recruitment is a critical life-history stage. The effects of herbaceous vegetation on woody plant recruitment is variable, and multiple mechanisms can operate in complex ways to influence emergence and establishment (53, 54). Grasses may regulate woody plant recruitment directly (competition for light, water, nutrients) or indirectly (as fine fuel loads influence fire frequency and intensity). There is much evidence for a strong negative effect of grasses on trees still small enough to be within the grass layer or the flame zone of grass-layer fires. However, in the absence of grazing, browsing, and fire, changes from open- to closed-bush savanna can occur rapidly (135). Thus, while grasses may reduce emergence, growth, and survival of woody seedlings, the competitive reduction may not be large enough to cause high mortality or complete exclusion. For example, although survival of an evergreen tree (*Juniperus virginiana*) was highest in grazed pastures (57%), survival of seedlings in pastures that had not been grazed for >50 years was still 40% (145). In South African savannas, reductions in grass competition did little to affect seedling establishment of *Acacia karroo* (120). Results from field experiments indicate that in the absence of fire, woody plant encroachment in savannas can be high across a wide range of herbaceous composition and biomass (9). Tree recruitment may occur during periods of moisture availability, when competition from grasses is minimal. However, the fine fuels which accumulate during these periods of high moisture may also dispose the system to fire (166) and thereby limit tree recruitment (70). Where grazers consume grass and remove fine fuels, fire frequency declines (17, 138) and woody plant abundance increases (98). As woody plant abundance increases and further suppresses grass production, a positive feedback develops, making fire increasingly unlikely. Thus, while grass biomass may interact negatively with woody stand development in savannas, it may be a determinant to tree density and stature primarily as it influences the fire regime.

Many savanna trees are well-adapted to fire and readily regenerate vegetatively or from seed. For some species, tolerance to fire can be observed in very young (2–3 years of age—204) or small plants. In these cases, fire may suppress but not necessarily eliminate trees from savannas (77, 179). Many apparently treeless or sparsely tree-covered grasslands may contain many woody plants concealed among the grasses and these may be decades old with well-developed root systems (67, 69, 112). Reduction in browsing pressure (21) or suppression of grass growth and fires for a number of years, for instance by grazing and drought, allows suppressed trees to quickly escape the grass layer and achieve vertical dominance (36, 180).

GROWTH AND REPRODUCTION OF ADULT TREES   Few experiments have assessed the impact of grasses on the growth and reproduction of mature trees. Analysis by Hoffman et al (78) of a 17-year record of grass and shrub basal cover suggests that increased summer rain would increase grass cover and that increased grass cover would decrease shrub cover. In Patagonian steppe, removal of shrubs did not alter grass production, but removal of grasses increased deep soil water, shrub leaf water potential, and shrub production (133). At Nylsvley, South Africa, an experiment was done in two adjacent savannas with soils differing in nutrient content and water-holding capacity (93). The nutrient-rich savanna with fertile, fine-textured soil was dominated by widely spaced *Acacia tortilis* trees; the coarse-textured, nutrient-poor site was a savanna woodland dominated by *Burkea africana*. Grass removal had no significant effect on *Burkea* growth but led to increased *Acacia* trunk radial growth and twig extension. The lack of *Burkea* response suggests that on coarse-textured sites, soil moisture recharge to deeper portions of the soil profile being used by tree roots occurs whether or not grasses are present. Conversely, grasses were capable of limiting water recharge of deeper soils on the fine-textured site, thus reducing *Acacia* growth. Stuart-Hill et al (164), working on a heavier clay soil, noted an even stronger response in *Acacia karroo* tree growth to grass clearing (40–166% increase). Together, these studies suggest that the outcome of the interaction of tree and grass life-forms with contrasting rooting patterns is regulated by the interaction of rainfall amount, seasonality, soil texture, and grass cover. The fact that tree sizes and densities on savanna landscapes are often highest in intermittent drainages characterized by fine-textured soils seems contradictory to the trend for greater sizes and densities on coarse-textured uplands. However, in cases of intermittent drainage, the additional moisture and nutrients received as run-off from upland portions of the landscape apparently override soil texture constraints.

Grass species differ greatly in their capabilities for resource use and extraction. As a result, the intensity of grass effects on trees will relax or intensify as grass composition changes in response to climatic fluctuation, succession, or grazing. Where grasses are heavily grazed, woody plant longevity (197), resource acquisition (44) and growth, time to reproductive maturity, and seed output may increase relative to lightly grazed sites (9, 109). In addition, the inverse relationship between grass production and tree canopy cover may shift from linear to a negative exponential, as the capacity for grasses to sequester resources is diminished by grazing (110).

Tree-Tree Interactions

If competition within life-forms is stronger than between life-forms, coexistence of grasses and trees is theoretically possible. Tree-tree interactions likely determine density and pattern of woody plant distribution across savanna landscapes and, hence, patterns of grass biomass and distribution. Even so, there have been surprisingly few experimental studies investigating the mechanisms that influence how woody plants interact with each other in savannas.

Competition between trees in savannas has generally been inferred from field studies of the spacing and size of trees (68, 126, 158, 159). The sum of the size of neighboring savanna trees is generally positively correlated with the distance between them, suggesting a competitive interaction. Self-thinning in *Prosopis glandulosa* stands has been inferred from comparison of different-aged stands on similar soils. Young stands (median stem age is 27 years) were characterized by high densities of small plants, whereas older stands (median stem age is 42 yr) were characterized by low densities of large plants (194). Thinning may reflect competition among lateral roots, as has been observed in desert shrublands (99), or it may start as belowground competition for soil resources in high density patches and give way to competition for light, which leads to a regular pattern of dominant surviving trees (91).

Savanna tree distribution can be over-dispersed (a quasi-regular spacing, suggesting competition), clumped (suggesting facilitation), or random. Within a single savanna plot, some species may be over-dispersed while others are clumped, and there may be statistically significant associations or disassociations between species. Clumped tree dispersal also occurs in response to localized variation in topography (e.g. termite mounds), soil depth (10, 134, 136), and fire patchiness (77). Competition between trees can occur in most open savannas but is influenced by topoedaphic factors and is often counterbalanced by elements of facilitation, especially during the phase in which tree seedlings are established. Experimental investigations of tree-tree interactions are required if we are to increase our understanding of mechanisms influencing or controlling patterns of tree spacing and density in savannas.

Is the Tree-Grass Mix in Savannas Stable or Unstable?

The widespread and persistent occurrence of savannas suggests some form of stability. The phenomenon of bush-encroachment suggests meta-stability. The asymmetry of the competitive effect of trees on grass in relation to the effect of grass on trees implies instability.

These viewpoints are not necessarily incompatible because they are based on observations at different scales in space and time (157). Competition experiments are typically conducted on a few individual plants at a scale where the competitive interaction may be unstable, with favoring of mature trees over grasses, but grasses over immature trees. In a community patch a few hectares in extent, weakening the suppressive effect of the grass layer on young trees for a few years can lead to open savannas being converted to tree-dominated thickets. Once established, the thicket may take decades to revert to an open savanna, if it ever does. At the scale of a whole landscape, savannas can persist over periods of millennia, since the landscape consists of many patches in different states of transition between a grassy dominance and a tree dominance.

MODIFICATION OF THE ENVIRONMENT BY WOODY PLANTS

Environmental modifications caused by trees and shrubs have been widely investigated in arid and semiarid systems (74). These modifications take many forms, some of which have positive effects (facilitative) and others negative (competitive). Where the net outcome is positive, the overall interaction is said to be facilitation; where it is negative, the interaction is competitive. Since the strength of the modifications decrease with distance from the plant, a mixed tree-grass community consists of a spatial patchwork of different degrees of competition and facilitation (56, 183). Furthermore, the nature of the interaction may vary with time and climatic fluctuation (93, 198) or fire-herbivore interactions (55). Environmental conditions may favor trees in some years, grasses in other years. Trees may have a net facilitative effect on grasses some years and a net competitive effect in other years.

Water

All savannas are water-limited for at least part of the year. It has therefore been widely assumed that competition in savannas is for water. However, some grasses are better adapted to subcanopy environments than to environments away from trees (3). The input of rainfall to the subcanopy habitat via stemflow and throughfall will depend on the size and intensity of rainfall events and the size, bark characteristics, canopy architecture, and leaf area of the tree (74). Interception losses on the order of 5% to >50% of gross annual rainfall have been reported (94, 118, 169), but stemflow can concentrate significant amounts of moisture near the base of trees (128, 169). As a result, measurements of soil moisture on transects from the stem outward into the between-canopy zone generally show peaks of wetness close to the stem (due to stemflow) and at the canopy perimeter (due to edge drip), with a drier area beneath the main canopy (163, 165).

Proliferation of fine feeder roots may enable woody plants to monopolize near-surface soil moisture concentrated via stemflow (207). Deep percolation of stemflow water, facilitated by large root channels and low bulk densities associated with subcanopy soils, could also favor deeply rooted woody plants. Duff or litter layers accumulating under trees can reduce the amount of moisture reaching mineral soil and, in conjunction with reduced light levels, limit herbaceous production. In other cases, modest reductions in radiant energy associated with tree shading can lower soil temperatures, reduce evaporative demand and water stress on understory plants, and enhance subcanopy soil moisture storage, availability, and plant water-use efficiency (3, 25, 87, 94, 117). Through the process of hydraulic lift, deeply rooted trees and shrubs may increase soil moisture to shallow-rooted understory species (43, 52).

The interspersion of grasses and woody plants also alters the hydrology of the savanna at a landscape scale by influencing horizontal patterns of water distribution (86). On loamy and silty soils, infiltration of rainwater into the soil is sensitive to surface conditions. Soil surface exposed to direct raindrop impacts becomes sealed and sheds a large fraction of the rain. A dense grass cover improves infiltration (101), the extent depending upon the growth-form (92), but this cover is vulnerable to fire, drought, and grazing. In some cases, the presence of trees helps to maintain ground cover, since tree cover is less variable between years and less easily removed by herbivores or fire, and tree leaf litter is more persistent than grass litter. However, in cases where trees significantly depress herbaceous production, bare ground may increase, making sites more susceptible to losses of water and nutrients via surface flow. The trees and grasses in some savanna landscapes on gentle slopes and loamy soils segregate into alternating bands or patches known as tiger bush in Africa (202), mulga bands in Australia (97), or vegetation arcs in Mexico (114) and the Middle East (201). The bare or sparsely grass-covered strips shed water, which is captured in soils associated with the up-slope portion of tree-covered strips. Enhanced woody plant recruitment on the up-slope portion of the tree strip relative to the down-slope portion (102) may cause these bands to slowly move up the landscape.

Nutrient Accumulation Below Woody Plants

In addition to altering soil moisture and structure, both evergreen (18, 122, 143, 209) and deciduous trees typically enhance pools of soil nutrients (C, N, P, and cations) and their fluxes beneath their canopies (25, 47, 59, 60, 66, 82, 83, 90, 116, 170, 171, 172, 184, 195). Nutrient enrichment of soils may vary with species (19), and patches with leguminous trees may contain more soil nitrogen than patches with nonleguminous trees (31, 130, 185). In other cases, nitrogen pools or mineralization rates are comparable between legume and nonlegume tree patches (27, 64) and between deciduous and evergreen shrub patches (103). The higher carbon and nitrogen densities in subcanopy soils may reflect differences in abiotic conditions under and away from trees and the fact that trees have foliage with higher nutrient content, higher litter inputs, and/or lower decomposition rates than the plants in the herbaceous layer.

Nutrient enrichment occurs across a broad array of tree and shrub growth-forms and species inhabiting diverse climatic zones. Three mechanisms have been proposed to account for this phenomenon (184). Trees may act as nutrient pumps, drawing nutrients from deep horizons and laterally from areas beyond the canopy, depositing them mainly beneath the canopy via litterfall and canopy leaching (90, 146). The supporting evidence for this mechanism is circumstantial. As noted earlier, savanna tree roots may extend outward many times the canopy radius and penetrate more deeply into the soil than do grass roots. The second mechanism is that the tall, aerodynamically rough tree canopy acts as an effective trap for atmospheric dust (30, 58, 167). The dust contains nutrients, which wash off the leaves during rainstorms and drip into the subcanopy area. Though not well quantified, a third mechanism may be of importance where trees are sparse and may serve as focal points attracting roosting birds and mammals seeking shade or cover. Herbivores that take refuge in the shade of trees may enhance the local nutrient cycle (66). Perching birds may enrich soil nutrients (23) and deposit seeds of other trees and shrubs whose germination and establishment may be favored in the subcanopy environment (10).

While the "island of fertility" phenomenon has been widely recognized, little is known of the rates of nutrient enrichment in tree-dominated patches. Bernhard-Reversat (30) found good correlations between total C and N in soil under *Acacia senegal* and *Balanites aegyptiaca* tree canopies and tree girth ($R^2 = 0.62$ and 0.71, respectively), suggesting that soil nutrients change with time of woody plant occupancy of a patch. Barth (18) found significant correlations between piñon pine age and various soil properties. In savanna parklands of southern Texas, soil carbon storage increased linearly with tree stem age, ranging from 19 g C m^{-2} y^{-1} in uplands to 70 g C m^{-2} y^{-1} in moister lowlands (TW Boutton, SR Archer, unpublished information); rates of N storage ranged from 1.7 g N m^{-2} y^{-1} (uplands) to 6 g N m^{-2} y^{-1} (lowlands).

A consequence of the accumulation of nutrients in the subcanopy area may be impoverishment beyond the canopy zone (but see 47). The contrast in soil nutrients under vs. away from tree canopies may be further magnified by concurrent losses from herbaceous patches resulting from heavy grazing. This mechanism has been held responsible for the overall decrease in animal production in desert shrublands in the United States when shrubs invade grasslands (144). In Africa and Australia, nutrient accumulation in patches in the landscape is generally looked on favorably, since it raises the nutrient content of at least some of the grass above the critical threshold for digestion by ruminants (146).

Shading and Microclimate Modification

The tree leaf area index (LAI) in savannas is low compared to that of forests or grasslands (148). The LAI in the subcanopy zone may be two to six times higher than the average for the savanna as a whole (0.5 to 1.0 is typical for dry savannas), since the tree leaves typically are concentrated into a tree cover fraction of 15–50%. The savanna tree canopy reduces direct and indirect solar radiation reaching the shaded area by 25–90% (25, 66, 83, 94, 172). Since the semi-arid tropics where savannas predominate are among the sunniest places in the world, the amount of radiation reaching grasses below the tree canopy may still be sufficient for a relatively high rate of photosynthesis (117). The total energy budget in the subcanopy area is less attenuated than the photosynthetically active radiation because a large part of the energy is advected from the surrounding intercanopy areas in the form of heated air. Transpiration and leaf temperatures may therefore remain high in the subcanopy zone despite shading and reductions in soil temperature. The latter may be important for seed germination and seedling establishment.

HERBIVORES AND FIRE

Low-growing, thorny trees may provide a refuge in which palatable grasses can persist in heavily grazed environments (121, 193). Trees dispersed to and capable of establishing in grass-dominated zones can provide vertical structure attractive to perching birds that disseminate seeds of other woody species (13). Small mammals and insects attracted to trees or bush clumps may subsequently influence patch dynamics via granivory, burrowing, and seedling predation. Seedlings and small trees are particularly vulnerable to herbivory from insects (107), browsers, or by grazing herbivores, which trample them or consume them along with grass (34, 81). In some cases the tree seedlings associated with clumps of unpalatable grass may escape herbivory during this critical stage of their life cycle. In North America, the widespread eradication of prairie dogs in the early 1900s may have contributed to the release of trees and shrubs recorded in many Great Plains grasslands in recent history (194).

High grass biomass can affect tree biomass by fueling fires (89); grazing reduces the fuel load and hence affects fire frequency, intensity, or continuity of spread (17, 138). Browsing, on the other hand, helps to keep woody plants within the flame zone (and conversely, fires keep woody plants browsable). Thus, a strong grazer-browser-fire interaction influences tree-grass mixtures. For example, control of the rinderpest virus led to increases in Serengeti wildebeest populations; the resultant intensification of grazing pressure reduced fire and grass competition and released small trees (153). The increased small tree production allowed giraffe numbers to increase, while giraffe browsing limited tree growth. Woodland decline and grassland expansion in the Serengeti-Mara woodlands since the 1960s cannot be attributed to browsing alone but appears associated with the combined effects of browsing and fire (55). Similarly, the fire regime associated with wildebeest grazing cannot maintain the grassland state unless browsers are present in sufficient numbers. The fire mechanism can be subtle and complex, involving in one example interactions between frequent fires, windstorms, and basal bark-scars left by gnawing rodents (206). In savannas where grazing and browsing animals (domestic or wild) are of little consequence, fire may operate more directly to influence tree-grass mixtures and may slow, but not prevent, complete tree domination (77). The fact that such sites persist as savanna rather than forest or woodland may indicate that portions of the landscape do not provide physical or nutritive conditions required for tree establishment.

MODELS OF TREE-GRASS INTERACTION

The coexistence of apparent competitors can be accounted for in three ways. All three models have been proposed for savannas. Niche separation models posit that competitors avoid competition by using resources that are slightly different, obtained from different places, or obtained at different times. Balanced competition models admit interspecific competition but propose that it is weaker than intraspecific competition. Both niche separation and balanced competition can lead to a stable coexistence and are therefore known as equilibrium models. Disequilibrium models assume there is no stable equilibrium but that frequent changes in the environment (disturbances) prevent the extinction of either competitor by restarting the race or by biasing it alternately toward one or another competitor.

Niche Separation by Depth

The observation that the fibrous root systems of grasses intensively exploit a relatively small proportion of the soil profile, whereas woody plants extensively explore a larger volume, extending deeper, prompted Schimper (142) to suggest "moisture in the subsoil has little influence on the covering of grass; only moisture in the superficial soil is important to it" and that woody plants are favored by a moist subsoil "regardless of whether rain falls during a period of activity or rest." This perception formed the basis for Walter's explicit model for tree-grass coexistence in savannas (192). He assumed water was the limiting factor and hypothesized that trees had roots in both the topsoil and subsoil, while grasses rooted only in the topsoil. This difference would result in a stable equilibrium if grasses have a greater water use efficiency than trees. Stable tree-grass coexistence is theoretically possible even if there is not complete rooting depth separation (and there seldom is), as long as there is a sufficient degree of separation in relation to the pattern of water distribution in the profile and the relative water use efficiencies (188, 189).

Walter's two-layer hypothesis predicts that in any environment with a given climate and soil there should be a characteristic tree-grass ratio. Trees should be advantaged on soils of low water-holding capacity, such as sands, and under wetter climates, since both of these conditions lead to more water reaching the deeper soil layers. Conversely, shallow soils underlain by impervious layers, water-retaining clays, and arid environments should disfavor trees. Evidence for the supposition that grasses and woody plants use different water resources in the soil profile exists for shrub-steppe (96, 133, 161) and tree savannas (76, 88, 93, 156). In addition, it appears that tree species with rapid root development are capable of achieving this resource partitioning with grasses very early in their life cycle (37, 39). However, the two-layer hypothesis cannot account for the large variation in the tree-grass ratio within a single climate-soil combination (22).

The rooting depth distribution data for savannas only partly supports the two-layer hypothesis as well. Trees do tend to have a maximum and modal rooting depth deeper than that of grasses, but grass roots can be more abundant in absolute terms than tree roots to depths of up to a meter (148). Most roots of trees and grasses are in the upper soil horizons (84), which makes ecological sense in semi-arid environments because that is where water and nutrients co-occur. Some species of savanna trees send roots down to great depths or exploit fissures to gain access to resources accumulating beneath bedrock or claypans. However, it seems unlikely that all trees on a savanna landscape obtain the bulk of their water from these deep sources. Indeed, many woody species are quite shallow-rooted or plastic with respect to rooting patterns, and their physiological activity may be closely coupled to soil moisture availability in upper horizons (10, 50, 59, 76, 123, 152, 175). These shallow-rooted trees may increase in abundance with heavy grazing, apparently favored by an increase in water availability in the surface soils resulting from reductions in grass biomass (155). In humid savannas of West Africa (1000 mm annual rainfall; 2-month dry season), patterns of root distribution and soil water depletion are similar for trees and grasses, with each life-form obtaining water primarily from the top soil layers during both rainy and dry periods (95).

Niche Separation by Phenology

The characteristic climate associated with tropical savannas is alternating warm dry seasons with hot wet seasons. Might this seasonal pattern of growth opportunity provide a potential axis of niche separation for trees and grass? This model is conceptually similar to the model developed by Westoby (199) for coexistence of grasses, shrubs, and forbs in episodically wetted arid shrublands.

Deciduous savanna trees achieve full leaf expansion within weeks of the onset of the rainy season (148). In moist savannas, leaf expansion of the C_3 trees may precede the rains by up to several weeks. The peak leaf area of C_4 grasses is achieved only several months after the onset of the wet season. Trees that flush early do so, it is hypothesized, by using carbohydrates and nutrients carried over from the previous growing season and residual water deep in the soil. Grasses initiate growth from stored reserves, but full leaf growth appears more dependent on factors affecting current season photosynthesis. Thus, trees expand their leaves rapidly in a synchronous flush from preformed buds, whereas grasses initiate leaves sequentially and at staggered intervals over the growing season (14), the rates and dynamics being influenced by environmental conditions (161). Deciduous trees may also retain leaves for several weeks after grasses have senesced. As a result, trees can monopolize resources early and late in the growing season (132). Evergreen trees, which dominate many neotropical and temperate savannas, may have lower nutrient requirements, higher nutrient retention, and higher concentrations of secondary compounds than grasses and cast deeper shade year-round than deciduous trees. These features may more than offset their low photosynthetic rates and give them an advantage over grasses, especially on nutrient poor sites.

Using a model that includes the grass rooting-depth niche entirely within the tree niche, we would predict dominance by trees wherever seasonality is strong, protracted, and predictable. In arid unpredictable environments dominated by small rainfall events, the opportunistic strategy of grasses would be favored (161), while in continuously moist environments (such as those occupied by forests), a continuous growth, overtopping strategy is favored. Within its area of applicability, the model predicts greatest advantage for trees where there is greatest potential for between-season carryover of resources, which is in the moister savannas and on deeper, sandier soils. As in the rooting-depth niche separation model, these predictions are broadly consistent with what is observed, but many local exceptions occur.

Balanced Competition

If balanced competition occurs, the superior competitor becomes self-limiting at a biomass insufficient to eliminate the poorer competitor. It is reasonable to expect greater competitive overlap between trees than between trees and grasses, and, similarly, greater competitive overlap between grasses than between grasses and trees. Since mature trees generally dominate over grasses, the model predicts that all savannas should trend toward a woodland with a sparse understory of grasses. Those not currently in this state are unstable and presumably are held away from the equilibrium point by disturbances such as fire or browsing (7, 125).

The theory that a form of balanced competition would result if mature trees could outcompete grasses, but grasses could outcompete establishing trees, is supported by evidence from competition experiments. The model predicts two meta-stable states, one a dense woodland with little or no grass, and the other a dense grassland with no trees. Rapid and dramatic transitions from open grassland to thickets have been observed following heavy grazing and the exclusion of fire (140, 155, 156, 178, 182). Spontaneous reversions to grassland have rarely been observed, but that may be an artifact of the longevity of trees. It is possible to keep a cleared savanna treeless by repeated burning. Meta-stable models lend themselves to description as "state and transition" models (55, 200).

Spatially Explicit Models

These models are built on the observation that the interactions between coexisting organisms (trees and grasses in this example) are not spatially homogeneous. Whereas the other models consider trees and grasses as abstract classes that have a vertical structure (rooting and canopy layers, for instance) but no lateral structure, the spatially explicit models contend that where each individual plant occurs is as important as what its life-form is. In practice, because of the large-scale difference between trees and grasses, only the trees are usually treated spatially explicitly, and the grasses are combine into a zoned layer. The "ecological field theory" approach (205) describes a plant community as a mosaic of overlapping spheres of influence, each showing some pattern in relation to the plant at its center. The performance of the entire community or of any community component measured at a landscape patch scale is the integral of these spatially varying fields of growth opportunity.

Spatially explicit models (139) provide a plausible explanation for the non-linear relationship between grass production and tree cover described in the *Tree Density and Herbaceous Production* section. As a simplified example, imagine that every tree is the center of a circular area with a radius equivalent to the maximum root extension. The first tree in a grassland might suppress grass growth within this circle, causing the grass production on the landscape to decrease. A second tree would have an additional influence but would also have some small probability of partly overlapping with the area of influence of the first tree, so the incremental reduction in total area grass production would be slightly less. Extend this scenario to a large number of trees, and the result is an exponentially declining relationship between grass production and tree density very similar to that actually observed.

The model can be made more sophisticated with trees of different sizes occurring in nonrandom patterns and with spheres of influence having differing decay functions, giving rise to curves with different degrees of concavity. As would be expected in a high-fertility, moist site where the rooting radius could be small, the influence zone would be small, and the curve more nearly linear than on sites where the radius of influence is large. This concurs with field observations.

The anomalous convex tree-density–grass-production curves observed on some sites (151, 165) could also be generated with this approach, by including in the model zones of facilitation as well as zones of competition. By varying the relative strengths of competition and facilitation and their rate of decrease with distance from the tree stem, almost any shape of relationship between grass production and tree density could be simulated.

A Synthesis Model

All the models outlined above are at least partly correct, but no single model can account for the variety of phenomena at all savanna locations, or even the range of behaviors exhibited at one location in different seasons or stages of succession. A comprehensive model of tree-grass relationships in savannas must combine elements of all of them (168, 187).

It is further useful to divorce the issue of co-existence of trees and grasses in the long term from the issue of resource partitioning in the short term, since the two may bear little relationship to one another outside of computer simulations. The dynamics of savannas, and probably many other ecosystems, are driven to a large degree by factors such as climate that originate outside the tree-grass system (157). The vegetation structure is therefore only to a small degree the result of the cumulative effect of competition for resources. Co-existence of trees and grass in this view is largely a result of the interaction of a variety of stresses and disturbances, acting differentially on trees and grass and patchily in time and space. Equilibrium niche theory is useful for understanding patterns of resource partitioning in between these stress periods.

Niches of grass and tree archetypes differ in both rooting depth and phenology. Life-history and growth form within the archetypes also vary substantially. Deep-rootedness is very important for some tree species, but not all, and some grasses have a less opportunistic growth pattern than others. On average, there is probably more opportunity for phenological separation than for depth separation, simply because so much of the soil activity is concentrated in the topsoil that any plant without most of its roots there has little chance of capturing resources. Phenology and rooting depth are not unrelated because possession of a large and deep root system is one reason why trees can expand their leaves early and rapidly and retain them for longer into stress periods.

We suggest that the rapid, synchronous deployment of leaves and fine roots by trees allows them almost exclusive access to the nutrients (particularly nitrogen) mineralized in the early part of the growing season. The nutrients can be stored in the tree biomass for use throughout the season, and when the leaves are dropped, a large fraction of their nutrient content is reabsorbed into the permanent tissues (148, 175). Grasses do the same, but to a lesser degree because of their limited storage capacity, and therefore they are usually highly responsive to added nitrogen. Trees, especially the nitrogen-fixing species, may be limited much more by the availability of phosphorus than of nitrogen (79).

Nutrients are mineralized, transported to the root, and taken up only in the presence of water. Most of the observations advanced to suggest water is the main resource competed for by savanna vegetation (for instance, the near-linear relation between net primary production and annual rainfall) can apply equally to nutrients. A simplistic "law of the minimum" view of competition is not useful in this context. Multiple-factor limitation is the norm, either sequentially or simultaneously (48). The arrangement of various savanna types within the axes of plant-available moisture and plant-available nutrients has been the focus of recent discussion (22), but quantification of these variables remains elusive. In reality, savanna structure is hierarchically constrained: by climate at regional-to-continental scales; by topographic effects on rainfall and landscape water redistribution, and by geomorphic effects on soil and plant-available water at landscape-to-regional scales; and finally by water redistribution and disturbance at local and patch scales (51).

CONCLUSIONS

Tree-grass interactions in savannas cannot be predicted by a simple model. They include elements of competition and facilitation, varying complexly in both time and space. Coexistence is permitted by a combination of niche separation, stronger intra–life-form than inter–life-form competition toward the extremes of dominance, a balanced asymmetry of competition, and most importantly, frequent levelling disturbances, particularly fire.

Experimental evidence suggests all mixtures of mature trees and grass are unstable in savanna environments. In the absence of disturbances such as re-peated fires, clearing by humans, or feeding by large herbivores, the tree cover increases at the expense of grass production until it is limited by tree-tree com-petition. The amount of grass remaining in the system when this endpoint is reached depends on the match between the tree growth strategy and the growth opportunities offered by the environment. In moist, relatively predictable en-vironments the match is good, and a closed woodland with almost no grasses will result. In arid, unpredictable environments, substantial quantities of grass may occur in some years.

Tropical nonhydromorphic grassland states are meta-stable where fires are sufficiently frequent and intense to prevent trees from escaping the flame zone. How easily the trees can escape depends on factors that include intensity of browsing, rainfall, soil texture, and soil fertility. Moist fertile environments support a vigorous grass growth that, if not grazed, leads to frequent intense fires. These environments often support large herds of migratory grazers, which crop the grass very short and browse the small trees. In these environments, extensive grassland patches can occur (the Mitchell grasslands in Australia and the Serengeti grasslands in Africa may be examples), suggesting a relatively robust meta-stable state. They can nevertheless be invaded by trees if tree propagules are present and grass cover is reduced.

Semi-arid environments on sandy, low-fertility soils are seldom treeless, suggesting a very weak or absent meta-stable grassland state. Many semi-arid areas on relatively fertile loamy or clayey soils were relatively treeless in precolonial times but were encroached rapidly and apparently irreversibly when grazed continuously by cattle. The encroachment often occurred in relatively discrete episodes, following a prolonged period of drought during which no fires were possible.

Resource partitioning patterns do not explain coexistence of trees and grasses in savannas, but they do account for the observed patterns of primary production. In particular, horizontal spatial interactions in the rooting zone are the key to understanding the nonlinearities in the relationships between tree and grass primary production.

ACKNOWLEDGMENTS

Our collaboration on this paper stems from involvement in a series of meetings sponsored by the International Union of Biological Sciences/Man and the Biosphere program on Responses of Savannas to Stress and Disturbance (RSSD). We are particularly grateful to the RSSD coordinators (Malcom Hadley, Ernesto Medina, JC Menaut, Juan Sillva, Otto Solbrig, and Brian Walker) for funding our participation in the meetings and encouraging this endeavor. GR McPhernon made helpful comments and suggestions on earlier drafts of the manuscript. Preparation of this review was supported in part by NSF grant BSR-9109240, NASA-EOS grant NAGW-2662, and USDA-NRI grant 92-37101-7463.

Visit the *Annual Reviews home page* at
http://www.annurev.org.

Literature Cited

1. Adamoli J, Sennhauser E, Acero JM, Rescia A. 1990. Stress and disturbance: vegetation dynamics in the dry Chaco region of Argentina. *J. Biogeogr.* 17:491–500
2. Aguiar MR, Sala OE. 1994. Competition, facilitation, seed distribution and the origin on patches in a Patagonian steppe. *Oikos* 70:26–34
3. Amundson RG, Ali AR, Belsky AJ. 1995. Stomatal responsiveness to changing light intensity increases rain-use efficiency of below-crown vegetation in tropical

savannas. *J. Arid. Environ.* 29:139–53

4. Anderson RC, Fralish JS, Baskin JM, eds. 1997. *Savanna, Barren and Rock Outcrop Plant Communities of North America.* New York: Cambridge Univ. Press. In press

5. Ansley RJ, Jacoby PW, Cuomo GJ. 1990. Water relations of honey mesquite following severing of lateral roots: influence of location and amount of subsurface water. *J. Range Manage.* 43:436–42

6. Ansley RJ, Jacoby PW, Hicks RA. 1991. Leaf and whole plant transpiration in honey mesquite following severing of lateral roots. *J. Range Manage.* 44:577–83

7. Archer S. 1989. Have southern Texas savannas been converted to woodlands in recent history? *Am. Nat.* 134:545–61

8. Archer S. 1994. Woody plant encroachment into southwestern grasslands and savannas: rates, patterns and proximate causes. In *Ecological Implications of Livestock Herbivory in the West,* ed. M Vavra, W Laycock, R Pieper, pp. 13–68. Denver: Soc. Range Manage.

9. Archer S. 1995. Herbivore mediation of grass-woody plant interactions. *Trop. Grassl.* 29:218–35

10. Archer S. 1995. Tree-grass dynamics in a *Prosopis*-thornscrub savanna parkland: reconstructing the past and predicting the future. *Ecoscience* 2:83–99

11. Archer S. 1996. Assessing and interpreting grass-woody plant dynamics. In *The Ecology and Management of Grazing Systems,* ed. J Hodgson, A Illius, pp. 101–34. Wallingford, UK: CAB Int.

12. Archer S, Schimel DS, Holland EA. 1995. Mechanisms of shrubland expansion: land use, climate, or CO_2? *Clim. Change* 29:91–99

13. Archer S, Scifres CJ, Bassham CR, Maggio R. 1988. Autogenic succession in a subtropical savanna: conversion of grassland to thorn woodland. *Ecol. Monogr.* 58:111–27

14. Archer S, Tieszen LL. 1980. Growth and physiological responses of tundra plants to defoliation. *Arct. Alpine Res.* 12:531–52

15. Backéus I. 1992. Distribution and vegetation dynamics of humid savannas in Africa and Asia. *J. Veg. Sci.* 3:345–57

16. Bahn PG. 1996. Further back Down Under. *Nature* 383:577–78

17. Baisan CH, Swetnam TW. 1990. Fire history on a desert mountain range: Rincon Mountain Wilderness, Arizona, USA. *Can. J. For. Res.* 20:1559–69

18. Barth RC. 1980. Influence of pinyon pine trees on soil chemical and physical properties. *Soils Sci. Soc. Am. J.* 44:112–114

19. Barth RC, Klemmedson JO. 1978. Shrub-induced spatial patterns of dry matter, nitrogen, and organic carbon. *Soils Sci. Soc. Am. J.* 42:804–9

20. Bartolome JW, Allen-Diaz BH, Tietje WD. 1994. The effect of *Quercus douglasii* removal on understory yield and composition. *J. Range Manage.* 47:151–54

21. Belsky AJ. 1984. Role of small browsing mammals in preventing woodland regeneration in the Serengeti National Park, Tanzania. *Afr. J. Ecol.* 22:271–79

22. Belsky AJ. 1990. Tree/grass ratios in East African savannas: a comparison of existing models. *J. Biogeog.* 17:483–89

23. Belsky AJ. 1994. Influences of trees on savanna productivity: tests of shade, nutrients, and tree-grass competition. *Ecology* 75:922–32

24. Belsky AJ. 1996. Viewpoint: Western juniper expansion: Is it a threat to arid northwestern ecosystems? *J. Range Manage.* 49:53–59

25. Belsky AJ, Amundson RG, Duxberry RM, Riha SJ, Ali AR, Mwonga SM. 1989. The effects of trees on their physical, chemical and biological environments in a semi-arid savanna in Kenya. *J. Appl. Ecol.* 26:1004–24

26. Belsky AJ, Canham CD. 1994. Forest gaps and isolated savanna trees: an application of patch dynamics in two ecosystems. *BioScience* 44:77–84

27. Belsky AJ, Mwonga SM, Amundson RG, Duxbury JM, Ali AR. 1993. Comparative effects of isolated trees on their undercanopy environments in high- and low-rainfall savannas. *J. Appl. Ecol.* 30:143–55

28. Belsky AJ, Mwonga SM, Duxbury JM. 1993. Effects of widely spaced trees and livestock grazing on understory environments in tropical savannas. *Agrofor. Sys.* 24:1–20

29. Ben-Sharer R. 1992. The effects of bush clearance on African ungulates in a semi-arid nature reserve. *Ecol. Appl.* 2:95–101

30. Bernhard-Reversat F. 1982. Biogeochemical cycle of nitrogen in a semi-arid savanna. *Oikos* 38:321–32

31. Bernhard-Reversat F. 1988. Soil nitrogen mineralization under a *Eucalyptus* plantation and a natural *Acacia* forest in Senegal. *For. Ecol. Manage.* 23:233–44

32. Blackmore AC, Mentis MT, Scholes RJ. 1990. The origin and extent of nutrient-enriched patches within a nutrient-poor savannah in South Africa. *J. Biogeogr.* 17:463–70

33. Bojorquez-Tapia LA, Ffoliott PF, Guertin DP. 1990. Herbage production-forest overstory relationships in two Arizona ponderosa pine forests. *J. Range Manage.* 43:25–28

34. Borchert MI, Davis FW, Michaelsen J, Oyler LD. 1989. Interaction of factors affecting seedling recruitment of blue oak (*Quercus douglasii*) in California. *Ecology* 70:389–404

35. Bourliere F, ed. 1983. *Tropical Savannas.* Amsterdam: Elsevier. 730 pp.

36. Bragg TB, Hulbert LC. 1976. Woody plant invasion of unburned Kansas bluestem prairie. *J. Range Manage.* 29:19–23

37. Bragg WK, Knapp AK, Briggs JM. 1993. Comparative water relations of seedling and adult *Quercus* species during gallery forest expansion in tallgrass prairie. *For. Ecol. Manage.* 56:29–41

38. Brain SK, Sillen A. 1988. Evidence from the Swartkrans cave for the earliest use of fire. *Nature* 336:464–66

39. Brown JR, Archer S. 1990. Water relations of a perennial grass and seedlings vs adult woody plants in a subtropical savanna, Texas. *Oikos* 57:366–74

40. Burgess TL. 1995. Desert grassland, mixed shrub savanna, shrub steppe, or semidesert scrub? The dilemma of coexisting growth forms. In *The Desert Grasslands,* ed. MP McClaran, TR Van Devender, pp. 31–67. Tucson: Univ. Ariz. Press

41. Burrows WH, Carter JO, Scanlan JC, Anderson ER. 1990. Management of savannas for livestock production in north-east Australia: contrasts across the tree-grass continuum. *J. Biogeogr.* 17:503–12

42. Cable DR. 1977. Seasonal use of soil water by mature velvet mesquite. *J. Range Manage.* 30:4–11

43. Caldwell MM, Richards JH. 1989. Hydraulic lift: water efflux from upper roots improves effectiveness of water uptake by deep roots. *Oecologia* 79:1–5

44. Caldwell MM, Richards JH, Manwaring JH, Eissenstat DM. 1987. Rapid shifts in phosphate acquisition show direct competition between neighboring plants. *Nature* 327:615–16

45. Callaway RM. 1995. Positive interactions among plants. *Bot. Rev.* 61:306–49

46. Cameron DM, Rance SJ, Jones RM, Charles-Edwards DA, Barnes A. 1989. Project STAG: An experimental study in agroforestry. *Aust. J. Agric. Res.* 40:699–714

47. Campbell BM, Frost P, King JA, Mawanza M, Mhlanga L. 1994. The influence of trees on soil fertility on two contrasting semi-arid soil types at Matopos, Zimbabwe. *Agrofor. Sys.* 28:159–72

48. Chapin FS III, Bloom AJ, Field CB, Waring RH. 1987. Plant responses to multiple environmental factors. *BioScience* 37:49–57

49. Cole MM. 1986. *The Savannas: Biogeography and Geobotany.* Orlando, FL: Academic

50. Coughenour MB, Detling JK, Bamberg IE, Mugambi MM. 1990. Production and nitrogen responses of the African dwarf shrub *Indigofera spinosa* to defoliation and water limitation. *Oecologia* 83:546–52

51. Coughenour MB, Ellis JE. 1993. Landscape and climatic control of woody vegetation in a dry tropical ecosystem: Turkana District, Kenya. *J. Biogeogr.* 20:383–98

52. Dawson TE. 1993. Hydraulic lift and water use by plants—implications for water balance, performance and plant-plant interactions. *Oecologia* 95:565–74

53. De Steven D. 1991. Experiments on mechanisms of tree establishment in old-field succession: seeding survival and growth. *Ecology* 72:1076–88

54. De Steven D. 1991. Experiments on mechanisms of tree establishment in old-field succession: seedling emergence. *Ecology* 72:1066–75

55. Dublin HT, Sinclair ARE, McGlade J. 1990. Elephants and fire as causes of multiple stable states in the Serengeti-Mara woodlands. *J. Animal Ecol.* 49:1147–64

56. Dye KL, Ueckert DN, Whisenant SG. 1995. Redberry juniper-herbaceous understory interactions. *J. Range Manage.* 48:100–7

57. Engle DM, Stritzke JF, Claypool PL. 1987. Herbage standing crop around eastern redcedar trees. *J. Range Manage.* 40:237–39

58. Escudero A, Garcia B, Gomex JM, E. L. 1985. The nutrient cycling in *Quercus rotundifolia* and *Q. pyrenaica* ecosystems of Spain. *Oecol. Plant.* 6:73–86

59. Evans RD, Ehleringer JR. 1994. Water and nitrogen dynamics in an arid woodland. *Oecologia* 99:233–42

60. Frost WE, Edinger SB. 1991. Effects of tree canopies on soil characteristics of annual rangeland. *J. Range Manage.* 44:286–88

61. Fuhlendorf SD, Smeins FE, Taylor CA. 1997. Browsing and tree size influences on Ashe juniper understory. *J. Range Manage.* 50:507–12

62. Fulbright JE, Beasom SL. 1987. Long-term effects of mechanical treatment on

white-tailed deer browse. *Wildl. Soc. Bull.* 15:560–64

63. Gadgill M, Meher-Homji VM. 1985. Land use and productive potential of Indian savannas. In *Ecology and Management of the World's Savannas,* ed. JC Tothill, JJ Mott, pp. 107–13. Canberra: Aust. Acad. Sci.

64. Garcia-Moya E, McKell CM. 1970. Contribution of shrubs to the nitrogen economy of a desert-wash plant community. *Ecology* 51:81–88

65. Gardener GJ, McIvor JG, Williams J. 1990. Dry tropical rangelands: solving one problem and creating another. In *Australian Ecosystems: 200 Years of Utilization, Degradation and Reconstruction,* ed. DA Saunders, AJM Hopkins, RA How, pp. 279–86. Geraldton, West. Aust.: Ecol. Soc. Aust.

66. Georgiadis NJ. 1989. Microhabitat variation in an African savanna: effect of woody cover and herbivores in Kenya. *J. Trop. Ecol.* 5:93–108

67. Grubb PJ. 1977. The maintenance of species richness in plant communities: the importance of the regeneration niche. *Biol. Rev.* 52:107–45

68. Gutierrez JR, Fuentes ER. 1979. Evidence for intraspecific competition in the *Acacia caven* (Leguminosea) savanna of Chile. *Oecol. Plant.* 14:151–58

69. Hara M. 1987. Analysis of seedling banks of a climax beech forest: ecological importance of seedling sprouts. *Vegetatio* 71:67–74

70. Harrington GN. 1991. Effects of soil moisture on shrub seedling survival in a semi-arid grassland. *Ecology* 72:1138–49

71. Harrington GN, Johns GG. 1990. Herbaceous biomass in a *Eucalyptus* savanna woodland after removing trees and/or shrubs. *J. Appl. Ecol.* 27:775–87

72. Harris DR, ed. 1980. *Human Ecology in Savanna Environments.* London: Academic. 522 pp.

73. Haworth K, McPherson GR. 1994. Effects of *Quercus emoryi* on herbaceous vegetation in a semi-arid savanna. *Vegetatio* 112:153–59

74. Haworth K, McPherson GR. 1995. Effects of *Quercus emoryi* trees on precipitation distribution and microclimate in a semi-arid savanna. *J. Arid. Environ.* 31:153–70

75. Heitschmidt RK, Schultz RD, Scifres CJ. 1986. Herbaceous biomass dynamics and net primary production following chemical control of honey mesquite. *J. Range Manage.* 39:67–71

76. Hesla BI, Tieszen LL, Boutton TW. 1985. Seasonal water relations of savanna shrubs and grasses in Kenya, East Africa. *J. Arid. Environ.* 8:15–31

77. Hochberg ME, Menaut JC, Gignoux J. 1994. The influences of tree biology and fire in the spatial structure of the West African savannah. *J. Ecol.* 82:217–26

78. Hoffman MT, Barr GD, Cowling RM. 1990. Vegetation dynamics in the semi-arid eastern Karoo, South Africa: the effect of seasonal rainfall and competition on grass and shrub basal cover. *S Afr. J. Sci.* 86:462–63

79. Högberg P. 1986. Soil nutrient availability, root symbiosis and tree species composition in tropical Africa: a review. *J. Trop. Ecol.* 2:359–72

80. Huntley BJ, Walker BH, eds. 1982. *Ecology of Tropical Savannas.* New York: Springer-Verlag. pp.

81. Huntly N. 1991. Herbivores and the dynamics of communities and ecosystems. *Annu. Rev. Ecol. Syst.* 22:477–503

82. Isichei AO, Muoghalu JI. 1992. The effects of tree canopy cover on soil fertility in a Nigerian savanna. *J. Trop. Ecol.* 8:329–38

83. Jackson LE, Strauss RB, Firestone MK, Bartolome JW. 1990. Influence of tree canopies on grassland productivity and nitrogen dynamics in deciduous oak savanna. *Agric. Ecosyst. Environ.* 32:89–105

84. Jackson RB, Canadell J, Ehleringer JR, Mooney HA, Sala OE, Schulze ED. 1996. A global analysis of root distributions for terrestrial biomes. *Oecologia* 108:389–411

85. Jameson DA. 1967. The relationship of tree overstory and herbaceous understory vegetation. *J. Range Manage.* 20:247–49

86. Joffre R, Rambal S. 1993. How tree cover influences the water balance of Mediterranean rangelands. *Ecology* 74:570–82

87. Joffre R, Vacher J, De Los Llanos C, Long G. 1988. The dehesa: an agrosilvopastoral system of the Mediterranean region with special reference to the Sierra Morena area of Spain. *Agrofor. Sys.* 6:71–96

88. Johnson RW, Tothill JC. 1985. Definition and broad geographic outline of savanna lands. In *Ecology and Management of the World's Savannas,* ed. JC Tothill, JJ Mott, pp. 1–13. Canberra, ACT: Aust. Acad. Sci.

89. Kauffman JB, Cummings DL, Ward DE. 1994. Relationships of fire, biomass and nutrient dynamics along a vegetation gradient in the Brazilian cerrado. *J. Ecol.* 82:519–31

90. Kellman M. 1979. Soil enrichment by neotropical savanna trees. *J. Ecol.* 67:565–77

91. Kenkel NC. 1988. Pattern of self-thinning in jack pine: testing the random mortality hypothesis. *Ecology* 69:1017–24

92. Knight RW, Blackburn WH, Merrill LB. 1984. Characteristics of oak mottes, Edwards Plateau, Texas. *J. Range Manage.* 37:534–37

93. Knoop WT, Walker BH. 1985. Interactions of woody and herbaceous vegetation in southern African savanna. *J. Ecol.* 73:235–53

94. Ko LJ, Reich PB. 1993. Oak tree effects on soil and herbaceous vegetation in savannas and pastures in Wisconsin. *Am. Midl. Nat.* 130:31–42

95. Le Roux X, Bariac T, Mariotti A. 1995. Spatial partitioning of the soil water resource between grass and shrub components in a West African humid savanna. *Oecologia* 104:147–55

96. Lee CA, Lauenroth WK. 1994. Spatial distributions of grass and shrub root systems in the shortgrass steppe. *Am. Midl. Nat.* 132:117–23

97. Ludwig JA, Tongway DJ. 1995. Spatial organisation of landscapes and its function in semi-arid woodlands, Australia. *Landsc. Ecol.* 10:51–63

98. Madany MH, West NE. 1983. Livestock grazing-fire regime interactions within montane forests of Zion National Park, Utah. *Ecology* 64:661–67

99. Manning SJ, Barbour MG. 1988. Root systems, spatial patterns, and competition for soil moisture between two desert subshrubs. *Am. J. Bot.* 75:995–893

100. Marañón T, Bartolome JW. 1993. Reciprocal transplants of herbaceous communities between *Quercus agrifolia* woodland and adjacent grassland. *J. Ecol.* 81:673–82

101. Martin SC, Morton HL. 1993. Mesquite control increases grass density and reduces soil loss in southern Arizona. *J. Range Manage.* 46:170–75

102. Mauchamp A, Montana C, Lepart J, Rambal S. 1993. Ecotone dependent recruitment of a desert shrub, *Fluorensia cernua*, in vegetation stripes. *Oikos* 68:107–16

103. Mazzarino MJ, Oliva L, Nunez A, Nunez GB. 1991. Nitrogen mineralization and soil fertility in the dry chaco ecosystem (Argentina). *Soils Sci. Soc. Am. J.* 55:515–22

104. McClaran MP, Bartolome JW. 1989. Effect of *Quercus douglasii* (Fagacaeae) on herbaceous understory along a rainfall gradient. *Madroño* 36:141–53

105. McPherson G. 1997. *Ecology and Management of North American Savannas.* Tucson: Univ. Ariz. Press. In press

106. McPherson GR. 1992. Comparison of linear and non-linear overstory-understory models for ponderosa pine: a conceptual framework. *For. Ecol. Manage.* 55:31–34

107. McPherson GR. 1993. Effects of herbivory and herb interference on oak establishment in a semi-arid temperate savanna. *J. Veg. Sci.* 4:687–92

108. McPherson GR, Rasmussen GA, Wester DB, Masters RA. 1991. Vegetation and soil zonation associated with *Juniperus pinchotii* Sudw. Trees. *Gt. Basin Nat.* 51:316–23

109. McPherson GR, Wright HA. 1987. Factors affecting reproductive maturity of redberry juniper (*Juniperus pinchotii*). *For. Ecol. Manage.* 21:191–96

110. McPherson GR, Wright HA. 1990. Effects of cattle grazing and *Juniperus pinchotii* canopy cover on herb cover and production in western Texas. *Am. Midl. Nat.* 123:144–51

111. Menaut JC, Gignoux J, Prado C, Clobert J. 1990. Tree community dynamics in a humid savanna of the Cote d'Ivoire: modelling the effects of fire and competition with grass and neighbors. *J. Biogeogr.* 17:471–81

112. Merz RW, Boyce GS. 1956. Age of oak "seedlings". *J. For.* 54:774–75

113. Miller RF, Wigand PE. 1994. Holocene changes in semiarid pinyon-juniper woodlands. *BioScience* 44:465–74

114. Montaña C. 1992. The colonization of bare areas in two-phase mosaics of an arid ecosystem. *J. Ecol.* 80:315–27

115. Moore MM, Dieter DA. 1992. Stand density index as a predictor of forage production in northern Arizona pine forests. *J. Range Manage.* 45:267–71

116. Mordelet P, Abbadie L, Menaut JC. 1993. Effects of tree clumps on soil characteristics in a humid savanna of West Africa (Lamto, Cote d'Ivoire). *Plant Soil* 153:103–11

117. Mordelet P, Menaut J-C. 1995. Influence of trees on above-ground production dynamics of grasses in a humid savanna. *J. Veg. Sci.* 6:223–28

118. Návar J, Bryan R. 1990. Interception loss and rainfall redistribution by three semi-arid growing shrubs in northeastern Mexico. *J. Hydrol.* 115:51–63

119. Noble JC. 1997. *The delicate and noxious scrub: CSIRO research into native tree and shrub proliferation in the semi-arid woodlands of Eastern Australia.*

Canberra, ACT: CSIRO Div. Wildlife Ecol. In press

120. O'Connor TG. 1995. *Acacia karroo* invasion of grassland: environmental and biotic effects influencing seedling emergence and establishment. *Oecologia* 103: 214–23

121. O'Connor TG. 1995. Transformation of a savanna grassland by drought and grazing. *Afr. J. Range Forage Sci.* 12:53–60

122. Padien DJ, Lajtha K. 1992. Plant spatial pattern and nutrient distribution in pinyon-juniper woodlands along an elevational gradient in northern New Mexico. *Int. J. Plant Sci.* 153:425–33

123. Palaez DV, Distel RA, Boo RM, Elia OR, Mayor MD. 1994. Water relations between shrubs and grasses in the Patagonian steppe. *Oecologia* 81:501–5

124. Parker VT, Muller CH. 1982. Vegetational and environmental changes beneath isolated live oak trees (*Quercus agrifolia*) in a California annual grassland. *Am. Midl. Nat.* 107:69–81

125. Pellew RA. 1983. The impacts of elephant, giraffe and fire upon the *Acacia tortilis* woodlands of the Serengeti. *Afr. J. Ecol.* 41–74

126. Penridge LK, Walker J. 1986. Effect of neighboring trees on eucalypt growth in a semi-arid woodland in Australia. *J. Ecol.* 74:925–36

127. Pieper RD. 1990. Overstory-understory relationships in pinyon juniper woodlands in New Mexico. *J. Range Manage.* 43:413–15

128. Pressland AJ. 1973. Rainfall partitioning by an arid woodland (*Acacia aneura* F. Muell.) in southwestern Queensland. *Aust. J. Bot.* 21:235–45

129. Pressland AJ. 1975. Productivity and management of mulga in south-western Queensland in relation to tree structure and density. *Aust. J. Bot.* 23:965–76

130. Radwanski SA, Wickens GE. 1967. The ecology of *Acacia albida* on mantle soils in Zalingei Jebbel Marra, Sudan. *J. Appl. Ecol.* 4:569–79

131. Reid RS, Ellis JE. 1995. Impacts of pastoralists on woodlands in South Turkana, Kenya: livestock-mediated tree recruitment. *Ecol. Appl.* 5:978–92

132. Rutherford MC, Panagos MD. 1982. Seasonal woody plant growth in *Burkea africana-Ochna pulchra* savanna. *S. Afr. J. Bot.* 1:104–16

133. Sala OE, Golluscio RA, Lauenroth WK, Soriano A. 1989. Resource partitioning between shrubs and grasses in the Patagonian steppe. *Oecologia* 81:501–5

134. San José JJ, Fariñas MR. 1983. Changes in tree density and species composition in a protected Trachypogon savanna, Venezuela. *Ecology* 64:447–53

135. San José JJ, Fariñas MR. 1991. Temporal changes in the structure of a Trachypogon savanna protected for 25 years. *Acta Oecol.* 12:237–47

136. San José JJ, Fariñas MR, Rosales J. 1991. Spatial patterns of trees and structuring factors in a Trachypogon savanna of the Orinoco Llanos. *Biotropica* 23:114–23

137. Sarmiento G. 1984. *The Ecology of Neotropical Savannas.* Cambridge, MA: Harvard Univ. Press. 235 pp.

138. Savage M, Swetnam TW. 1990. Early 19th-century fire decline following sheep pasturing in a Navajo ponderosa pine forest. *Ecology* 71:2374–78

139. Scanlan JC. 1992. A model of woody-herbaceous biomass relationships in eucalypt and mesquite communities. *J. Range Manage.* 45:75–80

140. Scanlan JC, Archer S. 1991. Simulated dynamics of succession in a North American subtropical *Prosopis* savanna. *J. Veg. Sci.* 2:625–34

141. Scanlan JC, Burrows WH. 1990. Woody overstory impact on herbaceous understory in *Eucalyptus* spp. communities in central Queensland. *Aust. J. Ecol.* 15:191–97

142. Schimper AFW. 1903. *Plant Geography on a Physiological Basis.* Oxford, UK: Clarendon

143. Schlesinger WH, Raikes JA, Hartley AE, Cross AF. 1996. On the spatial pattern of soil nutrients in a desert ecosystem. *Ecology* 72:364–74

144. Schlesinger WH, Reynolds JF, Cunningham GL, Huenneke LF, Jarrell WM, et al. 1990. Biological feedbacks in global desertification. *Science* 247:1043–48

145. Schmidt TL, Stubbendieck J. 1993. Factors influencing eastern redcedar seedling survival on rangeland. *J. Range Manage.* 46:448–51

146. Scholes RJ. 1990. The influence of soil fertility on the ecology of African savannas. *J. Biogeog.* 17:417–19

147. Scholes RJ, Hall DO. 1996. The carbon budget of tropical savannas, woodlands, and grasslands. In *Modelling Terrestrial Ecosystems,* ed. A Breymeyer, D Hall, J Melillo, G Agren, pp. 69–100. Chichester: Wiley

148. Scholes RJ, Walker BH. 1993. *An African savanna: synthesis of the Nylsvley Study.* Cambridge, UK: Cambridge Univ. Press

149. Scifres CJ. 1980. *Brush Management: Principles and Practices for Texas and the*

150. Scifres CJ, Hamilton WT, Koerth BH, Flinn RC, Crane RA. 1988. Bionomics of patterned herbicide application for wildlife habitat enhancement. *J. Range Manage.* 41:317–21

151. Scifres CJ, Mutz JL, Whitson RE, Drawe DL. 1982. Interrelationships of huisache canopy cover with range forage on the coastal prairie. *J. Range Manage.* 35:558–62

152. Seghieri J, Floret C, Pontanier R. 1995. Plant phenology in relation to water availability: herbaceous and woody species in the savannas of northern Cameroon. *J. Trop. Ecol.* 11:237–54

153. Sinclair ARE. 1979. Dynamics of the Serengeti ecosystem. In *Serengeti: Dynamics of an Ecosystem,* ed. ARE Sinclair, M Norton-Griffiths, pp. 1–30. Chicago: Univ. Chicago Press

154. Sinclair ARE, Fryxell JM. 1985. The Sahel of Africa: ecology of a disaster. *Can. J. Zool.* 63:987–94

155. Skarpe C. 1990. Shrub layer dynamics under different herbivore densities in an arid savanna, Botswana. *J. Appl. Ecol.* 27:873–85

156. Skarpe C. 1990. Structure of the woody vegetation in disturbed and undisturbed arid savanna, Botswana. *Vegetatio* 87:11–18

157. Skarpe C. 1992. Dynamics of savanna ecosystems. *J. Veg. Sci.* 3:293–300

158. Smith TM, Goodman PS. 1986. The role of competition on the structure and dynamics of *Acacia* spp. savannas in southern Africa. *J. Ecol.* 74:1031–44

159. Smith TM, Grant K. 1986. The role of competition in the spacing of trees in a *Burkea africana-Terminalia sericea* savanna. *Biotropica* 18:219–23

160. Solbrig OT, Medina E, Silva JF, eds. 1996. *Biodiversity and Savanna Ecosystem Processes: A Global Perspective.* New York: Springer-Verlag. 233 pp.

161. Soriano A, Sala O. 1983. Ecological strategies on a Patagonian arid steppe. *Vegetatio* 56:9–15

162. Stewart OC. 1955. Fire as the first great force employed by man. In *Man's Role in Changing the Face of the Earth,* ed. WL Thomas, pp. 115–33. Chicago, IL: Univ. Chicago Press

163. Strang RM. 1969. Soil moisture relations under grassland and under woodland in the Rhodesian highveld. *Com. For. Rev.* 48:26–40

164. Stuart-Hill GC, Tainton NM. 1989. The competitive interaction between *Acacia*

165. Stuart-Hill GC, Tainton NN, Barnard HJ. 1987. The influence of an *Acacia karroo* tree on grass production in its vicinity. *J. Grassl. Soc. South. Afr.* 4:83–88

166. Swetnam TW, Betancourt JL. 1990. Fire—southern oscillation relations in the southwestern United States. *Science* 249:1010–20

167. Szott LT, Fernandes ECM, Sanchez PA. 1991. Soil-plant interactions in agroforestry systems. *For. Ecol. Manage.* 45:127–52

168. Teague WR, Smit GN. 1992. Relations between woody and herbaceous components and the effects of bush clearing in southern African savannas. *J. Grassl. Soc. South. Afr.* 9:60–71

169. Thurow TL, Blackburn WH, Taylor CAJ. 1987. Rainfall interception losses by midgrass, shortgrass, and live oak mottes. *J. Range Manage.* 40:455–60

170. Tiedemann AR, Klemmedson JO. 1973. Effect of mesquite on physical and chemical properties of the soil. *J. Range Manage.* 26:27–29

171. Tiedemann AR, Klemmedson JO. 1973. Nutrient availability in desert grassland soils under mesquite (*Prosopis juliflora*) trees and adjacent open areas. *Soils Sci. Soc. Am. J.* 37:107–10

172. Tiedemann AR, Klemmedson JO. 1977. Effect of mesquite trees on vegetation and soils in the desert grassland. *J. Range Manage.* 30:361–67

173. Tiedemann AR, Klemmedson JO. 1986. Long-term effects of mesquite removal on soil characteristics: I. Nutrients and bulk density. *Soils Sci. Soc. Am. J.* 50:472–75

174. Tiedemann AR, Klemmedson JO. 1986. Long-term effects of mesquite removal on soil characteristics: II. Nutrient availability. *Soils Sci. Soc. Am. J.* 50:472–75

175. Tolsma DJ, Ernst WHO, Verweij RA, Vooijs R. 1987. Seasonal variation of nutrient concentrations in a semi-arid savanna ecosystem in Botswana. *J. Ecol.* 75:755–70

176. Tongway DJ, Ludwig JA. 1990. Vegetation and soil patterning in semi-arid mulga lands of Eastern Australia. *Aust. J. Ecol.* 15:23–34

177. Tothill JC, Mott JJ, eds. 1985. *Ecology and Management of the World's Savannas.* Canberra: Aust. Acad. Sci. 384 pp.

178. Trapnell CG. 1959. Ecological results of

woodland burning in Northern Rhodesia. *J. Ecol.* 47:161–72

179. Trollope WSW. 1974. The role of fire in preventing bush encroachment in the Eastern Cape. *Proc. Grassl. Soc. South. Afr.* 15:173–77

180. Trollope WSW. 1980. Controlling bush encroachment with fire in savanna areas of South Africa. *Proc. Grassl. Soc. South. Afr.* 15:173–77

181. Uresk DW, Severson KE. 1989. Understory-overstory relationships in ponderosa pine forests, Black Hills, South Dakota. *J. Range Manage.* 42:203–8

182. Van Vegten JA. 1983. Thornbush invasion in a savanna ecosystem in eastern Botswana. *Vegetatio* 56:3–7

183. Vetaas OR. 1992. Micro-site effects of trees and shrubs in dry savannas. *J. Veg. Sci.* 3:337–44

184. Virginia RA. 1986. Soil development under legume tree canopies. *For. Ecol. Manage.* 16:69–79

185. Vitousek PM, Walker LR. 1989. Biological invasion by *Myrica faya* in Hawai'i: plant demography, nitrogen fixation, ecosystem effects. *Ecol. Monogr.* 59:247–65

186. Walker BH, ed. 1987. *Determinants of Tropical Savannas.* Oxford, UK: IRL. 156 pp.

187. Walker BH. 1987. A general model of savanna structure and function. In *Determinants of Savannas,* ed. BH Walker, pp. 1–12. Paris: IUBS

188. Walker BH, Ludwig D, Holling CS, Peterman RM. 1981. Stability of semi-arid savanna grazing systems. *J. Ecol.* 69:473–98

189. Walker BJ, Noy-Meir I. 1982. Aspects of stability and resilience of savanna ecosystems. In *Ecology of Tropical Savannas,* ed. BJ Huntley, BH Walker, pp. 577–90. Berlin: Springer-Verlag

190. Walker J, Moore RM, Robertson JA. 1972. Herbage response to tree and shrub thinning in *Eucalyptus populnea* shrub woodlands. *Aust. J. Agric. Res.* 23:405–10

191. Walker J, Robertson JA, Penridge LK, Sharpe PJH. 1986. Herbage response to tree thinning in a *Eucalyptus crebra* woodland. *Aust. J. Ecol.* 11:135–40

192. Walter H. 1971. *Ecology of Tropical and Subtropical Vegetation.* Edinburgh, UK: Oliver & Boyd.

193. Welsh RG, Beck RF. 1976. Some ecological relationships between creosote-bush and bush muhly. *J. Range Manage.* 29:472–75

194. Weltzin JF, Archer S, Heitschmidt RK. 1997. Small mammal regulation of vegetation structure in a temperate savanna. *Ecology* 78:751–63

195. Weltzin JF, Coughenour MB. 1990. Savanna tree influence on understory vegetation and soil nutrients in northwestern Kenya. *J. Veg. Sci.* 1:325–34

196. Werner PA. 1991. *Savanna Ecology and Management: Australian Perspectives and Intercontinental Comparisons.* London: Blackwell Sci.

197. West NE, Rea KH, Harniss RO. 1979. Plant demographic studies in sagebrush-grass communities in southeastern Idaho. *Ecology* 60:376–88

198. Western D, Praet CV. 1973. Cyclical changes in the habitat and climate of an East African ecosystem. *Nature* 241:104–6

199. Westoby M. 1979/1980. Elements of a theory of vegetation dynamics in arid rangelands. *Israel J. Bot.* 28:169–94

200. Westoby M, WAlker B, Noy-Meir I. 1989. Opportunistic management for rangelands not at equilibrium. *J. Range Manage.* 42:266–74

201. White LP. 1969. Vegetation arcs in Jordan. *J. Ecol.* 57:461–64

202. White LP. 1970. "Brousse tigree" patterns in southern Niger. *J. Ecol.* 58:549–53

203. Whittaker RH, Gilbert LE, Connell JH. 1979. Analysis of a two-phase pattern in a mesquite grassland, Texas. *J. Ecol.* 67:935–52

204. Wright HA, Bunting SC, Neunschwander LF. 1976. Effect of fire on honey mesquite. *J. Range Manage.* 29:467–71

205. Wu H, Sharpe PJH, Walker J, Penridge LK. 1985. Ecological field theory: a spatial analysis of resource interference among plants. *Ecol. Model.* 29:215–43

206. Yeaton RI. 1988. Porcupines, fires and the dynamics of the tree layer of the *Burkea africana* savanna. *J. Ecology* 76:1017–29

207. Young JA, Evans RA, Easi DA. 1984. Stemflow on western juniper trees. *Weed Sci.* 32:320–27

208. Young MD, Solbrig OT. 1993. *The World's Savannas: Economic Driving Forces, Ecological Constraints and Policy Options for Sustainable Land Use.* Carnforth, UK: Parthenon

209. Zinke PJ. 1962. The pattern of influence of forest trees on soil properties. *Ecology* 43:130–33

Annu. Rev. Ecol. Syst. 1997. 28:545–70

PLANT COMPETITION UNDERGROUND

Brenda B. Casper

Department of Biology, University of Pennsylvania, Philadelphia, Pennsylvania 19104-6018; e-mail: bcasper@sas.upenn.edu

Robert B. Jackson

Department of Botany, University of Texas at Austin, Austin, Texas 78713

KEY WORDS: nitrogen and phosphate uptake, root plasticity, soil diffusion, soil water partitioning, symmetry of competition, nutrient heterogeneity, root and shoot interaction

ABSTRACT

Belowground competition occurs when plants decrease the growth, survival, or fecundity of neighbors by reducing available soil resources. Competition belowground can be stronger and involve many more neighbors than aboveground competition. Physiological ecologists and population or community ecologists have traditionally studied belowground competition from different perspectives. Physiologically based studies often measure resource uptake without determining the integrated consequences for plant performance, while population or community level studies examine plant performance but fail to identify the resource intermediary or mechanism. Belowground competitive ability is correlated with such attributes as root density, surface area, and plasticity either in root growth or in the properties of enzymes involved in nutrient uptake. Unlike competition for light, in which larger plants have a disproportionate advantage by shading smaller ones, competition for soil resources is apparently more symmetric. Belowground competition often decreases with increases in nutrient levels, but it is premature to generalize about the relative importance of above- and belowground competition across resource gradients. Although shoot and root competition are often assumed to have additive effects on plant growth, some studies provide evidence to the contrary, and potential interactions between the two forms of competition should be considered in future investigations. Other research recommendations include the simultaneous study of root and shoot gaps, since their closures may not occur simultaneously, and improved estimates of the belowground neighborhood. Only

0066-4162/97/1120-0545$08.00

by combining the tools and perspectives from physiological ecology and population and community biology can we fully understand how soil characteristics, neighborhood structure, and global climate change influence or are influenced by plant competition belowground.

INTRODUCTION

Much of the competition among plants takes place underground. In contrast to aboveground competition which primarily involves a single resource, light, plants compete for a broad range of soil resources, including water and at least 20 essential mineral nutrients that differ in molecular size, valence, oxidation state, and mobility within the soil. Belowground competition often reduces plant performance more than does aboveground competition (141), and it is the principal form of competition occurring in arid lands or other systems with extremely low plant densities (47).

We review the mechanisms and ecological importance of belowground competition, emphasizing the certainties and uncertainties that have made it a productive area of research. We begin by describing the processes by which soil resources reach roots and consider plant traits and soil properties likely to affect competition for those resources. We then discuss how belowground competition is measured, describing current methods and their limitations. Next, we consider several questions related to the roles of belowground plant competition in population structure, community organization, and vegetation dynamics. Finally, we suggest a number of research directions for the future. We attempt to combine the dual perspectives of physiological ecology and population or community ecology; one of our goals is to promote the merging of tools and perspectives from these disparate fields to foster progress in the future.

DEFINITION OF BELOWGROUND COMPETITION

We apply Goldberg's definition of competition (50), which takes into account how plants affect the abundance of an intermediary and how other plants respond to the change in abundance (Figure 1). For resource-mediated competition to occur belowground, a plant must have a negative effect on the availability of some belowground resource to which another plant shows a positive response in growth, survival, or reproduction (50). In other words, the reduced level of the intermediary has a negative impact on the performance of competing plants measured per individual or per unit size. In reality, studies typically focus on only one aspect of the competitive interactions described in Figure 1.

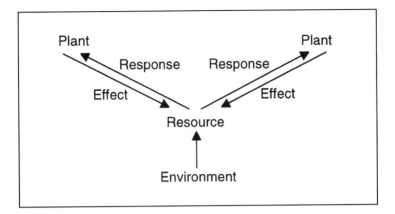

Figure 1 Plant competition as characterized by (50). In this framework plants must have an effect on the abundance of a resource and other plants must respond to the change. Both the effect and the response must be of appropriate sign for competition to occur.

Physiologically based studies commonly identify how resource uptake by one individual affects the quantity of resource taken up by another without determining the consequences for plant performance. Population or community level approaches rely on phenomenological responses of plants but usually fail to identify the intermediary resource.

Under our working definition of competition, the ability to take up soil resources and competitive ability are not necessarily correlated. For example, a plant may improve water uptake by growing a deeper root system and tapping a source of water unavailable to more shallow-rooted neighbors. Such habitat partitioning may not increase and could even decrease competition for water, although competition for mineral nutrients or light may increase as a consequence of more vigorous plant growth or increased plant densities. Likewise, drought may increase density-independent mortality of seedlings simply because their root systems are not well developed. Plants also differ in their ability to convert soil resources to biomass, referred to as water-use efficiency or nutrient-use efficiency (95), and these differences can affect relative plant growth rates across a soil resource gradient even in the absence of belowground interactions.

By limiting our discussion to resource-mediated plant competition, we ignore interference competition occurring through allelopathy (82), information-acquiring systems that allow plants to respond to neighbors prior to resource reduction (5a), and resource competition between plants and soil microbes (73). These are ecologically important but beyond the scope of this review.

RESOURCE UPTAKE AND MECHANISMS
OF COMPETITION

Soil resources reach the root surface through three general processes: 1. root interception, 2. mass flow of water and nutrients, and 3. diffusion (84). Root interception is the capture of water and nutrients as the root grows through the soil, physically displacing soil particles and clay surfaces. In general, root interception accounts for less than 10% of resource uptake by roots and is the least important of the three processes (84). Mass flow of water and dissolved mineral nutrients is driven by plant transpiration and is a function of the rate of water movement to the root and the concentration of dissolved nutrients in the soil solution. Diffusion of nutrients toward the root occurs when nutrient uptake exceeds the supply by mass flow, creating a local concentration gradient. Diffusion is especially important for nutrients with large fractions bound to the solid soil matrix, such as potassium and phosphate, whereas mass flow is often more important for nitrogen, particularly nitrate. Supply of the three major nutrients (N, P, and K) almost always depends on diffusion and mass flow working together (96), and the two processes are difficult, if not impossible, to separate experimentally in the field.

The size of the concentration gradient surrounding roots, referred to as the depletion zone, and the rate of ion diffusion depend on several factors (84). The effective diffusion coefficient (D_e) is the term applied to the mobility of nutrient ions; it depends on the ion's rate of diffusion in water, the volumetric water content of the soil, the impedance of the soil structure to ion movement through the aqueous fraction, and the ability of the soil matrix to release nutrients into the soil solution. Both the diffusion coefficient and the width of the depletion zone increase with soil water content. The width of the depletion zone also increases with overall nutrient concentration and with the root's ability to depress the nutrient concentration at its surface.

Among mechanisms of root interactions, competition via diffusion has received the most attention. Neighboring roots reduce nutrient uptake when nutrient depletion zones overlap (6, 96). For a given interroot distance, the degree of competition increases as effective diffusion increases, resulting in potentially greater competition for nitrate ions than for potassium or the relatively immobile phosphate ions. The relationship between the width of the diffusion zone and overall soil nutrient levels may imply that competition occurs at lower root densities in high nutrient soils than in low nutrient soils. The concept of overlapping diffusion zones is less applicable to water and dissolved nutrients that are primarily supplied to the root by mass flow. For those nutrients, competition must depend both on nutrient uptake and water uptake, driven by transpiration. Aboveground attributes such as maximum transpiration or stomatal conductance will both affect and be affected by rates of water uptake (113a).

TRAITS RELATED TO BELOWGROUND COMPETITIVE ABILITY

With a basic understanding of the processes involved in the acquisition of soil resources and the mechanisms by which competition for those resources takes place, we now consider morphological and physiological attributes likely to improve belowground competitive ability. Figure 2 provides a conceptual outline of important traits.

Root Surface Area and Rates of Resource Uptake

Of primary importance in belowground competition is the occupation of soil space. The ability to occupy space depends on several root characters, including relative growth rate, biomass, fine root density, and total surface area. As an example, Aerts et al (1) grew two evergreen shrubs and a perennial grass in a replacement series in the field with four competition treatments: no competition, aboveground only, belowground only, and above- and belowground together. The superior competitor in the experiment was the grass *Molinia caerulea*, which allocated three times the proportion of biomass to its roots as did either of the shrubs and was the only species to extend roots into the soil compartment of competitors. The authors concluded that its success was due in large part to

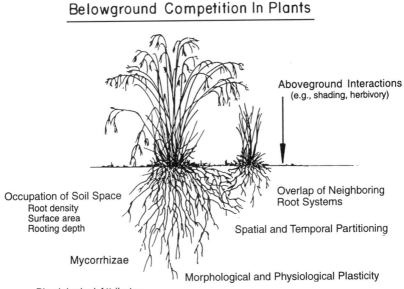

Figure 2   Plant traits that influence belowground competitive ability.

its high productivity and extensive root system. One of the only field experiments to measure root densities and nutrient uptake in competing root systems (20, 22) showed that root abundance alone was insufficient to explain relative nutrient uptake among three species in the sagebrush steppe. The non native tussock grass *Agropyron desertorum* had eight- to tenfold more roots in nutrient patches than did sagebrush, *Artemisia tridentata*, one week after the patches were created, and four- to sixfold more roots at three weeks. Despite this difference, the two species acquired the same amount of phosphate from the patches. The shrub also took up six to eight times more phosphate than did a native grass, *Pseudoroegneria spicata*, despite greater root densities for the grass.

Several factors may explain the lack of direct correspondence between root density and the outcome of belowground competition. First, competition may also occur among roots on the same plant, so the return per investment in new root growth may decline at higher root densities. Second, where and when roots are deployed may be just as important as average root density. A plant with much root surface area in one region of the soil might be poorly represented in a second region or less able to concentrate its roots in localized nutrient patches, or rooting density may vary temporally. Third, mycorrhizae play an important role but are frequently ignored in studies of nutrient acquisition, and fourth, physiological properties related to the rate of uptake are also crucial to competitive ability.

As a simplification, nutrient uptake by roots in most natural systems is governed by apparent Michaelis-Menten kinetics:

$$V = V_{max} C_l / (C_l + K_m) \qquad 1.$$

where V is the flux of ion into the root per unit time, V_{max} is the maximum such influx rate, K_m is the soil solution concentration where influx is $0.5 \times V_{max}$, and C_l is the soil solution concentration at the root surface (96). The equation is sometimes modified to include a C_{min} term, the soil solution concentration at which net influx into the root is zero (7). A species with more enzymes per root surface area (greater V_{max}), a higher ion affinity of enzymes (smaller K_m), or a greater ability to draw nutrients down to a low level (smaller C_{min}) will be at a competitive advantage, all else being equal.

Morphological and Physiological Plasticity

The ability to make morphological or physiological adjustments to the local soil environment may be critical to a plant's belowground competitive success.

Advantages of plasticity must be viewed both in terms of how much additional resource is taken up and how quickly because increasing the rate of uptake could be very important in the presence of competitors. Many plants respond to nutrient-enriched patches of soil by root proliferation (37, 38, 134). Proliferated roots tend to be smaller in diameter and greater in density than those found in the background soil, and consequently they have much greater surface area. In an experiment with eight British herbs, Campbell et al (25) showed that species with large rooting areas were less able to place roots selectively in high-nutrient patches, suggesting a trade-off between the ability to explore large soil volumes and that to exploit nutrient-rich patches. Beyond those experiments, we know little about scales of root foraging across broad groups of species. A factor rarely considered is fine root demography (59). When in situ root growth was examined for a hardwood forest in northern Michigan (100), roots were found to proliferate in response to localized water and nitrogen additions, and new roots in enriched patches lived significantly longer than new roots in control patches. Lengthening the lifespan of a root may be just as effective for a plant as growing a new cohort of roots and potentially less expensive (41).

Architectural adjustment is another type of morphological plasticity with the potential to increase resource capture (44, 45). While selective biomass allocation is architectural in a broad sense, we refer instead to local changes in topology, root length, or branching angles. Fitter (44) examined root architectural attributes of 11 herbaceous species. On average, roots in relatively high-nutrient patches had more of a herringbone branching pattern than did roots in low-nutrient patches. The herringbone pattern can increase the efficiency of nutrient uptake by concentrating higher order lateral roots in the enriched patch (45).

Physiological plasticity involves changes in uptake rates attributable to altered enzyme attributes or other physiological traits. In the case of mineral nutrients, plants in the laboratory and in the field both increase V_{max} and decrease K_m in response to local increases in nutrient concentration (37, 71, 79). For water, osmoregulation can lower cell water potential and maintain net uptake in the face of drying soils (15, 117). To show that plants are able to increase uptake by selectively altering physiological attributes is different from predicting when and if it is beneficial to do so. Another factor not usually considered is the cost involved in constructing or operating additional enzymes (40). For mineral nutrients, mobility within the soil is also important (16). When competition occurs through overlapping depletion zones, physiological plasticity should increase the uptake of relatively mobile nutrients (e.g. nitrate) more than the uptake of less mobile ones (e.g. ammonium, phosphate). Conversely, root proliferation may be less beneficial for the uptake of relatively mobile nutrients, since a single root depletes a broader volume of soil (96, 109). For extensive overviews of morphological and physiological plasticity and resource capture, see (33, 66, 108).

Spatial and Temporal Soil Partitioning

Some spatial and temporal rooting patterns seem to reduce belowground competition, thereby probably falling into the general category of niche separation. Although the majority of roots are found within the top 30 cm of soil, some reach great depths (70, 92, 106, 119). At least 50 species grow roots more than 5 m deep, 22 species reach below 10 m, and several desert species reach 50 m (26). These numbers almost certainly underestimate the importance and frequency of deep roots in many ecosystems because the study of deep soil

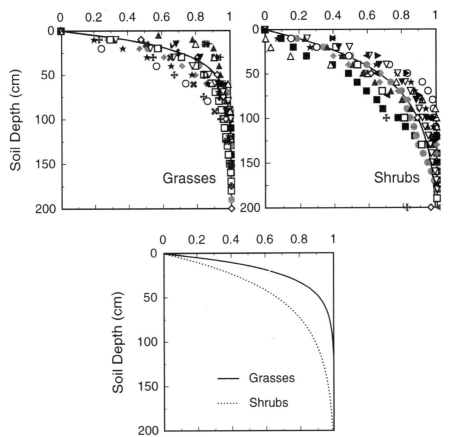

Figure 3 The distribution of grass and shrub roots as a function of soil depth. The data are from deserts, temperate grasslands, and tropical grasslands where the two growth forms potentially co-occur. See (70) for a key to the symbols.

has been largely neglected. Whether deep roots are engaged in belowground competition depends on the rooting depth of neighboring plants.

The lack of secondary thickening in roots of grasses and many herbs places a morphological limit on their ability to grow deep roots and provides an a priori rationale for separating plants into different belowground functional groups. Walter (131) first proposed the two-layer model of deep- and shallow-water partitioning between grasses and shrubs/trees in subtropical savannas. This simple model remains a useful way of subdividing vegetation by resource use (78, 105, 112, 129), and a recent comprehensive analysis (70) showed distinctly different rooting profiles for plants of the two groups (Figure 3). Analyses of hydrogen isotope ratios in plant tissues verified that the two groups acquire water from different depths (39). Some spatial and functional root overlap does occur between grasses and woody plants, however. Some grass roots are more than 5 m deep (134), and many woody plants are capable of taking up resources from both shallow and deep layers (32).

Spatial and temporal partitioning of soil resources can be related. Deep roots may allow plants access to a water source available after upper soil layers have dried out, enabling them to decouple the timing of growth from rainfall events, persisting after neighboring species have died or become dormant. Examples of temporal partitioning include early and late season annuals in the Mediterranean climate of California (88), shrub species of the Great Basin (42), and various trees (80).

MEASUREMENTS OF BELOWGROUND COMPETITION

Belowground competition is measured by quantifying the extent that root interactions reduce resource uptake, vegetative growth, or fecundity. Population or community level approaches generally estimate belowground competition from biomass increases when interactions with neighboring roots are prevented through the use of root exclusion tubes, trenching, or neighbor removal. Such methods often alter the soil environment and may even affect the availability of resources for which the plants are competing. Resource-based approaches usually involve less manipulation of the environment, but the integrated consequences of competition for plant performance are often unknown or require assumptions of scaling. Combining approaches is necessary to document both the resource intermediary and the ecological significance of belowground competition.

Root exclusion tubes (30, 103) are frequently employed in population or community level studies. Typically, these cylindrical steel or plastic partitions are inserted into the soil to separate the roots of target individuals, usually transplanted seedlings, from those of neighboring plants. Root competition is determined by comparing the growth or survival of target plants inside the partitions with those having roots systems that can interact freely with neighboring vegetation. Neighbors within partitions are killed by a fast-degrading herbicide, severed at the soil surface to remove shoots, or removed completely by excavating, sieving, and replacing the soil. By varying the length of the tubes, some studies have determined the soil depths at which the belowground competitive interactions take place (30, 103). Two related methods are neighbor removal without using exclusion tubes and trenching, which severs the roots of potential competitors so they do not extend into the rooting area of target plants. Trenching is often used for woody species with extensive root systems (10); a good recent example shows strong root competition between alder and spruce (29). Both neighbor removal and trenching attempt to eliminate belowground interactions without using a physical barrier (though some trenching experiments insert a root-restriction fabric to impede regrowth) and without much physical disturbance of the soil.

These methods all suffer from the possibility that soil resources are altered by the experimental protocol. A major worry is that roots left to decompose may release bursts of nutrients (11, 101) or result in nutrients being sequestered by rapidly growing populations of soil microorganisms (34). A greenhouse bioassay experiment (87) addressed this issue by comparing growth of two target plant species over three months in undisturbed field soil, soil that had been sieved to remove roots, and sieved soil to which roots had been added back. Root removal did not affect the performance of either species, but for the mycorrhizal species, soil disturbance lowered mycorrhizal infection and shoot growth. The results suggest that shoot removal alone or the application of herbicides may be preferable to removing roots physically.

Two additional concerns are related to potential differences in root growth rates in empty versus root-occupied soil and the extent that exclusion tubes create undesirable side effects. Starting with empty soil in treatments where neighbors' roots are allowed access to the space (49) may be comparable to greenhouse pot experiments, which Grubb (60) suggested could give fundamentally different measures of competition than field experiments in which root systems are already in place. He proposed that differences in the rates that roots fill vacant soil space may explain why greenhouse experiments, more often than field experiments, show an increase in root competition with increasing nutrient levels. Undesirable side effects of root exclusion tubes include restricting lateral movement of water and nutrients. Newer, more sophisticated "tubes" employ modern fabrics with small pore sizes that permit passage of water but not nutrients and exclude roots but not mycorrhizal hyphae (81). Since soil space itself may be a resource (85, 86), consideration should also be given to tube diameter and whether restricting lateral root spread has consequences for target plant biomass.

Physiologically based approaches identify the currency of the competition, examining the effect of belowground competitors on the acquisition of particular resources. In a study that documented interspecific differences in competitive ability, Caldwell et al (19) used a dual-labeling approach with ^{32}P and ^{33}P to examine competition in the sagebrush and grass system described earlier. By growing each grass species on opposite sides of the same sagebrush plants, and placing one label in the soil on each side of the shrub, they showed that the shrub took up 86% of its phosphate from the side shared with *P. spicata*. Although an elegant demonstration of interspecific effects on resource uptake, it is unclear how such short-term differences affect long-term plant performance.

EFFECTS OF THE SOIL ENVIRONMENT

Ecologists use the above methods to examine the role of belowground competition in natural systems, a subject that can be separated into two broad topics. The first deals with the influence of the soil environment on belowground competitive interactions, and the second examines how characteristics of neighboring plants figure in the interactions.

Productivity Gradients

One active line of research concerns whether the strength of belowground competition changes in a consistent way over gradients in habitat productivity. The research is motivated by two models with different predictions about the relationship between competition and soil resource levels. Grime (54, 55) suggested that more productive sites should exhibit greater overall importance in competition for both aboveground and belowground competition. Tilman's (120, 121) resource ratio model predicted no change in total competition as productivity increases but an increase in the ratio of aboveground to belowground components.

Experimental tests of these models require separating the effects of above- and belowground competition on plant growth. Root competition is measured as growth reduction in the presence of only neighboring roots, while total competition is measured as growth reduction in the presence of both roots and shoots of neighbors. Many studies do not measure shoot competition directly but calculate shoot competition as the difference between total competition and root competition. Most calculate competition as competitive intensity, where growth reduction is expressed on a percentage basis. This standardizes measurements to account for differences between sites in overall plant growth rates (51). Experimental systems include artificial productivity gradients created by adding fertilizer, and natural productivity gradients, where biomass standing crop is often used as an indicator of soil resource levels.

Experimental studies assessing competition as a function of habitat productivity often show a decrease in the magnitude of belowground competition with increased soil resource levels. It is premature, however, to generalize about the relative importance of above- and belowground competition under various soil resource conditions. In a Minnesota old field, the terrestrial system where Tilman has tested his model, nitrogen addition resulted in decreased belowground competition but no change in total competition for most species (142–144). Another study using tree seedlings planted under shrubs in three abandoned agricultural fields obtained similar results (102). This study measured aboveground and belowground competition independently and found that competition was largely belowground in the most xeric, nutrient-poor site, whereas shoot competition was stronger than root competition in the most productive site. Root and shoot competition acted together to reduce plant growth at the intermediate site. In contrast, a study comparing competition in two wetland habitats differing in standing biomass found that both the forb *Lythrum salicaria* and the sedge *Carex crinita* experienced greater aboveground and total competition in the more productive site, but the gradient in belowground competition differed between species (124). Belowground competition increased with productivity for *C. crinita* but decreased for *L. salicaria*.

Since standing biomass is often used as a proxy for resource levels, it is important to understand the relationship between neighbor biomass and competitive intensity. For example, Belcher et al (9) recognized that the relationship may be nonlinear, their explanation for why total competitive intensity did not increase over a soil depth gradient in shallow limestone-derived soils in Ontario. In that system, standing biomass was positively correlated with soil depth, but neither total competition nor root competition was correlated with standing biomass or soil depth; aboveground competition was insignificant. They suggested that competition may increase with neighbor biomass until some asymptotic level of competition is reached; additional biomass would have no additional effect on competition. An asymptotic relationship would mean that the inclusion of sites with relatively low standing crop would be more likely to detect a correlation between biomass and competition.

Belowground competition is thought to be the main component of competition in arid systems because low plant densities often result in minimal shading by neighbors. While we know of no studies specifically examining relative changes in root and shoot competition over aridity gradients, total competition has been compared among sites differing in soil moisture in desert or semi-desert habitats (61, 72, 98). Results often point to competition increasing as water becomes more abundant and productivity increases. Rainfall patterns can result in temporal soil moisture gradients, and the magnitude of density-dependent interactions also seems to decrease in dry years (8, 27, 46, 72). Without knowing the relative contributions of aboveground and belowground interactions, one cannot determine whether competition for water really increases with water availability—one explanation for these results. Under extreme aridity, shading can be beneficial by ameliorating temperatures and lowering evaporative water loss from plants and soil. These benefits may sometimes outweigh any detrimental effects of belowground or aboveground competition and result in facilitation between closely associated individuals (23, 53).

Belowground competition may not necessarily decrease with increasing levels of soil resources for at least two reasons. First, standing crop usually increases with soil resource levels, and many of the above studies found correlations between neighbor aboveground biomass and total competitive intensity. Of relevance to understanding the role of soil resources in affecting competition is whether the relationship between competition and standing biomass varies with soil resource levels; both water addition and nutrient addition have been found to change the slope of the regression relationship (72, 144). Second, understanding how nutrient levels affect the size of depletion zones leads to the prediction that competition will sometimes take place at lower root densities with increasing nutrients. Even though nutrient uptake per length of root may be greater under high nutrient conditions, the potential for uptake to be affected by neighboring roots may also be greater there. Some workers have argued that the development of depletion zones may result in strong competition even in fertile soils (24, 56, 57, 141), and the same could hold for water as well.

Heterogeneity in Soil Resources

Competition for soil resources may also change with their spatial and temporal distribution. The practical difficulties such variation presents for predicting crop yields were recognized early this century (133), and natural systems are just as variable at both fine and coarse scales (107, 118). While heterogeneity at scales as large as or larger than the rooting area of a plant is likely to affect relative plant performance and contribute to the maintenance of species diversity (123), fine-scale heterogeneity, occurring where root systems overlap, should have the greatest direct impact on belowground interactions. Campbell et al (25) suggested that small-scale heterogeneity may be even more important than average soil nutrient levels in determining competitive outcome. Likewise, Gross et al (59) suggested that small-scale heterogeneity, which increases in old fields over time, may be a factor contributing to successional vegetation changes in that habitat.

Numerous studies have documented fine-scale variation in nature and the ecological importance of its occurrence. In a classic study of *Trifolium repens*,

Snaydon (118) showed substantial variation in soil pH, calcium, and phosphate at scales of less than a meter in a Welsh grassland; documenting this variation was critical for understanding the local distribution of the species. In a later study, Jackson & Caldwell (68) found on average that soil ammonium and nitrate varied more than ten-fold in the rooting zone of individual plants in the sagebrush steppe. Phosphate at the same site varied three-fold around single plants, and soil pH differed up to 1.3 units in samples less than half a meter apart. A more recent study in the same system has shown that such spatial variability in mineral nutrients and water does not remain constant (111); both spatial patterning and the scale of heterogeneity changed over a single growing season.

Understanding how heterogeneity influences competitive outcome is an area that would especially benefit from integrating resource-based and higher level approaches. Much is known about the ability of roots to respond morphologically and physiologically to nutrient patches, but less is understood about how patchiness influences competition. There are two general possibilities. If co-occurring species differ simply in ability to harvest soil resources from patches, then heterogeneity may affect their relative performance, independent of competition. A second possibility is that heterogeneity directly alters the dynamics of root interactions. Since roots can proliferate in nutrient-rich patches, heterogeneity may result in spatial aggregation of competing root systems, potentially intensifying belowground competition.

Most information regarding the consequences of heterogeneity on plant performance come from studies of isolated plants grown in pots or hydroponic media. Plant growth is often enhanced when nutrients are patchily distributed in space and time (11a, 108). In experiments varying the spatial distribution of phosphate to potted crop plants plant growth increased as the same amount of phosphate was applied to smaller fractions of the soil volume (5, 14). This is thought to occur because a localized, more concentrated application of phosphate increases the amount in the soil solution proportionately more than it does the amount bound to the solid phase in forms unavailable to plants (75). While there are no similar data from field studies, model simulations based on measurements of spatial nutrient variability and root plasticity in the sagebrush steppe showed that spatial heterogeneity should result in more uptake of nitrate and phosphate than when the same quantities of nutrients are uniformly distributed (69).

At the population level, heterogeneity in the form of alternating 640 cm^3 patches of high and low nutrients was nearly inconsequential for the population structure of *Abutilon theophrasti* when compared to populations growing on a homogeneous mixture of the same two soil types (28). Heterogeneity increased productivity only at an intermediate planting density and slightly lowered mortality overall. Heterogeneity also did not affect plant size variation within the population, which would be expected if plants competed for patches, but did influence the ranking of plants within the population size hierarchy. Plants with stems located on high-nutrient patches were larger than individuals on low nutrient patches, probably because they grew faster as seedlings and gained dominance within the population. In populations of *Ocimum basilicum,* localization of nutrients within a thin horizontal layer increased local rooting densities, presumably increasing belowground competition, and lowered the intercept of the self-thinning line for both root biomass and shoot biomass (90). This occurred only when nutrients elsewhere were extremely limited.

Evidence that neighbors can affect access to resource patches is limited. Caldwell et al (21) removed frozen soil cores from interspaces between individuals of sagebrush and the same two grass species used in previous experiments and quantified the distribution and identity of fine roots within the cores. Roots of the sagebrush were more abundant when the soil space was shared with *P. spicata* than with *A. desertorum.* When nutrient patches were created between plants by fertilizing with a solution containing nitrate and phosphorus, both overall rooting densities and the proportionate representation of sagebrush roots increased. The proximate cause of these particular rooting distributions is not known. Other studies have found little evidence for species-specific root recognition among the same three species (65, 76). Species-specific root avoidance has been demonstrated, however, in at least one system (82).

THE BELOWGROUND COMPETITIVE NEIGHBORHOOD

In this section we explore belowground competition within the context of population and community structure. We discuss interactions between aboveground and belowground competition, the relationship between plant size and belowground competitive ability, the structure of the belowground neighborhood, and the potential importance of root gaps in community dynamics.

Interactions Between Root and Shoot Competition

The practice of calculating shoot competition from direct measurements of root competition and total competition is based on the assumption that aboveground competition and belowground competition are additive in their effects on plant growth. Some researchers acknowledge that an interaction between the two forms of competition is likely, that shoot competition may affect a plant's belowground competitive ability or vice versa, but they assume either that the interaction is not important or that the direction and magnitude of the interaction does not change over the resource gradients with which they work (9, 124, 142). Often cited is JB Wilson's review of root competition studies

(141), most conducted in the greenhouse, from which he concluded that non-additive interactions of root and shoot competition are rare.

Where the consequences for plant performance of root and shoot competition have been measured separately, results have suggested nonadditive effects of the two forms of competition for some species (35, 36, 48, 49, 102). In a study examining how interactions with two vine species affected saplings of *Liquidambar styraciflua*, clear nonadditive effects of aboveground and belowground competition were demonstrated for *Lonicera japonica* (35). Root interactions alone reduced sapling growth, but shoot interactions reduced growth only if root interactions were also present. A physiologically based study showed that shading reduced nutrient uptake in a perennial tussock grass (67); this is one way that above- and belowground competition could interact to affect plant growth, but there are others.

Field experiments have also provided evidence that interactions between root and shoot competition do not remain constant in different resource environments (102). JF Cahill (17a) measured separate effects of root and shoot competition and their combined effects on the growth of *Abutilon theophrasti* in an abandoned agricultural field. He found a strong interaction between below- and aboveground competition that changed in direction with the addition of nutrients. Where fertilizer had not been added, the sum of root competitive intensity and shoot competitive intensity measured separately was much greater than total competitive intensity measured as the combined effects of root and shoot interactions. With fertilizer addition, the sum of root competitive intensity and shoot competitive intensity was less than total competitive intensity. These results suggest that indirect measures of shoot competition may be inaccurate and that interactions between belowground and aboveground competition are physiologically complex.

The Symmetry of Belowground Competition

The extent to which plant size confers a competitive advantage may be an important difference between belowground and aboveground competition. The disproportionate advantage of size in competing for light occurs because larger plants shade smaller ones (137). The suppression of smaller plants results in the development of size hierarchies that become more pronounced over time. Plants that are larger aboveground do not seem to enjoy a similar disproportionate advantage in competing for belowground resources (137). Data come mostly from pot experiments in which partitions are installed to prevent the interaction of neighbor shoots (93, 136, 138, 141). Belowground competitive ability appears to be size-symmetric; root interactions with neighbors reduce plant growth but do not increase size variation among competing individuals. A recent study examining nitrate and ammonium uptake as a function of plant size provided corroborating physiological evidence. Within crowded populations of yellow birch seedlings were grown in tubs, individuals acquired ^{15}N tracer in direct proportion to several measures of root system size (132).

Fewer data regarding the relationship between size and belowground competitive ability are available for plants in field situations in which roots are not confined to pots and soil resources are more spatially heterogeneous. One field study varied the size of transplanted seedlings for six species used as target plants and found growth response to belowground competition to be in direct proportion to size of transplant (48). In considering situations where root competition might become asymmetric, Schwinning & Weiner (115) suggested the possibility that larger root systems would have a disproportionate advantage in a patchy soil environment because they should be more likely to encounter a high-nutrient patch (114). These authors also recognized that the relationship between root system size and resource uptake could depend on the mobility of the limiting resource.

Size of the Belowground Neighborhood

The amount of belowground competition experienced by an individual will be a function of both the sizes of neighbors and their numbers. How many plants make up the belowground neighborhood? Does root overlap and belowground competition primarily occur among nearest neighbors, or is lateral root spread so extensive that the belowground neighborhood includes many additional plants? Because of the difficulty of excavating roots, especially fine roots that are most involved in resource uptake, few realistic estimates of total rooting area or root system overlap exist.

When rooting area is larger or smaller than the spread of the canopy, aboveground neighborhoods and belowground neighborhoods may be different sizes. In arid and semiarid areas, the lateral spread of a plant belowground is almost always greater than the spread of the canopy aboveground (104). The belowground area occupied by dominant shrubs in a California chaparral, for example, was over 10 times the area occupied by their canopy (77). One study estimated the zone of nutrient uptake for the bunchgrass *Schizachyrium scoparium* in an old field habitat by killing all neighbors within a 2 m radius of target individuals and then measuring the distances over which target plants reduced soil nitrogen (122). A single clump of *S. scoparium* was found to forage for nitrogen over a 1–1.5 m^2 area, which is large enough to include hundreds of other plants.

Potential rooting area and the area occupied in the presence of neighbors are not necessarily the same, however. One population level study found the root systems of 32 excavated creosote shrubs to be shaped as irregular polygons, which seemed to reflect growth away from the greatest competitive pressure of adjacent plants (17). The irregular geometry of the roots filled soil space more effectively than would circular rooting areas, and root overlap occurred among four or more neighbors in only a small portion of the population. The study ignored roots smaller than 2 mm diameter, a factor that may not have affected measurements of geometry but certainly affected estimates of root overlap. Similarly, root systems in stands of loblolly pine and sweetgum were not circular in area, and structural roots also seemed to avoid those of other individuals (91).

Another consideration is how the distribution of neighboring root systems changes with the spatial distribution of nutrients. We expect spatial heterogeneity to affect the distribution of neighboring root systems based on evidence of root morphological plasticity within nutrient patches. In the study of loblolly pine and sweetgum, fine-root densities increased with local soil resource levels so that, at the population level, fine-root biomass and aboveground biomass were not correlated (91). Another study examined distributions of roots of the palm *Borassus aethiopum* in relation to nutrient patches associated with termite mounds and clumps of other trees in a humid savanna (89). Although rooting densities were highest within 2–3 m of the palm trunk, some roots extended laterally an impressive 20 m in accessing a nutrient-rich patch.

Root Gaps

Openings in the belowground neighborhood are likely to be important for plant establishment. The extent to which root gaps are associated with canopy openings and the role of root gaps in the dynamics of gap succession have only recently been considered. A series of studies have documented the importance of belowground gaps for seedling establishment in the shortgrass steppe (2, 3, 62, 63). Recruitment is apparently limited by water availability and is most successful in larger gaps, where rooting densities of neighboring adults are lowest. Root gaps formed by death of fine roots have also been documented in a hardwood forest (140); root death occurred after canopy openings were formed by felling trees to simulate bole breakage without disturbing the soil by forming tip-up mounds. Such gaps may be short-lived. Data from a similar study in a *Pinus menziesii* forest showed little evidence for the maintenance of fine root gaps six months after aboveground gaps as large as 50 m in diameter were formed (127). Fine roots were still lacking from mineral soil after that time, but they were found in other substrates such as buried pieces of dead wood, again suggesting that standing trees exploited nutrient-rich patches over great distances. Root competition is recognized as a potentially important factor limiting establishment of woody and herbaceous seedlings in forest communities (31, 64, 140). Root dynamics as well as nutrient dynamics (99, 125, 126) may be important determinants of the course of gap succession.

FUTURE DIRECTIONS

A full understanding of when and how belowground competition takes place and its ecological importance requires input from many fields of biology. While much is known about how roots respond to their soil environment, we are far from linking the physiological and growth responses of roots to the ways that plants affect each other. Furthermore, we are inadequately prepared to predict when and how strongly belowground competition occurs. Progress would be helped by experiments integrating the resource emphasis of physiological ecology and measurements of plant performance at the population or community level (58). Successful integrative approaches should identify the resources for which plants are competing, the mechanisms involved, and the ecological impact of the competitive interactions.

Mycorrhizae in Belowground Competition

In this review, we have given little consideration to mycorrhizae despite their obvious importance in the belowground community. The ways in which mycorrhizae alter root interactions are likely very complicated. Their presence can either increase competition (139) or cause nutrient sharing (94), and plants' reliance on mycorrhizae may be density-dependent (74). Mycorrhizae may also enable plants to capture quantities of nutrients that would otherwise be inaccessible (4), potentially increasing the pool of belowground resources available to competing individuals. Much is yet to be learned about these mechanisms and the role of mycorrhizae in affecting community composition and population structure (4).

Competition for Water

Although thousands of published studies deal with manipulating water availability to plants, we still have a poor understanding of how and under what circumstances competition for water occurs. Paying greater attention to the mechanisms of competition for water and measuring the strength of belowground competitive interactions under different conditions of water availability should determine whether competition really does increase across spatial or temporal gradients in soil moisture and the extent to which the increase is explained by correlated changes in plant growth or biomass. It is important to separate the phenomenon of water availability from plant competition for water; that water limitations may be greatest in arid systems does not necessarily mean that competition for water is greatest there.

Understanding Interactions Between Root and Shoot Competition

With the exception of arid systems, in which aboveground competition may be minimal, above- and belowground competition act together to affect plant performance. It should no longer be assumed that these forms of competition cause additive reductions in plant growth, given empirical evidence for non-additive interactions between the two. Experimental approaches should also allow for the possibility that neighbors could have opposing effects above- and belowground. For example, at least one old field species benefits from root interactions with neighbors (144), and it is reasonable to expect that belowground facilitation may sometimes co-occur with competition for light. Until more is understood about the frequency and causes of interactions between above- and belowground competition, their relative importance cannot be evaluated comprehensively. Here, too, combined research approaches could usefully identify the physiological bases of important ecological phenomena.

The relative importance of the two forms of competition should especially be considered is in the recovery of vegetation gaps. Because closure of root gaps and canopy gaps may not occur simultaneously, we suggest that succession within gaps should be examined separately above- and belowground. The degree that these processes occur independently and the degree to which root and shoot competition interact to affect plant growth have implications for the composition of successional vegetation.

Defining the Belowground Neighborhood

Determining the size of plant root systems and the area over which they take up resources is important for constructing neighborhood models of plant competition (13, 116, 130, 135). Built with the assumption that most interactions occur among nearby plants, these models are commonly used to predict size distributions in populations, population dynamics, or the outcome of interspecific encounters. The approach has been extended to examine how competitive outcomes may change with abiotic factors such as spatial heterogeneity in soil resources (12, 97), but the models are designed without much information about the structure of populations belowground and how that structure may change with plant density or variation in the soil environment. The application of radioactive tracers or stable nutrient analogs, tools used extensively in physiological studies (18, 83), could help define the area of nutrient uptake and thus the size of the belowground neighborhood.

Belowground Competition and Global Change

Whatever the current importance of belowground competition, it is likely to increase with predicted changes in the earth's atmosphere. The projected doubling in CO_2 by the end of the next century (113) should increase photosynthetic rates, at least for C_3 plants, and decrease carbon limitation for plants globally. Furthermore, root biomass generally increases in elevated CO_2 experiments (110), suggesting an increase in the importance of belowground competition for water and nutrients.

Progress in at least three belowground research areas would help in predicting vegetation responses to global change. First, no global rooting map exists for use by modelers, though recent progress has been made (26, 70, 119, 128). Such a map should provide a global description of total biomass, fine root biomass, and fine root surface area as a function of soil depth, as well as the maximum rooting depth for the dominant vegetation in each biome. This information is important if global models are to represent belowground phenomena in a realistic manner. Second, a better grasp of the interaction between aboveground and belowground competition would also improve predictions of community-level consequences of atmospheric changes (43). Third, progress is needed in the grouping of plants into meaningful belowground functional types and in determining their distribution worldwide. A better understanding of belowground competition is needed—now more than ever before—if we are to meet the challenge of predicting biotic responses to the altered environment.

ACKNOWLEDGMENTS

BBC dedicates this paper to the memory of her father, D Bowers, who passed away while it was being written.

We wish to thank M Caldwell, J Cahill, D Eissenstat, N Fowler, D Goldberg, C Hawkes, and J Weiner for reviewing and improving earlier drafts of the manuscript and J Cahill for helping to develop many of the ideas. We acknowledge the support of NSF grant IBN-95-27833 to BBC and NIGEC-DOE (TUL-038-95/97) and Andrew W Mellon Foundation grants to RBJ.

> Visit the *Annual Reviews home page* at
> http://www.annurev.org.

Literature Cited

1. Aerts R, Boot RGA, van der Aart PJM. 1991. The relation between above- and belowground biomass allocation patterns and competitive ability. *Oecologia* 87:551–59

2. Aguilera MO, Lauenroth WK. 1993. Seedling establishment in adult neighborhoods—intraspecific constraints in the regeneration of the bunchgrass *Bouteloua gracilis. J. Ecol.* 81:253–61

3. Aguilera MO, Lauenroth WK. 1995. Influence of gap disturbances and type of microsites on seedling establishment in *Bouteloua gracilis. J. Ecol.* 83:87–97

4. Allen EB, Allen MF. 1990. The mediation of competition by mycorrhizae in successional and patchy environments. See Ref. 52, pp. 367–89

5. Anghinoni I, Barber SA. 1980. Phosphorus application rate and distribution in the soil and phosphorus uptake by corn. *Am. J. Soil Sci. Soc.* 44:1041–44

5a. Aphalo PJ, Ballaré CL. 1995. On the importance of information-acquiring systems in plant interactions. *Funct. Ecol.* 9:5–14

6. Baldwin JP. 1976. Competition for plant nutrients in soil: a theoretical approach. *J. Agric. Sci.* 87:341–56

7. Barber SA. 1984. *Soil Nutrient Bioavailability.* New York: Wiley Intersci.

8. Bauder ET. 1989. Drought and competition effects on the local distribution of *Pogogyne abramsii. Ecology* 70:1083–89

9. Belcher JW, Keddy PA, Twolan-Strutt L.

1995. Root and shoot competition intensity along a soil depth gradient. *J. Ecol.* 83:673–82

10. Belsky AJ, Amundson RG, Duxbury JM, Riha SJ, Ali AR, Mwonga SM. 1989. The effects of trees on their physical, chemical, and biological environments in a semi-arid savanna in Kenya. *J. Appl. Ecol.* 26:1005–24

11. Berendse F. 1983. Interspecific competition and niche differentiation between *Plantago lanceolata* and *Anthoxanthum odoratum* in a natural hayfield. *J. Ecol.* 71:379–90

11a. Bilbrough CJ, Caldwell MM. 1997. Exploitation of springtime ephemeral N pulses by six Great Basin species. *Ecology* 78:231–43

12. Biondini ME, Grygiel CE. 1994. Landscape distribution of organisms and the scaling of soil resources. *Am. Nat.* 143:1026–54

13. Bonan GB. 1993. Analysis of neighborhood competition among annual plants: implications of a plant growth model. *Ecol. Model.* 65:123–36

14. Borkert CM, Barber SA. 1985. Soybean shoot and root growth and phosphorus concentration as affected by phosphorus placement. *Am. J. Soil Sci. Soc.* 49:152–55

15. Boyer JS. 1995. *Water Relations of Plants and Soils.* San Diego: Academic

16. Bray RH. 1954. A nutrient mobility concept of soil-plant relationships. *Soil Sci.* 78:9–22

17. Brisson J, Reynolds JF. 1994. The effect of neighbors on root distribution in a creotebush (*Larrea tridentata*) population. *Ecology* 75:1693–702

17a. Cahill JF. 1997. *Symmetry, intensity, and additivity: belowground interactions in an early successional field.* PhD diss. Univ. Penn, PA

18. Caldwell MM, Eissenstat DM. 1987. Coping with variability: examples of tracer use in root function studies. In *Plant Response to Stress—Functional Analysis in Mediterranean Ecosystems,* ed. JD Tenhunen, FM Catarino, OL Lange, WC Oechel, pp. 95–106. Berlin: Springer-Verlag

19. Caldwell MM, Eissenstat DM, Richards JH, Allen MF. 1985. Competition for phosphorus: differential uptake from dual-isotope-labeled soil interspaces between shrub and grass. *Science* 229:384–86

20. Caldwell MM, Manwaring JH, Durham SL. 1991. The microscale distribution of neighboring plant roots in fertile soil mi-

crosites. *Funct. Ecol.* 5:765–72

21. Caldwell MM, Manwaring JH, Durham SL. 1996. Species interactions at the level of fine roots in the field: influence of soil nutrient heterogeneity and plant size. *Oecologia* 106:440–47

22. Caldwell MM, Manwaring JH, Jackson RB. 1991. Exploitation of phosphate from fertile soil microsites by three Great Basin perennials when in competition. *Funct. Ecol.* 5:757–64

23. Callaway RM. 1995. Positive interactions among plants. *Bot. Rev.* 61:306–49

24. Campbell BD, Grime JP, Mackey JML, Jalili A. 1991. The quest of a mechanistic understanding of resource competition in plant communities: the role of experiments. *Funct. Ecol.* 5:241–53

25. Campbell BD, Grime JP, Mackey JML, Jalili A. 1991. A trade-off between scale and precision in resource foraging. *Oecologia* 87:532–38

26. Canadell J, Jackson RB, Ehleringer JR, Mooney HA, Sala OE, Schulze E-D. 1996. Maximum rooting depth for vegetation types at the global scale. *Oecologia* 108:583–95

27. Casper BB. 1996. Demographic consequences of drought in the herbaceous perennial *Cryptantha flava*: effects of density, associations with shrubs, and plant size. *Oecologia* 106:144–52

28. Casper BB, Cahill JF, Jr. 1996. Limited effects of soil nutrient heterogeneity on populations of *Abutilon theophrasti* (Malvaceae). *Am. J. Bot.* 83:333–41

29. Chapin FSI, Walker LR, Fastie CL, Sharman LC. 1994. Mechanisms of primary succession following deglaciation at Glacier Bay, Alaska. *Ecol. Monogr.* 64:149–75

30. Cook SJ, Ratcliff D. 1984. A study of the effects of root and shoot competition on the growth of green panic (*Panicum maximum* var. trichoglume) seedlings in an existing grassland using root exclusion tubes. *J. Appl. Ecol.* 21:971–82

31. Cristy EJ. 1986. Effect of root competition and shading on growth of suppressed western hemlock (*Tsuga heterophylla*). *J. Veg. Sci.* 65:21–28

32. Dawson TE, Pate JS. 1996. Seasonal water uptake and movement in root systems of Australian phraeatophytic plants of dimorphic root morphology: a stable isotope investigation. *Oecologia* 107:13–20

33. de Kroon H, Hutchings MJ. 1995. Morphological plasticity in clonal plants: the foraging concept reconsidered. *J. Ecol.* 83:143–52

34. Diaz S, Grime JP, Harris J, McPherson E. 1993. Evidence of a feedback mechanism limiting plant response to elevated carbon dioxide. *Nature* 364:616–17

35. Dillenburg LR, Whigham DF, Teramura AH, Forseth IN. 1993. Effects of below- and aboveground competition from vines *Lonicera japonica* and *Parthenocissus quinquefolia* on the growth of the tree host *Liquidambar styraciflua*. *Oecologia* 93:48–54

36. Donald CM. 1958. The interaction of competition for light and for nutrients. *Aust. J. Agric. Res.* 9:421–35

37. Drew MC, Saker LR. 1975. Nutrient supply and the growth of the seminal root system in barley. II. Localized, compensatory increases in lateral root growth and rates of nitrate uptake when nitrate supply is restricted to only part of the root system. *J. Exp. Bot.* 26:79–90

38. Duncan WG, Ohlrogge AJ. 1958. Principles of nutrient uptake from fertilizer bands. II. Root development in the band. *Agric. J.* 50:605–08

39. Ehleringer JR, Phillips SL, Schuster WSF, Sandquist DR. 1991. Differential utilization of summer rains by desert plants. *Oecologia* 88:430–34

40. Eissenstat DM. 1992. Costs and benefits of constructing roots of small diameter. *J. Plant Nutri.* 15:763–82

41. Eissenstat DM, Yanai RD. 1997. The ecology of root lifespan. *Adv. Ecol. Res.* 27:1–60

42. Fernandez OA, Caldwell MM. 1975. Phenology and dynamics of root growth of three cool semi-desert shrubs under field conditions. *J. Ecol.* 63:703–14

43. Field CB, Chapin FSI, Matson PA, Mooney HA. 1992. Responses of terrestrial ecosystems to the changing atmosphere: a resource-based approach. *Annu. Rev. Ecol. Syst.* 23:201–35

44. Fitter AH. 1994. Architecture and biomass allocation as components of the plastic response of root systems to soil heterogeneity. In *Exploitation of Environmental Heterogeneity by Plants: Ecophysiological Processes Above- and Belowground,* ed. MM Caldwell, RW Pearcy, pp. 305–23. San Diego: Academic

45. Fitter AH, Strickland TR, Harvey ML, Wilson GW. 1991. Architectural analysis of plant root systems. I. Architectural correlates of exploitation efficiency. *New Phytol.* 119:383–89

46. Fowler NL. 1986. Density-dependent population regulation in a Texas grassland community. *Ecology* 67:545–54

47. Fowler NL. 1986. The role of competition in plant communities in arid and semiarid regions. *Annu. Rev. Ecol. Syst.* 17:89–110

48. Gerry AK, Wilson SD. 1995. The influence of initial size on the competitive responses of six plant species. *Ecology* 76:272–79

49. Gill DS, Marks PL. 1991. Tree and shrub seedling colonization of old fields in central New York. *Ecol. Mono.* 61:183–205

50. Goldberg DE. 1990. Components of resource competition in plant communities. See Ref. 52, pp. 27–65

51. Grace JB. 1993. The effects of habitat productivity on competition intensity. *Trends Evol. Ecol.* 8:229–30

52. Grace JB, Tilman D. 1990. *Perspectives on Plant Competition* San Diego: Academic

53. Greenlee JT, Callaway RM. 1996. Abiotic stress and the relative importance of interference and facilitation in montane bunchgrass communities in western Montana. *Am. Nat.* 148:386–96

54. Grime JP. 1974. Vegetation classification by reference to strategies. *Nature* 250:26–31

55. Grime JP. 1979. *Plant Strategies and Vegetation Processes.* London: Wiley

56. Grime JP. 1993. Ecology sans frontières. *Oikos* 68:385–92

57. Grime JP. 1994. The role of plasticity in exploiting environmental heterogeneity. In *Exploitation of Environmental Heterogeneity by Plants,* ed. MM Caldwell, RW Pearcy, pp. 1–19. San Diego, CA: Academic

58. Grime JP. 1994. Defining the scope and testing the validity of C-S-R theory: a response to Midgley, Laurie, and Le Maitre. *Bull. Southern Afr. Inst. Ecologists Environ. Scientists* 13:4–7

59. Gross K, Peters A, Pregitzer KS. 1993. Fine root growth and demographic responses to nutrient patches in four old-field plant species. *Oecologia* 95:61–64

60. Grubb PJ. 1994. Root competition in soils of different fertility: a paradox resolved? *Phytocoenologia* 24:495–505

61. Gurevitch J. 1986. Competition and the local distribution of the grass *Stipa neomexicana. Ecology* 67:46–57

62. Hook PB, Lauenroth WK. 1994. Root system response of a perennial bunchgrass to neighborhood-scale soil water heterogeneity. *Funct. Ecol.* 8:738–45

63. Hook PB, Lauenroth WK, Burke IC. 1994. Spatial patterns of roots in a semiarid grassland: abundance of canopy

openings and regeneration gaps. *J. Ecol.* 82:485–94

64. Horn JC. 1985. Responses of understory tree seedlings to trenching. *Am. Midl. Natur.* 114:252–58

65. Huber-Sannwald E, Pyke DA, Caldwell MM. 1996. Morphological plasticity following species-specific recognition and competition in two perennial grasses. *Am. J. Bot.* 83:919–93

66. Hutchings MJ, de Kroon H. 1994. Foraging in plants: the role of morphological plasticity in resource acquisition. *Adv. Ecol. Res.* 25:159–238

67. Jackson RB, Caldwell MM. 1992. Shading and the capture of localized soil nutrients: nutrient contents, carbohydrates, and root uptake kinetics of a perennial tussock grass. *Oecologia* 91:457–62

68. Jackson RB, Caldwell MM. 1993. The scale of nutrient heterogeneity around individual plants and its quantification with geostatistics. *Ecology* 74:612–14

69. Jackson RB, Caldwell MM. 1996. Integrating resource heterogeneity and plant plasticity: modelling nitrate and phosphate uptake in a patchy soil environment. *J. Ecol.* 84:891–903

70. Jackson RB, Canadell J, Ehleringer JR, Mooney HA, Sala OE, Schulze E-D. 1996. A global analysis of root distributions for terrestrial biomes. *Oecologia* 108:389–411

71. Jackson RB, Manwaring JH, Caldwell MM. 1990. Rapid physiological adjustment of roots to localized soil enrichment. *Nature* 344:58–60

72. Kadmon R. 1995. Plant competition along soil moisture gradients: a field experiment with the desert annual *Stipa capensis. J. Ecol.* 83:253–62

73. Kaye JP, Hart SC. 1997. Competition for nitrogen between plants and soil microorganisms. *Trends Ecol. Evol.* 12:139–43

74. Koide RT. 1991. Density-dependent response to mycorrhizal infection in *Abutilon theorphrasti* Medic. *Oecologia* 85:389–95

75. Kovar JL, Barber SA. 1988. Phosphorus supply characteristics of 33 soils as influenced by seven rates of phosphorus addition. *Am. J. Soil Sci. Soc.* 52:160–65

76. Krannitz PG, Caldwell MM. 1995. Root growth responses of three Great Basin perennials to intra- and interspecific contact with other roots. *Flora* 190:161–67

77. Kummerow J, Krause D, Jow W. 1977. Root systems of chaparral shrubs. *Oecologia* 29:163–77

78. Le Roux X, Bariac T, Mariotti A. 1995. Spatial partitioning of the soil water resource between grass and shrub components in a West African humid savanna. *Oecologia* 104:147–55

79. Lee RB. 1982. Selectivity and kinetics of ion uptake by barley plants following nutrient deficiency. *Ann. Bot.* 50:429–49

80. Lyr H, Hoffmann G. 1967. Growth rates and growth periodicity of tree roots. *Inter. Rev. For. Res.* 1:181–236

81. Mäder P, Vierheilig H, Alt M, Wiemken A. 1993. Boundaries between soil compartments formed by microporous hydrophobic membranes (GORE-TEX^R) can be crossed by vesicular-arbuscular mycorrhizal fungi but not by ions in the soil solution. *Plant Soil* 152:201–06

82. Mahall BE, Callaway RM. 1992. Root communication mechanisms and intracommunity distributions of two Mojave desert shrubs. *Ecology* 73:2145–51

83. Mamolos AP, Elisseou GK, Veresoglou DS. 1995. Depth of root activity of coexisting grassland species in relation to N and P additions, measured using non-radioactive tracers. *J. Ecol.* 83:643–52

84. Marschner H. 1995. *Mineral Nutrition of Higher Plants.* London: Academic. 2nd ed.

85. McConnaughay KDM, Bazzaz FA. 1992. The occupation and fragmentation of space: consequences of neighboring roots. *Funct. Ecol.* 6:704–10

86. McConnaughay KMD, Newman EI. 1991. Is physical space a soil resource? *Ecology* 72:94–103

87. McLellan AJ, Fitter AH, Law R. 1995. On decaying roots, mycorrhizal colonization and the design of removal experiments. *J. Ecol.* 83:225–30

88. Mooney HA, Hobbs RJ, Gorman J, Williams K. 1986. Biomass accumulation and resource utilization in co-occurring grassland annuals. *Oecologia* 70:555–58

89. Mordelet P, Barot S, Abbadie L. 1996. Root foraging strategies and soil patchiness in a humid savanna. *Plant Soil* 182:171–76

90. Morris EC. 1996. Effect of localized placement of nutrients on root competition in self-thinning populations. *Ann. Bot.* 78:353–64

91. Mou P, Jones RH, Mitchell RJ, Zutter BR. 1995. Spatial distribution of roots in sweetgum and loblolly pine monocultures and relations with aboveground biomass and soil nutrients. *Funct. Ecol.* 9:689–99

92. Nepstad DC, de Carvalho CR, Davidson EA, Jipp PH, Lefebvre PA, et al. 1994. The role of deep roots in the hydrological and carbon cycles of Amazonian forests and pastures. *Nature* 372:666–69

93. Newberry DM, Newman EI. 1978. Competition between grassland plants of different sizes. *Oecologia* 33:361–80

94. Newman EI. 1988. Mycorrhizal links between plants: their functioning and ecological significance. *Adv. Ecol. Res.* 18:243–71

95. Nobel PS. 1991. *Physiochemical and Environmental Plant Physiology.* San Diego: Academic

96. Nye PH, Tinker PB. 1977. *Solute Movement in the Soil-Root System.* Berkeley: Blackwell

97. Pacala SW. 1987. Neighborhood models of plant population dynamics. 3. Models with spatial heterogeneity in the physical environment. *Theor. Pop. Biol.* 31:359–92

98. Pantastico-Caldas M, Venable DL. 1993. Competition in two species of desert annuals along a topographic gradient. *Ecology* 74:2192–203

99. Parsons WFJ, Knight DH, Miller SL. 1994. Root gap dynamics in lodgepole pine forest: nitrogen transformations in gaps of different sizes. *Ecol. Appl.* 4:354–62

100. Pregitzer KS, Hendrick RL, Fogel R. 1993. The demography of fine roots in response to patches of water and nitrogen. *New Phytol.* 125:575–80

101. Putwain PD, Harper JL. 1970. Studies in the dynamics of plant populations. III. The influence of associated species on populations of *Rumex acetosa* L. and *R. acetosella* L. in grassland. *J. Ecol.* 58:251–64

102. Putz FE, Canham CD. 1992. Mechanisms of arrested succession in shrublands: root and shoot competition between shrubs and tree seedlings. *For. Ecol. Manage.* 49:267–75

103. Reichenberger G, Pyke DA. 1990. Impact of early root competition on fitness components of four semiarid species. *Oecologia* 85:159–66

104. Richards J. 1986. Root form and depth distribution in several biomes. In *Mineral Exploration: Biological Systems and Organic Matter,* ed. D Carlisle, WL Berry, IR Kaplan, JR Watterson, pp. 82–96. Englewood Cliffs, NJ: Prentice-Hall

105. Richards JH, Caldwell MM. 1987. Hydraulic lift: substantial nocturnal water transport between soil layers by *Artemisia tridentata* roots. *Oecologia* 73:486–89

106. Richter DD, Markewitz D. 1995. How deep is soil? *BioScience* 45:600–09

107. Robertson GP, Klingensmith KM, Klug MJ, Paul EA, Crum JR, Ellis BG. 1997. Soil resources, microbial activity, and primary production across an agricultural ecosystem. *Ecol. Appl.* 7:158–70

108. Robinson D. 1994. The responses of plants to non-uniform supplies of nutrients. *New Phytol.* 127:635–74

109. Robinson D. 1996. Resource capture by localized root proliferation: Why do plants bother? *Ann. Bot.* 77:179–85

110. Rogers HH, Runion GB, Krupa SV. 1994. Plant response to CO_2 enrichment with emphasis on roots and the rhizosphere. *Environ. Pollut.* 83:155–89

111. Ryel RJ, Caldwell MM, Manwaring JH. 1996. Temporal dynamics of soil spatial heterogeneity in sagebrush-wheatgrass steppe during a growing season. *Plant Soil* 184:299–309

112. Sala OE, Golluscio RA, Lauenroth WK, Soriano A. 1989. Resource partitioning between shrubs and grasses in the Patagonian steppe. *Oecologia* 81:501–05

113. Schlesinger WH. 1991. *Biogeochemistry: an Analysis of Global Change.* San Diego: Academic

113a. Schulze E-D, Kelliher FM, Körner CH, Lloyd J, Leuning R. 1994. Relationships among maximum stomatal conductance, ecosystem surface conductance, carbon assimilation rate, and plant nitrogen nutrition: a global ecology scaling exercise. *Annu. Rev. Ecol. Syst.* 25:629–60

114. Schwinning S. 1996. Decomposition analysis of competitive symmetry and size structure. *Ann. Bot.* 77:47–57

115. Schwinning S, Weiner J. 1997. Mechanisms determining the degree of size-asymmetry in plant competition. *Oecologia.* In press

116. Silander JA, Jr., Pacala SW. 1985. Neighborhood predictors of plant performance. *Oecologia* 66:256–63

117. Slatyer RO. 1967. *Plant-Water Relationships.* London: Academic

118. Snaydon RW. 1962. Micro-distribution of *Trifolium repens* L. and its relation to soil factors. *J. Ecol.* 50:133–43

119. Stone E, Kalisz PJ. 1991. On the maximum extent of tree roots. *For. Ecol. Manage.* 46:59–102

120. Tilman D. 1982. *Resource Competition and Community Structure.* Princeton, NJ: Princeton Univ. Press

121. Tilman D. 1988. *Plant Strategies and the Dynamics and Structure of Plant*

Communities. Princeton, NJ: Princeton Univ. Press

122. Tilman D. 1989. Competition, nutrient reduction, and the competitive neighborhood of a bunchgrass. *Funct. Ecol.* 3:215–19

123. Tilman D, Pacala S. The maintenance of species richness in plant communities. In *Species Diversity in Ecological Communities: Historical and Geographical Perspectives*, ed. RE Ricklefs, D Schulter, pp. 13–25. Chicago, IL: Univ. Chicago Press

124. Twolan-Strutt L, Keddy PA. 1996. Above- and belowground competition intensity in two contrasting wetland plant communities. *Ecology* 77:259–70

125. Vitousek PM, Denslow JS. 1986. Nitrogen and phosphorus availability in treefall gaps of lowland tropical rainforest. *Ecology* 74:1167–78

126. Vitousek PM, Matson PA. 1985. Disturbance, nitrogen availability, and nitrogen losses in an intensively managed loblolly pine plantation. *Ecology* 66:1360–76

127. Vogt KA, Vogt DJ, Asbjornsen H, Dahlgren RA. 1995. Roots, nutrients and their relationship to spatial patterns. *Plant Soil* 168/169:113–23

128. Vogt KA, Vogt DJ, Boon P, O'Hara J, Asbjornsen H. 1996. Factors controlling the contribution of roots to ecosystem carbon cycles in boreal, temperate, and tropical forests. *Plant Soil.* 187:159–219

129. Walker BH, Ludwick D, Holling CS, Peterman RM. 1981. Stability of semiarid savanna grazing systems. *J. Ecol.* 69:473–98

130. Waller DM. 1981. Neighborhood competition in several violet populations. *Oecologia* 51:116–22

131. Walter H. 1954. Die verbuschung, eine erscheinung der subtropischen savannengebiete, und ihre ükologischen ursachen. *Vegetatio* 5/6:6–10

132. Wayne P, Berntson W. 1996. *Resource acquisition in crowded populations: linking mechanistic and phenomenological aspects of plant competition.* Presented at Ecol. Soc. Am. Ann. Comb. Meeting, 77th, Providence, RI

133. Waynick DD, Sharp LT. 1919. Variability in soils and its significance to past and future soil investigations. II. Variation in nitrogen and carbon in field soils and their relation to the accuracy of field trials. *Univ. Calif. Publ. Agric. Sci.* 4:121–39

134. Weaver JE. 1919. *The Ecological Relations of Roots.* Washington, DC: Carnegie Inst.

135. Weiner J. 1982. A neighborhood model of annual-plant interference. *Ecology* 63:1237–41

136. Weiner J. 1986. How competition for light and nutrients affects size variability in *Ipomoea tricolor* populations. *Ecology* 67:1425–27

137. Weiner J. 1990. Asymmetric competition in plant populations. *Trends Ecol. Evol.* 5:360–64

138. Weiner J, Wright DB, Castro S. 1997. Symmetry of below-ground competition between *Kochia scoparia* individuals. *Oikos.* 79:85–91

139. West HM. 1996. Influence of arbuscular mycorrhizal infection on competition between *Holcus lanatus* and *Dactylis glomera. J. Ecol.* 84:429–38

140. Wilczynski CJ, Pickett STA. 1993. Fine root biomass within experimental canopy gaps: evidence for a belowground gap. *J. Veg. Sci.* 4:571–74

141. Wilson JB. 1988. Shoot competition and root competition. *J. Appl. Ecol.* 25:279–96

142. Wilson SD, Tilman D. 1991. Components of plant competition along an experimental gradient of nitrogen availability. *Ecology* 72:1050–65

143. Wilson SD, Tilman D. 1993. Plant competition and resource availability in response to disturbance and fertilization. *Ecology* 74:599–611

144. Wilson SD, Tilman D. 1995. Competitive responses of eight old-field plant species in four environments. *Ecology* 76:1169–80

Annu. Rev. Ecol. Syst. 1997. 28:571–92

MALE AND FEMALE ALTERNATIVE REPRODUCTIVE BEHAVIORS IN FISHES: A New Approach Using Intersexual Dynamics

S. A. Henson and R. R. Warner

Department of Ecology, Evolution, and Marine Biology, University of California, Santa Barbara, California 93106; e-mail: henson@lifesci.lscf.ucsb.edu

KEY WORDS: alternative reproductive behaviors, sexual conflict, game theory, mating systems, fish

ABSTRACT

The study of alternative reproductive behaviors in fishes has contributed to our general understanding of reproductive strategies and mating systems. Despite extensive research on the mechanisms and patterns of alternatives, two important factors have not been addressed, and both may strongly influence the evolution of alternative reproductive behaviors. First, alternative female reproductive behaviors exist and should be considered in theoretical and empirical work. Second, interactions between the sexes will influence the evolution of alternative reproductive behaviors. In this review, we explore these two points and suggest the development of a more comprehensive theory of alternatives that will increase our ability to make predictions regarding the existence and expression of alternative reproductive behaviors in both sexes.

INTRODUCTION: WHAT DO WE MEAN BY ALTERNATIVE REPRODUCTIVE BEHAVIORS?

In fishes, an amazing diversity in mating patterns exists within and across species. Interspecific variation in mating systems ranges from pelagic group spawning to monogamy with biparental care (7, 67). Diversity within species is no less impressive. For example, in the ocellated wrasse, *Symphodus ocellatus*, four distinct male reproductive behaviors exist simultaneously within one

0066-4162/97/1120-0571$08.00

population (66, 81). In this species, large males build nests and provide parental care (43). The smallest males sneak into the nest to parasitize the mating between the nesting male and a female (66, 81). Meanwhile, intermediate-sized males defend another male's nest from sneakers and court females while sneaking spawns themselves. Finally, the largest males, called pirates, temporarily take over another male's nest by force (66). Such diversity is not unique to this species (65), and studies of variation in fish mating systems have played an important role in our general understanding of reproductive strategies (67, 78).

The relatively recent realization that differences within a species can represent evolutionarily adaptive patterns has led to a large number of studies on alternative reproductive patterns in a variety of taxa (46). A multitude of examples exist of both within- and between-individual differences in reproductive behavior, and variation may be more the rule than the exception. Separate populations of a species will often experience different habitats and selective pressures, and, as a result, may exhibit different reproductive strategies. Single populations may also exhibit changes in phenotype distributions through time due to environmental variation and selection. Finally, reproductive behavior may vary within a population at one point in time.

The study of alternative reproductive behaviors has focused mainly on discrete male alternatives within a population (e.g. 2, 15, 17, 22, 30, 33, 60, 65, 73). For a number of well-studied examples of male alternative reproductive behaviors in fishes, we know the pattern of expression, life-history pathways, relative fitness, and in some cases even the underlying physiology and genetic basis of these differences (e.g. 5, 6, 15, 16, 29, 30, 32, 61, 62, 68, 74, 78, 80). Although these well-known patterns are very striking, two other important aspects of alternative reproductive behaviors have been virtually ignored (but see 11, 57, 75). First, female alternative reproductive behaviors clearly exist and should be considered. Second, there is a need to treat the effect of interactions between the sexes on the evolution of alternative reproductive behaviors. One of our major goals is to include consideration of these factors, since they enrich the topic considerably and should enhance our understanding of the evolution of alternative reproductive behaviors in both sexes.

In order to have a meaningful discussion of alternative reproductive behaviors, it is necessary first to ask the question "What do we mean by alternatives?" For this review, we focus on situations in which distinct reproductive behaviors exist within a sex, within a population, at one point in time. While distinction is often made between the underlying rule (termed a strategy) and the expression of that rule (termed a tactic: 2, 14, 17, 30, 33), we avoid the use of this terminology and focus on observed differences in reproductive behavior within or between individuals.

We briefly review some general types of alternative reproductive behaviors in males and females, and we discuss whether there are consistent differences between the sexes in the types and patterns of alternatives. Although we discuss only a few illustrative examples, it is important to realize that numerous other species exhibit alternative reproductive behaviors. We then review the mechanisms that can maintain alternatives and ask what factors might determine the presence of alternative reproductive behaviors and their pattern of expression. Finally, we discuss some limitations of current theory and argue that not only do alternative female reproductive behaviors exist, but that females play an important role in the expression and evolution of male alternative behaviors. Males will play a corresponding role in the evolution and expression of female alternatives. In essence, we argue that interactions both within and between the sexes must be considered. This will increase our understanding of observed alternatives and our ability to make predictions about expected patterns of mating behavior in both sexes.

EXAMPLES OF OBSERVED VARIATION

Male Alternative Reproductive Behaviors

Male alternative reproductive behaviors in fishes have been well documented, as exemplified in the comprehensive review of Taborsky (65). This extensive diversity falls into three broad categories. Males are usually in competition for access to mates or resources, and existing alternatives often represent different solutions to the problem of obtaining mates (e.g. males may differ in their tendency to engage in male-male competition). Males may also vary in mating mode (e.g. males may mate singly with females or they may spawn en masse), and in degree of investment in parental care. Alternatives can exist in each of these three main components of the mating system. Although the exact details differ among species, a few examples demonstrate both the basic trends and the variation in these patterns.

Many species possess males that compete for territories as well as males that are nonterritorial. Although in some species nonterritorial males are not reproductively active, in others these males have mating success that is measurable, if not equal to that of territorial males (e.g. 30, 32, 62, 81). For example, in the bluegill sunfish, *Lepomis macrochirus*, large males build nests and defend territories (29, 30, 34). Other males in the population adopt alternative nonterritorial behaviors (15, 16, 29, 30, 32). Intermediate-sized males mimic females in order to gain access to the nest for spawning (29, 30). Smaller males hide nearby and use stealth to reach nests and to spawn with females ("sneaking": 29, 30). Although females will spawn only in nests, they do not appear to differentiate between male alternatives (30, 32). In this species, territorial and nonterritorial males may have completely separate life-history pathways with different growth rates and reproductive spans (15, 16, 29, 30, 32). Nonterritorial males mature early and reproduce at a lower rate for a longer time period, while territorial males delay maturity in exchange for higher eventual mating success. Despite all these differences, male alternative behaviors appear to have equal lifetime reproductive success on average (30, 32). In this species, male types vary in all three components: how they obtain mates, mating mode, and parental investment.

In other species, individual males adopt both territorial and nonterritorial behavior at different times of life. For example, in the bluehead wrasse, *Thalassoma bifasciatum*, young males on large reefs do not defend territories but spawn in groups at mating sites. Older males defend territories around other mating sites and are involved in pair-spawns with females (80). Males appear to switch from nonterritorial to territorial behavior when they are large enough to compete for a territory (36). An individual male, if it survives long enough, will be both nonterritorial and territorial. Although the mating success of territorial males is higher, males achieve a higher lifetime reproductive success by initially reproducing as nonterritorial males than by delaying sexual activity until they are large enough to compete for a territory (74, 80). Males of this species have two alternatives in access to mates and mating mode, but in neither alternative do males provide parental care.

In threespine sticklebacks, *Gasterosteus aculeatus*, most males build nests, defend territories, and provide parental care (4). However, nesting males often sneak at neighboring nests (37). Although a male's success is higher while spawning in its own nest, a male loses little by sneaking at nearby nests when the opportunity arises. An individual male switches between sneaking and territorial behavior throughout its lifetime, depending on mating success at its own nest. In this species, all males can provide parental care and defend territories, but the two alternatives differ in the way males obtain access to mates.

In many species of salmon, male size appears to be bimodally distributed (30, 31). Large males compete for the opportunity to guard females, while smaller males employ sneak matings. Males of these two types also appear to have separate life history pathways. Large males grow for many years before reproducing, while small males mature early. Individuals do not switch between types. It appears that males of the two types have approximately equal lifetime fitness, although success per breeding season is lower for small males (30, 31). This lower success is compensated for by earlier reproductive age and higher survival rate to reproduction. Males that grow quickly during the juvenile period become early-maturing, noncompetitive types, whereas males that were slower-growing juveniles eventually become males that compete directly for access to females; there may be a genetic component to these differences (30, 51). If early growth rate indicates general fitness, this comparison points out that larger males are not necessarily of better quality, or higher fitness.

Although we have focused our discussion on the territorial/nonterritorial male dichotomy, other patterns exist. Another common alternative is between males that court and those that coerce females into mating. This difference may depend on the female's choosiness, the male's quality, or even the local environment. For example, in guppies, *Poecilia reticulata*, individual males can either perform courtship displays or attempt to copulate without courtship (25, 26). Males adopt courtship more in the absence of predators and coercion more in the presence of predation risk (23, 26, 59). Alternatives also occur within species in degree of male parental investment; differences may represent differing life-history pathways, ontogenetic changes, or opportunistic switching. In many species, territorial males provide care for offspring while nonterritorial males do not (e.g. 29, 30, 43, 69, 70). In contrast, nonterritorial males of the cichlid *Lambrologus brichardi* help territorial males care for the offspring even if the nonterritorial male is unrelated to the young (63, 64).

Given the variety of basic patterns, it would be helpful to be able to predict the circumstances under which alternatives might evolve, and what the pattern of expression of alternatives might be. In order to do this it is necessary to understand what maintains the alternatives as well as the factors that determine pattern of expression. Classically, most male alternative reproductive behaviors have been assumed to arise as a consequence of "winner-take-all" situations in which a few individuals in the population accrue most of the mating success. In these cases, selection is expected to be very strong on other individuals to achieve matings through other means. For example, in mating systems in which only the few largest or oldest males win competitions for access to mates, smaller males will be selected to adopt reproductive behaviors that do not involve direct competition, such as sneaking. Although a "winner-take-all" situation clearly can lead to the evolution of alternatives, it is not the only or even the most common situation.

Female Alternative Reproductive Behaviors

Although the existence of alternative reproductive patterns is well established, there has been very little discussion of such alternatives in females. Females do not typically compete directly for resources or mates, and many of the classic male alternatives appear to be the result of direct competition between males. However, females are active participants in the mating process, and female choice and parental investment can be complex strategies. As in males, if female access to mates is limited, then we might expect behaviors to evolve to circumvent this limitation. Similarly, if costs and benefits change ontogenetically, we would expect behavior to differ with age or size. We present here some examples.

MATE CHOICE AND COPYING   Female choice of males before mating occurs in many species, and has gained attention as an important component of female reproductive strategies (1). Both theory and experimental studies suggest that some females may copy the mate choice of others in the population (18, 21, 38, 41, 45, 57, 73). This can be an evolutionarily adaptive behavior if females either differ in their ability to assess males or if assessment carries high costs (45, 57, 73), and younger or less experienced females do appear more prone to copy (20). Clearly, females can copy only the actual mate choices made by other females, and thus copying can exist only in the presence of another behavioral alternative. Copy or independent choice behaviors represent a good example of female alternative reproductive behaviors. In guppies, females copy the mate choice of other females when given the opportunity to observe another female's choice (18). Older females are not influenced by the decisions of younger females, while young females commonly copy the choice of older females (19). Individual females will both copy and choose depending on the circumstances, so both opportunity and age seem to influence female decisions.

RESISTANCE TO COERCION   Females may also differ in the degree to which they require male courtship or resist coercion. In guppies, females usually experience extensive courtship before they are willing to mate (24, 25, 49). Males often attempt to force copulations without courtship, which females usually resist (25, 26). However, in the presence of a predator, females are willing to copulate without courtship (27, 49). It has been argued that females avoid increased risk of predation by accepting copulations without courtship (56). Individual females within a population express both alternatives, and their tendency to require courtship is mediated by differences in predation risk.

MODE OF MATING   Females differ in the way they evaluate mates and may also choose between mating modes. Females of some species choose between pair- or group-mating modes. In the bluehead wrasse, both group-spawning and pair-spawning sites exist within a single population, and females are free to spawn at either type of site (74, 80). Some females tend to mate consistently at a group spawning site, where they mate with many males, while other females tend to visit a pair-spawning site, where they mate with only a territorial male (76, 77). While there appear to be no life-history differences between pair- and group-spawning females, nearly all females switch to pair-spawning when they reach large sizes (75). It is not known why some females choose one mode over the other, but these consistent differences persist between individual females within a single population.

In the damselfish *Chromis multilineata*, female sneaking occurs. Many individuals of this species carry the isopod ectoparasite *Anilocra multilineata*. A healthy male appears to deny parasitized females access to its nest. As a result, parasitized females rush into the nests of successful males to deposit their eggs and are quickly chased out (39). While these observations are preliminary, they indicate that sneaking may not be only a male behavior, and as in males, it is found in circumstances wherein access to mating is denied to a particular class of individuals.

PARENTAL CARE   Females may also differ with respect to their choice of post-mating care of eggs. A striking example of female alternatives is found in the peacock wrasse, *Symphodus tinca*, in which parental care is facultative (69, 70). A female tends to mate with territorial males that care for her eggs (43). However, the same females will also mate outside of territories with males that do not provide parental care (69, 82). While egg survival is always higher in a nest, for females that cannot find suitable nests in which to lay their eggs, the cost of searching can outweigh the lower survival of untended offspring (82). The relative abundance of the two behaviors changes through the season as the availability of nests and the chance of egg survival out of the nest change (82). Females may either mate with a parental male or non-parental male, and female choice varies both between and within individual females.

Thus, distinct alternatives may occur within and between individual females in the same population. As in males, alternatives appear in all three components of reproductive fitness: access to mates, mating mode, and parental care (summarized in Table 1). While males common differ in the degree to which they enter into male-male competition, territoriality, or parental care investment, females tend to differ more in exercising choice in pre-mating, mating, and post-mating situations. As with males, some female alternatives depend on the existence of another alternative in the same or opposite sex, while in other cases alternatives can exist independently. Male alternatives in these species would not exist independent of female alternatives. There is a need to develop theory relevant to the expression and maintenance of female alternatives and to consider the links between male and female alternative reproductive behaviors.

UNDERLYING MECHANISMS

Mechanisms for the Maintenance of Alternative Reproductive Behaviors

The variety of mechanisms that have been postulated to allow the stable coexistence of alternative reproductive behaviors (2, 10, 14, 17, 22, 30, 40, 44, 47, 52, 53) have been reviewed elsewhere (2, 8, 22, 30, 33, 60). Alternative behaviors

Table 1   Examples of alternative reproductive behaviors in males and females

| Species | Three main components of reproductive behaviors | | |
	Alternative access to mates	Mating mode alternatives	Parental investment
Oncorhynchus kisutch (coho salmon)	M: compete for access to females or sneak	none	none
	F: none	none	none
Lepomis macrochirus (bluegill sunfish)	M: territorial/nonterritorial	pair spawn or sneak spawn	care/no care
	F: none	none	none
Gasterosteus aculeatus (threespine stickleback)	M: none	spawn in own nest or neighbor's nest	none
	F: none	none	none
Poecilia reticulata (guppy)	M: coerce or court	coerce or court	none
	F: copy or choose	resist or accept coercion	none
Thalassoma bifasciatum (bluehead wrasse)	M: territorial/nonterritorial	group or pair spawn	none
	F: none	group or pair spawn	none
Symphodus ocellatus (ocellated wrasse)	M: territorial/nonterritorial	spawn in own nest or sneak spawn	care/no care
	F: none	none	none
Symphodus tinca (peacock wrasse)	M: territorial/nonterritorial	spawn in territory or out of territory	care/no care
	F: none	none	choose parental or nonparental male

classically have been divided into two categories. Discrete behavior differences that are the result of frequency dependence are often termed evolutionarily stable strategies (ESS: 50). Therefore, there is no single "best" behavior. For example, territorial and nonterritorial behaviors in the bluegill sunfish are postulated to be maintained by frequency-dependent fitness (32). Female choice may create frequency-dependence if females tend to choose rare males (42). If female fitness depends on the number of other females using a mating mode, the resulting frequency-dependence could give rise to alternative female behaviors. These behavior types are expected to have equal fitness at equilibrium (Figure 1; also 30, 60). The fitness of a frequency-dependent behavior is affected by the number of other individuals adopting the same behavior.

The other important mechanism facilitating the existence of alternatives is condition-dependent fitness (2, 17, 30). This occurs where the fitness of a behavior is dependent upon some characteristic of the individual or the environment. If two alternative behaviors each has a different relationship of fitness with condition, alternatives can be maintained (see Figure 2). For example, male behavior may depend on juvenile growth rate (30) in Pacific salmon, and female copying in guppies may depend on predation risk. Condition-dependence

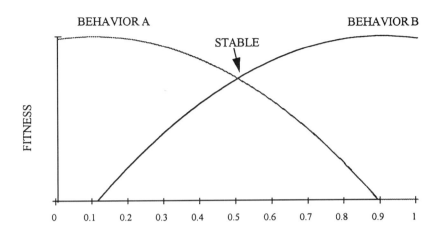

Figure 1   Negative frequency dependence can maintain alternative reproductive behaviors. At equilibrium (the ESS) alternatives have equal but suboptimal fitness. The proportion of alternatives in the population will be determined by the form of frequency-dependence. The stable solution occurs where the two fitness curves cross.

occurs if only individuals of high quality obtain most of the mating success. If individuals compete for access to mates or resources, and success in competition depends on age, experience, energy reserves, or size, then condition-dependent alternatives could result. For example, fecundity usually increases with size in fishes, and the cost of searching may decrease with experience.

A comprehensive theory does not currently exist regarding the relative importance of these two mechanisms in the maintenance of alternative behaviors. In reality, populations will often experience frequency-dependent selection and condition-dependent fitness simultaneously (33). Furthermore, complex interactions may occur among factors that influence the fitness of reproductive behaviors. For example, territorial males may compete for mating sites, while simultaneously competing with nonterritorial males using sperm competition tactics; territorial defense may be condition-dependent whereas sperm competition may be frequency-dependent. If males must allocate energy between sperm production and defense, maximizing fitness for one will not maximize it for the other. The best pattern of energy allocation may depend not only on social interactions, but also on total energy available and long-term tradeoffs between growth and reproduction. Female choice among males will affect the

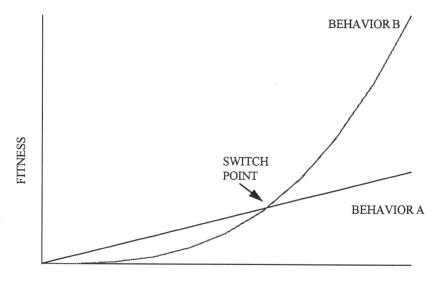

Figure 2 Condition-dependence can maintain alternatives. Two fitness curves are shown as a function of increasing condition. A particular individual is predicted to switch between behavior A and B at the point where the two functions cross. Different individuals are not predicted to have equal fitness.

fitness of each male reproductive tactic as well, and that choice may depend simultaneously on individual condition, male behavior, environmental state, and frequency of other females choosing the same site or male.

Genetic Basis of Alternatives

Alternative reproductive behaviors are sometimes classified with respect to underlying genetics. Alternatives maintained by frequency-dependence may be the result of either a genetic polymorphism or phenotypic plasticity. Condition-dependence can maintain alternatives only when variation in condition is independent of genotype.

In the swordtail, *Xiphophorus nigrensis*, size is sex-linked and determined by a single locus (61, 62, 83). Small males sneak behind a female to copulate, while large males display in front of a female before copulation, and intermediate-sized males adopt both behaviors. Male behavior is correlated with size, which is genetic in origin. In contrast, male guppies adopt both alternatives, and behavior choice appears to be determined by predation risk (59). Most of the examples of female alternatives mentioned above could occur within individuals and therefore are likely to result from conditional responses.

If alternatives are expected to evolve under predictable circumstances, is it also possible to predict their underlying genetics? As mentioned above, frequency-dependent alternatives can result from either a genetic polymorphism or phenotypic plasticity. Since these two mechanisms can lead to the same phenotypic distributions, they may in some circumstances be equivalent. However, if individuals switch between alternatives, then only behavioral plasticity can be responsible. While simple genetic polymorphisms may evolve easily and adjust to changes in the equilibrium frequency, the fitness of the individual genotype will be greater if it can adapt to changes in the optimal frequency of each alternative.

Patterns of Expression

Alternative reproductive behaviors may occur either between or within individuals (Figure 3). In the bluegill sunfish, differences occur between males, which act either as nonterritorial or as nesting males. In contrast, male peacock wrasses facultatively switch between nesting behavior and spawning out of nests, and individual male sticklebacks switch between spawning in their own and their neighbor's nest. Clearly, if alternatives are expressed within individuals, the only possible mechanism is phenotypic plasticity, while alternatives occurring between individuals allow the possibility of a genetic polymorphism as well. Either frequency-dependence or condition-dependence can lead to alternatives occurring between and within individuals. For example, if bluegill alternatives are the result of frequency-dependence, then either genetic polymorphism or plasticity could be responsible. However, in peacock wrasse females, switching between alternatives depends on the availability of nests and thus likely does not represent a genetic polymorphism.

Do any factors exist that would predict one type of expression over another? Regardless of origin, the cost of switching will influence the pattern of expression. Although it has been suggested that, if alternatives involve morphological differences, they are likely to be irreversible, whereas solely behavioral variation may enable switching (5, 6, 8, 52). However, costs of switching between behavior types, as might occur if individuals must learn new behaviors or accrue territories, can be high. In the bluehead wrasse, males not only change anatomically by decreasing allocation to testes and changing color patterns, but they must also compete for a territory. Males in this species switch only once to territorial behavior (36). In peacock wrasse females, switching from searching for nests and spawning out of nests is relatively free of costs because nonterritorial males are always available to spawn with females of either inclination (82).

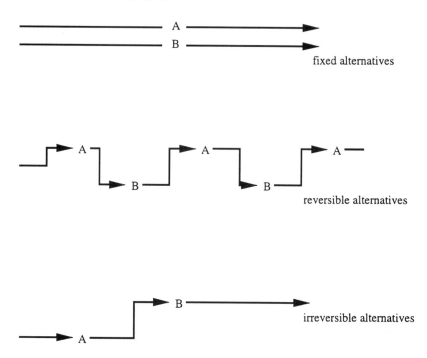

Figure 3 Alternative reproductive behavior patterns may be fixed, reversible, or irreversible within individuals. Redrawn with permission from Gross (30). Condition (age, size, or experience) increases from left to right.

If fitness is condition-dependent, the dynamics of the state of the individual or environment are expected to determine whether alternatives are fixed, reversible, or irreversible (Figure 3). For example, if fitness is size-dependent, but size is determined by factors in the juvenile stage, individuals may not be able to change size after maturity and thus can adopt only one behavior. In contrast, individuals may change behavior through ontogeny if growth continues after maturity. If absolute size determines behavior choice, switches need not be reversible. In contrast, relative size differences might select for reversible alternatives. If energy reserves or the current environment determine fitness, individuals may adopt reversible tactics. Stickleback males switch constantly between sneaking and nesting determined by mating opportunity (28), and female guppies switch between avoiding and accepting coercion based on predation risk (48, 49).

While Table 2 outlines some basic situations that will lead to one pattern over the other, predictive ability is clearly a weak link in our understanding

Table 2   Underlying mechanisms of alternative reproductive behaviors

	Frequency-dependence	Condition-dependence
Genetic basis	• genetic polymorphism or phenotypic plasticity	• phenotypic plasticity
Pattern of expression (fixed, reversible, or irreversible alternatives)	• any pattern possible • switches occur with a fixed probability • influenced by the cost of switching	• any pattern possible • condition determines switching • influenced by the cost of switching and the condition dynamics

of alternative reproductive behaviors in fishes. Future theory and experimental work should focus on the biological factors that determine the underlying mechanism, genetic basis, and pattern of expression of alternatives. Ideally, theory should allow both explanation of observed variation in mating system and prediction of the existence and pattern of alternative reproductive behaviors.

INTERSEXUAL DYNAMICS: A NEW VIEW OF THE EVOLUTION OF ALTERNATIVE REPRODUCTIVE BEHAVIORS

Traditionally, alternative reproductive behaviors have been thought to result from competition between males for access to mates or resources. We have argued that not only do females have alternative reproductive behaviors, but many factors interact to determine the evolution of alternatives in both sexes. An important factor that has been ignored is the effect that conflict between the sexes might have on the evolution of alternatives. Intersexual conflict is now recognized as an important factor in determining mating systems (e.g. 1, 9, 12, 13, 49), and a comprehensive theory on alternative reproductive behaviors must consider the simultaneous effect of both intra- and inter-sexual conflict interactions.

As an example, note that males often compete among themselves for territories, but the territory connected with the highest mating success will be determined by female mating strategies. The degree of female choosiness will dictate the distribution of success between males and thus the degree to which males must compete for territories. Correspondingly, female choosiness will be affected by the variation in male and territory quality. Female willingness to mate with a male may also be affected by the presence of other females on the territory. The existence of male alternatives depends on female willingness to mate with either male type, and female alternatives often differ in their choice between males. Thus, within-sex interactions may set the stage for between-sex conflict. Just as easily, mate choice can determine the degree of competition between males, and female choice can nullify predictions made, when only the interactions between males are considered. In order to have a complete understanding of the evolution of alternatives in either sex we must also consider the influence of the other sex on the existence of alternatives.

The Influence of Females on the Evolution of Male Alternatives

Females of some species exhibit a strong preference for one type of male, while females of other species show no choice between male behavior types. For example, females of the ocellated wrasse attempt to avoid nonterritorial males (71), while bluegill sunfish females seem indifferent (32). We would expect female choice to depend on whether male alternatives differentially affect female fitness. Furthermore, the cost of female choice and the ability to discriminate between males may also differ between species.

If male choices of alternatives have no direct effect on female fitness or offspring survival, then females should choose between alternatives only if indirect genetic benefits correlate with the alternative reproductive behaviors. However, mechanisms for the maintenance of genetic polymorphisms predict that at equilibrium male alternatives will have equal fitness. Thus, it is unlikely that females will choose between male alternatives maintained by frequency-dependence unless direct effects occur on female or offspring fitness. In support of this idea, females in the bluegill sunfish and Pacific salmon (in which male alternatives are proposed to be maintained by frequency-dependence) do not appear to choose actively between male alternatives (30, 32).

In contrast, females should exercise choice between male alternatives if direct fitness effects exist. For example, male alternatives might differ in their degree of investment in parental care and the resultant survival of offspring. In the peacock wrasse, *Symphodus tinca*, females prefer to spawn in nests where males provide parental care, spawning with nonparental males only if they are unable to find suitable nests within a limited period of searching (82). Equally, if females benefit from mate choice, they should prefer males that do not adopt coercive tactics. For example, in guppies, females actively avoid males that do not court, and they also benefit from choosing high-quality males (25, 58). Just as with the peacock wrasse, female choice between alternatives will depend on the costs, and in guppies, these costs seem to be mediated by predation risk (27, 49).

Clearly, if females choose directly between male alternatives, mate choice alone can either suppress or maintain alternatives in male behavior. Females can also influence the evolution of alternatives without actively choosing between male alternatives. For example, females can increase skew in reproductive success among males and thereby create a situation that favors the evolution of male alternatives. Thus, if females skew mating toward large males, sexual selection may lead to smaller males choosing other reproductive behaviors. In the end, this may not be to the female's benefit. One can imagine a situation in which females prefer to mate with males that are large because these males are able to obtain and defend the best territories. Smaller males may then be selected to attempt nonterritorial behaviors such as coercion of females or sneak spawning. If these small males are of lower quality, disrupt the mating, or decrease territorial male parental care through lowered paternity, female preference for large males may lead to a situation in which male alternatives actually lower female fitness.

The existence of alternatives in female choice could maintain male alternatives, or highly consistent female choice could lead to a situation in which a male alternative behavior has correspondingly consistent low fitness and does not persist. It is thus important to consider female behavior when attempting to make predictions about the evolution of male alternatives. Consider the basic predictions made by frequency- and condition-dependent fitness (Figures 1 and 2). By including female preference in the male fitness equation, we can ask what the equilibrium frequency of each alternative will be as a function of the degree of female preference for one alternative. Female choice can change the basic predictions drastically (Figure 4). The predicted frequency of male alternatives in the population increases as a function of female preference for that alternative. Sufficiently strong female choice can even suppress the existence of one alternative (Figure 4e). In the other extreme, female choice could conceivably maintain male alternative reproductive behaviors in the absence of either frequency- or condition-dependence. It is necessary to consider not only the benefits of a particular mate choice, but the costs as well, because these costs can create the opportunity for alternatives. Both female preferences and the ability to obtain a choice are important. For a complete understanding of male alternatives, male competition and female choice as well as the interactions between female choice and male behavior must be considered.

The Influence of Males on the Evolution of Female Alternatives

Female alternatives usually exist in terms of mate choice, and thus females often influence male fitness directly by determining male mating success. Males may

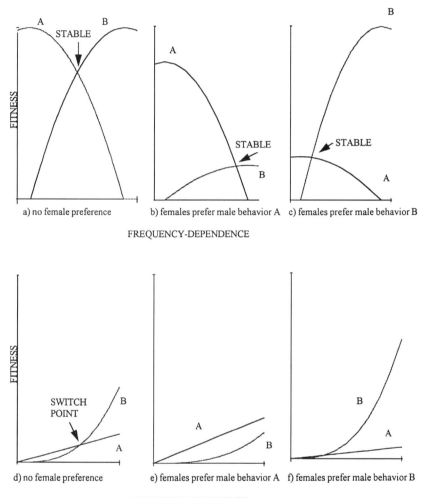

Figure 4 Female preference can influence predictions for male alternative reproductive patterns. Both frequency-dependence and condition-dependence can maintain alternatives within a sex. However, when female preference is included, predictions can change dramatically. Male distributions in the absence of female preference (left) can shift with female preference for male type A (center) or male type B (right). Female choice alone can change the frequency or even suppress the existence of either alternative.

not respond to female alternatives by being choosy themselves but may instead attempt to reduce a female's ability to choose through coercion or sneaking. Sneaking as an alternative to territorial behavior in many fishes may be the direct result of female choice to avoid spawning with small males.

As with females, males will influence female alternatives if they actively choose between females. This will occur when female alternatives have differential effects on male fitness and males are limited in their spawning rate (39, 55). Even if males are not limited, female choosiness could result in the appearance of a class of males that force copulations, which in turn leads to female alternatives between accepting coercion or attempting active choice (as in guppies, see above). If variation exists in the degree to which male alternatives invest in providing parental care, females may also adopt alternatives between demanding or doing without care of their offspring, as in the peacock wrasse. Male behavior may also affect female alternatives even in the absence of male alternatives. For example, high-quality females may choose high-quality males; if these males are limited in their mating rate, low-quality females have no option but to accept low-quality males (79). Admittedly, it is often impossible to identify cause and effect when both sexes have corresponding alternatives, and it is more instructive to model their joint evolution (see below).

Multiple Game Interactions and Coevolutionary Dynamics

The effect of males on females and of females on males both need to be considered in order to have a complete understanding of the evolution of alternative reproductive behaviors (35, 54). Females will determine, to a great extent, the mating success of various male behaviors, and skew in mating success due to female choice may lead to the evolution of male alternatives themselves. Male alternatives may, in turn, lead to the evolution of variation in female choice or other forms of variation in female reproductive behavior. Conflict between the sexes will be an important factor in determining patterns of mating behavior. This does not mean that competition or conflict within a sex is not also an important factor in mating systems. Instead, the existence and pattern of alternative reproductive behaviors in either sex will be the result of interactions within and between the sexes, and female behavior cannot be viewed separately from male behavior or vice versa. Considering the coevolution of male and female reproductive behavior by examining the links between male behavior and female behavior will lead to a better understanding of alternative reproductive behaviors and mating systems in general and may explain observations that seem counter-intuitive in light of present theory.

Imagine a species in which intrasexual competition leads a few strong males to all of the mating success, while most small males are not able to reproduce.

In another species, male-male competition is very weak. Standard theory would predict that only the former species should exhibit male alternative reproductive behaviors. However, strong and consistent female choice could lead to the absence of male alternatives in the species with male-male competition, and conditional female mate choice could lead to male alternatives in the species with no direct intrasexual competition. In parallel although theory might predict that female fitness would be higher in mating with one alternative, male coercion or sneaking might lead to the evolution of female alternative reproductive behaviors.

Ideally, theoretical predictions should include not only the effect of intersexual dynamics, but also the interactive effects of intra- and inter-sexual interactions. It has been suggested that dynamic programming games should be used to study the simultaneous effects of frequency- and condition-dependence on the evolution of alternative male reproductive behaviors (3). These models can be extended to model female behavior as well so that the dynamics of both intra- and inter-sexual interactions can be assessed, and even simple game theoretical models can demonstrate the importance of interactions between the sexes on the evolution of alternative reproductive behaviors (Henson, unpublished).

A consideration of intersexual dynamics can often suggest the basis for otherwise counter-intuitive behavior. For example, in the peacock wrasse, small sneaking males do not interfere with the mating of large males at the beginning of their nesting cycle, even though mating rates can be very high, and mating is performed in plain view of the small males (Warner, personal observation). Females in this species do not prefer to spawn with sneaker males (71) but will do so if a nest is very attractive. Since egg load in a nest increases its attractiveness to females (71, 82), small males may allow a nest to gain eggs before initiating alternative reproductive behaviors there.

In the ocellated wrasse, nests with high mating rates attract sneaker males, but mating rate decreases as sneaker numbers increase (71). Usually, nesting males court females and chase sneaker males away from the nest (66). However, at times, when the number of sneakers at the nest becomes very high, nesting males will refuse to mate with willing females (SA Henson, personal observation). This counter-intuitive observation may be explained by the interactions between male alternatives. If mating rate decreases at the nests, sneaker males leave, thus increasing the later mating success of the nesting male. Consideration of both female choice and interactions between male alternatives is needed to understand this otherwise surprising observation.

In general, females will choose between male alternatives when direct fitness effects occur. Male alternatives should exist only if female fitness is not affected or if male alternatives circumvent female choice, as with sneaking behavior or forced copulations. Similarly, female alternatives will exist either when male fitness is not affected by alternatives or when males have not been able to prevent females from exercising their options. Interactions between the sexes also can lead to counter-intuitive patterns of alternative reproductive behaviors, such as the absence of alternatives in a species where fitness is condition-dependent or the presence of alternatives in a system without any of the classical mechanisms for maintaining alternatives. To understand observed behavior and make predictions regarding the evolution and expression of alternatives, intersexual dynamics must be considered.

> **Visit the *Annual Reviews home page* at**
> **http://www.annurev.org.**

Literature Cited

1. Andersson MB. 1994. *Sexual Selection.* Princeton, NJ: Princeton Univ. Press. 599 pp.
2. Austad SN. 1984. A classification of alternative reproductive behaviors and methods for field-texting ESS models. *Am. Zool.* 24:309–19
3. Bednekoff PA, Clark CW. 1996. Dynamic games and conditional strategies. *Trends Ecol. Evol.* 11:383
4. Bell MA, Foster SA. 1994. *The Evolutionary Biology of the Threespine Stickleback.* Oxford: Oxford Univ. Press. 571 pp.
5. Brantley RK, Bass AH. 1994. Alternative male spawning tactics and acoustic signals in the plainfin midshipman fish *Porichthys notatus* Girard (Teleostei, Batrachoididae). *Ethology* 96:212–32
6. Brantley RK, Wingfield JC, Bass AH. 1993. Sex steroid levels in *Porichthys notatus*, a fish with alternative reproductive tactics and a review of the hormonal basis for male dimorphism among teleost fishes. *Horm. Behav.* 27:332–47
7. Breder CM, Rosen DE. 1966. *Modes of Reproduction in Fishes.* New York: Natl. Hist. Press. 941 pp.
8. Caro TM, Bateson P. 1986. Organization and ontogeny of alternative tactics. *Anim. Behav.* 34:1483–99
9. Clutton-Brock TA, Parker GA. 1995. Coercion in animal societies. *Anim. Behav.* 49:1345–65
10. Cooper WS, Kaplan RH. 1982. Adaptive coin-flipping: a decision-theoretic examination of natural selection for random individual variation. *J. Theor. Biol.* 94:135–51
11. Crowley PH, Travers SE, Linton MC, Cohn SL, Sih A, Sargent RC. 1991. Mate density,

predation risk and the seasonal sequence of mate choice: a dynamic game. *Am. Nat.* 137:567–96
12. Davies NB. 1989. Sexual conflict and the polygamy threshold. *Anim. Behav.* 38:226–34
13. Davies NB. 1992. *Dunnock Behavior and Social Evolution.* Oxford, UK: Oxford Univ. Press. 272 pp.
14. Dawkins R. 1980. Good strategy or evolutionarily stable strategy? In *Sociobiology: Beyond Nature/Nurture,* ed. GW Barlow, J Silverberg, pp. 331–67. Boulder: Westview. 627 pp.
15. Dominey WJ. 1980. Female mimicry in male bluegill sunfish—a genetic polymorphism? *Nature* 284:546–48
16. Dominey WJ. 1981. Maintenance of female mimicry as a reproductive strategy in bluegill sunfish, *Lepomis macrochirus.* *Environ. Biol. Fish.* 6:59–64
17. Dominey WJ. 1984. Alternative mating tactics and evolutionarily stable strategies. *Am. Zool.* 24:385–96
18. Dugatkin LA. 1992. Sexual selection and imitation: females copy the mate choice of others. *Am. Nat.* 139:1384–89
19. Dugatkin LA, Godin J-GJ. 1992. Reversal of female mate choice by copying in the guppy *Poecilia reticulata. Proc. R. Soc. London Ser. B* 249:179–84
20. Dugatkin LA, Godin J-GJ. 1993. Female mate copying in the guppy (*Poecilia reticulata*): age-dependent effects. *Behav. Ecol.* 4:289–92
21. Dugatkin LA, Hoglund J. 1995. Delayed breeding and the evolution of mate copying in lekking species. *J. Theor. Biol.* 174:261–67
22. Dunbar RIM. 1983. Life history tactics and

alternative strategies of reproduction. In *Mate Choice*, ed. P Bateson, pp. 423–47. Cambridge, UK: Cambridge Univ. Press. 472 pp.

23. Endler JA. 1987. Predation, light intensity, and courtship behaviour in *Poecilia reticulata* (Pisces: Poeciliidae). *Anim. Behav.* 35:1376–85

24. Farr JA. 1977. Male rarity or novelty, female choice behavior and sexual selection in the guppy, *Poecilia reticulata. Evolution* 31:162–68

25. Farr JA. 1980. Social behavior patterns as determinants of reproductive success in the guppy, *Poecilia reticulata* (Pisces: Poeciliidae): an experimental study of the effects of intermale competition, female choice and sexual selection. *Behaviour* 74:38–91

26. Farr JA. 1980. The effects of sexual experience and female receptivity on courtship-rape decisions in male guppies, *Poecilia reticulata* (Pisces: Poeciliidae). *Anim. Behav.* 28:1195–1201

27. Godin J-GJ, Briggs SE. 1996. Female mate choice under predation risk in the guppy. *Anim. Behav.* 51:117–30

28. Goldschmidt TS, Foster A, Sevenster P. 1992. Inter-nest distance and sneaking in the three-spined stickleback. *Anim. Behav.* 44:793–95

29. Gross MR. 1982. Sneakers, satellites, and parental males: polymorphic mating strategies in North American sunfishes. *Z. Tierpsychol.* 60:1–26

30. Gross MR. 1984. Sunfish, salmon, and the evolution of alternative reproductive strategies and tactics in fishes. In *Fish Reproduction: Strategies and Tactics*, ed. GW Potts, RJ Wootton, pp. 55–75. London: Academic. 410 pp.

31. Gross MR. 1985. Disruptive selection for alternative life histories in Salmon. *Nature* 313:47–48

32. Gross MR. 1991. Evolution of alternative reproductive strategies: frequency-dependent selection in male bluegill sunfish. *Philos. Trans. R. Soc. London Ser. B* 332:59–66

33. Gross MR. 1996. Alternative reproductive strategies and tactics: diversity within the sexes. *Trends Ecol. Evol.* 11:92–98

34. Gross MR, Charnov EL. 1980. Alternative male life histories in bluegill sunfish. *Proc. Natl. Acad. Sci. USA* 11:6937–40

35. Hammerstein P, Parker GA. 1987. Sexual selection: games between the sexes. In *Sexual Selection: Testing the Alternatives*, ed. JW Bradbury, MB Andersson, pp. 119–42. Chichester: Wiley. 308 pp.

36. Hoffman SG, Schildhauer MP, Warner RR.

1985. The cost of changing sex and the ontogeny of males under contest competition for mates. *Evolution* 39:915–22

37. Jamieson IG, Colgan PW. 1992. Sneak spawning and egg stealing by male three-spine sticklebacks. *Can. J. Zool.* 70:963–67

38. Jamieson I. 1995. Do female fish prefer to spawn in nests with eggs for reasons of mate choice copying or egg survival? *Am. Nat.* 145:824–32

39. Johnston BA. 1996. *The pathological and ecological consequences of parasitism by a cymothoid isopod (Anilocra chromis) for its damselfish host* (Chromis multilineata). MS thesis. Univ. Calif., Santa Barbara. 81 pp.

40. Kaplan RH, Cooper WS. 1984. The evolution of developmental plasticity in reproductive characteristics: an application of the adaptive coin-flipping principle. *Am. Nat.* 123:393–410

41. Kirkpatrick M, Dugatkin LA. 1994. Sexual selection and the evolutionary effects of copying mate choice. *Behav. Ecol. Sociobiol.* 34:443–49

42. Knoppien P. 1985. Rare male mating advantage: a review. *Biol. Rev.* 60:81–117

43. Lejeune P. 1985. Étude écoéthologique des comportements reproducteurs et sociaux des Labridés méditerranéens des genres *Symphodus* (Rafinesque 1810) et *Coris* (Lacepede 1802). *Cah. Ethol. Appl.* 5:1–208

44. Levene H. 1953. Genetic equilibrium when more than one ecological niche is available. *Am. Nat.* 87:331–33

45. Losey GSF, Stanton FG, Telecky M, and the Zoology 691 Graduate Seminar Class. 1986. Copying others: an evolutionary stable strategy for mate choice: a model. *Am. Nat.* 128:653–64

46. Lott DF. 1991. *Intraspecific Variation in the Social Systems of Wild Vertebrates*. Cambridge, UK: Cambridge Univ. Press. 238 pp.

47. Magurran AE. 1993. Individual differences and alternative behaviours. In *Behaviour of Teleost Fishes*, ed. TJ Pitcher, pp. 441–77. London: Chapman & Hall. 715 pp. 2nd ed.

48. Magurran AE, Seghers BH. 1990. Risk sensitive courtship behavior in the guppy, *Poecilia reticulata. Behaviour* 112:194–201

49. Magurran AE, Seghers BH. 1994. Sexual conflict as a consequence of ecology: evidence from guppy, *Poecilia reticulata,* populations in Trinidad. *Proc. R. Soc. London Ser. B* 255:31–36

50. Maynard-Smith J. 1982. *Evolution and*

Theory of Games. Cambridge, UK: Cambridge Univ. Press. 224 pp.

51. Metcalfe NB, Huntingford FA, Graham WD, Thorpe JE. 1989. Early social status and the development of life-history strategies in Atlantic salmon. *Proc. R. Soc. London Ser B* 236:7–19

52. Moore MC. 1991. Application of organization-activation theory to alternative male reproductive strategies: a review. *Horm. Behav.* 25:154–79

53. Moran NA. 1992. The evolutionary maintenance of alternative phenotypes. *Am. Nat.* 139:971–89

54. Parker GA. 1979. Sexual selection and sexual conflict. In *Sexual Selection and Reproductive Competition in Insects*, ed. MS Blum, NA Blum, pp. 123–66. New York: Academic. 463 pp.

55. Parker GA. 1983. Mate quality and mating decisions. In *Mate Choice*, ed. P Bateson, pp. 141–66. Cambridge, UK: Cambridge Univ. Press. 463 pp.

56. Pocklington P, Dill LM. 1995. Predation on females or males: Who pays for bright male traits? *Anim. Behav.* 49:1122–24

57. Pruett-Jones S. 1992. Independent versus nonindependent mate choice: Do females copy each other? *Am. Nat.* 140:1000–9

58. Reynolds JD, Gross MR. 1992. Female mate preference enhances offspring growth and reproduction in a fish, *Poecilia reticulata. Proc. R. Soc. London Ser. B* 250:57–62

59. Reynolds JD, Gross MR, Coombs MJ. 1993. Environmental conditions and male morphology determine alternative mating behaviour in Trinidadian guppies. *Anim. Behav.* 45:145–52

60. Rubenstein DI. 1980. On the evolution of alternative mating strategies. In *Limits to Action: The Allocation of Individual Behaviour*, ed. JER Staddon, pp. 65–100. London: Academic. 308 pp.

61. Ryan MJ, Causey BA. 1989. Alternative mating behavior in the swordtails *Xiphophorus nigrensis* and *Xiphophorus pygmaeus* (Pisces: Poeciliidae). *Behav. Ecol. Sociobiol.* 24:341–48

62. Ryan MJ, Pease CM, Morris MR. 1992. A genetic polymorphism in the swordtail, *Xiphophorus nigrensis*: testing the prediction of equal fitnesses. *Am. Nat.* 139:21–31

63. Taborsky M. 1984. Broodcare helpers in the cichlid fish *Lamprologus brichardi*: their costs and benefits. *Anim. Behav.* 32:1236–52

64. Taborsky M. 1985. Breeder-helper conflict in a cichlid fish with broodcare helpers: an experimental analysis. *Behaviour* 95:45–75

65. Taborsky M. 1994. Sneakers, satellites, and helpers: parasitic and cooperative behavior in fish reproduction. *Adv. Study Behav.* 21:1–100

66. Taborsky M, Hudde B, Wirtz P. 1987. Reproductive behaviour and ecology of *Symphodus (Crenilabrus) ocellatus*, a European wrasse with four types of male behaviour. *Behaviour* 102:82–118

67. Thresher RE. 1984. *Reproduction in Reef Fishes.* Neptune City, NJ: TFH. 399 pp.

68. Travis J, Woodward BD. 1989. Social context and courtship flexibility in male sailfin mollies, *Poecilia latipinna* (Pisces: Poeciliidae). *Anim. Behav.* 38:1001–11

69. van den Berghe E. 1990. Variable parental care in a labrid fish: How care might evolve. *Ethology* 84:319–33

70. van den Berghe E. 1992. Parental care and the cost of reproduction in a Mediterranean fish. *Behav. Ecol. Sociobiol.* 30:373–78

71. van den Berghe E, Wernerus F, Warner RR. 1989. Female choice and the mating cost of peripheral males. *Anim. Behav.* 38:875–84

72. Wade MJ, Pruett-Jones SG. 1990. Female copying increases the variance in male mating success. *Proc. Natl. Acad. Sci. USA* 87:5749–53

73. Waltz EC. 1982. Alternative mating tactics and the law of diminishing returns: the satellite threshold model. *Behav. Ecol. Sociobiol.* 10:75–83

74. Warner RR. 1984. Deferred reproduction as a response to sexual selection in a coral reef fish: a test of the life historical consequences. *Evolution* 38:148–62

75. Warner RR. 1985. Alternative mating behaviors in a coral reef fish: a life-history analysis. *Proc. Fifth Int. Coral Reef Conf.*, Tahiti. pp. 145–50

76. Warner RR. 1987. Female choice of sites versus mates in a coral reef fish, *Thalassoma bifasciatum. Anim. Behav.* 35:1470–78

77. Warner RR. 1990. Male versus female influences on mating-site determination in a coral reef fish. *Anim. Behav.* 39:540–48

78. Warner RR. 1991. The use of phenotypic plasticity in coral reef fishes as tests of theory in evolutionary ecology. In *The Ecology of Fishes on Coral Reefs*, ed. PF Sale, pp. 387–98. San Diego: Academic. 754 pp.

79. Warner RR, Harlan RH. 1982. Sperm competition and sperm storage as determinants of sexual dimorphism in the dwarf surfperch, *Micrometrus minimus. Evolution* 36:44–55

80. Warner RR, Hoffman SG. 1980. Local population size as a determinant of mating sys-

tem and sexual composition in two tropical marine fishes (*Thallasoma* spp.). *Evolution* 34:508–18

81. Warner RR, Lejeune P. 1985. Sex change limited by parental care: a test using four Mediterranean labrid fishes, genus *Symphodus. Mar. Biol.* 87:89–99

82. Warner RR, Wernerus F, Lejeune P, van den Berghe E. 1995. Dynamics of female choice for parental care in a fish species where care is facultative. *Behav. EcoL* 6:73–81

83. Zimmerer EJ, Kallman KD. 1989. Genetic basis for alternative reproductive tactics in the pygmy swordtail *Xiphophorus nigrensis. Evolution* 43:1298–1307

Annu. Rev. Ecol. Syst. 1997. 28:593–619

THE ROLE OF HYBRIDIZATION AND INTROGRESSION IN THE DIVERSIFICATION OF ANIMALS

Thomas E. Dowling and Carol L. Secor

Department of Biology, Arizona State University, Tempe, Arizona 85287-1501;
e-mail: Thomas.Dowling@ASU.EDU, Secor@ASU.EDU

KEY WORDS: speciation, evolutionary independence, historical stability, gene exchange, reticulate evolution

ABSTRACT

Although hybridization and introgression have been considered important in generation of plant diversity, their role in evolutionary diversification of animals remains unclear. In this review, we reconsider the significance of introgressive hybridization in evolution and diversification of animals to determine if the generally negative assessment of these processes is warranted. Unlike the situation for plants, hybrid animal taxa appear to be relatively rare. This could, however, be due to negative attitudes toward hybridization and difficulty in detecting such forms. Hybridization has been responsible for instantaneous creation of several unique complexes of polyploid and unisexual animals. Allopolyploidy has allowed for diversification, whereas unisexual taxa have acted as conduits of gene exchange among related sexual species. Many instances of diploid, bisexual taxa of hybrid origin have been put forward, but few have been carefully tested. Changing attitudes toward hybrids and technological advances should allow for careful consideration of hypothesized hybrid taxa and will undoubtedly increase the number of known animal hybrid taxa.

INTRODUCTION

Around the time of the neo-Darwinian synthesis, considerable effort was directed toward understanding the origin of taxonomic diversity. This work led to several central principles, such as the importance of geographic isolation (82, 83) and selection for reproductive isolation (33). Several of these tenets have since come under scrutiny (e.g. 15, 19, 40, 63, 77, 91, 156), leading

0066-4162/97/1120-0593$08.00

to renewed interest in the processes generating and maintaining organismal diversity. Gene exchange among animal species has been more common than previously believed, opening the door to several new avenues of evolutionary research. Hybrid zones have been used for study of evolutionary processes (54, 55, 59), providing insight into processes responsible for patterns of geographic variation among taxa and maintenance of their distinctiveness. The genetics of reproductive isolation is still poorly understood, but experimental hybridization studies have allowed for identification of specific genetic changes responsible for species differences and reduced fitness of hybrids (25, 114, 158).

While hybridization and introgression have been deemed important by botanists (7a, 113, 114), zoologists have not seriously considered the significance of these factors in the evolutionary process. Anderson & Stebbins (6) summarized views toward the significance of hybridization in plant evolution. They hypothesized that introgression and hybridization could transfer blocks of genes among stabilized, adapted groups, permitting rapid reshuffling of varying adaptations and complex modifier systems. In this way, levels of variation would be greatly increased, and selection would be able to act upon segregating blocks of genic material derived from different adaptive systems instead of one or two new alleles generated by mutation. This set of circumstances would be particularly advantageous where new ecological niches are created by changing environments, allowing evolution to proceed at "maximum rate."

One possible outcome of horizontal transfer of genetic variation among lineages would be creation of taxonomic diversity. Although this process has been considered important in generation of plant diversity (53), the role of hybridization in evolutionary diversification of animals remains unclear. Grant (53) noted that "several generations of zoologists have concluded that hybridization does not play an important role in animal evolution," with this perspective based upon "a store of background knowledge that should not be dismissed lightly" (p. 161).

With this caution in mind, we reconsider the significance of introgressive hybridization in the evolution and diversification of animals to determine if the generally negative assessment of these processes is warranted. This is achieved by placing perceptions of hybridization and introgression in historical perspective, followed by an assessment of hybrid origins of animal taxa and evaluation of the potential role for these factors in evolutionary diversification. It is impossible for such a review to be exhaustive because many suggestions of hybrid origin are presented as digressions in papers dedicated to taxonomic treatments or studies of hybridization. Therefore, we address conceptual issues to stimulate further consideration of the role for hybridization in the diversification of animals.

DEFINITIONS

We choose to focus on hybridization and introgression as processes, bypassing conceptual difficulties associated with defining categorical units (e.g. species) involved (139). Hybridization is defined as "the interbreeding of individuals from two populations, or groups of populations, which are distinguishable on the basis of one or more heritable characters" (55, p. 5), and introgression is "the permanent incorporation of genes from one set of differentiated populations into another, i.e. the incorporation of alien genes into a new, reproductively integrated population system" (115, p. 71).

Definition of hybrid taxa is difficult, as selection of criteria for placing groups into specific taxonomic categories (e.g. subspecies, species, genera) will suffer from the same difficulties associated with nonhybrid species, reflecting the bias of the user in determining the quantity and type of differences necessary for taxonomic recognition (40). This is particularly problematic for hybrid taxa as boundaries defining species are semipermeable (54), and the extent and persistence of introgression may vary among genes. In addition, recent technological advances (61) allow for fine-scaled examination with many loci, identifying unusual patterns of introgression (e.g. 7a, 36, 42, 79, 115). Therefore, the extent of introgressive hybridization at any locality might occur anywhere along a continuous distribution of possibilities, ranging from introgression at one locus [e.g. mitochondrial DNA (mtDNA)] to enough loci that populations be recognized as distinct taxonomic entities.

Given these difficulties, we define hybrid taxa in terms of the processes generating them: the derivation of features by horizontal transfer from multiple independent lineages at some point in their history. Under this guise, a hybrid taxon is an independently evolving, historically stable population or group of populations possessing a unique combination of heritable characteristics derived from interbreeding of representatives from two or more discrete units (e.g. races, subspecies, species, etc). Historical stability implies that the mosaic of characters inherited from independent lineages is retained in the population, passed on from parent to offspring.

Stability and evolutionary independence are essential for distinguishing taxa of hybrid origin from instances of ongoing hybridization among taxa. Unfortunately, these features are difficult to evaluate. Stability can be based only on human perceptions of longevity, typically from collections that cover 300 years at most. Where taxa are allopatric, evolutionary independence is often based on degree of divergence. Thus, these criteria are open to subjective interpretation, producing a situation analogous to evaluation of specific status for allopatric forms.

HISTORICAL PERSPECTIVE

The significance of hybridization and introgression for animals has not been the subject of any major review; however, the topic has received cursory coverage in a variety of texts and articles (e.g. 15, 113). This limited coverage seems to result largely from the ideas and perspectives of prominent researchers working at the time of the modern synthesis. This sentiment is summarized by Fisher (44): "The grossest blunder in sexual preference, which we can conceive of an animal making, would be to mate with a species different from its own . . ." (p. 144).

Dobzhansky (32) recognized the importance of allopolyploidy (i.e. elevation of chromosome numbers due to hybridization of two or more species) to evolution, but he noted that the rarity of allopolyploids "constitutes the greatest known difference between the evolutionary patterns in the two kingdoms" (32, p. 219). Introgressive hybridization was considered to occur more frequently in plants than in animals, possibly due to greater attention paid to this problem by botanists. He also noted that "hybrid swarms" may reflect inheritance of shared polymorphism from a variable common ancestor, not exchange of genes among species. In a later work, Dobzhansky (34) discussed reasons for differences in the frequency of introgressive hybridization between plants and animals. Vegetative and asexual reproduction found in many plants provides for longer life and increased spreading capacity of individuals (or clones), whereas sexual reproduction maintains evolutionary plasticity. In addition, the greater complexity of tissues and organ systems in animals may require that the adaptive value of genotypes depends upon integration of a larger constellation of genes than for plants (with their simplified tissues and open systems of growth). These features would constrain the number of adaptive recombinants in animals, reducing the frequency of hybridization and increasing the importance of reproductive isolation.

One of the most outspoken and influential opponents of a significant role for hybridization and introgression in evolution was Mayr (82, 83). He acknowledged that animal species could arise instantaneously through allopolyploidy, but the rarity of hybridization in animals made this process much less important than for plants. Transformation or fusion of species by introgression was also viewed as implausible, especially when the parental species continue to exist. Since hybrids would exhibit less reproductive isolation than their progenitors, they would have to remain geographically isolated until they could maintain their integrity. In light of this perspective, Mayr argued that combinations of characters from putative parental species and morphological intermediacy are inadequate for identification of hybrid species because of both the polygenic nature of these characters and the developmental stability of interactions among genes resulting in the phenotype. A more plausible explanation was the evolution of such intermediate forms from a single, polymorphic ancestor.

Mayr also provided several objections to the view that introgression significantly increased genetic variability in animal populations. Hybrid animals are rare in nature, even in groups that have been extensively studied. Where F_1 hybrids occur, they tend to be sterile or to exhibit reduced fertility. Hybrid and backcross individuals would also suffer reductions in fitness because gene exchange breaks up "internally balanced chromosome sections," resulting in elimination of these individuals and "severe selection against introgression." In his view, "the total weight of the available evidence contradicts the assumption that hybridization plays a major evolutionary role among higher animals" (83, p. 133).

Recently, some zoologists have become more open to a significant role for hybridization in the evolutionary process, including the formation of new species. Stebbins (140) contrasted the variation in perceived significance of hybridization in the evolution of plants and animals. Because the proportion of successful progeny segregating from hybrids is much lower for animals (due to complex patterns of development controlled by more intricate and integrated complexes of genes than found in plants), zoologists have downplayed the significance of introgression. Stebbins noted that rare gene combinations can establish rapidly; therefore, hybridization may have played a larger role in the evolution of animals than is recognized.

In his evaluation of mechanisms of speciation, Templeton (143) contemplated the production of new species by hybridization. Genetic and structural incompatibilities associated with unisexual reproduction and polyploidization were considered important means for maintenance of hybrid taxa. Conditions for stabilization of hybrid recombinants were also described. Under his scenario, hybridization is followed by inbreeding and hybrid breakdown. If recombinants with the greatest viability and fertility are able to survive, they could become reproductively isolated from both parental forms and become distinct entities. Hybridization followed by inbreeding could enhance mutation rates in recombinants, leading to increased divergence from parental forms, especially where the population of recombinants is geographically isolated. Divergence is required for coexistence; otherwise the new form could be swamped out by gene flow or eliminated through competition. On the basis of several experimental studies, Templeton suggested that this mode of speciation can occur and does not require the evolution of postmating barriers between recombinants and parental taxa.

The potential significance of introgressive hybridization in animal evolution was supported experimentally by Lewontin & Birch (73). They hypothesized that dramatic range expansion of an Australian fruit fly (*Dacus tryoni*) occurred through adaptation to extreme temperatures, with genetic variation obtained through introgression with a closely related species, *D. humeralis*. This possibility was tested by maintaining "pure" and "hybrid" populations initiated from *D. tryoni* and F_1 hybrids between *D. tryoni* and *D. humeralis*, respectively, at a series of optimal and extreme temperatures. Hybrid populations initially produced fewer pupae than did pure lines of *D. tryoni*, but there was no difference in this feature or morphological characteristics by the end of the experiment, around two years later. Despite these similarities, hybrid lineages were found to increase more rapidly at higher temperatures, indicating that introgression of alleles allowed for adaptation to extreme warmth. As Lewontin & Birch note, these results do not prove that *D. tryoni* became adapted through introgressive hybridization, only that such a series of events could have been involved.

Several ornithologists have indicated that introgressive hybridization has been important in avian evolution. Estimates of hybridization among bird species range from 9% worldwide (50) to 15% for the Nearctic fauna (126). Short (125) noted that introgressive hybridization preserved genetic variation of each progenitor in hybrids, making "such forms 'preadapted' by virtue of introgressive hybridization." In most instances dispersal has made it difficult to ascertain the fate of hybrids; however, detailed studies of Darwin's finches allowed Grant & Grant (51, 52) to conclude that introgressive hybridization plays a greater role in the evolution of animals than previously believed.

MODES OF HYBRID ORIGIN

Both hybridization and introgression have been hypothesized to enhance animal diversity, but differences between these processes require they be considered separately. Hybridization may directly produce distinct taxa, either through polyploidization or generation of clonally reproducing unisexual lineages. Alternatively, introgression could eventually lead to stable, independent lineages definable by unique combinations of characteristics. In either case, taxonomic diversification is best viewed as an incidental by-product, resulting from accumulation of genetic differences that confer distinctiveness and evolutionary independence.

Given problems associated with defining taxa of hybrid origin, assessment of a possible role for hybridization and introgression in the generation of taxonomic diversity is difficult. Nevertheless, reticulate origins have been invoked for many taxa, and it is important to determine the extent to which gene exchange has played a role in generation of taxonomic diversity.

Diversification by Hybridization

Hybridization can instantaneously produce distinct animal taxa in two ways, through increase in chromosome number (allopolyploidy) or through conversion to an essentially all-female (unisexual) mode of reproduction. These two pathways can be difficult to separate as most animal allopolyploids are hermaphroditic or unisexual while many all-female complexes include polyploids.

All-female reproductive systems occur in three general forms, parthenogenesis, gynogenesis, and hybridogenesis (27). In parthenogenesis, females reproduce progeny essentially identical to themselves (barring mutation) without involvement of males. Gynogenetic and hybridogenetic forms require sperm to initiate egg development. In gynogenesis, fertilization only serves to stimulate development, with sperm nuclei excluded. In hybridogenesis, true fertilization occurs and paternal genes are expressed; however, only the maternal lineage is transmitted to the next generation.

Hybrid lineages have typically been identified by morphological intermediacy, increased heterozygosity at nuclear gene loci (usually allozymes), and/or excess amounts of DNA in multiples of the standard complement (e.g. 3N, 4N, etc.) obtained through standard karyotypic methods or measurement of DNA content. Mitochondrial DNA variation has been used to identify maternal lineages; however, it does not allow for recognition of hybrids unless applied in conjunction with other characters (e.g. allozymes, karyotypes, morphology). Application of these methods has provided a number of excellent examples in many groups of animals, each with their own unique features. Instead of recounting extensive reviews of polyploid and unisexual taxa (11, 14, 29, 76, 122, 156), important concepts are summarized below and new cases and updated references are provided in Table 1.

THEORETICAL CONSIDERATIONS   The role of hybridization in the production of polyploid animal lineages has sometimes been difficult to determine. Allopolyploids have typically been distinguished from autopolyploids (in which multiple sets of chromosomes are derived from the same ancestral species) by the lack of multivalent sets of chromosomes or disomic inheritance of allozymes. Unfortunately, the distinction between these categories is not always clear. White (156) noted that multivalents, e.g. structures formed by association of more than two chromosomes during meiosis, have been identified in diploid organisms, possibly due to translocation heterozygosity. In addition, autotetraploids become functionally diploid over time (105). Therefore, older autopolyploids could exhibit fewer multivalents (11) and multisomic loci than expected; however, salmoniform fishes are ancient autotetraploids that have not become diploidized (4). Conversely, one could imagine a scenario in which hybridization among closely related but distinct taxa could result in polyploids that form multivalents and have multisomic inheritance, reducing the effectiveness of these characteristics for discriminating among these modes of origin.

Allopolyploid plants have been thought to be more common than autopolyploids because of increased fitness attained through their hybrid constitution,

Table 1 Hypothesized examples of asexual and polyploid taxa of hybrid origin

Taxonomic complex	Ploidy	Mode of reproduction[a]	References[b]
Platyhelminths			
Paragonimus westermani	3N, 4N	?	1, 3
Fasciola sp.	2N, 3N	P	2
Mollusks			
Lasaea species complex	3N-6N	G	104
Bulinus truncatus complex	2N, 4N, 8N	P or S?	47
Campeloma decisum	?	P	66
Ancylus fluviatilis	4N	P or S?	138
Insects			
Bacillus species complexes	2N, 3N	P, H	13, 145
Curculionidae	2N-10N	P, G/H?	101, 120
Simuliidae	3N	P	118
Crustaceans			
Daphnia pulex complex	2N, 4N	P	30
Cyprinotus incongruens	3N	P	146
Trichoniscus pusillus pusillus	3N	P	144
Fishes			
Fundulus heteroclitusXdiaphanus	2N	G	28
Phoxinus eos-neogeaus complex	2N, 2N/3N, 3N	G	45, 46
Poecilia complex	2N, 3N	G,?	9, 121, 132
Poeciliopsis complex	2N, 3N	G, H	109, 152
Rutilus alburnoides complex	2N, 3N	?	5
Family Catostomidae	4N	B	128, 147
Cobitis complexes	3N, 4N	G	124, 148, 149
Amphibians			
Ambystoma complexes	2N-5N	G?, H?	58, 69, 70, 134, 135
Bufo danatensis	4N	B	88
Hyla versicolor	4N	B	80, 108, 112, 116
Rana esculenta complexes	2N, 3N	H	62
Reptiles			
Cnemidophorous complexes	2N, 3N	P	20, 96, 127, 154
Gymnophthalmus underwoodi	2N	P	21, 22, 160
Hemidactylus garnotti	3N	P	94, 111
Heteronotia binoei	3N	P	92
Kentropyx borckiana	2N	P	23
Lacerta complex	2N, 2N/3N, 3N	P	95
Lepidodactylus lugubris	2N, 3N	P	94, 110, 111, 150
Nactus pelagicus	2N	P	35

[a]Abbreviations for modes of reproduction are: G, gynogenetic; H, hybridogenetic; P, parthenogenetic; S, self-fertilizing hermaphrodite; B, bisexual; ?, uncertain.

[b]Additional references may be found within those cited.

allowing for adaptation to a wider range of environments provided by multiple sets of genes. Using this information, White (156) reasoned that allopolyploid animals were also more likely to establish than autopolyploids. In any case, he concluded that polyploids are far less common in animals than plants.

The rarity of allopolyploid animals has been attributed to their systems of sex determination, with polyploidization directly interfering with sex determination (97) or indirectly affecting fitness through disruption of dosage compensation of sex-linked genes (106). White (156) concluded that the occurrence of polyploidy may also be constrained by modes of reproduction, particularly in those groups not exhibiting heterogametic sex determination. Many animals are obligate outcrossers, with polyploid progeny of hybridization events typically tetraploid. Such hybrids would most likely mate with diploids, producing sterile triploids. Therefore, polyploidy would be most likely found in groups exhibiting hermaphroditic or all-female modes of reproduction.

White (156) estimated that only 1 of every 1000 "species" of animals possessed all-female reproductive systems, many of which resulted from hybridization. The relative infrequency of hybrid asexual taxa has been attributed to severe ecological and genetic constraints on the origin and maintenance of such lineages. Moritz et al (93) argued that the chance of a hybrid founding a unisexual lineage is determined by a balance of genetic factors affecting the disruption of meiosis and the remainder of the developmental program, reducing the window in which hybrids can become unisexual lineages. As parental taxa diverge, genes regulating meiosis change. Combination of different alleles in F_1 hybrids may result in disruption of meiosis, increasing the proportion of unreduced ova generated in these individuals and possibly allowing for unisexual reproduction. At some point, parental taxa become too divergent, with genetic changes dramatically reducing fecundity and viability of hybrid offspring and also reducing the likelihood of establishment of unisexual lineages.

Once unisexual hybrids surmount genetic barriers to formation, a variety of genetic, ecological, and evolutionary constraints must be overcome (151, 152). Clonal transmission will result in accumulation of deleterious mutations (98) leading to "mutational meltdown" and relatively rapid extinction of such lineages (78). Pseudogamous (i.e. gynogenetic, hybridogenetic) taxa require contribution of sperm from compatible host taxa. Since these lineages are clonal, unisexual populations will be locked into certain niches that may place them at risk through competition with their bisexual progenitors or an inability to adapt to changing environments.

INFERENCE FROM DIPLOID AND POLYPLOID UNISEXUAL LINEAGES   Polyploidy is typically associated with asexual reproduction and self-fertilization, especially in invertebrates (76, 156), likely due to increased probability of establishment for such forms. In vertebrates, virtually all unisexual taxa have documented hybrid origins (153). Polyploid and unisexual lineages are often more widely distributed than their parental taxa, possibly indicative of advantages provided by their increased variability and mode of reproduction (151).

Allopolyploid and hybrid unisexual origins are more common than previously believed. Molecular methods have documented multiple origins for many polyploid and unisexual lineages, indicating that these phenomena are not accidents but reflect some general aspects of the interaction of parental genomes. Multiple origins are often attributable to several hybridization events, and reciprocal crosses may produce different results. Hybridization of male *Daphnia pulicaria* and female *D. pulex* produces diploid parthenogenetic lineages, whereas tetraploid parthenogens are produced by the reciprocal cross (38a). In some instances, only one species may serve as the maternal parent, indicating specific combinations may be necessary for production of such lineages.

Diverse complexes of clonal lineages often develop, due to genetic differences among bisexual parents. Hybridization of polyploids or unisexuals with bisexual diploids (sometimes a third taxon) has led to variation in ploidy level and/or mode of reproduction as well as allelic differences (e.g. *Bacillus, Bulinus, Cobitis, Phoxinus, Ambystoma, Cnemidophorus*). This diversity is likely to increase the probability of establishment and persistence as demonstrated by studies of unisexual forms of topminnow, *Poeciliopsis* (151, 152). This complex includes diploid hybridogenetic and triploid gynogenetic groups of clones produced by a variety of hybridization events involving *P. monacha* and some combination of four other species: *P. latidens, P. lucida, P. occidentalis,* and *P. viriosa.* Genetically distinct clonal lineages have different life-history, physiological, and ecological characteristics, allowing multiple clones to coexist and occupy at least part of the potential niche of the sexual species.

In some cases, hybrid origins have been attributed to other factors. Johnson (66) examined the evolution of parthenogenesis in the hermaphroditic snail *Campeloma decisum.* Unisexual reproduction was thought to have evolved spontaneously in some populations due to a parasitic trematode that ingests or blocks sperm. Fixed homozygosity at 19 allozyme loci supported this hypothesis for a subset of populations; however, all individuals from several localities were heterozygous at six loci that were fixed or nearly fixed allelic differences between eastern and western populations, implicating a hybrid origin for a subset of parthenogenetic populations. Dufresne & Hebert (38a) found that homozygous parthenogenetic clones of *Daphnia pulex* thought to be autopolyploid possessed mtDNA of *D. middendorffiana.* They concluded that these clones originated through hybridization with expulsion of the nuclear genome of the female parent while retaining mtDNA.

Asexual lineages have been considered dead-ends, not persisting long enough to contribute to the evolutionary process; however, recent studies have called this view into question. DNA variation has been used to suggest that several unisexual lineages are older than previously believed (e.g. 58, 109, 134). Unfortunately, difficulties in accurate dating of divergence have an impact on the utility of such estimates. Failure to include the maternal progenitor will inflate estimates, requiring extensive samples to maximize chances of including the maternal parent and accurate characterization of geographic variation. For mtDNA, most estimates of divergence time are based upon a standard calibration. These are not likely accurate due to tremendous variation in rates of evolution across groups (e.g. 8, 80). Worst of all, divergence of alleles could predate (or postdate) speciation events (100), reducing the utility of such estimates.

The most significant contribution of unisexual lineages may be their ability to act as conduits for gene exchange among bisexual forms (e.g. *Ambystoma, Bacillus, Phoxinus, Rana*). Some members of the *Phoxinus eos-neogaeus* complex exhibit unusual modes of reproduction that could lead to gene exchange among parental taxa (46). The two triploid individuals examined produced haploid eggs that were fertilized by *P. eos*. Progeny were indistinguishable from *P. eos* on the basis of allozymes and external morphological traits, but mtDNA of all (KA Goddard, personal communication) and gut morphology of some was like *P. neogaeus*. Sex ratios of resulting progeny were approximately 1:1, and males appeared normal. Unfortunately, their ability to successfully reproduce was not determined. If such males are fertile, their contribution could lead to increased genetic diversity in the bisexual species.

Bogart (12) discussed evidence for transfer of alleles among four species of *Ambystoma* and reconstitution of parental forms. The existence of certain genotypic combinations was consistent with reconstitution; however, analysis of such individuals from a single egg mass indicated reconstituted individuals may suffer from increased mortality due to karyotypic anomalies (142). Bogart concluded that circumstantial evidence supported a potentially significant role for unisexuals in the evolutionary process. Unisexuality increases the opportunity for selection to act upon various recombinant gene combinations by maintenance of these variants for longer periods of time than otherwise possible. He further hypothesized that the ultimate outcome could be a population of bisexuals with mosaic genotypes superior to the parental and hybrid forms. This prediction has come to fruition in *Poeciliopsis,* in which a clone of the hybridogen *P. monacha-occidentalis* has reverted to sexuality. This new bisexual species contains a mosaic of genetic features from its parental contributors and occupies a unique niche (152), implicating a significant role for unisexual taxa in the evolutionary process.

INFERENCE FROM BISEXUAL POLYPLOID LINEAGES   Despite expected difficulties of formation associated with outcrossing, several bisexual polyploid lineages have been generated by hybridization. Identification of polyploid lineages has not always been simple, and some hypothesized cases have been controversial. White (156) rejected most instances of polyploidy in obligate sexual lineages, indicating that the best examples are provided by fishes and anurans. Among fishes, bisexual polyploids have been reported from five orders, with two extensively studied (16, 122). While salmoniforms are considered autotetraploid (4), allopolyploidy has been hypothesized for three cypriniform families: Catostomidae, Cobitidae, and Cyprinidae. The catostomids are a widely distributed group of more than 60 tetraploid species exhibiting disomic inheritance (41, 128). Their success has been obtained through diversification and adaptation to a variety of habitats, with representatives found in most streams, rivers, and lakes of North America and parts of Asia. Uyeno & Smith (147) hypothesized that, given the extent of polyploidy, the entire family descended from a cyprinid-like lineage approximately 50 Mya.

In cobitids, tetraploids are found in three separate groups, but only indirect evidence supports their hybrid origin (149). Ferris & Whitt (43) theorized that, based on levels of duplicate gene expression in tetraploid cobitids from the genus *Botia*, the polyploidization event was likely more recent than that giving rise to the catostomids. In their review of cyprinids, Buth et al (17) noted that 52 taxa were polyploid. Most of these are members of the subfamily Cyprininae, identified through analyses of karyotypes and genome sizes. Chromosome counts in nearly all cyprinid polyploids occur in multiples or combinations of the most common karyotypes (48 or 50 chromosomes); thus it was speculated that tetraploids (96, 98, or 100 chromosomes) and hexaploids (148 and 150 chromosomes) arose through hybridization involving pairs of diploid taxa or diploid and tetraploid taxa, respectively.

Smith (128) identified several morphological characters uniting catostomids and cyprinines, consistent with a single origin of tetraploidy for these two groups. Collares-Pereira & Coelho (24) used evidence of polyploidy in all three families to suggest that the entire order was derived from a single polyploid ancestor; however, this is inconsistent with the distribution of polyploidy throughout the families and relative timing of events obtained from levels of duplicate gene expression.

The only amphibian bisexual polyploids (3N-8N) are anurans, with incidence more phylogenetically restricted than for fishes (11). Hybridization has been implicated in several instances, but not without debate. A prime example of this difficulty is provided by North American tree frogs. Diploid *Hyla chrysoscelis* is subdivided into eastern and western populations by the tetraploid *H. versicolor*. These two species are generally allopatric, but co-occur in several narrow contact zones throughout the eastern United States. *Hyla versicolor* has been considered autopolyploid due to the presence of quadrivalents (11) and lack of allozymic variation consistent with specific status of the presumed progenitors, eastern and western *H. chrysoscelis* (112, 116); however, immuno-logical and allozymic data have also been used to support a hybrid origin (11, 81). Ptacek et al (108) provided mtDNA evidence supporting multiple origins of *H. versicolor*; thus, available evidence is not consistent with hybrid origins for this form.

Diversification by Introgression

Formation of new taxa by introgressive hybridization involves different cir-cumstances than those described above. Gene exchange among taxa produces groups of recombinant individuals that eventually stabilize to form an evo-lutionarily independent, sexually reproducing taxon. This process has been hypothesized to include a variety of geographic scales, ranging from complete fusion of taxa to stabilization of local hybrid populations. Diversification is not instantaneous as recombinant lineages must be geographically isolated for long enough to evolve genetic differences that maintain independence and allow coexistence with parental forms.

Stable recombinant taxa have been hypothesized on the basis of several types of characters, including morphology, chromosomes, allozymes, and DNA. Hy-brids have been expected to be intermediate to parental forms for complex fea-tures such as morphological variation. However, F_1 hybrids are not always mor-phologically intermediate between their progenitors (99, 117), whereas not all morphological intermediates are hybrids (156, 157). For example, morpholog-ical characteristics of recent year classes of the endangered sucker *Chasmistes cujus* were intermediate between older fish and *Catostomus tahoensis*, leading to concerns over human-induced hybridization. Analysis of allozymic varia-tion failed to detect introgressed alleles in the younger *C. cujus* or to identify significant differences among younger and older representatives of this species, indicating that morphological differences reflect ontogenetic variation (18).

For more simplistic characters, (e.g. allozymes, karyotypes), the popula-tion should consist of an equilibrium distribution of variants contributed by all parental forms. Because the proportion of specific variants will be influenced by selection and drift, parental contributions may not persist for all character sets but could be represented by fixation for different diagnostic traits (e.g. alternate diagnostic alleles at two or more allozyme loci, mtDNA from one parental taxon and allozymic variants from the other). Therefore, hybrid taxa are most easily identified by examination of multiple independent character sets. Tests of hybrid origin are most readily obtained through phylogenetic analysis of each set of characters (e.g. morphology, allozymes, mtDNA, nuclear loci), with reticulate origins indicated by discordance among resulting topologies. Arnold (7) demonstrated the general utility of this approach, implicating introgressive hybridization in several animal and plant groups.

Technological advances have simplified identification of mosaic distributions of characters. Since primers for amplification and sequencing of many mitochondrial and nuclear genes have been developed, it is possible to closely examine patterns of genetic variation for a variety of independent loci. This approach has advantages over use of phenotypic characters such as morphology or allozymes because it allows for phylogenetic assessment of the source taxon for each allele. Careful examination of results derived from each data set is critical because most parsimonious topologies obtained from different character sets or loci may appear inconsistent when they are actually not well supported. Methods for assessment of phylogenetic trees are considered by Hillis et al (60) and Swofford et al (141).

THEORETICAL CONSIDERATIONS   Previous objections to hybrid taxa may have been influenced by misconceptions concerning geographic scale and the structure of hybrid zones. Traditionally, gene flow has been considered to be strong enough to maintain cohesion of each species (39), and hybridization was thought to occur in continuous zones in which characters graded clinally from one form to the other. Under such circumstances, formation of hybrid taxa would be severely constrained by constant gene flow from parental taxa, preventing populations of recombinants from stabilizing and establishing independence.

Studies of population structure and hybrid zones, however, have indicated that this view of geographic structure and hybridization is oversimplified (39, 56). Organisms have a tendency to track specific favorable environments, fragmenting distributions. Because of this behavior, hybridization is often found in tracts of intermediate habitat or where conditions favoring two taxa are found in proximity. Since such conditions tend to be patchily distributed, hybridization is often better represented as a mosaic of potentially distinctive interactions among taxa, each proceeding along an independent evolutionary trajectory. In addition, environments are not stable, with resulting changes in selection pressures causing temporal shifts in the genetic composition of populations. Environmental heterogeneity and temporal instability are conducive to formation of hybrid taxa, especially where patches of hybrids have become isolated from both parental species.

Rarity of recombinant animal taxa has also influenced perceptions of significance. Difficulty detecting such forms is partly responsible for their rarity, with probability of successful identification dependent upon a variety of factors. Production of hybrid taxa should exhibit a negative correlation with levels of divergence among parental forms, with more divergent forms more likely to produce inviable or sterile hybrids. When the level of divergence among parental taxa is low, horizontal transfer will be difficult to discriminate from ancestral polymorphism. In origins involving gene exchange between common and rare taxa, as is often the case (64), there are two potential difficulties. Influence of the less frequent taxon may be difficult to detect because it contributes proportionally fewer alleles to the recombinant taxon. In addition, rare taxa are more likely to go extinct, making it impossible to identify the source of some variants.

Probability of detection will also be reduced by time since origin of recombinant lineages. Soon after initiation, recombinant taxa are readily identifiable by mosaic combinations of character states. As time passes, however, such taxa will evolve their own unique characteristics, making it more difficult to discriminate between horizontally transmitted and convergent character states. Over time, internodal branches become relatively shorter and recombination among parental alleles becomes more likely, making ancient events more difficult to resolve with allele phylogenies.

Given all of these factors, only a subset of hybrid taxa will be detectable. Unfortunately, it is not possible to estimate probabilities of origination and detection, but it seems likely that these factors will severely limit identification of hybrid origins, making stable recombinant taxa appear much more rare than they actually are.

INFERENCE FROM STABILIZED RECOMBINANT LINEAGES   The frequency of diploid bisexual taxa of hybrid origin is difficult to assess. Hypothesized instances of hybrid origin are often included as anecdotes in systematic studies and are not amenable to recovery with standard library search procedures, whereas many others have not been published and are known only to specialists. Of those cases identified (Table 2), many have not been examined with multiple sets of characters and remain conjecture based on morphological intermediacy. Additional instances of reticulate evolution have been identified (e.g. 7, 7a, 37, 102a); however, taxonomic implications have not been considered.

Tests of hybrid origin can fail to discriminate among alternate hypotheses. Smith et al (130) proposed a hybrid origin for the sucker *Catostomus (Pantosteus) discobulus yarrowi* on the basis of allozymic and morphological variation. Crabtree & Buth (26), however, found only limited evidence for contribution from one of the putative parental taxa, *C. plebius*. Morphological evidence was interpreted to represent shared primitive traits retained in *C. d. yarrowi*, and it was considered a distinctive form of the other putative progenitor, *C. discobolus*. Based on intermediacy of morphological and allozymic characters, Menzel (87) proposed that *Luxilus albeolus* was a taxon generated by hybridization of *L. cornutus* and *L. cerasinus*. Meagher & Dowling (85) examined morphological, allozymic, and mtDNA variation for these three taxa to test this hypothesis. Allozyme alleles supposedly contributed by *L. cerasinus* to *L. albeolus* were also found in local allopatric populations of *L. cornutus*,

Table 2 Hypothesized examples of bisexual diploid taxa of hybrid origin

Taxon	Evidence[a]	Reference[b]
Mollusks		
Mercenaria campechiensis texana	M, A, mt	103
Cerion "columna"-like	M, A	49, 50
Cerion "rubicundum"-like	M	49
Insects		
Andrena montrosensis	M	71, 72
Papilio joanae	M, mt	133
Papilio brevicauda	M, mt	133
Crustaceans		
Bosmina coregoni/longispina	M	74
Daphnia wankeltae	M, A	57
Daphnia cucullata procurva	M, A	75, 123
Fishes		
Brachymystax sp.	M, A	90, 107
Catostomus discobolus yarrowi	M, A	26, 130
Chasmistes brevirostris	M	89
Chasmistes liorus mictus	M	89
Gila robusta jordani	M, A, mt	37
Gila seminuda	M, A, mt	31, 131
Luxilus albeolus	M, A, mt	85, 87
Mimagoniates microlepis	M	86
Pararhynichthys bowersi	M, A	48, 136, 137
Amphibians		
Ambystoma tigrinum stebbinsi	M, A, mt	68
Reptiles		
Pseudemoia cryodroma	M, A	65
Birds		
Passer italiae	M	67
Mammals		
Canis rufus	M, A, mt, u	119,155
Mus musculus molossinus	M, A, mt	159

[a]Abbreviations for character sets are: A, allozymes; M, morphology; mt, mtDNA; u, microsatellites.
[b]Additional references may be found within those cited.

preventing discrimination of hybrid origin from convergence of allozyme alleles or shared ancestral polymorphism. Reconstruction of allele phylogenies will provide more power for discrimination among these alternatives.

Analysis of proposed hybrid taxa has also yielded support for hybrid origins. DeMarais et al (31) used morphological, allozymic, and mtDNA data to examine the hypothesized hybrid origin of the minnow *Gila seminuda* (131). Taken individually, each character set yielded well-resolved differences between the putative parental taxa, *G. elegans* and *G. robusta*. *Gila seminuda* was intermediate to *G. robusta* and *G. elegans* in phylogenetic analysis of morphological and allozymic characters, but it exhibited mtDNA essentially identical to *G. elegans*. Phylogenetic analysis of additional *Gila* species (37) identified conflict between allozymic and mtDNA topologies, patterns best explained by past episodic introgression.

Several groups of fishes from the western United States exhibit evidence of past introgression (31, 37, 129), implicating a general causative factor. Ecosystems of this region have gone through dramatic environmental change, including considerable tectonic activity and progressive aridification (10). These changes had a severe impact on aquatic ecosystems, likely producing cycles of isolation and sympatry. Divergent, isolated forms would have been forced together by desertification, allowing for gene exchange. Later periods of isolation would allow for stabilization and independent evolution of hybrid derivatives, resulting in taxonomic diversification.

The fossil record has provided historical perspective of introgressive origins for several taxa. Variation in *Cerion*, a speciose group of West Indian pulmonate land snails, has long perplexed taxonomists and evolutionary biologists (50). Goodfriend & Gould (49) described two cases in *Cerion* where introgressive hybridization has yielded temporally stable, morphologically distinct populations of hybrid origin on Great Inagua, Bahamas. In the first, a snail with a distinctive flat-topped morphology (*C. dimidiatum*) invaded over 13,000 ya ago, followed by subsequent transition of the local population to a morphology intermediate to the invader and the native species, *C. columna*. Morphological and genetic analysis of the extant population (50) indicates that the influence of introgression has been retained long after the disappearance of *C. dimidiatum* from the fossil record. A second, older introgression event between invading *C. rubicundum* and native *C. excelsior* produced snails with intermediate morphotypes (49). A series of dated fossil shells identified progressive shifts in morphotype for over 13,000 ya, changing from intermediate to that of *C. rubicundum*. Although morphological characters are good indicators of hybridization, they do not always retain evidence of past introgression. These populations are indistinguishable from *C. rubicundum*, but evidence of past hybridization likely persists at the genetic level.

Not all cases of hybrid origin are simple to intepret. Cladocerans hybridize extensively, yielding complexes of intermediate and parental morphotypes (123). Because of their ability to reproduce asexually as well as sexually, hybrid populations can be established in isolated ponds from small numbers of individuals. Distribution of such populations has been attributed to postglacial vicariant events or recent dispersal, making identification of such populations as distinct taxa controversial.

Humans have induced introgressive hybridization among taxa, directly through introduction of exotic species (e.g. *Felis concolor coryi*) or modification of habitats (e.g. *Chasmistes liorus mictus*). While the impact varies from case to case, sometimes sufficient introgression occurs to change entire taxa. It can be difficult, however, to discriminate human influence from natural effects. For example, the red wolf (*Canis rufus*) has been hypothesized to have originated through hybridization between grey wolf and coyote, stimulated by human agricultural activity (119, 155). However, available data are equally consistent with an earlier origin, possibly due to habitat changes associated with the Pleistocene glaciation.

While not hypothesized to be hybrid taxa, Darwin's finches provide a strong example of the significance of introgressive hybridization for evolution and diversification (51, 52). Long-term field studies of life-history parameters of entire island populations are possible due to their isolation, allowing for assessment of the frequency of hybridization among species, fates of hybrid individuals, and their contribution to future generations. In the case of hybridization among *Geospiza fortis*, *G. scandens*, and *G. fuliginosa* on Daphne Major, one cohort of F_1 hybrids and backcrosses analyzed over a four-year period was found to have higher average fitnesses than parental types. Because climatic stochasticity in the Galapagos Islands causes large population fluctuations in these finches, Grant & Grant (51, 52) hypothesized that hybridization could introduce genetic variation and reduce inbreeding depression during periods of small population size. Fitness of hybrids likely fluctuates with normal variation in climatic conditions, resulting in a long-term balance between hybridization and selection. Introduction of new variation by hybridization may allow selection to shift finch populations among adaptive peaks (52). The outcome of hybridization among isolated islands could vary, with each having its own local complex of interacting species.

CONCLUSIONS

Incidence of Hybrid Taxa

Hybridization and introgression increase genetic diversity through production of new recombinant genotypes, probably more rapidly than is possible by mutation. Enhanced levels of variability could allow organisms to more readily track environmental change, leading to increased rates of evolution. A possible outcome of these processes would be origination of new species through combination of preexisting characters in other taxa.

Despite the rationale of this perspective, animal hybrid taxa appear to be relatively rare. Misperception and negative attitudes have contributed to an apparent scarcity. Bullini & Nascetti (14) noted that many invertebrate hybrid taxa were not recognized, with parthenogens and polyploids assumed to be derivatives of bisexual species even when hybridization provided a more parsimonious explanation. Such attitudes likely stem from the fact that generations of zoologists have been taught that hybridization disrupts coadapted gene complexes, yielding inferior progeny. Thus, natural selection will favor those individuals that do not hybridize, with the final stage of speciation requiring perfection of mechanisms that prevent the wastage of gametes. The inferiority of hybrids and reinforcement of reproductive isolation have recently been called into question (e.g. 7a, 19, 91), casting doubt on the negative assessment of hybridization and introgression.

Rarity is also partly due to difficulty of detection. Identification of hybrid taxa requires examination of several sets of characters, with some of the best approaches only recently technologically accessible. Even the best technology available may not allow for discrimination among alternate hypotheses as probability of detection is limited by factors beyond scientific control. To distinguish spontaneous from hybrid origins (e.g. *Hyla versicolor*) for polyploid and unisexual taxa, progenitors must be sufficiently different to increase levels of variation in hybrid progeny. Hybrid origin of bisexual diploid taxa is likely even more difficult to verify. Events must be recent enough to exclude convergence as an explanation for observed patterns of variation, and parental taxa must be sufficiently divergent to rule out ancestral polymorphism (e.g. *Catostomus discobolus yarrowi, Luxilus albeolus*).

Rarity of hybrid species could also result from differences in modes of production, with the pathway to evolutionary independence for hybrid taxa possibly more difficult than those taken by typical species. While hybrid unisexual and polyploid taxa are instantaneously isolated from their progenitors, they must surmount major barriers to formation and establishment. Production of chromosomal and meiotic conditions that give these forms autonomy will develop only under a limited set of conditions (93). Once produced, these forms must survive genetic and ecological constraints such as the accumulation of deleterious mutations, inability to find appropriate mates, and competition with progenitors (151, 156).

Bisexual taxa of hybrid origin do not face the same limitations; however, evolution of independence will likely be constrained by the homogenizing effects of gene exchange with progenitors (83). Production and stabilization of recombinants require a set of conditions that seem best exemplified by patchy and fluctuating environments (6). In such circumstances, species forced together in patches may hybridize due to rarity of mates (64) or breakdown of premating isolation, and resulting progeny may be as fit as parental taxa (7a).

In these isolated pockets, populations will evolve unique features through combinations of characteristics inherited from both parental taxa, and, if isolated long enough, attain evolutionary independence.

Implications

Even if hybrid taxa are uncommon, rarity should not be equated with insignificance of hybridization and introgression in the evolution and diversification of animals. Unisexual taxa have acted as conduits of gene exchange among related sexual species and may allow for reconstitution as new sexual taxa. Allopolyploidy has been important in diversification, as exemplified by diversity of catostomid fishes. Bogart (11) noted that perceptions of unisexual and polyploid hybrids are changing, with such taxa proving to be more significant for speciation than previously believed.

Stabilized recombinant taxa are indicative of a more significant role for gene exchange in the evolutionary process than generally believed. Introgressive hybridization among taxa will quickly increase levels of variation, allowing for rapid response to environmental change. The potential impact of gene exchange, however, will be limited by levels of premating reproductive isolation and reduction in fitness of recombinants relative to progeny of homospecific matings. The balance of these factors will provide a subset of circumstances under which introgressive hybridization will be important, likely determined by the degree of temporal and spatial variability of the biotic and abiotic environment and levels of divergence among taxa.

While it is difficult to predict the association between levels of divergence and fitness of hybrids, variability in habitat is readily observable. Extensive introgressive hybridization is often associated with habitats disturbed by anthropogenic activity (6, 64). More dramatic environmental changes occurred prior to human influence that would also have been conducive to introgressive hybridization (10) as indicated by evidence for extensive introgression in the evolution of several groups of western fishes (31, 37, 129). Association between habitat and reticulate evolution could be tested, with hybrid taxa expected to be more common in areas with a history of disturbance.

Several ancillary effects of introgressive hybridization require consideration. Events leading to hybrid taxa can be very difficult to reconstruct as genes will have different phylogenetic histories. These problems can extend to phylogenetic analyses involving hybrid and nonhybrid taxa, reducing resolution (129) especially where progenitors of hybrids are not closely related (84). Estimates of divergence time will also be compromised, providing dating of alleles and not taxonomic diversification (129).

Increased concern over biodiversity requires consideration of the role of hybridization in evolution. Extensive introgressive hybridization involving introduction of exotic species and habitat disturbance indicates how humans can influence the natural balance. Instead of simply viewing hybrids as detrimental and expendable, the impact of introgression must be appraised on a case-by-case basis (38, 102). Careful consideration should be provided to biological systems in which hybridization has played a key role in the evolution of a taxonomic complex (e.g. *Gila*). Such systems are not readily amenable to captive propagation and reintroduction and will require creative management solutions to preserve this mode of evolution.

While many zoologists have come to appreciate the potential significance of such events for adaptation, few have considered hybridization and introgression as creative forces. Unlike the situation for plants, it is still too early to evaluate the actual incidence of hybrid animal taxa. Many instances have been put forward, but few have been carefully tested, particularly those involving bisexual diploid populations. Closer examination of insects has led Bullini & Nascetti (14) to conclude that hybrid speciation has been more common than previously believed. Changing attitudes toward hybrids and technological advances should allow for careful consideration of hypothesized hybrid taxa and will undoubtedly increase the number of known animal hybrid taxa.

ACKNOWLEDGMENTS

We thank Anne Gerber and Alana Tibbets for critical review; Don Buth, Jim Collins, John Gold, W L Minckley, Bill Moore for discussion. This work was supported by funds from the National Science Foundation and Bureau of Reclamation.

Visit the *Annual Reviews home page* at
http://www.annurev.org.

Literature Cited

1. Agatsuma T, Ho L, Jian H, Habe S, Terasaki K, et al. 1992. Electrophoretic evidence of a hybrid origin for tetraploid *Paragonimus westermani* discovered in northeastern China. *Parasitol. Res.* 78:537–38
2. Agatsuma T, Terasaki K, Yang L, Blair D. 1994. Genetic variation in the triploids of Japanese *Fasciola* species, and relationships with other species in the genus. *J. Helminthol.* 68:181–86
3. Agatsuma T, Yang L, Kim D, Yonekawa H. 1994. Mitochondrial DNA differentiation of Japanese diploid and triploid *Paragonimus westermani*. *J. Helminthol.* 68:7–11
4. Allendorf FW, Thorgaard GH. 1984. Tetraploidy and the evolution of sal-

monid fishes. In *Evolutionary Genetics of Fishes*, ed. BJ Turner, pp. 1–53. New York: Plenum
5. Alves MJ, Coelho MM, Collares-Pereira MJ. 1997. The *Rutilus alburnoides* complex (Cyprinidae): evidence for hybrid origin. *J. Zool. Syst. Evol. Res.* 35:1–10
6. Anderson E, Stebbins GL. 1954. Hybridization as an evolutionary stimulus. *Evolution* 8:378–88
7. Arnold ML. 1992. Natural hybridization as an evolutionary process. *Annu. Rev. Ecol. Syst.* 23:237–61
7a. Arnold ML. 1997. *Natural Hybridization and Evolution.* Oxford, UK: Oxford Univ. Press. 215 pp.
8. Avise JC, Bowen BW, Lamb T, Meylan

AB, Bermingham E. 1992. Mitochondrial DNA evolution at a turtle's pace: evidence for low genetic variability and reduced microevolutionary rate in Testudines. *Mol. Biol. Evol.* 9:457–73

9. Avise JC, Trexler JC, Travis J, Nelson WS. 1991. *Poecilia mexicana* is the recent female parent of the unisexual fish *P. formosa. Evolution* 45:1530–33

10. Axelrod DI. 1979. Age and origin of Sonoran desert vegetation. *Occas. Pap. Calif. Acad. Sci* 132:1–74

11. Bogart JP. 1980. Evolutionary implications of polyploidy in amphibians and reptiles. In *Polyploidy: Biological Relevance,* ed. WH Lewis, pp. 341–78. New York: Plenum

12. Bogart JP. 1989. A mechanism for interspecific gene exchange via all-female salamander hybrids. See Ref. 29, pp. 170–79

13. Bullini L. 1994. Origin and evolution of animal hybrid species. *Trends Ecol. Evol.* 9:422–6

14. Bullini L, Nascetti G. 1990. Speciation by hybridization in phasmids and other insects. *Can. J. Zool.* 68:1747–60

15. Bush GL. 1975. Modes of animal speciation. *Annu. Rev. Ecol. Syst.* 6:339–64

16. Buth DG. 1983. Duplicate isozyme loci in fishes: origins, distribution, phyletic consequences, and locus nomenclature. *Isozymes: Curr. Topics Biol. Med. Res.* Vol. 10. *Genetics and Evolution.* pp. 381–400

17. Buth DG, Dowling TE, Gold JR. 1991. Molecular and cytological investigations. In *The Biology of Cyprinid Fishes,* ed. I Winfield, J Nelson, pp. 83–126. London: Chapman & Hall

18. Buth DG, Haglund TR, Minckley WL. 1992. Duplicate gene expression and allozymic divergence diagnostic for *Catostomus tahoensis* and the endangered *Chasmistes cujus* in Pyramid lake, Nevada. *Copeia* 1992:935–41

19. Butlin R. 1989. Reinforcement of premating isolation. In *Speciation and Its Consequences,* ed. D Otte, JA Endler, pp. 158–79. Sunderland, MA: Sinauer Assoc.

20. Cole CJ, Dessauer HC. 1993. Unisexual and bisexual whiptail lizards of the *Cnemidophorus lemniscatus* complex (Squamata: Teiidae) of the Guiana region, South America, with descriptions of new species. *Am. Mus. Nov.* 3081:1–30

21. Cole CJ, Dessauer HC, Markezich AL. 1993. Missing link found: the second ancestor of *Gymnophthalmus underwoodi* (Squamata: Teiidae), a South American unisexual lizard of hybrid origin. *Am. Mus. Nov.* 3055:1–13

22. Cole CJ, Dessauer HC, Townsend CR, Arnold MG. 1990. Unisexual lizards of the genus *Gymnophthalmus* (Reptilia: Teiidae) in the Neotropics: genetic, origin, and systematics. *Am. Mus. Nov.* 2994:1–29

23. Cole CJ, Dessauer HC, Townsend CR, Arnold MG. 1995. *Kentropyx borckiana* (Squamata: Teiidae): a unisexual lizard of hybrid origin in the Guiana region, South America. *Am. Mus. Nov.* 3145:1–23

24. Collares-Pereira MJ, Coelho MM. 1989. Polyploidy versus diploidy: a new model for the karyological evolution of Cyprinidae. *Arq. Mus. Boc.* Nova Série 1(26):375–83

25. Coyne JA. 1992. Genetics and speciation. *Nature* 355:511–15

26. Crabtree CB, Buth DG. 1987. Biochemical systematics of the catostomid genus *Catostomus:* assessment of *C. clarki, C. plebeius* and *C. discobolus* including the Zuni sucker, *C. d. yarrowi. Copeia* 1987:843–54

27. Dawley RM. 1989. An introduction to unisexual vertebrates. See Ref. 29, pp. 1–18

28. Dawley RM. 1992. Clonal hybrids of the common laboratory fish *Fundulus hetroclitus. Proc. Natl. Acad. Sci. USA* 89:2485–88

29. Dawley RM, Bogart JP. 1989. *Evolution and Ecology of Unisexual Vertebrates.* Albany: NY State Mus. Bull. 466. 302 pp.

30. Deleted in proof

31. DeMarais BD, Dowling TE, Douglas ME, Minckley WL, Marsh PC. 1992. Origin of *Gila seminuda* (Teleostei: Cyprinidae) through introgressive hybridization: implications for evolution and conservation. *Proc. Natl. Acad. Sci. USA* 89:2747–51

32. Dobzhansky T. 1937. *Genetics and the Origin of Species.* New York: Columbia Univ. Press. 364 pp. 1st ed.

33. Dobzhansky T. 1940. Speciation as a stage in evolutionary divergence. *Am. Nat.* 74:312–21

34. Dobzhansky, T. 1951. *Genetics and the Origin of Species.* New York: Columbia Univ. Press. 364 pp. 3rd ed.

35. Donnellan SC, Moritz C. 1995. Genetic diversity of bisexual and parthenogenetic populations of the tropical

gecko *Nactus pelagicus. Herpetologica* 51:140–54

36. Dowling TE, Hoeh WR. 1991. The extent of introgression outside the hybrid zone between *Notropis cornutus* and *Notropis chrysocephalus* (Teleostei: Cyprinidae). *Evolution* 45:944–56

37. Dowling TE, DeMarais BD. 1993. Evolutionary significance of introgressive hybridization in cyprinid fishes. *Nature* 362:444–46

38. Dowling TE, Minckley WL, Douglas ME, Marsh PC, DeMarais BD. 1992. Response to Wayne, Nowak, and Phillips and Henry: use of molecular characters in conservation biology. *Conserv. Biol.* 6:600–3

38a. Dufresne F, Hebert PDN. 1994. Hybridization and origins of polyplopidy. *Proc. R. Soc. Lond. B* 258:141–46

39. Ehrlich PR, PH Raven. 1969. Differentiation of populations. *Science* 165:1228–32

40. Endler JA. 1989. Conceptual and other problems in speciation. See Ref. 19, pp. 625–48

41. Ferris SD. 1984. Tetraploidy and the evolution of catostomid fishes. See Ref. 4, pp. 55–93

42. Ferris SD, Sage RD, Huang C-M, Nielsen JT, Ritte U, et al. 1983. Flow of mitochondrial DNA across a species boundary. *Proc. Natl. Acad. Sci. USA* 80:2290–94

43. Ferris SD, Whitt GS. 1977. Duplicate gene expression in diploid and tetraploid loaches (Cypriniformes, Cobitidae). *Biochem. Genet.* 15:1097–112

44. Fisher RA. 1958. *The Genetical Theory of Natural Selection.* New York: Dover. 291 pp. 2nd rev. ed.

45. Goddard KA, Dawley RM. 1990. Clonal inheritance of a diploid nuclear genome by a hybrid freshwater minnow (*Phoxinus eos-neogaeus*, Pisces: Cyprinidae). *Evolution* 44:1052–65

46. Goddard KA, Schultz RJ. 1993. Aclonal reproduction by polyploid members of the clonal hybrid species *Phoxinus eos-neogaeus* (Cyprinidae). *Copeia* 1993:650–60

47. Goldman MA, LoVerde PT, Chrisman, CL. 1983. Hybrid origin of polyploidy in freshwater snails of the genus *Bulinus* (Mollusca: Planorbidae). *Evolution* 37:592–600

48. Goodfellow WL Jr, Hocutt CH, Morgan RP, Stauffer JR Jr. 1984. Biochemical assessment of the taxonomic status of "*Rhinichthys bowersi*" (Pisces: Cyprinidae). *Copeia* 1984:652–59

49. Goodfriend GA, Gould SJ. 1996. Paleontology and chronology of two evolutionary transitions by hybridization in the Bahamian land snail *Cerion. Science* 274:1894–97

50. Gould SJ, Woodruff DS. 1990. History as a cause of area effects: an illustration from *Cerion* on Great Inagua, Bahamas. *Biol. J. Linn. Soc.* 40:67–98

51. Grant PR, Grant BR. 1992. Hybridization of bird species. *Science* 256:193–97

52. Grant PR, Grant BR. 1994. Phenotypic and genetic effects of hybridization in Darwin's finches. *Evolution* 48:297–316

53. Grant V. 1971. *Plant Speciation.* New York: Columbia Univ. Press. 435 pp.

54. Harrison RG. 1990. Hybrid zones: windows on evolutionary process. In *Oxford Surveys in Evolutionary Biology,* ed. D Futuyma, J Antonovics, 7:69–128. Oxford: Oxford Univ. Press.

55. Harrison RG. 1993. Hybrids and hybrid zones: historical perspective. In *Hybrid Zones and the Evolutionary Process,* ed. RG Harrison, pp. 3–12. Oxford, UK: Oxford Univ. Press.

56. Harrison RG, Rand DM. 1989. Mosaic hybrid zones and the nature of species boundaries. See Ref. 19, pp. 111–33

57. Hebert PDN. 1985. Interspecific hybridization between cyclic parthenogens. *Evolution* 39:216–20

58. Hedges SB, Bogart JP, Maxson LR. 1992. Ancestry of unisexual salamanders. *Nature* 356:708–10

59. Hewitt GM. 1988. Hybrid zones—natural laboratories for evolutionary studies. *Trends Ecol. Evol.* 3:158–67

60. Hillis DM, Mable BK, Moritz C. 1996. Applications of molecular systematics: the state of the field and a look to the future. See Ref. 61, pp. 515–543

61. Hillis DM, Moritz C, Mable BK. 1996. *Molecular Systematics.* Sunderland, MA: Sinauer Assoc. 655 pp. 2nd ed.

62. Hotz H, Beerli P, Spolsky C. 1992. Mitochondrial DNA reveals formation of nonhybrid frogs by natural matings between hemiclonal hybrids. *Mol. Biol. Evol.* 9:610–20

63. Howard DJ. 1993. Reinforcement: origin, dynamics, and fate of an evolutionary hypothesis. See Ref. 55, pp. 46–69

64. Hubbs CL. 1955. Hybridization between fish species in nature. *Syst. Zool.* 4:1–20

65. Hutchinson MN, Donnellan SC. 1992. Taxonomy and genetic variation in the Australian lizards of the genus *Pseudemoia* (Scincidae: Lygosominae). *J. Nat. Hist.* 26:215–64

66. Johnson, SG. 1992. Spontaneous and hybrid origins of parthenogenesis in *Campeloma decisum* (freshwater prosobranch snail). *Heredity* 68:253–61

67. Johnston RF. 1969. Taxonomy of house sparrows and their allies in the Mediterranean basin. *Condor* 71:129–39

68. Jones TR, Routman EJ, Begun DJ, Collins JP. 1995. Ancestry of an isolated subspecies of salamander, *Ambystoma tigrinum stebbinsi* Lowe: the evolutionary significance of hybridization. *Mol. Phyl. Evol.* 4:194–202

69. Kraus F, Ducey PK, Moler P, Miyamoto MM. 1991. Two new triparental unisexual *Ambystoma* from Ohio and Michigan. *Herpetologica* 47:429–39

70. Kraus F, Miyamoto MM. 1990. Mitochondrial genotype of a unisexual salamander of hybrid origin is unrelated to either of its nuclear haplotypes. *Proc. Natl. Acad. Sci. USA* 87:2235–38

71. Lanham UN. 1981. Some Colorado *Andrena* of the subgenus *Scaphandrena* of presumed hybrid origin, with special reference to the tarsal claws (Hymenoptera: Apoidea). *J. Kans. Entomol. Soc.* 54:537–46

72. Lanham UN. 1984. The hybrid swarm of *Andrena* (Hymenoptera: Apoidea) in Western North America: a possible source for the evolutionary origin of a new species. *J. Kans. Entomol. Soc.* 57:197–208

73. Lewontin RC, Birch LC. 1966. Hybridization as a source of variation for adaptation to new environments. *Evolution* 20:315–36

74. Lieder U. 1983. Introgression as a factor in the evolution of polytypical plankton cladocera. *Int. Revue ges Hydrobiol.* 68:269–84

75. Lieder U. 1987. The possible origin of *Daphnia cucullata procurva* Poppe 1887 in the lakes of the Pomeranian Lakeland by hybridization in the past. *Hydrobiologia* 145:201–11

76. Lokki J, Saura A. 1980. Polyploidy in insect evolution. See Ref. 11, pp. 277–312

77. Lynch JD. 1989. The gauge of speciation: on the frequencies of modes of speciation. See Ref. 19, pp. 527–53

78. Lynch M, Gabriel W. 1990. Mutation load and the survival of small populations. *Evolution* 44:1725–37

79. Marchant AD, Arnold ML, Wilkinson P. 1988. Gene flow across a chromosomal tension zone. I. Relicts of ancient hybridization. *Heredity* 61:321–28

80. Martin AP, GJP Naylor, SR Palumbi. 1992. Rates of mitochondrial DNA evolution in sharks are slow compared to mammals. *Nature* 357:153–55

81. Maxson L, Pepper E, Maxson RD. 1977. Immunological resolution of a diploid-tetraploid species complex of treefrogs. *Science* 197:1012–13

82. Mayr E. 1942. *Systematics and the Origin of Species.* New York: Columbia Univ. Press. 334 pp.

83. Mayr E. 1963. *Animal Species and Evolution.* Cambridge: Harvard Univ. Press. 797 pp.

84. McDade LA. 1992. Hybrids and phylogenetic systematics. II. The impact of hybrids on cladistic analysis. *Evolution* 46:1329–46

85. Meagher S, Dowling TE. 1991. Hybridization between the cyprinid fishes *Luxilus albeolus, L. cornutus* and *L. cerasinus*, with comments on the hybrid origin of *Luxilus albeolus. Copeia* 1991:979–91

86. Menezes NA, Weitzman SH. 1990. Two new species of *Mimagoniates* (Teleostei: Characidae: Glandulocaudinae), their phylogeny and biogeography and a key to the glandulocaudin fishes of Brazil and Paraguay. *Proc. Biol. Soc. Wash.* 103:380–426

87. Menzel BW. 1977. Morphological and electrophoretic identification of a hybrid cyprinid fish, *Notropis cerasinus* × *Notropis c. cornutus*, with implications on the evolution of *N. albeolus. Comp. Biochem. Physiol.* 57B:215–18

88. Mezhzherin SV, Pisanets EM. 1995. Genetic structure and origin of a tetraploid toad species *Bufo danatensis* Pisanetz, 1978 (Amphibia, Bufonidae) from central Asia: Description of biochemical polymorphism and comparison of heterozygosity levels in diploid and tetraploid species. *Russ. J. Genetics* 31:34–43

89. Miller RR, Smith GR. 1981. Distribution and evolution of *Chasmistes* (Pisces: Catostomidae) in western North America. *Occ. Pap. Mus. Zool. Univ. Mich.* 696:1–46

90. Mina MV. 1992. An interpretation of diversity in the Salmonid genus *Brachymystax* (Pisces, Salmonidae) as the possible result of multiple hybrid speciation. *J. Ichthyol.* 32:117–22

91. Moore WS. 1977. An evaluation of narrow hybrid zones in vertebrates. *Q. Rev. Biol.* 52:263–77

92. Moritz C. 1993. The origin and evolution of parthenogenesis in *Heteronotia*

binoei (Gekkonidae): synthesis. *Genetica* 90:269–80

93. Moritz C, Brown WM, Densmore LD, Wright JW, Vyas D, et al. 1989. Genetic diversity and the dynamics of hybrid parthenogenesis in *Cnemidophorus* (Teiidae) and *Heteronotia* (Gekkonidae). See Ref. 29, pp. 87–112

94. Moritz C, Case TJ, Bolger DT, Donnellan S. 1993. Genetic diversity and the history of Pacific island house geckos (*Hemidactylus* and *Lepidodactylus*). *Biol. J. Linn. Soc.* 48:113–33

95. Moritz C, Uzzell T, Spolsky C, Hotz H, Darevsky I, et al. 1992. The maternal ancestry and approximate age of parthenogenetic species of Caucasian rock lizards (Lacerta: Lacertidae). *Genetica* 87:53–62

96. Moritz C, Wright JW, Brown WM. 1992. Mitochondrial DNA analyses and the origin and relative age of parthenogenetic *Cnemidophorus*: phylogenetic constraints on hybrid origins. *Evolution* 46:184–92

97. Muller HJ. 1925. Why is polyploidy rarer in animals than plants? *Am. Nat.* 59:346–53

98. Muller HJ. 1964. The relation of mutation to mutational advance. *Mutat. Res.* 1:2–9

99. Neff NA, Smith GR. 1978. Multivariate analysis of hybrid fishes. *Syst. Zool.* 28:176–96

100. Neigel J, Avise JC. 1986. Phylogenetic relationships of mitochondrial DNA under various demographic models of speciation. In *Evolutionary Processes and Theory,* ed. S Karlin, E Nevo, pp. 515–34. New York: Academic. 786 pp.

101. Normark BB. 1996. Phylogeny and evolution of parthenogenetic weevils (Coleoptera: Curculionidae: Naupactini): evidence from mitochondrial DNA sequences. *Evolution* 50:734–45

102. O'Brien SJ, Mayr E. 1991. Bureaucratic mischief: recognizing endangered species and subspecies. *Science* 251:1187–88

102a. Oderico DM, Miller DJ. 1997. Variation in the ribosomal internal transcribed spacers and 5.85 rDNA among five species of *Acropora* (Cnidaria; Scleractinia); patterns of variation consistent with reticulate evolution. *Mol. Biol. Evol.* 14:465–73

103. O'Foighil D, Hilbish TJ, Showman RM. 1996. Mitochondrial gene variation in *Mercenaria* clam sibling species reveals a relict secondary contact zone in the western Gulf of Mexico. *Mar. Biol.* 126:675–83

104. O'Foighil D, Smith MJ. 1995. Evolution of asexuality in the cosmopolitan marine clam *Lasaea. Evolution* 49:140–50

105. Ohno S. 1970. *Evolution by Gene Duplication.* New York: Springer-Verlag. 160 pp.

106. Orr HA. 1990. "Why polyploidy is rarer in animals than plants" revisited. *Am. Nat.* 136:759–70

107. Osinov AG. 1993. Countercurrent dispersal, secondary contact, and speciation in lenoks of the genus *Brachymystax* (Salmonidae: Salmoniformes). Genetika 29:654–69

108. Ptacek MB, Gerhardt HC, Sage RD. 1994. Speciation by polyploidy in treefrogs: multiple origins of the tetraploid, *Hyla versicolor. Evolution* 48:898–908

109. Quattro JM, Avise JC, Vrijenhoek RC. 1992. An ancient clonal lineage in the fish genus *Poeciliopsis* (Atheriniformes: Poeciliidae). *Proc. Natl. Acad. Sci. USA* 89:348–52

110. Radtkey RR, Becker B, Miller RD, Riblet R, Case TJ. 1996. Variation and evolution of Class I MHC in sexual and parthenogenetic geckos. *Proc. R Soc. Lond. B* 263:1023–32

111. Radtkey RR, Donnellan SC, Fisher RN, Moritz C, Hanley KA, et al. 1995. When species collide: the origin and spread of an asexual species of gecko. *Proc. R Soc. Lond. B* 259:145–52

112. Ralin DB, Romano MA, Kirkpatrick CW. 1983. The tetraploid treefrog *Hyla versicolor*: evidence for a single origin from the diploid *H. chrysocelis. Herpetologica* 39:212–25

113. Rensch B. 1959. *Evolution above the Species Level.* New York: Columbia Univ. Press. 419 pp.

114. Rieseberg LH. 1997. Hybrid origins of plant species. *Annu. Rev. Ecol. Syst.* 28:359–89

115. Rieseberg LH, Wendel JF. 1993. Introgression and its consequences in plants. See Ref. 55, pp. 70–109

116. Romano MA, Ralin DB, Guttman SI, Skillings JH. 1987. Parallel electromorph variation in the diploid-tetraploid gray treefrog complex. *Am. Nat.* 130:864–78

117. Ross MR, Cavender TM. 1981. Morphological analyses of four experimental intergeneric cyprinid hybrid crosses. *Copeia* 1981:377–87

118. Rothfels K. 1989. Speciation in black flies. *Genome* 32:500–9

119. Roy MS, Geffen E, Smith D, Ostrander EA, Wayne RK. 1994. Patterns of differentiation and hybridization in North American wolflike canids, revealed by analysis of microsatellite loci. *Mol. Biol. Evol.* 11:553–70

120. Saura A, Lokki J, Soumalainen E. 1993. Origin of polyploidy in parthenogenetic weevils. *J. Theor. Biol.* 163:449–56

121. Schartl M, Wilde B, Schlupp I, Parzefall J. 1995. Evolutionary origin of a parthenoform, the Amazon molly *Poecilia formosa*, on the basis of a molecular genealogy. *Evolution* 49:827–35

122. Schultz RJ. 1980. Role of polyploidy in the evolution of fishes. See Ref. 11, pp. 313–40

123. Schwenk K, Spaak P. 1995. Evolutionary and ecological consequences of interspecific hybridization in cladocerans. *Experientia* 51:465–81

124. Sezaki K, Watabe S, Ochiai Y, Hashimoto K. 1994. Biochemical genetic evidence for a hybrid origin of spiny loach, *Cobitis taenia taenia*, in Japan. *J. Fish Biol.* 44:683–91

125. Short LL. 1965. Hybridization in the flickers (*Colaptes*) of North America. *Bull. Am. Mus. Nat. Hist.* 129:307–428

126. Short LL. 1972. Hybridization, taxonomy and avian evolution. *Ann. Miss. Bot. Gar.* 59:447–53

127. Sites JW Jr, Peccinini-Seale DM, Moritz C, Wright JW, Brown WM. 1990. The evolutionary history of parthenogenetic *Cnemidophorus lemniscatus* (Sauria: Teiidae). I. Evidence for hybrid origin. *Evolution* 44:906–21

128. Smith GR. 1992. Phylogeny and biogeography of the Catostomidae, freshwater fishes of North America and Asia. In *Systematics, Historical Ecology, and North American Freshwater Fishes*, ed. RL Mayden, pp. 778–826. Stanford, CA: Stanford Univ. Press.

129. Smith GR. 1992. Introgression in fishes: significance for paleontology, cladistics, and evolutionary rates. *Syst. Biol.* 41:41–57

130. Smith GR, Hall JG, Koehn RK, Innes DJ. 1983. Taxonomic relationships of the Zuni mountain sucker *Catostomus discobolus yarrowi. Copeia* 1983:37–48

131. Smith GR, Miller RR, Sable WD. 1979. Species relationships among fishes of the genus *Gila* in the upper Colorado River drainage. *Proc. First Conf. Sci. Res. Natl. Parks* 1:613–23

132. Sola L, Iaselli V, Rossi AR, Rasch EM, Monaco PJ. 1992. Cytogenetics of bisexual/unisexual species of *Poecilia*. III.

The karyotype of *Poecilia formosa*, a gynogenetic species of hybrid origin. *Cytogenet. Cell Genet.* 60:236–40

133. Sperling FAH, Harrison RG. 1994. Mitochondrial DNA variation within and between species of the *Papilio machaon* group of swallowtail butterflies. *Evolution* 48:408–22

134. Spolsky C, Phillips CA, Uzzell T. 1992. Antiquity of clonal salamander lineages revealed by mitochondrial DNA. *Nature* 356:706–8

135. Spolsky C, Phillips CA, Uzzell T. 1992. Gynogenetic reproduction in hybrid mole salamanders (Genus *Ambystoma*). *Evolution* 46:1935–44

136. Stauffer JR Jr, Hocutt CH, Denoncourt RF. 1978. Status and distribution of the hybrid *Nocommis micropogon* × *Rhinichthys cataractae*, with a discussion of hybridization as a viable mode of vertebrate speciation. *Am. Midl. Nat.* 101:355–65

137. Stauffer JR Jr, Hocutt CH, Mayden RL. 1997. *Pararhinichthys*, a new monotypic genus of minnows (Teleostei: Cyprinidae) of hybrid origin from eastern North America. *Ichthyol. Explor. Freshwaters* 7:327–36

138. Städler T, Loew M, Streit B. 1993. Genetic evidence for low outcrossing rates in polyploid freshwater snails (*Ancylus fluviatilis*). *Proc. R. Soc. Lond. B* 251:207–13

139. Stebbins GL. 1959. The role of hybridization in evolution. *Proc. Am. Philos. Soc.* 103:231–51

140. Stebbins GL. 1966. *Processes of Organic Evolution.* Inglewood Cliffs, NJ: Prentice-Hall. 191 pp.

141. Swofford DL, Olsen GJ, Waddell PJ, Hillis DM. 1996. Phylogenetic inference. See Ref. 61, pp. 407–514

142. Taylor AS. 1992. Reconstitution of diploid *Ambystoma jeffersonianum* (Amphibia: Caudata) in a hybrid, triploid egg mass with lethal consequences. *J. Hered.* 83:361–66

143. Templeton AR. 1981. Mechanisms of speciation—a population genetic approach. *Annu. Rev. Ecol. Syst.* 12:23–48

144. Theisen BF, Christensen B, Arctander P. 1995. Origin of clonal diversity in triploid parthenogenetic *Trichoniscus pusillus pusillus* (Isopoda, Crustacea) based upon allozyme and nucleotide sequence data. *J. Evol. Biol.* 8:71–80

145. Tinti F, Scali V. 1995. Allozymic and cytological evidence for hemiclonal, allpaternal, and mosaic offspring of the hybridogenetic stick insect *Bacillus*

rossius-grandii grandii. J. Exp. Zool 273:149–59

146. Turgeon J, Hebert PDN. 1994. Evolutionary interactions between sexual and all-female taxa of *Cyprinotus* (Ostracoda: Cyprididae). *Evolution* 48:1855–65

147. Uyeno T, Smith GR. 1972. Tetraploid origin of the karyotype of catostomid fishes. *Science* 175:644–46

148. Vasil'yev VP, Osinov AG, Vasil'yeva YeD. 1991. Reticulate species formation in vertebrates: the diploid-triploid-tetraploid complexes in the genus *Cobitis* (Cobitidae). V. Origin of even-polyploid species. *J. Ichthyol.* 31:22–36

149. Vasil'yev VP, Vasil'yeva YeD, Osinov AG. 1990. Reticulate evolution in vertebrates: the diploid-triploid-tetraploid complexes in the genus *Cobitis* (Cobitidae). III. Origin of the triploid form. *J. Ichthyol.* 30:22–31

150. Volobouev V, Pasteur G, Ineich I, Dutrillaux B. 1993. Chromosomal evidence for a hybrid origin of diploid parthenogenetic females from the unisexual-bisexual *Lepidodactylus lugubris* complex (Reptilia, Gekkonidae). *Cytogenet. Cell Genet.* 63:194–9

151. Vrijenhoek RC. 1989. Genetic and ecological constraints on the origins and establishment of unisexual vertebrates. See Ref. 29, pp. 24–31

152. Vrijenhoek RC. 1993. The origin and evolution of clones versus the maintenance of sex in *Poeciliopsis. J. Hered.* 84:388–95.

153. Vrijenhoek RC, Dawley RM, Cole CJ, Bogart JP. 1989. A list of the known unisexual vertebrates. See Ref. 29, pp. 19–23

154. Vyas DK, Moritz C, Peccinini-Seale D, Wright JW, Brown WM. 1990. The evolutionary history of parthenogenetic *Cnemidophorus lemniscatus* (Sauria: Teiidae). II. Maternal origin and age inferred from mitochondrial DNA analyses. *Evolution* 44:922–32

155. Wayne RK, Jenks SM. 1991. Mitochondrial DNA analysis implying extensive hybridization of the endangered red wolf, *Canis rufus. Nature* 351:565–68

156. White MJD. 1978. *Modes of Speciation.* San Francisco: WH Freeman. 455 pp.

157. Wilson P. 1992. On inferring hybridity from morphological intermediacy. *Taxon* 41:11–23

158. Wu C-I, Palopoli MF. 1994. Genetics of postmating reproductive isolation in animals. *Annu. Rev. Genet.* 27:283–308

159. Yonekawa H, Moriwaki K, Gotoh O, Miyashita N, Matsushima Y, et al. 1988. Hybrid origin of Japanese mice "*Mus musculus molossinus*": evidence from restriction analysis of mitochondrial DNA. *Mol. Biol. Evol.* 5:63–78

160. Yonenaga-Yassuda Y, Vanzolini PE, Rodrigues MT, de Carvalho CM. 1995. Chromosome banding patterns in the unisexual microteiid *Gymnophthalmus underwoodi* and in two related sibling species (Gymnophthalmidae, Sauria). *Cytogenet. Cell Genet.* 70:29–34

Annu. Rev. Ecol. Syst. 1997. 28:621–58

THE ECOLOGY OF INTERFACES:
Riparian Zones

Robert J. Naiman and Henri Décamps

School of Fisheries, Box 357980, University of Washington, Seattle, Washington
98195 USA, naiman@u.washington.edu and Centre National de la Recherche
Scientifique, Centre d'Ecologie des Systèmes Aquatiques Continentaux, 29, rue
Jeanne Marvig, 31055 Toulouse, France; e-mail: decamps@cemes.cemes.fr

KEY WORDS: interfaces, ecotones, riparian, buffer, corridor

ABSTRACT

Riparian zones possess an unusually diverse array of species and environmental
processes. The ecological diversity is related to variable flood regimes, geographically unique channel processes, altitudinal climate shifts, and upland influences
on the fluvial corridor. The resulting dynamic environment supports a variety of
life-history strategies, biogeochemical cycles and rates, and organisms adapted
to disturbance regimes over broad spatial and temporal scales. Innovations in
riparian zone management have been effective in ameliorating many ecological
issues related to land use and environmental quality. Riparian zones play essential
roles in water and landscape planning, in restoration of aquatic systems, and in
catalyzing institutional and societal cooperation for these efforts.

INTRODUCTION

First described nearly a century ago (36), interfaces between environmental
patches occur where structural or functional system properties change discon-
tinuously in space or time.

The terms transition zone, ecotone, and boundary, which describe interfaces
(91, 182), are used synonymously in this review. Interfaces between adjacent
ecological systems have a set of characteristics uniquely defined by space and
time scales and by the strength of interactions between the adjacent ecological
systems (102). Thus, interfaces possess specific physical and chemical attributes, biotic properties, and energy and material flow processes, but they are

621

unique in their interactions with adjacent ecological systems (183, 231). The strength of these interactions, which vary over wide temporal and spatial scales, is controlled by the contrast between adjacent resource patches or ecological units. In general, an interface may be thought of as being analogous to a semipermeable membrane regulating the flow of energy and material between adjacent environmental patches (183). Interfaces have resources, control energy and material flux, are potentially sensitive sites for interactions between biological populations and their controlling variables, have relatively high biodiversity, maintain critical habitat for rare and threatened species, and are refuge and source areas for pests and predators (91, 102, 183, 231). Some interfaces may also be sites for longitudinal migration (e.g. along windbreaks or riparian zones) and genetic pools or sites for active microevolution (e.g. forest/agricultural interfaces—76). Interfaces between terrestrial and freshwater ecosystems are particularly sensitive to environmental change (155, 182). Examples include riparian forests, marginal wetlands, littoral lake zones, floodplain lakes and forests, and areas with groundwater–surface water exchanges. This article reviews riparian zones associated with streams and rivers because they encompass most of the characteristics important for interfaces.

Natural riparian zones are some of the most diverse, dynamic, and complex biophysical habitats on the terrestrial portion of the planet (184, 185). This article provides an overview of important characteristics of riparian zones, describes physical effects on adjacent environments, summarizes ecological characteristics, and discusses consequences of environmental alterations on ecosystem form and function.

THE RIPARIAN ZONE

Riparius is a Latin word meaning "of or belonging to the bank of a river" (Webster's New Universal Unabridged Dictionary 1976). The anglicized term *riparian* refers to biotic communities on the shores of streams and lakes. Riparian zones are an unusually diverse mosaic of landforms, communities, and environments within the larger landscape, and they serve as a framework for understanding the organization, diversity, and dynamics of communities associated with fluvial ecosystems (53, 84, 183, 184, 185). A variety of natural disturbances creates a spatial and temporal environmental mosaic with few parallels in other systems.

Defining and Delineating Riparian Zones

The spatial extent of the riparian zone may be difficult to delineate precisely because the heterogeneity is expressed in an array of life-history strategies and successional patterns, while the functional attributes depend on community composition and the environmental setting. The riparian zone encompasses the stream channel between the low and high water marks and that portion of the terrestrial landscape from the high water mark toward the uplands where vegetation may be influenced by elevated water tables or flooding and by the ability of the soils to hold water (184, 185; exact definitions differ among researchers). The width of the riparian zone, the level of control that the streambed vegetation has on the stream environment, and the diversity of functional attributes (e.g. information flow, biogeochemical cycles) are related to the size of the stream, the position of the stream within the drainage network, the hydrologic regime, and the local geomorphology (53, 126, 182, 236, 238). The riparian zone may be small in the numerous headwater streams that are almost completely embedded in the forest. In mid-sized streams, the riparian zone is larger, being represented by a distinct band of vegetation whose width is determined by long-term (>50 years) channel dynamics and the annual discharge regime. Riparian zones of large streams are characterized by well-developed but physically complex floodplains with long periods of seasonal flooding, lateral channel migration, oxbow lakes in old river channels, a diverse vegetative community, and moist soils (155, 240). Vegetation outside the zone that is not directly influenced by hydrologic conditions but that contributes organic matter (e.g. leaves, wood, dissolved materials) to the floodplain or channel, or that influences the physical regime of the floodplain or channel by shading, may be considered part of riparian zones (26, 84). These attributes suggest that riparian zones are key systems for regulating aquatic-terrestrial linkages (182, 271) and that they may provide early indications of environmental change (52, 183, 230).

Defining riparian zones is important for both ecological and managerial reasons. Riparian buffer zones, a defined distance from a stream where land use activities are restricted for stream protection purposes, are becoming an increasingly common management tool (20, 80, 136, 242). Definitions generally incorporate ecological characteristics such as the spatial extent of herbaceous plants adapted to wetted soils, production of nutritional resources for aquatic systems, local geomorphology, and area of sediment generation (84, 136, 242). Increasingly, geographic information systems (GIS) and digital elevation models (DEM) are being used to provide an initial estimate of riparian zone area and distribution in drainage networks (13, 274). Where the riparian vegetation has been removed, methods are being developed to determine the spatial potential for its regeneration (136, 210).

Life-History Strategies

Streams are nonequilibrium systems that have strong effects on the biotic characteristics of riparian communities. The active channel and floodplain are harsh environments for the establishment of plants and animals (185). Here we focus on the morphological, physiological, and reproductive strategies of the vegetation even though an equally informative review could be made for the animal community (see 225).

Seasonal variations in discharge and wetted areas create environmental conditions that challenge even the most tolerant species. Nearly every year, most riparian plants are subjected to floods, erosion, abrasion, drought, freezing, and occasionally toxic concentrations of ammonia in addition to the normal biotic challenges; the life-history strategies of most riparian plants are such that extreme conditions are either endured, resisted, or avoided (4, 86, 185).

In general, riparian plant communities are composed of specialized and disturbance-adapted species within a matrix of less-specialized and less-frequently disturbed upland forest (185). The classification of plants into four broad categories of functional adaptations is useful for understanding processes leading to riparian forest succession and distributional patterns.

1. Invader—produces large numbers of wind- and water-disseminated propagules that colonize alluvial substrates.

2. Endurer—resprouts after breakage or burial of either the stem or roots from floods or after being partially eaten.

3. Resister—withstands flooding for weeks during the growing season, moderate fires, or epidemics.

4. Avoider—lacks adaptations to specific disturbance types; individuals germinating in an unfavorable habitat do not survive.

Some widely distributed species illustrate the variety of life-history strategies. Sitka willow (*Salix sitchensis*) and Scouler's willow (*Salix scouleriana*) are pioneer plants well adapted to living under a number of disturbance regimes. Individuals of these species invade post-fire landscape or resprout if the root system remains intact in a fire (185). They are well suited as invaders, endurers, or resisters depending upon local environmental conditions. In contrast, Sitka spruce (*Picea sitchensis*), which can colonize woody debris and mineral substrata on the floodplain (94, 165) and resist both flooding and sediment deposition, does not persist where fire is a regular occurrence.

MORPHOLOGICAL AND PHYSIOLOGICAL ADAPTATIONS   Flood tolerance in trees includes both morphological and physiological adaptations. In general, morphological adaptations such as adventitious roots, stem buttressing, and root flexibility in riparian plants are a response to either soil anoxia or unstable substrata. Morphological adaptations to anoxia by vascular plants in periodically flooded areas include air spaces (aerenchyma) in the roots and stems that allow for the diffusion of oxygen from aerial portions of the plant to the roots and adventitious roots that grow above the anaerobic zone enabling oxygen absorption by the plant. The development of these structures is mediated by increased levels of ethylene, the production of which is initiated by anaerobic soil conditions (17). Root and stem aerenchyma are common in species of families Cyperaceae and Juncaceae, which are normally found on poorly drained floodplains. Adventitious roots occur in a variety of tree genera (e.g. *Populus, Salix, Alnus, Sequoia*) living in riparian environments where sediment deposition and wetted soils are common (17, 248, 252).

Flooding also mechanically disturbs plants by eroding substrata and by abrasion. Stem flexibility among woody genera (i.e. *Populus, Salix, Lanes*) imparts endurance and resistance to potentially high levels of shear stress accompanying seasonal floods. Floods often occur during periods when the vegetation is without leaves, further reducing potential damage (73).

Anoxic conditions are challenging to plants not only because they need oxygen, but also because anoxic conditions mobilize soluble reduced ions (such as manganese) that are toxic (171). Rhizosphere oxygenation reduces this threat by moving oxygen from the root to the adjacent soil to form a very small but effective oxidized zone. However, riparian species show a large variety of responses to flooding. For example, in the tropical gallery forests of Brazil, *Sebastiana klotzchyana* accelerates glycolysis with ethanol as the major end product of anaerobic metabolism but without detectable oxygen diffusion to the root system (123), whereas *Hymenaea courbaril* maintains aerobic root metabolism (with a 50% decrease in metabolism) through oxygen diffusion from the aboveground stem system. *Chorisia speciosa* develops hypertrophic lenticels to improve the aeration of the root system, although it does not reach full metabolism. A fourth species, *Schyzolobium parahyba*, which does not accelerate glycolysis enough to maintain the rate of energy production required by the roots, does not grow in that environment (123).

REPRODUCTIVE ADAPTATIONS   Plant life-history strategies include a suite of co-adapted characters that enhance reproductive success in specific environments (9). The primary reproductive characteristics of riparian plants are tradeoffs between sexual and asexual reproduction, seed size, timing of dormancy, timing of seed dispersal, seed dispersal mechanisms, and longevity. For example, several plants (e.g. *Populus, Salix*) disperse seeds in phase with the seasonal retreat of floodwaters, ensuring moist seedbeds for successful germination and plant colonization (120, 248). Many species use transport by flowing water, a phenomenon known as hydrochory, for dispersal of seeds as well as vegetative fragments to new sites (116, 196, 198). Bald cypress (*Taxodium distichum*) and water tupelo (*Nyssa aquatica*) in a South Carolina swamp forest rely on water more than wind to disperse seeds (244). Further, small increases in water levels tend to introduce seeds from adjacent plant communities. Dispersal by animals (zoochory) and by wind (anemochory) may be even more important, but little empirical data exist for comparison.

The mechanism for propagule dispersal also structures riparian flora and may help explain species distribution patterns. In Sweden, Johansson et al (116) found a positive relationship between floating capacity of the diaspores and species occurrence in the riparian vegetation. Seeds of water-dispersed species floated better than other seeds. Floating time did not differ between seeds dispersed by animals or wind.

Many riparian plants reproduce vegetatively (e.g. asexually). Otherwise healthy branch tips, of cottonwood, for example, which are shed in the process of branch abscission, or cladoptosis (59, 81), may develop adventitious roots and grow into genetically identical trees (164).

Successional and Vegetative Patterns

PHYSICAL CONTROLS   Complex interactions among hydrology, geomorphology, light, temperature, and fire influence the structure, dynamics, and composition of riparian zones (23, 155). The literature suggests that hydrology (and its interactions with local geology) is the most important factor. For example, in riparian floodplains having a ridge-and-swale topography, vegetative patch types alternate between those on topographic lows adapted to long hydroperiods and those on topographic highs with species also found in mesic uplands (23, 168, 240).

Brinson's (23) conceptual model is that power and frequency of inundation are inversely proportional and exist in a continuum from high-power, low-frequency floods that affect the whole floodplain to low-power, high-frequency floods that influence only the area adjacent to the wetted channel. The former create large geographic features that persist for hundreds to thousands of years (e.g. oxbow lakes, relic levees). Medium-power, intermediate-frequency floods determine patterns of ecosystem structure that have lifetimes of tens to hundreds of years. Tree community zonation is influenced at this scale because many tree species have similar generation times (8, 96, 109). The low-power floods that occur annually determine short-term patterns such as seed germination and seedling survival (8, 137, 138). For example, in southwestern Colorado, seedlings of cottonwood (*Populus angustifolia*) are most abundant in years with cool winters, wet springs, and cool, wet autumns. Both good seedling years and stand-origin years are associated with winter blocking of storms in the North Pacific, but a persistent late-summer Arizona monsoon is needed for survival (8). Good seedling years occur more frequently (about every 3–4 years) than stand-origin years (about every 10–15 years).

The ability of soils and sediments to hold water and the existence of tributary and groundwater flows are equally important in determining vegetative distribution (23, 50, 109, 131, 262). Distance from the river and microtopographic variations determine lag times between rising discharge in the main channel and arrival of water on-site. Once water has arrived, the composition of the soil and the alluvium (as well as the rate of evapotranspiration) determine how long the substrata remains saturated. External water sources can allow vegetation to persist largely independent of the flood regime (14, 50, 131, 262). Indeed, older cottonwood, river red gum (*Eucalyptus camaldulensis*), and most mature riparian trees use groundwater rather than nearby creek water, presumably because groundwater is a more reliable source (50, 155).

The geomorphic template upon which the riparian forest develops is constantly undergoing change induced by the discharge regime (23, 180). The drainage network, from headwaters to the estuary, represents a mosaic of sites that may be aggrading, degrading, or maintaining a steady state. Even sites in a steady state, where the downstream movement of deposited materials is balanced by the transport of alluvium from upstream, the stream channels will continue to meander laterally and down-valley so that the physical features of the riparian zone continue to change (68). Levees support riparian gallery forests that may flood frequently, but the coarse deposits normally result in rapid drainage when water levels drop. Oxbow lakes are the most hydric of the riparian habitats, supporting species adapted to constant flooding and anaerobic soils.

A rich and comprehensive literature describes fluvial geomorphology (68). Two aspects particularly important for understanding patterns and processes in riparian vegetation are site-specific erosion and deposition, and lateral channel migration. Lateral channel migration may be slow (cm yr^{-1}) to fast (10^2 m yr^{-1}), depending on the type of stream and channel hydraulics (12, 23, 239); this substantially influences the composition and demography of the vegetative communities (111, 113, 127).

Sediment supply depends upon land use, climate, and tectonic activity. Rates of erosion and deposition range from a few millimeters to several meters annually. Tectonic activity, which has been occurring in the Andes Mountains for millennia, appears to be a central speciation mechanism for riparian forest communities in the floristically rich upper Amazonian basin (226, 239).

The physical influences of light and temperature on the vegetative community are less well investigated than either hydrology or geomorphology. Understory light tends to be highest at the riparian forest edge but rapidly declines toward the forest interior (unless gaps have been created by fallen trees), where levels are often <2% of full sunlight. However, seedling densities and diversity are not correlated with variations in understory light intensity, suggesting that other factors are more important in germination and establishment (152, 219, 220). The temperature regime is markedly different from upland forests, but no studies have attempted to link this difference to the vegetation (26). Compass orientation does influence light and temperature regimes, and significant floristic differences appear between riparian zones on north- and south-facing slopes (256).

Fire occurs rarely in the riparian zones of humid regions where most vegetation cannot withstand even mild fires (185) but plays a significant role in drier climates (4). In the lower Colorado River, nearly 40% of the riparian vegetation burned within 12 years, with halophytic shrubs recovering faster than mesophytic trees (30).

BIOTIC PATTERNS   As corridors within watersheds, riparian zones have a unique longitudinal pattern that exerts substantial controls on the movements of water, nutrients, sediment, and species (76, 155). Forman (76) recognized eight common shapes for riparian corridors, ranging from strictly linear (mostly uplands) to highly variable (low-elevation valley bottoms). Cross-sectional profiles that provide an important third dimension depend upon the local geographic setting. The simplest tend to be upstream and the more complex ones downstream (76, 235).

Despite strong hydrologic and geographic influences on riparian vegetation, ecological influences (such as competition, herbivory, soils, and disease) are significant in shaping communities. Even though competition is probably reduced because of frequent disturbance, a competitive hierarchy does exist. There is evidence that some species could exist in environments beyond their present range, specifically in the direction of less stress, except for the presence of competitors (129).

In regions with intact animal populations, herbivory exerts strong influences on vegetative characteristics (33, 190). The physical and trophic activities associated with herbivory have ecosystem-level consequences that go far beyond the requirements of individuals for food and habitat (121, 124, 179). The habitat is further modified by the activities of larger animals as they burrow and wallow in soils and build dams on streams, among other activities. The net result is that the heterogeneity of riparian habitats (or resource patches) is increased, and the distribution and cycling rates of elements (such as N and P) are modified.

Soil conditions (especially the degree of saturation) also coincide with patterns of sediment grain size and microtopography, affecting plant distribution from the river to the uplands (250) and are discussed later in this review. Disease as an agent influencing biotic patterns is little studied, although pests (such as budworm caterpillar, *Choristoneura fumiferana*) are known to spread rapidly along riparian corridors (76).

In general, the basal area of riparian forests is as great or greater than that of upland forests (23). Although values vary widely within regions, largely due to stand age, the variation is usually less than an order of magnitude. Riparian forests in the southeastern United States and the humid tropics tend toward greater stem density and basal area than those in more arid regions and more northern latitudes (23).

The aboveground biomass of riparian forests ranges between 100 and 300 tons ha^{-1} with few exceptions (23). Leaves represent 1–10% of the total. In general, belowground biomass tends to be less than aboveground biomass, ranging from 5% to 120% of it. Riparian forests have relatively high rates of production in comparison with upland forests. Much of the variation apparent in belowground biomass—12 to 190 tons ha^{-1}—may be due to sampling methods and site-specific conditions. Published values for aboveground production range from 6.5 to 21.4 t ha^{-1} yr^{-1}; litter fall averages 47% of the annual production. The limited data do not reveal latitudinal or successional gradients. Rates of belowground production have received virtually no attention in studies of riparian forests. The data suggest that there are no strong limiting factors associated with water or nutrients that would result in unusually low production.

Spatial zonation often exists as a transverse gradient perpendicular to the wetted channel that is made complex by vegetative responses to local variations in topography and susceptibility to flood (236). Vegetative patterns typically follow predictable patterns in physical features (273) and disturbance patterns (185). Décamps et al (54) presented a model of riparian forest succession for the Garonne River, France, in which cyclical successional processes occur within the floodplain where flood-induced erosion and deposition are common. However, on the higher terraces without repeated flooding, the successional dynamics are not reversible and internal autogenic forces dominate.

The earliest studies of vegetation dynamics in riparian zones did not refer to the concept of succession (75) but nevertheless illustrated the process. Many successional patterns in riparian areas are primary succession, but an equal number of successional patterns begin with plant fragments, propagules, or biomass remaining from previous communities (155). Avalanche, flood, wind, fire, drought, disease, herbivory, and other physical influences on the vegetation leave unique biotic legacies that are displayed in various successional patterns (185, 201). Many riparian plants possess adaptations allowing them to recover and reproduce by root suckering, adventitious root development on plant fragments, and stem flexibility. The amount of biotic material remaining to initiate succession and its viability depend on the type, intensity, frequency, and duration of the disturbance (53). Disturbances also prepare the site for invasion by additional species favored by the new conditions.

PHYSICAL FUNCTIONS OF RIPARIAN ZONES

Mass Movements of Materials and Channel Morphology

Material supplied to streams comes from erosion of stream banks, a process influenced by root strength and resilience (85), as well as from the uplands (178). Stream banks largely devoid of riparian vegetation are often highly unstable and subject to mass wasting, which can widen channels by several to tens of meters annually (111, 239). Major bank erosion is 30 times more prevalent on nonvegetated banks exposed to currents as on vegetated banks (11).

Riparian vegetation also modifies sediment transport either by physically entrapping materials, which appears to be most important in relatively low-gradient environments, or by altering channel hydraulics. In experimental channels, Kentucky bluegrass (*Poa pratensis*) entrains sediment at the base of the vegetation, with 30–70% retained dependent on blade length (3). Accretion of sediment and organic matter by vegetation can be substantial, especially during floods (107, 146). Sediment deposition from 1880 to 1979 on a coastal plain river in the southeastern United States averaged 35–52 Mg ha^{-1} yr^{-1} (146). Alteration of channel hydraulics is accomplished either by roots or by large woody debris in the channel at low flows and by stems at high flows. All provide physical structure that slows water, decreases stream power, and holds materials in place. Experimental manipulations involving the removal of large woody debris have resulted in dramatically increased erosion rates (16, 151, 191).

Erosional and depositional events shaping channel morphology are the subject of a large literature (68). In general, spatial heterogeneity introduced into the channel by either the vegetation or the large woody debris produced by the vegetation shapes channel morphology by redirecting flows of water and sediment, sorting sediments, and either retaining or moving materials (110).

Wood in Streams and Riparian Zones

Woody debris plays important biophysical roles at the land-water interface (95, 157). In the riparian forest, on exposed alluvial substrates, and in streams, it accumulates during floods in discrete and conspicuous piles. Each pile usually includes at least one large piece of dimensionally complex wood (i.e. a key member) which can resist most flow and physically capture smaller pieces of wood, making the pile even larger. In temperate forests, densities of up to 160 woody piles per km of stream bank may be found (251), but in tropical regions, where termites are prevalent, the density of woody debris piles is reduced substantially (190). Further, some of this wood, especially in the larger pieces, may be quite old. Nanson et al (192) found wood >17,000 years old in the floodplain of a Tasmanian stream; ages >300 years have been measured for large wood in streams of the Pacific coastal rain forest (1). Woody debris piles dissipate energy, trap moving materials, and form habitat. Depending on size, position in the channel, and geometry, they can resist and redirect water currents, causing the erosive power of water to become spatially heterogeneous, thereby creating a mosaic of erosional and depositional patches in the riparian corridor (173, 191). In steeper channels, the spatial arrangement of pools is independent of the amount of woody debris (173), so redirected water currents widen the channel and capture erosional materials (157). Such processes have been examined extensively by geomorphologists in alluvial rivers (228).

Woody debris also results in longer water residence times (69), and the temporary storage of materials can be substantial. Experimental removal of wood allows sediment and organic matter export rates in the first year to exceed baseline conditions by several hundred percent (16, 151). Analogous experiments with the addition of organic materials have produced similar insights. Leaves and small pieces of wood added to a stream with woody debris move only 65% and 8%, respectively, of the distance traveled by leaves and small wood in streams without large woody debris (69); 80% of the salmon carcasses added to nutrient-poor coastal rainforest streams are retained by woody debris within 200 m of the release sites (34).

Woody debris provides habitat for fish and macroinvertebrates within the stream channel (5, 95, 157, 167); its role as habitat within the terrestrial portion of the riparian corridor is only now being investigated. Woody debris physically retains plant propagules (seeds and plant fragments) and further protects them from erosion, abrasion, and, in some cases, drought and herbivory (74, 110). On exposed cobble bars, most seedling germination and survivorship are associated with woody debris, which provides a protective and relatively moist, nutrient-rich microenvironment. Woody debris also affords protection for small mammals and birds; the diversity and abundance of small mammals such as shrews, voles, and mice are significantly greater in areas with woody debris accumulations, while several bird species preferentially use woody debris for perching and feeding (61, 251).

Microclimate

Riparian forests exert strong controls on the microclimate of streams, but there are few comprehensive studies of the forest microclimate itself. Stream water temperatures are highly correlated with riparian soil temperatures, and strong microclimatic gradients appear in air, soil, and surface temperatures and in relative humidity but not in short-wave solar radiation or wind speed (26). Riparian forests, especially in warmer cimates and seasons, influence stream discharge through evapotranspiration. Reduced streamflow causes physiological difficulties for organisms preferring cooler temperatures (100).

Riparian Zones As Ecological Corridors

Riparian zones, as networks distributed over large areas, are key landscape components in maintaining biological connections along extended and dynamic environmental gradients (184, 218, 220). Perhaps the best evidence for plants using riparian zones as corridors comes from exotic invasions. Exotic plants rapidly move both up and down riparian corridors in preference to overland routes (58). However, it is not clear that riparian zones function as dispersal corridors in all cases. Certainly, adaptations of many plants allow vegetative fragments and seeds to float for various distances (196, 198, 241), while many other riparian species are dispersed by wind or animals (especially in the feces of birds). Schneider & Sharitz (244), for example, found that seeds of *Taxodium distichum* and *Nyssa aquatica* could float for 42 and 85 days, respectively, covering downstream distances of up to 2 km.

Few data document riparian zones as corridors for terrestrial animals, despite the common assumption in models. Two exceptions are the use of the riparian forest along the Garonne River, France, for birds moving between the Central Massif and the Pyrenees (55), and a three-year field experiment in Alberta, Canada, demonstrating enhanced movements of juvenile birds through riparian strips before and after harvest of adjacent forest (153).

ECOLOGICAL FUNCTIONS OF RIPARIAN ZONES

Sources of Nourishment: Allochthonous Inputs and Herbivory

Organic matter from riparian vegetation is a source of nourishment for aquatic organisms (112, 261). In temperate zones, values vary from about 200–900 g AFDM (ash-free dry mass) of litter m^{-2} in small and medium streams (1st to 3rd order) to about 20–50 g AFDM m^{-2} in larger rivers (35, 39, 272). As a general trend, the proportion of coarse particulate organic matter (CPOM; >1 mm diameter) decreases as river size increases. For example, in eastern Quebec, annual litterfall declines exponentially from 307–539 g AFDM m^{-2} in a 1st-order stream to 15–17 g AFDM m^{-2} in a sixth-order stream (39). Lateral inputs from the soil surface are not related to stream size but are strongly influenced by riparian structure and entrapment of organic matter during spring flooding (272).

Similar local effects also are found in intermittent prairie streams in Kansas where total annual input of CPOM is lowest in the headwater reaches (90). Prior to the wet season, storage of benthic CPOM in the dry channel and on the bank is 320–341 g AFDM m^{-2} in the upstream reaches and 999 g AFDM m^{-2} in the 4th- and 5th-order gallery forest reaches. The storage of CPOM increases during the wet season in headwater channels where retention is high, with the result that these reaches have more CPOM than do downstream reaches, although bank storage is always highest in downstream reaches.

Riparian structure appears to be the main factor influencing litter entering streams either directly or transported laterally from the forest floor. Depending on the vegetative cover of the riparian zone, annual inputs range from 52 to 295 g AFDM m^{-2} in Alaskan streams (64) and from 63 to 474 g AFDM m^{-2} in a Moroccan stream (150). In an Australian rainforest stream, laterally transported litter forms 6.8% of the total annual input, varying in response to bank slope and microtopography (15). The proportion of direct and laterally transported litter entering streams may significantly influence in-stream community dynamics as a consequence of input quality and timing (45). Even though laterally transported litter may not exceed 10% of the total litter input, it may be qualitatively important as a source of nourishment, due to a higher nitrogen concentration than that of leaves falling directly into the stream (15). However, the organic material may be rearranged only during high discharge periods and not during the leaf fall period (156), and forest floor litter may require the cumulative effects of several floods to move measurable amounts of litter laterally to the channel (172).

In addition to particulate organic matter, riparian zones contribute substantial amounts of dissolved organic matter (DOM; <0.5 μm) to river ecosystems. Soil water DOM may originate directly from unsaturated regions of riparian zones during floods or indirectly from the saturated through-flow at medium discharge rates (193). Soil water DOM originating in the riparian zone also can influence stream communities through macropore transfer of subsurface water.

At the scale of the Amazon basin, McClain & Richey (158) identified five transfer pathways of terrestrial organic matter to streams: direct litterfall, blow-in from the soil surface (e.g. lateral movement), groundwater baseflow, stormflow, and seepage from fringing wetlands. Direct litterfall from overhanging canopies and blow-in contributions were similar in mass and elemental composition among all landforms, amounting to 700 g m^{-2} yr^{-1}. Fresh and labile organic material dominated in all riparian zones examined. Groundwater baseflow DOM concentrations and proportions of hydrophobic organic acids were strongly correlated with soil type (characterized by old and refractory molecules). Stormflow contributions were dominated by saturated overland flow originating in riparian areas, which transferred a wide spectrum of dissolved and particulate organic matter, largely fresh and labile material. Fringing wetlands contributed high concentrations of DOM, particularly in the lowlands. In contrast to the other sites, the organic matter of wetlands is dominated by refractory hydrophobic dissolved compounds within a compositionally diverse molecular array. Although somewhat preliminary, the results and conceptual model of McClain & Richey (158) may prove useful as a framework for quantifying organic matter concentrations and compositions between contrasting trophic pathways.

Living riparian vegetation is a source of nourishment for many animals, from insects to mammals, that can considerably alter system function by their feeding activities. Outbreaks of defoliating insects can alter riparian forest production and thereby alter water yield, nutrient cycling, and streamwater chemistry (254, 263). Through selective browsing, large animals such as moose (*Alces alces*) can shift the riparian plant community from deciduous trees to conifers, altering soil formation and nutrient cycling and ultimately affecting plant productivity and moose population dynamics (206). Beaver (*Castor canadensis*) also exert a substantial impact on the structure and function of riparian systems, enhancing floodplain complexity and multiplying vegetative successional pathways, some of which affect the landscape for centuries (186).

Riparian Zones as Nutrient Filters

Karr & Schlosser's (128) demonstration that the land-water interface reduces nutrient movements to streams led to understanding the role played by riparian zones in controlling nonpoint sources of pollution by sediment and nutrients in agricultural watersheds (114, 147, 208).

Important biogeochemical processes that affect streamside as well as aquatic systems occur within the riparian zone, which is influenced on one hand by watershed hydrology and on the other hand by channel hydraulics (264, 265). The subsurface transfer of water and materials is mostly unidirectional toward the channel across the terrestrial boundary. In contrast, it is bidirectional across the aquatic boundary, where oxidized hyporheic water from the streambed mixes with (often reduced) interstitial water coming from the riparian zone. Both boundaries appear to be major locations for regulating and diminishing the transfer of inorganic nitrogen and phosphorus from subsurface water to stream water (159).

PHYSICAL BUFFERS   Sediments and sediment-bound pollutants carried in surface runoff are deposited effectively in mature riparian forests as well as in streamside grasses. Sediment trapping is facilitated by sheet flow runoff, which allows deposition of sediment particles and prevents channelized erosion of accumulated sediments. Riparian areas remove 80–90% of the sediments leaving agricultural fields in North Carolina (42, 47). Sediment deposition may be substantial in the long term, with coarse sediments deposited within a few meters of the field-forest boundary, and finer sediments deposited further into the forest and near the stream where they mix with coarse sediments deposited in overbank flows (42, 146).

Finer sediments carry higher concentrations of labile nutrients and adsorbed pollutants; their removal from the runoff occurs as a consequence of the interactive processes of deposition and erosion, infiltration, dilution, and adsorption/desorption reactions with forest soil and litter. In forested watersheds with relatively low nutrient concentrations, riparian zones can be sources or sinks for nutrients, depending on oxidation-reduction conditions. For example, the riparian zone of a small deciduous forest stream in eastern Tennessee was a net source of inorganic phosphorus when dissolved oxygen concentrations in riparian groundwater were low, but a sink when dissolved oxygen concentrations were high (174).

Phosphorus dynamics may be particularly complex. Although riparian zones may act effectively as physical traps (sinks) for incoming particulate phosphorus, they may enrich runoff waters in available soluble phosphorus (60). Significant amounts of phosphorus may first accumulate in riparian zones, then be transported to aquatic ecosystems in a different form via shallow groundwater flow, possibly as a result of increased decomposition of organic matter (71, 268).

Grassy riparian areas trap more than 50% of sediments from uplands when overland water flows are <5 cm deep (60, 154). Grassy areas influence the uniformity of runoff by transforming channelized flows into expanded shallow flows, which are more likely to deposit sediment. However, the performance of grassy vegetation seems to be highly variable and of short duration when several floods occur within a limited period. For example, sediment trapping efficiency may decrease from 90% in a first rainfall simulation event to 5% in a sixth rainfall simulation event (60).

BIOLOGICAL BUFFERS   Plant uptake, an important mechanism for nutrient removal in riparian forests (43, 72, 88, 208), results in a short-term accumulation of nutrients in nonwoody biomass and a long-term accumulation in woody biomass. Riparian forests are especially important sites for biotic accumulations of nutrients because transpiration may be quite high, increasing the mass flow of nutrient solutes toward root systems, and because morphological and physiological adaptations of the many flood-tolerant species facilitate nutrient uptake under low-oxygen conditions. In some species, such as water tupelo, saturated conditions enhance nutrient uptake and growth (103). Peterjohn & Correll (208) estimated vegetation uptakes of 77 and 10 $kg\ ha^{-1}\ yr^{-1}$ for N and P, respectively, rates that are comparable with upland rates. However, potential N uptake rates may be much higher, as shown by Cole (37): Poplar (*Populus nigra*) assimilated 213 $kg\ N\ ha^{-1}\ yr^{-1}$ when fertilized with a nutrient-rich effluent at a rate of 400 $kg\ N\ ha^{-1}\ yr^{-1}$ for three years, but sites not receiving nutrient effluent assimilated only 16 $kg\ N\ ha^{-1}\ yr^{-1}$. Further, due to nitrogen saturation

(2), phosphorus may become the limiting factor for tree growth, particularly in wetlands (260), making vegetation an effective phosphorus sink.

The importance of plants as nutrient filters may be reduced by restricted accessibility to water, by the seasonal phenology of uptake and release of nutrients, and by the saturation of mature forests. Water is accessible to plants only if the water table is high or if transpiration demand moves water and solutes into the root zone. During intense rain storms, concentrated surface flow and macropore-dominated percolation may not be available to plants (115). Nutrient uptake declines or stops during the winter, precisely when high discharges occur. In addition, litter decomposition releases nutrients to forest soils stored during the growing season. Finally, the ability to sequester nutrients in woody biomass may decline as trees mature, leading to saturation (2, 88). The contribution of individual riparian tree species to nutrient retention remains to be elucidated.

Microbial uptake of nutrients, similar to plant uptake, initially results in the immobilization of dissolved nutrients followed by cell growth, death, decomposition, and eventual nutrient release. Nitrogen, in contrast to other nutrients, has an alternate pathway of major importance in most riparian forests. Denitrification (43, 88, 114, 125, 212, 214) depends on the presence of nitrate, a suitable carbon substrate, and the absence of oxygen. Soil temperature, moisture, and the type of carbon influence the reaction rate. Soil pH affects whether N_2O or N_2 is produced. In riparian zones, anaerobic microsites associated with decomposing organic matter fragments allow denitrification in otherwise well-drained soils. Within a riparian zone, denitrification rates of 30–40 kg ha^{-1} yr^{-1} have been recorded; the fastest rates occur at the riparian-stream boundary where nitrate-enriched water enters organic surface soil (40).

Denitrification of groundwater-borne nitrate is less well established. Carbon availability usually limits subsurface microbial activity, preventing anaerobic conditions from developing (99). Since denitrification is concentrated in the upper 12–15 cm of the soil (which is only occasionally part of the shallow aquifer), nitrate disappearance from shallow groundwater may require the riparian vegetation to play a primary role (144). Groundwater-derived nitrate may be eventually denitrified in surface soil after plant uptake of the nitrate from groundwater, litter decomposition, and ammonium release, followed by nitrification (88, 92, 93). Therefore, microbial denitrification interacts with vegetation nitrogen uptake and organic carbon availability via litterfall and root decay to remove nitrate. Such an interaction varies within and between riparian forests under the influence of subsurface water, plant cover, and soil characteristics (101).

BUFFER VARIABILITY   Subsurface water flow paths have strong effects on nutrient characteristics. In several locations of the Coastal Plain of the Chesapeake

Bay watershed, average annual terrestrial boundary nitrate concentrations of 7 to 14 mg NO_3-N L^{-1} decrease to 1 mg NO_3-N $^{L-}1$ or less in shallow groundwater near streams (145). However, in the same area, a single site with a nitrate concentration of 25 mg NO_3-N L^{-1} at depth had a concentration of 18 mg NO_3-N L^{-1} in shallow groundwater at the stream. Lowrance et al (148), who estimated annual denitrification rates to average 31 kg N ha^{-1} yr^{-1} in the top 50 cm of soil, measured denitrification rates between 1.4 kg N ha^{-1} yr^{-1} in a riparian zone adjacent to an old field (which received no fertilizer) to 295 kg N ha^{-1} yr^{-1} under conditions of high nitrogen and carbon subsidies. Such results illustrate the potential for denitrification in surface soils as well as the high variability to be expected in field measurements. Most of this variability is driven by fine-scale differences between rooted and nonrooted soil layers as well as between anoxic and oxic conditions, and it depends on subsurface flow paths. On the whole, enrichment of riparian zones may lead to significantly higher soil inorganic-nitrogen concentrations, litter nitrogen contents, and potentials for net nitrogen mineralization and nitrification; all suggest nitrogen saturation. Nevertheless, high rates of denitrification and storage usually maintain enriched riparian zones as sinks for upland-derived nitrate (92).

Plant cover also influences the efficiency of riparian zones in filtering nutrients and pesticides. A riparian zone vegetated with poplar is more effective for winter nitrate retention than one vegetated with grass (97). Some trees are better than others in filtering nitrate: *Populus x canadensis* effectively removes nitrate from saturated soils with a subsequent accumulation of nitrogen in root biomass (203). Roots of alder, willow, and poplar seem to favor colonization by proteolytic and ammonifying microorganisms and, particularly for alder roots, to inhibit nitrifying microorganisms (221). Changing plant cover may affect water quality: In a set-aside riparian zone in New Zealand 12 yr after retirement from grazing, dominant vegetation returned to native tussock (*Poa cita*), leading to a zone likely to be a sink for sediment-bound nutrients and dissolved nitrogen but a source for dissolved phosphorus (41).

Soil characteristics influence redox conditions and the availability of dissolved nitrogen to plant roots. Pinay et al (215) calculated that sandy riparian forest soils retained 32% of the total organic nitrogen flux during a flood, but 70% was retained on loamy riparian soils. McDowell et al (162) compared riparian nitrogen dynamics in two geomorphologically different tropical rain forest sites. At one site, a deep layer of coarse sand conducted subsurface water to the adjacent stream below most plant roots, through oxic and anoxic zones from upslope to downslope, respectively, whereas at the other site, a dense clay layer impeded infiltration, and subsurface water rapidly moved through a shallow and variably oxidized rooting zone. Although intense biotic activity controlled hydrologic export of nitrogen at both sites, soil differences strongly modified interactions. Such interactions have important consequences at watershed and landscape scales whenever the transient retention of ammonium or nitrate in riparian sediments influences biotic nitrogen cycling, thereby altering the timing and form of dissolved inorganic nitrogen export (266).

VEGETATIVE PATTERNS OF DIVERSITY   Riparian forests provide insights into how plant species richness varies at regional scales with respect to natural disturbances (218). Profiles along river courses are especially effective in evaluating the intermediate disturbance hypothesis for explaining biodiversity patterns (199). Plant species richness is not always highest in median sections of river courses where flood disturbances are at intermediate levels of intensity and duration (8, 57, 199, 259).

In northern Sweden, post-glacial history of the landscape results in species diversity peaking where rivers begin to downcut into sediments deposited during a higher coastal stage (9200 yr BP; 198). In southern Spain, semi-arid rivers show irregular patterns of species richness along their courses, as a consequence of irregular water availability (259). Main channels and their tributaries may differ as in the Adour basin of France, where exotic species are more numerous in the main channel and display different longitudinal patterns than native species (57, 216). In the Vindel basin, northern Sweden, the main channel also has a higher species richness than the tributaries; in addition, species richness is greatest at mid-altitude in the main channel but least in the tributaries (195). The distributions of native and exotic flora (58, 217, 257, 258, 259), or long-lived ruderal species (195, 196, 199, 200) differ markedly in longitudinal profiles of richness.

It remains to be demonstrated whether riparian forests maintain biodiversity through continuous upstream movement of species from other catchments or from extension of species distributions downstream during favorable environmental periods (195). Clearly, propagule dispersal by water has a role in structuring the riparian flora (116), but dispersion mechanisms such as anemochory and ornithochory deserve consideration (169), as do the different stages of a plant's life cycle, which are important for successful establishment and growth (56).

Black cottonwood (*Populus trichocarpa*) from Pacific river banks in North America respond to climatic selection pressures at regional as well as local scales, with significant variations in survival, growth, and photosynthesis in high light regimes; additionally, leaf and crown traits respond to whether the habitat is mesic or xeric (65, 66, 67). Genetic discontinuities among riparian populations of *P. trichocarpa* along the same river coincide with upstream-downstream changes in atmospheric moisture levels (66, 67). Such discontinuities occur even though black cottonwood can disperse sexual and asexual propagules over wide areas.

NATURAL DISTURBANCES   Floods create heterogeneity within the riparian zone and thereby create distinct regeneration niches that facilitate the coexistence of congeneric species. For example, six species of *Salix* co-occur, despite similar adult ecology, along the Sorachi River, Hokkaido, Japan (194). Dispersal periods of these species overlap as water levels decrease following spring floods. The dominant species, *S. sacchalinensis,* establishes on a wide range of soil textures. The subordinate species *S. rorida* and *S. subfragilis* coexist with *S. sacchalinensis* on the finest and coarsest soils, respectively, where the dominant species does not grow as efficiently. The occurrence of the three other rarer subordinate species also is related to the fitness of early stages to flooding and soil characteristics.

Periodic flood disturbances of various intensities are also critical for maintaining the four dominant tree species of the lowland floodplain podocarp forest in New Zealand (63)—*Dacrycarpus dacrydioides, Dacrydium cupressinum, Prumnopitys ferruginea* (conifers), and *Weinmannia racemosa* (angiosperm). The two former are upper-canopy trees, and the two latter are sub-canopy trees. Intense floods denude large areas, allowing *D. dacrydioides* to establish on silt substrates and *D. cupressinum* to establish on elevated microsites covered with organic debris. Less intense floods expose debris, providing sites more suitable for *D. cupressinum* and occasionally for *P. ferruginea*. Minor floods do not disturb the canopy but provide opportunities for establishment of *P. ferruginea* under canopy cover and *W. racemosa* in small canopy gaps caused by nonflood mortality of the mature conifers.

At local scales, floods affect species diversity of herbaceous plants through physical heterogeneities created by the erosion and deposition of litter and silt. In northern Sweden, leaf litter accumulations of approximately 150 g m^{-2} result in a maximum number and diversity of adult vascular plant species at the upper elevational limit of riparian zones (197). Likewise, in southern France, organic matter accumulations of 150 to 300 g m^{-2} result in high germination rates and decreased mortality rates, whereas accumulations of 600 to 1200 g m^{-2} decrease germination rates. Around 150 g m^{-2} of litter corresponds to maximal density and species richness (137). Similarly, silt accumulations of about 500 g m^{-2} correspond to maximal density and species richness of seedlings (138).

INVASION BY EXOTIC SPECIES   The richest communities also have the greatest proportion of exotics, along the rivers as well as within specific sites. This suggests that the richest communities in riparian corridors may be the most invasible because of the substantial environmental heterogeneity created by moderate floods. Comparing the riparian plant communities of the Adour River, France, with the Mackenzie River, Oregon, Planty-Tabacchi et al (217) discovered about 1400 (24% exotic) and 850 (30% exotic) constituent species, respectively. More woody species are exotic along the Adour (46%) than along the Mackenzie (17%). In contrast, herbaceous communities of the Mackenzie are more invasible (32%) than those of the Adour (21%). Although these differences parallel more intensive forest management in Oregon than in France, the ecological mechanisms explaining these similarities remain obscure.

Several interactive processes appear to control establishment of exotic species in riparian zones (56). Although common in nature, biological invasions have been accelerated through human activities (142). Life-history characteristics of invaders control the various stages of establishment, stabilization, and expansion. For example, Pysek & Prach (222) reported that the shorter the life span, the higher the rate of invasion among four species alien to central Europe, *Impatiens glandulifera, Heracleum mantegazzianum, Reynoutria japonica*, and *R. sacchalinensis.* Landscape characteristics such as connectivity along rivers (116) and historical development of landscape patch structure (223) also exert control.

Natural environmental features may also slow the rate of invasion. For example, in the subarctic environment of northern Sweden, exotic plants from other continents are rare in riparian zones. In northern Australia, where the area occupied by the exotic *Mimosa nigra* doubles every 1.2 yr, the rate of expansion is related to the amount of rainfall in the previous wet season (143). At the regional scale, the areal doubling time is 6.7 yr, probably due to the spatial isolation of wetland habitats by eucalyptus savanna. Seed predation combined with folivores is also likely to slow the rate of expansion because the seeds of this species are dispersed by flotation.

Although exotic invasions can reduce native plant species diversity, there is no clear evidence it does. However, *Impatiens glandulifera,* the tallest annual plant in Europe, is expected to reduce species diversity and to out-compete native light-demanding species in riparian habitats (10, 216, 224). Similarly, *Tamarix* spp., invasive exotic woody plants in arid and semi-arid riparian habitats of western United States, are expected to replace or inhibit much of the native flora (25). However, clear data supporting or refuting these expectations do not appear to exist.

REFUGES FOR REGIONAL DIVERSITY   Riparian forest patches have acted as safe sites for regional flora during dry periods. Some present-day humid tropical zones appear to have experienced Pleistocene droughts, but there is no indication of mass extinctions, whereas there is indication of rapid species re-expansion during the early Holocene. Riparian forest patches may have been refuges for the maintenance of mesic plant diversity in Central America (166). The floristic attributes of the riparian forest are similar to those characterizing continuous forests in the area, while their stem density is higher and biomass lower, thus increasing their potential to maintain species richness. Moreover, riparian trees in a savanna matrix are often younger because more mature trees are removed by wind and fire as compared to continuous forests (132). Frequent removal of mature trees reduces the rate of competitive exclusion from the community while enhancing the potential for greater numbers of coexisting species.

Riparian forests in Central America average 52 species/ha (trees with > 10-cm diameter at breast height), a richness comparable to or slightly lower than for upland forests in the area (166). However, extremely diverse forests occur in other tropical rain forests such as in Amazonia (225 tree species ha^{-1}) and in southeast Asia (283 tree species ha^{-1}) where the refuge role of riparian forests may have been more limited (62).

MACROINVERTEBRATE COMMUNITIES   Stream macroinvertebrate community characteristics may be predicted from a knowledge of riparian vegetation (45). The presence or absence of riparian trees could be the single most important factor altered by human activity that affects these communities (255) since riparian vegetation affects macroinvertebrate diversity primarily through its effects on benthic habitat (21, 82, 140, 204, 227). In contrast, the effect of riparian vegetation as a source of nourishment for invertebrates is not so well established. Much of the woody debris is unpalatable, thus preventing a diverse consumer community from developing (202). Also, changes in species composition, rather than diversity, often follow changes in riparian composition. For example, a gatherer-collector community of chironomids may replace a shredder community when there is a reduction in slowly decomposing riparian litter in small woodland streams (89).

Habitat

With variations in flood duration and frequency, and concomitant changes in water table depth and plant succession, the environment is a complex of shifting habitats created and destroyed on different spatio-temporal scales (155). Most riparian zones are covered with a remarkable variety of woody vegetation from shrubs serving as refuges for small mammals to trees offering nesting and perching sites for birds. Also, sustained herbivory develops as a result of enhanced productivity and food quality, and fallen woody debris provides stability for terrestrial as well as aquatic invertebrate communities. Riparian forests act as refuges in adjacent areas and, in some cases, as corridors for migration and dispersal (24).

The occurrence of any species in riparian areas is probably due to several interrelated reasons. Juvenile wood turtles (*Clemmys insculpta*) prefer to remain near stream channels where they can move comparatively short distances to find appropriate thermal and moisture conditions and at the same time are less exposed to predation (22). Drinking water is obviously an important reason for mammals to visit riparian zones. However, small mammals may respond more to differences in tree communities. Higher capture rates for some small mammals have been reported in streamside habitat dominated by red alder (*Alnus rubra*) adjacent to uplands dominated by Douglas fir (*Pseudotsuga menziesii*) (160), but no difference was detected where upland and streamside habitats were similar in vegetative structure and composition (161). The white-tailed deer (*Odocoileus virginianus*) uses riparian zones almost twice as much as nonriparian areas, supposedly as an antipredation strategy (139).

Like mammals, more individuals and species of birds are found in riparian habitats than in adjacent ones. In the lower Mississippi River, more than 60 species of mammals, about 190 species of reptiles and amphibians, and about 100 species of birds are seasonally associated with riparian habitats (133). For example, 82% of the breeding birds of northern Colorado occur in riparian vegetation (135).

Bird assemblages of riparian zones and adjacent uplands are interdependent; the number of shared species varies seasonally and longitudinally along river courses with apparently greater interdependence at intermediate elevations (135, 141). However, bird individuals and species may not be more numerous where the riparian habitat is either similar to upland habitat or not clearly delineated (176). About 90% of the present bird fauna along the Platte River in northern Colorado has arrived since the development of a gallery forest within the last 90 years (134). In boreal forests, bird densities reportedly increase 30% to 70% in protected riparian forest strips the year after clear-cutting; they then decline during the following years while the adjacent clear-cut regenerates (48).

Bird communities are sensitive to the quality of riparian vegetation (44, 141). Destruction of riparian vegetation causes local extinction and also reduces the ability of some populations to recolonize sites (135). Along the Colorado River in the Grand Canyon, black-chinned hummingbirds (*Archilochus alexandri*) nest only in exotic tamarisk-dominated habitats that are greater than 0.5 ha in area (27). In disturbed areas, woody strips 2 m wide permit only portions of bird populations to occur; widths >25 m on each bank are necessary to maintain sensitive species (44). In a survey of 117 corridors ranging from 25 m to 800 m wide, Keller et al (130) concluded that the probability-of-occurrence increased most rapidly between 25 m and 100 m. An important conclusion of recent work on riparian habitat for birds is that conservation must be based on specific bird species and account for differences in behavior (for example, between generalist-opportunistic and riparian-obligate species). A drainage basin perspective is absolutely necessary (135).

The inhabitants of riparian zones can modify habitat structure and function. Through ponding water and storing sediments, beaver create wetlands and alter the vegetative composition of in-channel and riparian communities in temperate North America, strongly influencing riparian landscapes (186, 188, 189). Through selective cutting of trees, they change the composition of riparian communities. For example in Minnesota, under beaver influence, trembling aspen (*Populus tremuloides*) decreased in abundance, whereas alder (*Alnus rugosa*) and black spruce (*Picea glauca*) increased (122). Besides beaver, species such as moose, elk (*Cervus canadensis*), and brown bear (*Ursus arctos*) create and maintain tight networks of trails along river banks. In southern Africa, hippopotamus (*Hippopotamus amphibius*) modify riparian habitats, gathering in pools during the day where they stir up sediments and deepen aquatic habitats; during the night they create paths between pools and terrestrial grazing areas, thus maintaining connectivity between patches (190). Warthog (*Phacochoerus aethiopicus*) transform tens of hectares to ploughed fields by digging soils 10–15 cm deep to feed on underground plant storage organs in riparian forests and wetlands (229); consequently the replacement of perennial rhizomatous grasses by annual grasses and forbs is favored.

Feeding activities have long-term consequences for the structure and function of riparian forests. Zoochory, particularly ornithochory, facilitates the expansion of certain species through selective feeding and propagule transport. Moose browsing reportedly affects decomposition indirectly through changes in the quality of litterfall in North American riparian systems (163, 205). Similar effects occur in riparian corridors in Africa through bulk browsing by elephant (*Loxodonta africana*), bulk grazing by hippopotamus, and selective browsing by kudu (*Tragelaphus strepsicerous*), giraffe (*Giraffa camelopardalis*), and bushbuck (*Tragelaphus scriptus*) (190).

ENVIRONMENTAL ALTERATIONS

Human Alterations

Along European rivers, human-induced alterations include neolithic deforestation and land-clearing during Gallo-Roman and medieval periods (207). Civil engineering works in the nineteenth century and hydroelectric developments in the twentieth century accelerated these alterations (211). Similar events have occurred in North America on a reduced time scale since European settlement (246).

Flow variability and fluctuations in channel width, which are necessary for maintaining the biodiversity of riparian systems (78), have been dramatically decreased in many parts of the world through river impoundment, water management, and lowering of water tables. Substantial changes in riparian vegetation may occur without changing mean annual flow, as riparian vegetation is especially sensitive to changes in minimum and maximum flows (6). In many cases, hydrologic alterations result in shifts in riparian plant community composition as well as senescence of woody communities (18, 31, 54, 77, 104, 196, 198, 207, 232).

The rapid invasion of floodplains in the southwestern United States by *Tamarix ramossissima* has desiccated water courses (269), inducing other disturbances such as fire (32) and displacement of native species (e.g. *Salix gooddingii* and *Populus fremontii*; 253, 269). Along the Colorado River floodplain, ecophysiological tolerances and competition for moisture may be at the origin of shifts in riparian community structure from a gallery forest to riparian thickets (with *Tamarix* dominating or replacing *Salix*) and the disappearance of *Populus* (32).

Along the River Murray in Australia, flood frequency and duration historically prevented native red gum (*Eucalyptus camaldulensis*) from establishing on grass plains. By reducing depth and duration of flooding, river regulation has favored a red gum invasion that is expected to cause a complete extinction of the once-extensive grass plains in the near future (19). Alterations of water regimes also have contributed to dramatic declines and losses of cottonwood forests throughout western North America as a consequence of drought-induced mortality from abrupt flow reductions and lowering of water tables (233, 234). A possible mechanism for this decline is that a reduction of flooding diminishes the rate of recruitment of new stands (170).

In fact, the maintenance of phreatophytic riparian woodlands depends heavily on the relationship between river flow and life-history traits of cottonwood (108). For example, as growth conditions vary widely within and between years at the same site, an important factor for survival is the ability of seedlings to establish over a range of moisture conditions (247). Tyree et al (267) demonstrated species-specific responses to water regimes for several *Populus* species that differ in xylem water potentials and consequent cavitation.

Even though alterations to water regimes may result in declines of cottonwood forests, at least temporary increases occur in some other environmental situations (249). Along the meandering Missouri River, poplar-willow communities depend on flow peaks that erode outer banks and deposit sediments on inner banks (118, 119, 120). A reduction of peak flows after completion of the Garrison Dam in 1953 prevented meandering and thus the formation of suitable areas for poplar-willow establishment, resulting in a continuous decrease in pioneer stages and an increase in older stages (119). In contrast, poplar-willow communities have dramatically expanded along the Platte River after completion of dams on its two main tributaries, the North and South Platte rivers, reduced flow and exposed large areas suitable for seedling establishment and survival (120).

Invading riparian trees may have an impact on channel morphology, particularly when they replace formerly grassy areas (79, 237). Active investigations are needed on mechanisms linking water, landforms, and species in various landscape settings in order to predict the specific effects of manipulating flow on floodplain forests (53).

In northern Sweden, species-richness and the percentage vegetative cover are both lower per site in a regulated as compared to a nonregulated river (196, 198). The proportion of annual plus biennial species-richness is higher and perennial species-richness is lower along regulated rivers. Water level regime and mean annual discharge certainly are among the most important variables for maintaining species richness and plant cover.

The species composition of the riparian plant community is also important to consider in predicting responses to alterations (18, 196, 198, 207). Several scenarios may be expected for a plant community as a result of flow alterations: reproduction by on-site regeneration, colonization of other parts of the floodplain, or replacement by a new type of plant community (207). Alterations of riparian plant communities obviously affect aquatic macroinvertebrates and fishes as a consequence of modifying trophic pathways and in-stream habitat (49, 238) as well as the species diversity of amphibians, birds, and mammals (44, 130, 181).

Management and Restoration

Hydrological characteristics are of primary importance in managing riparian zones. Riparian buffer zones retain surface runoff pollutants as a result of their water storage capacity and infiltration (41). Riparian buffer zones intercept the dominant hydrological pathways that are dependent on soil type and permeability, adjacent land use, slope, potential run-off generation areas, and land drainage installations. For example, nitrate removal requires that the bulk of the water moves either across the surface or as shallow groundwater through biologically active soil zones (145), and sediment removal requires that surface runoff does not overwhelm the buffer system. Hydrological pathways are likely to change widely in space and time. Forest growth or weather variations may affect the degree of saturation of the riparian zone and the proximity of the water table to the soil surface (70, 98), thereby complicating the intricacy of groundwater routes and the ways water-borne nitrate encounters roots and soil microbes.

Increasing loading rates may affect riparian zones differentially according to the type of pollution (70, 145). For nitrate, higher rates of N-removal generally occur with higher loading rates as a consequence of denitrification and vegetative uptake. Nevertheless, buffering capacity may be limited by inefficient nitrate uptake rates, limited duration of anaerobic conditions, or organic carbon availability for microbial respiration (98). For other nutrients and metals, biological processes similar to denitrification are lacking, and higher rates of loading may result in excess release when the immobilization capacity of the riparian buffer is exhausted. In such cases it is necessary to manage riparian systems to facilitate sediment removal and infiltration so as to prevent these systems from becoming sources of pollutants. Also, flooding and erosion of riparian soils during winter may be a general limitation to buffering capacities for nutrients and metals (98).

An increasingly important managerial use of riparian zones is to control diffuse pollution. Riparian zones are more effective over the long term when upstream pollution has been limited through good agricultural practices at the catchment level (7, 51). The integrated effects of riparian zones on water quality will also differ according to stream order, smaller streams having a greater potential than larger ones to buffer against diffuse pollution (145). The control of water quality in headwater catchments is an effective management strategy because, once a river is contaminated, few inexpensive possibilities remain for improvement (98).

One model suggests that multi-species riparian buffer strips provide the best protection of streams against agricultural impact (106, 145, 245). This model uses three interactive zones that are in consecutive up-slope order from the stream: 1. a permanent forest about 10 m wide, 2. shrubs and trees up to 4 m wide (and managed so that biomass production is maximized), and 3. herbaceous vegetation up to 7 m wide. The first zone influences the stream environment (e.g. temperature, light, habitat diversity, channel morphology, food webs, and species richness). The second zone, which controls pollutants in subsurface flow and surface runoff, is where biological and chemical transformations, storage in woody vegetation, infiltration, and deposited sediments are maximized. The first two zones contribute to nitrogen, phosphorus, and sediment pollution removal. The third zone provides spreading of overland flow, thus facilitating deposition of coarse sediments. Clearly this basic model must be adapted to various catchment conditions and stream orders to provide effective management.

Long-term sustainability is likely to occur when managed systems imitate natural ones. For example, zone 1 of the multispecies riparian buffer strip (located near the stream) functions better if zone 2 is harvested infrequently; and zone 3, near the cropland, also functions better if accumulated sediment is removed and herbaceous vegetation is reestablished periodically (145). The literature, however, offers divergent examples ranging from efficient removal of nitrogen after 20 years of high nutrient loading (7) to exhaustible sinks (41). Whatever the example, improved land use practices within the catchment and the maintenance of riparian zones for interception of groundwater flows by vegetation in various stages of succession (which differ in absorption capacity) are key factors for the long-term vitality of buffer strips and streams (98, 270).

Other benefits obtained from creative management and restoration of riparian zones include provision of diversified habitat for terrestrial and aquatic wildlife, corridors for plant and perhaps animal dispersion, and input of organic matter to streams. Enhancement of the visual quality and increase in recreational value of the landscape are also important benefits. Management has been used positively to influence communities of aquatic animals in upland streams throughout Wales and Scotland (204), and managed and natural floodplain forests are recommended in British river and floodplain restoration projects (209).

Tools for the Future

Riparian systems are increasingly expected to fulfill ecological functions related to biodiversity, habitat, information flow, biogeochemical cycles, microclimate, and resistance and resilience to disturbance (187). They will be expected also to fulfill more social functions, including provision of resources, recreation, culture, and aesthetics. Clearly, no single riparian system will perform all these functions, but each will be likely to perform at least one.

An extensive linear approach to management, at the scale of river courses, is needed to delineate and classify riparian systems along streams. Significant progress has already been made in mapping riparian systems using remotely sensed data (175). Lowrance et al (145) recommended that linear forests be characterized at a resolution of 10 m to 20 m. In combination with hydrogeomorphic data, such precise maps help in assessing the potential for riparian systems to intercept surface- and subsurface-borne pollutants. Management-initiated investigations, such as those conducted on the poplars of the Platte River, demonstrate the need for considering entire river courses to understand various aspects of the dynamics of riparian systems (120), to restore the ecosystem integrity of rivers and floodplains (46), and to manage international greenways (28).

Both landscape and detailed site perspectives are required to judge whether planted trees will survive to reach the expected sizes. The landscape perspective may be attained through knowledge based on mapping past and present extents of floodplain plant communities, characterizing the ecology of appropriate species, and determining priorities at both landscape and site scales (105). Detailed site perspective may be attained through knowledge based on water regimes, suitable soil conditions, and long-term survival and growth rates as well as on the effects of variable water levels on tree metabolism (117).

An intensive site-specific approach integrates research, demonstration, and application of riparian zone buffers (145). Such an integration will aid in discovering the effects of vegetation type and management approach on the long-term control of nutrient and sediment pollution, the response of riparian zones to acute stresses such as large storms and extremes in temperature or growing season rainfall, the consequences of chronic stress leading to saturation of riparian zones by nutrients, and the processes controlling groundwater microbial dynamics (87, 145). Insights into these issues are requisite for developing models of risk assessment (149) and decision-making (106). Intensive site-specific studies also should improve the ability to model important functional issues such as the influence of soil wetness on key nutrient transformation processes (83). Finally, site-specific studies should improve the ability to evaluate the performances of restored riparian habitats through a better knowledge of hydrologic, geomorphic, and biologic conditions.

Research at the catchment scale, or at least at the representative hillslope scale, is essential to assess the effect of riparian systems on hydrologic inputs from uplands (145). As many buffer processes operate most effectively in headwater basins, downstream cumulative effects involving many small catchments must be given proper emphasis (29). The catchment is also the appropriate scale to improve hydrologic conditions within riparian zones (243) and to assess the potential of narrow riparian zones that are remnants of previously wider ones in most rural and urban temperate zone landscapes (54, 210).

Finally, considering riparian zones as management tools for the future requires the adoption of flexible, adaptive schemes to cope with surprises related to discontinuities and synergisms (177). Discontinuities in riparian systems may occur where nutrient accumulations reach disruptive thresholds that suddenly change the system from a sink to a source. Synergism also may occur from interactions between two chronic stresses (such as nutrient loading and global warming) or between a chronic and an acute stress (such as a large storm). In addition to such "anticipatable surprises," entirely unforeseeable future issues require increased efforts in research and management (187).

ACKNOWLEDGMENTS

Preparation of this article was supported by grants from the AW Mellon Foundation and the Pacific Northwest Research Station of the US Forest Service. We thank C Nilsson, T Coe, G Pinay, E Chauvet, and E Tabacchi for comments and suggestions that greatly improved the content, and Treva Coe and James Helfield for assistance with the bibliographic searches.

Visit the *Annual Reviews home page* at
http://www.annurev.org.

Literature Cited

1. Abbe TE, Montgomery DR. 1996. Large woody debris jams, channel hydraulics, and habitat formation in large rivers. *Reg. Riv.* 12:201–21
2. Aber JD, Nadelhoffer KJ, Steudler P, Melillo JM. 1989. Nitrogen saturation in northern forest ecosystems. *BioScience* 39:378–86
3. Abt SR, Clary WP, Thornton CI. 1994. Sediment deposition and entrapment in vegetated streambeds. *J. Irrig. Drain. Engin.* 120:1098–113
4. Agee JK. 1993. *Fire Ecology of Pacific Northwest Forests*. Washington, DC: Island
5. Anderson NH, Sedell JR. 1979. Detritus processing by macroinvertebrates in stream ecosystems. *Annu. Rev. Entomol.* 24:351–77
6. Auble GT, Friedman JM, Scott ML. 1994. Relating riparian vegetation to present and future streamflows. *Ecol. Appl.* 4:544–54
7. Baillie PW. 1995. Renovation of food-processing wastewater by a riparian wetland. *Environ. Manage.* 19:115–26
8. Baker, WL. 1990. Climatic and hydrologic effects on the regeneration of *Populus angustifolia* James along the Animas River, Colorado. *J. Biogeogr.* 17:59–73
9. Barbour MG, Burk JH, Pitts WA. 1987. *Terrestrial Plant Ecology*. Menlo Park, CA: Benjamin/Cummings
10. Beerling DJ, Perrins JM. 1993. Biological flora of British Isles: *Impatiens glandulifera* Royle (*Impatiens royeli* Walp.). *J. Ecol.* 81:367–82
11. Beeson CE, Doyle PF. 1995. Comparison of bank erosion and vegetated and non-vegetated channel bends. *Water Res. Bull.* 31:983–90
12. Begin Y, Lavoie J. 1988. Dynamique d'une bordure forestiere et variations recentes du niveau du Fleuve Saint-Laurent. *Can. J. Bot.* 66:1905–13
13. Belknap WC, Naiman RJ. 1998. GIS location, remote TIR detection, and mapping of wall-base channels in western Washington. *J. Environ. Manage.* In press
14. Bell DT, Johnson FL. 1974. Groundwater level in the flood plain and adjacent uplands of the Sangamon River. *Trans. Ill. State Acad. Sci.* 67:376–83
15. Benson LJ, Pearson RG. 1993. Litter inputs to a tropical Australian rainforest stream. *Aust. J. Ecol.* 18:377–83
16. Bilby, RE. 1981. Role of organic dams in regulating the export of dissolved and particulate matter from a forested watershed. *Ecology* 62:1234–43
17. Blom CWPM, Voesenek LACJ. 1996. Flooding: the survival strategies of plants. *Trends Evol. Ecol.* 11:290–95
18. Bravard JP, Amoros C, Pautou G. 1986. Impact of civil engineering works on the succession of communities in a fluvial system. *Oikos* 47:92–111
19. Bren LJ. 1992. Tree invasion of an intermittent wetland in relation to changes in the flooding frequency of the River Murray, Australia. *Aust. J. Ecol.* 17:395–408
20. Bren LJ. 1995. Aspects of the geometry of riparian buffer strips and its significance to forestry operations. *For. Ecol. Manage.* 75:1–10
21. Brewin PA, Newman TML, Ormerod SJ. 1995. Patterns of macroinvertebrate distribution in relation to altitude, habitat structure and land use in streams of the Nepalese Himalaya. *Arch. Hydrobiol.* 135:79–100
22. Brewster KN, Brewster CM. 1991. Movement and microhabitat use by juvenile wood turtles introduced into a riparian habitat. *J. Herpetol.* 25:379–82
23. Brinson MM. 1990. Riverine forests. In *Forested Wetlands*, ed. AE Lugo, MM Brinson, S Brown, 15:87–141. Amsterdam/New York: Elsevier
24. Brinson MM, Lugo AE, Browns S. 1981. Primary productivity, consumer activity, and decomposition in freshwater wetlands. *Annu. Rev. Ecol. Syst.* 12:123–61
25. Brock JH. 1994. *Tamarix* spp. (salt cedar), an invasive exotic woody plant in arid and semi-arid riparian habitats of western USA. In *Ecology and Management of Invasive Riverside Plants*, ed. LC de Waal, LE Child, PM Wade, JH Brock, pp. 27–44. Chichester: Wiley
26. Brosofske KD, Chen J, Naiman RJ, Franklin JF. 1997. Effects of harvesting on microclimate from small streams to uplands in western Washington. *Ecol. Appl.* In press
27. Brown BT. 1992. Nesting chronology, density and habitat use of black-chinned hummingbirds along the Colorado River, Arizona. *J. Field Ornithol.* 63:393–506
28. Burley JB. 1995. International greenways: a Red River Valley case study. *Landscape Urban Plan.* 33:195–210
29. Burt TP. 1997. The hydrological role of floodplains within the drainage basin system. In *Buffer Zones: Their Processes and Potential in Water Protection, Proc.*

of the Int. Conf. on Buffer Zones, Sept. 1996, ed. NE Haycock, TP Burt, KWT Golding, G. Pinay. Harpenden, pp. 21–31. UK: Quest Environ.

30. Busch DE. 1995. Effects of fire on southwestern riparian plant community structure. Southwest. Natur. 40:259–67

31. Busch DE, Ingraham NL, Smith SD. 1992. Water uptake in woody riparian phreatophytes of the southwestern US: stable isotope study. Ecol. Appl. 2:450–59

32. Busch DE, Smith SD. 1995. Mechanisms associated with decline of woody species in riparian ecosystems of the southwestern U.S. Ecol. Monogr. 65:347–70

33. Butler DR. 1995. Zoogeomorphology. Cambridge, UK: Cambridge Univ. Press

34. Cederholm CJ, Peterson NP. 1985. The retention of coho salmon (Oncorhynchus kisutch) carcasses by organic debris in small streams. Can. J. Fish. Aquat. Sci. 42:1222–25

35. Chauvet E, Jean-Louis AM. 1988. Production de litière de la ripisylve de la Garonne et apport au fleuve. Acta Oecolog., Oecolog. Gen. 9:265–79

36. Clements FE. 1905. Research Methods in Ecology. Lincoln, UK: Univ. Publ.

37. Cole DW. 1981. Nitrogen uptake and translocation by forest ecosystems. In Terrestrial Nitrogen Cycles, ed. FE Clark, T Rosswall. 33:219–32. Stockholm: Swedish Nat. Sci. Res. Counc.

38. Cole DW, Rapp M. 1980. Elemental cycling in forest ecosystems. In Dynamic Properties of Forest Ecosystems, ed. DE Reichle, pp. 341–409. New York: Cambridge Univ. Press

39. Conners ME, Naiman RJ. 1984. Particulate allochthonous inputs: relationships with stream size in an undisturbed watershed. Can. J. Fish. Aquat. Sci. 41:1473–84

40. Cooper AB. 1990. Nitrate depletion in the riparian zone and stream channel of a small headwater catchment. Hydrobiologia 202:13–26

41. Cooper AB, Smith CM, Smith MJ. 1995. Effects of riparian set-aside on soil characteristics in an agricultural landscape: implications for nutrient transport and retention. Agric. Ecosyst. Environ. 55:61–67

42. Cooper JR, Gilliam JW, Daniels RB, Robarge WP. 1987. Riparian areas as filters for agricultural sediment. Soil Sci. Soc. Am. Proc. 51:416–20

43. Correll DL, Weller DE. 1989. Factors limiting processes in freshwater wetlands: an agricultural primary stream riparian forest. In Freshwater Wetlands and Wildlife,

ed. R. Sharitz, J. Gibbons, pp. 9–23. Oak Ridge: US Dept. Energy

44. Croonquist MJ, Brooks RP. 1993. Effects of habitat disturbance on bird communities in riparian corridors. Soil Water Conserv. 48:65–70

45. Cummins KW, Wilzbach MA, Gates DM, Perry JB, Taliaferro WB. 1989. Shredders and riparian vegetation. BioScience 39:24–30

46. Dahm CN, Cummins KW, Valett HM, Coleman RL. 1995. An ecosystem view of the restoration of the Kissimmee River. Restoration Ecol. 3:225–38

47. Daniels RB, Gilliam JW. 1997. Sediment and chemical load reduction by grass and riparian filters. Soil Sci. Soc. Am. J. 60:246–51

48. Darveau M, Beauchesne P, Bélanger L, Huot J, Larue P. 1995. Riparian forest strips as habitat for breeding birds in boreal forest. J. Wildl. Manage. 59:67–78

49. Davies PE, Nelson M. 1994. Relationships between riparian buffer widths and the effects of logging on stream habitat invertebrate community composition and fish abundance. Aust. J. Mar. Freshwater Res. 45:1289–305

50. Dawson TE, Ehleringer JR. 1991. Streamside trees that do not use stream water. Nature 350:335–37

51. Debano LF, Schmidt LJ. 1990. Potential for enhancing riparian habitats in the southwestern United States with watershed practices. For. Ecol. Manage. 33/34:385–403

52. Décamps, H. 1993. River margins and environmental change. Ecol. Appl. 3:441–45

53. Décamps H. 1996. The renewal of floodplain forests along rivers: a landscape perspective. Verh. Int. Verein. Limnol. 26:35–59

54. Décamps H, Fortuné M, Gazelle F, Pautou G. 1988. Historical influence of man on the riparian dynamics of a fluvial landscape. Landscape Ecol. 1:163–73

55. Décamps H, Joachim J, Lauga J. 1987. The importance for birds of the riparian woodlands within the alluvial corridor of the River Garonne, S.W. France. Reg. Riv. 1:301–16

56. Décamps H, Planty-Tabacchi AM, Tabacchi E. 1995. Changes in the hydrological regime and invasions by plant species along riparian systems of the Adour River, France. Regul. Riv. 11:23–33

57. Décamps H, Tabacchi E. 1994. Species richness in vegetation along river margins. In Aquatic Ecology: Scale, Pattern and Process, ed. PS Giller, AG Hildrew,

DG Rafaelli, pp. 1–20. London: Blackwell

58. DeFerrari CM, Naiman RJ. 1994. A multiscale assessment of the occurrence of exotic plants on the Olympic Peninsula, Washington. *J. Veg. Sci.* 5:247–58

59. Dewit L, Reid DM. 1992. Branch abscission in balsam poplar (*Populous balsamifera*): characterization of the phenomenon and the influence of wind. *Int. J. Plant Sci.* 153:556–64

60. Dillaha TA, Reneau RB, Mostaghimi S, Lee D. 1989. Vegetative filter strips for agricultural nonpoint source pollution control. *Trans. Am. Soc. Agric. Eng.* 32:513–19

61. Doyle AT. 1990. Use of riparian and upland habitat by small mammals. *J. Mammal.* 71:14–23

62. Dumont JF, Lamotte S, Kahn F. 1990. Wetland and upland forest ecosystems in Peruvian Amazonia: plant species diversity in the light of some geological and botanical evidence. *For. Ecol. Manage.* 33/34:125–39

63. Duncan RP. 1993. Flood disturbance and the coexistence of species in a lowland podocarp forest, south Westland, New Zealand. *J. Ecol.* 81:403–16

64. Duncan WFA, Brusven MA. 1985. Energy dynamics of three low-order southeast Alaska streams: allochthonous processes. *J. Freshwater Ecol.* 3:233–48

65. Dunlap JM, Braatne JH, Hinckley TM, Stettler RF. 1993. Intraspecific variation in photosynthetic traits of *Populus trichocarpa. Can. J. Bot.* 71:1304–11

66. Dunlap JM, Heilman PE, Stettler RF. 1994. Genetic variation and productivity of *Populus trichocarpa* and its hybrids. VII. Survival and 2-year growth of native black cottonwood clones from four river valleys in Washington. *Can. J. For. Res.* 24:1439–549

67. Dunlap JM, Heilman PE, Stettler RF. 1995. Genetic variation and productivity of *Populus trichocarpa* and its hybrids. VIII. Leaf and crown morphology of native *P. trichocarpa* clones from four valleys in Washington. *Can. J. For. Res.* 25:1710–24

68. Dunne T, Leopold LB. 1979. *Water in Environmental Planning.* San Francisco: Freeman

69. Ehrman TP, Lamberti GA. 1992. Hydraulic and particulate matter retention in a 3rd-order Indiana stream. *J. North Am. Benth. Soc.* 11:341–49

70. Emmett BA, Hudson JA, Coward PA, Reynolds B. 1994. The impact of a riparian wetland on streamwater quality in a recently afforested upland catchment. *J. Hydrol.* 162:337–53

71. Fabre A, Pinay G, Ruffinoni C. 1996. Seasonal changes in inorganic and organic phosphorus in the soil of a riparian forest. *Biogeochemistry* 35:419–32

72. Fail JL Jr, Hamzah MN, Haines BL, Todd RL. 1986. Above and below ground biomass, production, and element accumulation in riparian forests of an agricultural watershed. In *Watershed Research Perspectives,* ed. DL Correll, pp. 193–224. Washington, DC: Smithsonian Inst.

73. Fetherston KL. 1998. *Temperate montane riparian forests: process and pattern in alluvial channels.* PhD diss. Coll. For. Resources, Univ. Wash., Seattle

74. Fetherston KL, Naiman RJ, Bilby RE. 1995. Large woody debris, physical process, and riparian forest development in montane river networks of the Pacific Northwest. *Geomorphology* 13:133–44

75. Fitzpatrick TJ, Fitzpatrick MFL. 1902. A study of the island flora of the Mississippi River near Sabula, Iowa. *Plant World* 5:198–201

76. Forman RTT. 1995. *Land Mosaics.* Cambridge, UK: Cambridge Univ. Press

77. Franklin JF, Shugart HH, Harmon ME. 1987. Tree death as an ecological process. *BioScience* 37:550–56

78. Friedman JM, Osterkamp WR, Lewis WM. 1996. Channel narrowing and vegetation development following a great plains flood. *Ecology* 77:2167–81

79. Friedman JM, Osterkamp WR, Lewis WM. 1996. The role of vegetation and bed-level fluctuations in the process of channel narrowing following a catastrophic flood. *Geomorphology* 14:341–51

80. Fry J, Steiner FR, Green DM. 1994. Riparian evaluation and site assessment in Arizona. *Landscape Urban Plan.* 28:179–99

81. Galloway G, Worrall J. 1979. Cladoptosis: a reproductive strategy in black cottonwood. *Can. J. For. Res.* 9:122–25

82. Glova GJ, Sagar PM. 1994. Comparison of fish and macroinvertebrate standing stocks in relation to riparian willows (*Salix* spp.) in three NZ streams. *NZ J. Mar. Freshwater Res.* 28:255–66

83. Gold A, Kellogg DQ. 1997. Modelling internal processes of buffer zones. See Ref. 29, pp. 192–207

84. Gregory SV, Swanson FJ, McKee WA, Cummins KW. 1991. An ecosystem perspective of riparian zones. *BioScience* 41:540–51

85. Griffiths GA. 1980. Stochastic estimation of bed load yield in pool-and-riffle

mountain streams. *Water Res. Bull.* 16: 931–37

86. Grime JP. 1979. *Plant Strategies and Vegetation Processes.* New York: Wiley & Sons

87. Groffman P. 1997. Contaminant effects on microbial functions in riparian buffer zones. See Ref. 29, pp. 83–92

88. Groffman PM, Gold AJ, Simmons RC. 1992. Nitrate dynamics in riparian forests: microbial studies. *J. Environ. Qual.* 21:666–71

89. Grubbs SA, Cummins KW. 1994. Processing and macroinvertebrate colonization of black cherry (*Prunus serotina*) leaves in two streams differing in summer biota, thermal regime and riparian vegetation. *Am. Midl. Nat.* 132:284–93

90. Gurtz ME, Marzolf GR, Killingbeck KT, McAuthur JV. 1988. Hydrologic and riparian influences on the import and storage of coarse particulate organic matter in a prairie stream. *Can. J. Fish. Aquat. Sci.* 45:655–65

91. Hansen AJ, di Castri F, eds. 1992. *Landscape Boundaries.* New York: Springer-Verlag

92. Hanson GC, Groffman PM, Gold AJ. 1994. Denitrification in riparian wetlands receiving high and low groundwater nitrate inputs. *J. Environ. Qual.* 23:917–22

93. Hanson GC, Groffman PM, Gold AJ. 1994. Symptoms of nitrogen saturation in a riparian forest. *Ecol. Appl.* 4:750–56

94. Harmon ME, Franklin JF. 1989. Tree seedlings on logs in Picea-Tsuga forests of Oregon and Washington. *Ecology* 70:48–59

95. Harmon ME, Franklin JF, Swanson FJ, Sollins P, Gregory SV et al. 1986. Ecology of coarse woody debris in temperate ecosystems. *Adv. Ecol. Res.* 15:133–302

96. Harris RR. 1987. Occurrence of vegetation on geographic surfaces in the active floodplain of a California alluvial stream. *Am. Midl. Natur.* 118:393–405

97. Haycock NE, Pinay G. 1993. Groundwater nitrate dynamics in grass and poplar vegetated riparian buffers during the winter. *J. Environ. Qual.* 22:273–78

98. Haycock NE, Pinay G, Walker C. 1993. Nitrogen retention in river corridors: European perspective. *Ambio* 22:340–46

99. Hedin LO, von Fischer JC, Ostrom NE, Kennedy BP, Brown MG, Robertson GP. 1997. Thermodynamic constraints on nitrogen transformations and other biogeochemical processes at soil-stream interfaces. *Ecology.* In press

100. Hicks BJ, Beschta RL, Harr RD. 1991. Long-term changes in streamflow following logging in western Oregon and associated fisheries implications. *Water Res. Bull.* 27:217–26

101. Hill AR. 1990. Groundwater flow paths in relation to nitrogen chemistry in the near stream zone. *Hydrobiologia* 206:39–52

102. Holland MM, Risser PG, Naiman RJ, eds. 1991. *Ecotones.* New York: Chapman & Hall

103. Hosner JF, Leaf AL, Dickson R, Hart JB Jr. 1965. Effects of varying soil moisture upon the nutrient uptake of four bottomland tree species. *Soil Sci. Soc. Am. Proc.* 29:313–16

104. Howe WH, Knopf FL. 1991. On the imminent decline of Rio Grande cottonwoods in central New Mexico. *Southwest. Natur.* 36:218–24

105. Howell J, Benson D, McDougall L. 1994. Developing a strategy for rehabilitating riparian vegetation of the Hawkesbury-Nepean River, Sydney, Australia. *Pacific Cons. Biol.* 1:257–69

106. Hubbard RK, Lowrance RR. 1994. Riparian forest buffer system research at the coastal plain experiment station, Tifton, GA. *Water Air Soil Pollut.* 77:409–32

107. Hubbard RK, Sheridan JM, Marti LR. 1990. Dissolved and suspended solids transport from Coastal Plain watersheds. *J. Environ. Qual.* 19:413–20

108. Hughes FMR. 1994. Environmental change, disturbance and regeneration in semi-arid floodplain forests. In *Environmental Change In Dry Lands: Biogeographical and Geomorphological Perspectives*, ed. AC Millington, K Pye, pp. 322–345. Chichester, UK: Wiley

109. Hupp CR, Osterkamp WR. 1985. Bottomland vegetation distribution along Passage Creek, Virginia, in relation to landforms. *Ecology* 66:670–81

110. Hupp CR, Osterkamp WR, Howard AD, eds. 1995. *Biogeomorphology, Terrestrial and Freshwater Systems.* Amsterdam: Elsevier Sci.

111. Hupp CR, Simon A. 1986. Vegetation and bank-slope development. *Proc. 4th Fed. Interagency Sedimentation Conf.,* March 24–27, 1986, Las Vegas, Nevada. Vol. II:583–592

112. Hynes HBN. 1963. Imported organic matter and secondary productivity in streams. *Proc. 16th Int. Congr. Zool.,* Vol. 4, pp. 324–329

113. Ishikawa S. 1991. Floodplain vegetation of the Ibi River in central Japan. II. Vegetation dynamics of the bars in the river course of the alluvial fan. *Jpn. J. Ecol.* 41:31–43

114. Jacobs TC, Gilliam JW. 1985. Riparian losses of nitrate from agricultural drainage waters. *J. Environ. Qual.* 14:472–78

115. Jaworski NA, Groffman PM, Keller A, Prager AC. 1992. A watershed-scale analysis of nitrogen loading: the Upper Potomac River. *Estuaries* 15:83–95

116. Johansson M, Nilsson C, Nilsson E. 1996. Do rivers function as corridors for plant dispersal? *J. Veg. Sci.* 7:593–98

117. Johnson RR, Mills GS, Carothers SW. 1990. Creation and restoration of riparian habitat in southwestern arid and semi-arid regions. In *Wetland Creation and Restoration: The Status of the Science*, pp. 351–66. Washington, DC: Island

118. Johnson WC. 1992. Dams and riparian forests: case study from the Upper Missouri River. *Rivers* 3:331–46

119. Johnson WC. 1993. Divergent responses of riparian vegetation to flow regulation on the Missouri and Platte rivers. *Proc. Symp. on Restoration Planning for the Rivers of the Mississippi River Ecosystem, Rep. 19*, ed. LW Hesse, CB Stalnaker, NG Benson, JR Zuboy, pp. 426–31. Washington, DC: US Dep. Interior

120. Johnson WC. 1994. Woodland expansion in the Platte River, Nebraska: patterns and causes. *Ecol. Monogr.* 64:45–84

121. Johnston CA. 1995. Effects of animals on landscape pattern. In *Mosaic Landscapes and Ecological Processes*, ed. R Hanson, L Fahrig, G Merriam, pp. 57–80. London: Chapman & Hall

122. Johnston CA, Naiman RJ. 1990. Browse selection by beaver: effects on riparian forest composition. *Can. J. For. Res.* 20:1036–43

123. Joly CA. 1991. Flooding tolerance in tropical trees. In *Plant Life Under Oxygen Deprivation*, ed. MB Jackson, DD Davies, H Lambers, pp. 23–34. The Hague: SPB Academic

124. Jones CG, Lawton JH, eds. 1995. *Linking Species to Ecosystems*. New York: Chapman & Hall

125. Jordan TE, Correll DL, Weller DE. 1993. Nutrient interception by a riparian forest receiving inputs from adjacent cropland. *J. Environ. Qual.* 22:467–73

126. Junk WJ, Bayley PB, Sparks RE. 1989. The flood pulse concept in river-floodplain systems. *Proc. Int. Large River Symp.*, Honey Harbor, Ontario, Canada, 1986, ed. DP Lodge. *Can. Spec. Publ. Fish. Aquat. Sci.* 106:110–27. Ottawa: Dep. Fish. Oceans

127. Kalliola R, Puhakka M. 1988. River dynamics and vegetation mosaicism: a case study of the River Kamajohka, northernmost Finland. *J. Biogeogr.* 15:703–19

128. Karr JR, Schlosser IJ. 1978. Water resources and the land-water interface. *Science* 201:229–34

129. Keddy, PA. 1989. *Competition, Population and Community Biology*, Ser. 6. London/New York: Chapman & Hall

130. Keller CME, Robbins CS, Hatfield JS. 1993. Avian communities in riparian forests of different widths in Maryland and Delaware. *Wetlands* 13:137–44

131. Keller EA, Kondolf GM, Hagerty DJ. 1990. Groundwater and fluvial processes; selected observations. In *Groundwater Geomorphology: The Role of Subsurface Processes in Earth-Surface Processes and Landforms*, pp. 319–40, Boulder: Geol. Soc. Am.

132. Kellman M, Tackaberry R. 1993. Disturbance and tree species coexistence in tropical riparian forest fragments. *Global Ecol. Biogeogr. Lett.* 3:1–9

133. Klimas CV, Martin CO, Teaford JW. 1981. *Impacts of flooding regime modification on wildlife habitats of bottomland hardwood forests in the lower Mississippi Valley. Tech. Rep. El-81–13.* US Army Engineers Water-ways Exp. Station, Vicksburg, Miss.

134. Knopf FL. 1986. Changing landscapes and the cosmopolitism of the eastern Colorado avifauna. *Wildl. Soc. Bull.* 14:132–42

135. Knopf FL, Samson FB. 1994. Scale perspectives on avian diversity in western riparian ecosystems. *Conserv. Biol.* 8:669–76

136. Kovalchik BL, Chitwood LA. 1990. Use of geomorphology in the classification of riparian plant associations in mountainous landscapes of central Oregon, USA. *For. Ecol. Manage.* 33–34:405–18

137. Langlade LR, Décamps O. 1994. Plant colonization on river gravel bars: the effect of litter accumulation. *CR Acad. Sci. Sér. III* 317:899–905

138. Langlade LR, Décamps O. 1995. Accumulation de limon et colonisation végétale d'un banc de galets. *CR Acad. Sci. Sér. III* 318:1073–82

139. Larue P, Bélanger L, Huot J. 1994. La fréquentation des peuplements riverains par le cerf de Virginie en hiver: sélection de site ou pure coincidence? *Ecoscience* 1:223–30

140. Lester PJ, Mitchell SF, Scott D. 1994. Effects of riparian willow trees (*Salix fragilis*) on macroinvertebrate densities in two small central Otago, NZ, streams. *NZ J. Mar. Freshwater Res.* 28:267–76

141. Lock PA, Naiman RJ. 1997. Effects of stream size on bird community structure in coastal temperate forests of the Pacific Northwest. *J. Biogeogr.* In press

142. Lodge DM. 1993. Biological invasions: lessons for ecology. *Trends Evol. Ecol.* 8:133–37

143. Lonsdale WM. 1993. Rates of spread of an invading species-*Mimosa pigra* in northern Australia. *J. Ecol.* 81:513–21

144. Lowrance R. 1992. Groundwater nitrate and denitrification in a coastal plain riparian soil. *J. Environ. Qual.* 21:401–05

145. Lowrance R, Altier LS, Newbold JD, Schnabel RR, Groffman PM, et al. 1995. Water quality functions of riparian forest buffer systems in the Chesapeake Bay Watershed. *Rep. Nutrient Subcommittee Chesapeake Bay Program.* Annapolis, MD: US Environ. Protect. Agency

146. Lowrance R, Sharpe JK, Sheridan JM. 1986. Long-term sediment deposition in the riparian zone of a coastal plain watershed. *J. Soil Water Conserv.* 41:266–71

147. Lowrance R, Todd RL, Asmussen LE. 1983. Waterborne nutrient budgets for the riparian zone of an agricultural watershed. *Agric. Ecosys. Environ.* 10:371–84

148. Lowrance RR, Todd RL, Fail J, Hendrickson O Jr, Leonard R, Asmussen L. 1984. Riparian forests as nutrient filters in agricultural watersheds. *BioScience* 34:374–77

149. Lowrance RR, Vellidis G. 1995. A conceptual model for assessing ecological risk to water quality function of bottomland hardwood forests. *Environ. Manage.* 19:239–58

150. Maamri A, Chergui H, Pattee E. 1994. Allochthonous input of coarse particulate organic matter to a Moroccan mountain stream. *Acta Oecologia* 15:495–508

151. Macdonald A, Keller EA. 1987. Stream channel response to the removal of large woody debris, Larry Damm Creek, northwestern California. In *Erosion and Sedimentation in the Pacific Rim,* pp. 405–406, *IAHS Pub. No. 165.* Washington, DC: Int. Assoc. Hydrolog. Sci.

152. MacDougall A, Kellman M. 1992. The understory light regime and patterns of tree seedlings in tropical riparian forest patches. *J. Biogeogr.* 19:667–75

153. Machtans CS, Villard MC, Hannon SJ. 1996. Use of riparian buffer strips as movement corridors by forest birds. *Conserv. Biol.* 10:1366–79

154. Magette WL, Brinsfield RB, Palmer RE, Wood JD. 1989. Nutrient and sediment removal by vegetated filter strips. *Trans. Am. Soc. Agric. Eng.* 32:663–67

155. Malanson GP. 1993. *Riparian Landscapes.* Cambridge, UK: Cambridge Univ. Press

156. Maridet L, Wasson JG, Philippe M, Amoros C. 1995. Benthic organic matter dynamics in three streams: riparian vegetation or bed morphology control? *Arch. Hydrobiol.* 132:415–25

157. Maser C, Sedell JR. 1994. *From the Forest to the Sea.* Delray Beach, FL: St. Lucie

158. McClain ME, Richey JE. 1996. Regional-scale linkages of terrestrial and lotic ecosystems in the Amazon basin: a conceptual model for organic matter. *Arch. Hydrobiol. Suppl.* 113(1/4):111–25

159. McClain ME, Richey JE, Pimentel TP. 1994. Groundwater nitrogen dynamics at the terrestrial-lotic interface of a small catchment in the Central Amazon Basin. *Biogeochemistry* 27:113–27

160. McComb WC, Chambers CL, Newton M. 1993. Small mammal and amphibian communities and habitat associations in red alder stands, Central Oregon coast range. *Northwest Sci.* 67:181–88

161. McComb WC, McGarigal K, Anthony RG. 1993. Small mammal and amphibian abundance in streamside and upslope habitats of mature Douglas-fir stands, western Oregon. *Northwest Sci.* 67:7–15

162. McDowell WH, Bowden WB, Asbury CE. 1992. Riparian nitrogen dynamics in two geomorphologically distinct tropical rain forest watersheds: subsurface solute patterns. *Biogeochemistry* 18:53–75

163. McInnes PF, Naiman RJ, Pastor J, Cohen Y. 1992. Effects of moose browsing on vegetation and litter of the boreal forest Isle Royale, Michigan, USA. *Ecology* 73:2059–75

164. McKay S. 1996. *The impact of river regulation on establishment processes of riparian black cottonwood.* Masters thesis. Coll. For. Resourc., Univ. Wash., Seattle

165. McKee A, Laroi G, Franklin JF. 1982. Structure, composition, and reproductive behavior of terrace forests, South Fork Hoh River, Olympic National Park. In *Ecological Research in National Parks of the Pacific Northwest,* ed. EE Starkey, JF Franklin, JW Matthews, pp. 22–29. Corvallis: Oregon State Univ. For. Res. Lab.

166. Meave J, Kellman M. 1994. Maintenance of rain forest diversity in riparian forests of tropical savannas: implications for species conservation during Pleistocene drought. *J. Biogeogr.* 21:121–35

167. Meehan WR, ed. 1991. *Influences of Forest and Rangeland Management on*

Salmonid Fishes and Their Habitats. Special Publ. 19. Bethesda, MD: Am. Fish. Soc.

168. Mertes LAK, Daniel DL, Melack JM, Nelson B, Martinelli LA, Forsberg BR. 1995. Spatial patterns of hydrology, geomorphology, and vegetation on the floodplain of the Amazon River in Brazil from a remote sensing perspective. *Geomorphology* 13:215–32

169. Metzger JP. 1995. *Structure du paysage et diversité des peuplements ligneux fragmentés du rio Jacaré-Pepira (Sud-Est du Brésil).* Thesis. Univ. Toulouse, Toulouse, France. 273 pp.

170. Miller JR, Schulz TT, Hobbs NT, Wilson KR, Schrupp DL, Baker WL. 1995. Changes in the landscape structure of a southeastern Wyoming riparian zone following shifts in stream dynamics. *Biol. Cons.* 72:371–79

171. Mitsch WJ, Gosselink JG. 1993. *Wetlands.* New York: Van Nostrand Reinhold

172. Molles MC Jr, Crawford CS, Ellis LM. 1995. Effects of an experimental flood on litter dynamics in the middle Rio Grande riparian ecosystem. *Regul. Riv.* 11:275–81

173. Montgomery DR, Buffington JM, Pess G. 1995. Pool spacing in forest channels. *Water Res. Res.* 31:1097–105

174. Mulholland PJ. 1992. Regulation of nutrient concentrations in a temperate forest stream: role of upland, riparian and instream processes. *Limnol. Oceanogr.* 37:1512–26

175. Muller E. 1995. Phénologie forestière révélée par l'analyse d'images Thematic Mapper. *C. R. Acad. Sci. Paris, Sér. III* 318:993–1003

176. Murray NL, Stauffer DF. 1995. Nongame bird use of habitat in central Appalachian riparian forests. *J. Wildl. Manage.* 59:78–88

177. Myers N. 1995. Environmental unknowns. *Science* 269:358–60

178. Myers TJ, Swanson S. 1992. Variation in stream stability with stream type and livestock bank damage in northern Nevada. *Water Res. Bull.* 28:743–54

179. Naiman RJ. 1988. Animal influences on ecosystem dynamics. *BioScience* 38:750–52

180. Naiman RJ, Beechie TJ, Benda LE, Berg DR, Bisson PA, et al. 1992. Fundamental elements of ecologically healthy watersheds in the Pacific Northwest coastal ecoregion. In *Watershed Management,* ed. RJ Naiman, pp. 127–188. New York: Springer-Verlag

181. Naiman RJ, Bilby RE, eds. 1997. *River Ecology and Management: Lessons from the Pacific Coastal Ecoregion.* New York: Springer-Verlag

182. Naiman RJ, Décamps H, eds. 1990. *The Ecology and Management of Aquatic-Terrestrial Ecotones.* Paris: UNESCO, Park Ridge: Parthenon

183. Naiman RJ, Décamps H, Pastor J, Johnston CA. 1988. The potential importance of boundaries to fluvial ecosystems. *J. North Am. Benth. Soc.* 7:289–306

184. Naiman RJ, Décamps H, Pollock M. 1993. The role of riparian corridors in maintaining regional biodiversity. *Ecol. Appl.* 3:209–12

185. Naiman RJ, Fetherston KL, McKay S, Chen J. 1997. Riparian forests. In *River Ecology and Management: Lessons from the Pacific Coastal Region,* ed. RJ Naiman, RE Bilby. New York: Springer-Verlag. In press

186. Naiman RJ, Johnston CA, Kelley JC. 1988b. Alteration of North American streams by beaver. *BioScience* 38:753–62

187. Naiman RJ, Magnuson JJ, McKnight DM, Stanford JA, eds. 1995. *The Freshwater Imperative. A Research Agenda.* Washington DC: Island. 165 pp.

188. Naiman RJ, Melillo JM, Hobbie JE. 1986. Ecosystem alteration of boreal forest streams by beaver. *Ecology* 67:1254–69

189. Naiman RJ, Pinay G, Johnston CA, Pastor J. 1994. Beaver influences on the long-term biogeochemical characteristics of boreal forest drainage networks. *Ecology* 75:905–21

190. Naiman RJ, Rogers KH. 1997. Large animals and the maintenance of system-level characteristics in river corridors. *BioScience.* In press

191. Nakamura F, Swanson FJ. 1993. Effects of coarse woody debris on morphology and sediment storage of a mountain stream system in western Oregon. *Earth Surf. Proc. Landforms* 18:43–61

192. Nanson GC, Barbetti M, Taylor G. 1995. River stabilisation due to changing climate and vegetation during the Late Quaternary in western Tasmania, Australia. *Geomorphology* 13(1/4):145–58

193. Neal C, Lock MA, Fiebig DM. 1990. Soil water in the riparian zone as a source of carbon for a headwater stream. *J. Hydrol.* 116:217–37

194. Niiyama K. 1990. The role of seed dispersal and seedling traits in colonization and coexistence of *Salix* species in a seasonally flooded habitat. *Ecol. Res.* 5:317–31

195. Nilsson C, Backe S, Carlberg B. 1994.

A comparison of species richness and traits of riparian plants between a main river channel and its tributaries. *J. Ecology* 82:281–95

196. Nilsson C, Ekblad A, Gardfjell M, Carlberg B. 1991. Long-term effects of river regulation on river margin vegetation. *J. Appl. Ecol.* 28:963–87

197. Nilsson C, Grelsson G. 1990. The effects of litter displacement on riverbank vegetation. *Can. J. Bot.* 68:735–41

198. Nilsson C, Grelsson G, Dynesius M, Johansson ME, Sperens U. 1991. Small rivers behave like large rivers: effects of postglacial history on plant species richness along riverbanks. *J. Biogeogr.* 18:533–41

199. Nilsson C, Grelsson G, Johansson M, Sperens U. 1989. Patterns of plant species richness along riverbanks. *Ecology* 70:77–84

200. Nilsson C, Jansson R. 1995. Floristic differences between riparian corridors of regulated and free-flowing boreal rivers. *Regul. Riv.* 11:55–66

201. Nilsson C, Nilsson E, Johansson ME, Dynesius M, Grelsson G, et al. 1993. Processes structuring riparian vegetation. *Curr. Top. Bot. Res.* 1:419–31

202. O'Connor NA. 1992. Quantification of submerged wood in a lowland Australian stream system. *Freshwater Biol.* 27:387–95

203. O'Neill GJ, Gordon AM. 1994. The nitrogen filtering capability of Carolina poplar in an artificial riparian zone. *J. Environ. Qual.* 23:1218–23

204. Ormerod SJ, Rundle SD, Lloyd EC, Douglas AA. 1993. The influence of riparian management of the habitat structure and macroinvertebrate communities of upland streams draining plantation forests. *J. Appl. Ecol.* 30:13–24

205. Pastor J, Dewey B, Naiman RJ, McInnes PF, Cohen Y. 1993. Moose browsing and soil fertility in the boreal forests of Isle Royale National Park. *Ecology* 74:467–80

206. Pastor J, Naiman RJ, Dewey B, McInnis P. 1988. Moose, microbes, and the boreal forest. *BioScience* 38:770–77

207. Pautou G, Girel J, Borel JL. 1992. Initial repercussions and hydroelectric developments in the French Upper Rhone Valley: a lesson for predictive scenarios propositions. *Environ. Manage.* 16:231–42

208. Peterjohn WT, Correll DL. 1984. Nutrient dynamics in an agricultural watershed: observations on the role of a riparian forest. *Ecology* 65:1466–75

209. Peterken GF, Hughes FMR. 1995. Restoration of floodplain forests in Britain. *Forestry* 68:187–202

210. Petersen RC Jr. 1992. The RCE: a riparian, channel, and environmental inventory for small streams in the agricultural landscape. *Freshwater Biol.* 27:295–306

211. Petts GE, Moller H, Roux AL, eds. 1989. *Historical Change of Large Alluvial Rivers: Western Europe.* Chichester: Wiley. 335 pp.

212. Pinay G, Décamps H. 1988. The role of riparian woods in regulating nitrogen fluxes between the alluvial aquifer and surface water: a conceptual model. *Regul. Riv.* 2:507–16

213. Deleted in proof

214. Pinay G, Roques L, Fabre A. 1993. Spatial and temporal patterns of denitrification in a riparian forest. *J. Appl. Ecol.* 30:581–91

215. Pinay G, Ruffinoni C, Fabre A. 1995. Nitrogen cycling in two riparian forest soils under different geomorphic conditions. *Biogeochemistry* 30:9–29

216. Planty-Tabacchi AM. 1993. *Invasions des corridors riverains fluviaux par des espèces végétales d'origine étrangère.* Thesis. Univ. Toulouse, Toulouse, France. 177 pp.

217. Planty-Tabacchi AM, Tabacchi E, Naiman RJ, DeFerrari C, Décamps H. 1996. Invasibility of species-rich communities in riparian zones. *Cons. Biol.* 10:598–607

218. Pollock MM. 1997. Biodiversity. See Ref. 185

219. Pollock MM, Naiman RJ, Erickson HE, Johnston CA, Pastor J, Pinay G. 1994. Beaver as engineers: influences on biotic and abiotic characteristics of drainage basins. In *Linking Species to Ecosystems,* ed. CG Jones, JH Lawton, pp. 117–26. New York: Chapman & Hall

220. Pollock MM, Naiman RJ, Hanley TA. 1997. An empirically based model for predicting plant diversity in forested and emergent wetlands. *Ecology.* In press

221. Pozuelo-Gonzalez JM, Gutierrez-Manero FJ, Probanza A, Acero N, Bermudez-de-Castro F. 1995. Effect of alder (*Alnus glutinosa* L. Gaertn.) roots on distribution of proteolytic, ammonifying and nitrifying bacteria in soil. *Geomicrobiol. J.* 13:129–38

222. Pysek P, Prach K. 1993. Plant invasions and the role of riparian habitats: a comparison of four species alien to central Europe. *J. Biogeogr.* 20:413–20.

223. Pysek P, Prach K. 1994. How important are rivers for supporting plant invasions. See Ref. 25, pp. 19–26

224. Pysek P, Prach K. 1995. Invasion dynamics of *Impatiens glandulifera*—a century of spreading reconstructed. *Biol. Conserv.* 74:41–48

225. Raedeke KJ, ed. 1988. *Streamside Management: Riparian Wildlife and Forestry Interactions. Contribution No. 59.* Seattle, WA: Inst. For. Resourc., Univ. Wash.

226. Räsänen ME, Salo JS, Kalliola R. 1987. Fluvial perturbance in the western Amazon basin: regulation by long-term sub-Andean tectonics. *Science* 238:1398–401

227. Richards C, Host G. 1994. Examining land use influences on stream habitats and macroinvertebrates: a GIS approach. *Water Resour. Bull.* 30:729–38

228. Richards K. 1982. *Rivers: Form and Process in Alluvial Channels.* London: Methuen

229. Rickard J. 1993. Warthog (*Phacochoerus aethiopicus*, Pallas) foraging patterns in stands of wild rice (*Oryza longistaminata*, A. Chev and Roehr) on the Nyl River floodplain. Masters Thesis. Univ. Witwatersrand, Johannesburg, S. Africa

230. Risser PG. 1990. The ecological importance of land-water ecotones, See Ref. 182, pp. 7–21

231. Risser PG, ed. 1993. Ecotones. *Ecol. Appl.* 3:369–445

232. Rood SB, Heinz-Milne S. 1989. Abrupt riparian forest decline following river damming in southern Alberta. *Can. J. Bot.* 67:1744–49

233. Rood SB, Mahoney JM. 1990. Collapse of riparian poplar forests downstream from dams in western prairies: probable causes and prospects for investigation. *Environ. Manage.* 14:451–64

234. Rood SB, Mahoney JM, Reid DE, Zilm L. 1995. Instream flows and the decline of riparian cottonwoods along the St. Mary River, Alberta. *Can. J. Bot.* 73:1250–60

235. Rosgen, DL. 1994. A classification of natural rivers. *Catena* 22:169–99

236. Rot BW, Naiman RJ, Bilby RE. 1997. Riparian succession, landform, and large woody debris in mature to old forests of the Pacific Northwest. *J. Biogeogr.* Submitted

237. Rowntree K. 1991. An assessment of the potential impact of alien invasive vegetation on the geomorphology of river channels in South Africa. *Southern Afr. J. Aquat. Sci.* 17:28–43

238. Salo EO, Cundy TW, eds. 1987. *Streamside Management: Forestry and Fishery Interactions. Contribution No. 57.* Seattle, WA: Inst. For. Resources, Univ. Wash.

239. Salo J. 1990. External processes influencing origin and maintenance of inland water-land ecotones. See Ref. 182, pp. 37–64

240. Salo J, Kalliola R, Häkkinen I, Mäkinen Y, Niemelä P, Puhakka M, Coley PD. 1986. River dynamics and the diversity of Amazon lowland forest. *Nature* 322:254–58

241. Sauer JD. 1988. *Plant Migration.* Berkeley: Univ. Calif. Press

242. Scatena PN. 1990. Selection of riparian buffer zones in humid tropical steeplands. In *Research Needs and Applications to Reduce Erosion and Sedimentation in Tropical Steeplands, IAHS Publ. No. 192*, pp. 328–337. Wallingford, CT: Int. Assoc. Hydrologic. Sci.

243. Schmidt LJ, Debano LF. 1990. Potential for enhancing riparian habitats in the south-western United States with watershed practices. *For. Ecol. Manage.* 33:385–403

244. Schneider RL, Sharitz RR. 1988. Hydrochory and regeneration in a bald cypress-water tupelo swamp forest. *Ecology* 69:1055–63

245. Schultz RC, Colletti JP, Isenhart TM, Simpkins WW, Mizc CW, Thompson ML. 1995. Design and placement of a multi-species riparian buffer strip system. *Agrofor. Syst.* 29:201–26

246. Sedell JR, Froggatt JL. 1984. Importance of streamside forests to large rivers: the isolation of the Willamette River, Oregon, USA, from its floodplain by snagging and streamside forest removal. *Verh. Int. Verein. Limnol.* 22:1824–34

247. Segelquist CA, Scott ML, Auble GT. 1993. Establishment of *Populus deltoides* under simulated alluvial groundwater declines. *Am. Midl. Nat.* 130:274–85

248. Sigafoos R. 1964. *Botanical evidence of floods and floodplain deposition. US Geol. Surv. Profess. Pap. 485A*

249. Snyder WD, Miller GC. 1991. Changes in plains cottonwoods along the Arkansas and South Platte rivers, eastern Colorado. *Prairie Nat.* 23:165–76

250. Sollers SC. 1973. Substrata conditions, community structure, and succession in a portion of the floodplain of Wissahickon Creek. *Bartonia* 42:24–42

251. Steel AE. 1993. *Woody debris piles: habitat for birds and small mammals in the riparian zone.* Masters thesis. Coll. For. Resources, Univ. Wash., Seattle.

252. Stone EC, Vasey RB. 1968. Preservation of coastal redwood on alluvial flats. *Science* 159:157–61

253. Stromberg JC, Patten DT, Richter BD. 1991. Flood flows and dynamics of Sonoran riparian forests. *Rivers* 2:221–35.

254. Swank WT, Wade JB, Crossley DA, Todd RL. 1981. Insect defoliation enhances nitrate export from forest ecosystems. *Oecologia* 51:297–99

255. Sweeney BW. 1993. Effects of streamside vegetation on macroinvertebrate communities of White Clay Creek in eastern North America. *Proc. Acad. Nat. Sci. Philadelphia* 144:291–340

256. Szaro RC. 1990. Southwestern riparian plant communities: site characteristics, tree species distribution, and size-class structures. *For. Ecol. Manage.* 33/3:315–34

257. Tabacchi E. 1995. Structural variability and invasions of pioneer plants community in riparian habitats of the middle Adour River. *Can. J. Bot.* 73:33–44

258. Tabacchi E, Planty-Tabacchi AM, Décamps O. 1990. Continuity and discontinuity of the riparian vegetation along a fluvial corridor. *Landscape Ecol.* 5:9–20

259. Tabacchi E, Planty-Tabacchi AM, Salinas MJ, Décamps H. 1996. Landscape structure and diversity in riparian plant communities: a longitudinal comparative study. *Regul. Riv.* 12:367–90

260. Taylor JR, Cardamone MA, Mitsch WJ. 1990. Bottomland hardwood wetlands: their function and values. In *Ecological Processes and Cumulative Impacts Illustrated by Bottomland Hardwood Wetland Ecosystems,* ed. JG Gosselink, LC Lee, TA Muir, pp. 13–86. Chelsea, MA: Lewis

261. Thienemann A. 1912. Der Bergbach des Sauerlandes. *Int. Rev. ges. Hydrobiol. Suppl.* 4:1–125

262. Thorburn PJ, Mensforth LJ, Walker GR. 1994. Reliance of creek-side river red gums on creek water. *Aust. J. Mar. Freshwater Res.* 45:1439–43

263. Torossian C, Roques L. 1989. Cycle biologique et écologique de Hyponomeuta rorellus Hubn. dans les Saulaies de la région toulousaine. *Acta Oecol., Oecol. Appl.* 10:47–63

264. Triska FJ, Duff JH, Avanzino RJ. 1993a. The role of water exchange between a stream channel and its hyporheic zone in nitrogen cycling at the terrestrial aquatic interface. *Hydrobiologia* 251:167–84

265. Triska FJ, Duff JH, Avanzino RJ. 1993b. Patterns of hydrological exchange and nutrient transformation in the hyporheic zone of a gravel bottom stream: examining terrestrial-aquatic linkages. *Freshwater Biol.* 29:259–74

266. Triska FJ, Jackman AP, Duff JH, Avanzino RJ. 1994. Ammonium sorption to channel and riparian sediments: a transient storage pool for dissolved inorganic nitrogen. *Biogeochemistry* 26:67–83

267. Tyree MT, Kolb KJ, Rood SB, Patino S. 1994. Vulnerability to drought-induced cavitation of riparian cottonwoods in Alberta: a possible factor in the decline of the ecosystem? *Tree Physiol.* 14:455–66

268. Vanek V. 1991. Riparian zone as a source of phosphorus for a groundwater-dominated lake. *Water Res.* 25:409–18

269. Vitousek PM. 1990. Biological invasions and ecosystem process: towards an integration of population biology and ecosystem studies. *Oikos* 57:7–13

270. Vitousek PM, Reiners WA. 1975. Ecosystem succession and nutrient retention: a hypothesis. *BioScience* 25:376–81

271. Ward JV. 1989. Riverine-wetland interactions. In *Freshwater Wetlands and Wildlife,* ed. RR Shartiz, JW Gibbons, pp. 385–400. Oak Ridge, TN: US Dept. Energy

272. Weigelhofer G, Waringer JA. 1994. Allochthonous input of coarse particulate organic matter (CPOM) in a first to fourth order Austrian forest stream. *Int. Rev. Ges. Hydrobiol.* 79:461–71

273. Welcomme RL. 1979. *Fisheries Ecology of Floodplain Rivers.* New York: Longman

274. Xiang WN. 1993. Application of a GIS-based stream buffer generation model to environmental policy evaluation. *Environ. Manage.* 17:817–27

Annu. Rev. Ecol. Syst. 1997. 28:659–87

ALLOMETRY FOR SEXUAL SIZE DIMORPHISM: Pattern and Process in the Coevolution of Body Size in Males and Females

D. J. Fairbairn

Department of Biology, Concordia University, 1455 de Maisonneuve Blvd., West, Montreal, Quebec, Canada H3G 1M8; e-mail: fairbrn@vax2.concordia.ca

KEY WORDS: sexual dimorphism, allometric constraint, sexual selection, correlational selection, genetic correlation

ABSTRACT

Sexual size dimorphism (SSD) is common in both plants and animals, and current evidence suggests that it reflects the adaptation of males and females to their different reproductive roles. When species are compared within a clade, SSD is frequently found to vary with body size. This allometry is detected as $\beta \neq 1$, where β is the slope of a model II regression of log(male size) on log(female size). Most frequently, β exceeds 1, indicating that SSD increases with size where males are the larger sex, but decreases with size where females are larger, a trend formalized as "Rensch's rule." Exceptions are uncommon and associated with female-biased SSD. These trends are derived from a sample of 40 independent clades of terrestrial animals, primarily vertebrates. Their extension to plants and aquatic animals awaits quantitative assessments of allometry for SSD within these groups. Many functional hypotheses have been proposed to explain the evolution of allometry for SSD, most featuring sexual selection on males or reproductive selection on females. Of these, the hypothesis that allometry evolves because of correlational selection between the sexes appears most promising as a general model but remains untested.

INTRODUCTION

Sexual dimorphism, defined as morphological differentiation of sexually mature males and females, is a conspicuous feature of dioecious plants and gonochoristic

659

0066-4162/97/1120-0659$08.00

animals (6, 29, 49, 50, 88, 100, 107, 140). Differentiation of reproductive roles leads directly to morphological differentiation of organs and structures associated with the production and release of gametes and, if fertilization is internal, with the formation and early development of the zygotes. Differences between the sexes in parental investment and in the relationship between mating frequency and reproductive success (sexual selection) also lead to dimorphism (6–8, 25, 88, 96, 100, 127, 150, 164, 165, 167), and such selection is frequently associated with hyperallometric growth of organs and structures associated with combat or display in males (22, 34, 53, 56, 77, 84, 87, 114, 115, 126, 134, 153, 165, 166). Although sexual selection and selection associated with differences in reproductive roles are often treated as independent explanations for the evolution of sexual dimorphism (29, 67, 120, 143), they can be viewed as components of a continuum determined by anisogamy, separation of the sexes, and the resulting specialization of reproductive roles (6, 24, 50, 88, 107, 124, 125, 127, 151, 164, 165). Hypotheses invoking selective processes along this continuum have the common property that the pattern (relative dimorphism of different traits), direction (which sex is larger), and magnitude of sexual dimorphism can be predicted from components of reproductive success (16, 24, 27, 31, 33, 36, 43, 47, 49, 50, 60, 66, 70, 71, 88, 92, 94, 98, 116–118, 125, 134, 135, 139, 141, 152, 159, 160, 161, 164, 166).

An alternative hypothesis, intersexual niche divergence or ecological sexual dimorphism, proposes that sexual dimorphism evolves to reduce intraspecific competition for food and is not associated directly with selection on reproductive traits (6, 67, 120, 134, 136, 143). Empirical tests of this hypothesis have produced both positive (89, 106, 134, 136, 141, 154) and negative conclusions (16, 27, 33, 46, 65, 75, 94, 103, 104, 118, 129, 137). Even where the results are positive, it is difficult to exclude the hypothesis that trophic dimorphism evolved as a consequence of pre-existing sexual dimorphism (6, 46, 120, 127, 136, 137). Further, although intersexual competition for food may lead to trophic dimorphism, it cannot explain the prevalence of dimorphism in reproductive and sexually-selected traits. Given the extensive evidence of the importance of reproductive and sexual selection, it seems unlikely that intersexual niche divergence plays more than a subsidiary role in the evolution and maintenance of common patterns of sexual dimorphism (6, 16, 46, 75, 88, 103, 118, 127, 136, 137, 164).

In many species, the selective processes producing sexual dimorphism result in dimorphism for overall body size (sexual size dimorphism or SSD). Male-biased size dimorphisms (i.e. males larger than females) tend to predominate in polycarpic seed-bearing plants (Gymnospermae and Angiospermae), while females tend to be the larger sex in other dioecious plants (49, 50, 57, 86, 88, 100,

101, 164, 165). In animals, female-biased SSD predominates among inverte-brates (5, 14, 50, 66, 72, 129, 142, 149, 152, 158, 161) and poikilothermic verte-brates (50, 51, 57, 135, 138, 139, 145). In these taxa, extreme female-biased SSD is not uncommon, with females of some species weighing several hun-dred times as much as their mates (6, 50, 52, 57, 59, 152). Male-biased SSD predominates only among birds and mammals, reaching its greatest extreme in the mammalian orders Primates, Pinnipedia, Proboscidea, and Artiodactyla, where males of larger species may weigh two to eight times as much as females (4, 13, 18, 57, 58, 112, 121, 134, 155).

Sexual size dimorphism varies within as well as among taxa, and it is not un-usual for SSD to range from female-biased to male-biased among species within a single family (Table 1 and 112, 124, 125, 138) or even a single genus (Table 1 and 7, 35, 138). These patterns of variation provide the necessary data for hy-pothesis testing, and quantitative comparative analyses have been widely used to test hypotheses concerning the evolution and adaptive significance of SSD. Such studies have provided much of the evidence for the importance of selec-tion on reproductive traits, particularly traits associated with sexual selection in males (4, 5, 13, 27, 32, 33, 38, 44, 45, 69, 75, 107, 109, 112, 135, 138, 139, 141, 151, 155, 158), fecundity selection in females (16, 18, 45, 66, 142, 151, 161), and patterns of parental investment (68, 85, 104, 108, 138, 141).

Comparative analyses of SSD also frequently reveal a strong statistical as-sociation between SSD and body size (Table 1 and 35, 107, 112, 120, 124, 125, 135, 146). In many studies, the effect of body size (allometry) is statisti-cally removed before hypotheses for the adaptive divergence of SSD are tested (13, 21, 26, 35, 84, 109, 137), the allometry itself being considered a functional constraint (cf 19, 42, 53, 63, 73, 147, 148). Other authors seek to understand the evolutionary processes responsible for the underlying allometry. Although many hypotheses have been proposed (26, 27, 37, 48, 57, 82, 96, 135, 157), none has proven sufficiently robust to emerge as a viable general hypothesis (6, 124, 125, 155). Thus, the functional and adaptive significance of allometry for SSD remains obscure.

In the following sections, I consolidate empirical and theoretical studies of allometry for SSD with the aim of discovering a general, functional hypothe-sis. The first section formally defines allometry and considers the problems of measuring allometry for SSD. The second section reviews the available quan-titative evidence of allometry for SSD and assesses its prevalence and pattern. The third section reviews and evaluates functional hypotheses for the evolution of allometry for SSD with reference to both the observed patterns of variation in allometry and knowledge of the adaptive significance of SSD, and the final section considers methods of testing these hypotheses.

Table 1 Quantitative evidence of interspecific allometry for sexual size dimorphism

Clade or taxon	N[a]	SSD[b] Mean	SSD[b] Range	RR[c]	β^d	r^e	Reference
MAMMALIA							
Primates	36	−0.32	−1.04 −0.08	++	1.15[f]	0.987[f]	2
Platyrrhini	52	−0.19	−0.76–0.17	++	1.05	0.992	44
Carnivora (excl. Mustelidae)	17	−0.25	−1.18–0.14	0	1.01[f]	0.991[f]	2
Mustelidae	26	−0.44	−1.33–0.03	0	1.05[f]	0.991[f]	2
Artiodactyla & Perisodactyla	27	−0.34	−0.77–0.00	0	1.02[f]	0.992[f]	2
Sciuridae (*Tamias*)	21	0.03	−0.01–0.06	0	1.07	0.977	78
AVES							
Passeriformes	65	−0.03	−0.21–0.03	++	1.13	0.990	13
Icterinae	35	−0.07	−0.23–0.03	++	1.20	0.964	155
Pipridae	30	−0.02	−0.11–0.06	++	1.14	0.983	112
Paradisaidae	42		−0.24–0.04	++			112
Pelicaniformes, Sphenisciformes	58	−0.08	−0.53–0.32	++	1.14	0.950	38
Procellariformes, Pelicaniformes	40	−0.06	−0.32–0.40	++	1.08[f]	0.985[f]	2
Galliformes	27	−0.24	−1.28–0.14	++	1.34[f]	0.961[f]	2
Trocholoformes	14	−0.03	−0.24–0.14	++	1.16[f]	0.950[f]	2
Falconiformes	22	0.36	0.04–1.17	0	1.00[f]	0.951[f]	2
Strigiformes	25	0.21	0.01–0.65	−−	0.91[f]	0.962[f]	2
Anseriformes	28	−0.13	−0.25–0.04	0	1.02[f]	0.985[f]	2
Charadriiformes[g]	65	0.04	−0.50–0.42	0	1.00[f]	0.953[f]	2
Haematopodidae	8	0.02	0–0.03	0	1.00	0.978	75
Charadriidae	25	0.01	−0.02–0.03	0	0.96	0.998	75
Scolopacidae	35	0.11	−0.50–0.32	0	1.01[f]	0.913[f]	2
Tringini	15	0.03	−0.01–0.06	0	1.02	0.993	75
Numeniini	9	0.03	−0.01–0.05	−	0.96	0.999	75
Calidridini	24	0.00	−0.21–0.05	+	1.10	0.972	75
Gallinagonini	12	0.01	−0.04–0.02	0	0.95	0.995	75
REPTILIA							
Squamata							
Elapidae							
Division A	19	0.16	−0.25–0.53	0	0.91[f]	0.762[f]	2
Division B	31	0.02	−0.19–0.32	++	1.21[f]	0.854[f]	2
Division C	16	−0.06	−0.22–0.09	0	1.05[f]	0.795[f]	2
Colubridae							
Natracinae	14	0.19	0.09–0.33	0	0.93[f]	0.924[f]	2
Colubrinae, Lycodontinae, Xenodontinae	18	0.05	−0.30–0.35	+	1.19[f]	0.814[f]	2
Viperidae	16	−0.09	−0.21–0.08	++	1.17[f]	0.915[f]	2
Lacertidae	8	0.00	−0.06–0.11	0	1.03	0.972	16

(Continued)

Table 1   *(Continued)*

Clade or taxon	N[a]	SSD[b] Mean	SSD[b] Range	RR[c]	β[d]	r[e]	Reference
Iguanidae	90	−0.12	−0.49–0.24	0	0.99[f]	0.902	2
Chelonia	77	0.05	−0.20–1.75	++	1.18	0.935	51
Kinosterninae	24		−0.35–0.11	++			74
AMPHIBIA							
Anura	51			0			124
Bufo	18			0			124
ARACHNIDA							
Araneae (excl. Araneidae)	44	0.23	−0.03–0.68	0	1.02[f]	0.750[f]	2
Araneidae	19	0.57	0.13–1.70	0	0.89	0.919	1
INSECTA							
Gerridae	33	0.13	0.00–0.32	++	1.17[f]	0.952[f]	2
Tephritidae	27	0.05	−0.17–0.19	++	1.12	0.989	142
Phasmatodea	152	>0		−−	0.84[h]	0.923	142

[a]Number of species compared.

[b]SSD estimated using the index of Lovich and Gibbons (90): SDI = ([mean size of largest sex]/[mean size of smallest sex])−1, arbitrarily defined as positive when females are larger and negative when males are larger.

[c]Rensch's rule: ++, statistically significant allometry consistent with Rensch's rule (P < 0.05); +, allometry consistent with Rensch's rule with 0.05 < P < 0.20; no allometric trend (P > 0.20); −, allometry inconsistent with Rensch's rule with 0.05 < P < 0.20; −, statistically significant allometry inconsistent with Rensch's rule (P < 0.05).

[d]Slope of the major axis regression of log(male size) on log(female size), estimated from the data provided in the original source.

[e]Pearson product moment correlation coefficient between log(male size) and log(female size).

[f]Corrected for phylogeny using independent contrasts analysis (2, 40, 110).

[g]Alcidae, Charadriidae, Jacanidae, Laridae, Recurvirosridae, Rostratulidae, Rynchopidae, and Stercoraridae.

[h]OLS slope from the original source.

ESTIMATING ALLOMETRY FOR SEXUAL SIZE DIMORPHISM

Allometry is formally defined as a departure from geometric similarity (53, 78, 133) or, more fully, as "differences in proportions correlated with changes in the absolute magnitude of the total organism" (53). By convention, allometric relationships are quantified using the power function, $y = \alpha x^\beta$, where x is body size, y is a measure of the trait of interest, and α and β are constants (19, 113, 125, 133, and see 53, 61, 73, 78 for justification of this model). If x and y are measured on the same scale, the ratio y/x will be a constant (α) when $\beta = 1$. This defines geometric similarity or isometry. By extension, allometry occurs when

$\beta \neq 1$. If $\beta > 1$, y increases faster than x, and y/x increases as x increases. This is termed positive allometry or hyperallometry (19, 53). Negative allometry or hypoallometry occurs when $\beta < 1$, indicating that y increases more slowly than x, and y/x declines as x increases. Quantifying allometry requires estimating the allometric coefficient β, and this is achieved by logarithmic transformation so that β becomes the slope of the linear regression, $\log(y) = \log(\alpha) + \beta \log(x)$ (53, 61, 78, 125).

Allometry for SSD has most frequently been estimated by regressing some measure of SSD against mean body size (27, 38, 51, 74, 139, 141), size of females (13, 44, 48, 84, 112, 155, 162), or size of males (16, 44). Unfortunately, the pattern of allometry revealed by these studies depends strongly on whether SSD is estimated as a difference or a ratio (48). For size ratio to remain constant as body size increases (i.e. isometry for size ratio), the difference in size between the sexes must increase with body size and thus must show positive allometry. Similarly, a constant size difference requires negative allometry for size ratio. In addition to this confusion, a regression of any index of SSD on body size violates the basic assumption of mathematical independence of the x and y variables (78, 131, 144). For example, in a regression of size ratio on mean size we expect x and y to covary because mean male size and mean female size are included in both variables. Because of the interdependence of y and x, the null hypothesis is not $\beta = 0$, and normal techniques of hypothesis testing are not valid.

Both of these problems are solved if allometry for SSD is estimated from a log/log plot of the size of one sex against the size of the other sex (2, 5, 27, 35, 37, 82, 94, 142, 155). As neither sex is, a priori, the independent variable, there is as yet no convention as to placement of the sexes on the x- and y-axes. Fairbairn & Preziosi (37) and Abouheif & Fairbairn (2) proposed assigning males to the x-axis. However, other authors have preferred to place females on this axis, both in regressions of SSD on size (13, 84, 112, 155, 162) and in regressions of the size of one sex against the size of the other sex (27, 48, 82, 94, 142, 155). To facilitate comparisons among studies, I adopt the convention of placing female size on the x-axis (Figure 1).

Because body size, x, is not fixed and is estimated with error, ordinary least squares regression (OLS, model I) will tend to underestimate both β and the confidence interval around β. Some form of model II regression is therefore more appropriate for both parameter estimation and hypothesis testing (53, 78, 97, 144). Given that male and female sizes are measured in the same manner and using the same scale, both measurement and intrinsic error should be very similar in the two sexes, especially when transformed to a logarithmic scale. Under these conditions, major axis regression (MA) provides an accurate estimate of β and allows statistical testing of the general null hypothesis,

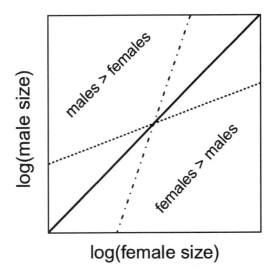

Figure 1   Regression model for estimating allometry for SSD. The diagonal *solid line* represents a size ratio of one (size of males = size of females). Male-biased size dimorphisms lie above the diagonal, female-biased dimorphisms below. The *dotted line* represents an allometric relationship in which female size varies more among species than male size and therefore the regression slope, β, is less than 1. The *dot-dashed line* represents an allometric relationship in which male size varies more among species than female size and therefore $\beta > 1$. Deviations of these lines from the 1:1 line illustrate the degree of SSD. When $\beta > 1$, SSD declines as size increases for female-biased SSD, but increases with size for male-biased SSD, as predicted by Rensch's rule (see text). If $\beta < 1$, the pattern of allometry for SSD is reversed and is inconsistent with Rensch's rule.

$\beta = 0$, as well as the null hypothesis of isometry, $\beta = 1$ (63, 78, 97, 144). Reduced major axis regression (RMA) would also provide an accurate estimate of β, but methods of testing the null hypothesis, $\beta = 0$, remain controversial for this technique (63, 78, 97, 144). Because of the high correlation generally found between male and female body sizes (Table 1), the two techniques should give almost identical estimates for β. To illustrate this, I compared slopes calculated by OLS, MA, and RMA regressions of log(male size) on log(female size) for ten independent data sets with sample sizes ranging from 14 to 65 species. The MA slopes ranged from 0.89 to 1.26. The mean difference between the MA and RMA slopes was only 0.0003 (SD 0.0040), while OLS slopes averaged 0.018 (SD 0.021) lower. Thus, choosing model II rather than model I regression may significantly influence the results, but the model II technique used is of little consequence for parameter estimation.

One problem remains for statistical assessments of allometry for SSD. As in any comparative analysis, the accuracy of both parameter estimation and hypothesis testing may be adversely influenced by phylogenetic autocorrelation among the taxa being compared (40, 55, 64, 95, 110, 119). Although several methods have been developed to remove phylogenetic effects from comparative data (reviewed in 64), the most efficient method for parameter estimation from species means is the method of pairwise independent comparisons or independent contrasts (40, 54, 110, 119). Simulation analyses indicate that the independent contrasts method performs better than standard comparisons across the "tips" of the phylogeny (i.e. comparisons among species, uncorrected for phylogeny) for both parameter estimation and hypothesis testing, even in the absence of a fully resolved phylogeny and branch length information (54, 95, 119). Abouheif (1) used the regression methods outlined above to estimate allometry for SSD in 21 independent clades and to compare the results of the standard analyses using species means ("tips analyses") with those from independent contrasts analyses. The independent contrasts analyses were less powerful than the tips analyses, with an average reduction in power of 10%, and five slopes judged significant (i.e. $P < 0.05$) according to the tips analysis were not significant after independent contrasts. This reduction in power reflects both the reduction in degrees of freedom associated with the independent contrasts analyses and the higher type I error rate of the tips analyses (1, 55, 95, 119). The allometric slopes estimated by the two methods were quite strongly correlated ($r_{19} = 0.71$, $P < 0.0005$) and did not differ systematically (paired t-test, $t_{20} = 0.607$, $P > 0.50$), indicating no relative bias. However, the coefficient of variation between the two estimates was only 51%, indicating considerable variance between methods. Thus, the slopes estimated from the standard tips regressions were unbiased but relatively inaccurate predictors of the slopes estimated from the independent contrasts (1). Because the latter are expected to be more accurate, with respect to both parameter values and statistical conclusions (40, 54, 64, 110, 119), independent contrasts analysis should be an integral part of assessments of allometric relationships (1, 2).

EMPIRICAL EVIDENCE OF ALLOMETRY FOR SSD

Numerous authors have stated that SSD generally increases with body size in taxa in which males are the larger sex (2, 27, 37, 112, 120, 124–126, 135, 155, 162). An overall trend is less apparent where females are the larger sex, but the most common pattern appears to be one of decreasing SSD with increasing size (2, 35, 37, 126, 157). Rensch formalized these trends into a general rule (2, 124–126). When male size is regressed against female size, allometry consistent with Rensch's rule yields a slope greater than one, regardless of which sex is larger (Figure 1). The major characteristic of this trend is that male size varies more among species than female size (i.e. $\Delta y > \Delta x$). In contrast, allometry inconsistent with Rensch's rule ($\beta < 1$ in Figure 1) would indicate greater interspecific variance in female size than in male size. Previous reviews of the quantitative evidence for Rensch's rule (2, 124, 125) indicate that it is indeed the most common allometric trend but that exceptions occur, particularly in taxa in which females are the larger sex.

The available quantitative evidence of allometry for SSD is consolidated in Table 1. I have restricted the data to interspecific comparisons because allometric patterns are often strongly influenced by taxonomic level (i.e. comparisons among species yield different slopes than comparisons among genera, families, etc; 2, 26, 63, 111). Allometry for SSD has also been recorded among populations within species, but these patterns may be poor indicators of evolutionary trends because of strong environmental effects (37, 146) and, if growth continues after sexual maturity, because of the confounding interaction between age and size distributions (146).

Although all of the studies in Table 1 are based upon comparisons among species, the constituent species have been grouped at various taxonomic levels, from a single genus to combinations of two related orders. In most cases the level of grouping may simply reflect the taxonomic scope of interest of the authors. However, Abouheif & Fairbairn (2) defined their 21 species groups objectively based on both phylogeny and homogeneity of allometric slopes, combining sister clades if the slopes were homogeneous, but analyzing them separately if the slopes were heterogeneous.

Table 1 includes only studies that assess allometry for overall body size measured as weight (16 analyses), length (19 analyses), or length of a body component known to be strongly correlated with overall size (9 analyses). In all examples, allometry was originally estimated from a bivariate plot of either SSD against body size, or the size of one sex against the size of the other sex. Where possible, I have used the data provided in the original sources to estimate the mean and range of SSD, the allometric coefficient (β), and the correlation coefficient (r) for log(body size) between the sexes. These standard statistics are then used in quantitative assessments of the overall pattern and prevalence of allometry.

Unfortunately, very few authors have corrected for phylogeny when assessing allometry for SSD (1, 2), and only 21 of the 42 analyses in Table 1 are based on independent contrasts. However, given that slopes estimated from standard tips regressions are unbiased estimators of slopes estimated from independent contrasts (1), the inclusion of both types of estimates should not bias our assessment of overall qualitative trends. However, the probability of a type I error may be higher than 0.05 in the tips analyses (54, 95, 119) and, conversely, the power is lower for the independent contrasts analyses if the underlying phylogeny is poorly resolved. Thus, the probability of concluding that the allometric trend is statistically significant is higher for tips analyses than for analyses based upon independent contrasts. Where this may influence interpretation of the data, it has been noted in the text.

In several cases, the same species have been included in more than one study, and thus the estimates of allometry cannot be considered independent. For example, significant allometry consistent with Rensch's rule has been demonstrated repeatedly for Primates, based on broadly overlapping data sets (2, 27, 44, 48, 82–84). Only the estimates from Abouheif & Fairbairn (2) and Ford (44) appear in Table 1. The former was chosen because it is based on independent contrasts (40, 110) and is therefore independent of phylogenetic effects. The latter was chosen because it is based primarily on Neotropical species not included in the other studies. Although Ford's analysis shares 10 species with that of Abouheif & Fairbairn (2), removal of these species did not affect the statistical conclusion and changed the major axis slope only from 1.052 to 1.058. Because of the large differences in both SSD and allometric slope between these two species sets (Table 1), I have included both as representative of patterns of allometry in different groups of Primates.

Three overlapping species sets occur in the bird data: Eight species of Ictinerae analyzed by Webster (155) are also included in Bjorklund's analysis of Passeriformes (13); seven species of Pelicaniformes occur in the data sets of both Fairbairn and Shine (12) and Abouheif & Fairbairn (2); and the subfamilies of Scolopacidae analyzed by Jehl & Murray (75) overlap with the Scolopacidae analyzed by Abouheif & Fairbairn (2). I have considered these data independent if at least 75% of species are unique to a given data set. By that criterion, only the Numeniini (75) are not independent, and these are therefore excluded from the quantitative analyses.

Among the reptiles, Abouheif & Fairbairn's analyses of six separate snake clades (2) supersede earlier analyses (41, 139) that combined snakes from clades differing markedly in behavior, ecology, reproductive biology, body size, SSD, and allometric slope (2, 138, 139). Similarly, the separate analyses of the lizard families Lacertidae (16) and Iguanidae (2) supersede earlier analyses (41, 145), and the recent analyses of allometry in turtles (51, 74) replace an earlier analysis (12) based on a much smaller sample size (eight species). The two turtle data sets share only five species and so are considered independent.

A final area of overlap occurs in between the two estimates of allometry in Amphibia. It is likely that some of the *Bufo* species used by Arak (unpublished data, cited in 124, 125) also occur in Crump's data set (28, cited in 124, 125). In the absence of evidence to the contrary, I do not consider them independent estimates.

For two clades, the Mustelidae and Araneae, the allometric pattern reported in Table 1 conflicts with patterns reported by previous authors. The mustelids have been used as an example of allometry inconsistent with Rensch's rule because SSD appears to decrease with body size even though males are larger than females (102, 122, 124, 125). However, controlling for the influence of phylogeny removes this apparent allometric trend (1, 2) resulting in $\beta \geq 1$ (Table 1). Allometry inconsistent with Rensch's rule has also been reported for the Araneae (66, 152). Female spiders are the larger sex, but SSD appears to increase with body size. However, both SSD and the pattern of allometry vary greatly among families within the Araneae (2, 66, 152), and as in the Mustelidae, removal of phylogenetic effects removed the apparent allometry among species (2), resulting in $\beta \geq 1$ for Aranae excluding the Araneidae (Table 1). The Araneidae (orb weavers) are clear outliers, with a strongly female-biased SSD, and an allometric slope less than one (1, 2, 66, Table 1). The interspecific variance in size is clearly much greater for females than for males in this family, but the correlation between the sexes is not sufficient to generate statistically significant allometry in our sample of 18 species (Table 1 and see 66). However, the slope for the Araneidae is lower than that reported for any other clade, and the hypothesis of significant allometry inconsistent with Rensch's rule in orb weaving spiders should not be rejected without further testing.

After consideration of nonindependence among data sets, there remain 40 independent estimates of allometry for SSD and 37 independent, quantitative estimates of the allometric and correlation coefficients. The correlations between the sexes are all high and statistically significant, indicating that strong covariation between the sexes is a conspicuous feature of interspecific variation in body size, regardless of the allometric pattern for SSD. Allometry for SSD is statistically significant ($\beta \neq 1$) in 18 independent data sets (10 of 19 tips analyses, and 8 of 13 independent contrasts analyses), across a wide range of clades. Sixteen of the estimates demonstrate statistically significant allometry consistent with Rensch's rule, and two estimates show the same trend with $0.20 > P > 0.05$. This pattern occurs in mammals, birds, reptiles, and insects, and in taxa in which females are the larger sex, as well as in taxa in which males are larger (Table 1, Figure 2a). Two clades, Strigiformes (owls) and Phasmatodea (walking sticks), show statistically significant allometry inconsistent with Rensch's rule, and a similar trend may also occur in the Araneidae. In all of these clades, females are the larger sex. Overall, 28 of the MA slopes are greater than one (consistent with Rensch's rule), and eight are less than one (Table 1, Figure 2).

The allometric slopes are negatively correlated with the index of SSD (Figure 2a, $r^{34} = -0.40$, $P < 0.02$). Since this index is negative when males are the larger sex (Table 1, and see 90 for a discussion of the statistical properties of this index), the negative correlation indicates that the slopes tend to decrease as SSD becomes more female-biased. Thus allometry consistent with Rensch's rule is more prevalent and more pronounced in clades with male-biased SSD.

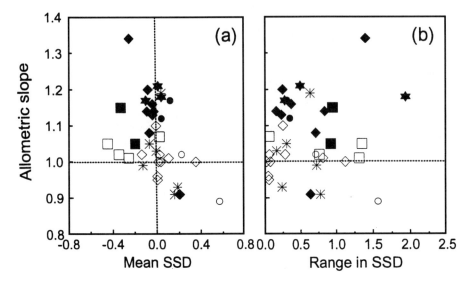

Figure 2   Scatter plots of allometric slopes against the (*a*) mean and (*b*) range in SSD, for 36 independent species groups (Table 1 and see text). The allometric slope is estimated as the slope of the major axis regression of log(male size) on log(female size). Sexual size dimorphism is estimated using the sexual dimorphism index of Lovich & Gibbons (90), which is negative when males are larger, zero for sexes of equal size, and positive when females are larger. The *horizontal* and *vertical dotted lines* indicate $\beta = 1$ (isometry), and SSD = 0 (no sexual size dimorphism), respectively. *Filled symbols* indicate slopes that differ significantly from 1.0. ■□, mammals; ◆◇, birds; ＊✳, reptiles; ●○, invertebrates.

Conversely, allometry tends to be weaker or inconsistent with Rensch's rule in clades with female-biased SSD. Abouheif & Fairbairn (2) reached similar conclusions based on a significant correlation between the effect sizes from their meta-analysis and SSD for their 21 clades.

The allometric slopes are not correlated with the within-clade range in SSD (Figure 2*b*, $r_{34} = 0.12$, P $= 0.49$). Thus, lack of significant allometry does not signal lack of variance in SSD. Several taxa of Charadriiformes and the *Tamias* species cluster at the far left and show neither allometry nor variance in SSD. However, some clades, notably the Araneidae, Carnivora, Mustelidae, and Falconiformes, show no allometry in spite of a broad range of size dimorphism. In these taxa, SSD varies but is not sufficiently correlated with overall size to generate a significant allometric slope.

The clades or taxa sampled have not been selected at random, and thus we cannot use these data to derive a quantitative estimate of the prevalence of allometry for SSD. In particular, plants are missing from our sample, invertebrates are severely underrepresented, and there are no aquatic organisms. However, the available evidence demonstrates that SSD is correlated with body size in many clades and is sufficiently common to warrant further attention. Allometry consistent with Rensch's rule is clearly the more common trend and tends to be both stronger and more prevalent in clades in which males are the larger sex. In contrast, allometry inconsistent with Rensch's rule is uncommon, at least among the available samples, and appears to be restricted to clades in which females are the larger sex. A final generalization from the empirical evidence is that even the most extreme allometry for SSD occurs in the presence of very strong covariation for body size between males and females. This implies that the pattern of evolutionary divergence leading to allometry among contemporary species consists primarily of correlated changes in size in males and females, allometry being produced because one sex (usually males) shows a greater magnitude of change than the other.

FUNCTIONAL HYPOTHESES FOR THE EVOLUTION OF ALLOMETRY FOR SSD

1. Allometry as a Functional Constraint

Allometry for SSD is merely a statistical description of the relationship between male and female body size and hence of the pattern of interspecific variation in SSD. The discovery of significant allometry for SSD neither establishes a causal relationship between body size and SSD nor suggests the evolutionary mechanism that might produce such a relationship. Nevertheless, body size is often treated as one of a suite of variables potentially affecting SSD, the other variables typically being indices of natural or sexual selection such as mating system, diet, or habitat. In some cases each variable is considered independently (35, 38, 51), but more often adaptive hypotheses are tested using residual variation after body size effects are statistically removed (13, 21, 35, 44, 83, 84, 166) or body size is entered as one variable in a simultaneous analysis including other variables representing adaptive hypotheses (48, 141, 155). All of these approaches assume that body size explains some subset of the variation in SSD not explained by the postulated mechanisms of natural or sexual selection.

The concept of considering allometry as distinct from adaptive patterns of divergence derives from the principle of functional similarity or size-required allometry (53, 113, 133). This principle refers to changes in the shape of structures or organs that must accompany changes in size to preserve the original function. For example, animals become increasingly stockier (i.e. bones become disproportionally thicker) as size increases. The allometric increase in bone cross-sectional area preserves the elastic strength of the bones so that the bones of larger animals are able to support the increased load without buckling (99). Such size-required allometry is clearly adaptive, but it describes changes in form required to maintain a given function. In this sense, it represents the functional constraints on shape imposed by changes in size. This is the meaning of the term "allometric constraint." Changes in the function of organs and body components are, by extension, revealed as deviations from this size-required allometry. Hence, adaptive divergence in function is often assessed by analyses of residuals from allometric relationships (10, 19, 53).

In the case of SSD, we have no a priori expectation of size-required allometry. What function is being preserved, and how does this translate into changes in SSD with size? Until these questions can be answered, it is premature to consider empirically derived allometric relationships for SSD as lines of functional equivalence, and hence, as representing "allometric constraints" (53). The observed divergence of SSD among species within a clade includes both the variance described by the allometric regression (i.e. the line of allometry) and the residual variance around this line; both components of variance require adaptive or functional explanation (35, 36, 53). Researchers who remove body size effects before testing functional hypotheses or include body size as one of a suite of functional variables predicting SSD may therefore be removing much of the variance that should be explained by their adaptive hypotheses.

2. Allometric Growth of Secondary Sexual Characters

Rensch (126) attributed allometry consistent with Rensch's rule to allometric growth of secondary sexual characters in males. This hypothesis is frequently cited (6, 27, 124, 125, 155), but to my knowledge has never been explained. The extent to which allometry for secondary sexual characters is associated with allometry for SSD has not been determined, but even if it proves to be general, the juxtaposition of two statistical relationships does not constitute a functional hypothesis. Recent studies indicate that the allometric growth patterns of secondary sexual traits respond to selection (34, 163), and patterns of interspecific allometry for such traits are expected to evolve in response to selection on the traits concerned (168). It therefore seems unlikely that the allometry of secondary sexual characters represents a functional constraint on the evolution of SSD. More probably, this allometry is simply one aspect of the adaptive divergence of SSD. A functional hypothesis attributing allometry for SSD to the allometric growth of secondary sexual characters in males would have to include both an adaptive explanation for the pattern of allometric growth and a mechanism that could produce the required correlation between this growth pattern in males and the body size of females. In the absence of these components, Rensch's hypothesis is only a description of allometric patterns, not a functional explanation for them.

3. Correlated Response of Females to Sexual Selection on Males

One of the earliest and most pervasive hypotheses for the evolution of allometry consistent with Rensch's rule is that female size evolves as a correlated response to sexual selection acting on male size (6, 37, 96, 112, 123, 155, 158). This expectation can be derived from basic quantitative genetic theory as follows. The direct response (R) of a trait to selection is given by $h^2 i \sigma$, where h^2 is the heritability of the trait (additive genetic variance/total phenotypic variance), i is the selection intensity in standard deviation units, and σ is the phenotypic standard deviation of the trait (39). If a trait is expressed in both sexes, the response of each sex is due to the direct response to selection in that sex plus the correlated response to selection on the other sex:

$$R_m = 0.5\left(h_m^2 i_m \sigma_m + r h_m h_f i_f \sigma_m\right) \tag{1.}$$

$$R_f = 0.5\left(h_f^2 i_f \sigma_f + r h_f h_m i_m \sigma_f\right), \tag{2.}$$

where r is the additive genetic correlation for that trait between the sexes and the subscripts m and f refer to male and female, respectively (39, 84, 123). If body size is measured on a log scale, as in the standard allometric model, allometry occurs whenever $R_m \neq R_f$ (i.e. the ratio of body sizes before and after selection is not the same). Since the correlated response model only specifies sexual selection acting on males and a positive genetic correlation between the sexes, we can simplify by setting $h_m^2 = h_f^2$, $\sigma_m = \sigma_f$, and $i_f = 0$. The ratio R_m/R_f then reduces to $1/r$, indicating allometry that is inversely proportional to the genetic correlation between the sexes. (See 79 for a similar analysis of expected allometry for two traits correlated within individuals, and 21 and 84 for a similar analysis based on SSD measured as the difference between the sexes, with allometry assessed as SSD/female size.)

In this scenario, female size changes only as a correlated response to selection on males. However, body size is unlikely to be a neutral trait (i.e. it is unlikely that $i_f = 0$), and most authors assume that female body size is at or near its natural selection optimum (80). As male size changes in response to sexual selection, natural selection on females is expected to slow and then reverse the correlated response of females, and females eventually return to their natural selection optimum while males stabilize at a size that balances the opposing forces of natural and sexual selection (80, Figure 3). Allometry consistent with Rensch's rule is expected only during the very early stages of evolutionary divergence (Figure 3). At equilibrium this model predicts no correlation between male and female body size and no allometry for SSD (33, 80, 168).

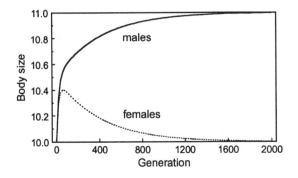

Figure 3 Typical theoretical trajectory for the evolution of male and female body size in response to sexual selection for increased size in males, based on Lande's quantitative genetic model (80, 123; J Reeve, unpublished data). With the exception of sexual selection, all parameters are the same for males and females. Natural (stabilizing) selection favors a body size of 10.0, the heritability of body size is 0.5, and the genetic correlation between the sexes is 0.9. Altering the model parameters alters the rate, duration, and maximum response to selection but does not change the basic shape of the trajectories.

Genetic correlations between the sexes are therefore unlikely to account for the widespread occurrence of allometry for SSD.

4. Variance Dimorphism

Cheverud et al (21) and Leutenegger & Cheverud (83, 84) used the same quantitative genetic model to derive the hypothesis that allometry for SSD can evolve solely as a consequence of differences between the sexes in the phenotypic variance for body size. Since the heritability of body size is set as equal in the males and females in the model, it is important to note that dimorphism in phenotypic variance implies dimorphism in additive genetic variance, a point that is insufficiently stressed in the original papers. To illustrate the effect of variance dimorphism, set $i_m = i_f$, and $h_m^2 = h_f^2$ in Equations 1 and 2, and define allometry as before as $R_m/R_f \neq 1$. The response ratio reduces to σ_m/σ_f , indicating that allometry now depends only upon the standard deviations of body size in each sex. Leutenegger & Cheverud (83, 84) used a difference rather than ratio measure of SSD and concluded that allometry is enhanced by both variance dimorphism and high genetic correlations between the sexes. However, the influence of the genetic correlation is not evident for allometry based on size ratios.

The key flaw in this model as a general explanation for allometry is that it predicts only transitory allometry as one sex responds faster than the other to the same direction and intensity of selection. No allometry is predicted at equilibrium. Further, while the sexes initially diverge in size, SSD will decline as the slower responding sex approaches equilibrium, producing a reversal of the allometric pattern along any given trajectory. Although Leutenegger and Cheverud (21, 83, 84) argue that variance dimorphism may be responsible for the strong allometric trend seen in primates, empirical evaluations of this hypothesis in other clades have produced negative results (35, 123, 155). Thus, it seems unlikely that variance dimorphism accounts for the evolution of allometry for SSD.

5. Large Size Facilitates the Evolution of SSD

Hypotheses in this category assume that natural selection acts similarly on males and females to produce both the interspecific divergence in mean body size and the covariance in body size between males and females. Allometry for SSD then evolves because large size itself facilitates the evolution of SSD by (a) promoting polygyny and thus increasing the intensity of sexual selection (82, 162); (b) facilitating response to sexual selection (27, 48); or (c) increasing the probability that sexual selection will act on body size rather than some other trait such as plumage, color, or vocalization (22, 44, 48, 135). The mechanism by which an increase in mean size promotes polygyny is not specified, and the first hypothesis therefore lacks theoretical justification. The second hypothesis derives from the assumption that energetic or competitive constraints on increasing size are relaxed in larger species (27, 48), an assumption that has not been supported analytically or empirically (124, 125). However, several arguments support the third hypothesis. Gaulin & Sailer (48) argue that the effectiveness of weapons and of contests of strength increases with body size because the energy that can be delivered by a blow increases faster with size than the ability to absorb the energy from that blow. Clutton-Brock (22) argues that, as size increases, strength increases but agility declines, leading larger species to rely more on tests of strength, and smaller species to rely more on agility and display. Finally, Shine (135) argues that large size reduces the risk of predation and thus increases the probability of sexual selection through male-male combat (the assumption being that combatants suffer increased predation risk). Therefore, it may be that as size increases within a clade, sexual selection is more likely to target size rather than some other male trait, resulting in allometry for SSD consistent with Rensch's rule.

6. Resource Distribution Influences Both Body Size and SSD

The central assumption of hypotheses in this category is that body size and SSD covary because both are influenced by the distribution of resources (4, 27, 155). For example, Clutton-Brock et al (27) suggested that dispersed, unpredictable food resources might favor both large size and intersexual niche divergence in primates. However, this hypothesis has been criticized on the grounds that it predicts greatest SSD in monogamous species and so is not consistent with the general relationship between SSD and polygygny (48, 124, 125). Alternatively, the key resource may be females rather than food. Mating systems associated with a clumped distribution of breeding females favor monopolization of many females by individual males, thus facilitating sexual selection for large males (4, 155). Webster (155) proposed that breeding aggregations may also favor large size in females, because larger females are more successful in intrasexual competition. This hypothesis predicts allometry for SSD only if female dispersion has a proportionally greater effect on male size than on female size, and it excludes many taxa in which allometry occurs in the absence of breeding aggregations (i.e. 2, 5, 13, 51, 74, 112, 139). It is therefore unlikely to provide a general explanation of allometry for SSD.

7. Stabilizing Selection on Females

Several authors have suggested that reproductive selection on females opposes further increases or decreases in female size at the extremes of size within a clade (6, 27, 58, 121, 155). Thus, in large species with very large males, selection should favor females that are small relative to their mates. At the other extreme, if males are very small, selection should favor females that are large relative to their mates. Because sexual selection on males is not expected to alter the equilibrium body size of females (80, 143), this hypothesis actually predicts covariation of male and female body sizes only if the diversifying selection is natural selection acting on both sexes. If the diversifying selection is sexual selection on males, no allometry is predicted. This contrasts with the predictions of most of the previous hypotheses and with the empirical evidence of a strong association between allometry and sexual selection (e.g. 2, 13, 27, 37, 44, 48, 112, 139, 155), and thus it seems unlikely that stabilizing selection alone accounts for the observed patterns of allometry for SSD.

8. Correlational Selection

All of the preceding hypotheses assume that, at equilibrium, the selective forces affecting body size in each sex are independent of the size of the other sex. This concept is implicit in the verbal hypotheses and explicit in the quantitative genetic models (21, 80, 83, 84, 123, 143). However, in many clades such independence seems unlikely (18). If one sex carries the other during courtship or mating, or if mating pairs remain physically joined for prolonged periods, as in many invertebrates, the mating success of males may depend significantly on their size relative to their mates (3, 11, 15, 30, 35, 93, 105). Similarly, if males must physically overcome female reluctance to mate, the mating success of males, the energetic cost of the struggle, and the risk of injury to both sexes are likely to be influenced by the relative sizes of the struggling pair (9, 132, 156).

The size ratio of mating pairs may also influence the proportion of eggs fertilized in a given clutch and so directly impact female fitness (11, 15). Even in the absence of correlational selection due to mating interactions, sexual selection favoring large males may be accompanied by selection favoring large females because larger females tend to produce larger sons (25, 155). The general correlation between mother's size and the size of eggs, neonates, and juveniles (6, 18, 113, 120, 130) suggests that any selection favoring larger males will ultimately favor larger size in mothers.

If, as suggested by these examples, the optimal body size in one sex is influenced by the mean size of the other sex, changes in the size of one sex will be accompanied by selection favoring correlated changes in the size of the other sex. This model is analogous to Zeng's quantitative genetic model for the evolution of allometry among traits within individuals, which predicts that, although patterns of genetic correlation among traits influence the evolutionary trajectories, the allometric pattern at equilibrium is determined only by the patterns of correlational selection (168). Although Zeng did not specifically consider the evolution of SSD, his model would predict allometry consistent with Rensch's rule whenever interspecific divergence in size was due to directional selection on males, with weaker, correlational selection on females. If the diversifying directional selection acts primarily on females (e.g. fecundity selection in spiders or butterflies; 66, 161), allometry inconsistent with Rensch's rule is expected, as males show the lower response to correlational selection.

The correlational selection hypothesis has the advantages of being consistent with quantitative genetic theory and at the same time predicting covariance in size between the sexes, greater interspecific variance in one sex than the other, and allometry for SSD at equilibrium. It is also one of the few hypotheses capable of predicting allometry inconsistent with Rensch's rule. It shows promise as a general, unifying hypothesis for the evolution of allometry for SSD but may be difficult to distinguish empirically from hypotheses in categories 5 and 7, which assume that the covariation between the sexes is caused by similar patterns of natural selection on males and females, rather than by correlational selection between sexes.

TESTING HYPOTHESES FOR THE EVOLUTION OF ALLOMETRY FOR SSD

Most of the functional hypotheses for the evolution of allometry for SSD suffer from a lack of both theoretical rigor and empirical validation, and they remain essentially post-hoc explanations for allometry within a given clade. Further, many of these hypotheses incorporate the same variables and differ mainly in the order of cause and effect. For example, hypotheses in category 5 assume that large size facilitates the evolution of SSD in response to sexual selection (i.e. sexual selection causes increased SSD only in species whose average size is initially large), while the correlated response hypothesis (category 3) assumes that sexual selection on males causes a correlated increase in both mean size and SSD. Is large average size a cause or a consequence of sexual selection? Similarly, does a clumped distribution of females promote polygyny and thus sexual selection for large size in males (category 6), or does large size of males promote polygyny (category 5[a]) Because these competing hypotheses were derived to explain similar patterns of correlation among mating systems, body size, and SSD, they cannot be distinguished by a purely correlative approach.

Directional comparative analyses that seek to determine the order of appearance of traits within a phylogeny (17, 64, 91, 128) hold more promise of distinguishing among the various hypotheses. Höglund (69) and Oakes (109) used this approach to test the hypothesis that lekking promotes the evolution of SSD in birds through increased sexual selection for large males (69, 112, 162). If this hypothesis is correct, cladistic analysis should reveal that the evolution of lekking is followed by the evolution of increased male size and SSD. Unfortunately, Höglund concluded that "lekking did not precede dimorphism more often than the reverse," while Oakes concluded that SSD is "more likely to evolve when birds lek" (109), and so the controversy is not yet resolved. However, Oakes provides an excellent comparison of the two studies, as well as explanations of the statistical problems associated with directional phylogenetic analyses. His study provides a good starting point for similar tests of hypotheses for the evolution of allometry for SSD within other clades.

A second promising approach to hypothesis testing in this area is construction of quantitative models to determine the plausibility of the proposed evolutionary mechanisms. For example, Reiss (124, 125) used simple analytical models to show that the degree to which the evolution of SSD is constrained by interspecific competition or by energetic requirements is expected to be independent of species size, contrary to the assumptions made by Clutton-Brock et al (27; category 5[b]). Similarly, the quantitative genetic models derived by Lande (80) and Cheverud et al (21) demonstrate that neither the correlated response hypothesis (category 3) nor the variance dimorphism hypothesis (category 4) are sufficient to generate persistant interspecific allometry for SSD. In these examples, quantitative modeling suggested that the prior verbal models were not plausible. In contrast, Zeng's quantitative genetic model for the evolution of allometry of correlated traits within individuals (168) provides the theoretical justification for the hypothesis that allometry for SSD is a function of correlational selection (category 8).

Quantitative models can also greatly facilitate hypothesis testing by providing estimates of constituent parameters, evolutionary trajectories, and equilibrium distributions that can then be compared to data collected from nature or from manipulative experiments. For example, although neither the correlated response hypothesis nor the variance dimorphism hypothesis predict allometry for SSD at equilibrium, they do predict allometry during the initial period of divergence of species from a common ancestor. Reeve & Fairbairn (123) used these models to predict the short-term evolution of mean size and SSD in *Drosophila melanogaster* in response to artificial selection on thorax width. Although body size responded well to selection, the response in SSD was consistently less than predicted by either model, and selection did not produce the predicted allometry for SSD among selected lines. Thus, these models are insufficient to account for even the short-term evolution of allometry.

Ultimately, our ability to test functional hypotheses for the evolution of allometry, whether through comparative analyses or the application of quantitative models, depends upon the accuracy with which the relevant parameters are estimated within species. Lack of accurate, quantitive estimates of the pattern and intensity of selection is a serious limitation throughout most of this literature. For example, sexual selection is typically reduced to a simple categorical classification of mating systems as monogamous (presumably low intensity of sexual selection) or polygynous (presumably high intensity of sexual selection) (5, 13, 21, 69, 75, 82, 83, 109, 112, 141). Other surrogate measures such as level of competition between males for access to mates (44), presence or absence of male combat (138, 139), number of matings per male (155), and socionomic sex ratio (27) have also been used. All but the latter two indices provide very little power because the presumed driving variable, sexual selection, is reduced to an ordinal variable with few classes. Further, using any of these indices as estimates of sexual selection on body size presupposes that sexual selection uniformly targets body size rather than other traits such as song, color, or vigor, an assumption that is unlikely to be correct (22, 44, 48, 135). The ability of such analyses to detect underlying functional relationships between sexual selection on body size and SSD is questionable, at best.

Fairbairn & Preziosi (37) provide an example of a more rigorous approach. To test the hypothesis that allometry for SSD among populations of the water strider, *Aquarius remigis*, is caused by sexual selection on males, they estimated sexual selection gradients (81) for male total length in 12 wild populations and found the predicted positive correlation between the magnitude of sexual selection for large size and mean male size across populations. This study illustrates the potential for more accurate comparative analyses of the relationships among selection intensity, body size, and SSD.

At present, we suffer from a surfeit of functional hypotheses for the evolution of allometry for SSD, and a clear deficiency of rigorous, quantitative tests of these hypotheses. While strictly correlational studies may still be valuable, particularly within less-studied clades, they will not be sufficient to sort out the morass of overlapping hypotheses. Significant progress in this direction is likely to be achieved only through a combination of more accurate parameter estimation, quantitative modeling, and directional comparative techniques that estimate the order of appearance of traits within a phylogeny.

SUMMARY AND CONCLUSIONS

Sexual size dimorphism varies greatly in direction and degree, both among and within clades. Given the importance of body size in the ecology, life history, and reproductive fitness of most organisms, patterns of interspecific variation in body size are expected to reflect patterns of adaptive divergence. This is true not only for the mean size of species, but also for the sizes of males and females, respectively. Quantitative genetic theory leads us to expect independent adaptation of males and females to the net selection acting on each sex. At equilibrium, therefore, each sex should be at its optimal size, and SSD should, in this sense, be adaptive. By extension, the patterns of variation in SSD among species should reflect differences in the patterns of adaptive divergence of males and females.

In many clades, interspecific variation in SSD is correlated with body size. When this variation is presented using the statistical methods of standard allometric analysis, it translates into slopes that deviate significantly from one when the size of one sex is plotted against the size of the other sex on a logarithmic scale. Such analyses reveal a predominant allometric pattern characterized by a strong correlation for body size between males and females and greater interspecific variance in male size than in female size. This pattern occurs in taxa as disparate as water striders and primates, increases in body size being associated with increasing SSD where males are the larger sex, and with decreasing SSD where females are the larger sex. Rensch (126) was the first to describe this general pattern, and it has been summarized as "Rensch's rule." Although found in some clades in which females are larger than males, allometry consistent with Rensch's rule is more common where males are the larger sex, and exceptions (i.e. allometry in which females vary more among species than do males) have been found exclusively in taxa with female-biased SSD.

We have no a priori reason to expect that allometry for SSD reflects patterns of functional constraint or nonadaptive divergence. In the absence of evidence to the contrary, we must assume that the observed allometric patterns reflect evolutionary divergence in response to the same selection that produces SSD within individual species. Therefore, selection associated with the different reproductive roles of the two sexes is likely to play a key role. Not surprisingly, almost all of the functional hypotheses for the evolution of allometry for SSD have invoked either sexual selection on males or reproductive selection on females, and it seems likely that the former is an important component of the evolutionary mechanism responsible for allometry consistent with Rensch's rule.

Many functional hypotheses have been proposed to explain the evolution of allometry for SSD, but few have been adequately tested. Many of these hypotheses incorporate the same variables (e.g. sexual selection, mating system) and differ mainly in the postulated order of cause and effect. They therefore cannot be distinguished by simple analyses of correlations among variables. Directional comparative analyses that estimate the order of appearance of traits within a clade hold more promise. Quantitative modeling of the proposed evolutionary mechanisms also holds promise as a tool for both assessment of the plausibility of the verbal hypotheses and generation of testable, quantitative predictions. However, in addition to the difficulty of rigorously testing and distinguishing among these hypotheses, many of them are specific to a given clade and lack sufficient generality to account for the observed prevalence and pattern of allometry for SSD.

A satisfactory general hypothesis must be applicable to a wide range of taxa, regardless of mean body size or direction of SSD, and must explain both the covariance of male and female size and the greater interspecific variance in one sex than the other. Hypotheses based on quantitative genetic models for the evolution of SSD (the correlated response and variance dimorphism hypotheses) hold the promise of generality, but they predict neither allometry at equilibrium nor a consistent pattern of allometry along the evolutionary trajectory. The correlational selection hypothesis derived from a quantitative genetic model for the evolution of interspecific allometry is more promising. According to this hypothesis, interspecific allometry for SSD evolves as a consequence of directional selection acting primarily on one sex (e.g. sexual selection on males or fecundity selection on females) combined with correlational selection on the other sex. This hypothesis appears to satisfy the requirements of a general model but has yet to be tested empirically.

Allometry for SSD is clearly a conspicuous component of patterns of variation in SSD among species. While factors influencing the evolution of SSD have intrinsic interest for many ecologists and evolutionary biologists, studies of allometry for SSD have an even broader appeal. As indicated by the plethora of hypotheses summarized above, attempts to understand the evolution of allometry for SSD force us to address many of the central issues of evolutionary biology such as adaptation versus constraint, equilibrium versus non-equilibrium models, and evolutionary mechanisms governing the correlated evolution of suites of traits. It is my hope that, by drawing together analytical techniques, empirical evidence, and current hypotheses, this review will encourage a transition from hypothesis generation to hypothesis testing, and so set the stage for greater understanding not only of the evolution of allometry for SSD, but also of these issues of broader significance to evolutionary biology.

ACKNOWLEDGMENTS

I thank J Reeve and A Abouheif for many lively and insightful discussions, for constructive comments on an earlier draft of the manuscript, and for allowing me access to unpublished data and analyses. The work was supported by the Natural Sciences and Engineering Research Council and Canada (individual research grant number OGP0000347).

> Visit the *Annual Reviews home page* at
> http://www.annurev.org.

Literature Cited

1. Abouheif E. 1995. *A comparative analysis of the allometry for sexual size dimorphism: testing Rensch's rule.* MSc thesis. Concordia Univ., Montreal, Canada. 95 pp.
2. Abouheif E, Fairbairn DJ. 1997. A comparative analysis of allometry for sexual size dimorphism: assessing Rensch's rule. *Am. Nat.* 149:540–62
3. Adams J, Greenwood PJ. 1987. Loading constraints, sexual selection and assortative mating in peracarid Crustacea. *J. Zool. Lond.* 211:35–46
4. Alexander RD, Hoogland JL, Howard RD, Noonan KM, Sherman PW. 1979. Sexual dimorphisms and breeding systems in pinnipeds, ungulates, primates, and humans. In *Evolutionary Biology and Human Social Behavior: An Anthropological Perspective,* ed. NA Chagnon, W Irons, pp. 402–35. North Scituate, MA: Duxbury
5. Andersen NM. 1994. The evolution of sexual size dimorphism and mating systems in water striders (Hemiptera: Gerridae): a phylogenetic approach. *Ecoscience* 1:208–14
6. Andersson M. 1994. *Sexual Selection.* Princeton, NJ: Princeton Univ. Press. 599 pp.
7. Arak A. 1988. Sexual dimorphism in body size: a model and a test. *Evolution* 42:820–25
8. Arnold SJ, Duvall D. 1994. Animal mating systems: a synthesis based on selection theory. *Am. Nat.* 143:317–48

9. Arnqvist G. 1992. Pre-copulatory fighting in a water strider: inter-sexual conflict or mate assessment? *Anim. Behav.* 43:559–67
10. Barbault R. 1988. Body size, ecological constraints, and the evolution of life-history strategies. In *Evolutionary Biology,* ed. MK Hecht, B Wallace, 22:261–86. New York: Plenum
11. Bastos RP, Celio F, Haddad B. 1996. Breeding activity of the neotropical treefrog *Hyla elegans* (Anura, Hylidae). *J. Herpetol.* 30:355–60
12. Berry JF, Shine R. 1980. Sexual size dimorphism and sexual selection in turtles (order Testudines). *Oecologia* 44:185–91
13. Bjorklund M. 1990. A phylogenetic interpretation of sexual dimorphism in body size and ornament in relation to mating system in birds. *J. Evol. Biol.* 3:171–83
14. Blackith RE. 1957. Polymorphism in some Australian locusts and grasshoppers. *Biometrics* 13:183–96
15. Bourne GR. 1993. Proximate costs and benefits of mate acquisition at leks of the frog *Ololygon rubra. Anim. Behav.* 45:1051–59
16. Brana F. 1996. Sexual dimorphism in lacertid lizards: male head increase vs female abdomen increase? *Oikos* 75:511–23
17. Brooks DR, McLennan DA. 1991. *Phylogeny, Ecology and Behavior.* London, UK: Univ. Chicago Press
18. Cabana G, Frewin A, Peters RH, Randall L. 1982. The effect of sexual size dimor-

phism on variations in reproductive effort of birds and mammals. *Am. Nat.* 120:17–25

19. Calder WA. 1984. *Size, Function and Life History.* Cambridge, MA: Harvard Univ. Press

20. Campbell BC, ed. 1972. *Sexual Selection and the Descent of Man, 1871–1971.* Chicago, IL: Aldine. 378 pp.

21. Cheverud JM, Dow MM, Leutenegger W. 1985. The quantitative assessment of phylogenetic constraints in comparative analyses: sexual dimorphism in body weight among primates. *Evolution* 39:1335–51

22. Clutton-Brock TH. 1985. Size, sexual dimorphism, and polygyny in Primates. See Ref. 76, pp. 51–60

23. Clutton-Brock TH, ed. 1988. *Reproductive Success. Studies of Individual Variation in Contrasting Breeding Systems.* Chicago, IL: Univ. Chicago Press

24. Clutton-Brock TH. 1994. The costs of sex. See Ref. 140, pp. 347–62

25. Clutton-Brock TH, Albon SD, Guiness FE. 1988. Reproductive success in male red deer. See Ref. 23, pp. 325–43

26. Clutton-Brock TH, Harvey PH. 1984. Comparative approaches to investigating adaptation. In *Behavioral Ecology: An Evolutionary Approach.* ed. J Krebs, N Davies, pp. 7–29. New York: Sinauer. 2nd ed.

27. Clutton-Brock TH, Harvey PH, Rudder B. 1977. Sexual dimorphism, socionomic sex ratio and body weight in primates. *Nature* 269:797–800

28. Crump ML. 1974. Reproductive strategies in a tropical anuran community. *Misc. Publ. No. 61, Univ. Kans. Mus. Nat. Hist.* Lawrence, KS

29. Darwin C. 1874. *The Descent of Man and Selection in Relation to Sex.* New York: Humboldt

30. DeFrenza J, Kirner RJ, Maly EJ, van Leeuwen HC. 1986. The relationships of sex-size ratio and season to mating intensity in some Calanoid copepods. *Limnol. Oceangr.* 31:491–96

31. Dodson G. 1987. The significance of sexual dimorphism in the mating system of two species of tephritid flies (*Aciurina trixa* and *Valentibulla dodsoni*) (Diptera: Tephritidae). *Can. J. Zool.* 65:194–98

32. Elgar MA. 1991. Sexual cannibalism, size dimorphism, and courtship behavior in orb-weaving spiders (Aranidae). *Evolution* 45:444–48

33. Emerson SB. 1994. Testing pattern predictions of sexual selection. *Am. Nat.* 143:848–69

34. Emlen D. 1996. Artificial selection on horn length-body size allometry in the horned beetle *Onthophagus acuminatus* (Coleoptera: Scarabaeidae). *Evolution* 50:1219–30

35. Fairbairn DJ. 1990. Factors influencing sexual size dimorphism in temperate water striders. *Am. Nat.* 136:61–86

36. Fairbairn DJ. 1992. The origins of allometry: size and shape polymorphism in the common waterstrider, *Gerris remigis* Say (Heteroptera, Gerridae). *Biol. J. Linn. Soc.* 45:167–68

37. Fairbairn DJ, Preziosi RF. 1994. Sexual selection and the evolution of allometry for sexual size dimorphism in the water strider, *Aquarius remigis. Am. Nat.* 144:101–18

38. Fairbairn J, Shine R. 1993. Patterns of sexual size dimorphism in seabirds of the southern hemisphere. *Oikos* 68:139–45

39. Falconer DS. 1989. *Introduction to Quantitative Genetics.* London, Longman. 3rd ed.

40. Felsenstein J. 1985. Phylogenies and the comparative method. *Am. Nat.* 125:1–15

41. Fitch HS. 1981. Sexual size differences in reptiles. *Misc. Publ. No. 70, Univ. Kans. Mus. Nat. Hist.* Lawrence, KS.

42. Fleagle JG. 1985. Size and adaptation in Primates. See Ref. 76, pp. 1–19

43. Flemming IA, Gross MR. 1994. Breeding competition in a Pacific salmon (Coho: *Oncorhynchus kisutch*): measures of natural and sexual selection. *Evolution* 48:637–57

44. Ford SM. 1994. Evolution of sexual dimorphism in body weight in Platyrrhines. *Am. J. Primatol.* 34:221–44

45. Forsman A, Shine R. 1995. Sexual size dimorphism in relation to frequency of reproduction in turtles (Testudines: Emydidae). *Copeia* 1995:727–29

46. Freeman DC, Klikoff LG, Harper KT. 1976. Differential resource utilization by the sexes of dioecious plants. *Science* 193:597–99

47. Fukuyama K, Kusano T. 1989. Sexual size dimorphism in a Japanese stream-breeding frog, *Buergeria buergeri* (Rhacophoridae, Amphibia). *Curr. Herpetol. East* Asia, pp. 306–13

48. Gaulin SJC, Sailer LD. 1984. Sexual dimorphism in weight among the Primates: the relative impact of allometry and sexual selection. *Int. J. Primatol.* 5:515–35

49. Geber MA. 1995. Fitness effects of sexual dimorphism in plants. *Trends Ecol. Evol.* 10:222–23

50. Ghiselin MT. 1974. *The Economy of Nature and the Evolution of Sex.* Berkeley: Univ. Calif. Press

51. Gibbons WJ, Lovich JE. 1990. Sexual dimorphism in turtles with emphasis on the slider turtle (*Trachemys scripta*). *Herpetol. Monogr.* 4:1–29
52. Gotelli NJ, Spivey HR. 1992. Male parasitism and intersexual competition in a burrowing barnacle. *Oecologia* 91:474–80
53. Gould SJ. 1966. Allometry and size in ontogeny and phylogeny. *Biol. Rev.* 41:587–640
54. Grafen A. 1989. The phylogenetic regression. *Philos. Trans. R. Soc. Lond. B Biol. Sci.* 326:119–57
55. Grafen A. 1992. The uniqueness of the phylogenetic regression. *J. Theor. Biol.* 156:405–23
56. Green AJ. 1992. Positive allometry is likely with mate choice, competitive display and other functions. *Anim. Behav.* 43:170–72
57. Greenwood PJ, Adams J. 1987. *The Ecology of Sex*. London, UK: Edward Arnold. 74 pp.
58. Greenwood PJ, Wheeler P. 1985. The evolution of sexual size dimorphism in birds and mammals: a hot-blooded hypothesis. In *Evolution: Essays in Honour of John Maynard Smith*, ed. PH Harvey, M Slatkin, pp. 287–99. Cambridge, UK: Cambridge Univ. Press
59. Hanken J, Wake DB. 1993. Miniaturization of body size: organismal consequences and evolutionary significance. *Annu. Rev. Ecol. Syst.* 24:501–19
60. Harvey AW. 1990. Sexual differences in contemporary selection acting on size in the hermit crab *Clibanarius digueti*. *Am. Nat.* 136:292–304
61. Harvey PH. 1982. On rethinking allometry. *J. Theor. Biol.* 95:37–41
62. Harvey PH, Clutton-Brock TH. 1983. The survival of the theory. *New Sci.* 5:313–15
63. Harvey PH, Mace GM. 1982. Comparisons between taxa and adaptive trends: problems of methodology. In *Current Problems in Sociobiology*, ed. King's College Sociobiology Group, pp. 343–61. Cambridge, UK: Cambridge Univ. Press
64. Harvey PH, Pagel MD. 1991. *The Comparative Method in Evolutionary Biology*. New York: Oxford Univ. Press
65. Harvey PH, Ralls K. 1985. Homage to the null weasel. In *Evolution: Essays in Honour of John Maynard Smith*, ed. MK Hecht, B Wallace, pp. 155–171. Cambridge, UK: Cambridge Univ. Press
66. Head G. 1995. Selection on fecundity and variation in the degree of sexual size dimorphism among spider species (class Aranae). *Evolution* 49:776–81
67. Hedrick AV, Temeles EJ. 1989. The evolution of sexual dimorphism in animals: hypotheses and tests. *Trends Ecol. Evol.* 4:136–38
68. Helms KR. 1994. Sexual size dimorphism and sex ratios in bees and wasps. *Am. Nat.* 143:418–34
69. Höglund J. 1989. Size and plumage dimorphism in lek-breeding birds: a comparative analysis. *Am. Nat.* 134:72–87
70. Holtby LB, Healey MC. 1990. Sex-specific life-history tactics and risk-taking in Coho salmon. *Ecology* 71:678–90
71. Howard RD. 1988. Reproductive success in two species of anurans. See Ref. 23, pp. 99–113
72. Hurlbutt B. 1987. Sexual size dimorphism in parasitoid wasps. *Biol. J. Linn. Soc.* 30:63–89
73. Huxley JS. 1932. *Problems of Relative Growth*. London, UK: Methuen. 276 pp.
74. Iverson JB. 1991. Phylogenetic hypotheses for the evolution of modern kinosternine turtles. *Herpetol. Monogr.* 5:1–27
75. Jehl JR Jr, Murray BG Jr. 1986. The evolution of normal and reverse sexual size dimorphism in shorebirds and other birds. *Curr. Ornithol.* 3:1–86
76. Jungers WL, ed. 1985. *Size and Scaling in Primate Biology*. New York: Plenum. 491 pp.
77. Kawano K. 1995. Horn and wing allometry and male dimorphism in giant rhinoceros beetles (Coleoptera: Scarabaeidae) of tropical Asia and America. *Ann. Entomol. Soc. Am.* 88:92–99
78. LaBarbera M. 1989. Analyzing body size as a factor in ecology and evolution. *Annu. Rev. Ecol. Syst.* 20:97–117
79. Lande R. 1979. Quantitative genetic analysis of multivariate evolution applied to brain:body size allometry. *Am. Nat.* 33:402–16
80. Lande R. 1980. Sexual dimorphism, sexual selection and adaptation in polygenic characters. *Evolution* 34:292–307
81. Lande R, Arnold SJ. 1983. The measurement of selection on correlated characters. *Evolution* 37:1210–26
82. Leutenegger W. 1978. Scaling of sexual dimorphism in body size and breeding system in primates. *Nature* 272:610–11
83. Leutenegger W, Cheverud J. 1982. Correlates of sexual dimorphism in primates: ecological and size variables. *Int. J. Primatol.* 3:387–402
84. Leutenegger W, Cheverud J. 1985. Sexual dimorphism in Primates. The effects of size. See Ref. 76, pp. 33–60

85. Levenson H. 1990. Sexual size dimorphism in chipmunks. *J. Mammal.* 71:161–70

86. Levitan DR, Petersen C. 1995. Sperm limitation in the sea. *Trends Ecol. Evol.* 10:228–31

87. Lincoln GA. 1994. Teeth, horns and antlers: the weapons of sex. See Ref. 140, pp. 131–58

88. Lloyd DG, Webb CJ. 1977. Secondary sex characters in plants. *Bot. Rev.* 43:177–216

89. Longland WS. 1989. Reversed sexual size dimorphism: its effect on prey selection by the great horned owl, *Bubo virginianus. Oikos* 54:395–98

90. Lovich JE, Gibbons JW. 1992. A review of techniques for quantifying sexual dimorphism. *Growth Dev. Aging* 56:269–81

91. Maddison WP. 1990. A method for testing the correlated evolution of two binary characters: Are gains or losses concentrated on certain branches of a phylogenetic tree? *Evolution* 44:539–57

92. Madsen T, Shine R. 1994. Costs of reproduction influence the evolution of sexual size dimorphism in snakes. *Evolution* 48:1389–97

93. Marden JH. 1989. Effects of load-lifting constraints on the mating system of a dance fly. *Ecology* 70:496–502

94. Martin RD, Willner LA, Dettling A. 1994. The evolution of sexual size dimorphism in primates. See Ref. 140, pp. 159–200

95. Martins EP, Garland T. 1991. Phylogenetic analysis of the correlated evolution of continuous characters: a simulation study. *Evolution* 45:534–57

96. Maynard Smith J. 1977. Parental investment: a prospective analysis. *Anim. Behav.* 25:1–9

97. McArdle BH. 1988. The structural relationship: regression in biology. *Can. J. Zool.* 66:2329–39

98. McLachlin AJ. 1986. Sexual dimorphism in midges: strategies for flight in the rain-pool dweller *Chironomous imicola* (Diptera: Chironomidae). *J. Anim. Ecol.* 55:261–67

99. McMahon. T. 1973. Size and shape in biology. *Science* 179:1201–4

100. Meagher TR. 1992. The quantitative genetics of sexual dimorphism in *Silene latifolia* (Caryophyllaceae). I. Genetic variation. *Evolution* 46:445–57

101. Meagher TR. 1994. The quantitative genetics of sexual dimorphism in *Silene latifolia* (Caryophyllaceae). II. Response to sex-specific selection. *Evolution* 48:939–51

102. Moors PJ. 1980. Sexual dimorphism in the body size of mustelids (Carnivora): the roles of food habits and breeding systems. *Oikos* 34:147–58

103. Mueller HC. 1986. The evolution of reversed sexual dimorphism in owls: an empirical analysis of possible selective factors. *Wilson Bull.* 98:387–406

104. Mueller HC, Meyer K. 1985. The evolution of reversed sexual dimorphism in size. A comparative analysis of the Falconiformes of the western Palearctic. *Curr. Ornithol.* 2:65–101

105. Naylor C, Adams J. 1987. Sexual dimorphism, drag constraints and male performance in *Gammarus duebeni* (Amphipoda). *Oikos* 48:23–27

106. Nudds TD, Kaminski RM. 1984. Sexual size dimorphism in relation to resource partitioning in North American dabbling ducks. *Can. J. Zool.* 62:2009–102

107. Nylin S, Wedell N. 1994. Sexual size dimorphism and comparative methods. In *Phylogenetics and Ecology*, ed. P Eggleton, R Vane-Wright, pp. 253–309. London: Academic

108. O'Neill KM. 1985. Egg size, prey size, and sexual size dimorphism in digger wasps (Hymenoptera: Sphecidae). *Can. J. Zool.* 63:2187–93

109. Oakes EJ. 1992. Lekking and the evolution of sexual dimorphism in birds: comparative approaches. *Am. Nat.* 140:665–84

110. Pagel MD. 1992. A method for the analysis of comparative data. *J. Theor. Biol.* 156:431–42

111. Pagel MD, Harvey PH. 1988. The taxon-level problem in the evolution of mammalian brain size: facts and artifacts. *Am. Nat.* 132:344–59

112. Payne RB. 1984. Sexual selection, lek and arena behavior, and sexual dimorphism in birds. *Orthithol. Monogr. 33.* Washington, DC: Am. Ornithol. Union

113. Peters RH. 1983. *The Ecological Implications of Body Size.* New York: Cambridge Univ. Press

114. Petrie M. 1988. Intraspecific variation in structures that display competitive ability: Large animals invest relatively more. *Anim. Behav.* 36:1174–79

115. Petrie M. 1992. Are all secondary sexual display structures positively allometric and, if so, why? *Anim. Behav.* 43:173–75

116. Preziosi RF, Fairbairn DJ. 1996. Sexual size dimorphism and selection in the wild in the waterstrider *Aquarius remigis*: body size, components of body size and male mating success. *J. Evol. Biol.* 9:317–36

117. Preziosi RF, Fairbairn DJ. 1997. Sexual size dimorphism and selection in the wild in the waterstrider *Aquarius remigis*: lifetime fecundity selection on female total length and its consequences. *Evolution* 51:467–74

118. Price TD. 1984. The evolution of sexual size dimorphism in Darwin's finches. *Am. Nat.* 123:500–18

119. Purvis A, Gittleman JL, Luh HK. 1994. Truth or consequences: effects of phylogenetic accuracy on two comparative methods. *J. Theor. Biol.* 167:293–300

120. Ralls K. 1976. Mammals in which females are larger than males. *Q. Rev. Biol.* 51:245–76

121. Ralls K. 1977. Sexual dimorphism in mammals: avian models aand unanswered questions. *Am. Nat.* 111:917–38

122. Ralls K, Harvey PH. 1985. Geographic variation in size and sexual dimorphism of North American weasels. *Biol. J. Linn. Soc.* 25:119–67

123. Reeve JP, Fairbairn DF. 1996. Sexual size dimorphism as a correlated response to selection on body size: an empirical test of the quantitative genetic model. *Evolution* 50:1927–38

124. Reiss MJ. 1986. Sexual dimorphism in body size: are larger species more dimorphic? *J. Theor. Biol.* 121:163–72

125. Reiss MJ. 1989. *The Allometry of Growth and Reproduction.* Cambridge, UK: Cambridge Univ. Press

126. Rensch B. 1960. *Evolution above the Species Level.* New York: Columbia Univ. Press.

127. Reynolds JD, Harvey PH. 1994. Sexual selection and the evolution of sex differences. See Ref. 140, pp. 53–70

128. Ridley M. 1983. *The Explanation of Organic Diversity.* New York: Oxford Univ. Press

129. Rodriquez V, Jiménez F. 1990. Coexistance within a group of congeneric species of Acartia (Copepoda Calanoida): sexual dimorphism and ecological niche in *Acartia grani*. *J. Plankton Res.* 12:497–511

130. Roff DA. 1992. *The Evolution of Life Histories.* New York: Chapman & Hall

131. Roff DA, Fairbairn DJ. 1980. An evaluation of Gulland's method for fitting the Schaefer model. *Can. J. Fish. Aquat. Sci.* 37:1229–35

132. Rowe L. 1992. Convenience polyandry in a water strider: foraging conflicts and female control of copulation frequency and guarding duration. *Anim. Behav.* 44:189–202

133. Schmidt-Nielsen K. 1984. *Scaling. Why is Animal Size so Important?* New York: Cambridge Univ. Press

134. Selander RK. 1972. Sexual selection and dimorphism in birds. See Ref. 20, pp. 180–230

135. Shine R. 1979. Sexual selection and sexual dimorphism in the Amphibia. *Copeia* 2:297–306

136. Shine R. 1989. Ecological causes for the evolution of sexual dimorphism: a review of the evidence. *Q. Rev. Biol.* 64:419–61

137. Shine R. 1991. Intersexual dietary divergence and the evolution of sexual dimorphism in snakes. *Am. Nat.* 138:103–22

138. Shine R. 1994. Sexual size dimorphism in snakes revisited. *Copeia* 2:326–46

139. Shine R. 1994. Allometric patterns in the ecology of Australian snakes. *Copeia* 4:851–67

140. Short RV, Balaban E, ed. 1994. *The Differences between the Sexes.* Cambridge, UK: Cambridge Univ. Press, 479 pp.

141. Sigurjonsdottir H. 1981. The evolution of sexual size dimorphism in gamebirds, waterfowl and raptors. *Ornis Scand.* 12:249–60

142. Sivinski JM, Dodson G. 1992. Sexual dimorphism in *Anatrepha suspensa* (Loew) and other Tephritid fruit flies (Diptera: Tephritidae): possible roles of developmental rate, fecundity, and dispersal. *J. Insect Behav.* 5:491–506

143. Slatkin M. 1984. Ecological causes of sexual dimorphism. *Evolution* 38:622–30

144. Sokal RR, Rohlf FJ. 1995. *Biometry: The Principles and Practice of Statistics in Biological Research.* San Francisco, CA: Freeman. 3rd ed.

145. Stamps JA. 1983. Sexual selection, sexual dimorphism, and territoriality. In *Lizard Ecology,* ed. RB Huey, ER Pianka, TW Schoener, pp. 169–204. Cambridge, MA: Harvard Univ. Press

146. Stamps JA. 1993. Sexual size dimorphism in species with asymptotic growth after maturity. *Biol. J. Linn. Soc.* 50:123–45

147. Stearns SC. 1983. The influence of size and phylogeny on patterns of covariation among life-history traits in mammals. *Oikos* 41:173–87

148. Stearns SC. 1984. The effects of size and phylogeny on patterns of covariation in the life history traits of lizards and snakes. *Am. Nat.* 123:56–72

149. Stubblefield JW, Seger J. 1994. Sexual dimorphism in the hymenoptera. See Ref. 140, pp. 71–103

150. Trivers RL. 1972. Parental investment and sexual selection. See Ref. 20, pp. 136–79

151. Van den Assem J. 1976. Male courtship behaviour, female receptivity signal,

and size differences between the sexes in Pteromalinae (Hymen., Chalcidoidea, Pteromalidae), and comparative notes on other chalcidoids. *Netherlands J. Zool.* 26:535–48

152. Vollrath F, Parker GA. 1992. Sexual dimorphism and distorted sex ratios in spiders. *Nature* 360:156–59

153. Wallace B. 1987. Ritualistic combat and allometry. *Am. Nat.* 129:775–76

154. Weatherhead PJ. 1980. Sexual dimorphism in two savannah sparrow populations. *Can. J. Zool.* 58:412–15

155. Webster MS. 1992. Sexual dimorphism, mating system and body size in new world blackbirds (Icterinae). *Evolution* 46:1621–41

156. Weigensberg I, Fairbairn DJ. 1994. Conflicts of interest between the sexes: a study of mating interactions in a semi-aquatic bug. *Anim. Behav.* 48:893–90

157. Wheeler P, Greenwood PJ. 1983. The evolution of reversed sexual dimorphism in birds of prey. *Oikos* 40:145–49

158. Wickman P-O. 1992. Sexual selection and butterfly design—a comparative study. *Evolution* 46:1525–36

159. Wickman P-O, Karlsson B. 1989. Abdomen size, body size and the reproductive effort of insects. *Oikos* 56:209–14

160. Wiklund C, Kaitala A. 1995. Sexual selection for large male size in a polyandrous butterfly: the effect of body size on male versus female reproductive success in *Pieris napi. Behav. Ecol.* 6:6–13

161. Wiklund C, Karlsson B. 1988. Sexual size dimorphism in relation to fecundity in some Swedish Satyrid butterflies. *Am. Nat.* 131:132–38

162. Wiley RH. 1974. Evolution of social organization and life-history patterns among grouse. *Q. Rev. Biol.* 49:201–27

163. Wilkinson GS. 1993. Artificial selection alters allometry in the salk-eyed fly *Cyrtodiopsis dalmanni* (Diptera:Diopsidae). *Genet. Res. Camb.* 62:213–22

164. Willson MF. 1983. *Plant Reproductive Ecology.* New York: Wiley

165. Willson MF. 1991. Sexual selection, sexual dimorphism and plant phylogeny. *Evol. Ecol.* 5:69–87

166. Winquist T, Lemon RE. 1994. Sexual selection and exaggerated male tail length in birds. *Am. Nat.* 143:95–116

167. Ydenberg RC, Forbes LS. 1991. The survival-reproduction selection equilibrium and reversed size dimorphism in raptors. *Oikos* 60:115–19

168. Zeng Z-B. 1988. Long-term correlated response, interpopulation covariation, and interspecific allometry. *Evolution* 42:363–74

SUBJECT INDEX

A

Abelson, A., 317–36
Abouheif, E., 664, 667–68, 670
Abundance traits
  extinction vulnerability and,
    503–4
Abutilon theophrasti
  growth of
    root and shoot competition
      and, 560
  population structure of
    heterogeneity and, 558
Acacia karroo
  herbaceous production and,
    522
Acacia senegal
  nutrient accumulation below,
    529
Acacia tortilis
  grass removal and, 525
Acanthamoeba
  polyploidy in, 401
Acromyrmex versicolor
  pleometrotic associations of
    queens in
    relatedness in, 7
Adaptive radiations, 134
Aerosol filtration theory, 323
Aerts, R., 549
Age-structure
  population genetic models with,
    273–74
  population genetic models
    without, 272–73
Agropyron desertorum
  resource uptake by, 550
Albumin gene
  aligned DNA sequences of,
    452
Alces alces
  riparian systems and, 634
Alder
  riparian zones and, 643
Algal biomass
  Daphnia and, 472–73
Algal response factor (ARF), 472,
    479
Alleculids
  mouthparts of, 160
Allelic frequencies
  in haplodiploids, 57–58
Allometry
  sexual size dimorphism and,
    663–80

Allomyces
  polyploidy in, 401
Allopolyploidy, 400, 416, 596,
    604
  animals and, 601
Allozyme markers
  gene flow estimation and,
    109–10
Allozymes
  genetic studies of social insects
    and, 4
  variations in natural
    populations, 106
Alnus rubra
  riparian zones and, 642
Alnus rugosa
  riparian zones and, 643
Alpha diversity trends, 97–98
Altschul, S. F., 251
Alveolates, 399
  amphimixis and, 403–4
  polyploidy in, 401
Ambystoma
  allele transfer among, 603
American oyster
  gene flow in, 121
Amino acid sequencing
  echinoderm phyogenies and,
    220
Amitochondriate flagellates, 398
Amoeba dubia
  DNA amounts in, 248
Amoebae, 399
  polyploidy in, 400
Amoeboflagellates, 399
Amphibians
  polyploidy in, 401
Amphidiploidy, 416
Amphimictic cycles
  evolution of, 419–22
Amphimictic reproduction
  atypical, 408–9
Amphimixis, 396–97
  apomixis combined, 411–12
  distribution of, 402–6
Anadromous fish
  as terrestrial prey, 296
Anadromy, 297
Anderson, E., 366, 594
Anderson, W. B., 289–309
Andrena jacobi
  inbreeding in, 8
  low relatedness in, 8
Andrenid bees
  low relatedness in, 8

Anemochory, 626
Aneuploidy
  in eukaryotes, 400
Angiosperms
  allopolyploidy in, 401
  syngamy in, 406
Anilocra multilineata
  female alternative reproductive
    behaviors in, 577
Animal diversification, 593–613
  hybridization and, 598–605
  introgression and, 605–10
Animalia, 400
  amphimixis and, 406
  polyploidy in, 401
Anisogamy, 410, 420–21
Ants
  colonial sex ratios in, 14
  continuous populations of
    genetic structure of, 5
  female dispersal in, 6
  founding stage in
    kin groups and, 7
  inbreeding in, 63
  intracolonial relatedness in, 11
  microsatellite markers for, 4
  multi-nest aggregation in, 9
  relatedness asymmetry in, 14
  sex ratios in, 15
  sexual production in
    intraspecific parasitism and,
      10
Anurans
  polyploidy in, 604
Aoki, K., 111
Aphids
  eusociality in, 28
Apicomplexans
  reduced zygophase in, 403
Apis
  patrilines in, 6
Apis mellifera
  phylogeography in, 19–20
Apis mellifera capensis
  reproductive dominance in, 13
Apomixis, 396
  amphimixis combined, 411–12
  facultative, 411, 422
  obligate, 412, 421
Aquatic organisms
  nourishment for
    riparian vegetation and, 632
Aquatic systems
  nutrient flow in
    hydrologic cycle and, 292

CUMULATIVE INDEXES

CONTRIBUTING AUTHORS, VOLUMES 24–28

CHAPTER TITLES, VOLUMES 24–28